Multiples and Submultiples for SI Units			
Prefix	**Symbol**	**Mutiplier**	**Use**
tera	T	$1,000,000,000,000 = 10^{12}$	1 terameter (Tm)
giga	G	$1,000,000,000 = 10^{9}$	1 gigameter (Gm)
mega	M	$100,000 = 10^{6}$	1 megameter (Mm)
kilo	k	$1,000 = 10^{3}$	1 kilometer (km)
centi*	c	$0.01 = 10^{-2}$	1 centimeter (cm)
milli	m	$0.001 = 10^{-3}$	1 millimeter (mm)
micro	μ	$0.000001 = 10^{-6}$	1 micrometer (μm)
nano	n	$0.000000001 = 10^{-9}$	1 namometer
—	Å	$0.0000000001 = 10^{-10}$	1 angsstrom (Å)*
pico	p	$0.000000000001 = 10^{-12}$	1 picometer (pm)
* The use of the centimeter and the angstrom is discouraged but they are still widely used.			

The SI Base Units for Seven Fundamental Quantities and Two Supplemental Quantities		
Quantity	**Unit**	**Symbol**
Base Units		
Length	meter	m
Mass	kilogram	kg
Time	second	s
Electric current	ampere	A
Temperature	kelvin	K
Luminous intensity	candela	cd
Amount of substance	mole	mol
Supplemental units		
Plane angle	radian	rad
Solid angle	steradian	Sr

PHYSICS

SIXTH EDITION

Paul E. Tippens

Department Of Physics
Southern Polytechnic State University
Marietta, Georgia

 Glencoe
McGraw-Hill

New York, New York   Columbus, Ohio   Woodland Hills, California   Peoria, Illinois

Cover: Richard Magna/Fundamental Photographs

Part Openers:

Part 1, page 1: Paul L. Ruben
Part 2, page 359: Doug Martin
Part 3, page 511: Telegraph Colour Library/FPG
Part 4, page 833: Super Stock

To my wife, my friend, and my constant companion,
Joyce Scott Tippens

Library of Congress Cataloging-in-Publication Data
Tippens, Paul E.
  Physics/Paul E. Tippens.—6th ed.
    p. cm.
  Includes index.
  ISBN 0-07-820340-6
    1. Physics. I. Title

  QC21.2.T55 1999
  530—dc21
                      99-052055

Physics, Sixth Edition   ISBN 0-07-820340-6
Activities Manual   ISBN 0-07-820341-4
Instructor's Manual   ISBN 0-07-820342-2

Glencoe/McGraw-Hill

*A Division of The **McGraw·Hill** Companies*

Physics, Sixth Edition

Send all inquiries to:
Glencoe/McGraw-Hill
8787 Orion Place
Columbus, OH 43240-4027

ISBN 0-07-820340-6

2 3 4 5 6 7 8 9   027   06 05 04 03 02 01 00

Contents

Preface xi

PART 1   Mechanics   1

CHAPTER 1   Introduction 2
1-1   What Is Physics? 2
1-2   What Part Is Played by Mathematics? 3
1-3   How Should I Study Physics? 4

CHAPTER 2   Technical Mathematics 6
2-1   Signed Numbers 7
2-2   Algebra Review 10
2-3   Exponents and Radicals (Optional) 13
2-4   Scientific Notation 16
2-5   Graphs 17
2-6   Geometry 19
2-7   Right Triangle Trigonometry 23

CHAPTER 3   Technical Measurements and Vectors 38
3-1   Physical Quantities 38
3-2   The International System 40
3-3   Measurement of Length and Time 42
3-4   Significant Figures 44
3-5   Measuring Instruments 45
3-6   Unit Conversions 47
3-7   Vector and Scalar Quantities 50
3-8   Addition of Vectors by Graphical Methods 53
3-9   Force and Vectors 56
3-10 The Resultant Force 58
3-11 Trigonometry and Vectors 59
3-12 The Component Method of Vector Addition 63
3-13 Vector Difference 66

CHAPTER 4   Translational Equilibrium and Friction 73
4-1   Newton's First Law 73
4-2   Newton's Third Law 74
4-3   Equilibrium 75
4-4   Free-Body Diagrams 76
4-5   Solution of Equilibrium Problems 80
4-6   Friction 84

CHAPTER 5 Torque and
　　　　　　Rotational Equilibrium 100
5-1　Conditions for Equilibrium....... 100
5-2　The Moment Arm............. 101
5-3　Torque 103
5-4　Resultant Torque............. 105
5-5　Equilibrium 107
5-6　Center of Gravity 111

CHAPTER 6 Uniform Acceleration 120
6-1　Speed and Velocity 120
6-2　Acceleration 122
6-3　Uniform Acceleration 123
6-4　Other Useful Relations 125
6-5　Solutions to
　　　Acceleration Problems.......... 126
6-6　Sign Convention in
　　　Acceleration Problems.......... 129
6-7　Gravity and Free-Falling Bodies ... 131
6-8　Projectile Motion 136
6-9　Horizontal Projection 137
6-10 The More General
　　　Problem of Trajectories......... 140

CHAPTER 7 Newton's Second Law 150
7-1　Newton's Second Law of Motion... 150
7-2　The Relationship between Weight
　　　and Mass................... 153
7-3　Application of Newton's Second
　　　Law to Single-Body Problems 155
7-4　Problem-Solving Techniques 158

CHAPTER 8 Work, Energy, and Power ... 172
8-1　Work..................... 172
8-2　Resultant Work 174
8-3　Energy 176
8-4　Work and Kinetic Energy 177
8-5　Potential Energy 180
8-6　Conservation of Energy.......... 182
8-7　Energy and Friction Forces 184
8-8　Power 187

CHAPTER 9 Impulse and Momentum.... 196
9-1　Impulse and Momentum........ 196
9-2　The Law of Conservation
　　　of Momentum 199
9-3　Elastic and Inelastic Impacts 202

CHAPTER 10 Uniform Circular Motion.... 213
10-1　Motion in a Circular Path 213
10-2　Centripetal Acceleration 214
10-3　Centripetal Force............. 216
10-4　Banking Curves.............. 219
10-5　The Conical Pendulum 221
10-6　Motion in a Vertical Circle 223
10-7　Gravitation 225
10-8　The Gravitational Field
　　　 and Weight 227
10-9　Satellites in Circular Orbits 228
10-10 Kepler's Laws................. 232

CHAPTER 11 Rotation of Rigid Bodies 241
11-1　Angular Displacement........... 241
11-2　Angular Velocity 243
11-3　Angular Acceleration........... 244
11-4　Relationship between Rotational
　　　 and Linear Motion 247
11-5　Rotational Kinetic Energy:
　　　 Moment of Inertia 249
11-6　The Second Law of Motion
　　　 in Rotation 251
11-7　Rotational Work and Power 254
11-8　Angular Momentum 256
11-9　Conservation
　　　 of Angular Momentum.......... 258

CHAPTER 12 Simple Machines 267
12-1　Simple Machines and Efficiency ... 267
12-2　Mechanical Advantage.......... 269
12-3　The Lever 271
12-4　Applications of the
　　　 Lever Principle 273
12-5　The Transmission of Torque 275
12-6　The Inclined Plane 278
12-7　Applications of the
　　　 Inclined Plane............... 281

CHAPTER 13 Elasticity............. 290
13-1　Elastic Properties of Matter...... 290
13-2　Young's Modulus.............. 293
13-3　Shear Modulus 296
13-4　Volume Elasticity;
　　　 Bulk Modulus............... 298
13-5　Other Physical Properties
　　　 of Metals.................. 299

CHAPTER 14 Simple Harmonic Motion . . . 306
14-1   Periodic Motion 306
14-2   The Reference Circle 308
14-3   Velocity in Simple
       Harmonic Motion. 310
14-4   Acceleration in Simple
       Harmonic Motion. 312
14-5   The Period and Frequency 314
14-6   The Simple Pendulum. 316
14-7   The Torsion Pendulum 318

CHAPTER 15 Fluids 325
15-1   Density. 325
15-2   Pressure 328
15-3   Fluid Pressure 329
15-4   Measuring Pressure. 333
15-5   The Hydraulic Press 335
15-6   Archimedes' Principle 337
15-7   Fluid Flow. 341
15-8   Pressure and Velocity. 343
15-9   Bernoulli's Equation 345
15-10 Applications of Bernoulli's Equation. . 346

PART 2   Thermodynamics, Mechanical Waves, and Sound   359

CHAPTER 16 Temperature
                and Expansion 360
16-1   Temperature and Thermal Energy. . 360
16-2   The Measurement of Temperature. . 362
16-3   The Gas Thermometer 367
16-4   The Absolute Temperature Scale . . . 368
16-5   Linear Expansion 371
16-6   Area Expansion. 374
16-7   Volume Expansion 376
16-8   The Unusual Expansion of Water . . 378

CHAPTER 17 Quantity of Heat 384
17-1   The Meaning of Heat 384
17-2   The Quantity of Heat 385
17-3   Specific Heat Capacity. 387
17-4   The Measurement of Heat. 390
17-5   Change of Phase 394
17-6   Heat of Combustion 399

CHAPTER 18 Transfer of Heat. 405
18-1   Methods of Heat Transfer 406
18-2   Conduction. 406
18-3   Insulation: The *R*-value. 410
18-4   Convection 411
18-5   Radiation 413

CHAPTER 19 Thermal Properties of Matter . . 421
19-1   Ideal Gases and Boyle's Law 421
19-2   Gay-Lussac's Law. 424
19-3   General Gas Laws 425
19-4   Molecular Mass and the Mole 427

19-5   The Ideal Gas Law. 429
19-6   Liquefaction of a Gas 430
19-7   Vaporization 432
19-8   Vapor Pressure 433
19-9   Triple Point. 434
19-10 Humidity 435

CHAPTER 20 Thermodynamics 443
20-1   Heat and Work 443
20-2   The Internal Energy Function 444
20-3   The First Law of Thermodynamics. . 445
20-4   The *P-V* Diagram 446
20-5   The General Case for the First Law. . 448
20-6   Adiabatic Processes 448
20-7   Isochoric Processes 450
20-8   Isothermal Processes 451
20-9   The Second Law
       of Thermodynamics 452
20-10 The Carnot Cycle 454
20-11 The Efficiency of an Ideal Engine . . 455
20-12 Internal Combustion Engines 456
20-13 Refrigeration. 459

CHAPTER 21 Mechanical Waves 468
21-1   Mechanical Waves 468
21-2   Types of Waves 469
21-3   Calculating Wave Speed. 470
21-4   Periodic Wave Motion. 472
21-5   Energy of a Periodic Wave. 474
21-6   The Superposition Principle 475
21-7   Standing Waves. 476
21-8   Characteristic Frequencies. 478

CHAPTER 22 Sound.................. 486
22-1   Production of a Sound Wave...... 487
22-2   The Speed of Sound 488
22-3   Vibrating Air Columns 490
22-4   Forced Vibration and Resonance... 493

22-5   Audible Sound Waves 494
22-6   Pitch and Quality 498
22-7   Interference and Beats........ 499
22-8   The Doppler Effect 500

PART 3   Electricity, Magnetism, and Optics   511

CHAPTER 23 The Electric Force 512
23-1   The Electric Charge............ 512
23-2   The Electron 515
23-3   Insulators and Conductors 516
23-4   The Gold-Leaf Electroscope 517
23-5   Redistribution of Charge........ 519
23-6   Charging by Induction 520
23-7   Coulomb's Law 521

CHAPTER 24 The Electric Field 530
24-1   The Concept of a Field 531
24-2   Computing the Electric Intensity... 534
24-3   Electric Field Lines 538
24-4   Gauss' Law 539
24-5   Applications of Gauss' Law 541

CHAPTER 25 Electric Potential.......... 551
25-1   Electric Potential Energy 551
25-2   Calculating Potential Energy 554
25-3   Potential 556
25-4   Potential Difference........... 560
25-5   Millikan's Oil-Drop Experiment ... 562
25-6   The Electronvolt 564

CHAPTER 26 Capacitance 570
26-1   Limitations on Charging
       a Conductor 570
26-2   The Capacitor 573
26-3   Computing the Capacitance 575
26-4   Dielectric Constant; Permittivity... 577
26-5   Capacitors in Parallel
       and in Series 581
26-6   Energy of a Charged Capacitor ... 585

CHAPTER 27 Current and Resistance 592
27-1   The Motion of Electric Charge 592
27-2   The Direction of Electric Current .. 594
27-3   Electromotive Force 595
27-4   Ohm's Law; Resistance 597
27-5   Electric Power and Heat Loss 600
27-6   Resistivity 601
27-7   Temperature Coefficient
       of Resistance 604
27-8   Superconductivity 604

CHAPTER 28 Direct-Current Circuits 612
28-1   Simple Circuits; Resistors
       in Series 612
28-2   Resistors in Parallel........... 615
28-3   EMF and Terminal
       Potential Difference........... 618
28-4   Measuring Internal Resistance..... 619
28-5   Reversing the Current through
       a Source of EMF.............. 620
28-6   Kirchhoff's Laws 621
28-7   The Wheatstone Bridge.......... 626
28-8   Electric Conduction in Liquids 628
28-9   Electrolysis 628
28-10 DC Voltage Sources; the Lead
       Storage Battery 632
28-11 Capacity Rating of a Battery 633

CHAPTER 29 Magnetism
              and the Magnetic Field..... 641
29-1   Magnetism 641
29-2   Magnetic Fields............... 643
29-3   The Modern Theory of Magnetism.. 643
29-4   Flux Density and Permeability..... 645

29-5   Magnetic Field and Electric Current. . 648
29-6   The Force on a Moving Charge. . . . 649
29-7   Force on a Current-Carrying Wire . 652
29-8   Magnetic Field of a Long,
        Straight Wire. 654
29-9   Other Magnetic Fields. 655
29-10 Hysteresis 657

CHAPTER 30 Forces and Torques
                in a Magnetic Field **667**
30-1   Force and Torque on a Loop 667
30-2   Magnetic Torque on a Solenoid. . . . 670
30-3   The Galvanometer. 670
30-4   The DC Voltmeter. 672
30-5   The DC Ammeter 674
30-6   The DC Motor 675

CHAPTER 31 Electromagnetic Induction. . **682**
31-1   Faraday's Law 682
31-2   EMF Induced by a Moving Wire . . . 686
31-3   Lenz's Law. 687
31-4   The AC Generator. 689
31-5   The DC Generator 693
31-6   Black EMF in a Motor. 693
31-7   Types of Motors 695
31-8   The Transformer. 697

CHAPTER 32 Alternating-Current Circuits . . **705**
32-1   The Capacitor. 705
32-2   The Inductor. 709
32-3   Alternating Currents 711
32-4   Phase Relation in AC Circuit. 713
32-5   Reactance 714
32-6   The Series AC Circuit 716
32-7   Resonance. 719
32-8   The Power Factor 719

CHAPTER 33 Light and Illumination. **727**
33-1   What Is Light?. 728
33-2   The Propagation of Light 730
33-3   The Electromagnetic Spectrum 732
33-4   The Quantum Theory. 734
33-5   The Speed of Light 736
33-6   Light Rays and Shadows 738

33-7   Luminous Flux 740
33-8   Luminous Intensity. 743
33-9   Illumination 744

CHAPTER 34 Reflection and Mirrors **753**
34-1   The Laws of Reflection 753
34-2   Plane Mirrors 755
34-3   Spherical Mirrors 757
34-4   Images Formed
        by Spherical Mirrors 760
34-5   The Mirror Equation. 762
34-6   Magnification 765
34-7   Spherical Aberration 767

CHAPTER 35 Refraction. **773**
35-1   Index of Refraction 774
35-2   The Laws of Refraction 775
35-3   Wavelength and Refraction 779
35-4   Dispersion. 780
35-5   Total Internal Reflection 780
35-6   Fiber Optics and Applications 782
35-7   Is Seeing the Same as Believing? . . . 785
35-8   Apparent Depth 787

CHAPTER 36 Lenses and
                Optical Instruments **793**
36-1   Simple Lenses 793
36-2   Focal Length and the
        Lensmaker's Equation 795
36-3   Image Formation by Thin Lenses . . 798
36-4   The Lens Equation
        and Magnification. 800
36-5   Combinations of Lenses 803
36-6   The Compound Microscope 804
36-7   Telescope. 806
36-8   Lens Aberrations. 806

CHAPTER 37 Interference, Diffraction,
                and Polarization **813**
37-1   Diffraction 813
37-2   Young's Experiment: Interference . . 814
37-3   The Diffraction Grating 818
37-4   Resolving Power of Instruments . . . 821
37-5   Polarization. 824

CHAPTER 38 Modern Physics
                and the Atom **834**
38-1   Relativity 834
38-2   Simultaneous Events:
        The Relativity of Time 836
38-3   Relativistic Length, Mass,
        and Time 837
38-4   Mass and Energy 841
38-5   Quantum Theory
        and the Photoelectric Effect 842
38-6   Waves and Particles 844
38-7   The Rutherford Atom 845
38-8   Electron Orbits 847
38-9   Atomic Spectra 848
38-10   The Bohr Atom 850
38-11   Energy Levels 853
38-12   Lasers and Laser Light 855
38-13   Modern Atomic Theory 857

CHAPTER 39 Nuclear Physics
                and the Nucleus **864**
39-1   The Atomic Nucleus 865
39-2   The Elements 866
39-3   The Atomic Mass Unit 870
39-4   Isotopes 871
39-5   The Mass Defect
        and Binding Energy 873

39-6   Radioactivity 877
39-7   Radioactive Decay 878
39-8   Half-Life 880
39-9   Nuclear Reactions 882
39-10   Nuclear Fission 883
39-11   Nuclear Reactors 885
39-12   Nuclear Fusion 886

CHAPTER 40 Electronics **893**
40-1   Thermionic Emission 893
40-2   Vacuum Tubes 894
40-3   The Cathode-Ray Tube 895
40-4   The X-Ray Tube 896
40-5   Semiconductors 897
40-6   N-Type and P-Type Semiconductors . . 899
40-7   PN Junction 900
40-8   Diode Applications 902
40-9   The Zener Diode 904
40-10   The Transistor 906
40-11   Transistor Amplification 908
40-12   Other Semiconductor Devices 912
40-13   Integrated Circuits 916

Appendix   Solution to
            Quadratic Equations 923

Index 925

Preface

The sixth edition of *Physics* is written for the noncalculus physics course designed for students who are pursuing technical or industrial careers or who desire a clear introduction to the applications of physical principles. A basic understanding of algebra and right-triangle trigonometry is usually assumed. Although many first-year physics students have excellent high school mathematics backgrounds, other students find it beneficial to review mathematics fundamentals as a part of the physics course. This new edition provides an excellent review of technical mathematics in the new second chapter. Instructors can assign this chapter, as needed, depending upon the abilities of their students.

ORGANIZATION OF THIS TEXT

This edition retains the successful four-part organization used in past editions, with much updated content and changes in the chapter-ending problems and questions. The treatment of statics now precedes the coverage of kinematics. Newton's first and third laws are covered early to provide qualitative understanding of force, but the second law is delayed until the concepts of free-body diagrams and static equilibrium are understood. This sequence allows each student to develop his or her understanding in a logical and continuous way. When statics is delayed until late chapters, a review of forces and vectors is often required. With the approach in this text, students' familiarity with free-body diagrams and static equilibrium makes it possible to provide detailed examples of Newton's second law.

The book's flexible organization allows instructors to rearrange topics to meet their course needs. Further, to facilitate flexibility in making assignments, there have been a number of important, yet subtle organizational changes. All projectile motion topics are included in Chap. 6. The two formerly separate chapters on fluids are now conveniently combined into a single chapter on fluid mechanics. The order of topics has been changed to reflect the traditional approach, in which optics and electromagnetic waves are covered after electricity and magnetism. This change was recommended so that the organization of the book would be consistent with many course syllabi in which the coverage of light and optical phenomena follows electricity and magnetism. The electronics chapter was carefully revised to be brought up to date with the latest solid-state theory. The material in each of the major topical areas has been written to allow maximum flexibility.

The goal of *Physics*, Sixth Edition, is to continue the record of excellence established with previous editions. It is noteworthy that *Physics*, Fifth Edition, won the prestigious McGuffey Award, an honor granted by the Text and Academic Authors' Association. This award pays tribute to textbooks that demonstrate long-term publishing excellence. As with previous editions, a major goal of this edition is to produce a friendly, easy-to-read, and accurate text and give a solid foundation in physics without discouraging students who have traditionally been intimidated by the subject.

This edition contains a number of significant enhancements and new features designed to facilitate the learning processes. Problem-solving strategies are enhanced and a number of conceptual example problems have been inserted. These conceptual examples differ slightly from the normal "number-crunching" examples because they are designed to give students a feeling or general understanding for the concept, as opposed to having them simply find correct numerical answers. For many students, it is difficult to know and understand the meaning of numerical answers. These conceptual examples should aid them in understanding the meaning of a correct answer.

CHANGES IN THE NEW EDITION

Perhaps the most extensive text changes can be seen with the chapter-ending pedagogy, which now consists of sections entitled Summary, Key Terms, Review Questions, Problems (organized and identified by sections), Challenge Problems, and Critical Thinking Questions. Key terms are cited to aid students in recalling key points discussed in the chapter. All the questions and problems were reexamined and reevaluated and then rewritten or revised.

Considerable effort has been given to the revision of problems. They have all been carefully studied for both content and order. More than 600 new problems have been added to give students more practice and instructors more options; 80 percent of all problems are new or have been revised. The problems were also reorganized and identified by section name and number for ease of making homework assignments. Within each of these sections, the problems are ordered according to level of difficulty. In addition, challenge problems covering one or more sections of the chapter are provided. As in the previous edition, the chapter-ending problem assignments that may be more difficult, require more time, or involve additional mathematical techniques are identified with an asterisk. All chapter-ending problems have been thoroughly checked and independently verified for accuracy and for their instructive value to the student.

Of particular significance, and new to this edition, is the addition of Critical Thinking Questions at the end of each chapter—totaling about 250 in the book. These problems, usually of moderate or greater difficulty, require significant reflection on the concepts and procedures. They are, in effect, learning examples that guide the students and build problem-solving skills. Students facing career challenges in the new millennium will be required to think logically and to reach sound and accurate solutions to ever more complex problems. These problems may be assigned in the usual way, or they may be assigned to teams of students. Professors and instructors may wish to add their own creative variations to these team problems.

The marginal features are an exciting addition. Math Tips, FYIs, Pocket Labs, Using a Calculator, and other items appear in margin and text areas to gain student interest and to show applications. Suggested Internet sites are provided to help students keep pace and to have access to new knowledge arising from the expansive growth of the Internet and the sheer explosion of information. The sites selected should encourage the readers to look beyond the text as they expand their own career paths. In turn, instructors may wish to create assignments based upon using the Internet.

Students will also notice the vignettes that help to demonstrate that we are surrounded daily by physical applications within our physical world. Whether the topics concern riding a roller coaster at an amusement park, or riding in the family sedan, van, or a sports utility vehicle, these everyday examples are based upon sound knowledge and technical application of physical principles and the laws of physics.

TEXTBOOK FEATURES

This new edition features the following:

- Clearly stated objectives begin each chapter.
- Summaries clearly identify each chapter's main points.
- Color is used to highlight pedagogical features.

- General problem-solving strategies are clearly outlined.
- Key terms or words are provided at the end of each chapter.
- FYIs are inserted in margins to generate interest and to stimulate further study.
- Marginal tips on the calculator, today's technology, Internet sites, and applications are included.
- Pocket experiments are inserted where appropriate to suggest simple experiments that students might perform to reinforce physical principles.
- Careful selection and use of more than 800 drawings illustrate concepts in a way that complements the homework assignments.
- The informative writing style enhances learning objectives.
- Detailed summaries of important concepts and formulas appear at the end of every chapter.
- More than 250 in-text examples provide immediate reinforcement of theory and provide easy-to-understand problem-solving techniques.
- More than 500 thought-provoking questions are offered at the end of chapters to stimulate and enhance conceptual understanding.
- More than 1750 carefully selected problems, ranging from simple to moderate to complex are provided, including nearly 250 new Critical Thinking Questions.

ACTIVITIES MANUAL

An activities manual is available. Containing two parts, the manual provides a concise overview of each chapter, including definitions, self-tests, and a sample of worked-out problems. Unit 1 of the manual includes chapter summaries, multiple-choice and true-false tests, and selected solutions to problems. Unit 2 provides laboratory experiments that are referenced to the text.

INSTRUCTOR'S MANAGEMENT SYSTEM

Physics, Sixth Edition, is supported by a comprehensive *Instructor's Management System.* It contains solutions to chapter-ending questions and problems, information on Tech Prep, an introduction on using the Internet in classroom teaching, and suggested answers to experiments contained in the Activities Manual.

This manual also contains a CD-ROM with a new PowerPoint presentation and a comprehensive Windows-based test bank. The PowerPoint slides provide instructors with a unique feature that identifies key chapter points chapter by chapter and section by section. This is an excellent way to introduce section or chapter content. The PowerPoint presentation can be projected on screen and used as a pretest review.

Paul E. Tippens

Acknowledgments

As with any work, many individuals contributed significantly to the process. The author is indebted to Dr. Russell Patrick, Southern College of Technology, Scott J. Tippens, Professor of Electrical Engineering Technology, Southern Polytechnic State University, Dr. Sam Nalley of Chattanooga State Technical Community College, Dr. Charles A. Schuler, California State University of Pennsylvania, and David Heiserman of Sweethaven Publishing Services for their assistance in preparing this edition. The author is also grateful to Shaikh Ali, City College of Fort Lauderdale, Miles L. Kirkhuff, Lincoln Technical Institute, Allentown, Pennsylvania, Henry Merrill, Fox Valley Technical College, Dr. Fred Einstein, County College of Morris, Randolph, New Jersey, Ajay Raychaudhuri, Seneca College of Arts and Technology, Northyork, Ontario, Ron Uhey, ITT Tech Institute, Fort Wayne, Indiana, Bob Tyndall, Forsyth Technical Community College, Winston-Salem, North Carolina, and Cliff Wurst, Motlow State Community College, Tullahoma, Tennessee.

Thanks also to John J. Beck, executive editor, and Linda D. Jefferson, production editor. As always, the author thanks the many users of previous editions for their critical reviews and comments. Comments, suggestions, and criticisms from instructors and students are always welcome.

PHYSICS

Mechanics

Introduction

A knowledge of physics is essential to an understanding of our world. No other science has been as active in revealing the causes and effects of natural events. A casual glance at our past demonstrates a continuum of experiment and discovery ranging from early measurements of gravity to later conquests of space. By studying objects at rest and in motion, scientists have been able to derive fundamental laws for many applications in mechanical engineering. The investigation of electricity and magnetism produced new sources of energy and new methods of distributing power for the use of humanity. An understanding of the physical principles that govern the production of heat, light, and sound has added countless applications that have served to make us more comfortable and more able to cope with our environment.

It is difficult to imagine a single product available today that does not represent an application of some physical principle. This means that regardless of your career choice, you will need to understand physics in some way. Granted, there are some occupations and professions that do not require the depth of understanding necessary for engineering applications, but all fields of work utilize and apply these concepts. With a thorough understanding of mechanics, heat, sound, and electricity, you carry with you the building blocks for almost any career. If you find it necessary or desirable to change careers either before or after graduation, you will be able to draw from a general base of science and mathematics. By taking this course seriously and by devoting an unusual amount of time and energy to it, you will have less trouble in the future. In your later coursework, and on the job, you will be riding the crest of the wave instead of merely staying afloat in an angry sea.

WHAT IS PHYSICS?

Even if you have previously taken courses in high school physics, you probably still have a rather "fuzzy" idea of what *physics* really means and how it might differ, for example, from *science*. For our purpose, the sciences can be divided between *biological* and *physical*. The biological sciences deal with living things. The physical sciences deal primarily with the nonliving sides of nature.

Physics can be defined as the science that investigates the fundamental concepts of matter, energy, and space and the relationships between them.

In terms of this broad definition, there are no clear boundaries between the many physical sciences. This is evident from the overlapping fields of biophysics, chemical physics, astrophysics, geophysics, electrochemistry, and so forth.

The goal of this textbook is to provide a basic introduction to the world of physics. The emphasis is on applications, and the broad field of physics will be narrowed to the essential concepts that underlie all technical knowledge. You will study mechanics, heat, light, sound, electricity, and atomic structure. The most basic of these topics, and probably the most important for beginning students, is mechanics.

Mechanics is concerned with the position (statics) and motion (dynamics) of matter in space. **Statics** represents the study of physics associated with bodies at rest. **Dynamics** is concerned with a description of motion and its causes. In each case, the engineer or technician is concerned with measuring and describing physical quantities in terms of their cause and effect.

An engineer, for example, uses physical principles to determine which type of bridge structure will be the most efficient for a given situation. The concern is for the *effect* of forces. If the completed bridge fails, the *cause* of failure must be analyzed to apply this knowledge to future construction. It is important to note that by *cause* the scientist means the sequence of physical events leading to an *effect*.

WHAT PART IS PLAYED BY MATHEMATICS?

Mathematics serves many purposes. It is philosophy, art, metaphysics, and logic. These values are subordinate, however, to its main value as a tool for the scientist, engineer, or technician. One of the rewards of a first course in physics is the growing awareness of the relevance of mathematics. A study of physics reveals specific applications of basic mathematics.

Suppose you wish to predict how long it takes to stop a car traveling at a given speed. You might record the initial speed, the change in speed, and the distance and time required to stop the car. When all these facts are recorded, the data might be used to establish a tentative relationship. We cannot do this without the tools of mathematics.

If you completed such an experiment, your measurements plus a knowledge of mathematics would lead you to the following relationship:

$$s = vt + \frac{1}{2}at^2$$

where s = stopping distance
v = initial speed
t = stopping time
a = rate of change in speed

This statement is a *workable hypothesis*. From this equation we can predict the stopping distance for other vehicles. When it has been used long enough for us to be reasonably sure that it is true, we call it a *scientific theory*. In other words, any scientific theory is just a workable hypothesis that has stood the test of time.

Thus, we can see that mathematics is useful in deriving formulas that describe physical events accurately. Mathematics plays an even larger role in solving such formulas for specific quantities.

For example, in the preceding formula, it would be a relatively simple matter to find values for s, v, and a when the other quantities are given, but the solution for t involves a more specialized knowledge of mathematics. How easily you derive or solve a theoretical relationship depends on your background in mathematics.

USING A CALCULATOR

Your calculator can give a wrong answer! A calculator performs only the operations in the order that you tell it to. Let's look at this problem.

$$Y = \frac{6 + 3}{2}$$

You might enter this into the calculator: ⑥ ⊞ ③ ⟋ ② ⊟. This would yield 7.5, which is incorrect. To get the *correct answer*, first group the ⑥ ⊞ ③ together with ⊂ ⊃. Entering this into the calculator like this ⊂ ⑥ ⊞ ③ ⊃ ⟋ ② ⊟ would give you 4.5. This is correct. Keep in mind when entering a problem into the calculator that fraction bars, radicals, and exponents must all be grouped correctly. Whenever possible, use parentheses to group terms in an expression. If your calculator does not have parentheses, calculate the expressions from within the group first and then perform the functions on that answer.

A review of the mathematics required for this text is presented in Chapter 2. If you are unfamiliar with any of the topics discussed, you should study this chapter carefully. Pay particular attention to the sections on powers of 10, literal equations, and trigonometry. Skill in applying the tools of mathematics will largely determine your success in any physics course.

HOW SHOULD I STUDY PHYSICS?

Reading technical material is different from other reading. Attention to specific meanings of words must be given if the material is to be understood. Graphs, drawings, charts, and photographs are often included in technical literature. They are always helpful and may even be essential to the description of physical events. You should study them thoroughly so that you understand the principles clearly.

Much of what you learn will be from classroom lectures and experiments. The beginning student often asks: How can I concentrate fully on the lecture and at the same time take accurate notes? Of course, it may not be possible to understand fully all the concepts presented and still take complete notes. You must learn to note only the significant portions of each lesson. Make sure you listen attentively to the explanation of various topics. Learn to recognize key words like *work*, *force*, *energy*, and *momentum*.

Adequate preparation before class should give you a good idea of which portions of the lecture are covered in the text and which are not. If a problem or definition is in the text, it is usually better to jot down a key word and concentrate fully on what the instructor is saying. These notes can be expanded later.

Each student who enters a beginning physics class has the prerequisites and ability to pass the course. There must be other reasons for failure, perhaps lack of motivation, heavy class loads, outside jobs, sickness, or personal problems. The following advice comes from seasoned instructors who have a history of success with beginning physics students.

- *The ultimate responsibility for learning rests with the student.* The instructor is only a facilitator, the college is only a campus, and the text is only a book. Come to class each day, on time and prepared for the scheduled instruction. Study the material in advance and jot down questions to ask the instructor.
- *Learning rarely occurs in the classroom.* For physics, learning most often comes from working problems. At least, make an attempt to work every assigned problem. Seek the help of others, but participate in the process. Work groups are effective only if participants have active roles.
- *Timely learning is efficient learning.* It is better to study an hour each day of the week, than to study for 20 hours every weekend. After each lecture or recitation, use the very next free period to reinforce what was taught by attempting example problems. The longer you wait, the more you forget, and the more you waste your time. If you wait until the weekend, often an hour or more is needed just to review and reconstruct the lecture from notes. Cramming for tests will not work. Instead, review problems that you have already worked and attempt to work similar problems in the textbook.

http://www2.physics.umd
.edu/~redish/
Did you know that countries around the world have put pictures of physicists on their money? To see a sampling, visit this Internet site.

- *Brush up on basic skills.* Chapter 2 (on technical mathematics) stresses skills that may be rusty or weak. Make sure that you understand these topics well.
- *Study the syllabus.* Know what will be covered on exams, when they will be given, and how they will be used to determine your grade.
- *Find a class partner and obtain this person's phone number.* Use a *buddy system* so that each of you can notify the other of class or lab activities that one of you has missed. Have this person pick up extra handouts and instructions given in your absence.
- *Organization is the key to effective learning.* Keep a loose-leaf filler notebook with tabbed sections: Handouts, Notes, Problems, Graded Exams, Graded Labs.

Technical Mathematics

OBJECTIVES

After completing this chapter, you should be able to:

1. Demonstrate your ability to add, subtract, multiply, and divide technical measurements.
2. Solve simple formulas for any quantity appearing in the formula and evaluate by substitution.
3. Solve simple problems that require operations with exponents and radicals.
4. Perform common mathematical operations in scientific notation.
5. Construct a graph from given technical data and interpret new information from the graph.
6. Apply the elementary rules of geometry to determine unknown angles in given situations.

It is often frustrating to open a physics book and see that it begins with mathematics. Naturally, you want to learn only those things for which you see a definite need. You want to make measurements, to work with machines or engines, to get your hands on something, or, at the very least, to know that your time is not wasted. Depending on your previous experience, you may omit much or all of the first chapter at the discretion of your instructor. Keep in mind that fundamentals are important and that some skills in mathematics are essential. You may have a complete understanding of force, mass, energy, and electricity as concepts. But you may not be able to apply the concepts to your work because of weaknesses in basic mathematics. Mathematics is the language of physics. In this book, an effort has been made to make that language as simple and as relevant as is necessary.

In any industrial or technical occupation, we are concerned with physical measurements of some kind. It might be the length of a board, the area of a sheet of metal, the number of bolts to be ordered, the stress on an aircraft wing, or the pressure in an oil tank. The only way we can make sense out of such data is through the use of numbers and symbols. Mathematics provides the necessary tools for organizing the data and for predicting outcomes. For example, the formula $F = ma$ expresses the relationship between an applied force F and the acceleration a that the force produces. The quantity m is a symbol that represents the mass of an object (a measure of the amount of matter it contains). Through appropriate mathematical steps, we can use formulas such as this to predict future events. However, a general knowledge of algebra and geometry is required in many instances. This chapter provides you with a review of a few essential concepts in mathematics. Chapter sections may be assigned or omitted at the discretion of your instructor.

2-1 SIGNED NUMBERS

Frequently, it is necessary to work with negative numbers as well as positive numbers. For example, a temperature of $-10°C$ means 10 degrees "below" a zero reference point, and $24°C$ refers to a temperature 24 degrees "above" zero (see Fig. 2-1). The numbers refer to the *magnitude* of the temperature, and the plus or minus sign refers to the *direction* from zero. The minus sign in $-10°C$ does not indicate a lack of temperature; it means that the temperature is less than zero. The number 10 in $-10°C$ describes how far the temperature is from zero; the minus sign is necessary to indicate the direction from zero.

The value of a number without its sign is called its *absolute value*. In other words, if we disregard the signs of $+7$ and -7, the value is the same. Each number is 7 units from zero. The absolute value of a number is indicated by vertical bar symbols. The number $+7$ does not equal the number -7, but $|+7|$ does equal $|-7|$. When arithmetic operations are performed with signed numbers, the absolute values are used.

Plus and minus signs are also used to indicate arithmetic operations, for example:

$7 + 5$ means "add the number $+5$ to the number $+7$"
$7 - 5$ means "subtract the number $+5$ from the number $+7$"

If we wish to indicate the addition or subtraction of negative numbers, parentheses are helpful:

$(+7) + (-5)$ means "add the number -5 to the number $+7$"
$(+7) - (-5)$ means "subtract the number -5 from the number $+7$"

When signed numbers are added, it is useful to recall the following rule:

Addition rule: To add two numbers of like sign, we add the absolute values of the numbers and give the sum the common sign. To add two numbers of unlike sign, we find the difference of their absolute values and give the result the sign of the number of larger value.

Figure 2-1

Consider the following examples:

$$(+6) + (+2) = + (6 + 2) = +8$$
$$(-6) + (-2) = - (6 + 2) = -8$$
$$(+6) + (-2) = + (6 - 2) = +4$$
$$(-6) + (+2) = - (6 - 2) = -4$$

Now, let's examine the procedure for subtraction. Whenever a number has another subtracted from it, we change the sign of the second number and then add it to the first, using the addition rule. In the expression $7 - 5$, the number $+5$ is to be subtracted from the number $+7$. The subtraction is accomplished by changing $+5$ to -5 and then adding the two numbers of unlike sign: $(+7) + (-5) = +(7 - 5) = +2$.

> **Subtraction rule: To subtract one signed number *b* from another signed number *a*, we change the sign of *b* and then add it to *a*, using the addition rule.**

Consider the following examples:

$$(+8) - (+5) = 8 - 5 = 3$$
$$(+8) - (-5) = 8 + 5 = 13$$
$$(-8) - (+5) = - 8 - 5 = -13$$
$$(-8) - (-5) = - 8 + 5 = -3$$

EXAMPLE 2-1

The velocity of an object is considered positive when it is moving upward and negative when it is moving downward. What is the change in velocity of a ball if it strikes the floor at 12 meters per second (m/s) and rebounds upward at 7 m/s? Refer to Fig. 2-2.

+7 m/s

−12 m/s

+

Figure 2-2

Solution

The initial velocity is -12 m/s because the ball is moving downward. Later its velocity is $+7$ m/s because it is moving upward. We find the change in velocity by subtracting the initial velocity from the final velocity. Thus,

$$\text{Change in velocity} = \text{final velocity} - \text{initial velocity}$$
$$= (+7 \text{ m/s}) - (-12 \text{ m/s})$$
$$= 7 \text{ m/s} + 12 \text{ m/s} = 19 \text{ m/s}$$

Without understanding signed numbers, we might have guessed that the change in speed was only 5 m/s (12 − 7). A moment's thought, however, makes us realize that the speed must first decrease to zero (a change of 12 m/s) and then attain a speed of 7 m/s in the opposite direction (an additional change of 7 m/s).

When two or more numbers are to be multiplied, each number is called a *factor* and the result is the **product.** We can now state the multiplication rule for signed numbers:

> **Multiplication rule. If two factors have like signs, their product is positive. If two factors have unlike signs, their product is negative.**

Examples are

$$(+2)(+3) = +6 \qquad (-3)(-4) = +12$$
$$(-2)(+3) = -6 \qquad (-3)(+4) = -12$$

An extension of the multiplication rule is often helpful for products resulting from several factors. Rather than multiplying a series of factors two at a time, we might recall that

> **The product will be positive if all factors are positive or if there is an even number of negative factors. The product will be negative if there is an odd number of negative factors.**

Consider the following examples:

$$(-2)(+2)(-3) = +12 \qquad \text{(two negative factors—even)}$$
$$(-2)(+4)(-3)(-2) = -48 \qquad \text{(three negative factors—odd)}$$
$$(-3)^3 = (-3)(-3)(-3) = -27 \qquad \text{(three negative factors—odd)}$$

Notice that in the last example a superscript 3 was used to indicate the number of times the number − 3 was to be taken as a factor. The superscript 3 written in this manner is called an *exponent.*

When two numbers are to be divided, the number being divided is called the *dividend.* The number divided into the dividend is called the *divisor.* The result of division is called the *quotient.* The rule for dividing signed numbers is as follows:

> **Division rule. The quotient of two numbers of like sign is positive, and the quotient of two numbers of unlike sign is negative.**

For example,

$$(+2) \div (+2) = +1 \qquad (-4) \div (-2) = +2$$

$$\frac{+4}{-2} = -2 \qquad\qquad \frac{-4}{+2} = -2$$

If either the numerator or the denominator of a fraction contains two or more factors, the following rule is also useful:

The quotient is negative if the total number of negative factors is odd; otherwise, the quotient is positive.

For example,

$$\frac{(-4)(3)}{2} = -6 \qquad \text{odd}$$

$$\frac{(-2)(-2)(-3)}{(2)(-3)} = +2 \qquad \text{even}$$

You should practice the application of all the rules in this entire section. It is a serious mistake to assume that you understand these concepts without adequate proof. A major source of errors in physics problems can be traced to signed numbers.

2-2 ALGEBRA REVIEW

physics.nist.gov/cuu/
Constants/index.html
For a listing of the values of
the constants and related
information, see this Internet
site (but do not type www.
before the site).

Algebra is really a generalization of arithmetic in which letters are used to replace numbers. For example, we will learn that the space occupied by some objects (their volume) can be calculated by multiplying the length by the width by the height. By assigning letters to each of these measurements, we can establish a general *formula*, such as

$$\text{Volume} = \text{length} \times \text{width} \times \text{height}$$
$$V = l \cdot w \cdot h \qquad\qquad \text{2-1}$$

The advantage of formulas is that they work for any number of situations. Given the length, width, and height of any rectangular solid, we can use Eq. (2-1) to calculate its volume. If we wish to determine the volume of a rectangular metal block, we need only *substitute* the proper numbers into the formula.

EXAMPLE 2-2

Calculate the volume of a solid whose length is 6 inches (in.), whose width is 4 in., and whose height is 2 in.

Solution
Substitution yields

$$V = lwh$$
$$= (6 \text{ in.})(4 \text{ in.})(2 \text{ in.})$$
$$= 48 \ (\text{in.} \times \text{in.} \times \text{in.}) = 48 \text{ in.}^3$$

The treatment of units that results in volume being expressed as cubic inches is discussed later. For now, you should concentrate on the substitution of numbers.

When numbers are substituted for letters in a formula, it is very important to insert the proper sign of the number. Consider the following formula:

$$P = c^2 - ab$$

Suppose that $c = +2$, $a = -3$, and $b = +4$. Remember that plus and minus signs in formulas do not apply to any of the numbers that might be substituted. In this example, we have

$$
\begin{aligned}
P &= (c)^2 - (a)(b) \\
&= (+2)^2 - (-3)(+4) \\
&= 4 + 12 = 16
\end{aligned}
$$

It is easy to see how confusing a sign in the formula with the sign of a substituted number can result in an error.

It is frequently necessary to solve a formula or an equation for some letter that is only a part of the formula. For example, suppose we wanted to give a formula for the length of a rectangular solid in terms of its volume, height, and width. The letters in $V = lwh$ would need to be rearranged so that the letter l would appear by itself on the left side. Formula rearrangement is not difficult if we recall several rules for working with equations.

Basically, an equation is a mathematical statement that two expressions are equal. For example,

$$2b + 4 = 3b - 1$$

is an equation. In this case, it is evident that the letter b represents the *unknown* quantity. If we substitute $b = 5$ into both sides of this equation, we obtain $14 = 14$. Thus, $b = 5$ is the *solution* to the equation.

We can obtain solutions for equalities by performing the same operations on each side of an equation. Consider the equality $4 = 4$. If we add, subtract, multiply, or divide the number 2 into both sides, we do not change the fact that the two sides are equal. We *do* increase or decrease the magnitude of each side, but the equality remains. (You should verify this statement for the equality $4 = 4$.) Note also that taking the square or square root of each side does not disturb the equality. By performing a series of identical operations to each side of an equation, we can eventually obtain an equality with one letter by itself on the left side. In this case, we have *solved* the equation for that letter.

EXAMPLE 2-3

Solve the following equation for m:

$$3m - 5 = m + 3$$

Solution

The point is to get m by itself on one side of the equals sign and a number by itself on the other side. We first add $+5$ to both sides, then subtract m from both sides:

$$3m - 5 + 5 = m + 3 + 5$$
$$3m = m + 8$$
$$3m - m = m + 8 - m$$
$$2m = 8$$

Finally, we divide both sides by 2:

$$\frac{2m}{2} = \frac{8}{2}$$
$$m = 4$$

To check this answer, we substitute $m = 4$ into the original equation and obtain $7 = 7$. This shows that $m = 4$ is the solution.

In formulas, the solution to an equation may also be expressed in terms of letters. For example, the literal equation

$$ax - 5b = c$$

might be solved for x in terms of a, b, and c. In cases such as this one, we decide in advance which letter is to be the "unknown." In our example, we will choose x. The remaining letters are treated as though they were known numbers. Adding $5b$ to both sides, we have

$$ax - 5b + 5b = c + 5b$$
$$ax = c + 5b$$

Now, we can divide both sides by a to obtain

$$\frac{ax}{a} = \frac{c + 5b}{a}$$

$$x = \frac{c + 5b}{a}$$

which is the solution for x. The values for a, b, and c in a given situation can be substituted to find a particular value for x.

EXAMPLE 2-4

The volume of a right circular cone is expressed by the formula

$$V = \frac{\pi r^2 h}{3} \qquad \text{2-2}$$

What is the height of such a cone of radius $r = 3$ in. and $V = 81$ cubic inches (in.^{3})? (Assume that $\pi = 3.14$.)

Solution

First, we solve the formula for h in terms of r and V. Multiplying both sides by 3 gives

$$3V = \pi r^2 h$$

Dividing both sides by πr^2 yields

$$\frac{3V}{\pi r^2} = h \qquad \text{or} \qquad h = \frac{3V}{\pi r^2}$$

Substitution for V, π, and r gives

$$h = \frac{3(81 \text{ in.}^3)}{(3.14)(3 \text{ in.})^2} = \frac{243 \text{ in.}^3}{28.26 \text{ in.}^2} = 8.60 \text{ in.}$$

The height of the cylinder is 8.60 in.

2-3 EXPONENTS AND RADICALS (OPTIONAL)

It is often necessary to multiply the same quantity a number of times. A shorthand method of indicating the number of times a quantity is taken as a factor uses a superscript number called an **exponent.** This notation works according to the following scheme:

For any number a:

$$a = a^1$$
$$a \times a = a^2$$
$$a \times a \times a = a^3$$
$$a \times a \times a \times a = a^4$$

For the number 2:

$$2 = 2^1$$
$$2 \times 2 = 2^2$$
$$2 \times 2 \times 2 = 2^3$$
$$2 \times 2 \times 2 \times 2 = 2^4$$

The powers of the number a are read as follows: a^2 is read "a squared," a^3 is read "a cubed," and a^4 is read "a to the fourth power." More generally, we speak of a^n as "a to the nth power." In these examples, the letter a is called the **base,** and the superscript numbers 1, 2, 3, 4, and n are called the **exponents.**

We will review several rules to follow in working with exponents.

Rule 1. When two quantities of the same base are multiplied, their product is obtained by adding the exponents algebraically:

$$(a^m)(a^n) = a^{m+n} \qquad \text{Multiplication Rule} \quad \textbf{2-3}$$

Examples:

$$(2^4)(2^3) = 2^{4+3} = 2^7$$
$$y^8 y^6 = y^{14}$$
$$x^2 x^5 y^3 x^3 = x^{2+5+3} y^3 = x^{10} y^3$$

Rule 2. When a is not zero, a negative exponent may be defined by either of the following expressions:

$$a^{-n} = \frac{1}{a^n} \qquad \text{and} \qquad a^n = \frac{1}{a^{-n}} \qquad \text{Negative Exponent} \quad \textbf{2-4}$$

USING A CALCULATOR

To calculate a root using a calculator, find the symbol $\sqrt[x]{\ }$. First, press the number that should be where the x symbol is. Then press the $\sqrt[x]{\ }$ button. Then press the number that goes under the radical. Then press equals. To be sure that you've got it right, try it out with the fourth root of 16.

The exact procedure varies with different calculators.

Examples:

$$3^{-4} = \frac{1}{3^4} = \frac{1}{81} \qquad 10^2 = \frac{1}{10^{-2}}$$

$$a^{-5} = \frac{1}{a^5} \qquad \frac{x^{-3}y^2}{a^{-4}b^3} = \frac{a^4y^2}{x^3b^3}$$

Rule 3. Any quantity raised to the power of zero is equal to 1:

$$a^0 = 1 \qquad\qquad \text{Zero Exponent} \quad \textbf{2-5}$$

Examples:

$$x^3y^0 = x^3 \qquad (x^3y^2)^0 = 1$$

Rule 4. The quotient of two nonzero quantities of the same base is found by taking the algebraic difference of their exponents:

$$\frac{a^m}{a^n} = a^{m-n} \qquad\qquad \text{Division} \quad \textbf{2-6}$$

Examples:

$$\frac{2^3}{2} = 2^{3-1} = 2^2 \qquad \frac{2^5}{2^7} = 2^{5-7} = 2^{-2} = \frac{1}{2^2}$$

$$\frac{a^{-3}}{a^{-5}} = a^{-3-(-5)} = a^{-3+5} = a^2$$

Rule 5. When a quantity a^m is raised to the power n, the exponents are multiplied:

$$(a^m)^n = a^{mn} \qquad\qquad \text{Power of a Power} \quad \textbf{2-7}$$

Examples:

$$(2^2)^3 = 2^{2 \cdot 3} = 2^6 \qquad (2^{-3})^2 = 2^{-6} = \frac{1}{2^6}$$

$$(a^2)^4 = a^8 \qquad (a^2)^{-4} = a^{-8} = \frac{1}{a^8}$$

Rule 6. The power of a product and of a quotient is obtained by applying the exponent to each of the factors:

$$(ab)^n = a^n b^n \qquad \left(\frac{a}{b}\right)^n = \frac{a^n}{b^n} \qquad\qquad \textbf{2-8}$$

Examples:

$$(2 \cdot 3)^2 = 2^2 \cdot 3^2 = 4 \cdot 9 = 36$$

$$(ab)^3 = a^3 b^3$$

$$(ab^2)^3 = a^3(b^2)^3 = a^3 b^6$$

$$\left(\frac{ax^3}{y^2}\right)^4 = \frac{a^4 x^{12}}{y^8}$$

If $a^n = b$, then not only is b equal to the nth power of a, but also by definition a is said to be the nth *root* of b. In general, this fact is expressed by using a *radical* ($\sqrt{}$):

$$\sqrt[n]{b} \qquad n\text{th root of } b$$

Consider the following statements:

$$2^2 = 4 \text{ means that 2 is the } \textit{square root} \text{ of 4, or } \sqrt{4} = 2$$
$$2^3 = 8 \text{ means that 2 is the } \textit{cube root} \text{ of 8, or } \sqrt[3]{8} = 2$$
$$2^5 = 32 \text{ means that 2 is the } \textit{fifth root} \text{ of 32, or } \sqrt[5]{32} = 2$$

A radical may also be expressed by using a fractional exponent. In general, we may write

$$\sqrt[n]{b} = b^{1/n}$$

For example,

$$\sqrt[3]{8} = 8^{1/3} \qquad \text{or} \qquad \sqrt{10} = 10^{1/2}$$

There are two additional rules to follow when you are working with radicals.

Rule 7. The nth root of a product is equal to the product of the nth roots of each factor:

$$\sqrt[n]{ab} = \sqrt[n]{a}\,\sqrt[n]{b} \qquad \text{Roots of a Product} \quad \textbf{2-9}$$

Examples:

$$\sqrt{4 \cdot 16} = \sqrt{4}\,\sqrt{16} = 2 \cdot 4 = 8$$
$$\sqrt[5]{ab} = \sqrt[5]{a}\,\sqrt[5]{b}$$

Rule 8. The roots of a power are found by using the definition of fractional exponents:

$$\sqrt[n]{a^m} = a^{m/n} \qquad \text{Roots of Powers} \quad \textbf{2-10}$$

Examples:

$$\sqrt[3]{2^9} = 2^{9/3} = 2^3 = 8$$

$$\sqrt{10^{-4}} = 10^{-4/2} = 10^{-2} = \frac{1}{10^2}$$

$$\sqrt{4 \times 10^8} = \sqrt{4}\,\sqrt{10^8} = 2(10)^{8/2} = 2 \times 10^4$$

$$\sqrt[3]{8 \times 10^{-6}} = \sqrt[3]{8}(10)^{-6/3} = 2 \times 10^{-2}$$

Most of the problems in this textbook will require only a limited understanding of the above rules. Squares, cubes, square roots, and cube roots are encountered most often. However, a good understanding of the rules for exponents and radicals is helpful.

2-4 SCIENTIFIC NOTATION

In scientific work, one frequently encounters very large or very small numbers. For example, a machinist may measure the thickness of a thin metal sheet to be 0.00021 in. Similarly, an engineer may encounter an area of 1,200,000 ft^2 for an airport runway. It is convenient to be able to express these numbers, respectively, as 2.1×10^{-4} in. and 1.2×10^6 ft^2. Powers of 10 are used to keep track of the decimal point, and we are not forced to carry a large number of zeros along with our calculations. The expression of any number as a number between 1 and 10 times an integral power of 10 is called *scientific notation.*

Electronic calculators are often equipped with an $\boxed{\text{EE}}$ or $\boxed{\text{EXP}}$ button which allows even beginning students to use scientific notation in many computations. You are virtually certain to encounter scientific notation, even if your job does not require frequent use of numbers expressed in this form. Review the manual which comes with your calculator to learn how to work with powers of 10 on the calculator.

Consider the following multiples of 10 and examples of their use in scientific notation:

$$
\begin{aligned}
0.0001 &= 10^{-4} & 2.34 \times 10^{-4} &= 0.000234 \\
0.001 &= 10^{-3} & 2.34 \times 10^{-3} &= 0.00234 \\
0.01 &= 10^{-2} & 2.34 \times 10^{-2} &= 0.0234 \\
0.1 &= 10^{-1} & 2.34 \times 10^{-1} &= 0.234 \\
1 &= 10^0 & 2.34 \times 10^0 &= 2.34 \\
10 &= 10^1 & 2.34 \times 10^1 &= 23.4 \\
100 &= 10^2 & 2.34 \times 10^2 &= 234.0 \\
1000 &= 10^3 & 2.34 \times 10^3 &= 2340.0 \\
10,000 &= 10^4 & 2.34 \times 10^4 &= 23,400.0
\end{aligned}
$$

To write a number larger than 1 in scientific notation, you must determine the number of times the decimal point must be moved to the left in order to arrive at the shorthand notation. Examples are

$$
\begin{aligned}
467 &= 4{\,}6{\,}7. = 4.67 \times 10^2 \\
30 &= 3{\,}0. = 3.0 \times 10^1 \\
35,700 &= 3{\,}5{\,}7{\,}0{\,}0. = 3.57 \times 10^4
\end{aligned}
$$

Any decimal number less than 1 can be written as a number between 1 and 10 times a *negative* power of 10. The negative exponent in this case is the number of times the decimal point is moved to the right. This is always one more than the number of zeros that separate the first digit from the decimal. Examples are

$$
\begin{aligned}
0.24 &= 0.2{\,}4 = 2.4 = 10^{-1} \\
0.00327 &= 0.0{\,}0{\,}3{\,}2{\,}7 = 3.27 \times 10^{-3} \\
0.0000469 &= 0.0{\,}0{\,}0{\,}0{\,}4{\,}6{\,}9 = 4.69 \times 10^{-5}
\end{aligned}
$$

To transfer from scientific notation to decimal notation, the procedure is simply reversed.

By recalling the laws of exponents, scientific notation can be used in multiplication and division of very small or very large numbers. When two numbers are multiplied, the exponents of 10 are added. For example, 200×4000 may be written $(2 \times 10^2)(4 \times 10^3) = (2)(4) \times (10^2)(10^3) = 8 \times 10^5$. Other examples are

$$2200 \times 40 = (2.2 \times 10^3)(4 \times 10^1) = 8.8 \times 10^4$$
$$0.0002 \times 900 = (2.0 \times 10^{-4})(9.0 \times 10^2) = 18 \times 10^{-2}$$
$$1002 \times 3 = (1.002 \times 10^3)(3 \times 10^0) = 3.006 \times 10^3$$

Similarly, when one number is divided by another number, the exponent of 10 in the denominator is subtracted from the exponent of 10 in the numerator. Examples are

$$\frac{7000}{35} = \frac{7 \times 10^3}{3.5 \times 10^1} = \frac{7.0}{3.5} \times 10^{3-1} = 2.0 \times 10^2$$

$$\frac{1200}{0.003} = \frac{1.2 \times 10^3}{3.0 \times 10^{-3}} = \frac{1.2}{3.0} \times 10^{3-(-3)} = 0.4 \times 10^6$$

$$\frac{0.008}{400} = \frac{8 \times 10^{-3}}{4 \times 10^2} = \frac{8}{4} \times 10^{-3-2} = 2.0 \times 10^{-5}$$

When two numbers in scientific notation are added, care must be taken to adjust all numbers to be added so that they have identical powers of 10. Examples are

$$2000 + 400 = 2 \times 10^3 + 0.4 \times 10^3 = 2.4 \times 10^3$$
$$0.006 - 0.0008 = 6 \times 10^{-3} - 0.8 \times 10^{-3} = 5.2 \times 10^{-3}$$
$$4 \times 10^{-21} - 6 \times 10^{-20} = 0.4 \times 10^{-20} - 6 \times 10^{-20} = -5.6 \times 10^{-20}$$

Scientific calculators automatically make the necessary adjustments when adding and subtracting such numbers.

Scientific notation and powers of 10 take on an important and special significance when you work with metric units. In the next chapter, you will see that multiples of 10 are used to define a number of units in the metric system. For example, one kilometer is defined as a thousand (1×10^3) meters, and one millimeter is defined as one-thousandth (1×10^{-3}) of a meter.

2-5 GRAPHS

Frequently, it is desirable to show the relationship between two quantities in the form of a graph. For example, we know that a car traveling at constant speed covers the same distances every minute (min) that it travels. We might record the distance traveled in feet at particular times as follows:

Distance, ft	200	400	600	800	1000
Time, min	1	2	3	4	5

Along the bottom of a sheet of graph paper, we might establish a time scale, maybe letting each division equal 1 min. On the left side of the paper, we can establish a distance scale. The scale selected should fill the graph paper. (This makes it easier to locate points on the graph.) Simple scale divisions are 1 division = 1, 2, or

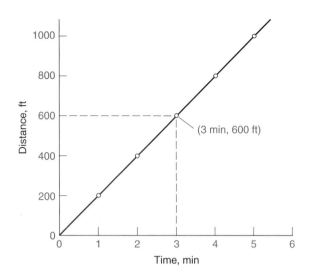

Figure 2-3 A graph of distance as a function of time (a *direct* relationship).

Math Tip

When graphing collected data, make sure that you have enough data points to accurately draw the curve. For example, when a voltage output gives you the following points, what would be the simplest graph for the data?

Seconds	Volts
1	1
2	0
3	1
4	0

Would the graph be a smooth sine wave or a sawtooth wave? Using only the data we are given, we cannot answer this question. We must take more voltage readings at finer time intervals until a clear pattern is established. For the final graph, always smooth out the curve that best fits the data. Do not simply connect the dots.

5 times some power of 10. Good examples are 1 division $= 1 \times 10^3 = 1000$, or 1 division $= 2 \times 10^0 = 2$, or 1 division $= 5 \times 10^{-2} = 0.05$. Awkward scale divisions such as 3 divisions $= 100$ ft should be avoided because they hinder the location of points. In our example, we let each division represent 100 ft. The data may then be plotted on the graph, as shown in Fig. 2-3. Each point on the horizontal line has a corresponding point on the vertical line. For example, the distance traveled after 3 min is 600 ft. Notice that when the points are connected, a straight line results.

When a graph of one quantity versus another results in a straight line, there is a *direct relationship*. In this example, the distance traveled is directly proportional to the time. When one quantity increases, the other increases by the same proportion. Doubling the time elapsed doubles the distance traveled.

Indirect or *inverse relationships*, in which an increase of one quantity results in a proportional decrease in another quantity, are also encountered. If we were to decrease the speed of an automobile, we would find that greater and greater time intervals would be required to cover the same distance. Suppose we measure the time in seconds (s) required to travel a distance of 1 kilometer (km) [0.621 mile (mi)] at speeds of 20, 40, 60, 80, and 100 kilometers per hour (km/h). The following data are recorded:

Speed, km/h	20	40	60	80	100
Time, s	180	90	60	45	36

A graph of these data is shown in Fig. 2-4. Notice that the graph of an inverse relationship is not a straight line but a curved one.

A graph can be used to obtain information that was not available before the graph was constructed. For example, in Fig. 2-4, we can learn that a time of 120 s would be required to cover the distance if our speed were 30 km/h.

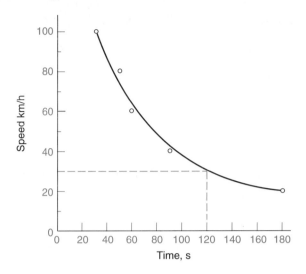

Figure 2-4 A graph of the time required to cover a distance of 1 km as a function of speed (an *indirect* relationship).

2-6 GEOMETRY

In this brief review, it is assumed that you already have an understanding of the concept of a point and a line. Other important concepts are reviewed only to the extent that they may be needed in physics problems. A lengthy review of the many possible theorems in this field is unnecessary. We will begin with angles and lines.

The **angle** between two straight lines is defined by drawing a circle with its center at the point of intersection (see Fig. 2-5a). The magnitude of angle A is proportional to the fraction of a complete circle which lies between the two lines. Angles are measured in **degrees,** as defined in Fig. 2-5b. One degree (°) is a portion of a circle equal to 1/360 of a full revolution (rev). Thus, there are 360° in 1 rev:

$$1° = \frac{1}{360} \text{ rev} \qquad 1 \text{ rev} = 360° \qquad \qquad \textbf{2-11}$$

A special name is given to the angle which forms one-fourth of 1 rev, or 90°. It is called a **right angle** and is illustrated in Fig. 2-6a. When two straight lines meet

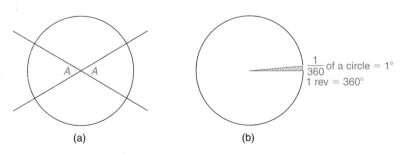

Figure 2-5 An angle is a fraction of a complete circle. One degree is a portion of a circle equal to 1/360 of a full revolution.

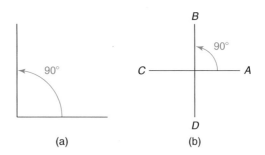

(a)   (b)

Figure 2-6   (a) A right angle is one-fourth of a circle. (b) Lines which cross at right angles are said to be perpendicular.

so that the angle between them is a right angle, they are said to be ***perpendicular.*** Line *CA* in Fig. 2-6b is perpendicular to line *BD*. This may be written

$$CA \perp BD$$

where $\perp$ means "is perpendicular to."

Two straight lines are said to be ***parallel*** if the extension of their sides will never intersect. In Fig. 2-7, line *AB* is parallel to line *CD*. We write

$$AB \parallel CD$$

where $\parallel$ means "is parallel to."

The application of geometry usually requires only a few general rules. We will describe three of the most important.

> **Rule 1. When two straight lines intersect, they form opposing angles that are equal (Fig. 2-8).**

> **Rule 2. When a straight line intersects (cuts across) two parallel lines, the alternate interior angles are equal (Fig. 2-9).**

Figure 2-7   Parallel lines extended indefinitely will never intersect (*AB* $\parallel$ *CD*).

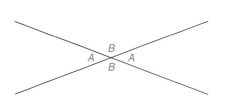

Figure 2-8   When two straight lines intersect, the opposing angles are equal.

Figure 2-9   When a straight line intersects two parallel lines, the alternate interior angles are equal.

Note from Fig. 2-9 that the angles *A* are on alternate sides of the intersecting line and are inside the two parallel lines. According to Rule 2, these *alternate interior* angles are equal. (The remaining two interior angles are also equal.)

EXAMPLE 2-5

Two studs at a construction site are braced by a cross member, as shown in Fig. 2-10. Determine angle *C*, using geometry.

Figure 2-10

Solution

Angle *A* is 60° by Rule 1, angle *B* is 60° by Rule 2, and angle *C* is 60° by Rule 1. From this example, it is seen that *alternate exterior* angles are also equal, but no new rule is necessary.

A *triangle* is a plane closed figure with three sides. Figure 2-11 shows a general triangle with sides *a*, *b*, and *c* and angles *A*, *B*, and *C*. Such a triangle in which no two sides are equal and no two angles are equal is called a *scalene triangle*.

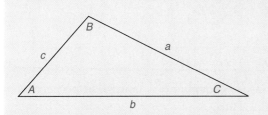

Figure 2-11   A scalene triangle.

Figure 2-12   A right triangle in which one of the interior angles is a right angle.

A special triangle of interest to us is a right triangle, as shown in Fig. 2-12. A right triangle has one angle equal to 90° (two of the sides are perpendicular). The side opposite the 90° angle is called the *hypotenuse.*

Rule 3. For any triangle, the sum of the interior angles is equal to 180°.

$$A + B + C = 180°$$

Corollary. For any right triangle ($C = 90$), the sum of the two smaller angles is equal to 90°.

$$A + B = 90°$$

EXAMPLE 2-6

Apply the rules of geometry to determine the unknown angles for the arrangement shown in Fig. 2-13.

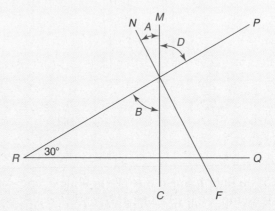

Figure 2-13

Solution

Since the line MC is perpendicular to line RQ, we have a right triangle in which the smallest angle is 30°. From Rule 3:

$$30° + B = 90°$$

and

$$B = 90° - 30° = 60°$$

Since opposite angles are equal, D is also equal to 60°. Line NF is perpendicular to line RP so that $A + D = 90°$. Thus,

$$A + 60° = 90°$$
$$A = 90° - 60° = 30°$$

The only additional rule you will need is the *pythagorean theorem*. This rule states that, for right triangles, the square of the hypotenuse is equal to the sum of the squares of the other two sides. We will discuss this rule in the following section.

2-7 RIGHT TRIANGLE TRIGONOMETRY

Often it is necessary to determine lengths and angles from three-sided figures known as *triangles*. By learning a few principles that apply to all right triangles, you can significantly improve your ability to work with vectors. Moreover, hand-held calculators make many of the calculations relatively simple.

First, let's review some of the things we already know about right triangles. We will follow the convention that uses Greek letters for angles and Roman letters for sides. Commonly used Greek symbols are

α alpha $\quad\quad$ β beta $\quad\quad$ γ gamma
θ theta $\quad\quad$ ϕ phi $\quad\quad$ δ delta

In the right triangle drawn as Fig. 2-14, the symbols R, x, and y refer to the side dimensions, and θ, ϕ, and 90° are the angles. Recall that the sum of the smaller angles in a right triangle is 90°:

$$\phi + \theta = 90° \quad\quad\quad \text{Right Triangle}$$

There is also a relationship between the sides, which is known as the pythagorean theorem:

Pythagorean theorem. The square of the hypotenuse is equal to the sum of the squares of the other two sides.

$$R^2 = x^2 + y^2 \quad\quad \text{Pythagorean Theorem} \quad \textbf{2-12}$$

The *hypotenuse* is defined as the longest side. It is conveniently located as that side directly opposite the right angle—the line joining the two perpendicular sides.

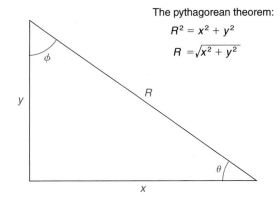

The pythagorean theorem:
$$R^2 = x^2 + y^2$$
$$R = \sqrt{x^2 + y^2}$$

Figure 2-14

EXAMPLE 2-7

What length of guy wire is needed to stretch from the top of a 12-m telephone pole to a ground stake located 8 m from the foot of the pole?

Solution

Draw a rough sketch, such as Fig. 2-15. Identify the length R of cable wire as the hypotenuse of a right triangle; then from the pythagorean theorem:

$$R^2 = (12 \text{ m})^2 + (8 \text{ m})^2$$
$$= 144 \text{ m}^2 + 64 \text{ m}^2 = 208 \text{ m}^2$$

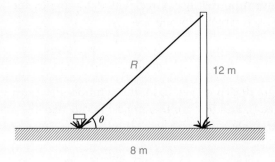

Figure 2-15

Taking the square root of both sides gives

$$R = \sqrt{208 \text{ m}^2} = 14.4 \text{ m}$$

Remember to give your answer in three significant figures. In this text we assume that all measurements have three significant digits. In other words, the height of the pole is 12.0 m and the base of the triangle is 8.00 m even though they are given as 12 m and 8 m for convenience.

In general, to find the hypotenuse we could express the pythagorean theorem as

$$R = \sqrt{x^2 + y^2} \qquad \qquad \text{Hypotenuse} \quad \textbf{2-13}$$

On some electronic calculators, the sequence of entries might be as follows:

$$x \quad \boxed{x^2} \quad \boxed{+} \quad y \quad \boxed{x^2} \quad \boxed{=} \quad \boxed{\sqrt{x}} \qquad \textbf{2-14}$$

In this instance, x and y are the values of the shorter sides, and the boxed symbols are operation keys on the calculator. You should verify the solution to the previous problem, using $x = 8$ and $y = 12$. (The input procedure varies depending on the make of calculator.)

Of course, the pythagorean theorem can also be used to find either of the shorter sides if the remaining sides are known. Solution for x or for y yields

$$x = \sqrt{R^2 - y^2} \qquad \qquad y = \sqrt{R^2 - x^2} \qquad \textbf{2-15}$$

Trigonometry is the branch of mathematics that takes advantage of the fact that similar triangles are proportional in size. In other words, for a given angle the ratio of any two sides is the same regardless of the overall dimensions of the triangle. For the three triangles in Fig. 2-16, the ratios of corresponding sides are equal as long as the angle is 37°. From Fig. 2-16, it is seen that

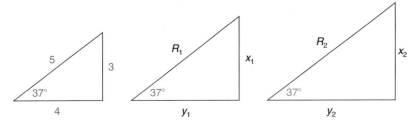

Figure 2-16 All right triangles with the same internal angles are similar; that is, their sides are proportional.

$$\frac{3}{4} = \frac{x_1}{y_1} = \frac{x_2}{y_2}$$

and

$$\frac{4}{5} = \frac{y_1}{R_1} = \frac{y_2}{R_2}$$

Once an angle is identified in a right triangle, the sides *opposite* and *adjacent* to that angle may be labeled. The meanings of opposite, adjacent, and hypotenuse are given in Fig. 2-17. You should study this figure until you understand fully the meaning of these terms. Verify that the side opposite to θ is y and that the side adjacent to θ is x. Also notice that the sides described by "opposite" and "adjacent" change if we refer to angle ϕ.

In a right triangle, there are three side ratios that are important. They are the *sine*, the *cosine*, and the *tangent*, defined as follows for angle θ:

$$\sin \theta = \frac{\text{opp } \theta}{\text{hyp}}$$

$$\cos \theta = \frac{\text{adj } \theta}{\text{hyp}} \qquad\qquad \text{(2-16)}$$

$$\tan \theta = \frac{\text{opp } \theta}{\text{adj}}$$

Hypotenuse, R

y, adjacent to ϕ
opposite to θ

ϕ

θ

x, adjacent to θ
opposite to ϕ

Figure 2-17

To make sure that you understand these definitions, you should verify the following for the triangles in Fig. 2-18:

$$\sin \theta = \frac{9}{15} \qquad \cos \gamma = \frac{m}{H} \qquad \tan \alpha = \frac{y}{x}$$

$$\sin \alpha = \frac{y}{R} \qquad \cos \beta = \frac{n}{H} \qquad \tan \phi = \frac{12}{9}$$

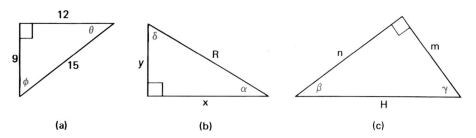

(a) (b) (c)

Figure 2-18

You should first identify the right angle and then label the longest side (opposite to the 90° angle) as the hypotenuse. Then, for a particular angle, the sides opposite and adjacent may be identified.

The constant values for the trigonometric functions of all angles between 0 and 90° have been calculated. They may be found from tables or, more conveniently, with a small calculator. Suppose we wish to know the cosine of 47°, for example. In a table of cosines we would see the number 0.682 adjacent to the angle 47°, and we would write

$$\cos 47° = 0.682$$

With some calculators, we would enter the angle 47, then strike the $\boxed{\cos}$ key to obtain the same result. From either a table or your calculator, you should verify the following:

$$\tan 38° = 0.781 \qquad \cos 31° = 0.857$$
$$\sin 22° = 0.375 \qquad \tan 65° = 2.147$$

To find the angle whose tangent is 1.34 or to find the angle whose sine is 0.45, we would reverse the above process. On a calculator, for example, we would enter the number 1.34, then we would look for one of the following sequences, depending on the calculator: $\boxed{\text{INV}}$ $\boxed{\tan}$, $\boxed{\text{ARC}}$ $\boxed{\tan}$, or $\boxed{\tan^{-1}}$. Any of these will give the angle whose tangent is the entered value. In the above examples, we find that

$$\tan \theta = 1.34 \qquad \theta = 53.3°$$
$$\sin \theta = 0.45 \qquad \theta = 26.7°$$

You should now be able to apply trigonometry to find unknown angles or sides in a right triangle. The following problem-solving procedure will be helpful:

Problem-Solving Strategy

Application of Trigonometry

1. Draw the right triangle from the stated conditions of the problem. (Label all sides and angles with either their known values or a symbol for an unknown value.)
2. Isolate an angle for study; if an angle is known, choose that one.
3. Label each side according to whether its relation to the chosen angle is opp, adj, or hyp.
4. Decide which side or angle is to be found.

5. Recall the definitions of the trigonometric functions:

$$\sin = \frac{opp}{hyp} \qquad \cos = \frac{adj}{hyp} \qquad \tan = \frac{opp}{adj}$$

6. Choose that trigonometric function which involves (a) the unknown quantity and (b) no other unknown quantity.
7. Write the trigonometric equation and solve for the unknown.

CONCEPTUAL EXAMPLE 2-8

What is the length of the rope segment x in Fig. 2-19?

20 m, HYP

40°

x, OPP y, ADJ

Figure 2-19

Solution

The sketch is already drawn, so we proceed with step 2. The 40° angle is selected for reference, and *opp*, *adj*, and *hyp* are labeled on the figure. The decision is made to solve for x (*opp*). Since the sine function involves *opp* and *hyp*, we choose that function and write the equation

$$\sin 40° = \frac{x}{20 \text{ m}}$$

Solving for x by multiplying both sides by 20 m, we obtain

$$x = (20 \text{ m})(\sin 40°)$$

If we use tables, we determine that $\sin 40° = 0.643$, and then

$$x = (20 \text{ m})(0.643) = 12.9 \text{ m}$$

On some calculators, we would calculate x as follows:

$$(20 \text{ m})(\sin 40°) = 20 \quad \boxed{\times} \quad 40 \quad \boxed{\sin} \quad \boxed{=} \quad = 12.9 \text{ m}$$

It is easy to see what an advantage the calculator is in the application of trigonometry. As an additional exercise, you might show that $y = 15.3$ m.

EXAMPLE 2-9

Refer to Fig. 2-15 (repeated below). What is the angle θ that the guy wire makes with the ground?

Solution

Note the location of the angle θ; then label *opp*, *adj*, and *hyp*. For the angle θ, we may write

$$\tan \theta = \frac{opp}{adj} = \frac{40 \text{ ft}}{60 \text{ ft}}$$

The tangent function was the only one that involved the two *known* sides. If tables are to be used, then θ is the angle whose tangent is $^{40}\!/_{60}$, or 0.667. In this case, we find that

$$\tan \theta = 0.667 \qquad \text{and} \qquad \theta = 33.7°$$

The sequence on some calculators might be as follows:

$$40 \quad \boxed{\div} \quad 60 \quad \boxed{=} \quad \boxed{\text{INV}} \quad \boxed{\tan}$$

Of course, the procedure varies. Study your manual and work with your calculator until you understand the process thoroughly.

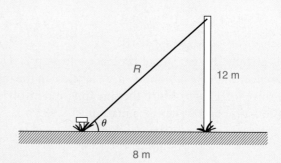

Summary and Review

Summary

This chapter was intended as a review of technical mathematics. Now that you have completed the chapter, go back and review the chapter objectives. If you feel uneasy about any of the objectives, perhaps another review would be helpful. Make sure that you understand the important topics in this chapter before applying the concepts of physics in later chapters. Remember the following points:

- When adding numbers of like sign, we add the absolute values of the numbers and give to the sum the common sign. To add numbers of unlike sign, we find the difference of their absolute values and affix the sign of the larger.
- To subtract a number b from a number a, we change the sign of number b and then add it to number a, using the addition rule.
- When we multiply or divide a group of signed numbers, the result is negative if the total number of negative factors is odd; otherwise, the result is positive.
- Formulas can be rearranged to solve for a particular unknown by performing similar operations (addition, subtraction, multiplication, division, etc.) to both sides of an equality.
- The following rules apply for exponents and radicals (optional):

$$a^m a^n = a^{m+n} \qquad a^{-n} = \frac{1}{a^n}$$

$$\frac{a^m}{a^n} = a^{m-n} \qquad (a^m)^n = a^{mn}$$

$$(ab)^n = a^n b^n \qquad \left(\frac{a}{b}\right)^n = \frac{a^n}{b^n}$$

$$\sqrt[n]{ab} = \sqrt[n]{a}\,\sqrt[n]{b} \qquad a^m = a^{m/n}$$

- Scientific notation uses positive or negative powers of 10 to express large or small numbers in shorthand notation.
- Graphs are used to give a continuous description of the relationship between two variables, based on observed data.
- When two straight lines intersect, they form opposing angles which are equal.
- When a straight line intersects two parallel lines, the alternate interior angles are equal.
- For any triangle, the sum of the interior angles is 180°; for a right triangle, the sum of the smaller angles is 90°.
- Application of the pythagorean theorem and basic trigonometric functions is essential when studying physics.

$$R^2 = x^2 + y^2 \qquad \text{Sin } \theta = \frac{\text{Opp}}{\text{Hyp}}$$

$$\text{Cos } \theta = \frac{\text{Adj}}{\text{Hyp}} \qquad \text{Tan } \theta = \frac{\text{Opp}}{\text{Adj}}$$

Key Terms

1. formula
2. scientific notation
3. exponent
4. base
5. radical
6. parallel
7. perpendicular
8. triangle
9. right triangle

Review Questions

2-1. The sum of two numbers is always greater than their difference. Is this statement true? Justify your answer by giving examples.

2-2. If you subtract the number (-8) from the number $(+4)$, what is the result? What if instead you subtracted the second number from the first? What is their sum?

Summary and Review

2-3. On a cold winter day, the temperature changes from $-5°C$ to $+10°C$. What is the change in temperature? What is the change if the temperature cools back to $-5°C$? Explain the difference.

2-4. Is it true that a negative number raised to an odd power will always be negative?

2-5. Distinguish clearly between -9^2 and $(-9)^2$. Are they the same? Why or why not?

2-6. Discuss two ways in which positive and negative signs are used when working with formulas. When you substitute signed numbers into formulas that contain the operations of addition and subtraction, what precautions should you take?

2-7. When you transpose a term from one side of an equation to the other side, the sign is changed. Explain how this procedure works and why it works.

2-8. Cross-multiplication is sometimes used in formula rearrangement in which one fraction is equal to another fraction. For example,

$$\frac{a}{b} = \frac{c}{d} \qquad \text{becomes} \qquad ad = bc$$

Explain why this procedure works and discuss the dangers involved.

2-9. A common error in formula rearrangement is canceling terms instead of factors. The following are *not* permitted:

$$\frac{x+y}{x} \neq y \qquad \frac{x^2 + y^2}{x+y} \neq \frac{x+y}{1}$$

2-10. If the graph of two variables (x, y) is a straight line, can we say that when x increases by 10 units, y must also increase by 10 units? Can we say that if the value of x doubles, the value of y must also double?

2-11. When two parallel lines are crossed by a transversal, the alternate interior angles are equal. Are the alternate exterior angles also equal?

2-12. One of the angles of a right triangle is $33°$. What are the other two angles?

2-13. A window is 6 ft tall. A diagonal two-by-four, 9 ft in length, fits exactly from the top corner to the bottom corner. How wide is the window?

2-14. The complement ϕ of an angle θ is that angle ϕ such that $\phi + \theta = 90°$. Show that the sine of an angle is equal to the cosine of its complement.

2-15. If angles ϕ and θ are complementary, show that $\tan \theta$ is the reciprocal of $\tan \phi$.

Problems

Section 2-1—Algebra Review

In items 2-1 through 2-26, perform the indicated operation.

2-1. $(+2) + (+5)$		Ans. $+7$
2-2. $(-2) + (6)$		
2-3. $(-4) - (-6)$		Ans. $+2$
2-4. $(+6) - (+8)$		
2-5. $(-3) - (+7)$		Ans. -10
2-6. $(-15) - (+18)$		
2-7. $(-4) - (+3) - (-2)$		Ans. -5
2-8. $(-6) + (-7) - (+4)$		
2-9. $(-2)(-3)$		Ans. $+6$
2-10. $(-16)(+2)$		
2-11. $(-6)(-3)(-2)$		Ans. -36

2-12. $(-6)(+2)(-2)$		
2-13. $(-3)(-4)(-2)(2)$		Ans. -48
2-14. $(-6)(2)(3)(-4)$		
2-15. $(-6) \div (-3)$		Ans. $+2$
2-16. $(-14) \div (+7)$		
2-17. $(+16) \div (-4)$		Ans. -4
2-18. $(+18) \div (-6)$		
2-19. $\dfrac{-4}{-2}$		Ans. $+2$
2-20. $\dfrac{+16}{-4}$		
2-21. $\dfrac{(-2)(-3)(-1)}{(-2)(-1)}$		Ans. -3

2-22. $\dfrac{(-6)(+4)}{(-2)}$

2-23. $\dfrac{(-16)(4)}{2(-4)}$ Ans. +8

2-24. $\dfrac{(-1)(-2)^2(12)}{(6)(2)}$

2-25. $(-2)(+4) - \dfrac{(-6)}{(+2)} - (-5)$ Ans. 0

2-26. $(-2)(-2)^2 + \dfrac{(-3)(-2)(-8)}{(-4)(1)} - (-6)^3$

In items 2-27 through 2-30, solve the given problem.

2-27. Distances above the ground are positive, and distances below the ground are negative. If an object is dropped from 20 ft above the ground into a 12-ft hole, what is the difference between the initial position and the final position?   **Ans. 32 ft**

2-28. In physics, work is measured in joules (J) and may be positive or negative depending on the direction of the force which does the work. What is the total work done if the works of the forces are 20 J, −40 J, and −12 J?

2-29. The temperature of a bolt is −12°C. (a) If the temperature increases by 6C°, what is the resultant temperature? (b) If the temperature decreases by 5C°, what is the resultant temperature? (c) If the temperature is multiplied by a factor of −3, what is the resultant temperature?
   Ans. (a) −6°C, (b) −17°C, (c) 36°C

3-30. A metal expands when heated and contracts when cooled. Assume that the length of a rod changes by 2 millimeters (mm) for each 1C° change in temperature. What is the total change in the length of this rod as the temperature changes from −5 to −30°C?

Section 2-3—Exponents and Radicals

In items 2-31 through 2-46, determine the value for x when $a = 2$, $b = -3$, and $c = -2$.

2-31. $x = a + b + c$ Ans. −3

2-32. $x = a - b - c$

2-33. $x = b + c - a$ Ans. −7

2-34. $x = b(a - c)$

2-35. $x = \dfrac{b - c}{a}$ Ans. $-\dfrac{1}{2}$

2-36. $x = \dfrac{a + b}{c}$

2-37. $x = b^2 - c^2$ Ans. +5

2-38. $x = \dfrac{-b}{ac}$

2-39. $x = \dfrac{a}{bc}(a - c)$ Ans. $+\dfrac{4}{3}$

2-40. $x = a^2 + b^2 + c^3$

2-41. $x = \sqrt{a^2 + b^2 + c^2}$ Ans. $\sqrt{17}$

2-42. $x = ab(c - a)^2$

2-43. $2ax - b = c$ Ans. $-\dfrac{5}{4}$

2-44. $ax + bx = 4c$

2-45. $3ax = \dfrac{2ab}{c}$ Ans. +1

2-46. $\dfrac{4ac}{b} = \dfrac{2x}{b} - 16$

In items 2-47 through 2-56, solve the equations for the unknown letter.

2-47. $5m - 16 = 3m - 4$ Ans. $m = 6$

2-48. $3p = 7p - 16$

2-49. $4m = 2(m - 4)$ Ans. $m = -4$

2-50. $3(m - 6) = 6$

2-51. $\dfrac{x}{3} = (4)(3)$ Ans. $x = 36$

2-52. $\dfrac{P}{3} = \dfrac{2}{6}$

2-53. $\dfrac{96}{x} = 48$ Ans. $x = 2$

2-54. $14 = 2(b - 7)$

2-55. $R^2 = (4)^2 + (3)^2$ Ans. $R = +5$

2-56. $\dfrac{1}{2} = \dfrac{1}{P} + \dfrac{1}{6}$

In items 2-57 through 2-70, solve the formulas for the letter indicated.

2-57. $V = IR$, R Ans. $R = \dfrac{V}{I}$

2-58. $PV = nRT$, T

2-59. $F = ma$, a Ans. $a = \dfrac{F}{m}$

2-60. $s = vt + d$, d

2-61. $F = \dfrac{mv^2}{R}$, R Ans. $R = \dfrac{mv^2}{F}$

2-62. $s = \frac{1}{2}at^2$, a

2-63. $2as = v_f^2 - v_o^2$, a Ans. $a = \dfrac{v_f^2 - v_o^2}{2s}$

Summary and Review

2-64. $C = \dfrac{Q^2}{2V}$, V

2-65. $\dfrac{1}{R} = \dfrac{1}{R_1} + \dfrac{1}{R_2}$, R Ans. $R = \dfrac{R_1 R_2}{R_1 + R_2}$

2-66. $MV = Ft$, t

2-67. $mv_2 - mv_1 = Ft$, v_2 Ans. $V_2 = \dfrac{Ft + mv_1}{m}$

2-68. $\dfrac{P_1 V_1}{T_1} = \dfrac{P_2 V_2}{T_2}$, T_2

2-69. $v = v_o + at$, a Ans. $a = \dfrac{v - v_o}{t}$

2-70. $c^2 = a^2 + b^2$, b

In items 2-71 through 2-92, simplify the given expressions, using the laws of exponents and radicals.

2-71. $2^5 \cdot 2^7$ Ans. 2^{12}

2-72. $3^2 \cdot 2^3 \cdot 3^3$

2-73. $x^7 x^3$ Ans. x^{10}

2-74. $x^7 x^{-5} x^3$

2-75. $a^{-3} a^2$ Ans. $\dfrac{1}{a}$

2-76. $a^3 a^{-2} b^{-3} b$

2-77. $\dfrac{2^3}{2^5}$ Ans. $\dfrac{1}{2^2}$

2-78. $\dfrac{2a^3 b}{2ab^3}$

2-79. $\dfrac{2x^{17}}{x^{12}}$ Ans. $2x^5$

2-80. $(ab)^{-2}$

2-81. $(m^{-3})^{-2}$ Ans. m^6

2-82. $(n^3 c^{-2})^{-2}$

2-83. $(4 \times 10^2)^3$ Ans. 64×10^6

2-84. $(6 \times 10^{-2})^{-2}$

2-85. $\sqrt[3]{64}$ Ans. 4

2-86. $\sqrt[4]{81}$

2-87. $\sqrt[5]{x^{15}}$ Ans. x^3

2-88. $\sqrt{a^4 b^6}$

2-89. $\sqrt{4 \times 10^4}$ Ans. 2×10^2

2-90. $\sqrt[3]{8 \times 10^{-27}}$

2-91. $\sqrt[5]{32a^{10}}$ Ans. $2a^2$

2-92. $\sqrt{(x + 2)^2}$

Section 2-4—Scientific Notation

In items 2-93 through 2-100, convert the decimal numbers to scientific notation.

2-93. 40,000   Ans. 4×10^4

2-94. 67

2-95. 480   Ans. 4.80×10^2

2-96. 497,000

2-97. 0.0021   Ans. 2.1×10^{-3}

2-98. 0.789

2-99. 0.087   Ans. 8.7×10^{-2}

2-100. 0.000967

In items 2-101 through 2-108, convert the numbers to decimal notation.

2-101. 4×10^6 Ans. 4,000,000

2-102. 4.67×10^3

2-103. 3.7×10^1 Ans. 37

2-104. 1.4×10^5

2-105. 3.67×10^{-2} Ans. 0.0367

2-106. 4×10^{-1}

2-107. 6×10^{-3} Ans. 0.006

2-108. 4.17×10^{-5}

In items 2-109 through 2-132, simplify and express as a single number written in scientific notation.

2-109. $400 \times 20,000$ Ans. 8×10^6

2-110. 37×2000

2-111. $(4 \times 10^{-3})(2 \times 10^5)$ Ans. 8×10^2

2-112. $(3 \times 10^{-1})(6 \times 10^{-8})$

2-113. $(6.7 \times 10^3)(4.0 \times 10^5)$ Ans. 2.68×10^9

2-114. $(3.7 \times 10^{-5})(200)$

2-115. $(4 \times 10^{-3})^2$ Ans. 1.60×10^{-5}

2-116. $(3 \times 10^6)^3$

2-117. $(6000)(3 \times 10^{-7})$ Ans. 1.8×10^{-3}

2-118. $(4)(300)(2 \times 10^{-2})$

2-119. $7000 \div (3.5 \times 10^{-3})$ Ans. 2.00×10^6

2-120. $60 \div 30,000$

2-121. $(6 \times 10^{-5}) \div (3 \times 10^4)$ Ans. 2×10^{-9}

2-122. $(4 \times 10^{-7}) \div (7 \times 10^{-7})$

2-123. $\dfrac{4600}{0.02}$ Ans. 2.3×10^5

2-124. $\dfrac{(1600)(4 \times 10^{-3})}{1 \times 10^{-2}}$

2-125. $4.0 \times 10^2 + 2 \times 10^3$ Ans. 2.40×10^3

2-126. $6 \times 10^{-5} - 4 \times 10^{-6}$

2-127. $6 \times 10^{-3} - 0.075$ Ans. -6.90×10^{-2}

2-128. $0.0007 - 4 \times 10^{-3}$

2-129. $\dfrac{4 \times 10^{-6} + 2 \times 10^{-5}}{4 \times 10^{-2}}$ Ans. 6×10^{-4}

2-130. $\dfrac{6 \times 10^3 + 4 \times 10^2}{1 \times 10^{-3}}$

2-131. $\dfrac{600 - 3000}{0.0003}$ **Ans.** -8×10^6

2-132. $(4 \times 10^{-3})^2 - 2 \times 10^{-5}$

Section 2-5—Graphs

2-133. Plot a graph for the following data recorded for an object falling freely from rest.

Speed, ft/s	32	63	97	129	159	192	225
Time, s	1	2	3	4	5	6	7

What do you expect the speed to be after 4.5 s? How much time is required for the object to attain a speed of 100 ft/s?
Ans. $V = 144$ **ft/s,** $t = 3.1$ **s**

2-134. The advance of a right-hand screw is proportional to the number of complete turns. For a particular screw, the following data are recorded:

Advance, in.	0.5	1.0	1.5	2.0	2.5	3.0
Number of turns	16	32	48	64	80	96

Plot a graph with the number of turns as the horizontal divisions and the advance in inches as the vertical divisions. What number of turns is required to advance the screw 2.75 in.?

2-135. Plot a graph showing the relationship between frequency and wavelength of electromagnetic waves. The following data are given:

Frequency, kilohertz (kHz)	150	200	300	500	600	900
Wavelength, meters (m)	2000	1500	1000	600	500	333

What are the wavelengths of electromagnetic waves with frequencies of 350 kHz and 800 kHz?   **Ans. 10.4 W, 64.8 W**

2-136. The electric power loss in a resistor varies directly with the square of the current. In a particular experiment, the following data are recorded:

Current, amperes (A)	1.0	2.5	4.0	5.0	7.0	8.5
Power, watts (W)	1.0	6.5	16.2	25.8	50.2	72.0

Plot a graph and from the curve determine the power loss when the current is (a) 3.2 A and (b) 8.0 A.

Section 2-6—Geometry

2-137. What are the magnitudes of the angles in Fig. 2-20 by your estimation?
Ans. 90°, 180°, 270°, and 45°

2-138. Use a ruler and a protractor to measure lines and angles. Draw two parallel lines AB and CD that are separated by 2 cm. Now draw a third line EF that crosses each of the other lines at any angle other than 90°. Verify Rules 1 and 2 by measuring the angles formed by the transversal EF. Now draw another transversal line GH slanted in the other direction that intersects line AB at the same point as for line EF. Verify Rule 3 for the triangle you have just formed.

2-139. Determine angles A and B for each of the arrangements drawn in Fig. 2-21.
Ans. (a) $A = 17°$, $B = 35°$, **(b)** $A = 50°$, $B = 40°$

2-140. Determine angles A and B in Fig. 2-22.

(a)   (b)

(c)   (d)

Figure 2-20

Figure 2-21

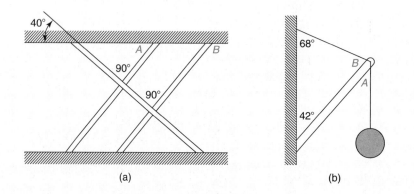

Figure 2-22

Section 2-7—Right Triangle Trigonometry

In items 2-161 through 2-178, use your calculator or tables to evaluate each example.

2-141.	sin 67°	**Ans. 0.921**
2-142.	cos 48°	
2-143.	tan 59°	**Ans. 1.66**
2-144.	sin 34°	
2-145.	cos 29°	**Ans. 0.875**
2-146.	tan 15°	
2-147.	20 cos 15°	**Ans. 19.3**
2-148.	400 sin 21°	
2-149.	600 tan 24°	**Ans. 267**
2-150.	170 cos 79°	
2-151.	240 sin 78°	**Ans. 235**
2-152.	1400 tan 60°	

2-153. $\dfrac{200}{\sin 17°}$ **Ans. 684**

2-154. $\dfrac{300}{\sin 60°}$

2-155. $\dfrac{167}{\cos 78°}$ **Ans. 803**

2-156. $\dfrac{256}{\cos 16°}$

2-157. $\dfrac{670}{\tan 17°}$ **Ans. 2191**

2-158. $\dfrac{2000}{\tan 51°}$

In items 2-179 through 2-187, determine the unknown angles, using your tables or a calculator.

2-159.	sin θ = 0.811	**Ans. 54.2°**
2-160.	sin θ = 0.111	
2-161.	tan θ = 1.2	**Ans. 50.2°**
2-162.	tan θ = 0.511	
2-163.	cos β = 0.228	**Ans. 76.8°**
2-164.	cos θ = 0.81	

2-165. cos θ = $\dfrac{400}{500}$ **Ans. 36.9°**

2-166. $\tan \theta = \dfrac{16}{4}$

2-167. $\sin \phi = \dfrac{140}{270}$ **Ans. 31.2°**

In items 2-188 through 2-195, solve the triangles for the unknown angles and sides.

2-168.

2-169.

Ans. $R = 721$ m, $\phi = 6.3°$

2-170.

2-171.

Ans. $x = 458$ km, $\phi = 23.6°$

2-172.

2-173.

Ans. $x = 164$ in., $y = 202$ in.

2-174.

2-175.

Ans. $x = 150$ m, left, $y = 260$ m

Challenge Problems

2-176. Early one morning a barometer reads 30.21. When an afternoon storm occurs, the reading decreases by 0.59 in. What is the later reading?

2-177. A thermometer reads 29.0°C. After some time in a freezer, it reads $-15°$C. What is the change in temperature?   **Ans. -44**

2-178. The change in temperature of an object is $-34°$C. If the original temperature was 20°C, what was the final temperature?

2-179. A length of wood is cut into six pieces, each 28 cm long. Each cut wastes 1 mm of wood. What was the original length of the board in inches?   **Ans. 28.8 in.**

2-180. The volume V of a right circular cylinder is the area of the bottom (πr^2) times the height h. Given the radius, write a formula to find the height h.

Summary and Review

 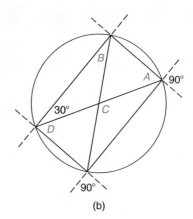

(a)   (b)

Figure 2-23

2-181. The centripetal force F_C is found by multiplying the mass m by the square of its velocity v and then dividing by the radius r of the circle. Write the formula and solve it for the radius r.

$$\text{Ans. } F_C = \frac{mv^2}{R}, \quad R = \frac{mv^2}{F_C}$$

2-182. Solve the equation $xb + cd = a(x + 2)$ for x and find the value of x when $a = 2$, $b = -2$, $c = 3$, and $d = -1$.

2-183. Solve for b when $c^2 = a^2 + b^2$ and find the value of b when $a = 50$ and $b = 20$.

Ans. 53.9

2-184. Newton's law of gravitation is written as $F = Gm_1m_2/R^2$. The following numerical values are given: $G = 6.67 \times 10^{-11}$, $m_1 = 4 \times 10^{-8}$, $m_2 = 3 \times 10^{-7}$, and $R = 4 \times 10^{-2}$. What is the value of F?

2-185. The length of a rod is initially $L_0 = 21.41$ cm at $t_0 = 20°C$. It is heated to a final temperature $t = 100°C$. The new length is given by:

$$L = L_0 + \alpha L_0(t - t_o)$$

where $\alpha = 2 \times 10^{-3}/°C$. What is the value of L? **Ans. $L = 24.84$ cm**

2-186. Plot a graph of the function $y = 2x$. From the graph, verify that $x = 3.5$ when $y = 7$.

2-187. Determine the unknown angles in Fig. 2-23a and b. **Ans. (a) $A = 30°$, $B = 60°$, $C = 60°$; (b) $A = 60°$, $B = 30°$, $C = 120°$, $D = 60°$**

Critical Thinking Questions

2-188. Subtract -4 cm from -8 cm to get length A. Add -6 cm to $+14$ cm to get a length B. What is the length $C = A - B$? Is $A - B$ the same value as $B - A$? Consider a number line along the x axis. Draw a line from the origin to point C. Then draw a line from the origin to the point $B - A$. How far are these points apart? **Ans. -12 cm, no, 24 cm**

2-189. The period T of a pendulum is the time required to make a complete oscillation (out and back). The period is given by the following equation:

$$T = 2\pi \sqrt{\frac{L}{g}}$$

where L is the length of the pendulum and g is the acceleration due to gravity. (a) Solve for the length L. (b) If the length L of the pendulum is quadrupled, how much larger will the period be? (c) If the period of a pendulum on Earth is 2.0 s, what would be the period on the Moon, where g is one-sixth of its Earth value?

*2-190. The length of a tiny chip is 3.45×10^{-4} m and its width is 9.77×10^{-5} m. (a) Find the area and perimeter of the chip. (b) If the width is doubled and the length is cut in half, what is the change in the area and what is the change in the perimeter?

Ans. (a) 3.37×10^{-8} m^2, 8.85×10^{-4} m;

(b) 0, -1.5×10^{-4} m

*2-191. The pressure in a storage tank is dependent on the temperature. The following measurements are recorded:

Temperature, K	300	350	400	450	500	550
Pressure, lb/in.^{2}	400	467	535	598	668	733

Plot a graph of these data. What is the slope of this graph? Can you write a description of the relationship between pressure and temperature on the basis of this information? What would you expect the pressure to be at the temperatures of 420 and 600 degrees kelvin (K)?

*2-192. The voltage (V) in volts and the electric current (I) in milliamperes is given for a particular resistor. The recorded data follows:

Voltage, V	10	20	30	40	50	60
Current, mA	145	289	435	581	724	870

Plot a graph of these data. What would you expect the current in the resistor to be for applied voltages of 26 and 48 V?

Ans. 377 mA, 696 mA

Technical Measurement and Vectors

OBJECTIVES

After completing this chapter, you should be able to:
1. Write the base units for mass, length, and time in SI and U.S. Customary System (USCS) units.
2. Define and apply the SI prefixes that indicate multiples of base units.
3. Convert from one unit to another unit for the same quantity when given the necessary definitions.
4. Define a vector quantity and a scalar quantity and give examples of each.
5. Determine the components of a given vector.
6. Find the resultant of two or more vectors.

The application of physics, whether in the shop or in a technical laboratory, always requires measurements of some kind. An automobile mechanic might be measuring the diameter or *bore* of an engine cylinder. Refrigeration technicians might be concerned with volume, pressure, and temperature measurements. Electricians use instruments that measure electric resistance and current, and mechanical engineers are concerned about the effects of forces whose magnitudes must be accurately determined. In fact, it is difficult to imagine any occupation that is not involved with the measurement of some physical quantity.

In the process of physical measurement, we are often concerned with the direction as well as the magnitude of a particular quantity. The length of a wooden rafter is determined by the angle it makes with the horizontal. The direction of an applied force determines its effectiveness in producing a displacement. The direction in which a conveyor belt moves is often just as important as the speed with which it moves. Such physical quantities as *displacement, force,* and *velocity* are often encountered in industry. In this chapter, the concept of *vectors* is introduced to permit the study of both the magnitude and the direction of physical quantities.

3-1 PHYSICAL QUANTITIES

The language of physics and technology is universal. Facts and laws must be expressed in an accurate and consistent manner if everyone is to mean exactly the same thing by the same term. For example, suppose an engine is said to have a piston displacement of 3.28 liters (200 cubic inches). Two questions must be answered if this statement is to be understood: (1) How is the *piston displacement* measured, and (2) what is the *liter?*

Piston displacement is the volume that the piston displaces, or "sweeps out," as it moves from the bottom of the cylinder to the top. It is really not a displacement in the usual sense of the word; it is a volume. A standard measure for volume that is easily recognized throughout the world is the liter. Therefore, when an engine has a label on it that reads "piston displacement = 3.28 liters," all mechanics will give the same meaning to the label.

In the above example, the piston displacement (volume) is an example of a *physical quantity.* Notice that this quantity was defined by describing the procedure for its measurement. In physics, all quantities are defined in this manner. Other examples of physical quantities are length, weight, time, speed, force, and mass.

A physical quantity is measured by comparison with some known standard. For example, we might need to know the length of a metal bar. With appropriate instruments we might determine the length of the bar to be 4 meters. The bar does not contain four things called "meters"; it is merely compared with the length of some standard known as a "meter." The length could also be represented as 13.1 ft or 4.37 yards if we used other known measures.

The *magnitude* of a physical quantity is given by a *number* and a *unit* of measure. Both are necessary because either the number or the unit by itself is meaningless. Except for pure numbers and fractions, it is necessary to include the unit with the number when listing the magnitude of any quantity.

> **The magnitude of a physical quantity is specified completely by a number and a unit, for example, 20 meters or 40 liters.**

Since there are many different measures for the same quantity, we need a way of keeping track of the exact size of particular units. To do this, it is necessary to establish standard measures for specific quantities. A **standard** is a permanent or easily determined physical record of the size of a unit of measurement. For example, the standard for measuring electrical resistance, the *ohm,* might be defined by comparison with a standard resistor whose resistance is accurately known. Thus, a resistance of 20 ohms would be 20 times as great as that of a standard 1-ohm resistor.

Remember that every physical quantity is defined by telling how it is measured. Depending on the measuring device, each quantity can be expressed in a number of different units. For example, some distance units are *meters, kilometers, miles,* and *feet,* and some speed units are *meters per second, kilometers per hour, miles per hour,* and *feet per second.* Regardless of the units chosen, however, distance must be a *length* and speed must be a *length* divided by *time.* Thus, *length* and *length/time* are the *dimensions* of the physical quantities *distance* and *speed.*

Note that speed is defined in terms of two more fundamental quantities (length and time). It is convenient to establish a small number of base quantities such as length and time from which all other physical quantities can be derived. Thus, we might say that speed is a *derived* quantity and that length or time is a *base* quantity. If we reduce all physical measurements to a small number of quantities with standard base units, there will be less confusion in their application.

3-2 THE INTERNATIONAL SYSTEM

**physics.nist.gov/cuu/units
/index.html**

For an in-depth look at the
International System of Units
(SI), see this Internet site.

The international system of units is called *Système International d'Unités* (SI) and is essentially the same as what we have come to know as the *metric system*. The International Committee on Weights and Measures has established seven base quantities and has assigned official base units to each quantity. A summary of these quantities, their base units, and the symbols for the base units is given in Table 3-1.

TABLE 3-1 The SI Base Units for Seven Fundamental Quantities and Two Supplemental Quantities		
Quantity	**Unit**	**Symbol**
Base units		
Length	meter	m
Mass	kilogram	kg
Time	second	s
Electric current	ampere	A
Temperature	kelvin	K
Luminous intensity	candela	cd
Amount of substance	mole	mol
Supplemental units		
Plane angle	radian	rad
Solid angle	steradian	sr

Each of the units listed in Table 3-1 has a specific measurable definition that can be duplicated anywhere in the world. Of these base units only one, the *kilogram,* is currently defined in terms of a single physical sample. This standard specimen is kept at the International Bureau of Weights and Measures in France. Copies of the original specimen have been made for use in other nations. All other units are defined in terms of reproducible physical events and can be accurately determined at a number of locations throughout the world.

We can measure many quantities, such as volume, pressure, speed, and force, that are combinations of two or more fundamental quantities. However, no one has ever encountered a measurement that cannot be expressed in terms of length, mass, time, current, temperature, luminous intensity, or amount of substance. Combinations of these quantities are referred to as *derived* quantities, and they are measured in derived units. Several common derived units are listed in Table 3-2.

The SI units are not fully implemented in many industrial applications. The United States is making progress toward the adoption of SI units. However, wholesale conversions are costly, particularly in many mechanical and thermal applications, and total conversion to the international system will require some time. For this reason it is necessary to be familiar with older units for physical quantities. The U.S. Customary System (USCS) units for several important quantities are listed in Table 3-3.

TABLE 3-2 Derived Units for Common Physical Quantities

Quantity	Derived Units	Symbol	
Area	square meter	m^2	
Volume	cubic meter	m^3	
Frequency	hertz	Hz	s^{-1}
Mass density (density)	kilogram per cubic meter	kg/m^3	
Speed, velocity	meter per second	m/s	
Angular velocity	radian per second	rad/s	
Acceleration	meter per second squared	m/s^2	
Angular acceleration	radian per second squared	rad/s^2	
Force	newton	N	$kg \cdot m/s^2$
Pressure (mechanical stress)	pascal	Pa	N/m^2
Kinematic viscosity	square meter per second	m^2/s	
Dynamic viscosity	newton-second per square meter	$N \cdot s/m^2$	
Work, energy, quantity of heat	joule	J	$N \cdot m$
Power	watt	W	J/s
Quantity of electricity	coulomb	C	
Potential difference, electromotive force	volt	V	J/C
Electric field strength	volt per meter	V/m	
Electric resistance	ohm	Ω	V/A
Capacitance	farad	F	C/V
Magnetic flux	weber	Wb	$V \cdot s$
Inductance	henry	H	$V \cdot s/A$
Magnetic flux density	tesla	T	Wb/m^2
Magnetic field strength	ampere per meter	A/m	
Magnetomotive force	ampere	A	
Luminous flux	lumen	lm	$cd \cdot sr$
Luminance	candela per square meter	cd/m^2	
Illuminance	lux	lx	lm/m^2
Wave number	1 per meter	m^{-1}	
Entropy	joule per kelvin	J/K	
Specific heat capacity	joule per kilogram kelvin	$J/(kg \cdot K)$	
Thermal conductivity	watt per meter kelvin	$W/(m \cdot K)$	
Radiant intensity	watt per steradian	W/sr	
Activity (of a radioactive source)	1 per second	s^{-1}	

TABLE 3-3 U.S. Customary System Units		
Quantity	SI unit	USCS unit
Length	meter (m)	foot (ft)
Mass	kilogram (kg)	slug (slug)
Time	second (s)	second (s)
Force (weight)	newton (N)	pound (lb)
Temperature	kelvin (K)	degree Rankine (R)

It should be noted that, even though the foot, pound, and other units are frequently used in the United States, they have been redefined in terms of the SI standard units. Thus, all measurements are currently based on the same standards.

3-3 MEASUREMENT OF LENGTH AND TIME

The standard SI unit of length, the *meter* (m), was originally defined as one ten-millionth of the distance from the North Pole to the equator. For practical reasons, this distance was marked off on a standard platinum-iridium bar. In 1960, the standard was changed to allow greater access to a more accurate measure of the meter based on an atomic standard. One meter was said to be equal to exactly 1,650,763.73 wavelengths of orange-red light from krypton 86. The number was chosen so that the new standard would be close to the older standard. However, even this standard was not without its problems. The wavelength of light emitted from krypton was uncertain because of the processes occurring within the atom during emission. Also, the development of stabilized lasers made it possible to measure a wavelength much more accurately in terms of time and the velocity of light. In 1983, the latest (and probably the final) standard for the meter was adopted:

One meter is the length of path traveled by a light wave in a vacuum in a time interval of 1/299,792,458 second.

The new standard for the meter is more precise, but it also has other advantages. Its definition depends on the standard for time (s) as given below, and it hinges on a standard value for the velocity of light. The velocity of light is now exactly:

$$c = 2.99792458 \times 10^8 \text{ m/s} \qquad \text{(exact by definition)}$$

A standard value for the speed of light makes sense because, according to Einstein's theory, the speed of light is a fundamental constant. Moreover, any future refinements to the standard for measuring time will automatically improve the standard for length. Of course, we do not usually need to know the precise definition of length to make accurate practical measurements. Many tools, such as simple meter sticks and calipers, are calibrated to agree with the standard measure.

TABLE 3-4 Multiples and Submultiples for SI Units

Prefix	Symbol	Multiplier	Use
tera	T	$1{,}000{,}000{,}000{,}000 = 10^{12}$	1 terameter (Tm)
giga	G	$1{,}000{,}000{,}000 = 10^{9}$	1 gigameter (Gm)
mega	M	$1{,}000{,}000 = 10^{6}$	1 megameter (Mm)
kilo	k	$1{,}000 = 10^{3}$	1 kilometer (km)
centi	c	$0.01 = 10^{-2}$	1 centimeter (cm)*
milli	m	$0.001 = 10^{-3}$	1 millimeter (mm)
micro	μ	$0.000001 = 10^{-6}$	1 micrometer (mm)
nano	n	$0.000000001 = 10^{-9}$	1 nanometer (nm)
—	Å	$0.0000000001 = 10^{-10}$	1 angstrom (Å)*
pico	p	$0.000000000001 = 10^{-12}$	1 picometer (pm)

*The use of the centimeter and the angstrom is discouraged, but they are still widely used.

The original definition of time depended on the idea of a solar day as the time interval between two successive passages of the sun over a given meridian on the earth. One second was then 1/86,400 of a mean solar day. It's not hard to imagine the difficulties and the inconsistencies associated with such a standard. In 1976 the SI standard for time was defined as follows:

One second is the time needed for 9,192,631,770 vibrations of a cesium atom.

Thus the atomic standard for one second is the period of vibration of an atom of cesium. The best cesium clocks are so accurate that they gain or lose no more than 1 second in 300,000 years.

Because this measure of time tends to drift ahead of mean solar time, the National Bureau of Standards periodically inserts a *leap second,* usually once a year on December 31. Therefore, the last minute of each year often contains 61 seconds, rather than 60 seconds.

A distinct advantage of the metric system over other systems of units is the use of prefixes to indicate multiples of the base unit. Table 3-4 defines the accepted prefixes and demonstrates their use to indicate multiples and subdivisions of the meter. From the table, you can determine that

$$1 \text{ meter (m)} = 1000 \text{ millimeters (mm)}$$
$$1 \text{ meter (m)} = 100 \text{ centimeters (cm)}$$
$$1 \text{ kilometer (km)} = 1000 \text{ meters (m)}$$

The relationship between the centimeter and the *inch* can be seen in Fig. 3-1 on page 44. By definition, 1 inch is equal to exactly 25.4 millimeters. This definition and other useful definitions are as follows (symbols for the units are in parentheses):

$$1 \text{ inch (in.)} = 25.4 \text{ millimeters (mm)}$$
$$1 \text{ foot (ft)} = 0.3048 \text{ meter (m)}$$
$$1 \text{ yard (yd)} = 0.914 \text{ meter (m)}$$
$$1 \text{ mile (mi)} = 1.61 \text{ kilometers (km)}$$

1 in. = 2.54 cm

inches
centimeters

Figure 3-1 A comparison of the inch to the centimeter as a measure of length.

In reporting data, it is preferable to use the prefix that will allow the number to be expressed in the range from 0.1 to 1000. For example, 7,430,000 meters should be expressed as 7.43×10^6 m, and then it should be reported as 7.43 megameters, abbreviated 7.43 Mm. It would not usually be desirable to write this measurement as 7430 kilometers (7430 km) unless the distance is being compared with other distances measured in kilometers. In the case of the quantity 0.00064 ampere, it is proper to write either 0.64 milliampere (0.64 mA) or 640 microamperes (640 μA). Normally, prefixes are chosen for multiples of a thousand.

3-4 SIGNIFICANT FIGURES

Some numbers are exact, and others are approximate. If we select 20 bolts from a bin and use only one-fourth of them, the numbers 20 and $\frac{1}{4}$ are considered to be exact quantities. If we measured the length or the width of a rectangular plate, however, the accuracy of the measurement would be determined by the precision of the instrument and by the skill of the observer. Suppose we measure the width of such a plate with a vernier caliper and find it to be 3.42 cm. The last digit is estimated and is therefore susceptible to error. The actual width is between 3.40 cm and 3.50 cm. To write this width as 3.420 cm would imply greater accuracy than was justified. We say that the number 3.42 has three *significant* figures, and we are careful not to write more numbers or zeros than are meaningful.

All physical measurements are assumed to be approximate, with the last significant digit having been determined by an estimation of some kind. When writing such numbers, we often include some zeros to indicate properly the location of the decimal point. Except for these zeros, however, all other digits are considered significant digits. For example, the distance 76,000 m has only two significant digits. We assume that the three zeros following the 6 were needed to locate the decimal unless there is information to the contrary. Other examples are:

4.003 cm	4 significant figures
0.34 cm	2 significant figures
60,400 cm	3 significant figures
0.0450 cm	3 significant figures

Zeros that are not specifically needed to locate the decimal are significant (as in the last two examples).

With the widespread use of calculators, students often report their results with a greater accuracy than is warranted. For example, suppose we measure the length of a rectangular plate to be 9.54 cm and its width to be 3.4 cm. The area of such a plate might be calculated as 32.436 cm^2 (five significant figures). But a chain is only as strong as its weakest link. Since the width was accurate to only two significant figures, we cannot report a greater accuracy in our result. The area should be given as 32 cm^2. The number in our calculator gives a false indication of the accuracy and is misleading to others who did not participate in the measurement.

Rule 1 When approximate numbers are multiplied or divided, the number of significant digits in the final answer is the same as the number of significant digits in the least accurate of the factors.

Another problem arises when approximate numbers are added or subtracted. In such cases consideration should be given to the precision indicated by each measurement. The sum of a group of measurements may have more significant digits than some of the individual measurements, but it cannot be more *precise*. For example, suppose we determine the perimeter of the rectangular plate. We might write:

$$9.54 \text{ cm} + 3.4 \text{ cm} + 9.54 \text{ cm} + 3.4 \text{ cm} = 25.9 \text{ cm}$$

The least precise measurement was to the nearest tenth of a centimeter; therefore, the perimeter should be given to the nearest tenth of a centimeter (even though it has three significant figures).

Rule 2 When approximate numbers are added or subtracted, the number of decimal places in the result should equal the smallest number of decimal places of any term in the sum.

Throughout this text, we will generally assume that the data given are precise enough to give answers rounded to three significant figures. Thus, an 8-in. rod would be treated as though it were 8.00 in. long. If converted to metric units, the length would be reported as 203 mm. This practice is common since it avoids the necessity of cluttering the text with zeros.

3-5 MEASURING INSTRUMENTS

The choice of a measuring instrument is determined by the accuracy required and by the physical conditions surrounding the measurement. A basic choice for the mechanic or machinist is most often the steel rule, such as the one shown in Fig. 3-2 (page 46). This rule is often accurate enough when you are measuring openly accessible lengths. Steel rules may be graduated as fine as thirty-seconds or even sixty-fourths of an inch. Metric rules are usually graduated in millimeters.

For the measurement of inside and outside diameters, calipers such as those shown in Fig. 3-3 (page 46) may be used. The caliper itself cannot be read directly and therefore must be set to a steel rule or a standard size gauge.

The best accuracy possible with a steel rule is determined by the size of the smallest graduation and is on the order of 0.01 in. or 0.1 mm. For greater accuracy the machinist often uses a standard micrometer caliper, as in Fig. 3-4 (page 46), or a standard vernier caliper, as in Fig. 3-5 (page 47). These instruments make

physics.nist.gov/cuu/ Uncertainty/index.html
When expressing a quantity, you need to know the number of significant figures. This will ensure that you know how believable your answer is. For guidelines on measurement uncertainty, see this Internet site.

ts.nist.gov
All measuring instruments need to be calibrated. Information on calibration can be found at Technology Services Information sites.

When a micrometer caliper is calibrated, thin rectangles of uniform thickness are joined to make a known width. Grime, dust, or even tarnish could change the measurement. The calibration bars are rubbed clean and carefully stored away from moisture.

(a)

(b)

Figure 3-2 Some 6-in. (15-cm) steel scales. (a) Scales 1/32 in. and 0.5 mm. (*The L. S. Starrett Company.*) (b) Scales 1/1000 and 1/50 in. (*The L. S. Starrett Company.*)

Figure 3-3 Using calipers to measure an inside diameter.

Figure 3-4 A micrometer caliper, showing a reading of 5.78 mm. (*The L. S. Starrett Company.*)

Figure 3-5 The vernier caliper. (*The L. S. Starrett Company.*)

Figure 3-6 Measuring the depth of a shoulder with a micrometer depth gauge.

use of sliding scales to record very accurate measurements. Micrometer calipers can measure to the nearest ten-thousandth of an inch (0.002 mm), and vernier calipers are used to measure within 0.001 in. or 0.02 mm.

The depth of blind holes, slots, and recesses is often measured with a micrometer depth gauge. Fig. 3-6 shows such a gauge used to measure the depth of a shoulder.

3-6 UNIT CONVERSIONS

Because so many different units are required for a variety of jobs, it is often necessary to convert a measurement from one unit to another. For example, suppose that a machinist records the outside diameter of a pipe as $1\frac{3}{16}$ in. To order a fitting for the pipe, the machinist may need to know this diameter in millimeters. Such conversions can be accomplished easily by treating units algebraically and using the principle of cancellation.

In the above case, the machinist should first convert the fraction to a decimal.

$$1\frac{3}{16} \text{ in.} = 1.19 \text{ in.}$$

Next, the machinist should write down the quantity to be converted, giving both the number and the unit (1.19 in.). The definition that relates inches to millimeters is recalled:

$$1 \text{ in.} = 25.4 \text{ mm}$$

Since this statement is an equality, we can form two ratios, each equal to 1.

$$\frac{1 \text{ in.}}{25.4 \text{ mm}} = 1 \qquad \frac{25.4 \text{ mm}}{1 \text{ in.}} = 1$$

Note that the number 1 does not equal the number 25.4, but the *length* of 1 in. is equal to the *length* of 25.4 mm. Thus, if we multiply some other length by either of these ratios, we will get a new number, but we will not change the length. Such ratios are called **conversion factors.** Either of the above conversion factors may be multiplied by 1.19 in. without changing the length represented. Multiplication by the first ratio does not give a meaningful result. Note that units are treated as algebraic quantities.

$$(1.19 \text{ in.}) \left(\frac{1 \text{ in.}}{25.4 \text{ mm}} \right) = \left(\frac{1.19}{25.4} \right) \left(\frac{\text{in.}^2}{\text{mm}} \right) \qquad \text{Wrong!}$$

Multiplication by the second ratio, however, gives the following results:

$$(1.19 \text{ in.}) \left(\frac{25.4 \text{ mm}}{1 \text{ in.}} \right) = \frac{(1.19)(25.4)}{(1)} \text{ mm} = 30.2 \text{ mm}$$

Therefore, the outside diameter of the pipe is 30.2 mm.

Sometimes it is necessary to work with quantities that have multiple units. For example, **speed** is defined as *length* per unit *time* and may have units of *meters per second* (m/s), *feet per second* (ft/s), or other units. The same algebraic procedure can help with conversion of multiple units.

Problem-Solving Strategy

Procedure for Converting Units

1. Write down the quantity to be converted.
2. Define each unit appearing in the quantity to be converted in terms of the desired unit(s).

3. For each definition, form two conversion factors, one being the reciprocal of the other.
4. Multiply the quantity to be converted by those factors that will cancel all but the desired units.

EXAMPLE 3-1

Convert a speed of 60 km/h to units of meters per second.

Solution

We recall two definitions that might result in four possible conversion factors.

$$1 \text{ km} = 1000 \text{ m} \begin{cases} \dfrac{1000 \text{ m}}{1 \text{ km}} \\[2ex] \dfrac{1 \text{ km}}{1000 \text{ m}} \end{cases}$$

$$1 \text{ h} = 3600 \text{ s} \begin{cases} \dfrac{1 \text{ h}}{3600 \text{ s}} \\[8pt] \dfrac{3600 \text{ s}}{1 \text{ h}} \end{cases}$$

We write down the quantity to be converted and then choose conversion factors that will cancel nondesired units.

$$60 \; \frac{\cancel{\text{km}}}{\cancel{\text{h}}} \left(\frac{1000 \text{ m}}{1 \; \cancel{\text{km}}} \right) \left(\frac{1 \; \cancel{\text{h}}}{3600 \text{ s}} \right) = 16.7 \; \frac{\text{m}}{\text{s}}$$

Additional examples of the procedure are as follows:

$$30 \; \frac{\cancel{\text{mi}}}{\cancel{\text{h}}} \left(\frac{5280 \text{ ft}}{1 \; \cancel{\text{mi}}} \right) \left(\frac{1 \; \cancel{\text{h}}}{60 \; \cancel{\text{min}}} \right) \left(\frac{1 \; \cancel{\text{min}}}{60 \text{ s}} \right) = 44 \text{ ft/s}$$

$$20 \; \frac{\cancel{\text{lb}}}{\text{in.}^2} \left(\frac{1550 \; \cancel{\text{in}^2}}{1 \text{ m}^2} \right) \left(\frac{4.448 \text{ N}}{1 \; \cancel{\text{lb}}} \right) = 1.38 \times 10^5 \text{ N/m}^2$$

If the required definitions are not available in this chapter, they can be found on the inside front cover of this book.

When you are working with technical formulas, it is always helpful to substitute units as well as numbers. For example, the formula for speed v is

$$v = \frac{s}{t}$$

where s is the distance traveled in a time t. Thus, if a car travels 400 m in 10 s, its speed will be

$$v = \frac{400 \text{ m}}{10 \text{ s}} = 40 \; \frac{\text{m}}{\text{s}}$$

Notice that the units of velocity are meters per second, written m/s.

Whenever velocity appears in a formula, it must always have units of *length* divided by *time*. These are said to be the **dimensions** of velocity. There may be many different units for a given physical quantity, but the dimensions result from a definition, and they do not change.

When you are working with formulas, it will be useful to remember two rules concerning dimensions:

Rule 1 If two quantities are to be added or subtracted, they must be of the same dimensions.

Rule 2 The quantities on both sides of an equals sign must be of the same dimensions.

EXAMPLE 3-2

Show that the following formula is dimensionally correct.

$$s = v_0 t + \frac{1}{2}at^2$$

The symbol s represents the distance traveled in a time t while accelerating at a rate a from an initial speed v_0. Assume that acceleration has units of meters per second squared (m/s^2).

Solution

Since the units of a are given, the unit of length must be meters (m) and the unit of time must be seconds (s). Ignoring the factor $\frac{1}{2}$, which has no dimensions, we substitute units for the quantities in the equation.

$$\text{m} = \frac{\text{m}}{\cancel{s}}(\cancel{s}) + \frac{\text{m}}{\cancel{s^2}}(\cancel{s})^2$$

Notice that both Rule 1 and Rule 2 are satisfied. Therefore, the equation is dimensionally correct.

The fact that an equation is dimensionally correct is a valuable check. Such an equation still may not be a *true* equation, but at least it is consistent dimensionally.

3-7 VECTOR AND SCALAR QUANTITIES

Some quantities can be described totally by a number and a unit. Only the *magnitudes* are of interest in an area of 12 m^2, a volume of 40 ft^3, or a distance of 50 km. Such quantities are called **scalar quantities.**

> A scalar quantity is specified completely by its magnitude—a number and a unit. Examples are speed (15 mi/h), distance (12 km), and volume (200 cm^3).

Scalar quantities that are measured in the same units may be added or subtracted in the usual way. For example,

$$14 \text{ mm} + 13 \text{ mm} = 27 \text{ mm}$$
$$20 \text{ ft}^2 - 4 \text{ ft}^2 = 6 \text{ ft}^2$$

Some physical quantities, such as force and velocity, have direction as well as magnitude. In such cases, they are called **vector quantities.** The direction must be a part of any calculations involving such quantities.

> A vector quantity is specified completely by a magnitude and a direction. It consists of a number, a unit, and a direction. Examples are displacement (20 m, N) and velocity (40 mi/h, 30° N of W).

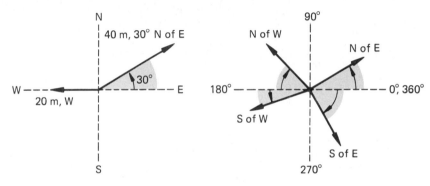

Figure 3-7 Indicating the direction of a vector by reference to north (N), south (S), east (E), and west (W).

The direction of a vector may be given by reference to conventional north, east, west, and south directions. Consider, for example, the vectors 20 m, W and 40 m at 30° N of E, as shown in Fig. 3-7. The expression "north of east" indicates that the angle is formed by rotating a line northward from the easterly direction.

Another method of specifying direction that will be particularly useful later on is to make reference to perpendicular lines called *axes*. These imaginary lines are usually chosen to be horizontal and vertical, but they may be oriented along other directions as long as the two lines remain perpendicular. An imaginary horizontal line is usually called the *x* axis, and an imaginary vertical line is called the *y* axis. See Fig. 3-8. Directions are given by angles measured counterclockwise from the positive *x* axis. The vectors 40 m at 60° and 50 m at 210° are shown in the figure.

Assume that a person travels by car from Atlanta to St. Louis. The *displacement* from Atlanta can be represented by a line segment drawn to scale from Atlanta to St. Louis (see Fig. 3-9 on page 52). An arrowhead is drawn on the St. Louis end to denote the direction. It is important to note that the displacement, represented by the vector D_1, is completely independent of the actual path or the mode of transportation. The odometer would show that the car had actually traveled a scalar distance s_1 of 541 mi. The magnitude of the displacement is only 472 mi.

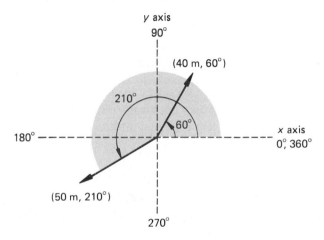

Figure 3-8 Indicating the direction of a vector as an angle measured from the positive *x* axis.

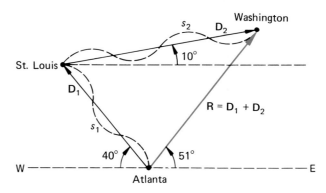

Figure 3-9 Displacement is a vector quantity; its direction is indicated by a solid arrow. Distance is a scalar quantity, indicated above by a dotted line.

Another important difference between a vector displacement and a scalar displacement is that the vector component has a constant direction of 140° (or 40° N of W). The direction of the car at any instant on the trip is not important, however, when considering the scalar distance.

Now, let us suppose that our traveler continues the drive to Washington. This time the vector displacement D_2 is 716 mi at a constant direction of 10° N of E. The corresponding ground distance s_2 is 793 mi. The total distance traveled for the entire trip from Atlanta is the arithmetic sum of the scalar quantities s_1 and s_2.

$$s_1 + s_2 = 541 \text{ mi} + 793 \text{ mi} = 1334 \text{ mi}$$

The *vector sum* of the two displacements D_1 and D_2 must take note of direction, however, as well as magnitudes. The question now is not the distance traveled but the resultant displacement from Atlanta. This vector sum is represented in Fig. 3-9 by the symbol **R,** where

$$\mathbf{R} = \mathbf{D}_1 + \mathbf{D}_2$$

Methods we will discuss in the next section will allow us to determine the magnitude and direction of **R.** Using a ruler and a device for measuring angles, we would see that

$$\mathbf{R} = 545 \text{ mi}, 51°$$

Remember that, in performing vector additions, both the magnitude and the direction of the displacements must be considered. The additions are geometric instead of algebraic. It is possible for the magnitude of a vector sum to be less than the magnitude of either of the component displacements.

A vector is usually denoted in print by boldface type. For example, the symbol D_1 denotes a displacement vector in Fig. 3-9. A vector can be denoted conveniently in handwriting by underscoring the letter or by putting an arrow over it. In print, the magnitude of a vector is usually indicated by italics; thus, D denotes the magnitude of the vector **D.** A vector is often specified by a pair of numbers (R, θ). The first number and unit give the magnitude, and the second number gives the angle measured counterclockwise from the positive x axis. For example,

$$\mathbf{R} = (R, \theta) = (200 \text{ km}, 114°)$$

Note that the magnitude R of a vector is always positive. A negative sign before the symbol of a vector merely reverses its direction; in other words, it interchanges the arrow tip without affecting the length. If $\mathbf{A} = (10 \text{ m}, \text{E})$, then $-\mathbf{A}$ would be $(10 \text{ m}, \text{W})$.

3-8 ADDITION OF VECTORS BY GRAPHICAL METHODS

In this section we discuss two common graphical methods for finding the geometric sum of vectors. The **polygon method** is the more useful one, since it can be readily applied to more than two vectors. The **parallelogram method** is useful for the addition of two vectors at a time. In each case the magnitude of a vector is indicated to scale by the length of a line segment. The direction is denoted by an arrow tip at the end of the line segment.

EXAMPLE 3-3

A ship travels 100 km due north on the first day of a voyage, 60 km northeast on the second day, and 120 km due east on the third day. Find the resultant displacement by the polygon method.

Solution
A suitable scale might be 20 km = 1 cm, as in Fig. 3-10 on page 54. Using this scale, we note that

$$100 \text{ km} = 100 \text{ km} \times \frac{1 \text{ cm}}{20 \text{ km}} = 5 \text{ cm}$$

$$60 \text{ km} = 60 \text{ km} \times \frac{1 \text{ cm}}{20 \text{ km}} = 3 \text{ cm}$$

$$120 \text{ km} = 120 \text{ km} \times \frac{1 \text{ cm}}{20 \text{ km}} = 6 \text{ cm}$$

By measuring with a ruler, we find from the scale diagram that the arrow for the resultant is 10.8 cm long. Therefore, the magnitude is

$$10.8 \text{ cm} = 10.8 \text{ cm} \times \frac{20 \text{ km}}{1 \text{ cm}} = 216 \text{ km}$$

Measuring the angle θ with a protractor shows the direction to be 41°. The resultant displacement is therefore

$$\mathbf{R} = (216 \text{ km}, 41°)$$

Math Tip

Here are some tips to remember when checking to see whether your answer makes sense. A minus times a minus equals a plus. An even times an even equals an even. An even times an odd equals an even. An odd times an odd equals an odd. Two numbers between 0 and 1 that are multiplied will result in a smaller number.

Math Tip

Some numbers represented as fractions can be larger than 1. For example, 4/3, $\pi/2$, and 12/7 are each larger than 1.

Figure 3-10 The polygon method of vector addition.

Note that the order in which the vectors are added does not change the resultant in any way. We could have begun with any of the three distances traveled by the ship in the previous example.

Graphical methods can be used to find the resultant of all kinds of vectors. They are not restricted to measuring displacement, and they are particularly useful in finding the resultant of a number of *forces*. For now, consider a force to be defined as a push or pull that tends to produce motion. The force vector is also specified by a number, unit, and angle, as with displacements, and they can be added in the same manner as for displacement vectors.

Problem-Solving Strategy

The Polygon Method of Vector Addition

1. Choose a scale and determine the length of the arrows that correspond to each vector.
2. Draw to scale an arrow representing the magnitude and direction of the first vector.
3. Draw the arrow of the second vector so that its tail is joined to the tip of the first vector.
4. Continue the process of joining tail to tip until the magnitude and direction of all vectors have been represented.
5. Draw the resultant vector with its tail at the origin (starting point) and its tip joined to the tip of the last vector.
6. Measure with ruler and protractor to determine the magnitude and direction of the resultant vector.

What is the resultant force on the donkey?

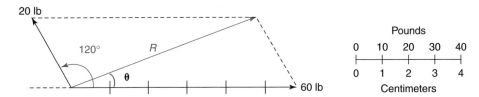

Figure 3-11 The parallelogram method of vector addition.

In the next example we determine the resultant force on a donkey being pulled in different directions with two ropes (see Fig. 3-11). This time we will apply the *parallelogram method,* which is useful only for two vectors at a time. Each vector is drawn to scale, and both tails are at a common origin. The two arrows then form two adjoining sides of a parallelogram. The other two sides are constructed by drawing parallel lines of equal length. The resultant is represented by the diagonal of the parallelogram included between the two vector arrows.

EXAMPLE 3-4

Find the resultant force on the donkey in Fig. 3-11 if the angle between the two ropes is 120°. One end is pulled with a force of 60 lb, and the other with a force of 20 lb.

Solution
Using a scale of 1 cm = 10 lb gives

$$60 \text{ lb} \times \frac{1 \text{ cm}}{10 \text{ lb}} = 6 \text{ cm} \qquad 20 \text{ lb} \times \frac{1 \text{ cm}}{10 \text{ lb}} = 2 \text{ cm}$$

A parallelogram is constructed in Fig. 3-11 by drawing the two forces to scale from a common origin. Use a protractor to make sure that the angle between the vectors is 120°. Completing the parallelogram allows the resultant to be drawn as a diagonal from the origin. Measurement of R and θ with a ruler and protractor and converting as before gives 52.9 lb for the magnitude and 19.1° for the direction.

$$\mathbf{R} = (52.9 \text{ lb}, 19.1°)$$

A second look at the parallelogram will show that the same answer would result from applying the polygon method and attaching the 20-lb vector to the tip of the 60-lb vector.

3-9 FORCE AND VECTORS

Pocket Lab

Ask two students to hold onto the ends of a strong 10-m rope, one student at each end. Have each student pull on his or her end. The rope will seem straight. Then have a third student push down with one finger on the middle of the rope. Can the students pulling at each end of the rope keep it from sagging? Explain the results. (*Answer:* It is impossible to prevent the third student from causing the rope to sag because the students pulling on the rope cannot create enough upward force to counteract the downward force of the finger pushing on the rope.)

As you saw in the previous section, *force* vectors can be added graphically, just as was done earlier for displacements. Because forces are such an important part of the study of mechanics, it is wise to develop vector skills by using force applications in addition to displacement applications. A stretched spring exerts forces on the objects to which its ends are attached, compressed air exerts forces on the walls of its container, and a tractor exerts a force on the trailer it is pulling. Probably the most familiar force is that of gravitational attraction exerted on every body by the earth. This force is called the **weight** of the body. A definite force exists even though there is no contact between the earth and the bodies it attracts. Weight as a vector quantity is directed toward the center of the earth.

The SI unit of force is the newton (N), which will be properly defined in a later chapter. For the moment, consider its relationship to the pound:

$$1 \text{ N} = 0.225 \text{ lb} \qquad 1 \text{ lb} = 4.45 \text{ N}$$

A 120-lb woman has a weight of 534 N. If the weight of a pipe wrench were 20 N, it would weigh about 4.5 lb in USCS units. Until all industries have converted completely to SI units, the pound will still be with us, and frequent conversions are necessary. We will use both units of force in this chapter as we learn to work with vector quantities.

Two of the measurable effects of forces are (1) changing the dimensions or shape of a body and (2) changing a body's motion. If in the first case there is no resultant displacement of the body, the push or pull causing the change in shape is called a **static force.** If a force changes the motion of a body, it is called a **dynamic force.** Both types of forces are conveniently represented by vectors, as in Example 3-4.

The effectiveness of any force depends on the direction in which it acts. For example, it is easier to pull a sled along the ground with an inclined rope, as shown in Fig. 3-12, than to push it. In each case, the applied force is producing more than a single effect. That is, the pull on the cord is both lifting the sled and moving it forward. Similarly, pushing the sled would have the effect of adding to the weight of the sled. We are thus led to the idea of **components** of a force: the effective values of a force in directions other than that of the force itself. In Fig. 3-12 the force **F** can be replaced by its horizontal and vertical components, $\mathbf{F}_x$ and $\mathbf{F}_y$.

Figure 3-12

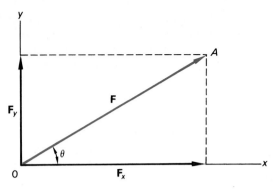

Figure 3-13 Graphical representation of the x and y components of **F**.

If a force is represented graphically by its magnitude and an angle (R, θ), its components along the x and y directions can be determined. A force **F** acting at an angle θ above the horizontal is drawn in Fig. 3-13. The meaning of the x and y components, F_x and F_y, can be seen in this diagram. The segment from O to the perpendicular dropped from A to the x axis is called the x component of **F** and is labeled $\mathbf{F}_x$. The segment from O to the perpendicular line from A to the y axis is called the y component of **F** and is labeled $\mathbf{F}_y$. By drawing the vectors to scale, we can determine the magnitudes of the components graphically. These two components acting together would have the same effect as the original force **F**.

EXAMPLE 3-5

A lawn mower is pushed downward with a force of 40 N at an angle of 50° with the horizontal. What is the magnitude of the horizontal effect of this force?

Figure 3-14 Finding the components of a force by the graphical method.

Solution

A sketch is drawn, as in Fig. 3-14a, to translate the word problem into a picture. This approach often aids in understanding the problem. The 40-N force is transmitted through the handle to the body of the lawn mower. A vector diagram is shown in Fig. 3-14b. A ruler and protractor are used to draw the diagram accurately. A scale of 1 cm = 10 N is convenient for this example. The horizontal effect of the 40-N force is the *x* component, as labeled in the figure. Measurement of this line segment gives

$$F_x = 2.57 \text{ cm}$$

Since 1 cm = 10 N, we obtain

$$F_x = 2.57 \text{ cm} \frac{10 \text{ N}}{1 \text{ cm}} = 25.7 \text{ N}$$

Notice that the effective force is quite a bit less than the applied force. As an additional exercise, you should show that the magnitude of the *downward* component of the 40-N force is $F_y = 30.6$ N.

3-10 THE RESULTANT FORCE

When two or more forces act at the same point on an object, they are said to be *concurrent forces.* The combined effect of such forces is called the *resultant force.*

> **The resultant force is the single force that will produce the same effect in both magnitude and direction as two or more concurrent forces.**

Resultant forces may be calculated graphically by representing each concurrent force as a vector. The polygon or parallelogram method of vector addition will then give the resultant force.

Often forces act in the same line, either together or in opposition to each other. If two forces act on a single object in the same direction, the resultant force is equal to the sum of the magnitudes of the forces. The direction of the resultant would be the same as that of either force. Consider, for example, a 15-N force and a 20-N force acting in the same easterly direction. Their resultant is 35 N, E, as demonstrated in Fig. 3-15a.

If the same two forces act in opposite directions, the magnitude of the resultant force is equal to the *difference* of the magnitudes of the two forces, and it acts in the direction of the larger force. Suppose the 15-N force in our example were changed so that it pulled to the west. As seen in Fig. 3-15b, the resultant would be 5 N, E.

If forces act at an angle between 0 and 180° to each other, their resultant is the vector sum. The polygon method or the parallelogram method of vector addition may be used to find the resultant force. In Fig. 3-15b, our two forces of 15 and 20 N act at an angle of 60° with each other. The resultant force, calculated by the parallelogram method, is found to be 30.4 N at 34.7°.

(a) Forces in the same direction.

(b) Forces acting in opposite directions.

(c) Forces acting at an angle of 60° with respect to each other.

Figure 3-15 The effect of direction on the resultant of two forces.

3-11 TRIGONOMETRY AND VECTORS

Graphical treatments of vectors are effective for visualizing forces, but they are usually not too accurate. A much more useful approach is to take advantage of simple right-triangle trigonometry, and today's calculators have simplified the process considerably. Familiarity with the *pythagorean* theorem and some experience with the *sine, cosine,* and *tangent* functions is all you will need for this unit of study.

Trigonometric methods can improve your accuracy and speed in determining the resultant vector or in finding the components of a vector. In most cases, it is helpful to use imaginary *x* and *y* axes when working with vectors in an analytical way. Any vector can then be drawn with its tail at the center of these imaginary lines. Components of the vector might be seen as effects along the *x* and *y* axis.

USING A CALCULATOR

Although most angles in this book are given in degrees, some may be in radians. Keep this in mind when performing calculations involving angles. Your calculator has three angle settings: DEG (degrees), RAD (radians), and GRAD (gradients). Check the manual that came with your calculator to see how to change from one setting to another. Make sure that the calculator and the problem angle types agree.

EXAMPLE 3-6

What are the *x* and *y* components of a force of 200 N at an angle of 60°?

Solution

A diagram is drawn placing the tail of the 200-N vector at the center of the *x* and *y* axes (see Fig. 3-16 on page 60).

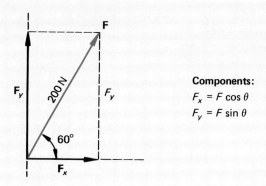

Components:
$$F_x = F \cos \theta$$
$$F_y = F \sin \theta$$

Figure 3-16 Using trigonometry to find the x and y components of a vector.

We first compute the x component, F_x, by noting that it is the side adjacent. The 200-N vector is the hypotenuse. Using the cosine function, we obtain

$$\cos 60° = \frac{F_x}{200 \text{ N}}$$

from which

$$F_x = (200 \text{ N}) \cos 60° = 100 \text{ N}$$

For purposes of calculation, we recognize that the side opposite to 60° is equal in length to F_y. Thus, we may write

$$\sin 60° = \frac{F_y}{200 \text{ N}}$$

or

$$F_y = (200 \text{ N}) \sin 60° = 173.2 \text{ N}$$

In general, we may write the x and y components of a vector in terms of its magnitude F and direction θ.

$$F_x = F \cos \theta$$
$$F_y = F \sin \theta$$

Components of a Vector   **3-1**

where θ is the angle between the vector and the positive x axis, measured in a counterclockwise direction.

The sign of a given component can be determined from a vector diagram. The four possibilities are shown in Fig. 3-17. The magnitude of the component can be found by using the small angle φ when angle θ is greater than 90°.

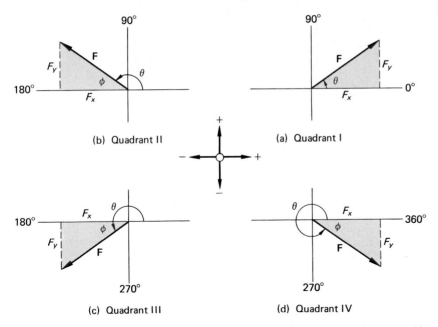

90°

(b) Quadrant II

90°

(a) Quadrant I

180°

180°

270°

(c) Quadrant III

270°

(d) Quadrant IV

360°

Figure 3-17 (a) In the first quadrant, angle θ is between 0° and 90°; both F_x and F_y are positive. (b) In the second quadrant, angle θ is between 90° and 180°; F_x is negative and F_y is positive. (c) In the third quadrant, angle θ is between 180° and 270°; F_x and F_y are negative. (d) In the fourth quadrant, angle θ is between 270° and 360°; F_x is positive and F_y is negative.

EXAMPLE 3-7

Find the x and y components of a 400-N force at an angle of 220° from the positive x axis.

Solution

Refer to Fig. 3-17c, which describes this problem where $\theta = 220°$. The small angle ϕ is found by reference to 180°.

$$\phi = 220° - 180° = 40°$$

From the figure, both F_x and F_y will be negative.

$$F_x = -|F \cos \phi| = -(400 \text{ N}) \cos 40°$$
$$= -(400 \text{ N})(0.766) = -306 \text{ N}$$
$$F_y = -|F \sin \phi| = -(400 \text{ N}) \sin 40°$$
$$= -(400 \text{ N})(0.643) = -257 \text{ N}$$

Note that the signs were determined from Fig. 3-17. With many electronic calculators, both the magnitude and the sign of F_x and F_y can be found directly from Eq. (3-1) using $\theta = 220°$. You should verify this fact.

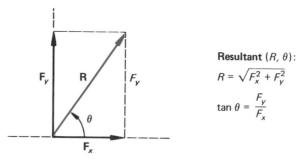

$$\textbf{Resultant } (R, \theta):$$

$$R = \sqrt{F_x^2 + F_y^2}$$

$$\tan \theta = \frac{F_y}{F_x}$$

Figure 3-18 The resultant of perpendicular vectors.

Trigonometry is also useful in calculating the resultant force. In the special case for two forces $\mathbf{F}_x$ and $\mathbf{F}_y$ at right angles to each other, as in Fig. 3-18, the resultant (R, θ) may be found from

$$R = \sqrt{F_x^2 + F_y^2} \qquad \tan \theta = \frac{F_y}{F_x} \qquad\qquad \textbf{3-2}$$

If either F_x or F_y is negative, it is usually easier to determine the small angle ϕ, as described in Fig. 3-17. The sign (or direction) of the forces F_x and F_y determines which of the four quadrants is used. Then, Eq. (3-2) becomes

$$\tan \phi = \left| \frac{F_y}{F_x} \right|$$

Only the absolute values of F_x and F_y are needed. If desired, the angle θ from the positive x axis may be determined. In any case, the direction must be clearly identified.

EXAMPLE 3-8

What is the resultant of a 5-N force directed horizontally to the right and a 12-N force directed vertically downward?

Solution

Label the two components $F_x = 5$ N and $F_y = -12$ N (downward). Draw a diagram for the situation described by Fig. 3-17d. The magnitude of the resultant R is found from Eq. (3-2):

$$R = \sqrt{F_x^2 + F_y^2} = \sqrt{(5 \text{ N})^2 + (-12 \text{ N})^2}$$

$$= \sqrt{25 \text{ N}^2 + 144 \text{ N}^2} = \sqrt{169 \text{ N}^2} = 13 \text{ N}$$

To find the direction of R, we first find the small angle ϕ:

$$\tan \phi = \left| \frac{-12 \text{ N}}{5 \text{ N}} \right| = 2.4$$

$$\phi = 67.4° \text{ below positive } x \text{ axis}$$

The angle θ measured counterclockwise from the positive x axis is

$$\theta = 360° - 67.4° = 292.6°$$

The resultant force is 13 N at 292.6°.

3-12 THE COMPONENT METHOD OF VECTOR ADDITION

Often a body is acted on by a number of forces having different magnitudes, directions, and points of application. Forces that intersect at a common point or have the same point of application are said to be *concurrent forces.* When such forces are not at right angles with each other, calculating the resultant can be more difficult. The vectors would not fall along either the x or the y axis in all cases. The component method of addition of vectors is needed for these more general cases. Consider the vectors **A, B,** and **C** in Fig. 3-19. The resultant **R** is the vector sum **A** + **B** + **C**. However, **A** and **B** are not along an axis and cannot be added in the usual manner.

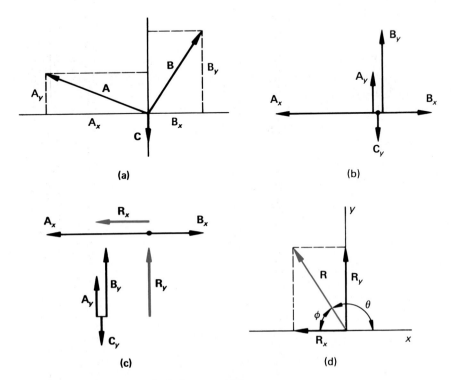

Figure 3-19 The component method of vector addition.

Problem-Solving Strategy

The Component Method of Vector Addition

(These steps are shown graphically in Fig. 3-19.)

1. Draw each vector from the center of imaginary x and y axes.
2. Find the x and y components of each vector.
3. Find the x component of the resultant by adding the x components of all vectors. (Components to the right are positive and those to the left are negative.)

$$R_x = A_x + B_x + C_x$$

4. Find the y component of the resultant by adding the y components of all vectors. (Upward components are positive and downward components are negative.)

$$R_y = A_y + B_y + C_y$$

5. Find the magnitude and direction of the resultant from its perpendicular components R_x and R_y.

$$R = \sqrt{R_x^2 + R_y^2} \qquad \tan \theta = \frac{R_y}{R_x}$$

EXAMPLE 3-9

Three ropes are tied to a stake, and the following forces are exerted: $\mathbf{A} = 20$ lb, E; $\mathbf{B} = 30$ lb, 30° N of W; and $\mathbf{C} = 40$ lb, 52° S of W. Determine the resultant force.

Solution

Follow the steps described in the problem-solving strategy.

1. Draw a figure representing each force (Fig. 3-20a). Two things should be noticed from the figure: (a) all angles are determined from the x axis, and (b) the components of each vector are labeled opposite and adjacent to known angles.
2. Find the x and y components of each vector. (Refer to Fig. 3-20b, c, and d.) Note that the force A has no y component. Care must be taken to obtain the correct sign for each component. For example B_x, C_x, and C_y are each negative. The results are listed in Table 3-5.
3. Add the x components to obtain R_x.

$$R_x = A_x + B_x + C_x$$
$$= 20 \text{ lb} - 26 \text{ lb} - 24.6 \text{ lb} = -30.6 \text{ lb}$$

4. Add the y components to obtain R_y.

$$R_y = A_y + B_y + C_y$$
$$= 0 + 15 \text{ lb} - 31.5 \text{ lb} = -16.5$$

5. Now we find R and θ from R_x and R_y (see Fig. 3-21):

$$R = R_x^2 + R_y^2 = \sqrt{(-30.6)^2 + (-16.5)^2}$$
$$= \sqrt{936.4 + 272.2} = 34.8 \text{ lb}$$

$$\tan \phi = \left| \frac{R_y}{R_x} \right| = \left| \frac{-16.5}{-30.6} \right| = 0.539$$

$$\phi = 28.3° \text{ S of W (or 208.3°)}$$

Thus, the resultant force is 34.8 lb at 28.3° S of W.

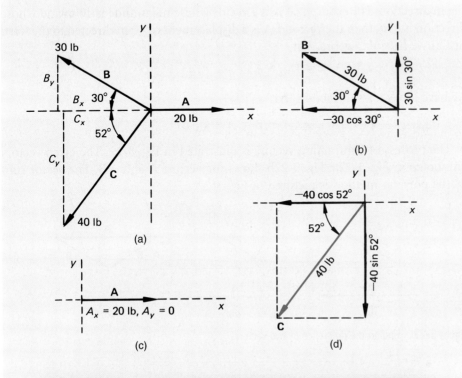

(a)

(b)

(c)

(d)

Figure 3-20 Calculating the x and y components of all vectors.

TABLE 3-5 A Table of Components

Force	ϕ_x	x component	y component
$A = 20$ lb	0°	$A_x = 20$ lb	$A_y = 0$
$B = 30$ lb	30°	$B_x = -(30 \text{ lb})(\cos 30°)$ $= -26.0$ lb	$B_y = (30 \text{ lb})(\sin 30°)$ $= 15$ lb
$C = 40$ lb	52°	$C_x = (40 \text{ lb})(\cos 52°)$ $= -24.6$ lb	$C_y = (-40 \text{ lb})(\sin 52°)$ $= -31.5$ lb
		$R_x = \sum F_x = -30.6$ lb	$R_y = \sum F_y = -16.5$ lb

Figure 3-21

3-13 VECTOR DIFFERENCE

When we study relative velocity, acceleration, and certain other quantities, we must be able to find the difference of two vector quantities. The difference between two vectors is obtained by adding one vector to the negative of the other. The negative of a vector is found by constructing a vector equal in magnitude but opposite in direction. For example, if **A** is a vector whose magnitude is 40 m and whose direction is east, then the vector −**A** is a displacement of 40 m directed to the west. Just as we have in algebra that

$$a - b = a + (-b)$$

we have in vector subtraction that

$$\mathbf{A} - \mathbf{B} = \mathbf{A} + (-\mathbf{B})$$

The process of subtracting vectors is illustrated in Fig. 3-22. The given vectors are shown in Fig. 3-22a; Fig. 3-22b shows the vectors **A** and −**B**. The vector sum by the polygon method is pictured in Fig. 3-22c.

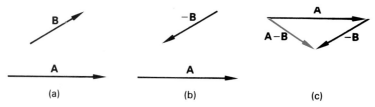

(a)   (b)   (c)

Figure 3-22   Finding the difference of two vectors.

ABOUT PHYSICS

Are Crash Test Dummies History?
At BMW's design facilities in Munich, Germany, advanced supercomputers and high-powered workstations are performing crash test simulations to help design safer vehicles. The calculations these engineers program into the computers are based on the methods of vectorial addition of forces. Although computer simulations are not likely to put crash test dummies out of a job, the computer tests can make it less likely for surprises in the crash test to send the designers back to the drawing board.

Summary and Review

Summary

Technical measurement is essential for the application of physics. You have learned that there are seven fundamental quantities and that each has a single approved SI unit. In mechanics, the three fundamental quantities of length, mass, and time are the only ones required for most applications. Some of these applications involve *vectors* and some involve *scalars*. Since vector quantities have direction, they must be added and subtracted by special methods. The following points summarize this unit of study:

- The SI prefixes used to express multiples and subdivisions of the base units are given below:

giga (G) = 10^9	milli (m) = 10^{-3}
mega (M) = 10^6	micro (μ) = 10^{-6}
kilo (k) = 10^3	nano (n) = 10^{-9}
centi (c) = 10^{-2}	pico (p) = 10^{-12}

- To convert one unit to another.

  a. Write down the quantity to be converted (number and unit).
  b. Recall the necessary definitions.
  c. Form two conversion factors for each definition.
  d. Multiply the quantity to be converted by those conversion factors that cancel all but the desired units.

- The *polygon method* of vector addition: The **resultant vector** is found by drawing each vector to scale, placing the tail of one vector to the tip of another until all vectors are drawn. The resultant is the straight line drawn from the starting point to the tip of the last vector (Fig. 3-23).
- The **parallelogram method** of vector addition: The resultant of two vectors is the diagonal of a parallelogram formed by the two vectors as adjacent sides. The direction is away from the common origin of the two vectors (Fig. 3-24).
- The *x* and *y components* of a vector (R, θ):

$$R_x = R \cos \theta \qquad R_y = R \sin \theta$$

- The *resultant* of two perpendicular vectors (R_x, R_y):

$$R = \sqrt{R_x^2 + R_y^2} \qquad \tan \phi = \frac{R_y}{R_x}$$

- The **component method** of vector addition:

$$R_x = A_x + B_x + C_x + \cdots$$
$$R_y = A_y + B_y + C_y + \cdots$$
$$R = \sqrt{R_x^2 + R_y^2}$$

$$\tan \phi = \frac{R_y}{R_x}$$

Figure 3-23

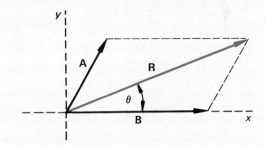

Figure 3-24

Summary and Review

Key Terms

1. fundamental quantity
2. SI units
3. conversion factor
4. dimensions
5. vector quantity
6. scalar quantity
7. components
8. resultant vector
9. concurrent forces
10. polygon method
11. parallelogram method
12. component method

Review Questions

3-1. Express the following measurements in proper SI form using the appropriate prefixes. The symbol for the base unit is given in parentheses:

  a. 298,000 meters (m)
  b. 7600 volts (V)
  c. 0.000067 amperes (A)
  d. 0.0645 newtons (N)
  e. 43,000,000 grams (g)
  f. 0.00000065 farads (F)

3-2. What three fundamental quantities appear in the definition of most laws of mechanics? Name the three fundamental units associated with each quantity in the SI and USCS units.

3-3. A unit of specific heat capacity is cal/g · C°. How many definitions are needed to convert these units to their corresponding units in the USCS, where the units are Btu/lb · F°? Show by a series of products how you would perform such a conversion.

3-4. Given that the units of s, v, a, and t are meters (m), meters per second (m/s), meters per second squared (m/s²), and seconds (s), respectively, what are the dimensions of each quantity? Accept or reject the following equations on the basis of dimensional analysis:

  a. $s = vt + \frac{1}{2}at^2$ **c.** $v_f = v_0 + at^2$
  b. $2as = \frac{2}{f} - v_0^2$ **d.** $s = vt + 4at^2$

3-5. Distinguish between vector and scalar quantities, and give examples of each. Explain the difference between adding vectors and adding scalars. Is it possible for the sum of two vectors to have a magnitude less than either of the original vectors?

3-6. What are the minimum and maximum resultants of two forces of 10 N and 7 N if they act on the same object?

3-7. Look up a section on rectangular and polar coordinates in a math book. What are the similarities between components of a vector and the rectangular and polar coordinates of a point?

3-8. If a vector has a direction of 230° from the positive x axis, what are the signs of its x and y components? If the ratio R_y/R_x is negative, what are the possible angles for R as measured from the positive x axis?

Note: In this chapter and others the assumption is that all numbers are accurate to three significant digits unless otherwise indicated. Answers are given to odd-numbered problems and to selected Critical Thinking Questions.

Problems

Section 3-6—Unit Conversions

3-1. A soccer field is 100 m long and 60 m across. What are the length and width of the field in feet?   **Ans. 328 ft, 197 ft**

3-2. A wrench has a handle 8 in. long. What is the length of the handle in centimeters?

3-3. A 19-in. computer monitor has a viewable area that measures 18 in. diagonally. Express this distance in meters.   **Ans. 0.457 m**

3-4. The length of a notebook is 234.5 mm and the width is 158.4 mm. Express the surface area in square meters.

3-5. A cube has 5 in. on a side. What is the volume of the cube in SI units and in USCS units? **Ans. 0.00205 m³, 0.0723 ft³**

3-6. The speed limit on an interstate highway is posted at 75 mi/h. (a) What is this speed in kilometers per hour? (b) In feet per second?

3-7. A Nissan engine has a piston displacement (volume) of 1600 cm³ and a bore diameter of 84 mm. Express these measurements in cubic inches and inches.

Ans. 97.6 in.³, 3.31 in.

3-8. An electrician must install an underground cable from the highway to a home located 1.20 mi into the woods. How many feet of cable will be needed?

3-9. One U.S. gallon is a volume equivalent to 231 in.³. How many gallons are needed to fill a tank that is 18 in. long, 16 in. wide, and 12 in. high? **Ans. 15.0 gal**

3-10. The density of brass is 8.89 g/cm³. What is the density in kilograms per cubic meter?

Section 3-8—Addition of Vectors by Graphical Methods

3-11. A woman walks 4 km east and then 8 km north. (a) Use the polygon method to find her resultant displacement. (b) Verify the result by the parallelogram method.

Ans. 8.94 km, 63.4° N of E

3-12. A land rover on the surface of Mars moves a distance of 38 m at an angle of 180°. It then turns and moves a distance of 66 m at an angle of 270°. What is the displacement from the starting position?

3-13. A surveyor starts at the southeast corner of a lot and charts the following displacements: $A = 600$ m, N; $B = 400$ m, W; $C = 200$ m, S; and $D = 100$ m, E. What is the net displacement from the starting point? **Ans. 500 m, 126.9°**

3-14. A downward force of 200 N acts simultaneously with a 500-N force directed to the left. Use the polygon method to find the resultant force.

3-15. The following three forces act simultaneously on the same object. $A = 300$ N, 30° N of E; $B = 600$ N, 270°; and $C = 100$ N due east. Find the resultant force using the polygon method. **Ans. 576 N, 51.4° S of E**

3-16. A boat travels west a distance of 200 m, then north for 400 m, and finally 100 m at 30° S of E. What is the net displacement?

3-17. Two ropes A and B are attached to a mooring hook so that an angle of 60° exists between the two ropes. The tension in rope A is 80 lb and the tension in rope B is 120 lb. Use the parallelogram method to find the resultant force on the hook. **Ans. 174 lb**

3-18. Two forces A and B act on the same object and produce a resultant force of 50 lb at 36.9° N of W. The force $A = 40$ lb due west. Find the magnitude and direction of force B.

Section 3-11—Trigonometry and Vectors

3-19. Find the x and y components of (a) a displacement of 200 km at 34°, (b) a velocity of 40 km/h at 120°, and (c) a force of 50 N at 330°. **Ans. 166 km, 112 km; −20 km/h, 34.6 km/h; 43.3 N, −25 N**

3-20. A sled is pulled with a force of 540 N at an angle of 40° with the horizontal. What are the horizontal and vertical components of this force?

3-21. The hammer in Fig. 3-25 applies a force of 260 N at an angle of 15° with the vertical. What is the upward component of the force on the nail? **Ans. 251 N**

260 lb

15°

Figure 3-25

Summary and Review

3-22. A jogger runs 2.0 mi west and then 6.0 mi north. Find the magnitude and direction of the resultant displacement.

3-23. A river flows south with a velocity of 20 km/h. A boat has a maximum speed of 50 km/h in still water. In the river, at maximum throttle, the boat heads due west. What are the resultant speed and direction of the boat? **Ans. 53.9 km/h, 21.8° S of W**

3-24. A rope that makes an angle of 30° with the horizontal drags a crate along the floor. What must be the tension in the rope, if a horizontal force of 40 lb is required to drag the crate?

3-25. A vertical lift of 80 N is needed to lift a window. A long pole is used to lift the window. What force must be exerted along the pole if it makes an angle of 34° with the wall?
Ans. 96.5 N

3-26. The resultant of two forces **A** and **B** is 400 N at 210°. If force **A** is 200 N at 270°, what are the magnitude and direction of force **B**?

Section 3-12—The Component Method of Vector Addition

3-27. Find the resultant of the following perpendicular forces: (a) 400 N, 0°, (b) 820 N, 270°, and (c) 500 N, 90°. **Ans. 512 N, 321.3°**

3-28. Four ropes, all at right angles to each other, pull on a ring. The forces are 40 lb, E; 80 lb, N; 70 lb, W; and 20 lb, S. Find the resultant force on the ring.

3-29. Two forces act on the car as shown in Fig. 3-26. Force **A** is equal to 120 N, west and force **B** is equal to 200 N at 60° N of W. What are the magnitude and direction of the resultant force on the car?
Ans. 280 N, 38.2° N of W

3-30. Suppose that the direction of force **B** in Prob. 3-29 is reversed (+180°) and other parameters are unchanged. What is the new resultant? (This result is the vector difference **A − B**.)

3-31. Determine the resultant force on the bolt in Fig. 3-27. **Ans. 69.6 lb, 154.1°**

Figure 3-26

Figure 3-27

3-32. Determine the resultant of the following forces by the component method of vector addition: **A** = (200 N, 30°), **B** = (300 N, 330°), and **C** = (400 N, 250°).

3-33. Three boats exert forces on a mooring hook as shown in Fig. 3-28. Find the resultant of these three forces. **Ans. 853 N, 101.7°**

Figure 3-28

Challenge Problems

3-34. Find the horizontal and vertical components of the following vectors: $A = (400$ N, $37°)$, $B = 90$ m, $320°)$, and $C = (70$ km/h, $150°)$.

3-35. A cable is attached to the end of a beam. What pull at an angle of $40°$ with the horizontal is needed to produce an effective horizontal force of 200 N? **Ans. 261 N**

3-36. A fishing dock runs north and south. What must be the speed of a boat heading at an angle of $40°$ E of N if its velocity component along the dock is to be 30 km/h?

3-37. Find the resultant $\mathbf{R} = \mathbf{A} + \mathbf{B}$ for the following pairs of forces: (a) $\mathbf{A} = (520$ N, south), $\mathbf{B} = 269$ N, west, (b) $\mathbf{A} = 18$ m/s, north, $\mathbf{B} = 15$ m/s, west.
Ans. 585 N, 242.6°; 23.4 m/s, 129.8°

3-38. Determine the vector difference $(\mathbf{A} - \mathbf{B})$ for the pairs of forces in Prob. 3-37.

3-39. A traffic light is attached to the midpoint of a rope so that each segment makes an angle of $10°$ with the horizontal. The tension in each rope segment is 200 N. If the resultant force at the midpoint is zero, what must be the weight of the traffic light?
Ans. 69.5 N

3-40. Determine the resultant of the forces shown in Fig. 3-29.

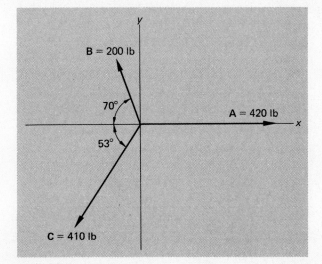

Figure 3-29

3-41. Determine the resultant of the forces shown in Fig. 3-30. **Ans. 225 N, 124.6°**

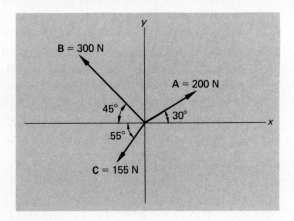

Figure 3-30

3-42. A 200-N block rests on a $30°$ inclined plane. If the weight of the block acts vertically downward, what are the components of the weight down the plane and perpendicular to the plane?

3-43. Find the resultant of the following three displacements: $\mathbf{A} = 220$ m, $60°$; $\mathbf{B} = 125$ m, $210°$; and $\mathbf{C} = 175$ m, $340°$.
Ans. 180 m, 22.3°

Critical Thinking Questions

*3-44. Consider three vectors: $\mathbf{A} = 100$ m, $0°$; $\mathbf{B} = 400$ m, $270°$; and $\mathbf{C} = 200$ m, $30°$. Choose an appropriate scale and show graphically that the order in which these vectors is added does not matter; that is, $\mathbf{A} + \mathbf{B} + \mathbf{C} = \mathbf{C} + \mathbf{B} + \mathbf{A}$. Is this also true for subtracting vectors? Show graphically how $\mathbf{A} - \mathbf{C}$ differs from $\mathbf{C} - \mathbf{A}$.

3-45. Two forces $A = 30$ N and $B = 90$ N can act on an object in any direction desired. What is the maximum resultant force? What is the minimum resultant force? Can the resultant force be zero? **Ans. 120 N, 60 N, no**

3-46. Consider two forces $A = 40$ N and $B = 80$ N. What must be the angle between these two forces in order to produce a resultant force of 60 N?

*3-47. What third force **F** must be added to the following two forces so that the resultant force is zero? **A** = 120 N, 110° and **B** = 60 N, 200°?   **Ans. 134 N, 316.6°**

*3-48. An airplane needs a resultant heading of due west. The speed of the plane is 600 km/h in still air. If the wind has a speed of 40 km/h and blows in a direction of 30° S of W, in what direction should the aircraft be pointed and what will be its speed relative to the ground?

*3-49. What are the magnitude F and direction θ of the force needed to pull the car of Fig. 3-31 directly east with a resultant force of 400 lb?   **Ans. 223 lb, 17.9°**

Figure 3-31

Translational Equilibrium and Friction

Forces may act in such a manner as to cause motion or to prevent motion. Large bridges must be designed so that the overall effect of forces is to prevent motion. Every truss, girder, beam, and cable must be in *equilibrium.* In other words, the resultant forces acting at any point on the entire structure must be balanced. Shelves, chain hoists, hooks, lifting cables, and even large buildings must be constructed so that the effects of forces are controlled and understood. In this chapter we will continue our study of forces by studying objects at rest. The friction force that is so essential for equilibrium in many applications will also be introduced in this chapter as a natural extension of our work with all forces.

4-1 NEWTON'S FIRST LAW

We know from experience that a stationary object remains at rest unless acted on by some outside force. A can of oil will stay on a workbench until someone tips it over. A suspended weight will hang until it is released. We know that forces are necessary to cause anything to move if it is originally at rest.

Less obvious is the fact that an object in motion will continue in motion until an outside force changes the motion. For example, a steel bar that slides on the shop floor soon comes to rest because of its interaction with the floor. The same bar would slide much farther on ice before stopping. This is because the horizontal interaction, called *friction,* between the floor and the bar is much greater than the friction between the ice and the bar. This leads to the idea that a sliding bar on a perfectly frictionless horizontal plane would stay in motion forever. These ideas are a part of Newton's first law of motion.

www.iisa.org
Different types of roller blade brakes slow down a roller blader's motion. To read about these brakes, see this Web site.

Newton's first law **A body at rest remains at rest and a body in motion remains in uniform motion in a straight line unless acted on by an external unbalanced force.**

Because of the existence of friction, no actual body is ever completely free from external forces. But there are situations in which it is possible to make the resultant force zero or approximately zero. In such cases, the body will behave in accordance with the first law of motion. Since we recognize that friction can never be eliminated completely, we also recognize that Newton's first law is an expression of an *ideal* situation. A flywheel rotating on lubricated ball bearings tends to keep on spinning, but even the slightest friction will eventually bring it to rest.

Newton called the property of a particle that allows it to maintain a constant state of motion or rest *inertia.* His first law is sometimes called the *law of inertia.* When an automobile is accelerated, the passengers obey this law by tending to remain at rest until the external force of the seat compels them to move. Similarly, when the automobile stops, the passengers continue in motion with constant speed until they are restrained by their seat belts or through their own efforts. All matter has inertia. The concept of *mass* is introduced later as a measure of a body's inertia.

A person trained in karate can break a 3.8-cm-thick concrete block with a bare hand. The hand moves at 11 m/s; this creates 3069 N of force. The bones in the hand are able to withstand forces up to 40 times that amount.

4-2 NEWTON'S THIRD LAW

There can be no force unless two bodies are involved. When a hammer strikes a nail, it exerts an "action" force on the nail. But the nail must also "react" by pushing back against the hammer. In all cases there must be an *acting* force and a *reacting* force. Whenever two bodies interact, the force exerted by the second body on the first (the **reaction force**) is equal in magnitude and opposite in direction to the force exerted by the first body on the second (the action force). This principle is stated in *Newton's third law:*

Newton's third law **To every action there must be an equal and opposite reaction.**

Therefore, there can never be a single isolated force. Consider the examples of action and reaction forces in Fig. 4-1.

Note that the acting and reacting forces do not cancel each other. They are equal in magnitude and opposite in direction, but they act on *different* objects. For two forces to cancel, they must act on the same object. It might be said that the action forces create the reaction forces.

For example, someone starting to climb a ladder begins by putting one foot on the rung and pushing on it. The rung must exert an equal and opposite force on the foot to prevent collapse. The greater the force exerted by the foot on the rung, the greater the reaction against the foot must be. Of course the rung cannot create a reaction force until the force of the foot is applied. The action force acts on the object, and the reacting force acts on the agent that applies the force.

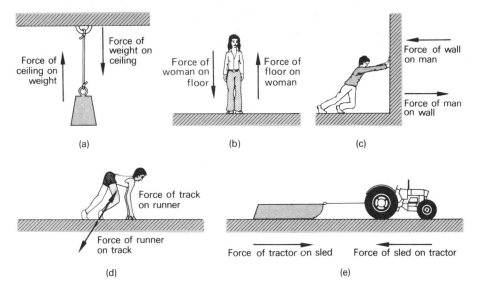

Figure 4-1 Examples of action and reaction forces.

4-3 EQUILIBRIUM

The resultant force was defined as a single force whose effect is the same as a given system of forces. If the tendency of a number of forces is to cause motion, the resultant will also produce this tendency. A condition of equilibrium exists where the resultant of all external forces is zero. This is the same as saying that each external force is balanced by the sum of all the other external forces when equilibrium exists. Therefore, according to Newton's first law, a body in equilibrium must be either at rest or in motion with constant velocity since there is no unbalanced force.

Consider the system of forces in Fig. 4-2a. The vector polygon solution shows that, regardless of the sequence in which the vectors are added, their resultant is always zero. The tip of the last vector lands on the tail of the first vector (see Sec. 3-7).

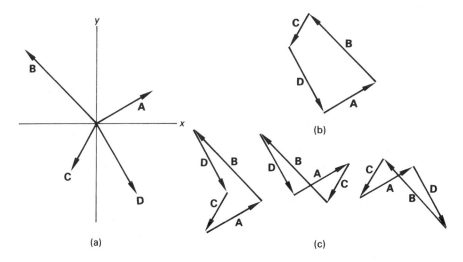

Figure 4-2 Forces in equilibrium.

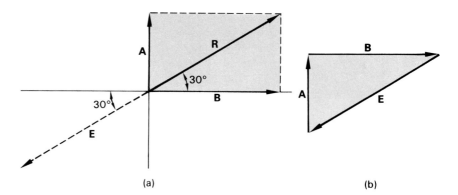

(a)   (b)

Figure 4-3 The equilibrant.

A system of forces not in equilibrium can be put in equilibrium by replacing their resultant force with an equal but opposite force called the **equilibrant.** For instance, the two forces **A** and **B** in Fig. 4-3a have a resultant **R** in a direction 30° above the horizontal. If we add **E,** which is equal in magnitude to **R** but which has an angle 180° greater, the system will be in equilibrium, as shown in Fig. 4-3b.

We have shown in the previous chapter that the magnitudes of the x and y components of any resultant **R** are given by

$$R_x = \sum F_x = A_x + B_x + C_x + \ldots$$

$$R_y = \sum F_y = A_y + B_y + C_y + \ldots$$

When a body is in equilibrium, the resultant of all forces acting on it is zero. In this case, both R_x and R_y must be zero; hence, for a body in equilibrium,

$$\sum F_x = 0 \qquad \sum F_y = 0 \qquad \qquad \text{4-1}$$

These two equations represent a mathematical statement of the *first condition for equilibrium,* which can be stated as follows:

A body is in translational equilibrium if and only if the vector sum of the forces acting on it is zero.

The term ***translational equilibrium*** is used to distinguish the first condition from the second condition for equilibrium, which involves rotational motion, discussed in Chap. 5.

4-4 FREE-BODY DIAGRAMS

www.skatepatrol.org
To see the effects of forces on a braking roller blader, go to this Web site.

Before applying the first condition for equilibrium to the solution of physical problems, you must be able to construct vector diagrams. Consider, for example, the 40-lb weight suspended by ropes shown in Fig. 4-4a. There are three forces acting on the knot—those exerted by the ceiling, the wall, and the earth (weight). If each of these forces is labeled and represented as a vector, we can draw a vector diagram such as the one in Fig. 4-4b. Such a diagram is called a ***free-body diagram.***

A free-body diagram is a vector diagram that describes all forces acting on a particular body or object. Note that in the case of concurrent forces, all vectors point away from the center of the x and y axes, which cross at a common origin.

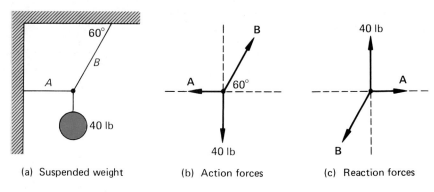

(a) Suspended weight   (b) Action forces   (c) Reaction forces

Figure 4-4   Free-body diagrams showing action and reaction forces.

In drawing free-body diagrams, it is important to distinguish between action and reaction forces. In our example, there are forces on the knot, but there are also three equal and opposite reaction forces exerted *by* the knot. From Newton's third law, the reaction forces exerted *by* the knot *on* the ceiling, wall, and earth are shown in Fig. 4-4c. To avoid confusion, it is important to pick a point at which all forces are acting and draw those forces that act *on* the body at that point.

Problem-Solving Strategy

How to Construct a Free-Body Diagram

1. Draw a sketch and label the conditions of the problem. Be sure to indicate all known and unknown forces and angles.
2. Isolate each body of the system to be studied. Do this either mentally or by drawing a light circle around the point where all forces are applied.
3. Construct a force diagram for each body to be studied. The forces are represented as vectors with their tails placed at the center of a rectangular coordinate system. (See examples in Figs. 4-5 and 4-7.)
4. Represent the *x* and *y* axes with dotted lines. These axes need not necessarily be drawn horizontally and vertically, as we shall see.
5. Dot in rectangles corresponding to the *x* and *y* components of each vector, and determine known angles from given conditions.
6. Label all known and unknown components opposite and adjacent to known angles.

Although this process may appear laborious, it is helpful and sometimes necessary for a clear understanding of a problem. As you gain practice drawing free-body diagrams, their use will become routine.

The two types of forces that act on a body are *contact forces* and *field forces*. Both must be considered in the construction of a force diagram. For example, the gravitational attraction on a body by the earth, called its **weight,** does not have a point of contact with the body. Nevertheless, it exerts a real force and must be considered an important factor in any force problem. The direction of the weight vector should always be assumed to be downward.

EXAMPLE 4-1

A block of weight W hangs from a cord that is knotted to two other cords, A and B, fastened to the ceiling. If cord B makes an angle of $60°$ with the ceiling and cord A forms a $30°$ angle, draw the free-body diagram of the knot.

Solution

Following the procedure above, we construct the diagram shown in Fig. 4-5. This free-body diagram is valid and workable, but choosing the x and y axes along the vectors **B** and **A,** instead of horizontally and vertically, simplifies the diagram. Thus, in Fig. 4-6 we need only resolve one force **W** into components since **A** and **B** would now lie entirely along a particular axis.

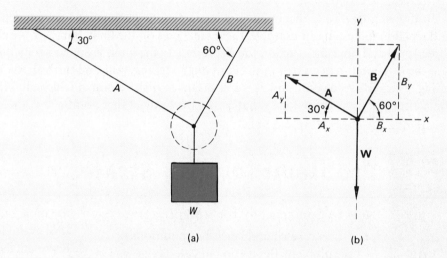

(a) (b)

Figure 4-5 (a) A sketch is drawn to clarify the problem. (b) A free-body diagram is constructed.

Figure 4-6 Rotation of x and y axes to coincide with the perpendicular vectors **A** and **B**.

Hint: Whenever possible, choose x and y axes so that as many forces as possible are completely specified along them.

Probably the most difficult part of constructing vector diagrams is the visualization of forces. In drawing free-body diagrams, it is helpful to imagine that the forces are acting on *you.* Become the knot in a rope or the block on a table and try to see the forces you would experience. Two additional examples are shown in Fig. 4-7. Note that the force exerted by the light boom in Fig. 4-7a is outward and not toward the wall. This is because we are interested in forces exerted *on* the end of the boom and not those exerted *by* the end of the boom. We pick a point at the end of the boom where the two ropes are attached. The 60-N weight and the tension **T** are action forces exerted by the ropes at this point. If the end of the boom is not to move, these forces must be balanced by a third force—the force exerted by the wall (through the boom). This third force **B,** acting on the end of the boom, must not be confused with the inward **reaction force** that acts *on* the wall.

The second example (Fig. 4-7b) also shows action forces acting on two weights connected by a light cord. Friction forces, which are discussed later, are not included in these diagrams. The tension in the cord on either side is shown as **T,** and the normal forces $\mathcal{N}_1$ and $\mathcal{N}_2$ are perpendicular forces exerted by the plane on the blocks. If these forces were absent, the blocks would swing together. (Note the choice of axes in each diagram.)

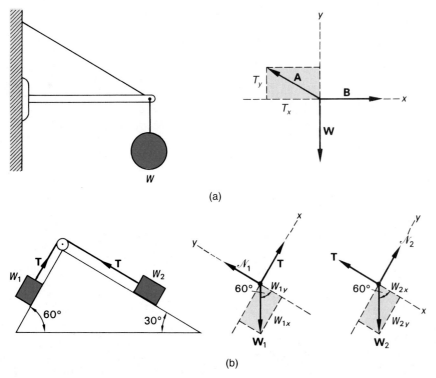

(a)

(b)

Figure 4-7 Examples of free-body diagrams. Note that the components of vectors are labeled opposite and adjacent to known angles.

In Chap. 3 we discussed a procedure for finding the resultant of a number of forces by rectangular resolution. A similar procedure can be used to add forces that are in equilibrium. In this case, the first condition for equilibrium tells us that the resultant is zero, or

$$R_x = \sum F_x = 0 \qquad\qquad R_y = \sum F_y = 0 \qquad\qquad \textbf{4-2}$$

Thus we have two equations that can be used to find unknown forces.

Problem-Solving Strategy

Translational Equilibrium

1. Draw a sketch and label the conditions of the problem.
2. Draw a free-body diagram. (See Sec. 4-4.)
3. Resolve all forces into their x and y components, even though they may contain unknown factors, such as $A \cos 60°$ or B

sin 60°. (You may wish to construct a force table like Table 4-1.)

4. Use the first condition for equilibrium [Eq. (4-1)] to set up two equations in terms of the unknown forces.
5. Solve algebraically for the unknown factors.

EXAMPLE 4-2

A 100-N ball suspended by a rope A is pulled aside by a horizontal rope B and held so that rope A forms an angle of 30° with the vertical wall. (See Fig. 4-8.) Find the tensions in ropes A and B.

Figure 4-8 Forces acting on the knot are represented in a free-body diagram.

Solution

We solve by following the above steps illustrated in Fig. 4-8.

1. Draw a sketch (Fig. 4-8a).
2. Draw a free-body diagram (Fig. 4-8b).
3. Resolve all forces into their components (Table 4-1). Note from the figure that A_x and W_y are negative.

TABLE 4-1

Force	θ_x	x component	y component
A	60°	$A_x = -A \cos 60°$	$A_y = A \sin 60°$
B	0°	$B_x = B$	$B_y = 0$
W	−90°	$W_x = 0$	$W_y = -100$ N
		$\sum F_x = B - A \cos 60°$	$\sum F_y = A \sin 60° - 100$ N

4. We now apply the first condition for equilibrium. Summing the forces along the x axis yields

$$\sum F_x = B - A \cos 60° = 0$$

from which we obtain

$$B = A \cos 60° = 0.5A \qquad \textbf{4-3}$$

since $\cos 60° = 0.5$. A second equation results from summing the y components.

$$\sum F_y = A \sin 60° - 100 \text{ N} = 0$$

from which

$$A \sin 60° = 100 \text{ N} \qquad \textbf{4-4}$$

5. Finally, we solve for the unknown forces. Since $\sin 60° = 0.866$, we have from Eq. (4-4)

$$0.866A = 100 \text{ N}$$

or

$$A = \frac{100 \text{ N}}{0.866} = 115 \text{ N}$$

Now that the value of A is known, Eq. (4-3) can be solved for B as follows:

$$B = 0.5A = (0.5)(115 \text{ N})$$
$$= 57.5 \text{ N}$$

EXAMPLE 4-3

A 200-N ball hangs from a cord knotted to two other cords, as shown in Fig. 4-9. Find the tensions in ropes A, B, and C.

Solution

Since a sketch is already provided, the first step is to construct a free-body diagram, as illustrated in Fig. 4-9b. The x and y components for each vector as calculated from the figure are given in the following table:

Math Tip

To reduce rounding errors, perform calculations as close to the end of a problem as possible. A wooden roof is sloped 20°. What is the coefficient of static friction between the sole of a shoe worn by a 75-kg man and the roof if the man is not to slip? Start with the equation for the force of friction:

$$\mathscr{F}_s = \mu_s \mathscr{N}$$

Solve for μ_s by dividing by $\mathscr{N}$:

$$\frac{\mathscr{F}_s}{\mathscr{N}} = \mu_s$$

From the problem:

$$\mathscr{F}_s = mg \sin \theta \text{ and}$$
$$\mathscr{N}_n = mg \cos \theta$$

Dividing the first equation by the second and substituting μ_s yields:

$mg \sin \theta / mg \cos \theta$
$= \sin \theta / \cos \theta = \tan \theta = \mu_s$

Now we substitute $\theta = 20°$ to find μ_s.

$$\mu_s = \tan 20° = 0.364$$

Notice that the mass of the man was not a factor in this problem.

x component	y component
$A_x = -A \cos 60°$	$A_y = A \sin 60°$
$B_x = B \cos 45°$	$B_y = B \sin 45°$
$C_x = 0$	$C_y = -200\ \text{N}$

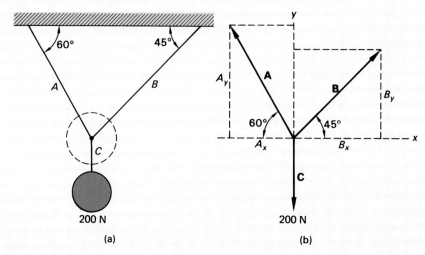

(a) (b)

Figure 4-9

Summing the forces along the x axis, we obtain

$$\sum F_x = -A \cos 60° + B \cos 45° = 0$$

which can be simplified by substituting known trigonometric functions. Hence,

$$-0.5A + 0.707B = 0 \qquad \text{4-5}$$

More information is needed to solve this equation. We obtain a second equation by summing the forces along the y axis, giving

$$0.866A + 0.707B = 200\ \text{N} \qquad \text{4-6}$$

Eqs. (4-5) and (4-6) are now solved simultaneously for A and B by the process of substitution. Solving for A in Eq. (4-5) gives

$$A = \frac{0.707B}{0.5} \qquad \text{or} \qquad A = 1.414B \qquad \text{4-7}$$

Now we substitute this equality into Eq. (4-6), obtaining

$$0.866(1.414B) + 0.707B = 200\ \text{N}$$

which can be solved for B as follows:

$$1.225B + 0.707B = 200\ \text{N}$$
$$1.93B = 200\ \text{N}$$
$$B = \frac{200\ \text{N}}{1.93} = 104\ \text{N}$$

The tension A can now be found by substituting $B = 104$ N into Eq. (4-7):

$$A = 1.414(104 \text{ N}) = 146 \text{ N}$$

The tension in cord C is, of course, 200 N, since it must be equal to the weight.

EXAMPLE 4-4

A 200-lb block rests on a frictionless inclined plane of slope angle 30°. A cord attached to the block passes over a frictionless pulley at the top of the plane and is attached to a second block. What must the weight of the second block be if the system is to be in equilibrium? (Neglect the weight of the cord.)

| (a) | (b) | (c) |

Figure 4-10 A free-body diagram is drawn for each body in the problem.

Solution

After the system is sketched, a free-body diagram is constructed for each body, as shown in Fig. 4-10. Applying the first condition for equilibrium to the second block (Fig. 4-10c), we find that

$$T - W = 0$$

or

$$T = W$$

Since the rope is continuous and the system is frictionless, the tension in Fig. 4-10b for the 200-lb block must also be equal to the weight W.

Considering the diagram for the 200-lb block, we determine the components of each force as follows:

x component	y component
$T_x = T = W$	$T_y = 0$
$N_x = 0$	$N_y = N$
$(200 \text{ lb})_x = (-200 \text{ lb})(\sin 30°)$	$(200 \text{ lb})_y = (-200 \text{ lb})(\cos 30°)$

Applying the first condition yields

$$\sum F_x = 0 \qquad T - (200 \text{ lb})(\sin 30°) = 0 \qquad \text{4-8}$$

$$\sum F_y = 0 \qquad \mathcal{N} - (200 \text{ lb})(\cos 30°) = 0 \qquad \text{4-9}$$

From Eq. (4-8) we obtain

$$T = (200 \text{ lb})(\sin 30°) = 100 \text{ lb}$$

and, therefore, $W = 100$ lb since $T = W$. Thus, a 100-lb weight is required to maintain equilibrium.

The normal force exerted by the plane on the 200-lb block can be found from Eq. (4-9), although this calculation was not necessary to determine the weight W. Hence,

$$\mathcal{N} = (200 \text{ lb})(\cos 30°) = (200 \text{ lb})(0.866)$$
$$= 173 \text{ lb}$$

4-6 FRICTION

Whenever a body moves while it is in contact with another object, friction forces oppose the relative motion. These forces are caused by the adhesion of one surface to the other and by the interlocking of irregularities in the rubbing surfaces. It is friction that holds a nail in a board, allows us to walk, and makes automobile brakes work. In all these cases, friction has a desirable effect.

In many other instances, however, friction must be minimized. For example, it increases the work necessary to operate machinery, it causes wear, and it generates heat, which often causes additional damage. Automobiles and airplanes are streamlined to decrease air friction, which is large at high speeds.

Whenever one surface moves past another, the frictional force exerted by each body on the other is parallel or tangent to the two surfaces and acts in such a manner as to oppose relative motion. It is important to note that these forces exist not only when there is relative motion but even when one object only *tends* to slide past another.

Suppose a force is exerted on a block that rests on a horizontal surface, as shown in Fig. 4-11. At first the block will not be moved because of the action of a force called the *force of static friction* $\mathcal{F}_s$. But as the applied force is increased, motion eventually occurs, and the friction force exerted by the horizontal surface while the block is moving is called the *force of kinetic friction* $\mathcal{F}_k$.

(a)   (b)

Figure 4-11 (a) In static friction, motion is impending. (b) In kinetic friction, the two surfaces are in relative motion.

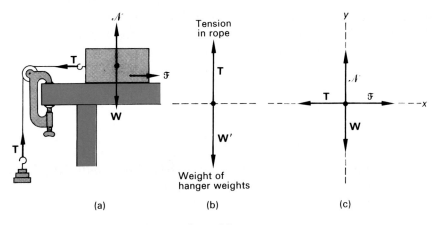

(a) (b) (c)

Figure 4-12 Experiment to determine the force of friction.

The laws governing friction forces can be determined experimentally in the laboratory by using an apparatus similar to the one shown in Fig. 4-12a. A box of weight W is placed on a horizontal table, and a string attached to the box is passed over a light, frictionless pulley and attached to a weight hanger. All forces acting on the box and hanger are shown in their corresponding free-body diagrams (Fig. 4-12b and c).

Let us consider that the system is in equilibrium, which requires the box to be stationary or moving with a constant velocity. In either case, we may apply the first condition for equilibrium. Consider the force diagram of the box as shown in Fig. 4-12c.

$$\sum F_x = 0 \qquad \mathscr{F} - T = 0$$
$$\sum F_y = 0 \qquad \mathscr{N} - W = 0$$

or

$$\mathscr{F} = T \quad \text{and} \quad \mathscr{N} = W$$

Thus, the force of friction is equal in magnitude to the tension in the string, and the normal force exerted by the table on the box is equal to the weight of the box. Note that the tension in the string is determined by the weight of the hanger plus the weights placed on the hanger.

We begin the experiment by slowly adding weights to the hanger, thus gradually increasing the tension in the string. As the tension is increased, the equal but oppositely directed force of static friction is also increased. If T is increased sufficiently, the box will start to move, indicating that T has overcome the *maximum* force of static friction $\mathscr{F}_{s,\text{max}}$. Thus, although the force of static friction $\mathscr{F}_s$ will vary according to values of tension in the string, there exists a single maximum value $\mathscr{F}_{s,\text{max}}$.

To continue the experiment, suppose we add weights to the box, thereby increasing the normal pressure between the box and the table. Our normal force will now be

$$\mathscr{N} = W + \text{added weights}$$

Repeating the above procedure will show that a *proportionately* larger value of T will be necessary to overcome $\mathscr{F}_s$. In other words, if we double the normal force between two surfaces, the maximum force of static friction that must be overcome is also doubled. If $\mathcal{N}$ is tripled, $\mathcal{F}_s$ is tripled, and so it will be for other factors. Therefore, it can be said that the maximum force of static friction is directly proportional to the normal force between the two surfaces. We can write this proportionality as

$$\mathcal{F}_s \propto \mathcal{N}$$

which can be stated as an equation:

$$\mathcal{F}_s = \mu_s \mathcal{N} \qquad \qquad \text{4-10}$$

where μ_s is a proportionality constant called the ***coefficient of static friction.*** Since μ_s is the constant ratio of two forces, it is a dimensionless quantity.

In the above experiment it will be noticed that after $\mathcal{F}_s$ has been overcome, the box will increase its speed, or accelerate, until it is stopped by the pulley. This indicates that a lesser value of **T** would be necessary to keep the box moving with a constant speed. Thus, the force of kinetic friction $\mathcal{F}_k$ must be smaller than the value of $\mathcal{F}_s$ for the same surfaces. In other words, it requires more force to start a block moving than it does to keep it moving with constant speed. In the latter case, the first condition for equilibrium is still satisfied; hence, the same reasoning that led to Eq. (4-10) for static friction will yield the following proportionality for kinetic friction:

$$\mathcal{F}_k \propto \mathcal{N}$$

which can be stated as an equation. As before,

$$\mathcal{F}_k = \mu_k \mathcal{N} \qquad \qquad \text{4-11}$$

where μ_k is a proportionality constant called the ***coefficient of kinetic friction.***

The proportionality coefficients μ_s and μ_k can be shown to depend on the roughness of the surfaces but not on the area of contact between the two surfaces. It can be seen from the equations above that μ depends only on the frictional force $\mathcal{F}$ and the normal force $\mathcal{N}$ between the surfaces. Of course, it must be realized that Eqs. (4-10) and (4-11) are not fundamentally rigorous, like other physical equations. Many variables interfere with the general application of these formulas. No one who has experience in automobile racing, for instance, will believe that the friction force is *completely* independent of the contact area. Nevertheless, the equations are useful tools for estimating resistive forces in specific cases.

Table 4-2 shows some representative values for the coefficients of static and kinetic friction for different types of surfaces. These values are approximate and depend on the condition of the surfaces.

TABLE 4-2 Approximate Coefficients of Friction		
Material	μ_s	μ_k
Wood on wood	0.7	0.4
Steel on steel	0.15	0.09
Metal on leather	0.6	0.5
Wood on leather	0.5	0.4
Rubber on concrete, dry	0.9	0.7
wet	0.7	0.57

Problem-Solving Strategy

Considerations for Problems Involving Friction

1. Friction forces are parallel to the surfaces and directly oppose motion or impending motion.
2. The maximum force of static friction is usually larger than the force of kinetic friction for the same materials.
3. In drawing free-body diagrams, it is usually better to choose the x axis along the direction of motion and the y axis normal to the direction of motion or impending motion.

4. The first condition for equilibrium can be applied to set up two equations representing forces along and perpendicular to the plane of motion.
5. The relations $\mathcal{F}_s = \mu_s \mathcal{N}$ and $\mathcal{F}_k = \mu_k \mathcal{N}$ can be applied to solve for the desired quantity.
6. The *normal force* should never be assumed to be equal to the weight. Determine its magnitude by summing forces along the normal axis.

EXAMPLE 4-5

A 50-N block rests on a horizontal surface. A horizontal pull of 10 N is required just to start the block moving. After motion is started, only a 5-N force is required to move the block with a constant speed. Find the coefficients of static and kinetic friction.

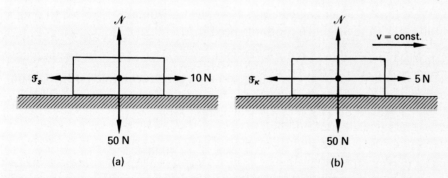

Figure 4-13 (a) A force of 10 N is required to overcome the maximum force of static friction. (b) A force of only 5 N is required to move the block with constant speed.

Solution

The key words that should be recognized are *just to start moving* and *move with a constant speed*. The first phrase implies static friction, whereas the second implies kinetic friction. In each case a condition of equilibrium exists. The correct free-body diagrams are shown in Fig. 4-13a and b. Let us first consider the force that overcomes static friction. Applying the first condition for equilibrium to Fig. 4-13a gives

$$\sum F_x = 0 \qquad 10 \text{ N} - \mathscr{F}_s = 0$$

$$\sum F_y = 0 \qquad \mathscr{N} - 5 \text{ N} = 0$$

from which we note that

$$\mathscr{F}_s = 10 \text{ N} \qquad \mathscr{N} = 50 \text{ N}$$

Thus, we can find the coefficient of static friction from Eq. (4-10).

$$\mu_s = \frac{\mathscr{F}_s}{\mathscr{N}} = \frac{10 \text{ N}}{50 \text{ N}} = 0.2$$

The force that overcomes kinetic friction is 5 N. Hence, summing the forces along the x axis yields

$$5 \text{ N} - \mathscr{F}_k = 0$$

or

$$\mathscr{F}_k = 5 \text{ N}$$

The normal force is still 50 N, and so

$$\mu_k = \frac{\mathscr{F}_k}{\mathscr{N}} = \frac{5 \text{ N}}{50 \text{ N}} = 0.1$$

EXAMPLE 4-6

What force **T** at an angle of 30° above the horizontal is required to drag a 40-lb block to the right at constant speed if $\mu_k = 0.2$?

Solution

We shall first sketch the problem and then construct a free-body diagram, as shown in Fig. 4-14. Applying the first condition for equilibrium yields

$$\sum F_x = 0 \qquad T_x - \mathscr{F}_k = 0 \qquad \text{4-12}$$

$$\sum F_y = 0 \qquad \mathscr{N} + T_y - 40 \text{ lb} = 0$$

Figure 4-14 The force **T** at an angle above the horizontal reduces the normal force, resulting in a smaller force of friction.

The latter equation shows the normal force to be

$$\mathcal{N} = 40 \text{ lb} - T_y \qquad \qquad \textbf{4-13}$$

It should be noted that the normal force is decreased by the y component of $\mathbf{T}$. Substituting $\mathcal{F}_k = \mu_k \mathcal{N}$ into Eq. (4-12) gives

$$T_x - \mu_k \mathcal{N} = 0$$

But $\mathcal{N} = 40 \text{ lb} - T_y$ from Eq. (4-13), and so

$$T_x - \mu_k(40 \text{ lb} - T_y) = 0 \qquad \qquad \textbf{4-14}$$

From the free-body diagram it is noted that

$$T_x = T \cos 30° = 0.866T$$

and

$$T_y = T \sin 30° = 0.5T$$

thus, recalling that $\mu_k = 0.2$, we can write Eq. (4-14) as

$$0.866T - (0.2)(40 \text{ lb} - 0.5T) = 0$$

which can be solved for T as follows:

$$0.866T - 8 \text{ lb} + 0.1T = 0$$
$$0.966T - 8 \text{ lb} = 0$$
$$0.966T = 8 \text{ lb}$$
$$T = \frac{8 \text{ lb}}{0.966} = 8.3 \text{ lb}$$

Therefore, a force of 8.3 lb is required to pull the block with constant speed if the rope makes an angle of 30° above the horizontal.

EXAMPLE 4-7

A 100-N block rests on a 30° inclined plane. If $\mu_k = 0.1$, what push $\mathbf{P}$ parallel to the plane and directed up the plane will cause the block to move (a) up the plane with constant speed and (b) down the plane with constant speed?

Solution (a)

The general problem is sketched in Fig. 4-15a. For motion up the plane, the force of friction is directed down the plane, as shown in Fig. 4-15b. Applying the first condition for equilibrium, we obtain

$$\sum F_x = 0 \qquad P - \mathcal{F}_k - W_x = 0 \qquad \qquad \textbf{4-15}$$

$$\sum F_y = 0 \qquad \mathcal{N} - W_y = 0 \qquad \qquad \textbf{4-16}$$

From the figure, the x and y components of the weight are

$$W_x = (100 \text{ N})(\sin 30°) = 50 \text{ N}$$
$$W_y = (100 \text{ N})(\cos 30°) = 86.6 \text{ N}$$

Figure 4-15 Friction on the inclined plane.

Substitution of the latter into Eq. (4-16) allows us to solve for the normal force. Hence,

$$\mathcal{N} - 86.6 \text{ N} = 0$$
$$\mathcal{N} = 86.6 \text{ N}$$

The push required to move up the plane is, from Eq. (4-15),

$$P = \mathcal{F}_k + W_x$$

But $\mathcal{F}_k = \mu_k \mathcal{N}$, so that

$$P = \mu_k \mathcal{N} + W_x$$

Substituting known values for μ_k, $\mathcal{N}$, and W_x, we obtain

$$P = (0.1)(86.6 \text{ N}) + 50 \text{ N}$$
$$= 58.7 \text{ N}$$

Note that the push up the plane in this case must overcome both the frictional force of 8.66 N and the 50-N component of the weight down the plane.

Solution (b)

Now we must consider the push **P** required to retard the downward motion. The only difference between this problem and the problem in part (a) is that the friction force is now directed up the plane. The normal force does not change, and the components of the weight do not change. Therefore, if we sum the forces along the *x* axis in Fig. 4-15c, we have

$$\sum F_x = 0 \qquad P + \mathcal{F}_k - W_x = 0$$

from which

$$P = W_x - \mathcal{F}_k$$

or

$$P = 50 \text{ N} - 8.66 \text{ N}$$
$$= 41.3 \text{ N}$$

the force of 41.3 N directed up the plane retards the downward motion of the block so that its speed is constant. If this force **P** were not exerted, the block would accelerate down the plane of its own accord.

EXAMPLE 4-8

What is the maximum slope angle θ for an inclined plane such that a block of weight W will not slide down the plane?

Figure 4-16   The limiting angle of repose.

Solution

A sketch and free-body diagram are constructed as shown in Fig. 4-16. The maximum value of θ would be the value that overcomes static friction $\mathcal{F}_s = \mu_s \mathcal{N}$. Applying the first condition for equilibrium gives

$$\sum F_x = 0 \qquad \mathcal{F}_s - W_x = 0$$

$$\sum F_y = 0 \qquad \mathcal{N} - W_y = 0$$

from which we transpose to obtain

$$\mathcal{F}_s = W_x \qquad \mathcal{N} = W_y \qquad \text{4-17}$$

From Fig. 4-16 we note that θ is an angle whose tangent is W_x/W_y; hence, from Eq. (4-17), we have

$$\tan \theta = \frac{W_x}{W_y} = \frac{\mathcal{F}_s}{\mathcal{N}}$$

But $\mathcal{F}_s/\mathcal{N}$ is equal to the coefficient of static friction μ_s. Hence,

$$\tan \theta = \mu_s$$

Thus, a block, regardless of its weight, will remain at rest on an inclined plane unless $\tan \theta$ equals or exceeds μ_s. The angle θ in this case is called the *limiting angle* or the *angle of repose*.

It is left as an exercise for you to show that a block will slide down the plane with constant speed if $\tan \theta = \mu_k$.

Summary and Review

Summary

In this chapter, we have defined objects that are at rest or in motion with constant speed to be in equilibrium. Through the use of vector diagrams and Newton's laws, we have found it possible to determine unknown forces for systems that are known to be in equilibrium. The following items will summarize the more important concepts to be remembered:

- *Newton's first law of motion* states that an object at rest and an object in motion with constant speed will maintain the state of rest or constant motion unless acted on by a resultant force.
- *Newton's third law of motion* states that every action must produce an equal and opposite reaction. The action and reaction forces do not act on the same body.
- Free-body diagrams: From the conditions of the problem, a neat sketch is drawn and all known quantities are labeled. Then a force diagram indicating all forces and their components is constructed. All information such as that given in Fig. 4-17 should be part of the diagram.
- Translational equilibrium: A body in translational equilibrium has no resultant force acting on it. In such cases, the sum of all the x components is zero, and the sum of all the y components is zero. This is known as the first condition for equilibrium and is written

$$R_x = \sum F_x = 0 \qquad R_y = \sum F_y = 0$$

- Applying these conditions to Fig. 4-17, for example, we obtain two equations in two unknowns:

$$B \cos 45° - A \cos 60° = 0$$
$$B \sin 45° + A \sin 60° - 200 \text{ N} = 0$$

These equations can be solved to find A and B.

- Static friction exists between two surfaces when motion is impending; kinetic friction occurs when the two surfaces are in relative motion. In either case, the friction forces are proportional to the normal force. They are given by

$$\mathcal{F}_s = \mu_s \mathcal{N} \qquad \mathcal{F}_k = \mu_k \mathcal{N}$$

These friction forces must often be considered in equilibrium problems.

Figure 4-17

Key Terms

Define the following terms:

1. inertia
2. reaction force
3. equilibrium
4. equilibrant
5. free-body diagram
6. friction force
7. coefficient of friction
8. normal force
9. angle of repose

Review Questions

4-1. A popular stunt consists of placing a coin on a card and the card on the top of a glass. The edge of the card is flipped briskly with the forefinger, causing the card to fly off the top of the glass as the coin drops into the glass. Explain. What law does this illustrate?

4-2. When the head of a hammer becomes loose, you can reseat it by holding the hammer vertically and tapping the base of the handle against the floor. Explain. What law does this illustrate?

4-3. Explain the part played by Newton's third law of motion in the following activities: (a) walking, (b) rowing, (c) rocket launching, and (d) parachuting.

4-4. Can a moving body be in equilibrium? Give several examples.

4-5. According to Newton's third law of motion, every force has an equal and oppositely directed reaction force. Therefore, the concept of a resultant, unbalanced force must be an illusion that does not hold up under close examination. Do you agree with this statement? Give the reasons for your answer.

4-6. A brick is suspended from the ceiling by a light string. A second identical string is attached to the bottom of the brick and hangs within the reach of a student. When the student pulls the lower string slowly, the upper string breaks, but when the lower string is jerked, it breaks. Explain.

4-7. A long steel cable is stretched between two buildings. Show by diagrams and discussion why it is not possible to pull the cable so taut that it will be perfectly horizontal with no sag in the middle.

4-8. We have seen that it is always advantageous to choose the x and y axes so that as many forces as possible are completely specified along an axis. Suppose that no two forces are perpendicular to each other. Will there still be an advantage to rotating axes to align an unknown force with an axis, as opposed to aligning a known force? Test this approach by applying it to one of the text examples.

4-9. Discuss a few beneficial uses of the force of friction.

4-10. Why do we speak of a *maximum* force of static friction? Why do we not discuss a maximum force of kinetic friction?

4-11. Why is it easier to pull a sled at an angle than it is to push a sled at the same angle? Draw free-body diagrams to show what the normal force would be in each case.

4-12. Is the normal force acting on a body always equal to its weight?

4-13. When walking across a frozen pond, should you take short steps or long ones? Why? If the ice were completely frictionless, would it be possible for you to get off the pond? Explain.

Problems

Note: For all of the problems at the end of this chapter, the rigid booms or struts are considered to be of negligible weight. All forces are considered to be concurrent forces.

Section 4-4—Free-Body Diagrams

4-1. Draw a free-body diagram for the arrangements shown in Fig. 4-18a and b. Isolate a point where the important forces are acting, and represent each force as a vector. Determine the reference angle and label the components.

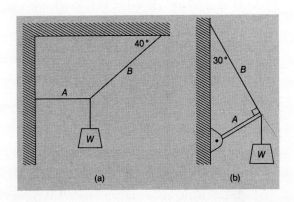

Figure 4-18

4-2. Study each force acting at the end of the light strut in Fig. 4-19. Draw the appropriate free-body diagram.

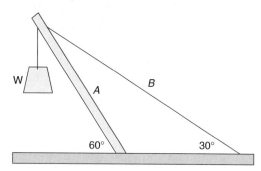

Figure 4-19

Section 4-5—Solution of Equilibrium Problems

4-3. Three identical bricks are strung together with cords and hung from a scale that reads a total of 24 N. what is the tension in the cord that supports the lowest brick? What is the tension in the cord between the middle brick and the top brick? **Ans. 8 N, 16 N**

4-4. A single chain supports a pulley whose weight is 40 N. Two identical 80-N weights are then connected with a cord that passes over the pulley. What is the tension in the supporting chain? What is the tension in each cord?

4-5. If the weight of the block in Fig. 4-18a is 80 N, what are the tensions in ropes *A* and *B*? **Ans. A = 95.3 N, B = 124 N**

4-6. If rope *B* in Fig. 4-18a will break for tensions greater than 200 lb, what is the maximum weight *W* that can be supported?

4-7. If *W* = 600 N in Fig. 4-18b, what is the force exerted by the rope on the end of the boom *A* in Fig. 4-18B? What is the tension in rope *B*? **Ans. A = 300 N, B = 520 N**

4-8. If rope *B* in Fig. 4-18a will break if its tension exceeds 400 N, what is the maximum weight *W*?

4-9. What is the maximum weight *W* for Fig. 4-18b if the rope can sustain a maximum tension of only 800 N? **Ans. 924 N**

4-10. A 70-N block rests on a 35° inclined plane. Determine the normal force and find the friction force that keeps the block from sliding.

4-11. A wire is stretched between two poles 10 m apart. A sign is attached to the midpoint of the line causing it to sag vertically a distance of 50 cm. If the tension in each line segment is 2000 N, what is the weight of the sign? **Ans. 398 N**

4-12. An 80-N traffic light is supported at the midpoint of a 30-m length of cable between two poles. Find the tension in each cable segment if the cable sags a vertical distance of 1 m.

*4-13. The ends of three 8-ft studs are nailed together, forming a tripod with an apex that is 6 ft above the ground. What is the compression in each of these studs if a 100-lb weight is hung from the apex? **Ans. 44.4 lb**

4-14. A 20-N picture is hung from a nail as in Fig. 4-20, so that the supporting cords make an angle of 60°. What is the tension of each cord segment?

Figure 4-20

Section 3-6—Friction

4-15. A horizontal force of 40 N will just start an empty 600-N sled moving across packed snow. After motion is begun, only 10 N is needed to keep motion at constant speed. Find the coefficients of static and kinetic friction.   **Ans. 0.0667, 0.0167**

4-16. Suppose 200-N of supplies are added to the sled in Prob. 4-15. What new force is needed to drag the sled at constant speed?

4-17. Assume surfaces where $\mu_s = 0.7$ and $\mu_k = 0.4$. What horizontal force is needed to just start a 50-N block moving along a wooden floor? What force will move it at constant speed?   **Ans. 35 N, 20 N**

4-18. A dockworker finds that a horizontal force of 60 lb is needed to drag a 150-lb crate across the deck at constant speed. What is the coefficient of kinetic friction?

4-19. The dockworker in Prob. 4-18 finds that a smaller crate of similar material can be dragged at constant speed with a horizontal force of only 40 lb. What is the weight of this crate?   **Ans. 100 lb.**

4-20. A steel block weighing 240 N rests on a level steel beam. What horizontal force will move the block at constant speed if the coefficient of kinetic friction is 0.12?

4-21. A 60-N toolbox is dragged horizontally at constant speed by a rope making an angle of 35° with the floor. The tension in the rope is 40 N. Determine the magnitude of the friction force and the formal force.   **Ans. 32.8 N, 37.1 N**

4-22. What is the coefficient of kinetic friction for the example in Prob. 4-21?

*4-23. The coefficient of static friction for wood on wood is 0.7. What is the maximum angle for an inclined wooden plane if a wooden block is to remain at rest on the plane?   **Ans. 35°**

*4-24. A roof is sloped at an angle of 40°. What is the maximum coefficient of static friction between the sole of a shoe and the roof to prevent slipping?

*4-25. A 200-N sled is pushed along a horizontal surface at constant speed with a 50-N force that makes an angle of 28° below the horizontal. What is the coefficient of kinetic friction?   **Ans. 0.198**

*4-26. What is the normal force on the block in Fig. 4-21? What is the component of the weight acting down the plane?

*4-27. What push **P** directed up the plane will cause the block in Fig. 4-21 to move up the plane with constant speed?   **Ans. 54.1 N**

*4-28. If the block in Fig. 4-21 is released, it will overcome static friction and slide rapidly down the plane. What push **P** directed up the incline will retard the downward motion until the block moves at constant speed?

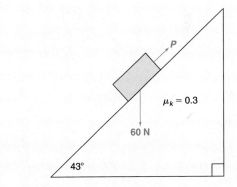

Figure 4-21

Challenge Problems

4-29. Determine the tension in rope A and the compression B in the strut for Fig. 4-22.

Ans. A = 231 N, B = 462 N

Figure 4-22

4-30. If the breaking strength of cable A in Fig. 4-23 is 200 N, what is the maximum weight that can be supported by this apparatus?

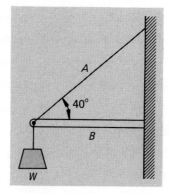

Figure 4-23

4-31. What is the minimum push **P** parallel to a 37° inclined plane if a 90-N wagon is to be rolled up the plane at constant speed? Ignore friction. **Ans. 54.2 N**

4-32. A horizontal force of only 8 lb moves a cake of ice with constant speed across a floor ($\mu_k = 0.1$). What is the weight of the ice?

4-33. Find the tension in ropes A and B for the arrangement shown in Fig. 4-24a.

Ans. 170 N, 294 N

4-34. Find the tension in ropes A and B in Fig. 4-24b.

4-35. A cable is stretched horizontally across the top of two vertical poles 20 m apart. A 250-N sign suspended from the midpoint causes the rope to sag a vertical distance of 1.2 m. What is the tension in each cable segment? **Ans. 1049 N**

4-36. Assume that the cable in Prob. 4-35 has a breaking strength of 1200 N. What is the maximum weight that can be supported at the midpoint?

4-37. Find the tension in the cable and the compression in the light boom for Fig. 4-25a.

Ans. A = 43.2 lb, B = 34.5 lb

4-38. Find the tension in the cable and the compression in the light boom for Fig. 4-25b.

4-39. Determine the tension in ropes A and B for Fig. 4-26a. **Ans. A = 1410 N, B = 1150 N**

*4-40. Find the forces in the light boards of Fig. 4-26b and state whether the boards are under tension or compression.

Figure 4-24

Figure 4-25

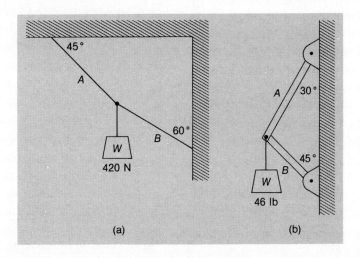

Figure 4-26

Critical Thinking Questions

4-41. Study the structure drawn in Fig. 4-27 and analyze the forces acting at the point at which the rope is attached to the light poles. What is the direction of the forces acting *on* the ends of the poles? What is the direction of the forces exerted *by* the poles at that point? Draw the appropriate free-body diagram.

*4-42. Determine the forces acting *on* the ends of the poles in Fig. 4-27 if $W = 500$ N.

*4-43. A 2-N eraser is pressed against a vertical chalkboard with a horizontal push of 12 N. If $\mu_s = 0.25$, find the horizontal force

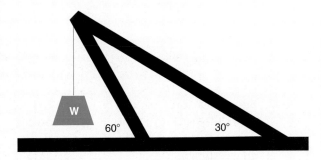

Figure 4-27

required to start motion parallel to the floor. What if you want to start its motion up or down? Find the vertical forces required to just start motion up the board and then down the board.

Ans. 3.00 N, up = 5 N, down = 1 N

*4-44. It is determined experimentally that a 20-lb horizontal force will move a 60-lb lawn mower at a constant speed. The handle of the mower makes an angle of 40° with the ground. What push along the handle will move the mower at a constant speed? Is the normal force equal to the weight of the mower? What is the normal force?

*4-45. Suppose the lawn mower of Question 4-44 is to be moved backward. What pull along the handle is required to move with constant speed? What is the normal force in this case? Discuss the differences between this example and the one in the previous problem. **Ans. 20.4 N, 46.9 N**

*4-46. A truck is pulled out of mud by a line attached to the truck and to a tree. When the angles are as shown in Fig. 4-28, a force of 40 lb is exerted at the midpoint of the line. What force is exerted on the truck?

Figure 4-28

*4-47. Suppose a force of 900 N is required to move the truck in Fig. 4-28. What force is required at the midpoint of the line for the angles shown? **Ans. 616 N**

4-48. A 70-N block of steel is at rest on a 40° incline. What is the magnitude of the static friction force directed up the plane? Is this necessarily the maximum force of

static friction? What is the normal force at this angle?

*4-49. Determine the compression in the center strut B and the tension in the rope A for the situation described by Fig. 4-29. Distinguish clearly the difference between the compression force in the strut and the force indicated on your free-body diagram. **Ans. A = 643 N, B = 940 N**

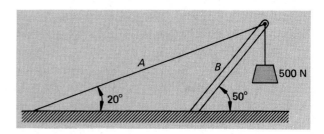

Figure 4-29

*4-50. What *horizontal* push P is required to just prevent a 200-N block from slipping down a 60° inclined plane where $\mu_s = 0.4$? Why does it take a lesser force if P acts *parallel* to the plane? Is the friction force greater, less, or the same for these two cases?

*4-51. Find the tension in each cord of Fig. 4-30 if the suspended weight is 476 N. **Ans. A = 476 N, B = 275 N, C = 275 N**

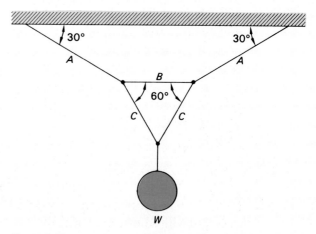

Figure 4-30

*4-52. Find the force required to pull a 40-N sled horizontally at a constant speed by exerting a pull along a pole that makes a 30° angle with the ground ($\mu_k = 0.4$). Now find the force required if you push along the pole at the same angle. What is the major factor that changes in these cases?

*4-53. Two weights are hung over two frictionless pulleys as shown in Fig. 4-31. What weight W will cause the 300-lb block to just start moving to the right? Assume $\mu_s = 0.3$. *Note:* The pulleys merely change the direction of the applied forces. **Ans. 108 lb**

*4-54. Find the maximum weight that can be hung at point O in Fig. 4-32 without upsetting the equilibrium. Assume that $\mu_s = 0.3$ between the block and table.

Figure 4-32

Figure 4-31

Torque and Rotational Equilibrium

In previous chapters, we have discussed forces that act at a single point. Translational equilibrium exists when the vector sum of forces is zero. There are many cases, however, in which the forces acting on an object do not have a common point of application. Such forces are said to be *nonconcurrent.* For example, a mechanic exerts a force on the handle of a wrench to tighten a bolt. A carpenter uses a long lever to pry the lid from a wooden box. The engineer considers twisting forces that tend to snap a beam attached to a wall. The steering wheel of an automobile is turned by forces that do not have a common point of application. In such cases, there may be a *tendency to rotate* that we will define as **torque.** If we learn to measure or predict the torques produced by certain forces, we can obtain desired rotational effects. If no rotation is desired, there must be no resultant torque. This leads naturally to a condition for **rotational equilibrium** that is important for industrial and engineering applications.

5-1 CONDITIONS FOR EQUILIBRIUM

When a body is in equilibrium, it is either at rest or in uniform motion. According to Newton's first law, only the application of a resultant force can change this condition. We have seen that, if all forces acting on such a body intersect at a single point and their vector sum is zero, the system must be in equilibrium. When a body is acted on by forces that do not have a common **line of action,** it may be in translational equilibrium but not in rotational equilibrium. In other words, it may not move to the right or left or up or down, but it may still rotate. In studying equilibrium, we must consider the point of application of each force as well as its magnitude.

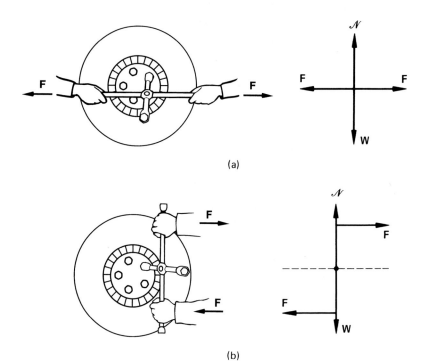

(a)

(b)

Figure 5-1   (a) Equilibrium exists because the forces have the same line of action. (b) Equilibrium does not exist because opposing forces do not have the same line of action.

Consider the forces exerted on the lug wrench in Fig. 5-1a. Two equal opposing forces **F** are applied to the right and to the left. The first condition for equilibrium tells us that the vertical and horizontal forces are balanced. Hence, the system is said to be in equilibrium. If the same two forces are applied as shown in Fig. 5-1b, however, the wrench has a definite tendency to rotate. This is true even though the vector sum of the forces is still zero. Clearly, we need a second condition for equilibrium to cover rotational motion. A formal statement of this condition will be given later. First, we need to define some terms.

In Fig. 5-1b, the forces **F** do not have the same *line of action*.

> **The line of action of a force is an imaginary line extended indefinitely along the vector in both directions.**

When the lines of action of forces do not intersect at a common point, rotation may occur about a point called the **axis of rotation.** In our example, the axis of rotation is an imaginary line passing through the stud perpendicular to the page.

5-2 THE MOMENT ARM

The perpendicular distance from the axis of rotation to the line of action of a force is called the **moment arm** of that force. It is the moment arm that determines the effectiveness of a given force in causing rotational motion. For example, if we exert a force **F** at increasing distances from the center of a large wheel, it becomes easier and easier to rotate the wheel about its center. (See Fig. 5-2.)

> **The moment arm of a force is the perpendicular distance from the line of action of the force to the axis of rotation.**

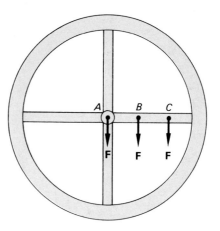

Figure 5-2 The unbalanced force **F** has no rotational effect at point *A* but becomes increasingly effective as the moment arm lengthens.

If the line of action of a force passes through the axis of rotation (point *A* of Fig. 5-2), the moment arm is zero. No rotational effect is observed, regardless of the magnitude of the force. In this simple example, the moment arms at points *B* and *C* are simply the distance from the axis of rotation to the point of application of the force. Note, however, that the line of action of a force is a mere geometrical construction. The moment arm is drawn perpendicular to this line. It may be equal to the distance from the axis to the point of application of a force, but this is true only when the applied force is directed perpendicular to this distance. In the examples of Fig. 5-3, *r* represents the moment arm and *O* represents the axis of rotation. Study each example, observing how the moment arms are drawn and reasoning whether the rotation is clockwise or counterclockwise about *O*.

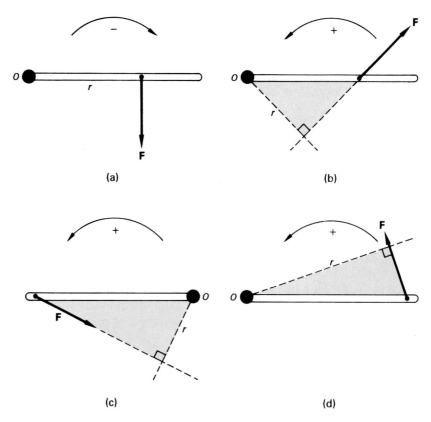

Figure 5-3 Some examples of moment arms.

Force has been defined as a push or pull that tends to cause motion. **Torque** τ can be defined as the tendency to produce a change in rotational motion. It is also called the *moment of force* in some textbooks. As we have seen, rotational motion is affected by both the magnitude of a force F and its moment arm r. Thus, we will define torque as the product of a force and its moment arm.

$$Torque = force \times moment\ arm$$

$$\tau = Fr \qquad\qquad 5\text{-}1$$

It must be understood that r in Eq. (5-1) is measured perpendicular to the line of action of the force **F**. The units of torque are the units of force times distance, for example, *newton-meters* (N · m) and *pound-feet* (lb · ft).

Earlier, we established a sign convention to indicate the direction of forces. The direction of torque depends on whether it tends to produce clockwise (cw) or counterclockwise (ccw) rotation. We shall follow the same convention we used for measuring angles. If the force **F** tends to produce counterclockwise rotation about an axis, the torque will be considered positive. Clockwise torques will be considered negative. In Fig. 5-3, all the torques are positive (ccw) except for that in Fig. 5-3a.

EXAMPLE 5-1

A force of 20 N is exerted on a cable wrapped around a drum that has a diameter of 120 mm. What is the torque produced about the center of the drum? (See Fig. 5-4.)

0.12 m

20 N

Figure 5-4 The tangential force exerted by a cable wrapped around a drum.

Solution

Notice that the line of action of the 20-N force is perpendicular to the diameter of the drum. The moment arm, therefore, is equal to the radius of the drum. Converting the diameter to meters (0.12 m), the radius is 0.06 m. The torque is found from Eq. (5-1):

$$\tau = Fr = -(20 \text{ N})(0.06 \text{ m}) = -1.20 \text{ N} \cdot \text{m}$$

The torque is negative because it tends to cause clockwise rotation.

EXAMPLE 5-2

A mechanic exerts a 20-lb force at the end of a 10-in. wrench, as shown in Fig. 5-5. If this pull makes an angle of 60° with the handle, what is the torque produced on the nut?

(a)   (b)

Figure 5-5 Calculating torque.

Solution

First, we draw a neat sketch, extend the line of action of the 20-lb force, and draw in the moment arm as shown. Note that the moment arm r is perpendicular both to the line of action of the force and to the axis of rotation. You must remember that the moment arm is a geometrical construction and may or may not be along some physical structure such as the handle of the wrench. From the figure we obtain

$$r = (10 \text{ in.}) \sin 60° = 8.66 \text{ in.}$$
$$\tau = Fr = (20 \text{ lb})(8.66 \text{ in.}) = 173 \text{ lb} \cdot \text{in.}$$

If desired, this torque can be converted to 14.4 lb · ft.

In some applications, it is more useful to work with the *components* of a force to obtain the resultant torque. In the preceding example, for instance, we could have resolved the 20-lb vector into its horizontal and vertical components. Instead of finding the torque of a single force, we would then need to find the torque of two component forces. As shown in Fig. 5-6, the 20-lb vector has components F_x and F_y, which are found from trigonometry:

$$F_x = (20 \text{ lb})(\cos 60°) = 10 \text{ lb}$$
$$F_y = (20 \text{ lb})(\sin 60°) = 17.3 \text{ lb}$$

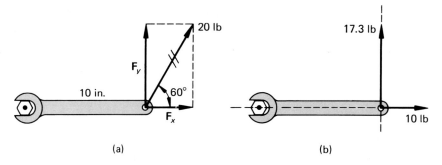

Figure 5-6 The component method of calculating torque.

Notice from Fig. 5-6b that the line of action of the 10-lb force passes through the axis of rotation. It does not produce any torque because its moment arm is zero. The entire torque is, therefore, due to the 17.3-lb component perpendicular to the handle. The moment arm of this force is the length of the wrench, and the torque is

$$\tau = Fr = (17.3 \text{ lb})(10 \text{ in.}) = 173 \text{ lb} \cdot \text{in.}$$

Note that the same result is obtained using this method. No more calculations are required because the horizontal component has a zero moment arm. If we choose the components of a force along and perpendicular to a known distance, we need only concern ourselves with the torque of this perpendicular component.

5-4 RESULTANT TORQUE

In Chap. 3, we demonstrated that the resultant of a number of forces could be obtained by adding the x and y components of each force to find the components of the resultant.

$$R_x = A_x + B_x + C_x + \cdots \qquad R_y = A_y + B_y + C_y + \cdots$$

This procedure applies to forces that have a common point of intersection. Forces that do not have a common line of action may produce a resultant torque in addition to a resultant translational force. When the applied forces act in the same plane, the resultant torque is the algebraic sum of the positive and negative torques due to each force.

$$\tau_R = \sum \tau = \tau_1 + \tau_2 + \tau_3 + \cdots \qquad \text{5-2}$$

Remember that counterclockwise torques are positive and clockwise torques are negative.

One essential element of effective problem-solving techniques is organization. The following procedure is useful for calculating resultant torque.

www.explorescience.com/
The Explore Science Web site has simple shock wave demos to help explain the concept of torque.

Problem-Solving Strategy

Calculating Resultant Torque

1. Read the problem and then draw and label a figure.
2. Construct a free-body diagram showing all forces, distances, and the axis of rotation.
3. Extend the lines of action of each force using dotted lines.
4. Draw and label the moment arms for each force.

5. Calculate the moment arms if necessary.
6. Calculate the torques due to each force independent of other forces; make sure to affix the proper sign (ccw = + and cw = −).
7. The resultant torque is the algebraic sum of the torques due to each force. See Eq. (5-2).

CONCEPTUAL EXAMPLE 5-3

A piece of angle iron is hinged at point A as shown in Fig. 5-7. Determine the resultant torque at A due to the 60- and 80-N forces.

Figure 5-7

Solution

A free-body diagram is drawn and the moment arms r_1 and r_2 are constructed as in Fig. 5-7b. The lengths of the moment arms are:

$$r_1 = (12 \text{ cm}) \sin 50° = 9.19 \text{ cm}$$
$$r_2 = (10 \text{ cm}) \sin 70° = 9.40 \text{ cm}$$

Considering A as the axis of rotation, the torque due to $\mathbf{F}_1$ is negative (cw) and the torque due to $\mathbf{F}_2$ is positive (ccw). The resultant torque is found as follows:

$$\begin{aligned}
\tau_R = \tau_1 + \tau_2 &= F_1 r_1 + F_2 r_2 \\
&= -(60\ \text{N})(9.19\ \text{cm}) + (80\ \text{N})(9.40\ \text{cm}) \\
&= -552\ \text{N} \cdot \text{cm} + 752\ \text{N} \cdot \text{cm} \\
&= 200\ \text{N} \cdot \text{cm}
\end{aligned}$$

The resultant torque is 200 N · cm, counterclockwise. This answer is best expressed as 2.00 N · m in SI units.

5-5 EQUILIBRIUM

We are now ready to discuss the necessary condition for rotational equilibrium. The condition for translational equilibrium was stated in equation form as

$$\sum F_x = 0 \qquad \sum F_y = 0 \qquad \text{5-3}$$

If we are to ensure that the rotational effects are also balanced, we must stipulate that there is no resultant torque. Hence, the second condition for equilibrium is:

The algebraic sum of all the torques about any axis must be zero.

$$\sum \tau = \tau_1 + \tau_2 + \tau_3 + \cdots = 0 \qquad \text{5-4}$$

The second condition for equilibrium simply tells us that the clockwise torques are balanced exactly by the counterclockwise torques. Moreover, since rotation is not occurring about any point, we may choose whatever point we wish as an axis of rotation. As long as the moment arms are measured to the same point for each force, the resultant torque will be zero. Choosing the axis of rotation at the point of application of an unknown force is a way to simplify problems. If a particular force has a zero moment arm, it does not contribute to torque, regardless of its magnitude.

Hint for Success

Prior to graduation, use your networking skills. Tell friends, family, and acquaintances the type of career you are looking for and ask them whether they know of any jobs or of other people you should contact. If someone gives you a contact name in a particular company, mention that person when you apply for the job.

Problem-Solving Strategy

Rotational Equilibrium

1. Draw and label a sketch with all given information.
2. Draw a free-body diagram (if needed) with distances given between forces.
3. Choose an axis of rotation at the point where the least information is given, for example, at the point of application of an unknown force.

4. Sum the torques due to each force about the chosen axis of rotation and set the result equal to zero.

$$\tau_R = \tau_1 = \tau_2 + \tau_3 + \cdots = 0$$

5. Apply the first condition for equilibrium to obtain two additional equations.

$$\sum F_x = 0 \qquad \sum F_y = 0$$

6. Solve for the unknown quantities.

CONCEPTUAL EXAMPLE 5-4

Consider the arrangement shown in Fig. 5-8. A uniform beam that weighs 200 N is held up by two supports A and B. Given the distances and forces listed in this figure, what are the forces exerted by the supports?

(a) Sketch of system

(b) Free-body diagram

Figure 5-8 The solution of torque problems is easier if a rough sketch of the system is drawn. A free-body diagram can then be drawn to indicate forces and distances.

Solution

A free-body diagram is drawn to show clearly all forces and distances between forces. Note that all the weight of the uniform beam is considered as acting at the center of the board. Next, we will apply the first condition for equilibrium, Eq. (5-3):

$$\sum F_y = A + B - 300 \text{ N} - 200 \text{ N} - 400 \text{ N} = 0$$

or

$$A + B = 900 \text{ N}$$

Since this equation has two unknowns, we must have more information. Therefore, we will apply the second condition for equilibrium.

First, we must select an axis about which we can measure moment arms. The logical choice is at the point of application of some unknown force. Choosing the axis at B gives this force a zero moment arm. The sum of the torques about B results in the following equation.

$$\sum \tau_B = -A(12 \text{ m}) + (300 \text{ N})(10 \text{ m}) + (200 \text{ N})(4 \text{ m}) - (400 \text{ N})(4 \text{ m}) = 0$$

Note that the 400-N force and the force **A** tend to produce clockwise rotation about B. (Their torques were negative.) Simplifying gives

$$-(12 \text{ m})A + 3000 \text{ N} \cdot \text{m} - 1600 \text{ N} \cdot \text{m} + 800 \text{ N} \cdot \text{m} = 0$$

Adding $(12 \text{ m})A$ to both sides and simplifying, we obtain

$$2200 \text{ N} \cdot \text{m} = (12 \text{ m})A$$

Dividing both sides by 12 m, we find that

$$A = 183 \text{ N}$$

Now, to find the force exerted by support B, we can return to the equation obtained from the first condition for equilibrium.

$$A + B = 900 \text{ N}$$

Solving for B, we obtain

$$
\begin{aligned}
B &= 900 \text{ N} - A = 900 \text{ N} - 183 \text{ N} \\
&= 717 \text{ N}
\end{aligned}
$$

As a check on this solution, we could choose the axis of rotation at A and then apply the second condition for equilibrium to find B.

EXAMPLE 5-5

A uniform 200-lb boom, 24 ft long, is supported by a cable as shown in Fig. 5-9. The boom is hinged at the wall, and the cable makes a 30° angle with the boom, which is horizontal. If a load of 500 lb is hung from the right end, what is the tension T in the cable? What are the horizontal and vertical components of the force exerted by the hinge?

Solution

Let us consider the boom as the object in equilibrium. Of the two unknown forces **F** and **T**, we know least about the force **F**. Therefore, it is logical to choose this hinge as our axis of rotation when summing torques. In this manner, the unknown force **F** will have zero moment arm, rendering its torque about point A zero also. (Do not make the mistake of assuming that the force exerted by the hinge is entirely along the boom.) We can determine the cable tension from the second condition of equilibrium.

$$
\begin{aligned}
\sum \tau_A &= F(0) - (200 \text{ lb})(12 \text{ ft}) - (500 \text{ lb})(24 \text{ ft}) + T_x(0) + T_y(24 \text{ ft}) = 0 \\
&= 0 - 2400 \text{ lb} \cdot \text{ft} - 12{,}000 \text{ lb} \cdot \text{ft} + T_y(24 \text{ ft}) = 0 \\
T_y(24 \text{ ft}) &= 14{,}400 \text{ lb} \cdot \text{ft}
\end{aligned}
$$

From Fig. 5-9b,

$$T_y = T \sin 30° = 0.5T$$

Figure 5-9 Forces on a horizontal boom.

so that

$$(0.5T)(24 \text{ ft}) = 14{,}400 \text{ lb} \cdot \text{ft}$$
$$12t = 14{,}400 \text{ lb}$$
$$T = 1200 \text{ lb}$$

To find the horizontal and vertical components of **F**, we can apply the first condition for equilibrium. The horizontal component is found by summing forces along the x axis.

$$\sum F_x = 0 \qquad F_x - T_x = 0$$

from which

$$F_x = T_x = T \cos 30°$$
$$= (1200 \text{ lb})(\cos 30°) = 1040 \text{ lb}$$

The vertical component is found by summing the forces along the y axis.

$$\sum F_y = 0 \qquad F_y + T_y - 200 \text{ lb} - 500 \text{ lb} = 0$$

Solving for F_y, we obtain

$$F_y = 700 \text{ lb} - T_y$$

or

$$F_y = 700 \text{ lb} - (1200 \text{ lb})(\sin 30°)$$
$$= 700 \text{ lb} - 600 \text{ lb} = 100 \text{ lb}$$

It is left as an exercise for you to show that the magnitude and direction of the force **F** from its components is 1045 lb at 5.5° above the boom.

5-6 CENTER OF GRAVITY

Every particle on the earth has at least one force in common with every other particle—its *weight*. In the case of a body made up of many particles, these forces are essentially parallel and directed toward the center of the earth. Regardless of the shape and size of the body, there exists a point at which the entire weight of the body may be considered to be concentrated. This point is called the ***center of gravity*** of the body. Of course, the weight does not in fact all act at this point. But we would calculate the same torque about a given axis if we considered the entire weight to act at that point.

The center of gravity of a regular body, such as a uniform sphere, cube, rod, or beam, is located at its geometric center. This fact was used in the examples of the previous section, where we considered the weight of an entire beam as acting at its center. Although the center of gravity is a fixed point, it does not necessarily lie within the body. For example, a hollow sphere, a circular hoop, and a rubber tire all have centers of gravity outside the material of the body.

From the definition of the center of gravity, it is recognized that any body that is suspended at this point will be in equilibrium. This is true because the weight vector, which represents the sum of forces acting on each portion of the body, will have a zero moment arm. Thus, we can compute the center of gravity of a body by determining the point at which an upward force will produce rotational equilibrium.

EXAMPLE 5-6

Compute the center of gravity of the two spheres in Fig. 5-10 if they are connected by a 30-in. rod of negligible weight.

Solution

We first draw an upward vector indicating the force at the center of gravity that would balance the system. Suppose we choose to locate this vector at a distance from the center of the 16-lb sphere. The distance x would be drawn and labeled on the figure. Since the upward force must equal the sum of the downward forces, the first condition for equilibrium tells us that

$$F = 16 \text{ lb} + 8 \text{ lb} = 24 \text{ lb}$$

Now we will choose our axis of rotation at the center of the 16-lb sphere. This is the best choice because the distance x is measured from this point. The second condition for equilibrium is applied as follows:

$$\sum \tau = (24 \text{ lb})x - (8 \text{ lb})(30 \text{ in.}) = 0$$
$$(24 \text{ lb})x = 240 \text{ lb} \cdot \text{in.}$$
$$x = 10 \text{ in.}$$

Figure 5-10 Computing the center of gravity.

If the rod joining the two spheres is suspended from the ceiling at a point 10 in. from the center of the 16-lb sphere, the system will be in equilibrium. This point is the center of gravity. You should show that the same conclusion follows if you sum torques about the right end or about any other point.

Summary and Review

Summary

When forces acting on a body do not have the same line of action or do not intersect at a common point, rotation may occur. In this chapter, we introduced the concept of torque as a measure of the tendency to rotate. The major concepts are summarized below:

- The *moment arm* of a force is the perpendicular distance from the line of action of the force to the axis of rotation.
- The *torque* about a given axis is defined as the product of the magnitude of a force and its moment arm:

$$Torque = force \times moment\ arm$$

It is positive if it tends to $\oplus$ ccw produce counterclockwise motion and negative $\ominus$ cw if the motion produced is clockwise.

$$\tau = Fr$$

- The *resultant torque* τ_R about a particular axis A is the algebraic sum of the torques produced by each force. The signs are determined by the above convention.

$$\tau_R = \sum \tau_A = F_1 r_1 + F_2 r_2 + F_3 r_3 + \cdots$$

- *Rotational equilibrium:* A body in rotational equilibrium has no resultant torque acting on it. In such cases, the sum of all the torques about *any* axis must equal zero. The axis may be chosen anywhere because the system is not tending to rotate about any point. This is called the second condition for equilibrium and may be written

The sum of all torques about any point is zero.

$$\sum \tau = 0$$

- *Total equilibrium* exists when the first and second conditions are satisfied. In such cases, three independent equations can be written:

$$\text{(a)}\ \sum F_x = 0 \quad \text{(b)}\ \sum F_y = 0 \quad \text{(c)}\ \sum \tau = 0$$

By writing these equations for a given situation, unknown forces, distances, or torques can be found.

- The *center of gravity* of a body is the point through which the resultant weight acts, regardless of how the body is oriented. For applications involving torque, the entire weight of the object may be considered as acting at this point.

Key Terms

1. line of action
2. axis of rotation
3. moment arm
4. torque
5. rotational equilibrium
6. center of gravity

Review Questions

5-1. You lift a heavy suitcase with your right hand. Describe and explain the position of your body.

5-2. A parlor trick consists of asking you to stand against a wall with your feet together so that the side of your right foot rests against the wall. You are then asked to raise your left foot off the floor. Why can't you do this without falling?

5-3. Why is a minivan more likely to turn over than a Corvette or some other sports car?

5-4. If you know the weight of a brick to be 6 lb, describe how you could use a meterstick and a pivot to measure the weight of a baseball.

5-5. Describe and explain the arm and leg motions used by a tightrope walker to maintain balance.

Summary and Review

5-6. Discuss the following items and their use of the principle of torque: (a) screwdriver, (b) wrench, (c) pliers, (d) wheelbarrow, (e) nutcracker, and (f) crowbar.

Problems

Section 5-2—The Moment Arm

5-1. Draw and label the moment arm of the force F about an axis at point A in Fig. 5-11a. What is the magnitude of the moment arm?   **Ans. 0.845 ft**

5-2. Find the moment arm about axis B in Fig. 5-11a.

5-3. Draw and label the moment arm if the axis of rotation is at point A in Fig. 5-11b. What is the magnitude of the moment arm?
                                Ans. 1.73 m

5-4. Find the moment arm about axis B in Fig. 5-11b.

Section 5-3—Torque

5-5. If the force F in Fig. 5-11a is equal to 80 lb, what is the resultant torque about axis A (neglecting the weight of the rod)? What is the resultant torque about axis B?
                  Ans. −67.5 lb · ft, 101 lb · ft

5-6. The force F in Fig. 5-11b is 400 N, and the angle iron is of negligible weight. What is the resultant torque about axis A and about axis B?

5-7. A leather belt is wrapped around a pulley 20 cm in diameter. A force of 60 N is applied to the belt. What is the torque at the center of the shaft?   **Ans. 6 N · m**

5-8. The light rod in Fig. 5-12 is 60 cm long and pivoted about point A. Find the magnitude and sign of the torque due to the 200-N force if angle θ is (a) 90°, (b) 60°, (c) 30°, and (d) 0°.

Figure 5-12

5-9. A person who weighs 650 N rides a bicycle. The pedals move in a circle of radius 40 cm. If the entire weight acts on each downward-moving pedal, what is the maximum torque?   **Ans. 260 N · m**

5-10. A single belt is wrapped around two pulleys. The drive pulley has a diameter of 10 cm, and the output pulley has a diameter of 20 cm. If the top belt tension is essentially 50 N at the edge of each pulley, what are the input and output torques?

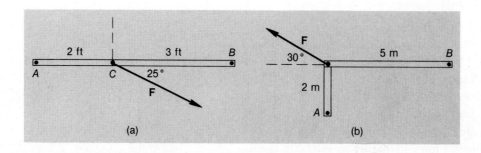

(a)

(b)

Figure 5-11

Section 5-4—Resultant Torque

5-11. What is the resultant torque about point A in Fig. 5-13? Neglect the weight of the bar.
Ans. 90 N · m

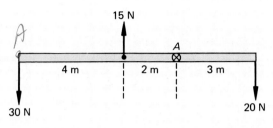

Figure 5-13

5-12. Find the resultant torque in Fig. 5-13 if the axis is moved to the left end of the bar.

5-13. What horizontal force must be exerted at point A in Fig. 5-11b to make the resultant torque about point B equal to zero when the force $F = 80$ N? **Ans. 100 N**

5-14. Two wheels of diameters 60 cm and 20 cm are fastened together and turn on the same axis as in Fig. 5-14. What is the resultant torque about a central axis for the shown weights?

5-15. Suppose you remove the 150-N weight from the small wheel in Fig. 5-14. What new weight can you hang to produce zero resultant torque? **Ans. 400 N**

5-16. Determine the resultant torque about the corner A for Fig. 5-15.

5-17. Find the resultant torque about point C in Fig. 5-15. **Ans. −16.0 N · m**

Figure 5-14

*5-18. Find the resultant torque about axis B in Fig. 5-15.

Section 5-5—Equilibrium

5-19. A uniform meterstick is balanced at its midpoint with a single support. A 60-N weight is suspended at the 30-cm mark. At what point must a 40-N weight be hung to balance the system? **Ans. At the 80-cm mark**

5-20. Weights of 10 N, 20 N, and 30 N are placed on a meterstick at the 20 cm, 40 cm, and 60 cm marks, respectively. The meterstick is balanced by a single support at its midpoint. At what point may a 5-N weight be attached to produce equilibrium?

Figure 5-15

5-21. An 8-m board of negligible weight is supported at a point 2 m from the right end, where a 50-N weight is attached. What downward force must be exerted at the left end to produce equilibrium? **Ans. 16.7 N**

5-22. A 4-m pole is supported at each end by hunters carrying an 800-N deer which is hung at a point 1.5 m from the left end. What are the upward forces required by each hunter?

5-23. Assume that the bar in Fig. 5-16 is of negligible weight. Find the forces F and A provided the system is in equilibrium.

Ans. A = 26.7 N, F = 107 N

Figure 5-16

5-24. For equilibrium, what are the forces F_1 and F_2 in Fig. 5-17? Neglect the weight of the bar.

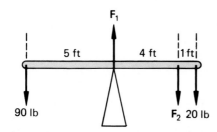

Figure 5-17

5-25. Consider the light bar supported as shown in Fig. 5-18. What are the forces exerted by the supports A and B?

Ans. A = 50.9 N, B = 49.1 N

Figure 5-18

5-26. A V-belt is wrapped around a pulley 16 in. in diameter. If a resultant torque of 4 lb ft is required, what force must be applied along the belt?

5-27. A bridge whose total weight is 4500 N is 20 m long and supported at each end. Find the forces exerted at each end when a 1600-N tractor is located 8 m from the left end.

Ans. 2890 N, 3210 N

5-28. A 10-ft platform weighing 40 lb is supported at each end by stepladders. A 180-lb painter is located 4 ft from the right end. Find the forces exerted by the supports.

*5-29. A horizontal, 6-m boom weighing 400 N is hinged at the wall as shown in Fig. 5-19. A

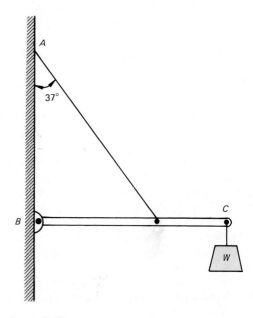

Figure 5-19

cable is attached at a point 4.5 m away from the wall, and a 1200-N weight is attached to the right end. What is the tension in the cable? **Ans. 2340 N**

*5-30. What are the horizontal and vertical components of the force exerted by the wall on the boom? What are the magnitude and direction of this force?

Section 5-6—Center of Gravity

5-31. A uniform 6-m bar has a length of 6 m and weighs 30 N. A 50-N weight is hung from the left end and a 20-N force is hung at the right end. How far from the left end will a single upward force produce equilibrium?
Ans. 2.10 m

5-32. A 40-N sphere and a 12-N sphere are connected by a light rod 200 mm in length. How far from the middle of the 40-N sphere is the center of gravity? 46, 2 mm

5-33. Weights of 2, 5, 8, and 10 N are hung from a 10-m light rod at distances of 2, 4, 6, and 8 m from the left end. How far from the left in is the center of gravity? **Ans. 6.08 m**

5-34. Compute the center of gravity of a sledge-hammer if the metal head weighs 12 lb and the 32-in. supporting handle weighs 2 lb. Assume that the handle is of uniform construction and weight.

_r = 0.0462 m_ (handwritten, top)

r = .14 ft, 2.28 inch. (handwritten)

Challenge Problems

5-35. What is the resultant torque about the hinge in Fig. 5-20? Neglect the weight of the curved bar. **Ans. −3.42 N · m**

5-36. What horizontal force applied to the left end of the curved bar in Fig. 5-20 will produce rotational equilibrium? The bar forms a right angle.

5-37. Weights of 100, 200, and 500 lb are placed on a light board resting on two supports, as shown in Fig. 5-21. What are the forces exerted by the supports? **Ans. 375 N, 425 N**

5-38. An 8-m steel metal beam weighs 2400 N and is supported 3 m from the right end. If a 9000-N weight is placed on the right end, what force must be exerted at the left end to balance the system?

*5-39. Find the resultant torque about point A in Fig. 5-22. **Ans. −3.87 N · m**

*5-40. Find the resultant torque about point B in Fig. 5-22.

Figure 5-21

Figure 5-20

Figure 5-22

Summary and Review

Critical Thinking Questions

*5-41. A 30-lb box and a 50-lb box are on opposite ends of a 16-ft board that is supported only at its midpoint. How far from the left end should a 40-lb box be placed to produce equilibrium? Would the result be different if the board weighed 90 lb? Why, or why not? **Ans. 4.00 ft, no**

5-42. On a lab bench you have a small rock, a 4-N meterstick, and a single knife-edge support. Explain how you can use these three items to find the weight of the small rock.

5-43. Find the forces F_1, F_2, and F_3 such that the system drawn in Fig. 5-23 is in equilibrium. **Ans. F_1 = 213 lb, F_2 = 254 lb, F_3 = 83.9 lb**

5-44. (a) What weight W will produce a tension of 400 N in the rope attached to the boom in Fig. 5-24? (b) What would be the tension in the rope if W = 400 N? Neglect the weight of the boom in each case.

5-45. Suppose the boom in Fig. 5-24 has a weight of 100 N and the suspended weight W is equal to 400 N. What is the tension in the cord? **Ans. T = 1169 N**

5-46. For the conditions set in Question 5-45, what are the horizontal and vertical components of the force exerted by the floor hinge on the base of the boom?

Figure 5-23

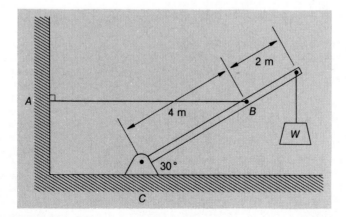

Figure 5-24

*5-47. What is the tension in the cable for Fig. 5-25? The weight of the boom is 300 N, but its length is unknown. **Ans. 360 N**

*5-48. What are the magnitude and direction of the force exerted by the wall on the boom in Fig. 5-25? Again, assume that the weight of the board is 300 N.

*5-49. An automobile has a distance of 3.4 m between front and rear axles. If 60 percent of the weight rests on the front wheels, how far is the center of gravity located from the front axle? **Ans. 1.36 m**

Figure 5-25

Uniform Acceleration

6-1 SPEED AND VELOCITY

The simplest kind of motion an object can experience is uniform motion in a straight line. If the object covers the same distances in each successive unit of time, it is said to move with constant speed. For example, if a train covers 8 m of track every second that it moves, we say that it has a ***constant speed*** of 8 m/s. Whether the speed is constant or not, the average speed of a moving object is defined by

$$Average\ speed = \frac{distance\ traveled}{time\ elapsed}$$

$$\bar{v} = \frac{s}{t} \qquad\qquad 6\text{-}1$$

The bar over the symbol v means that the speed represents an average value for the time interval t.

Remember that the dimension of speed is the ratio of a length to a time interval. Hence, the units of miles per hour, feet per second, meters per second, and centimeters per second are all typical units of speed.

EXAMPLE 6-1

A golfer sinks a putt 3 s after the ball leaves the club face. If the ball traveled with an average speed of 0.8 m/s, how long was the putt?

Solution

Solving Eq. (6-1) for *s*, we have

$$s = \bar{v}t = (0.8 \text{ m/s})(3 \text{ s})$$

Therefore, the distance of the putt is

$$s = 2.4 \text{ m}$$

It is important to recognize that speed is a scalar quantity that is completely independent of direction. In Example 6-1, it was not necessary for us to know either the speed of the golf ball at any instant or the nature of its path. Similarly, the **average speed** of a car traveling from Atlanta to Chicago is a function only of the distance registered on its odometer and the time required to make the trip. It makes no difference as far as computation is concerned whether the driver of the car took the direct or scenic route or even if he or she stopped for meals.

We must make a clear distinction between the scalar quantity *speed* and its directional counterpart **velocity.** This is best done by recalling the difference between *distance* and *displacement,* as discussed in Chap. 3. Suppose, in Fig. 6-1, an object moves along the broken path from *A* to *B.* The actual distance traveled is denoted by *s,* whereas the displacement is represented by the polar coordinates

$$\mathbf{D} = (D, \theta)$$

As an example, suppose the distance *s* in Fig. 6-1 is 500 km, and the displacement is 350 km at 45°. If the actual traveling time is 8 h, the average speed is

$$\bar{v} = \frac{s}{t} = \frac{500 \text{ km}}{8 \text{ h}} = 62.5 \text{ km/h}$$

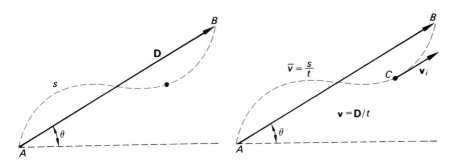

Figure 6-1 Displacement and velocity are vector quantities, whereas distance and speed are independent of direction: *s*, distance; **D**, displacement; **v**, velocity; *t*, time.

The average *velocity*, however, must consider the displacement magnitude and direction. The average velocity is given by

$$\bar{\mathbf{v}} = \frac{\mathbf{D}}{t} = \frac{350 \text{ km}, 45°}{8 \text{ h}}$$
$$\bar{\mathbf{v}} = 43.8 \text{ km/h}, 45°$$

Therefore, if the path of the moving object is curved, the difference between speed and velocity is one of magnitude as well as direction.

Automobiles cannot always travel at constant speeds for long periods of time. In traveling from point A to B, we may be required to slow down or speed up because of road conditions. For this reason, it is sometimes useful to talk of ***instantaneous speed*** or ***instantaneous velocity***.

The instantaneous speed is a scalar quantity representing the speed at the instant the car is at an arbitrary point C. It is, therefore, the time rate of change in distance.

The instantaneous velocity is a vector quantity representing the velocity v_i at any point C. It is the time rate of change in displacement.

In this chapter we shall be concerned with motion along a straight line so that the magnitudes of speed and velocity are the same at any instant. If the direction does not change, the instantaneous speed is the scalar part of the the instantaneous velocity. It is good practice, however, to reserve the term *velocity* for the more complete description of motion. As we shall see in the coming sections, a change in velocity may result in a change of direction as well. In such cases, the terms *velocity* and *displacement* are more appropriate than the terms *speed* and *distance*.

6-2 ACCELERATION

In most cases, the velocity of a moving object changes as motion continues. Motion in which the magnitude or direction changes with time is called ***acceleration.*** For example, suppose we observe the motion of a sprinter for a period of time *t*. The initial velocity $\mathbf{v}_0$ of an object is defined as the velocity at the beginning of the time interval (usually $t = 0$). The final velocity is defined as the velocity at the end of the time interval (when $t = t$). Thus, if we are able to measure initial and final velocities of a moving object, we can say that its acceleration is given by

$$Acceleration = \frac{change \text{ } in \text{ } velocity}{time \text{ } interval}$$

$$\mathbf{a} = \frac{\mathbf{v}_f - \mathbf{v}_0}{t} \qquad \text{6-2}$$

Acceleration written in the above manner is a vector quantity and thus depends on changes in direction as well as changes in magnitude. If the direction does not change and the motion is in a straight line, only the *speed* of the object changes.

However, if the path is curved, there must be acceleration, even if the speed is not changed. For motion in a perfect circle and constant speed, the acceleration will be at right angles to the velocity. Such uniform circular motion will be treated later.

6-3 UNIFORM ACCELERATION

The simplest kind of acceleration is motion in a straight line in which the speed changes at a constant rate. This special kind of motion is generally called **uniform acceleration** or **constant acceleration.** Since the direction does not change, the vector difference in Eq. (6-2) becomes simply the difference between the signed values of final and initial velocities. However, it must be remembered that velocity is still a *vector* quantity and that signs affixed to velocity denote *direction* rather than *magnitude.* For constant acceleration we can write

$$a = \frac{v_f - v_0}{t} \qquad \text{6-3}$$

As an example, consider the car in Fig. 6-2, which moves with constant acceleration from an initial speed of 12 m/s to a final speed of 22 m/s. Considering the right direction as positive, the car's velocity at A is $+12$ m/s and its final velocity at B is $+22$ m/s. If the increase in velocity requires 5 s, the acceleration can be found from Eq. (6-2).

$$a = \frac{v_f - v_0}{t} = \frac{22 \text{ m/s} - 12 \text{ m/s}}{5 \text{ s}}$$

$$= \frac{10 \text{ m/s}}{5 \text{ s}} = 2 \text{ m/s}^2$$

buphy.bu.edu
For an excellent demonstration of the effect of gravity on a feather and a penny, with and without air present, see this Web site (but do not type www.).

Figure 6-2   Uniformly accelerated motion.

The answer is read as *two meters per second per second* or as *two meters per second squared.* The meaning is that each second the car increases its speed by 2 m/s. Since the car began at 12 m/s, its speed after times of 1, 2, and 3 s, would be 14, 16, and 18 m/s, respectively.

EXAMPLE 6-2

A train reduces its velocity from 80 km/h to 20 km/h in a time of 8 s. Find the acceleration in SI units.

Solution

We first convert the velocity from kilometers per hour to the SI units of m/s.

$$60 \, \frac{km}{h} = \frac{1000 \text{ m}}{1 \text{ m}} = \frac{1 \text{ h}}{3600 \text{ s}} = 16.7 \text{ m/s}$$

Similarly, it is determined that 20 km/h is equal to 5.56 m/s. Substitution into Eq. (6-3) gives

$$a = \frac{v_f - v_0}{t} = \frac{5.56 \text{ m/s} - 16.7 \text{ m/s}}{8 \text{ s}}$$

$$a = -1.39 \text{ m/s}$$

Since the original direction of the train was considered positive, the negative sign for acceleration means that the train reduced its speed by 1.39 m/s every second. Such motion is sometimes called *deceleration,* but this term is problematic because $a = -1.39$ m/s^2 really means that the velocity *becomes more negative* by that amount every second. If the speed increases in a negative direction, the acceleration is also negative. Acceleration refers to the *change* in velocity, which can mean either an increase or a decrease in speed.

Many times the same equation is used to solve for different quantities. You should therefore solve each equation literally for each symbol in the equation. A convenient form for Eq. (6-3) arises when it is solved explicitly for the final velocity. Thus,

Final velocity = initial velocity + change in velocity

$$v_f = v_0 + at \qquad\qquad 6\text{-}4$$

EXAMPLE 6-3

An automobile maintains a constant acceleration of 8 m/s^2. If its initial velocity was 20 m/s, what will its velocity be after 6 s?

Solution

The final velocity is obtained from Eq. (6-4).

$$v_f = v_0 + at = 20 \text{ m/s} + (8 \text{ m/s}^2)(6 \text{ s})$$

or

$$v_f = 20 \text{ m/s} + 48 \text{ m/s}$$

Hence, the final velocity is

$$v_f = 68 \text{ m/s}$$

Now that the concept of initial and final velocities is understood, let us return to the equation for average velocity and express it in terms of initial and final states. The average velocity of an object moving with uniform acceleration is found just like the arithmetic mean of two numbers. Given an initial velocity and a final velocity, the average velocity is simply

$$\bar{v} = \frac{v_f + v_0}{2} \qquad \text{6-5}$$

Utilizing this relationship in Eq. (6-1) gives us a more useful expression for computing distance:

$$s = \bar{v}t = \frac{v_f + v_0}{2}t \qquad \text{6-6}$$

EXAMPLE 6-4

A moving object increases its velocity uniformly from 20 to 40 m/s in 2 min. What is its average velocity, and how far did it travel in the 2 min?

Solution

The average velocity is found by direct substitution into Eq. (6-5).

$$\bar{v} = \frac{v_f + v_0}{2} = \frac{40 \text{ m/s} + 20 \text{ m/s}}{2}$$

or

$$\bar{v} = \frac{60 \text{ m/s}}{2} = 30 \text{ m/s}$$

The distance traveled is then found from Eq. (6-1).

$$s = \bar{v}t = (30 \text{ m/s})(2 \text{ min})$$

The units of time are inconsistent, but since 2 min = 120 s, we have

$$s = (30 \text{ m/s})(120 \text{ s}) = 3600 \text{ m}$$

6-4 OTHER USEFUL RELATIONS

Thus far we have presented two fundamental relations. One arises from the definition of velocity and the other from the definition of acceleration. They are

$$s = \bar{v}t = \frac{v_f + v_0}{2}t \qquad \text{6-6}$$

Math Tip

In the equation
$$s = vt + \frac{1}{2}at^2,$$
when t is the only unknown, the solution sometimes requires the quadratic equation. First get the equation equal to zero by subtracting s from both sides.

$$s - s = vt + \frac{1}{2}at^2 - s$$

Rearranging terms gives us

$$0 = \frac{1}{2}at^2 + vt - s$$

From here $a = \frac{1}{2}a$, $b = v$, $c = -s$ for the quadratic formula.

$$t = \frac{-b \pm \sqrt{b^2 - 4ac}}{2a}$$

and

$$v_f = v_0 + at \qquad \text{6-4}$$

Although these are the only formulas necessary to attack the many problems presented in this chapter, three other useful relationships can be obtained from them. The first is derived by eliminating the final velocity from Eqs. (6-6) and (6-4). Substituting the latter into the former yields

$$s = \frac{(v_0 + at) + v_0}{2}t$$

Simplifying gives

$$s = v_0t + \frac{1}{2}at^2 \qquad \text{6-7}$$

A similar equation is obtained by eliminating v_0 from the same two equations:

$$s = v_ft - \frac{1}{2}at^2 \qquad \text{6-8}$$

The third additional equation comes from the elimination of time t from the basic equations. A little algebra yields

$$2as = v_f^2 - v_0 \qquad \text{6-9}$$

Although these equations add no new information, they are useful for solving problems in which three of the parameters are given and one of the other two is to be found.

6-5 SOLUTIONS TO ACCELERATION PROBLEMS

Although the solution to problems involving constant acceleration depends primarily on choosing the correct formula and substituting known values, there are several suggestions to help the beginning student. Problems frequently occur with motion that either started from rest or was brought to a stop from some initial velocity. In either case, the formulas discussed can be simplified by the substitution of either $v_0 = 0$ or $v_f = 0$, as the case may be. Table 6-1 summarizes the general formulas.

A close look at five general equations will reveal a total of five parameters: s, v_0, v_f, a, and t. Given any three of these quantities, the remaining two can be found from the general equations. Therefore, a starting point in solving any problem is to read it thoroughly with a view to establishing the three quantities required for solution. It is

Table 6-1 Summary of Acceleration Formulas

(1) $s = \dfrac{v_0 + v_v}{2}t$	(4) $s = v_ft - \dfrac{1}{2}at^2$
(2) $v_f = v_0 + at$	(5) $2as = v_f^2 - v_0^2$
(3) $s = v_0t + \dfrac{1}{2}at^2$	

also important to choose a direction to call positive and apply it consistently to velocity, displacement, and acceleration when inserting the values into equations.

If you have difficulty in deciding which equation should be used, it may help to recall the conditions such an equation must satisfy. First, it must contain the unknown parameter. Second, all other parameters that appear in the equation must be known. For example, if a problem gives you values for v_f, v_0, and t, you may solve for a in Eq. (2) of Table 6-1.

Problem-Solving Strategy

Constant Acceleration Problems

1. Read the problem and then draw and label a rough sketch.
2. Indicate the consistent positive direction.
3. Establish three given parameters and two that are unknown. Make sure the signs and units are consistent.

   Given: _____   Find: _____
   _____   _____
   _____

4. Select the equation that contains one of the unknown parameters but not the other one.

$$s = \left(\frac{v_0 + v_f}{2}\right)t \qquad 2as = v_f^2 - v_0^2$$

$$v_f = v_0 + at \qquad s = v_f t - \frac{1}{2}at^2$$

$$s = v_0 t + \frac{1}{2}at^2$$

5. Substitute known quantities and solve the equation.

The examples that follow are abbreviated and do not include figures, but they do illustrate the approach.

CONCEPTUAL EXAMPLE 6-5

A motorboat starting from rest attains a velocity of 50 km/h in 15 s. What was its acceleration, and how far did it travel?

Solution

Given: $v_0 = 0$ Find: $a = ?$

$v_f = 50$ km/h $= 13.9$ m/s $s = ?$

$t = 0$

In solving for acceleration, we must choose an equation that contains a but not s. Equation (2) in Table 6-1 can be used, where $v_0 = 0$. Hence,

$$v_f = 0 + at$$

from which

$$a = \frac{v_f}{t} = \frac{13.9 \text{ m/s}}{15 \text{ s}}$$

$$= 0.926 \text{ m/s}^2$$

The displacement can be found from Eq. (1) in Table 6-1, as follows:

$$s = \frac{v_0 + v_f}{2}t = \frac{0 + 13.9 \text{ m/s}}{2}(15 \text{ s})$$

$$a = 104 \text{ m}$$

EXAMPLE 6-6

An airplane lands on a carrier deck at 200 mi/h and is brought to a stop in 300 ft. Find the acceleration and the stopping time.

Solution

Given: $v_0 = 200 \text{ mi/h} = 294 \text{ ft/s}$ Find: $a = ?$

$$v_f = 0 \qquad\qquad\qquad t = ?$$

$$s = 300 \text{ ft}$$

Looking at Table 6-1, we must select Eq. (5) as the one that contains a and not t:

$$2as = v_f^2 - v_0^2$$

$$a = \frac{v_f^2 - v_0^2}{2s} = \frac{(0)^2 - (294 \text{ m/s})^2}{2(300 \text{ ft})}$$

$$a = -144 \text{ ft/s}^2$$

A person undergoing such acceleration would experience a stopping force equal to about four and a half times their weight. Next we find the stopping time by selecting the equation that contains t and not a. Solving Eq. (2) explicitly for t, we have

$$t = \frac{v_f - v_0}{a} = \frac{(0) - (294 \text{ ft/s})}{-144 \text{ ft/s}^2}$$

$$t = 2.04 \text{ s}$$

EXAMPLE 6-7

A train traveling initially at 16 m/s is under constant acceleration of 2 m/s². How far will it travel in 20 s? What will be its final velocity?

Solution

Given: $v_0 = 16 \text{ m/s}$ Find: $s = ?$

$$a = 2 \text{ m/s}^2 \qquad\qquad v_f = ?$$

$$t = 20 \text{ s}$$

Selecting Eq. (3) from Table 6-1 as the one that contains s and not v_f, we have

$$s = v_0 t + \frac{1}{2}at^2$$

$$= (16 \text{ m/s})(20 \text{ s}) + \frac{1}{2}(2 \text{ m/s}^2)(20 \text{ s})^2$$

$$= 320 \text{ m} + 400 \text{ m} = 720 \text{ m}$$

The final velocity is found from Eq. (2)

$$v_f = v_0 + at$$
$$= (16 \text{ m/s} + (2 \text{ m/s}^2)(20 \text{ s}) = 56 \text{ m/s}$$

The train travels for a distance of 720 m and attains a velocity of 56 m/s.

6-6 SIGN CONVENTION IN ACCELERATION PROBLEMS

The signs of acceleration, displacement, and velocity are mutually independent and are each determined by different criteria. Probably no other point is more confusing for beginning students. Whenever the direction of motion changes, as with an object tossed into the air or an object attached to an oscillating spring, the signs of displacement and acceleration are particularly difficult to visualize. Only the sign of velocity is determined by the direction of motion. The sign of displacement depends on the location or position of the object, and the sign of acceleration is determined by the force that causes the velocity to change.

Consider for example a baseball tossed upward as shown in Fig. 6-3. The ball moves upward along a straight line until it stops and returns in a downward path along the same line. We will take the point of release as zero for displacement ($s = 0$). Now, the sign of displacement will be positive at any point *above* the release point, and it will be negative at any point *below* the release point. Note that it doesn't matter whether the ball is moving upward or downward; it is only the location (the y coordinate of its position) that determines the sign of displacement. For example, the value of s may be $+1$ m on the way up and $+1$ m on the way down. Its displacement will become negative only when the ball is physically located below its release point.

Now, observe the signs of the velocity during the flight of the ball. Assuming the upward direction to be positive, we assert that the velocity of the ball is positive any time its motion is directed upward and negative any time its motion is directed downward. It doesn't matter that the velocity is changing with time; nor does its location in space matter.

Finally, consider the acceleration of the ball during its flight. The only force acting on the ball during the entire flight is its weight, which is always directed downward. Thus, the sign of the acceleration is negative (downward) during the entire motion. Note that the acceleration is negative when the ball is moving up and also as it moves downward. In essence, the velocity is always becoming more negative. To determine whether the acceleration of an object is positive or negative, we must

Math Tip

When you are working through a problem with many steps, it is possible to lose sight of the meaning of the final answer. When the answer is negative, some students assume that it is incorrect. However, a negative can often indicate the direction of a vector (such as velocity or acceleration) in the direction opposite to the one we assigned as positive.

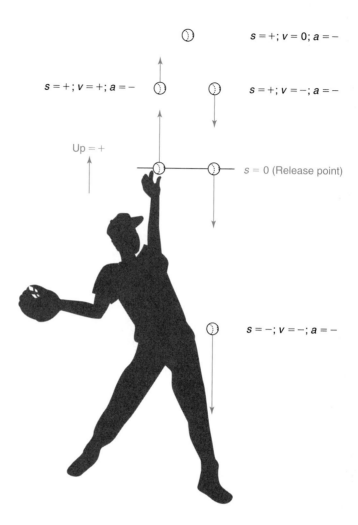

$s = +; v = 0; a = -$

$s = +; v = +; a = -$

$s = +; v = -; a = -$

Up = +

$s = 0$ (Release point)

$s = -; v = -; a = -$

Figure 6-3

not look at its location or its direction of motion; we must look, instead, at the direct of the *force* that is causing the velocity to change. In this example, that force is its weight.

Once a choice is made for positive direction, the following conventions will determine the signs of velocity, displacement, and acceleration:

> **Displacement is positive or negative depending on the location or coordinate of position of the object relative to its zero position.**

> **Velocity is positive or negative depending on whether the direction of motion is with or against the chosen positive direction.**

> **Acceleration is positive or negative depending on whether the resultant force is with or against the positive direction.**

6-7 GRAVITY AND FREE-FALLING BODIES

Much of our knowledge about the physics of falling bodies originated with the Italian scientist Galileo Galilei (1564–1642). He was the first to deduce that in the absence of friction all bodies, large or small, heavy or light, fall to the earth with the same acceleration. This was a revolutionary idea, for it contradicted what a person might expect. Until the time of Galileo, people followed the teachings of Aristotle—that heavy objects fall proportionally faster than lighter objects. The classic explanation for the paradox rests with the fact that heavier bodies are proportionately more difficult to accelerate. This resistance to motion is a property of a body called its *inertia.* Thus, in a vacuum, a feather will fall at the same rate as a steel ball because the larger inertial effect of the steel ball compensates exactly for its larger weight. (See Fig. 6-4.)

Figure 6-4 All bodies fall with the same acceleration in a vacuum.

In the treatment of falling bodies given in this chapter, the effects of air friction are neglected entirely. Under these circumstances, **gravitational acceleration** is uniformly accelerated motion. At sea level and 45° latitude, this acceleration has been measured to be 32.17 ft/s², or 9.806 m/s², and is denoted by g. For our purposes, the following values will be sufficiently accurate:

$$g = 32 \text{ ft/s}^2$$
$$g = 9.8 \text{ m/s}^2$$

6-10

Since gravitational acceleration is constant acceleration, the same general equations of motion apply. One of the parameters is known in advance, however, and need not be stated in the problem. If the constant g is inserted into the general equations (Table 6-1) the following modified forms will result:

(1a) $s = \dfrac{v_f + v_0}{2t}$ $\qquad\qquad$ $s = \bar{v}t$

(2a) $v_f = v_0 + gt$

(3a) $s = v_0 t + \dfrac{1}{2}gt^2$

(4a) $s = v_f t - \dfrac{1}{2}at^2$

(4a) $2gs = v_f^2 - v_0^2$

Before these equations are used, a few general comments are in order. In problems dealing with free-falling bodies, it is extremely important to choose a direction to call positive and to follow through consistently in the substitution of known values. The sign of the answer is necessary to determine the location of a point or the direction of the velocity at specific times. For example, the distance s in the above equations represents the displacement above or below the origin. If the upward direction is chosen as positive, a positive value for s indicates a displacement above the starting point; if s is negative, it represents a displacement below the starting point. Similarly, the signs of v_0, v_f, and g indicate their directions.

EXAMPLE 6-8

A rubber ball is dropped from rest, as shown in Fig. 6-5. Find its velocity and position after 1, 2, 3, and 4 s.

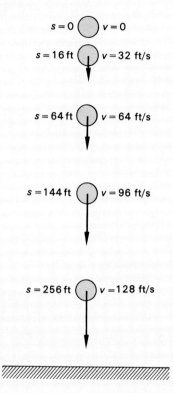

Figure 6-5 A freely falling body has a constant downward acceleration of 32 ft/s².

Solution

Since all parameters will be measured downward, it will be more convenient to choose the downward direction as positive. Organizing the data, we have

Given: $v_0 = 0$ Find: $v_f = ?$

$g = +32$ ft/s² $s = ?$

$t = 1, 2, 3,$ and 4 s

The velocity as a function of time is given by Eq. (2a), where $v_0 = 0$.

$$v_f = v_0 + gt = gt$$
$$= (32 \text{ ft/s}^2)t$$

After 1 s we have

$$v_f = (32 \text{ ft/s}^2)(1 \text{ s}) = 32 \text{ ft/s} \qquad \text{(downward)}$$

Similar substitution of $t = 2, 3,$ and 4 s will yield final velocities of 64, 96, and 128 ft/s, respectively. All these velocities are directed downward since that direction was chosen as positive.

The position as a function of time is calculated from Eq. (3a). Since the initial velocity is zero, we write

$$s = v_0 t + \frac{1}{2}gt^2 = \frac{1}{2}gt^2$$

from which

$$s = \frac{1}{2}(32 \text{ ft/s}^2)t^2 = (16 \text{ ft/s}^2)t^2$$

After 1 s, the body will fall a distance as given by

$$s = (16 \text{ ft/s}^2)(1 \text{ s})^2 = (16 \text{ ft/s}^2)(1 \text{ s}^2)$$
$$= 16 \text{ ft}$$

After 2 s,

$$s = (16 \text{ ft/s}^2)(2 \text{ s})^2 = (16 \text{ ft/s}^2)(4 \text{ s}^2)$$
$$= 64 \text{ ft}$$

Similarly, calculations give 144 and 256 ft for the positions after 3 and 4 s, respectively. The above results are summarized in Table 6-2.

Table 6-2		
Time t, s	Velocity at the end of time t, ft/s	Position at the end of time t, ft
0	0	0
1	32	16
2	64	64
3	96	144
4	128	256

CONCEPTUAL EXAMPLE 6-9

Assuming a ball is projected upward with an initial velocity of 96 ft/s, explain without using equations how its upward motion is just the reverse of its downward motion.

Solution

We shall assume the upward direction to be positive, making the acceleration due to gravity equal to -32 ft/s^2. The negative sign indicates that an object projected vertically will have its velocity reduced by 32 ft/s every second it rises. (Refer to Fig. 6-6). If its initial velocity is 96 ft/s, its velocity after 1 s will be reduced to 64 ft/s. After 2 s its velocity will be 32 ft/s, and after 3 s its velocity will be reduced to zero. When the velocity becomes zero, the ball has reached its maximum height and begins to fall freely from rest. Now, however, the

$3\,s$ $v_3 = 0$

$v_2 = 32$ ft/s, $2\,s$ $4\,s$, $v_4 = -32$ ft/s

$v_1 = 64$ ft/s, $1\,s$ $5\,s$, $v_5 = -64$ ft/s

$v_0 = 96$ ft/s

$v_6 = -96$ ft/s

Figure 6-6 A ball thrown vertically upward returns to the ground with the same speed.

velocity of the ball will be *increasing* by 32 ft/s every second since both the direction of motion and the acceleration of gravity are in the negative direction. Its velocity after 4, 5, and 6 s will be -32, -64, and -96 ft/s, respectively. Except for the sign, which indicates the direction of motion, the velocities are the same at equal heights above the ground.

EXAMPLE 6-10

A baseball thrown vertically upward from the roof of a tall building has an initial velocity of 20 m/s. (a) Calculate the time required to reach its maximum height. (b) Find the maximum height. (c) Determine its position and velocity after 1.5 s. (d) What are its position and velocity after 5 s? (See Fig. 6-7.)

Solution (a)

Let us choose the upward direction as positive since the initial velocity is directed upward. At the highest point, the final velocity of the ball will be zero. Organizing the known data, we have

Given: $v_0 = 20$ m/s   Find: $t = ?$

$v_f = 0$ $s = ?$

$g = -9.8$ m/s^2

The time required to reach the maximum height can be determined from Eq. (2a):

$$t = \frac{v_f - v_0}{g} = -\frac{v_0}{g}$$

$$= \frac{-20 \text{ m/s}}{-9.8 \text{ m/s}^2} = 2.04 \text{ s}$$

Solution (b)

The maximum height is found by setting $v_f = 0$ in Eq. (1a).

$$s = \frac{v_f + v_0}{2}t = \frac{v_0}{2}t$$

$$= \frac{20 \text{ m/s}}{2}(2.04 \text{ s}) = 20.4 \text{ m}$$

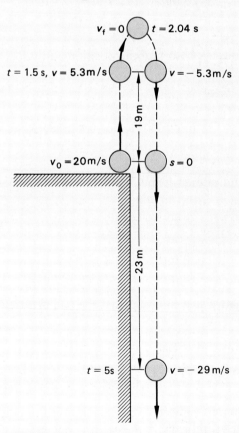

Figure 6-7   A ball projected vertically upward rises until its velocity is zero; then it falls with increasing downward velocity.

Solution (c)

To find the position and velocity after 1.5 s, we must establish new conditions.

Given: $v_0 = 20$ m/s   Find: $s = ?$

$g = -9.8$ m/s^2 $v_f = ?$

$t = 1.5$ s

We can now calculate the position as follows:

$$s = v_0 t + \frac{1}{2}gt^2$$

$$= (20 \text{ m/s})(1.5 \text{ s}) + \frac{1}{2}(-9.8 \text{ m/s}^2)(1.5 \text{ s})^2$$

$$= 30 \text{ m} - 11 \text{ m} = 19 \text{ m}$$

The velocity after 1.5 s is given by

$$v_f = v_0 + gt$$
$$= 20 \text{ m/s} + (-9.8 \text{ m/s}^2)(1.5 \text{ s})$$
$$= 20 \text{ m/s} - 14.7 \text{ m/s} = 5.3 \text{ m/s}$$

Solution (d)

The same equations apply to find the position and velocity after 5 s. Thus,

$$s = v_0 t + \frac{1}{2}gt^2$$

$$= (20 \text{ m/s})(5 \text{ s}) + \frac{1}{2}(-9.8 \text{ m/s}^2)(5 \text{ s})^2$$

$$= 100 \text{ m} - 123 \text{ m} = -23 \text{ m}$$

The negative sign indicates that the ball is located 23 m below the point of release.

The velocity after 5 s is given by

$$v_f = v_0 + gt$$
$$= 20 \text{ m/s} + (-9.8 \text{ m/s}^2)(5 \text{ s})$$
$$= 20 \text{ m/s} - 49 \text{ m/s} = -29 \text{ m/s}$$

In this case, the negative sign indicates that the ball is traveling downward.

**www3.adnc.com/
~topquark/fun**

For an applet showing the effect of launch angle on projectile motion, click on Java for this site.

6-8 PROJECTILE MOTION

We have seen that objects projected vertically upward or downward or dropped from rest are accelerated uniformly in the earth's gravitational field. Now we consider the more general problem of an object projected freely into a gravitational field in a nonvertical direction, as in Fig. 6-8, in which a football is kicked into space. In the absence of friction, such motion is another example of uniform or constant acceleration.

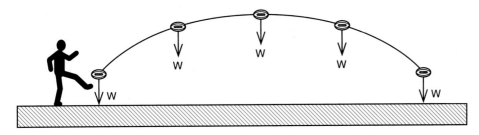

Figure 6-8 A kicked football is a projectile launched freely into space under the influence of gravity alone. If we neglect air resistance, the only force acting on such an object is its weight.

An object launched into space without motive power of its own is called a *projectile.* Neglecting air resistance, the only force acting on a projectile is its weight **W,** which causes its path to deviate from a straight line. It receives constant downward acceleration because of the gravitational force acting toward the center of the earth. But it differs from the motion previously studied in that the acceleration is usually not in line with the direction of initial velocity. Since no force acts horizontally to change the velocity, the horizontal acceleration is zero; this results in constant horizontal velocity. The downward force of gravity, on the other hand, causes the vertical velocity to change uniformly. Thus the motion of a projectile is normally in two dimensions, and it must be studied that way.

6-9 HORIZONTAL PROJECTION

If an object is projected horizontally, it motion can best be described by considering its horizontal and vertical motion separately. For example, in Fig. 6-9 an electronic device is rigged to simultaneously project one ball horizontally while

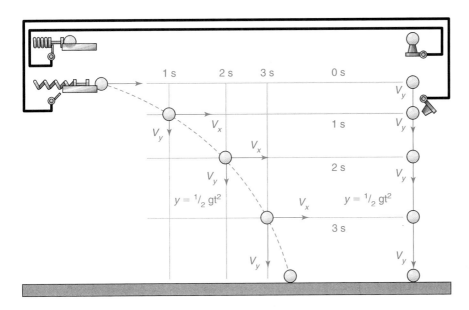

Figure 6-9

releasing another from rest at the same height. The horizontal velocity of the projected ball is unchanged, as indicated by the arrows of the same length throughout its trajectory. The vertical velocity, on the other hand, is initially zero and increases uniformly, in accordance with equations derived earlier for one-dimensional motion. The balls will strike the floor at the same instant, even though one is also moving horizontally. Thus, problems are simplified greatly by finding separate solutions for horizontal and vertical components.

The same general equations listed in Table 6-1 for uniform acceleration also apply for projectile motion. However, we know in advance that if the particle is projected near the earth, the vertical acceleration is equal to 9.8 m/s^2 or 32 ft/s^2 and that it is always directed downward. Thus, if the upward direction is chosen as positive, the acceleration of a projectile will be negative and equal to the gravitational acceleration. It is probably better to replace the symbol s used for distance along any line with the symbol x or y depending on whether the displacement is horizontal or vertical. Similarly, we can indicate the components of velocity with appropriate subscripts. For example, if we wish to express the vertical motion as a function of time, we might write:

$$y = v_{0y}t + \frac{1}{2}gt^2$$

Note that y represents the vertical displacement, y_0 the initial vertical velocity, and g the acceleration due to gravity (normally -9.8 m/s^2).

Let's first consider the motion of a projectile that is launched horizontally in a gravitational field. In this case, we note that

$$v_{0x} = v_x \qquad v_{0y} = 0 \qquad a_x = 0 \qquad a_y = g = -9.8 \text{ m/s}^2$$

since the horizontal velocity is constant and the vertical velocity is zero initially. Thus, the horizontal and vertical displacements of such a projectile can be found from:

$$x = v_{0x}t \qquad \textit{Horizontal Displacement}$$

$$y = \frac{1}{2}gt^2 \qquad \textit{Vertical Displacement}$$

6-11

Similar reasoning from the fact that $v_f = v_0 + gt$, we can find the horizontal and vertical components of the final velocity from:

$$v_x = v_{0x} \qquad \textit{Horizontal Velocity}$$
$$v_y = v_{0y} + gt \qquad \textit{Vertical Velocity}$$

6-12

The final displacement and velocity of a horizontally projected particle can be found from their components. In all of the above equations, the symbol g may be taken as the acceleration due to gravity, which is always directed downward. Thus, g will be negative when up is chosen as the positive direction.

Table 6-3 provides an excellent summary of how the general formulas for uniform acceleration may be modified for applications to projectiles.

Table 6-3 Uniform Acceleration and Projectiles

General acceleration formulas	Modified for projectile motion	
(1) $s = \bar{v}t + \dfrac{v_0 + v_f}{2}t$	$x = v_{0x}t$	$y = \dfrac{v_{0y} + v_y}{2}t$
(2) $v_f = v_0 + at$	$v_x = v_{0x}$	$v_y = v_{0y} + gt$
(3) $s = v_0 t + \dfrac{1}{2}at^2$	$x = v_{0x}t$	$y = v_{0y}t + \dfrac{1}{2}gt^2$
(4) $s = v_f t - \dfrac{1}{2}at^2$	$x = v_{0x}t$	$y = v_y t - \dfrac{1}{2}gt^2$
(5) $2as = v_f^2 - v_0^2$		$2gy = v_y^2 - v_{0y}^2$

CONCEPTUAL EXAMPLE 6-11

A skier leaves the jump horizontally with an initial velocity of 25 m/s as shown in Fig. 6-10. The initial height is 80 m above the landing position. (a) How long is the skier in the air? (b) How far does the skier travel horizontally? (c) What are the horizontal and vertical components of the final velocity?

Figure 6-10 A skier leaves the jump with an initial horizontal velocity of 25 m/s. His position and velocity can be found as a function of time.

Solution (a)

The time spent in the air is a function only of vertical parameters. The initial velocity in the y direction is zero, and the skier must fall a vertical distance of 80 m. Choosing the downward direction as positive, we have

$$y = \frac{1}{2}gt^2$$

Solving for time and substituting known values, we obtain

$$t = \sqrt{\frac{2y}{g}} = \sqrt{\frac{2(80 \text{ m})}{9.8 \text{ m/s}^2}}$$

$$t = 4.04 \text{ s}$$

It takes a time of 4.04 s for the skier to reach the landing point.

Solution (b)

Since the horizontal velocity is constant, the **range** is determined only by the time of flight.

$$x = v_{0x}t = (25 \text{ m/s})(4.04 \text{ s}) = 101 \text{ m}$$

Solution (c)

The horizontal component of the velocity is unchanged and, therefore, equal to 25 m/s at the landing point. The final vertical component is given by

$$v_y = gt = (9.8 \text{ m/s}^2)(4.04 \text{ s}) = 39.6 \text{ m/s}$$

The horizontal component is 25 m/s to the right and the vertical component is 39.6 m/s directed *downward*. Remember that the downward direction was chosen as positive for convenience. It is left as an exercise for you to show that the final velocity is 46.8 m/s at an angle of 57.7° below the horizontal. This will be the velocity the instant before striking the ground.

6-10 THE MORE GENERAL PROBLEM OF TRAJECTORIES

The more general case of projectile motion occurs when the projectile is thrown at an angle. This problem is illustrated in Fig. 6-11, where the motion of a projectile thrown at an angle θ with an initial velocity v_0 is compared with the motion of an object projected vertically upward. Once again, it is easy to see the advantage of treating the horizontal and vertical motions separately. In this case the equations listed in Table 6-3 may be used, and we should consider the upward direction as positive. Thus, if the vertical position y is above the origin, it will be positive; if it is below the origin, it will be negative. Similarly, upward velocities will be positive. Since the acceleration is always directed downward, we must substitute a negative value for g.

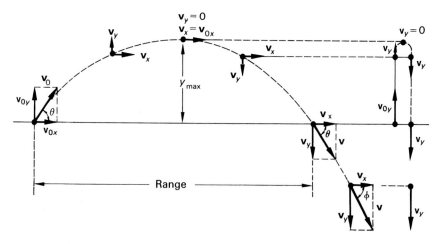

Figure 6-11   The motion of a projectile thrown at an angle is compared with the motion of an object thrown vertically upward.

Problem-Solving Strategy

Projectile Motion

1. Resolve the initial velocity v_0 into its x and y components:

$$v_{0x} = v_0 \cos\theta \qquad v_{0y} = v_0 \sin\theta$$

2. The horizontal and vertical components of the displacement at any instant are given by

$$x = v_{0x}t$$

$$y = v_{0y}t + \frac{1}{2}gt^2$$

3. The horizontal and vertical components of the velocity at any instant are given by

$$v_x = v_{0x}$$
$$v_y = v_{0y} + gt$$

4. The final position and velocity can then be obtained from their components.

5. Make sure that you use the appropriate signs and consistent units. Remember that gravity g may be positive or negative depending on your initial choice.

CONCEPTUAL EXAMPLE 6-12

A projectile is fired with an initial velocity of 400 ft/s at an angle of 30° above the horizontal. Find (a) its position and velocity after 8 s, (b) the time required to reach its maximum height, and (c) the horizontal range R as indicated in Fig. 6-11.

Solution (a)

The horizontal and vertical components of the initial velocity are

$$v_{0x} = v_0 \cos\theta = (400 \text{ ft/s})(\cos 30°) = 346 \text{ ft/s}$$
$$v_{0y} = v_0 \sin\theta = (400 \text{ ft/s})(\sin 30°) = 200 \text{ ft/s}$$

The x component of its position after 8 s is

$$x = v_{0x}t = (346 \text{ ft/s})(8 \text{ s}) = 2770 \text{ ft}$$

The y component of its position at this time is given by

$$y = v_{0y}t + \frac{1}{2}gt^2$$

from which

$$y = (200 \text{ ft/s})(8 \text{ s}) + \frac{1}{2}(-32 \text{ ft/s}^2)(8 \text{ s})^2$$

$$= 1600 \text{ ft} - 1024 \text{ ft} = 576 \text{ ft}$$

Hence, the position of the projectile after 8 s is 2770 ft down range and 576 ft above its initial position.

In computing its velocity at this point, we first recognize that the x component of the velocity does not change. Thus,

$$v_x = v_{0x} = 346 \text{ ft/s}$$

USING A CALCULATOR

When you calculate complex equations involving multiplication and powers, it is wise to use sets of parentheses for any calculation whose input might be misinterpreted by the calculator. If necessary, redo the problem, calculating the multiplication and power of two functions first, and then add the results. Here is how to do a calculation in one step, using the parentheses buttons on your calculator.

$$y = v_{0y}t + \frac{1}{2}gt^2$$

(2 0 0 × 8) + (.
5 × 3 2 +/- × (8 x^2)
) = 576

The y component must be computed from

$$v_y = v_{0y} + gt$$

so that

$$v_y = 200 \text{ ft/s} + (-32 \text{ ft/s})(8 \text{ s})$$
$$= 200 \text{ ft/s} - 256 \text{ ft/s}$$
$$= -56 \text{ ft /s}$$

The minus sign indicates that the projectile is on its way down. Finally, we can compute the resultant velocity from its components, as shown in Fig. 6-12.

Figure 6-12

The angle ϕ is calculated from

$$\tan \phi = \frac{v_y}{v_x} = \frac{56 \text{ ft/s}}{346 \text{ ft/s}} = 0.162$$

from which

$$\phi = 9.2°$$

The magnitude of the velocity is

$$v = \frac{v_y}{\sin \phi} = \frac{56 \text{ ft/s}}{\sin 9.2°} = 350 \text{ ft/s}$$

Solution (b)

At the maximum point in the projectile's ***trajectory,*** the y component of its velocity is zero. Hence, the time to reach this height can be computed from

$$v_y = v_{0y} + gt$$

Substituting given values for v_{0y} and g, we obtain

$$0 = 200 \text{ ft/s} + (-32 \text{ ft/s})t$$

from which

$$t = \frac{200 \text{ ft/s}}{32 \text{ ft/s}^2} = 6.25 \text{ s}$$

As an exercise, you should use this result to show that the minimum height in the trajectory is 625 ft.

Solution (c)

The range of the projectile can be computed by recognizing that the total time t' of the entire flight is equal to twice the time to reach the highest point. Hence,

$$t' = 2t = (2)(6.25 \text{ s}) = 12.5 \text{ s}$$

and the range is

$$R = v_{0x}t' = (346 \text{ ft/s})(12.5 \text{ s})$$
$$= 4325 \text{ ft}$$

Therefore, the projectile rises for 6.25 s to a height of 625 ft and then falls to the ground at a horizontal distance of 4325 ft from the point of release.

Summary and Review

Summary

A convenient way of describing objects in motion is to discuss their *velocity* or their acceleration. In this chapter, a number of applications have been presented that involve these physical quantities.

- *Average velocity* is the distance traveled per unit of time, and *average acceleration* is the change in velocity per unit of time.

$$\bar{v} = \frac{s}{t} \qquad a = \frac{v_f - v_0}{t}$$

- The definitions of velocity and acceleration result in the establishment of five basic equations involving uniformly accelerated motion:

$$s = \left(\frac{v_0 + v_f}{s}\right)t$$

$$v_f = v_0 + at$$

$$s = v_0 t + \frac{1}{2}at^2$$

$$s = v_f t - \frac{1}{2}at^2$$

$$2as = v_t^2 - v_0^2$$

Given any three of the five parameters (v_0, v_f, a, s, and t), the other two can be determined from one of these equations.

- To solve acceleration problems, read the problem with a view to establishing the three given parameters and the two that are unknown. You might set up columns like this:

Given: $a = 4$ m/s^2 Find: $v_f = ?$
$s = 500$ m $v_0 = ?$
$t = 20$ s

This procedure helps you choose the appropriate equation. Remember to choose a direction as positive and to apply it consistently throughout the problem.

- *Gravitational acceleration:* Problems involving gravitational acceleration can be solved like other acceleration problems. In this case, one of the parameters is known in advance to be

$$a = g = 9.8 \text{ m/s}^2 \text{ or } 32 \text{ ft/s}^2$$

The sign of gravitational acceleration is $+$ or $-$ depending on whether you choose up or down as the positive direction.

- *Projectile motion:* The key to problems involving projectile motion is to treat the horizontal motion and the vertical motion separately. Most projectile problems are solved using the following approach:
  - Resolve the initial velocity v_0 into its x and y components:

$$v_{0x} = v_0 \cos \theta \qquad v_{0y} = v_0 \sin \theta$$

  - The horizontal and vertical components of its position at any instant are given by

$$x = v_{0x}t \qquad y = v_{0y}t + \frac{1}{2}gt^2$$

  - The horizontal and vertical components of its velocity at any instant are given by

$$v_x = v_{0x} \qquad v_y = v_{0y} + gt$$

  - The final position and velocity can then be obtained from their components.

The important point to remember in applying these equations is to be consistent throughout with units and sign conversion.

Key Terms

1. constant speed
2. average speed
3. velocity
4. acceleration
5. instantaneous speed
6. instantaneous velocity
7. uniform acceleration
8. acceleration due to gravity
9. projectile
10. trajectory
11. range

Review Questions

6-1. Distinguish clearly between the terms *speed* and *velocity*. A stock car racer drives 500 laps around a 1-mi track in a time of 5 h. What was the average speed? What was the average velocity?

6-2. A bus driver travels a distance of 300 km in 4 h. At the same time a tourist travels the same 300 km in a car but stops for two 30-min breaks. Still, the tourist arrives at the destination at the same instant as the bus driver. Compare the average speeds of each driver.

6-3. Give some examples of motion in which the speed is constant but the velocity is not.

6-4. A bowling ball rolls down a 30° incline for a short distance and then levels out for a few seconds before starting up another 30° incline. Discuss the acceleration of the ball as it rolls along each segment of track. Is the acceleration the same going up the incline as it is going down the incline? Suppose the ball is given an initial push at the top of the first incline and then released to travel at greater speeds. Will the acceleration be any different on any of the segments?

6-5. A rock is thrown vertically upward. It moves to the maximum height, returns to its release point, and continues downward. Discuss the signs of its displacement, velocity, and acceleration at each point in its path. What is its acceleration when the velocity reaches zero at the highest point?

6-6. In the absence of air resistance, the same time is required for a projectile to reach its highest point as is required for it to return. Will this be true if air resistance is not neglected? Draw free-body diagrams for each situation.

6-7. Is the motion of a projectile fired at an angle an example of uniform acceleration? What is it if it is fired vertically upward or vertically downward? Explain.

6-8. At what angle should a baseball be thrown to achieve its maximum range? At what angle must it be thrown to attain the maximum height?

6-9. A hunter shoots an arrow directly at a squirrel on a tree limb, and the squirrel drops at the instant the arrow leaves the bow. Will the arrow strike the squirrel? Draw the trajectories that might be expected. Under what conditions will the arrow not strike the squirrel?

6-10. A child drops a ball from the window of a car traveling at a constant speed of 60 km/h. What is the initial velocity of the ball relative to the ground? Describe the motion.

6-11. A toy car is pulled across the floor with uniform speed. A spring attached to the car projects a marble vertically upward. Describe the motion of the marble relative to the floor and relative to the car.

6-12. Explain the adjustment of sights on a rifle for increasing target distances.

6-13. Explain the reasoning behind the use of high or low trajectories for punting in a football game.

Problems

Section 6-1—Speed and Velocity

6-1. A car travels a distance of 86 km at an average speed of 8 m/s. How many hours were required for the trip?   **Ans. 2.99 h**

6-2. Sound travels at an average speed of 340 m/s. Lightning from a distant thundercloud is seen almost immediately. If the sound of thunder reaches the ear 3 s later, how far away is the storm?

6-3. A small rocket leaves its pad and travels a distance of 40 m vertically upward before returning to the earth 5 s after it was launched. What was the average velocity for the trip?   **Ans. 16 m/s**

6-4. A car travels along a U-shaped curve for a distance of 400 m in 30 s. Its final location, however, is only 40 m from the starting position. What is the average speed, and what is the magnitude of the average velocity?

6-5. A woman walks for 4 min directly north with an average velocity of 6 km/h; then she walks eastward at 4 km/h for 10 min. What is her average speed for the trip?

Ans. 4.57 km/h

6-6. What is the average velocity for the entire trip described in Prob. 6-5?

6-7. A car travels at an average speed of 60 mi/h for 3 h and 20 in. What was the distance traveled?

Ans. 200 mi

6-8. How long will it take to travel 400 km if the average speed is 90 km/h?

6-9. A marble rolls up an inclined ramp a distance of 5 m and then stops and returns to a point 5 m below its starting point. The entire trip took only 2 s. What was the average speed, and what was the average velocity?

Ans. 7.5 m/s, −2.5 m/s

Section 6-3—Uniform Acceleration

6-10. The tip of a robot arm is moving to the right at 8 m/s. Four seconds later, it is moving to the left at 2 m/s. What is the change in velocity, and what is the acceleration?

6-11. An arrow accelerates from zero to 40 m/s in the 0.5 s it is in contact with the bow string. What is the average acceleration?

Ans. 80 m/s^2

6-12. A car traveling initially at 50 km/h accelerates at a rate of 4 m/s^2 for 3 s. What is the final speed?

6-13. A truck traveling at 60 mi/h brakes to a stop in 180 ft. What were the average acceleration and stopping time?

Ans. −21.5 ft/s^2, 4.09 s

6-14. An arresting device on a carrier deck stops an airplane in 1.5 s. The average acceleration was 49 m/s^2. What was the stopping distance? What was the initial speed?

6-15. In a braking test, a car traveling at 60 km/h is stopped in a time of 3 s. What were the acceleration and stopping distance?

Ans. −5.56 m/s^2, 25.0 m

6-16. A bullet leaves a 28-in. rifle barrel at 2700 ft/s. What were its acceleration and time in the barrel?

6-17. The ball in Fig. 6-13 is given an initial velocity of 16 m/s at the bottom of an inclined plane. Two seconds later it is still moving up the plane, but with a velocity of only 4 m/s. What is the acceleration?

Ans. −6.00 m/s^2

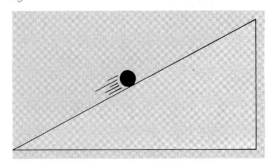

Figure 6-13

6-18. For Prob. 6-17, what is the maximum displacement from the bottom and what is the velocity 4 s after leaving the bottom?

6-19. A monorail train traveling at 80 km/h must be stopped in a distance of 40 m. What average acceleration is required, and what is the stopping time? **Ans. −6.17 m/s^2, 3.6 s**

Section 6-7—Gravity and Free-Falling Bodies

6-20. A ball is dropped from rest and falls for 5 s. What are its position and velocity at that instant?

6-21. A rock is dropped from rest. When will its displacement be 18 m below the point of release? What is its velocity at that time?

Ans. 1.92 s, −18.8 m/s

6-22. A woman drops a weight from the top of a bridge while a friend below measures the time it takes for the object to strike the water below. What is the height of the bridge if the time is 3 s?

6-23. A brick is given an initial downward velocity of 6 m/s. What is its final velocity after falling a distance of 40 m?

Ans. 28.6 m/s, downward

6-24. A projectile is thrown vertically upward and returns to its starting position in 5 s.

What was its initial velocity and how high did it rise?

6-25. An arrow is shot vertically upward with an initial velocity of 80 ft/s. What is its maximum height?   **Ans. 100 ft**

6-26. In Prob. 6-25, what are the position and velocity of the arrow after 2 s and after 6 s?

6-27. A hammer is thrown vertically upward to the top of a roof 16 m high. What minimum initial velocity was required?
   Ans. 17.7 m/s

Section 6-9—Horizontal Projection

6-28. A baseball leaves a bat with a horizontal velocity of 20 m/s. In a time of 0.25 s, how far will it have traveled horizontally and how far will it have fallen vertically?

6-29. An airplane traveling at 70 m/s drops a box of supplies. What horizontal distance will the box travel before striking the ground 340 m below?   **Ans. 583 m**

6-30. At a lumber mill, logs are discharged horizontally at 15 m/s from a greased chute that is 20 m above a mill pond. How far do the logs travel horizontally?

6-31. A steel ball rolls off the edge of a tabletop 4 ft above the floor. If it strikes the floor 5 ft from the base of the table, what was its initial horizontal speed?   **Ans. $v_{0x} = 10.0$ ft/s**

6-32. A bullet leaves the barrel of a weapon with an initial horizontal velocity of 400 m/s. Find the horizontal and vertical displacements after 3 s.

6-33. A projectile has an initial horizontal velocity of 40 m/s at the edge of a rooftop. Find the horizontal and vertical components of its velocity after 3 s.   **Ans. 40 m/s, −29.4 m**

Section 6-10—The More General Problem of Trajectories

6-34. A stone is given an initial velocity of 20 m/s at an angle of 58°. What are its horizontal and vertical displacements after 3 s?

6-35. A baseball leaves the bat with a velocity of 30 m/s at an angle of 30°. What are the horizontal and vertical components of its velocity after 3 s?   **Ans. 26.0 m/s, −14.4 m/s**

6-36. For the baseball in Prob. 6-35, what is the maximum height and what is the range?

6-37. An arrow leaves the bow with an initial velocity of 120 ft/s at an angle of 37° with the horizontal. What are the horizontal and vertical components of its displacement two seconds later?
   Ans. x = 208 ft, y = 56.0 ft

*6-38. In Prob. 6-37, what are the magnitude and direction of arrow's velocity after 2 s?

*6-39. A golf ball in Fig. 6-14 leaves the tee with a velocity of 40 m/s at 65°. If it lands on a green located 10 m higher than the tee, what was the time of flight, and what was the horizontal distance to the tee?
   Ans. 7.11 s, 120 m

*6-40. A projectile leaves the ground with a velocity of 35 m/s at an angle of 32°. What is the maximum height attained?

*6-41. The projectile in Prob. 6-40 rises and falls, striking a billboard at a point 8 m above the ground. What was the time of flight, and how far did it travel horizontally?
   Ans. 3.29 s, 97.6 m

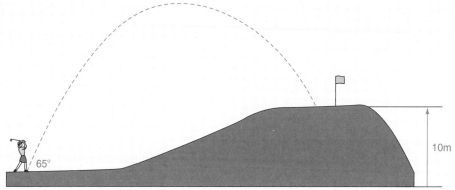

65°

10m

Figure 6-14

Summary and Review

Challenge Problems

6-42. A rocket travels in space at 60 m/s before it is given a sudden acceleration. If its velocity increases to 140 m/s in 8 s, what was its average acceleration and how far did it travel in this time?

6-43. A railroad car starts from rest and coasts freely down an incline. With an average acceleration of 4 ft/s^2, what will be the velocity after 5 s? What distance does it travel? **Ans. 20 ft/s, 50 ft**

*6-44. An object is projected horizontally at 20 m/s. At the same time, another object located 12 m down range is dropped from rest. When will they collide, and how far are they located below the release point?

6-45. A truck moving at an initial velocity of 30 m/s is brought to a stop in 10 s. What was the acceleration of the car and what was the stopping distance? **Ans. −3.00 m/s, 150 m**

6-46. A ball is thrown vertically upward with an initial velocity of 23 m/s. What are its position and velocity after 2 s, after 4 s, and after 8 s?

6-47. A stone is thrown vertically downward from the top of a bridge. Four seconds later it strikes the water below. If the final velocity was 60 m/s, what was the initial velocity of the stone and how high was the bridge? **Ans. v_0 = 20.8 m/s, downward; y = 162 m**

6-48. A ball is thrown vertically upward with an initial velocity of 80 ft/s. What are its position and velocity after **(a)** 1 s; **(b)** 3 s; and **(c)** 6 s?

6-49. An aircraft flying horizontally at 500 mi/h releases a package. Four seconds later, the package strikes the ground below. What was the altitude of the plane? **Ans. −256 ft**

6-50. In Prob. 6-49, what was the horizontal range of the package and what are the components of its final velocity?

6-51. A putting green is located 240 ft horizontally and 64 ft vertically from the tee. What must be the magnitude and direction of the initial velocity if a ball is to strike the green at this location after a time of 4 s? **Ans. 100 ft/s, 53.1°**

Critical Thinking Questions

6-52. A long strip of pavement is marked off in 100-m intervals. Students use stopwatches to record the times at which a car passes each mark. The following data is listed:

Distance, m	0	10	20	30	40	50
Time, s	0	2.1	4.3	6.4	8.4	10.5

Plot a graph with distance along the y axis and time along the x axis. What is the significance of the slope of this curve? What is the average speed of the car? At what instant in time is the distance equal to 34 m? What is the acceleration of the car? **Ans. Slope is v, 4.76 m/s, 7.14 s, 0**

6-53. An astronaut tests gravity on the Moon by dropping a tool from a height of 5 m. The following data are recorded electronically.

Height, m	5.00	4.00	3.00	2.00	1.00	0
Time, s	0	1.11	1.56	1.92	2.21	2.47

Plot the graph of this data. Is it a straight line? What is the average speed for the entire fall? What is the acceleration? How would you compare this with gravity on Earth?

6-54. A car is traveling initially north at 20 m/s. After traveling a distance of 6 m, the car passes point A, where its velocity is still northward but is reduced to 5 m/s. (a) What are the magnitude and direction of the acceleration of the car? (b) What time was required? (c) If the acceleration is held constant, what will be the velocity of the car when it returns to point A?

Ans. (a) 31.2 m/s², south; (b) 0.48 s; (c) −5 m/s

*6-55. A ball moving up an incline is initially located 6 m from the bottom of an incline and has a velocity of 4 m/s. Five seconds later, it is located 3 m from the bottom. Assuming constant acceleration, what was the average velocity? What is the meaning of a negative average velocity? What are the average acceleration and final velocity?

*6-56. The acceleration due to gravity on a distant planet is determined to be one-fourth its value on Earth. Does this mean that a ball dropped from a height of 4 m above this planet will strike the ground in one-fourth the time? What are the times required on the planet and on Earth?

Ans. $t_p = 1.81$ s, $t_e = 0.904$ s

*6-57. Consider the two balls A and B shown in Fig. 6-15. Ball A has a constant acceleration of 4 m/s² directed to the right, and ball B has a constant acceleration of 2 m/s² directed to the left. Ball A is initially traveling to the left at 2 m/s, and ball B is traveling to the left initially at 5 m/s. Find the time t at which the balls collide. Also, assuming x = 0 at the initial position of ball A, what is their common displacement when they collide? Ans. $t = 2.0$ s, $x = +4$ m

*6-58. Initially, a truck with a velocity of 40 ft/s is located a distance of 500 ft to the right of

a car. If the car begins at rest and accelerates at 10 ft/s², when will it overtake the truck? How far is the point from the initial position of the car?

*6-59. A ball is dropped from rest at the top of a 100-m-tall building. At the same instant, a second ball is thrown upward from the base of the building with an initial velocity of 50 m/s. When will the two balls collide, and at what distance above the street?

Ans. 2.00 s, 80.4 m

*6-60. A balloonist rising vertically with a velocity of 4 m/s releases a sandbag at the instant when the balloon is 16 m above the ground. Compute the position and velocity of the sandbag relative to the ground after 0.3 s and 2 s. How many seconds after its release will it strike the ground?

*6-61. An arrow is shot vertically upward with a velocity of 40 m/s. Three seconds later, another arrow is shot upward with a velocity of 60 m/s. At what time and position will they meet? Ans. 4.54 s, 80.6 m

*6-62. Someone wishes to strike a target that has a horizontal range of 12 km. What must be the velocity of an object projected at an angle of 35° if it is to strike the target? What is the time of flight?

*6-63. A wild boar charges directly at a hunter with a constant speed of 60 ft/s. At the instant the boar is 100 yd away, the hunter fires an arrow at 30° with the ground. What must be the velocity of the arrow if it is to strike its target? Ans. 76.2 ft/s

Figure 6-15

Newton's Second Law

OBJECTIVES

After completing this chapter, you should be able to:

1. Describe the relationships among force, mass, and acceleration and give the consistent units for each in the metric and U.S. customary systems of units.
2. Define the units *newton* and *slug* and explain why they are derived units rather than fundamental units.
3. Demonstrate by definition and example your understanding of the distinction between mass and weight.
4. Determine mass from weight and weight from mass at a point at which the acceleration due to gravity is known.
5. Draw a free-body diagram for objects in motion with constant acceleration, set the resultant force equal to the total mass times the acceleration, and solve for unknown parameters.

According to Newton's first law of motion, an object will undergo a change in its state of rest or motion *only* when acted on by a resultant, unbalanced force. We now know that a change in motion, for example, a change in speed, results in *acceleration*. In many industrial applications, we need to be able to predict the acceleration that will be produced by a given force. For example, the forward force required to accelerate a car from rest to a speed of 60 km/h in 8 s is of interest to the automotive industry. In this chapter, you will study the relationships among force, mass, and acceleration.

7-1 NEWTON'S SECOND LAW OF MOTION

Before studying the relationship between a resultant force and acceleration in a formal way, let us first consider a simple experiment. A linear air track is an apparatus for studying the motion of objects under conditions that approximate zero friction. Hundreds of tiny air jets create an upward force that balances the weight of the glider in Fig. 7-1. A string is attached to the front of the glider, and a spring scale of negligible weight is used to measure the applied horizontal force as shown. The acceleration the glider receives can be measured by determining the change in speed for a known time interval. The first applied force F_1 in Fig. 7-1a causes an acceleration a_1. Twice the force $2F_1$ will produce twice the acceleration $2a_1$, and three times the force $3F_1$ gives three times the acceleration $3a_1$.

Newton's Second Law

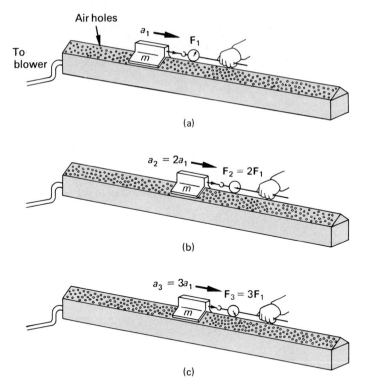

Figure 7-1   Variation of acceleration with force.

Such observations show that the acceleration of a given body is directly proportional to the applied force. This means that the ratio of force to acceleration is always constant:

$$\frac{\mathbf{F}_1}{\mathbf{a}_1} = \frac{\mathbf{F}_2}{\mathbf{a}_2} = \frac{\mathbf{F}_3}{\mathbf{a}_3} = \text{constant}$$

The constant is a measure of how effective a given force is in producing acceleration. We shall see that this ratio is a property of the body called its mass m, where

$$m = \frac{F}{a}$$

A mass of one kilogram (1 kg) was defined in Chap. 3 by comparison with a standard artifact. Keeping this definition, we can now define a new force unit such that it will give a unit mass a unit acceleration.

A force of one newton (1 N) is that resultant force that will give a 1-kg mass an acceleration of 1 m/s².

The *newton* is declared the SI unit of force. A resultant force of 2 N will produce an acceleration of 2 m/s², and a force of 3 N will give an acceleration of 3 m/s² to a 1-kg mass.

Now let's return to our air track experiment to see how acceleration is affected by increasing the mass. This time we shall keep the applied force **F** constant. The mass will be changed by adding in succession more gliders of equal size and weight.

F.Y.I.

Gravity varies over the surface of the earth, with latitude and altitude, and locally based on the earth's density.

www.exploratorium.edu/ronh
How much would you weigh on Mars? To find out, visit this site.

Figure 7-2 Variation of acceleration with mass.

Note from Fig. 7-2 that if the force is unchanged, an increase in mass will result in a proportionate *decrease* in the acceleration. Applying a constant force of 12 N in succession to 1-, 2-, and 3-kg masses will produce accelerations of 12 m/s², 6 m/s², and 4 m/s², respectively. These three cases are shown in Fig. 7-2a, b, and c.

From the above observations, we are now prepared to state Newton's second law of motion.

> **Newton's second law of motion** **Whenever an unbalanced force acts on a body, it produces in the direction of the force an acceleration that is directly proportional to the force and inversely proportional to the mass of the body.**

By using our newly defined unit, the newton, we can write this law as the following equation:

$$Resultant\ force = mass \times acceleration$$

$$F = ma \qquad Newton's\ Second\ Law \qquad \text{7-1}$$

Since this relationship depends on the definition of a new unit, we may substitute only units consistent with that definition. For example, if the mass is in kilograms (kg), the unit of force must be newtons (N) and the unit of acceleration must be meters per second squared (m/s²).

$$Force\ (\text{N}) = mass\ (\text{kg}) \times acceleration\ (\text{m/s}^2)$$

In USCS, a new mass unit is defined from the chosen units of *pound* (lb) for force and *feet per second squared* (ft/s^2) for acceleration. The new unit of mass is called the *slug* (from the sluggish or inertial property of mass).

A mass of one slug is that mass to which a resultant force of 1 lb will give an acceleration of 1 ft/s^2.

$$Force \text{ (lb)} = mass \text{ (slugs)} \times acceleration \text{ (ft/s}^2)$$

The SI unit of force is less than the USCS unit, and a mass of one slug is much greater than a mass of one kilogram. The following conversion factors might be helpful:

$$1 \text{ lb} = 4.448 \text{ N} \qquad 1 \text{ slug} = 14.59 \text{ kg}$$

A 1-lb bag of apples might contain four or five apples—each weighing about a newton. A person weighing 160 lb on earth would have a mass of 5 slugs or 73 kg.

It is important to recognize that the *F* in Newton's second law represents a *resultant* or unbalanced force. If more than one force acts on an object, it will be necessary to determine the resultant force *along the direction of motion*. The resultant force will always be along the direction of motion, since it is the *cause* of the acceleration. All components of forces that are perpendicular to the acceleration will be balanced. If the *x* axis is chosen along the direction of motion, we can determine the *x* component of each force and write

$$\sum F_x = ma_x \qquad\qquad \text{7-2}$$

A similar equation could be written for the *y* components if the *y* axis were chosen along the direction of motion.

7-2 THE RELATIONSHIP BETWEEN WEIGHT AND MASS

Before we consider examples of Newton's second law, we must have a clear understanding of the difference between the *weight* of a body and its *mass*. Perhaps no other point is more confusing to the beginning student. The *pound* (lb), which is a unit of force, is often used as a mass unit, the pound-mass (lb$_m$). The kilogram, which is a unit of mass, is often used in industry as a unit of force, the kilogram-force (kgf). These seemingly inconsistent units result from the fact that there are many different systems of units in use. In this text, there should be less cause for confusion because we use only the SI and U.S. customary (British gravitational) units. Therefore, in this book the *pound* (lb) will always refer to *weight*, which is a force, and the unit *kilogram* (kg) will always refer to the *mass* of a body.

The weight of any body is the force with which the body is pulled vertically downward by gravity. When a body falls freely to the earth, the only force acting on it is its weight **W.** This net force produces an acceleration **g,** which is the same for all falling bodies. Thus, from Newton's second law we can write the relationship between a body's weight and its mass:

order.ph.utexas.edu/chaos/
For an explanation of Newton's laws, see this site, which is a five-part online physics course (but do not type www.).

$$W = mg \qquad \text{or} \qquad m = \frac{W}{g} \qquad\qquad 7\text{-}3$$

In either system of units (1) the mass of a particle is equal to its weight divided by the acceleration of gravity, (2) weight has the same units as the unit of force, and (3) the acceleration of gravity has the same units as acceleration.

Therefore, we can summarize as follows:

$$\text{SI: } W\,(\text{N}) = m\,(\text{kg}) \times g\,(9.8 \text{ m/s}^2)$$
$$\text{USCS: } W\,(\text{lb}) = m\,(\text{slug}) \times g\,(32 \text{ ft/s}^2)$$

The values for g, and hence the weights, in the above relationships apply only at points on earth near sea level, where g has these values.

Two things must be remembered to understand fully the difference between mass and weight:

Mass is a universal constant equal to the ratio of a body's weight to the gravitational acceleration due to its weight.

Weight is the force of gravitational attraction and varies depending on the acceleration of gravity.

Therefore, the mass of a body is only a measure of its inertia and is not in any way dependent on gravity. In outer space, a hammer has negligible weight, but it serves to drive nails just the same since its mass is unchanged.

In USCS units a body is usually described by stating its weight **W** in *pounds*. The mass, if desired, is computed from this weight and has the unit of *slugs*. In SI units a body is usually described in terms of its mass in *kilograms*. The weight, if desired, is computed from the given mass and has the units of *newtons*. In the following examples, all parameters are measured at points where $g = 32$ ft/s^2 or 9.8 m/s^2.

EXAMPLE 7-1

Find the mass of a person who weighs 150 lb.

Solution

$$m = \frac{W}{g} = \frac{150 \text{ lb}}{32 \text{ ft/s}^2} = 4.69 \text{ slugs}$$

EXAMPLE 7-2

Find the weight of an 18-kg block.

Solution

$$W = mg = 18 \text{ kg}(9.8 \text{ m/s}^2) = 176 \text{ N}$$

EXAMPLE 7-3

Find the mass of a body whose weight is 100 N.

Solution

$$m = \frac{W}{g} = \frac{100 \text{ N}}{9.8 \text{ m/s}^2} = 10.2 \text{ kg}$$

7-3 APPLICATION OF NEWTON'S SECOND LAW TO SINGLE-BODY PROBLEMS

The primary difference between the problems discussed in this chapter and earlier problems is that a net, unbalanced force is acting to produce an acceleration. Thus, after constructing the free-body diagrams that describe the situation, the first step is to determine the unbalanced force and set it equal to the product of mass and acceleration. The desired quantity can then be determined from the relation given as Eq. (7-1):

$$Resultant\ force = mass \times acceleration$$

$$F \text{ (resultant)} = ma$$

The following examples will serve to demonstrate the relationships among force, mass, and acceleration.

> **Hint for Success**
>
> Today's economy is driven by factors of world trade and economic expansion in our country and other countries. You must prepare for success by making education a lifelong priority.

EXAMPLE 7-4

What acceleration will a force of 20 N impart to a 10-kg body?

Solution

There is only one force acting, and so

$$F = ma \quad \text{or} \quad a = \frac{F}{m}$$

and

$$a = \frac{20 \text{ N}}{10 \text{ kg}} = 2 \text{ m/s}^2$$

EXAMPLE 7-5

What resultant force will give a 32-lb body an acceleration of 5 ft/s²?

Solution

To find the resultant force, we must first determine the mass of the body from its given weight.

$$m = \frac{W}{g} = \frac{32 \text{ lb}}{32 \text{ ft/s}^2} = 1 \text{ slug}$$

Then

$$F = ma = (1 \text{ slug})(5 \text{ ft/s}^2) = 5 \text{ lb}$$

EXAMPLE 7-6

What mass has a body if a force of 60 N will give it an acceleration of 4 m/s²?

Solution

Solving for m in Newton's law, we have

$$m = \frac{F}{a} = \frac{60 \text{ N}}{4 \text{ m/s}^2} = 15 \text{ kg}$$

In the preceding examples, the unbalanced forces were easily determined. As the number of forces acting on a body increases, however, the problem of determining the resultant force is less obvious. In these cases, perhaps it is wise to point out a few considerations.

According to Newton's second law, a resultant force always produces an acceleration *in the direction of the resultant force.* This means that the net force and the acceleration it causes are of the same algebraic sign, and they each have the same line of action. Therefore, if the direction of motion (acceleration) is considered positive, fewer negative factors will be introduced into the equation $F = ma$. For example, in Fig. 7-3b, it is preferable to choose the direction of motion (left) as positive since the equation

$$P - \mathcal{F}_k = ma$$

is preferable to the equation

$$\mathcal{F}_k - P = -ma$$

which would result if we chose the right direction as positive.

Another consideration that results from the above discussion is that the forces acting normal to the line of motion will be in equilibrium if the resultant force is constant. Thus, in problems involving friction, normal forces can be determined from the first condition for equilibrium.

(a) Net force $P - \mathcal{F}_K$(right)   **(b)** Net force $P - \mathcal{F}_K$(left)

Figure 7-3   The direction of acceleration should be chosen as positive.

In summary, we apply the following equations to acceleration problems:

$$\sum F_x = ma_x \qquad \sum F_y = ma_y = 0 \qquad \text{7-4}$$

where ΣF_x and a_x are taken as positive and along the line of motion and ΣF_y and a_y are taken normal to the line of motion.

EXAMPLE 7-7

A horizontal force of 20 N drags a 4-kg block across a level floor. If $\mu_k = 0.2$, find the acceleration of the block.

Figure 7-4

Solution

A free-body diagram is superimposed on the sketch in Fig. 7-4. We will choose right as positive. To avoid confusing mass with weight, it is often desirable to calculate each in advance. The mass (4 kg) is given, and the weight is found from $W = mg$.

$$m = 4 \text{ kg} \qquad W = (4 \text{ kg})(9.8 \text{ m/s}^2) = 39.2 \text{ N}$$

The resultant force is along the direction of motion. The vertical forces are balanced. Applying Newton's second law,

$$\textit{Resultant force} = \textit{mass} \times \textit{acceleration}$$
$$20 \text{ N} - \mathcal{F}_k = ma$$

Recalling that $\mathcal{F}_k = \mu_k \mathcal{N}$, we may write

$$20 \text{ N} - \mu_k \mathcal{N} = ma$$

Since the vertical forces are balanced, we see from Fig. 7-4 that

$$\mathcal{N} = W = 39.2 \text{ N}$$

Therefore, substituting $\mathcal{N} = 39.2$ N, $\mu_k = 0.2$, and $m = 4$ kg, we have

$$20 \text{ N} - (0.2)(39.2 \text{ N}) = (4 \text{ kg})\, a$$

$$a = \frac{12.2 \text{ N}}{4 \text{ kg}} = 3.04 \text{ m/s}^2$$

You should verify that *newtons per kilogram* is equivalent to *meters per second squared*.

Solving all physics problems requires an ability to organize the given data and to apply formulas in a consistent manner. Often a procedure is helpful to the beginning student, and this is particularly true for the problems in this chapter. A logical sequence of operations for problems involving Newton's second law is outlined in the following procedure.

Problem-Solving Strategy

Newton's Second Law of Motion

1. Read the problem and then draw and label a rough sketch.
2. Construct a free-body diagram with an axis along the direction of motion.
3. Indicate the positive direction for acceleration.
4. Distinguish between the mass and weight of each object.

$$W = mg \qquad m = \frac{W}{g}$$

5. From the free-body diagram, determine the resultant force along the assumed direction of motion (ΣF).
6. Determine the total mass ($m_t = m_1 + m_2 + m_3 \ldots$).
7. Set the resultant force (ΣF) equal to the total mass m_t times acceleration a:

$$\sum F = m_t a$$

8. Substitute known quantities and solve for the unknown.

USING A CALCULATOR

After calculating an answer, always ask yourself whether the answer makes sense. If your calculation of the distance of the sun from the earth is 930 miles, you know that the answer is not correct.

CONCEPTUAL EXAMPLE 7-8

A 2000-lb elevator is lifted upward with an acceleration of 4 ft/s². What is the tension in the supporting cable?

Solution

Read the problem, then draw a sketch from which a free-body diagram may be drawn (see Fig. 7-5). Notice that the positive direction of acceleration (upward) is indicated on the free-body diagram.

Figure 7-5 Upward acceleration in a gravitational field.

Now we determine the mass and the weight of the 2000-lb elevator. The weight, of course, is 2000 lb. The mass must be calculated from $m = W/g$.

$$W = 2000 \text{ lb} \qquad m = \frac{2000 \text{ lb}}{32 \text{ ft/s}^2} = 62.5 \text{ slugs}$$

Since the elevator is the only object moving, the 62.5 slugs represents the *total* mass m_t.

The resultant force from the free-body diagram is

$$\sum F = T - 2000 \text{ lb}$$

From Newton's second law, we now write

$$Resultant\ force = total\ mass \times acceleration$$
$$T - 2000 \text{ lb} = (62.5 \text{ slugs})(4 \text{ ft/s}^2)$$
$$T - 2000 \text{ lb} = 250 \text{ lb}$$

Finally, we solve for the unknown T by adding 2000 lb to both sides of the equation.

$$T = 250 \text{ lb} + 2000 \text{ lb}$$
$$T = 2250 \text{ lb}$$

EXAMPLE 7-9

A 100-kg ball is lowered by means of a cable with a downward acceleration of 5 m/s². What is the tension in the cable?

Figure 7-6 Downward acceleration.

Solution

As before, we construct a sketch and a free-body diagram (Fig. 7-6). Note that the downward direction is chosen as positive since that is the direction of motion.

This time the *mass* is given and the *weight* must be calculated from $W = mg$.

$$m = 100 \text{ kg} \qquad W = (100 \text{ kg})(9.8 \text{ m/s}^2) = 980 \text{ N}$$

The resultant force is the *net* downward force, or

$$\sum F \div W - T \qquad \text{(remember, down is positive)}$$

Now, from Newton's second law, we write

Net downward force = total mass × downward acceleration
$$W - T = ma$$

Substituting known quantities, we obtain

$$980 \text{ N} - T = (100 \text{ kg})(5 \text{ m/s}^2)$$
$$980 \text{ N} - T = 500 \text{ N}$$

from which we solve for T by adding T to both sides and subtracting 500 N from both sides:

$$980 \text{ N} - 500 \text{ N} = T$$
$$T = 480 \text{ N}$$

CONCEPTUAL EXAMPLE 7-10

An Atwood machine consists of a single pulley with masses suspended on each side. It is a simplified version of many industrial systems in which counterweights are used for balance. Assume that the mass on the right side is 10 kg and that the mass on the left side is 2 kg. (a) What is the acceleration of the system? (b) What is the tension of the cord?

(a)   (b)   (c)

Figure 7-7 Two masses suspended from a single pulley. Free-body diagrams are drawn; the positive direction of acceleration is chosen to be upward on the left and downward on the right.

Solution (a)

We first draw a sketch and a free-body diagram for each mass (Fig. 7-7). The weight and mass of each object are determined.

$$m_1 = 2 \text{ kg} \qquad W_1 = m_1g = (2 \text{ kg})(9.8 \text{ m/s}^2) \qquad \text{or} \qquad W_1 = 19.6 \text{ N}$$
$$m_2 = 10 \text{ kg} \qquad W_2 = m_2g = (10 \text{ kg})(9.8 \text{ m/s}^2) \qquad \text{or} \qquad W_2 = 98 \text{ N}$$

Now, the problem is to determine the net unbalanced force on the entire system. Note that the pulley merely changes the direction of the forces. The unbalanced force is, therefore, just the difference in the weights. This is just what we would expect from experience. Notice that the tension T is the same on either side, since there is only one rope. Thus, the tension cancels out and does not figure in the resultant force, which may be written as follows:

$$\sum F = W_2 - T + T - W_1$$
$$\sum F = W_2 - W_1$$

The total mass of the system is simply the sum of all the masses in motion.

$$m_t = m_1 + m_2 = 2 \text{ kg} + 10 \text{ kg}$$
$$m_t = 12 \text{ kg} \qquad \text{(total mass)}$$

From Newton's second law of motion, we have

$$\textit{Resultant force} = \textit{total mass} \times \textit{acceleration}$$
$$W_2 - W_1 = (m_1 + m_2)a$$

Substituting for W_2, W_1, m_1, and m_2, we have

$$98 \text{ N} - 19.6 \text{ N} = (2 \text{ kg} + 10 \text{ kg})a$$

From which we may solve for a as follows:

$$78.4 \text{ N} = (12 \text{ kg})a$$

$$a = \frac{78.4 \text{ N}}{12 \text{ kg}} = 6.53 \text{ m/s}^2$$

Solution (b)

To solve for the tension in the cord, we must consider either of the masses by itself, since considering the system as a whole would not involve cord tension. Suppose we consider the forces acting on m_1:

$$\textit{Resultant force} = \textit{mass} \times \textit{acceleration}$$
$$T - W_1 = m_1a$$

But $a = 6.53 \text{ m/s}^2$ and the mass and weight are known, so we have

$$T - 19.6 \text{ N} = (2 \text{ kg})(6.53 \text{ m/s}^2)$$
$$T - 19.6 \text{ N} = 13.06 \text{ N}$$
$$T = 32.7 \text{ N}$$

We would obtain the same value for the tension if we applied Newton's law to the second mass. You should demonstrate this fact as an additional exercise.

EXAMPLE 7-11

A 64-lb block rests on a frictionless tabletop. A rope attached to it passes over a light frictionless pulley and is attached to a weight W, as shown in Fig. 7-8. (a) What must the value of W be to give the system an acceleration of 16 ft/s²? (b) What is the tension in the rope?

Figure 7-8

Solution (a)

Draw free-body diagrams for each body in the system, as shown in Fig. 7-8b and c. Since the vertical forces on the 64-lb block are balanced, the net force on the entire system is simply the weight W. Hence, applying Newton's law yields

Resultant force on system = total mass × acceleration

$$W = \left(\frac{64 \text{ lb}}{g} + \frac{W}{g} \right) a$$

$$W = \frac{64 \text{ lb} + W}{g} a = (64 \text{ lb} + W) \frac{a}{g}$$

$$W = (64 \text{ lb} + W) \frac{16 \text{ ft/s}^2}{32 \text{ ft/s}^2}$$

$$W = \frac{64 \text{ lb} + W}{2}$$

$$2W = 64 \text{ lb} + W$$

$$2W - W = 64 \text{ lb}$$

$$W = 64 \text{ lb}$$

Solution (b)

To solve for the tension in the rope, we may choose either Fig. 7-8b or c since each diagram involves the unknown **T**. The better choice is the former because the net force on the 64-lb body is the tension **T**. Thus,

$$Resultant\ force = mass \times acceleration$$
$$T = \frac{64\ lb}{32\ ft/s^2}\ (16\ ft/s^2)$$
$$= 32\ lb$$

One more example will be given in this section to allow you to become more familiar with the reasoning process involved with more complex systems. Since the foundation has been laid in previous examples, we include only the significant steps in the solution.

EXAMPLE 7-12

Consider the masses $m_1 = 20$ kg and $m_2 = 18$ kg in the system represented by Fig. 7-9. If the coefficient of kinetic friction is 0.1 and the inclination angle θ is 30°, find (a) the acceleration of the system and (b) the tension in the cord joining the two masses.

(a)   (b)   (c)

Figure 7-9

Solution (a)

Using the symbols as defined in Fig. 7-9, we apply Newton's law to the system:

$$Resultant\ force\ on\ system = total\ mass \times acceleration$$
$$W_2 - W_{1x} - \mathscr{F}_k = (m_1 + m_2)a$$

The symbols on the left side are found as follows:

$$W_2 = m_2g = (18\ kg)(9.8\ m/s^2) = 176\ N$$
$$W_{1x} = m_1g\ \sin\ \theta = (20\ kg)(9.8\ m/s^2)(\sin\ 30°) = 98\ N$$
$$W_{1y} = m_1g\ \cos\ \theta = (20\ kg)(9.8\ m/s^2)(\cos\ 30°) = 170\ N$$
$$\mathscr{F}_k = \mu_k\ N = \mu_k\ W_{1y} = (0.1)(170\ N) = 17\ N$$

Substitution into the equation of motion yields

$$176 \text{ N} - 98 \text{ N} - 17 \text{ N} = (20 \text{ kg} + 18 \text{ kg})a$$

from which we obtain

$$a = 1.16 \text{ m/s}^2$$

Solution (b)

To find the tension in the cord, we apply Newton's law to the 18-kg mass, as shown in Fig. 7-9c:

$$\text{Resultant force} = mass \times acceleration$$
$$m_2 g - T = m_2 a$$
$$T = m_2 g - m_2 a = m_2(g - a)$$
$$= (18 \text{ kg})(9.8 \text{ m/s}^2 - 1.61 \text{ m/s}^2)$$
$$= 147 \text{ N}$$

Summary and Review

Summary

In this chapter, we have considered the fact that a resultant force will always produce an acceleration in the direction of the force. The magnitude of the acceleration is directly proportional to the force and inversely proportional to the mass, according to Newton's second law of motion. The following concepts are essential to applications of this fundamental law:

- The mathematical formula that expresses Newton's second law of motion may be written as follows:

$$Force = mass \times acceleration$$

$$F = ma \qquad m = \frac{F}{a} \qquad a = \frac{F}{m}$$

$$\text{In SI units: } 1 \text{ N} = (1 \text{ kg})(1 \text{ m/s}^2)$$
$$\text{In USCS units: } 1 \text{ lb} = (1 \text{ slug})(1 \text{ ft/s}^2)$$

- Weight is the force due to a particular acceleration g. Thus, weight W is related to mass m by Newton's second law:

$$W = mg \qquad m = \frac{F}{g} \qquad g = 9.8 \text{ m/s}^2 \text{ or } 32 \text{ ft/s}^2$$

For example, a mass of 1 kg has a weight of 9.8 N. A weight of 1 lb has a mass of $\frac{1}{32}$ slug. In a given problem you must determine whether weight or mass is given. Then you must determine what is needed in an equation. Conversions of mass to weight and weight to mass are common.

- Application of Newton's second law:

1. Construct a free-body diagram for each body undergoing an acceleration. Indicate on this diagram the direction of positive acceleration.
2. Determine an expression for the net resultant force on a body or a system of bodies.
3. Set the resultant force equal to the total mass of the system multiplied by the acceleration of the system.
4. Solve the resulting equation for the unknown quantity.

Key Terms

1. Newton's second law
2. mass
3. weight
4. slug
5. newton

Review Questions

7-1. Distinguish clearly between the mass of an object and its weight and give the appropriate units for each in the SI and USCS systems of units.

7-2. What exactly do we mean when we describe an athlete as a 160-lb person? What would be the mass of this person on the Moon?

7-3. A round piece of brass found in the laboratory is labeled 500 g. Is this its weight or its mass? How can you be sure?

7-4. A state of equilibrium is maintained on a force table by hanging masses from pulleys mounted at various locations on the circular edge. In calculating the equilibrant, we sometimes use grams instead of newtons. Are we justified in doing this?

7-5. When drawing free-body diagrams, why is it usually to our advantage to choose either the x or y axis along the direction of motion, even if it means rotated axes? Use the example of motion along an inclined plane as an illustration.

Summary and Review

7-6. In the example of an Atwood machine (Conceptual Example 7-10), we neglected the mass of the cord that connects the two masses. Discuss how this problem is altered if the mass of the cord is large enough to affect the motion.

7-7. In industry we often hear of a kilogram-force (kg_f) unit, which is defined as a force equivalent to the weight of a 1-kg mass near the earth's surface. In the United States, we also speak often of the pound-mass (lb_m) unit, which is the mass of an object that has a weight of 1 lb near the surface of the earth. Find the value of these quantities in appropriate SI units, and discuss the problems caused by their use.

Problems

Section 7-1—Newton's Second Law

7-1. A 4-kg mass is acted on by a resultant force of (a) 4 N, (b) 8 N, and (c) 12 N. What are the resulting accelerations?
Ans. (a) 1 m/s², (b) 2 m/s², (c) 3 m/s²

7-2. A constant force of 20 N acts on a mass of (a) 2 kg, (b) 4 kg, and (c) 6 kg. What are the resulting accelerations?

7-3. A constant force of 60 lb acts on each of three objects, producing accelerations of 4, 8, and 12 N. What are the masses?
Ans. 15, 7.5, and 5 slugs

7-4. What resultant force is necessary to give a 4-kg hammer an acceleration of 6 m/s²?

7-5. It is determined that a resultant force of 60 N will give a wagon an acceleration of 10 m/s². What force is required to give the wagon an acceleration of only 2 m/s²?
Ans. 12 N

7-6. A 1000-kg car moving north at 100 km/h brakes to a stop in 50 m. What are the magnitude and direction of the force?
Ans. 772 N, south

Section 7-2—The Relationship between Weight and Mass

7-7. What is the weight of a 4.8-kg mailbox? What is the mass of a 40-N tank?
Ans. 47.0 N, 4.08 kg

7-8. What is the mass of a 60-lb child? What is the weight of a 7-slug man?

7-9. A woman weighs 180 lb on earth. When she walks on the moon, she weighs only 30 lb. What is the acceleration due to gravity on the moon and what is her mass on the moon? On the earth?
Ans. 5.33 ft/s², 5.63 slugs both places

7-10. What is the weight of a 70-kg astronaut on the surface of the earth? Compare the *resultant* force required to give him or her an acceleration of 4 m/s² on the earth with the *resultant* force required to give the same acceleration in space where gravity is negligible.

7-11. Find the mass and the weight of a body if a resultant force of 16 N will give it an acceleration of 5 m/s². **Ans. 3.20 kg, 31.4 N**

7-12. Find the mass and weight of a body if a resultant force of 200 lb causes its speed to increase from 20 ft/s to 60 ft/s in a time of 5 s.

7-13. Find the mass and weight of a body if a resultant force of 400 N causes it to decrease its velocity by 4 m/s in 3 s.
Ans. 300 kg, 2940 N

Section 7-3—Application of Newton's Second Law to Single-Body Problems

7-14. What horizontal pull is required to drag a 6-kg sled with an acceleration of 4 m/s² if a friction force of 20 N opposes the motion?

7-15. A 2500-lb automobile is speeding at 55 mi/h. What resultant force is required to stop the car in 200 ft on a level road? What must be the coefficient of kinetic friction?
Ans. −1270 lb, 0.508

7-16. A 10-kg mass is lifted upward by a light cable. What is the tension in the cable if the acceleration is **(a)** zero, **(b)** 6 m/s² upward, and **(c)** 6 m/s² downward?

7-17. A 64-lb load hangs at the end of a rope. Find the acceleration of the load if the tension in the cable is **(a)** 64 lb, **(b)** 40 lb, and **(c)** 96 lb.   **Ans. 0, −12.0 ft/s², 16.0 ft/s²**

7-18. An 800-kg elevator is lifted vertically by a strong rope. Find the acceleration of the elevator if the rope tension is **(a)** 9000 N, **(b)** 7840 N, and **(c)** 2000 N.

7-19. A horizontal force of 100 N pulls an 8-kg cabinet across a level floor. Find the acceleration of the cabinet if $\mu_k = 0.2$.

Ans. 10.5 m/s²

7-20. In Fig. 7-10, an unknown mass slides down the 30° inclined plane. What is the acceleration in the absence of friction?

7-21. Assume that $\mu_k = 0.2$ in Fig. 7-10. What is the acceleration? Why did you not need to know the mass of the block?

Ans. 3.20 m/s², down the plane

*7-22. Assume that $m = 10$ kg and $\mu_k = 0.3$ in Fig. 7-10. What push **P** directed up and along the incline in Fig. 7-10 will produce an acceleration of 4 m/s² also up the incline?

*7-23. What force **P** down the incline in Fig. 7-10 is required to cause the acceleration *down* the plane to be 4 m/s²? Assume that $m = 10$ kg and $\mu_k = 0.3$.   **Ans. 16.5 N**

Section 7-3—Applications of Newton's Second Law to Multibody Problems

7-24. Assume zero friction for the system shown in Fig. 7-11. What is the acceleration of the system? What is the tension T in the connecting cord?

7-25. What force does block A exert on block B in Fig. 7-12?   **Ans. 15.0 N**

*7-26. What are the acceleration of the system and the tension in the connecting cord for the arrangement shown in Fig. 7-13?

Figure 7-10

Figure 7-11

Figure 7-12

Figure 7-13

*7-27. If the coefficient of kinetic friction between the table and the 4-kg block is 0.2 in Fig. 7-13, what is the acceleration of the system? What is the tension in the cord?

Ans. 5.10 m/s², 28.2 N

*7-28. Assume that the masses $m_1 = 2$ kg and $m_2 = 8$ kg are connected by a cord that passes over a light frictionless pulley as in Fig. 7-14. What are the acceleration of the system and the tension in the cord?

*7-29. The system described in Fig. 7-15 starts from rest. What is the acceleration assuming zero friction?

Ans. 2.69 m/s² down the incline

*7-30. What is the acceleration in Fig. 7-15 as the 10-kg block moves down the plane against friction? Assume $\mu_k = 0.2$.

*7-31. What is the tension in the cord for Prob. 7-30? **Ans. 22.2 N**

Figure 7-14

Figure 7-15

Challenge Problems

7-32. A 2000-lb elevator is lifted vertically with an acceleration of 8 ft/s². Find the minimum breaking strength of the cable pulling the elevator.

7-33. A 200-lb worker stands on weighing scales in the elevator of Prob. 7-32. What is the reading of the scales as he is lifted at 8 m/s²? **Ans. 2500 lb**

7-34. A 8-kg load is accelerated upward with a cord whose breaking strength is 200 N. What is the maximum acceleration?

7-35. For rubber tires on a concrete road $\mu_k = 0.7$. What is the minimum horizontal stopping distance for a 1600-kg truck traveling at 20 m/s? **Ans. 29.2 km**

*7-36. Suppose the 4- and 6-kg masses in Fig.

7-13 are switched so that the larger mass is on the table. What would be the acceleration and tension in the cord (neglecting friction)?

*7-37. Consider two masses A and B connected by a cord and hung over a single pulley. If mass A is twice that of mass B, what is the acceleration of the system? **Ans. 3.27 m/s²**

*7-38. A 5-kg mass rests on a 34° inclined plane where $\mu_k = 0.2$. What push up the incline will cause the block to accelerate at 4 m/s²?

*7-39. A 96-lb block rests on a table where $\mu_k = 0.4$. A cord tied to this block passes over a light, frictionless pulley. What weight must be attached to the free end if the system is to accelerate at 4 ft/s²? **Ans. 35.7 lb**

Critical Thinking Questions

7-40. In a laboratory experiment the acceleration of a small car is measured by the separation of spots burned at regular intervals in a paraffin-coated tape. Larger and larger weights are transferred from the car to a hanger at the end of a tape that passes over a light, frictionless pulley. In this manner, the mass of the entire system is kept constant. Since the car moves on a horizontal air track with negligible friction, the resultant force is equal to the weights at the end of the tape. The following data are recorded:

Weight, W	2	4	6	8	10	12
Acceleration, m/s²	1.4	2.9	4.1	5.6	7.1	8.4

Plot a graph of weight versus acceleration. What is the significance of the slope of this curve? What is the mass of the system?

7-41. In the experiment described in Prob. 7-40, the student places a constant weight of 4 N at the free end of the tape. Several runs are made, increasing the mass of the car each time by adding weights. What happens to the acceleration as the mass of the system is increased? What should the value of the product of the mass of the system and the acceleration be for each run? Is it necessary to include the mass of the constant 4-N weight in these experiments?

7-42. An arrangement similar to that described by Fig. 7-13 is set up except that the masses are replaced. What is the acceleration of the system if the suspended mass is three times that of the mass on the table and $\mu_k = 0.3$?

Ans. 6.62 m/s²

7-43. Three masses, 2 kg, 4 kg, and 6 kg, are connected (in order) by strings and hung from the ceiling with another string so that the largest mass is in the lowest position. What is the tension in each cord? If they are then detached from the ceiling, what must be the tension in the top string in order that the system accelerate upward at 4 m/s²? In the latter case what are the tensions in the strings that connect masses?

7-44. An 80-kg astronaut on a space walk pushes against a 200-kg solar panel that has become dislodged from a spacecraft. The force causes the panel to accelerate at 2 m/s². What acceleration does the astronaut receive? Do they continue to accelerate after the push? Why or why not?

Ans. −5.00 m/s², no

7-45. A 400-lb sled slides down a hill ($\mu_k = 0.2$) inclined at an angle of 60°. What is the normal force on the sled? What is the force of kinetic friction? What is the resultant force down the hill? What is the acceleration? Is it necessary to know the weight of the sled to determine its acceleration?

7-46. Three masses, $m_1 = 10$ kg, $m_2 = 8$ kg, and $m_3 = 6$ kg, are connected as shown in Fig. 7-16. Neglecting friction, what is the acceleration of the system? What are the tensions in the cord on the left and in the cord on the right? Would the acceleration be the same if the middle mass m_2 were removed?

*7-47. Assume that $\mu_k = 0.3$ between the mass m_2 and the table in Fig. 7-16. The masses m_2 and m_3 are 8 and 6 kg, respectively. What mass m_1 is required to cause the system to accelerate to the left at 2 m/s²?

*7-48. A block of unknown mass is given a push up a 40° inclined plane and then released. It continues to move up the plane (+) at an acceleration of −9 m/s². What is the coefficient of kinetic friction? **Ans. 0.360**

*7-49. Block A in Fig. 7-17 has a weight of 64 lb. What must be the weight of block B if block A moves up the plane with an acceleration of 6 ft/s²? Neglect friction.

Figure 7-16

60°

Figure 7-17

*7-50. The mass of block *B* in Fig. 7-17 is 4 kg. What must be the mass of block *A* if it is to move down the plane at an acceleration of 2 m/s²? Neglect friction.   **Ans. 7.28 kg**

*7-51. Assume that the masses *A* and *B* in Fig. 7-17 are 4 kg and 10 kg, respectively. The coefficient of kinetic friction is 0.3. Find the acceleration if **(a)** the system is initially moving up the plane, and **(b)** if the system is initially moving down the plane.

Work, Energy, and Power

The principal reason for the application of a resultant force is to cause a displacement. For example, a large crane lifts a steel beam to the top of a building; the compressor in an air conditioner forces a fluid through its cooling cycle; and electromagnetic forces move electrons across a television screen. Whenever a force acts through a distance, we will learn, *work* is done in a way that can be measured or predicted. The capacity for doing work will be defined as **energy**, and the rate at which it is accomplished will be defined as **power.** The control and use of energy is probably the major concern of industry today, and a thorough understanding of the three concepts of work, energy, and power is essential.

8-1 WORK

When we attempt to drag a block with a rope, as in Fig. 8-1a, nothing happens. We are exerting a force, but the block has not moved. On the other hand, if we continually increase our pull, eventually the block will be displaced. In this case we have actually accomplished something in return for our efforts. This accomplishment is defined in physics as **work.** The term *work* has an explicit, quantitative, operational definition. For work to be done, three things are necessary:

1. There must be an applied force.
2. The force must act through a certain distance, called the **displacement.**
3. The force must have a component along the displacement.

(a) Work = 0   (b) Work = $F \cos \theta \cdot s$

Figure 8-1   The work done by a force **F** undergoing a displacement *s*.

interNET
CONNECTION
buphy.bu.edu
How much work is performed by a pile driver? For a demonstration, see this Web site (but do not type www.).

Assuming that we are given these conditions, a formal definition of work may be stated:

> **Work is a scalar quantity equal to the product of the magnitudes of the displacement and the component of the force in the direction of the displacement.**

$$Work = Force\ component \times displacement$$

$$Work = F_x s \qquad\qquad 8\text{-}1$$

In this equation, F_x is the component of **F** along the displacement *s*. In Fig. 8-1, only F_x contributes to work. Its magnitude can be found from trigonometry, and work can be expressed in terms of the angle θ between **F** and *s*:

$$Work = (F \cos \theta)s \qquad\qquad 8\text{-}2$$

Quite often the force causing the work is directed entirely along the displacement. This happens when a weight is lifted vertically or when a horizontal force drags an object along the floor. In these simple cases, $F_x = F$, and the work is the simple product of force and displacement:

$$Work = Fs \qquad\qquad 8\text{-}3$$

Another special case occurs when the applied force is perpendicular to the displacement. In this instance, the work will be zero, since $F_x = 0$. An example is motion parallel to the earth's surface in which gravity acts vertically downward and is perpendicular to all horizontal displacements. Then the force of gravity does no work.

EXAMPLE 8-1

What work is done by a 60-N force in dragging the block of Fig. 8-1 a distance of 50 m when the force is transmitted by a rope making an angle of 30° with the horizontal?

Solution

We must first determine the component F_x of the 60-N force **F**. Only this component contributes to work. Graphically, this is accomplished by drawing the 60-N vector to scale at a 30° angle. Measuring F_x and converting to newtons gives

$$F_x = 52.0\ N$$

With trigonometry, we could accomplish the same calculation using the cosine function:

$$F_x = (60 \text{ N})(\cos 30°) = 52.0 \text{ N}$$

Now, applying Eq. (8-1), we obtain the work:

$$\text{Work} = F_x s = (52.0 \text{ N})(50 \text{ m})$$
$$= 2600 \text{ N} \cdot \text{m}$$

Note that the units of work are the units of force times distance. Thus, in SI units, work is measured in *newton-meters* (N · m). By agreement, this combination unit is renamed the *joule*, which is denoted by the symbol J.

> **One joule (1 J) is equal to the work done by a force of one newton in moving an object through a parallel distance of one meter.**

In Example 8-1, the work done in dragging the block would be written as 2600 J.

In the United States, work is sometimes also given in USCS units. When the force is given in *pounds* (lb) and the displacement is given in *feet* (ft), the corresponding work unit is called the *foot-pound* (ft · lb).

> **One foot-pound (1 ft · lb) is equal to the work done by a force of one pound in moving an object through a parallel distance of one foot.**

No special name is given to this unit.

The following conversion factors will be useful when comparing work units in the two systems:

$$1 \text{ J} = 0.7376 \text{ ft} \cdot \text{lb} \qquad 1 \text{ ft} \cdot \text{lb} = 1.356 \text{ J}$$

Math Tip

When solving work problems, be sure to find the force component directed along the line of motion. A person pulls a cart 24 m across a frictionless floor by a rope making a 60° angle with the floor. The tension in the rope is 8 N. What is the work done?

The force we are concerned with is that directed parallel to the floor.

$$F_p = F \cos 60°$$
$$W = Fd = F_p d = F \cos 60° d$$

Substitute in the numbers:

$$W = 8(\cos 60°)(24) =$$
$$8(0.5)24 = 96 \text{ N} \cdot \text{m parallel}$$
to the floor or 96 J

8-2 RESULTANT WORK

When we consider the work of several forces acting on the same object, it is often useful to distinguish between positive and negative work. In this text, we will follow the convention that the work of a particular force is positive if the force component is in the same direction as the displacement. Negative work is done by a force component that opposes the actual displacement. Hence, work done by a crane in lifting a load is positive, but the gravitational force exerted by the earth on the load is doing negative work. Similarly, when we stretch a spring, the work on the spring is positive; the work on the spring is negative when the spring contracts, pulling us back. Another important example of negative work is that performed by a frictional force that is opposite to the direction of displacement.

If several forces act on a body in motion, the **resultant work** (total work) is the algebraic sum of the works of the individual forces. This will also be equal to the work of the resultant force. The accomplishment of net work requires the existence of a resultant force. These ideas are clarified in Example 8-2.

EXAMPLE 8-2

A push of 80 N moves a 5-kg block up a 30° inclined plane, as shown in Fig. 8-2. The coefficient of kinetic friction is 0.25, and the length of the plane is 20 m. (a) Compute the work done by each force acting on the block. (b) Show that the net work done by these forces is the same as the work of the resultant force.

(a)   (b)

Figure 8-2 The work required to push a block up a 30° inclined plane.

Solution (a)

There are four forces acting on the block: $\mathcal{N}$, **P**, $\mathcal{F}_k$, and **W**. (See Fig. 8-2b). The normal force $\mathcal{N}$ does no work because it has no component along the displacement.

$$(\text{Work})_{\mathcal{N}} = 0$$

The push **P** is entirely along the displacement and in the direction of the displacement. Hence,

$$(\text{Work})_P = Ps = (80 \text{ N})(20 \text{ m}) = 1600 \text{ J}$$

To find the work of the friction force $\mathcal{F}_k$ and the work of the weight **W**, we must first determine the components of the weight along and perpendicular to the plane.

$$W = mg = (5 \text{ kg})(9.8 \text{ m/s}^2) = 49.0 \text{ N}$$
$$W_x = (49.0 \text{ N}) \sin 30° = 24.5 \text{ N}$$
$$W_y = (49.0 \text{ N}) \cos 30° = 42.4 \text{ N}$$

Now the friction force $\mathcal{F}_k = \mu_k \mathcal{N}$ and $\mathcal{N} = W_y$, so that

$$\mathcal{F}_k = \mu_k \mathcal{N} = \mu_k W_y$$
$$= -(0.25)(42.4 \text{ N}) = -10.6 \text{ N}$$

The minus sign indicates that the friction force is down the plane. The work will therefore be negative since the displacement is up the plane.

$$(\text{Work})_{\mathcal{F}} = \mathcal{F}_k s = (-10.6 \text{ N})(20 \text{ m}) = -212 \text{ J}$$

The weight W of the block also does negative work since its component W_x is directed opposite to the displacement.

$$(\text{Work})_W = -(24.5 \text{ N})(20 \text{ m}) = -490 \text{ J}$$

Solution (b)

The net work is obtained by summing the works of the individual forces.

$$\text{Net work} = (\text{work})_N = (\text{work})_P = (\text{work})_{\mathscr{F}} + (\text{work})_W$$
$$= 0 + 1600 \text{ J} - 212 \text{ J} - 490 \text{ J}$$
$$= 898 \text{ J}$$

To show that this is also the work of the resultant force, we first compute the resultant force. According to methods introduced in earlier chapters,

$$F_R = P - \mathscr{F}_k - W_x$$
$$= 80 \text{ N} - 10.6 \text{ N} - 24.5 \text{ N} = 44.9 \text{ N}$$

The work of F_R is therefore

$$\text{Net work} = F_R s = (44.9 \text{ N})(20 \text{ m}) = 898 \text{ J}$$

which compares with the value obtained by computing the work of each force separately.

It is important to distinguish between the *resultant* or *net* work and the work of an individual force. If we speak of the work required to move a block through a distance, the work done by the pulling force is not necessarily the resultant work. Work may be done by a friction force or by other forces. Resultant work is simply the work done by a resultant force. If the resultant force is zero, then the resultant work is zero, even though individual forces may be doing positive or negative work.

8-3 ENERGY

Energy may be thought of as *anything that can be converted into work*. When we say that an object has energy, we mean that it is capable of exerting a force on another object to do work on it. Conversely, if we do work on some object, we have added to it an amount of energy equal to the work done. The units of energy are the same as those for work: the *joule* and the *foot-pound*.

In mechanics we shall be concerned with two kinds of energy:

Kinetic energy E_k: Energy possessed by a body by virtue of its motion.

Potential energy E_p: Energy possessed by a system by virtue of position or condition.

One can readily think of many examples of each kind of energy. For instance, a moving car, a moving bullet, and a rotating flywheel all have the ability to do

(a)

(b)

Figure 8-3 Examples of (a) kinetic energy and (b) potential energy.

work because of their motion. Similarly, a lifted object, a compressed spring, and a stretched rubberband have the potential for doing work because of position. Several examples are provided in Fig. 8-3.

8-4 WORK AND KINETIC ENERGY

We have defined kinetic energy as the capability for performing work as a result of the motion of a body. To see the relationship between motion and work, let's consider a constant force **F** acting on the block in Fig. 8-4. Consider that the block has an initial velocity v_0 and that the force **F** acts through a distance s, causing the velocity to increase to a final value v_f. If the body has a mass m, Newton's second law tells us that it will gain speed, or accelerate, at a rate given by

$$a = \frac{F}{m} \qquad \text{8-4}$$

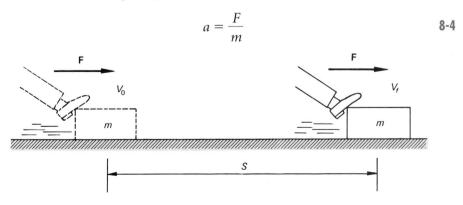

Figure 8-4 The work done by the force **F** results in a change in the kinetic energy of the mass m.

until it reaches the final velocity v_f. From Chap. 5, we recall

$$2as = v_f^2 - v_0^2$$

from which

$$a = \frac{v_f^2 - v_0^2}{2s}$$

Substitution of this into Eq. (8-4) yields

$$\frac{F}{m} = \frac{v_f^2 - v_0^2}{2s}$$

which can be solved for the product Fs to obtain

$$Fs = \frac{1}{2}mv_f^2 - \frac{1}{2}mv_0^2 \qquad \text{8-5}$$

The quantity on the left side of Eq. (8-5) represents the work done on the mass m. The quantity on the right side must be the change in kinetic energy as a result of this work. Thus, we can define kinetic energy E_k as

$$E_k = \frac{1}{2}mv^2 \qquad \text{8-6}$$

Following this notation, $\frac{1}{2}mv_f^2$ and $\frac{1}{2}mv_0^2$ would represent final and initial kinetic energies, respectively. The important result can be stated as follows:

The work of a resultant external force on a body is equal to the change in kinetic energy of the body.

A close look at Eq. (8-5) will show that an *increase* in kinetic energy ($v_f > v_0$) will result from *positive work*, whereas a *decrease* in kinetic energy ($v_f < v_0$) will result from *negative* work. In the special case in which zero work is done, the kinetic energy is a constant, given by Eq. (8-6).

EXAMPLE 8-3

Compute the kinetic energy of a 4-kg sledgehammer at the instant its velocity is 24 m/s.

Solution
Applying Eq. (8-6), we obtain

$$E_k = \frac{1}{2}mv^2 = \frac{1}{2}(4 \text{ kg})(24 \text{ m/s})^2$$
$$= 1152 \text{ N} \cdot \text{m} = 1152 \text{ J}$$

EXAMPLE 8-4

Compute the kinetic energy of a 3200-lb automobile traveling at 60 mi/h (88 ft/s).

Solution

We compute as before, except that we must determine the mass from the weight.

$$E_k = \frac{1}{2}mv^2 = \frac{1}{2}\left(\frac{W}{g}\right)v^2$$

Substituting given values for W and v, we have

$$E_k = \frac{1}{2}\left(\frac{3200 \text{ lb}}{32 \text{ ft/s}^2}\right)(88 \text{ ft/s})^2$$
$$= 3.87 \times 10^5 \text{ ft} \cdot \text{lb}$$

EXAMPLE 8-5

What average force **F** is necessary to stop a 16-g bullet traveling at 260 m/s as it penetrates into wood a distance of 12 cm?

Solution

The total work required to stop the bullet will be equal to the change in kinetic energy. (See Fig. 8-5.) Since the bullet is stopped, $v_f = 0$, and so Eq. (8-5) yields

$$Fs = -\frac{1}{2}mv_0^2$$

Substituting gives

$$F(0.12 \text{ m}) = -\frac{1}{2}(0.016 \text{ kg})(260 \text{ m/s})^2$$

Dividing by 0.12 m, we have

$$F = \frac{-(0.016 \text{ kg})(260 \text{ m/s})^2}{(2)(0.12 \text{ m})}$$
$$= -4510 \text{ N}$$

The minus sign indicates that the force was opposite to the displacement. It should be noted that this force is about 30,000 times the weight of the bullet.

Figure 8-5 The work done in stopping the bullet is equal to the change in kinetic energy of the bullet.

8-5 POTENTIAL ENERGY

The energy that systems possess by virtue of their positions or conditions is called *potential energy.* Since energy expresses itself in the form of work, potential energy implies that there must be a potential for doing work. For example, suppose the pile driver in Fig. 8-6 is used to lift a body of weight W to a height h above the ground stake. We say that the body-earth system has gravitational potential energy. When such a body is released, it will do work when it strikes the stake. If it is heavy enough and if it has fallen from a great enough height, the work done will result in driving the stake through a distance s.

The external force $\mathbf{F}$ required to lift the body must at least be equal to the weight W. Thus, the work done on the system is given by

$$\text{Work} = Wh = mgh$$

This amount of work can also be done *by* the body after it has dropped a distance h. Thus, the body has potential energy equal in magnitude to the external work required to lift it. This energy does not come from the earth-body system, but results from work done on the system by an external agent. Only *external* forces, such as $\mathbf{F}$ in Fig. 8-6 or friction, can add energy to or remove energy from the system made up of the body and the earth.

Note from the preceding discussion that potential energy E_p can be found from

$$E_p = Wh = mgh \qquad \text{\textit{Potential Energy}} \quad \textbf{8-7}$$

where W and m are the weight and the mass of an object located a distance h above some reference point.

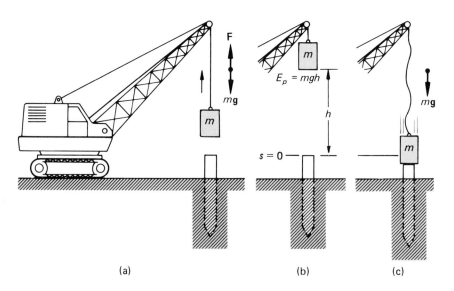

Figure 8-6 (a) Lifting a mass m to a height h requires the work mgh. (b) The body-earth system, therefore, has a potential energy $E_p + mgh$. (c) When the mass is released, it has the capacity for doing the work mgh on the stake.

The potential energy depends on the choice of a particular reference level. The gravitational potential energy for an airplane is quite different when measured with respect to a mountain peak, a skyscraper, or sea level. The capacity for doing work is much greater if the aircraft falls to sea level. Potential energy has physical significance only in the event that a reference level is established.

EXAMPLE 8-6

A 250-g wrench is held 200 mm above a workbench that is 1 m above the floor. Compute the potential energy relative to (a) the bench top and (b) the floor.

Solution (a)

The height h of the wrench above the bench is 200 mm (or 0.2 m), and the mass is 250 g (or 0.25 kg). Thus, the potential energy relative to the bench is

$$E_p = mgh = (0.25 \text{ kg})(9.8 \text{ m/s}^2)(0.2 \text{ m})$$
$$= 0.49 \text{ J}$$

Notice that kilograms, meters, and seconds are the only units of mass, length, and time that are consistent with the definition of a joule.

Solution (b)

The potential energy with reference to the floor is based on a different value of h.

$$E_p = mgh = (0.25 \text{ kg})(9.8 \text{ m/s}^2)(1.2 \text{ m})$$
$$= 2.94 \text{ J}$$

EXAMPLE 8-7

An 800-lb commercial air conditioner is lifted by a chain hoist until it is 22 ft above the floor. What is the potential energy relative to the floor?

Solution

Applying Eq. (8-7), we obtain

$$E_p = Wh = (800 \text{ lb})(22 \text{ ft}) = 17,600 \text{ ft} \cdot \text{lb}$$

We have stated that the potential for doing work is a function only of the weight and the height h above some reference point. The potential energy at a particular position is not dependent on the path taken to reach that position because the same work must be done against gravity regardless of the path. In Example 8-7, work of 17,600 ft · lb was required to lift the air conditioner through a vertical distance of 22 ft. If we choose to exert a lesser force by moving it up an incline, a greater distance would be required. In either case, the work done against gravity is 17,600 ft · lb because the end result is the placement of an 800-lb weight at a height of 22 ft.

8-6 CONSERVATION OF ENERGY

Quite often, at relatively low speeds, an interchange takes place between kinetic and potential energies. For example, suppose a mass m is lifted to a height h and dropped, as shown in Fig. 8-7. An external force has increased the energy of the system, giving it a potential energy $E_p = mgh$ at the highest point. This is the total energy available to the system, and it cannot change unless an external resistive force is encountered. As the mass falls, its potential energy decreases because its height above the ground is reduced. The lost potential energy reappears in the form of kinetic energy of motion. In the absence of air resistance, the total energy $(E_p + E_k)$ remains the same. Potential energy continues to be converted into kinetic energy until the mass reaches the ground $(h = 0)$. At this final position, the kinetic energy is equal to the total energy, and the potential energy is zero. The important point to be made is that the sum of E_p and E_k is the same at any point during the fall (see Fig. 8-7).

$$\text{Total energy} = E_p + E_k = \text{constant}$$

We say that mechanical energy is *conserved*. In our example, the total energy at the top is mgh and the total energy at the bottom is $\frac{1}{2}mv^2$, if we neglect air resistance. We are not prepared to state the principle of *conservation of mechanical energy*:

Conservation of Mechanical Energy **In the absence of air resistance or other dissipative forces, the sum of the potential and kinetic energies is a constant, provided that no energy is added to the system.**

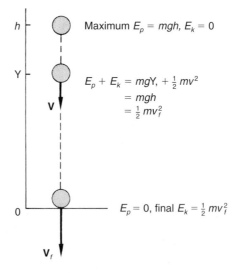

Figure 8-7 In the absence of friction, the total mechanical energy is constant. At any point, it is equal to either the potential energy at the top or the kinetic energy at the bottom.

In applying this principle it is useful to think of beginning and ending points for any process. At either point, if the velocity is not zero, there is kinetic energy, and if the height is not zero, there is potential energy. We can write

$$(E_p + E_k)_{\text{BEG}} = (E_p + E_k)_{\text{END}}$$

$$mgh_o + \frac{1}{2}mv_o^2 = mgh_f + \frac{1}{2}mv_f^2 \qquad \text{8-8}$$

The subscripts o and f refer to initial and final values. Equation (8-8), of course, applies when no friction forces are involved.

In the example of an object falling from rest at an initial position h_o, the total initial energy is $mgh_o(v_o = 0)$ and the total final energy is $\frac{1}{2}mv_f^2$ $(h = 0)$.

$$mgh_o = \frac{1}{2}mv_f^2$$

Solving this relationship for v_f gives a useful equation for determining the final velocity from energy considerations for an object falling from rest with no friction.

$$v_f = \sqrt{2gh_o}$$

A great advantage of this method is that the final velocity is determined from the initial and final energy states. The actual path taken does not matter in the absence of friction. For example, the same final velocity would result if the object followed a curved path from the same initial height h_o.

EXAMPLE 8-8

In Fig. 8-8, a 40-kg ball is pulled to one side until it is 1.6 m above its lowest point. Neglecting friction, what will be its velocity as it passes through its lowest point?

Figure 8-8 The velocity of a suspended mass as it passes through its lowest point can be found from energy considerations.

Solution

Conservation of total energy requires that $(E_p + E_k)$ be the same at the beginning and at the end. Thus,

$$mgh_o + 0 = 0 + \frac{1}{2}mv_f^2$$

from which we can divide out the mass m and obtain

$$v_f = \sqrt{2gh_o} = \sqrt{2(9.8 \text{ m/s}^2)(1.6 \text{ m})}$$
$$= 5.60 \text{ m/s}$$

As an additional example, you would show that the total energy at the beginning and at the end is 627 J.

8-7 ENERGY AND FRICTION FORCES

It is helpful to consider the conservation of mechanical energy as an accounting process in which one keeps track of what happens to the energy of a system from beginning to end. For example, suppose you withdraw $1000 from a bank, then pay $400 for an airline ticket to New York. You would have $600 left to spend on entertainment. The $400 was spent, but it still must be accounted for. Now consider a sled at the top of a hill. The initial total energy is 1000 J. If 400 J of energy was lost due to friction, the sled would arrive at the bottom with a total energy of only 600 J. By accounting for friction or other dissipative forces, we can give a more general statement of the conservation of energy principle:

Conservation of energy **The total energy of a system is always constant, although energy changes from one form to another may occur within the system.**

In real-world applications, it is not possible to remove external forces from consideration. Therefore, a more general statement of the principle of conservation of energy accounts for losses due to friction:

$$(E_p + E_k)_{\text{BEG}} = (E_p + E_k)_{\text{END}} + |\text{energy losses}| \qquad \textbf{8-9}$$

The absolute value signs attached to the energy-losses term are a reminder that we are not concerned with the sign of the work done against friction forces. We are simply accounting for the disposition of all the initial energy. If we represent the work of a friction force by the product $\mathcal{F}_k s$, we can write

$$mgh_o + \frac{1}{2}mv_o^2 = mgh_f + \frac{1}{2}mv_f^2 + |\mathcal{F}_k s| \qquad \textbf{8-10}$$

Of course, if an object begins at rest a height h_0 above its final position, this equation simplifies to

$$mgh_o = \frac{1}{2}mv_f^2 + |\mathcal{F}_k s| \qquad \textbf{8-11}$$

In working problems, it is useful to establish the sum of potential and kinetic energies at some beginning point. The absolute value of any energy losses must then be added to the total energy of the system at the ending point in such a manner that energy is conserved.

CONCEPTUAL EXAMPLE 8-9

A 20-kg sled rests at the top of a 30° slope 80 m in length, as shown in Fig. 8-9. If $\mu_k = 0.2$, what is the velocity at the bottom of the incline?

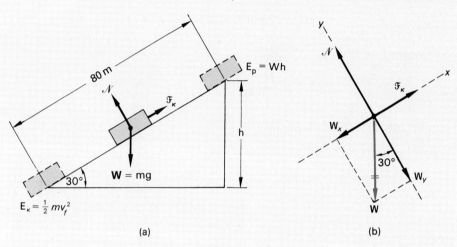

Figure 8-9 Some of the initial potential energy at the top of the incline is lost because of doing work against friction as the clock slides down.

Solution

The total energy at the beginning is potential energy since the initial velocity was zero. The height h_o is given by

$$h_o = (80 \text{ m}) \sin 30° = 40 \text{ m}$$

which allows us to compute the initial total energy:

$$\begin{aligned}
(E_p + E_k)_{\text{BEG}} &= mgh_o + 0 \\
&= (20 \text{ kg})(9.8 \text{ m/s}^2)(40 \text{ m}) \\
&= 7840 \text{ J}
\end{aligned}$$

Thus, we have 7840 J that must be accounted for as the sled moves to the bottom. To determine how much of this is lost in work against friction, we must first find the normal force $\mathcal{N}$ exerted by the plane on the block. From 8-9b,

$$\mathcal{N} = W_y = (20 \text{ kg})(9.8 \text{ m/s}^2) \cos 30° = 170 \text{ N}$$

from which the friction force is

$$\mathcal{F}_k = \mu_k \mathcal{N} = (0.2)(170 \text{ N}) = 34.0 \text{ N}$$

The absolute value of the work done by the friction force is

$$\mathcal{F}_k s = (34.0 \text{ N})(80 \text{ m}) = 2720 \text{ J}$$

We can now determine how much energy is left for velocity from Eq. (8-11):

$$mgh_o = \frac{1}{2}mv_f^2 + |\mathcal{F}_k s|$$

$$7840 \text{ J} = \frac{1}{2}mv_f^2 + 2720 \text{ J}$$

and

$$\frac{1}{2}mv_f^2 = 7840 \text{ J} - 2720 \text{ J} = 5120 \text{ J}$$

Substituting $m = 20$ kg, we have

$$\frac{1}{2}(20 \text{ kg})v_f^2 = 5120 \text{ J}$$

Finally, solving for v_f, we obtain

$$v_f = 22.6 \text{ m/s}$$

You should show that the final velocity would have been 28 m/s if there had been no friction forces present.

Problem-Solving Strategy

Conservation of Energy

1. Read the problem and then draw and label a rough sketch, identifying each object whose height or velocity changes.
2. Determine a reference point for measuring gravitational potential energy, for example, the bottom on an incline, the floor of a room, or the lowest point in the path of a particle.
3. For each object, make a note of the beginning and ending heights and velocities: h_o, v_o, h_f, and v_f. The heights are each measured from the chosen reference position and only the magnitudes are needed for velocities.
4. The total energy of the system at any instant is the sum of the kinetic and potential energies. Thus, the initial total energy E_o and the final total energy E_f are

$$E_o = mgh_o + \frac{1}{2}mv_o^2 \qquad E_f = mgh_f + \frac{1}{2}mv_f^2$$

5. Determine whether or not friction forces are present. If friction or air resistance is present, then the energy losses must be given or determined. Often, the energy loss in doing work against friction is the simple product of the friction force $\mathcal{F}$ and the displacement s. Remember that $\mathcal{F} = \mu \mathcal{N}$.
6. Write the conservation of energy equation and solve for the unknown.

$$mgh_o + \frac{1}{2}mv_o^2 = mgh_f + \frac{1}{2}mv_f^2 + |\text{energy losses}|$$

7. Remember to use the absolute value of the energy lost when applying the above relationship. The actual work against friction is always negative, but you are accounting for it here as a *loss*.

In our definition of work, *time* is not involved in any way. The same amount of work is done whether the task takes an hour or a year. Given enough time, even the weakest motor can lift an enormous load. If we wish to perform a task efficiently, however, the *rate* at which work is done becomes an important engineering quantity.

Power is the rate at which work is accomplished.

$$P = \frac{\text{work}}{t} \qquad \text{8-12}$$

The SI unit for power is the *joule per second*, which is renamed the *watt* (W). Thus, an 80-W light bulb burns energy at the rate of 80 J/s.

$$1 \text{ W} = 1 \text{ J/s}$$

In USCS units, we use the *foot-pound per second* (ft · lb/s). No special name is given to this unit of power.

The watt and the foot-pound per second are inconveniently small units for most industrial purposes. Therefore, the **kilowatt** (kW) and the **horsepower** (hp) are defined:

$$1 \text{ kW} = 1000 \text{ W}$$
$$1 \text{ hp} = 550 \text{ ft} \cdot \text{lb/s}$$

In the United States, the watt and kilowatt are used almost exclusively in connection with electric power; horsepower is reserved for mechanical power. This practice is purely a convention and by no means necessary. It is perfectly proper to speak of an 0.08-hp bulb or to brag about a 238-kW engine. The conversion factors are

$$1 \text{ hp} = 746 \text{ W} = 0.746 \text{ kW}$$
$$1 \text{ kW} = 1.34 \text{ hp}$$

Since work is frequently done in a continuous fashion, an expression for power that involves velocity is useful. Thus,

$$P = \frac{\text{work}}{t} = \frac{Fs}{t} \qquad \text{8-13}$$

from which

$$P = F\frac{s}{t} = Fv \qquad \text{8-14}$$

where v is the velocity of the body on which the parallel force F is applied.

EXAMPLE 8-10

A 40-kg load is raised to a height of 25 m. If the operation requires 1 min, find the power required. What is the power in units of horsepower?

Solution

The work done in lifting the load is

$$\text{work} = Fs = mgh = (40 \text{ kg})(9.8 \text{ m/s}^2)(25 \text{ m})$$
$$= 9800 \text{ J}$$

The power is then

$$P = \frac{\text{work}}{t} = \frac{9800 \text{ J}}{60 \text{ s}} = 163 \text{ W}$$

Since 1 hp = 746 W, the horsepower developed is

$$P = (163 \text{ W}) \frac{1 \text{ hp}}{746 \text{ W}} = 0.219 \text{ hp}$$

EXAMPLE 8-11

A 60-hp motor provides for the elevator of a hotel. If the weight of the elevator is 2000 lb, how much time is required to lift the elevator 120 ft?

Solution

The work required is given by

$$\text{Work} = Fs = (2000 \text{ lb})(120 \text{ ft})$$
$$= 2.4 \times 10^5 \text{ ft} \cdot \text{lb}$$

Since 1 hp = 550 ft · lb/s, the power developed is

$$P = (60 \text{ hp}) \frac{550 \text{ ft} \cdot \text{lb/s}}{1 \text{ hp}} = 3.3 \times 10^4 \text{ ft} \cdot \text{lb/s}$$

From Eq. (8-13),

$$P = \frac{Fs}{t}$$

so that

$$t = \frac{Fs}{P} = \frac{2.4 \times 10^5 \text{ ft} \cdot \text{lb}}{3.3 \times 10^4 \text{ ft} \cdot \text{lb/s}}$$
$$= 7.27 \text{ s}$$

Equation (8-12) can be solved for work: work = Pt. Therefore, the *kilowatthour* (kW · h) unit used by electric companies in billing is a unit of *power* (kilowatt) times *time* (hour), or a unit of work. Quite reasonably, the bill is for the amount of work that has been done. The price per kilowatthour may also be determined, however, by the peak power demand of the consumer.

Summary and Review

Summary

The concepts of work, energy, and power have been discussed in this chapter. The major points to remember are summarized below:

- The *work* done by a force **F** acting through a distance s is found from the following equations (refer to Fig. 8-1):

$$\text{Work} = F_x s \qquad \text{Work} = (F \cos \theta)s$$

SI unit: joule (J) USCS unit: foot-pound (ft · lb)

- *Kinetic energy* E_k is the capacity for doing work as a result of motion. It has the same units as work and is found from

$$E_k = \frac{1}{2}mv^2 \qquad E_k = \frac{1}{2}\left(\frac{W}{g}\right)v^2$$

- Gravitational *potential energy* is the energy that results from the position of an object relative to the earth. Potential energy E_p has the same units as work and is found from

$$E_p = Wh \qquad E_p = mgh$$

where W or mg is the weight of the object and h is the height above some reference position.

- Net work is equal to the change in kinetic energy.

$$Fs = \frac{1}{2}mv_f^2 - \frac{1}{2}mv_0^2$$

- Conservation of mechanical energy with no friction:

$$(E_p + E_k)_{\text{BEG}} = (E_p + E_k)_{\text{END}}$$

$$mgh_o + \frac{1}{2}mv_o^2 = mgh_f + \frac{1}{2}mv_f^2$$

- Conservation of energy including friction:

$$(E_p + E_k)_{\text{BEG}} = (E_p + E_k)_{\text{END}} + |\text{energy losses}|$$

$$mgh_o + \frac{1}{2}mv_o^2 = mgh_f + \frac{1}{2}mv_f^2 + |\mathcal{F}_k s|$$

- Power is the rate at which work is done:

$$P = \frac{\text{work}}{t} \qquad P = \frac{Fs}{t} \qquad P = Fv$$

SI unit: watt (W)   USCS unit: ft · lb/s
Other units: 1 kW = 10^3 W   1 hp = 550 ft · lb/s

Key Terms

1. work
2. joule
3. potential energy
4. kinetic energy
5. conservation of energy
6. power
7. horsepower
8. watt
9. kilowatt

Review Questions

8-1. Distinguish clearly between the physicist's concept of work and the general concept of work.

8-2. Two teams are engaged in a tug of war. Is work done? When?

8-3. Whenever net work is done on a body, will the body necessarily undergo acceleration? Discuss.

8-4. A diver stands on a board 10 ft above the water. What kind of energy results from this position? What happens to this energy as she dives into the water? Is work done? If so, what is doing the work, and on what is the work done?

8-5. Compare the potential energies for two bodies A and B if (a) A is twice as high as

B but of the same mass; (b) B is twice as heavy as A but at the same height; and (c) A is twice as heavy as B, but B is twice as high as A.

8-6. Compare the kinetic energies of two bodies A and B if (a) A has twice the speed of B, (b) A has half the mass of B, and (c) A has twice the mass and half the speed of B.

8-7. In stacking 8-ft boards, you lift an entire board at its center and lay it on the pile. Your helper lifts one end, rests it on the pile, and then lifts the other end. Compare the work done.

8-8. In the light of what you have learned about work and energy, describe the most efficient procedure for ringing the bell with a sledgehammer at the fair. What precautions should you take?

8-9. A roller coaster at the fair boasts "a maximum height of 100 ft with a maximum speed of 60 mi/h." Do you believe the advertisement? Explain.

8-10. A man mows his yard for years using a 4-hp mower. He then buys a 6-hp mower. After using the new mower for a while, it seems to him as if he has twice the power as before. Why do you think he is convinced of the increase in power?

Problems

Section 8-1—Work

8-1. What is the work done by a force of 20 N acting through a parallel distance of 8 m? What force will do the same work through a distance of 4 m? **Ans. 160 J, 40 N**

8-2. A worker lifts a 40-lb weight through a height of 10 ft. How many meters can a 10-kg block be lifted by the same amount of work?

8-3. A tugboat exerts a constant force of 4000 N on a ship, moving it a distance of 15 m. What work is done? **Ans. 60 kJ**

8-4. A 5-kg hammer is lifted to a height of 3 m. What is the minimum required work?

8-5. A push of 30 lb is applied along the handle of a lawn mower, producing a horizontal displacement of 40 ft. If the handle makes an angle of 30° with the ground, what work was done by the 30-lb force? **Ans. 1040 ft lb**

8-6. The trunk in Fig. 8-10 is dragged a horizontal distance of 24 m by a rope that makes an angle θ with the floor. If the rope tension is 8 N, what works are done for the following angles: 0°, 30°, 60°, 90°?

Figure 8-10

8-7. A horizontal force pushes a 10-kg sled along a driveway for a distance of 40 m. If the coefficient of sliding friction is 0.2, what work is done by the friction force? **Ans. −784 J**

8-8. A sled is dragged a distance of 12.0 m by a rope under constant tension of 140 N. The task requires 1200 J of work. What angle does the rope make with the ground?

Section 8-2—Resultant Work

8-9. An average force of 40 N compresses a coiled spring a distance of 6 cm. What is the work done by the 40-N force? What work is done by the spring? What is the resultant work? **Ans. 2.40 J, −2.40 J, 0**

8-10. A horizontal force of 20 N drags a small sled 42 m across the ice at constant speed. Find the work done by the pulling force and by the friction force. What is the resultant force?

8-11. A 10-kg block is dragged 20 m by a parallel force of 26 N. If $\mu_k = 0.2$, what is the resultant work and what acceleration results? **Ans. 128 J, 0.640 m/s²**

8-12. A rope making an angle of 35° drags a 10-kg toolbox a horizontal distance of 20 m. The tension in the rope is 60 N, and the constant friction force is 30 N. What work is done by the rope? What work is done by friction? What is the resultant work?

8-13. For the example described in Prob. 8-12, what is the coefficient of friction between the toolbox and the floor? **Ans. 0.472**

*8-14. A 40-kg sled is pulled horizontally for 500 m ($\mu_k = 0.2$). If the resultant work is 50 kJ, what was the parallel pulling force?

*8-15. Assume that $m = 8$ kg in Fig. 8-11 and $\mu_k = 0$. What minimum work is required by the force P to reach the top of the inclined plane? What work is required to lift the 8-kg block vertically to the same height? **Ans. 941 J, 941 J**

FF = FNμ

= mgμ.

Figure 8-11

*8-16. What is the minimum work by the force P to move the 8-kg block to the top of the incline if $\mu_k = 0.4$? Compare this with the work needed to lift it vertically to the same height.

*8-17. What is the resultant work when the 8-kg block slides from the top to the bottom of the incline in Fig. 8-11? Assume that $\mu_k = 0.4$ **Ans. 492 J**

Section 8-4—Work and Kinetic Energy

8-18. What is the kinetic energy of a 6-g bullet at the instant its speed is 190 m/s? What is the kinetic energy of a 1200-kg car traveling at 80 km/h?

8-19. What is the kinetic energy of a 2400-lb automobile when its speed is 55 mi/h? What is the kinetic energy of a 9-lb ball when its speed is 40 ft/s? **Ans. 244,000 ft-lb; 225 ft-lb**

8-20. What is the change in kinetic energy when a 50-g ball hits the pavement with a velocity of 16 m/s and rebounds with a velocity of 10 m/s?

8-21. A runaway 400-kg wagon enters a cornfield with a velocity of 12 m/s and eventually comes to rest. What work was done on the wagon? **Ans. −28.8 kJ**

8-22. A 2400-lb car increases its speed from 30 mi/h to 60 mi/h. What resultant work was required? What is the equivalent work in joules?

8-23. A 0.6-kg hammer head is moving at 30 m/s just before striking the head of a spike. Find the initial kinetic energy. What work can be done by the hammer head? **Ans. 270 J, 270 J**

8-24. A 12-lb hammer moving at 80 ft/s strikes the head of a nail moving it into the wall a distance of $\frac{1}{4}$ in. What was the average stopping force?

8-25. What average force is needed to increase the velocity of a 2-kg object from 5 m/s to 12 m/s over a distance of 8 m? **Ans. 14.9 N**

*8-26. Verify the answer to Prob. 8-25 by applying Newton's second law of motion.

*8-27. A 20-g projectile strikes a mud bank in Fig. 8-12, penetrating a distance of 6 cm before stopping. Find the stopping force F if the entrance velocity is 80 m/s. **Ans. 1070 N**

*8-28. A 1500-kg car is moving along a level road at 60 km/h. What work is required to stop

Summary and Review

Figure 8-12

the car? If $\mu_k = 0.7$, what is the stopping distance?

Section 8-5—Potential Energy

8-29. A 2-kg block rests on top of a table 80 cm from the floor. Find the potential energy of the block relative to (a) the floor, (b) the seat of a chair 40 cm from the floor, and (c) relative to the ceiling 3 m from the floor.

Ans. 15.7 J, 7.84 J, −43.1 J

8-30. A 1.2-kg brick is held a distance of 2 m above a manhole. The bottom of the manhole is 3 m below the street. Relative to the street, what is the potential energy at each location? What is the change in potential energy?

8-31. At a particular instant a mortar shell has a velocity of 60 m/s. If its potential energy at that point is one-half of its kinetic energy, what is its height above the earth?

Ans. 91.8 m

*8-32. A 20-kg sled is pushed up a 34° slope to a vertical height of 140 m. A constant friction force of 50 N acts for the entire distance. What external work was required? What was the change in potential energy?

*8-33. An average force of 600 N is required to compress a coiled spring a distance of 4 cm.

What work is done *by* the spring? What is the change in potential energy of the compressed spring?   **Ans. −24 J, +24 J**

Section 8-6—Conservation of Energy

8-34. A 64-lb weight is lifted to a height of 10 ft and then released to fall freely. What is the potential energy, the kinetic energy, and the total energy at (a) the highest point, (b) 3 ft above the ground, and (c) at the ground?

8-35. A 4-kg hammer is lifted to a height of 10 m and dropped. What are the potential and kinetic energies of the hammer when it has fallen to a point 4 m from the earth?

Ans. 157 J, 235 J

8-36. What will be the velocity of the hammer in Prob. 8-35 just before it strikes the ground? What is the velocity at the 4-m location?

8-37. What initial velocity must be given to a 5-kg mass if it is to rise to a height of 10 m? What is the total energy at any point in its path?   **Ans. 14 m/s, 490 J**

8-38. A simple pendulum 1 m long has an 8-kg bob. How much work is needed to move the pendulum from its lowest point to a horizontal position? From energy considerations find the velocity of the bob as it swings through the lowest point.

8-39. A ballistic pendulum is illustrated in Fig. 8-13. A 40-g ball is caught by a 500-g sus-

Figure 8-13

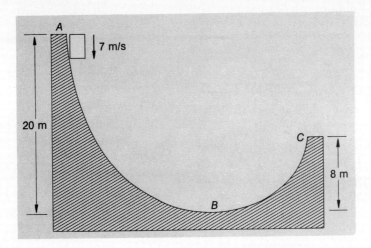

Figure 8-14

pended mass. After impact, the two masses rise a vertical distance of 45 mm. Find the velocity of the combined masses just after impact. **Ans. 93.9 cm/s**

*8-40. A 100-lb sled slides from rest at the top of a 37° inclined plane. The original height is 80 ft. In the absence of friction, what is the velocity of the sled when it reaches the bottom of the incline?

*8-41. An 8-kg block in Fig. 8-14 has an initial downward velocity of 7 m/s. Neglecting friction, find the velocity when it reaches point *B*. **Ans. 21.0 m/s.**

*8-42. What is the velocity of the 8-kg block at point *C* in Prob. 8-41?

*8-43. An 80-lb girl sits in a swing of negligible weight. If she is given an initial velocity of 20 ft/s, to what height will she rise?
Ans. 6.25 ft

Section 8-7—Energy and Friction Forces

8-44. A 60-kg sled slides to the bottom of a 25° slope of length 30 m. A 100-N friction force acts for the entire distance. What is the total energy at the top of the slope and at the bottom? What is the velocity of the sled at the bottom?

8-45. A 500-g block is released from the top of a 30° incline and slides 160 cm to the bot-

tom. A constant friction force of 0.9 N acts the entire distance. What is the total energy at the top? What work is done by friction? What is the velocity at the bottom?
Ans. 3.92 J, −1.44 J, 3.15 m/s

8-46. What initial velocity must be given to the 500-g block in Prob. 8-45 if it is to just reach the top of the same slope?

8-47. A 64-lb cart starts up a 37° incline with an initial velocity of 60 ft/s. If it comes to rest after moving a distance of 70 ft, how much energy was lost to friction?
Ans. 904 ft lb

*8-48. A 0.4-kg ball drops a vertical distance of 40 m and rebounds to a height of 16 m. How much energy was lost in collision with the floor?

*8-49. A 4-kg sled is given an initial velocity of 10 m/s at the top of a 34° slope. If $\mu_k = 0.2$, how far must the sled travel until its velocity reaches 30 m/s? **Ans. 104 m**

*8-50. Assume in Fig. 8-14 that the sliding mass is 6 kg and that 300 J of energy is lost doing work against friction. What is the velocity when the mass reaches point *C*?

*8-51. A bus slams on brakes to avoid an accident. The tread marks of the tires are 80 ft long. If $\mu_k = 0.7$, what was the speed before applying brakes? **Ans. 59.9 ft/s**

Summary and Review

Section 8-8—Power

8-52. A power station conveyor belt lifts 500 tons of ore to a height of 90 ft in 1 hr. What average horsepower is required?

8-53. A 40-kg mass is lifted through a distance of 20 m in a time of 3 s. What average power is employed? **Ans. 2.61 kW**

8-54. A 300-kg elevator is lifted vertical distance of 100 m in 2 min. What is the output power?

8-55. A 90-kW engine is used to lift a 1200-kg load. What is the average velocity of the lift? **Ans. 7.65 m/s**

8-56. To what height can a 400-W engine lift a 100-kg mass in 3 s?

8-57. An 800-N student runs up a flight of stairs rising 6 m in 8 s. What is the average power expended? **Ans. 600 W**

*8-58. A speedboat must develop 120 hp in order to move at a constant speed of 15 ft/s through the water. What is the average resistive force due to the water?

Challenge Problems

8-59. A worker lifts a 20-kg bucket from a well at constant speed and does 8 kJ of work. How deep is the well? **Ans. 40.8 m**

8-60. A horizontal force of 200 N pushes an 800-N crate horizontally for a distance of 6 m at constant speed. What work is done by the 200-N force? What is the resultant work?

*8-61. A 10-kg mass is lifted to a height of 20 m and released. What is the total energy of the system? What is the velocity of the mass when it is located 5 m from the floor? **Ans. 1960 J, 17.1 m/s**

8-62. A crate is lifted at a constant speed of 5 m/s by an engine whose output power is 4 kW. What is the mass of the crate?

8-63. A roller coaster boasts a maximum height of 100 ft. What is the maximum speed in miles per hour when it reaches its lowest point? **Ans. 54.4 mi/h**

8-64. A 20-N force drags an 8-kg block a horizontal distance of 40 m by a rope at an angle of 37° with the horizontal. Assume

$\mu_k = 0.2$ and that the time required is 1 min. What resultant work is done?

8-65. What is the velocity of the block in Prob. 8-64 at the end of the trip? What *resultant* power was expended? **Ans. 5.20 m/s, 1.80 W**

8-66. A 70-kg skier slides down a 30-m slope that makes an angle of 28° with the horizontal. Assume that $\mu_k = 0.2$. What is the velocity of the skier at the bottom of the slope?

*8-67. A 0.3-mg flea can jump to a height of about 3 cm. What must be the takeoff speed? Do you really need to know the mass of the flea? **Ans. 76.7 cm/s, no**

*8-68. A roller coaster goes through a low point and barely makes the next hill 15 m higher. What is the minimum speed at the bottom of the loop?

*8-69. The hammer of a pile driver weighs 800 lb and falls a distance of 16 ft before striking the pile. The impact drives the pile 6 in. deeper into the ground. What was the average force driving the pile? **Ans. 25,600 lb**

Critical Thinking Questions

*8-70. An inclined board is used to unload boxes of nails from the back of a truck. The height of the truck bed is 60 cm, and the

board is 1.2 m in length. Assume that $\mu_k = 0.4$ and the boxes are given an initial push to start sliding. What is their speed when they reach the ground below? What initial speed would they need at the bottom in order to slide back into the truck bed? In the absence of friction would these two questions have the same answer?

Ans. 1.90 m/s, 4.46 m/s, yes

*8-71. A 96-lb safe is pushed with negligible friction up a 30° incline for a distance of 12 ft. What is the increase in potential energy? Would the same change in potential energy occur if a 10-lb friction force opposed the motion up the incline? Why? Would the same work be required?

*8-72. A 2-kg ball is suspended from a 3-m cable attached to a spike in the wall. The ball is pulled out, so that the cable makes an angle of 70° with the wall and then released. If 10 J of energy is lost during the collision with the wall, what is the maximum angle between the cable and the wall after the first rebound?

Ans. 59.2°

*8-73. A 3-kg ball dropped from a height of 12 m has a velocity of 10 m/s just before hitting the ground. What is the average retarding force due to the air? If the ball rebounds from the surface with a speed of 8 m/s, what energy was lost on impact? How high will it rebound if the average air resistance is the same as before?

*8-74. Consider a roller coaster where the first hill is 34 m high? If the coaster loses only 8 percent of its energy between the first two hills, what is the maximum height possible for the second hill?

Ans. 31.3 m

*8-75. A 4-kg block is compressed against a spring at the bottom of the inclined plane in Fig. 8-15. A force of 4000 N was required to compress the spring a distance of 6 cm. If it is then released and the coefficient of friction is 0.4, how far up the incline will the block move?

Figure 8-15

Impulse and Momentum

Energy and work are scalar quantities that say absolutely nothing about direction. The law of conservation of energy describes only the relationship between initial and final states; it says nothing about how the energies are distributed. For example, when two objects collide, we can say that the total energy before collision must equal the energy after collision if we neglect friction and other heat losses. But we need a new concept if we are to determine how the total energy is divided between the objects or even their relative directions after impact. The concepts of impulse and momentum presented in this chapter add a vector description to our discussion of energy and motion.

9-1 IMPULSE AND MOMENTUM

When a golf ball is driven from the ground, as in Fig. 9-1, a large average force $\mathbf{F}$ acts on the ball during a short interval of time Δt, causing the ball to accelerate from rest to a final velocity $\mathbf{v}_f$. It is extremely difficult to measure either the force or its duration, but their product $\mathbf{F} \Delta t$ can be determined from the resulting change in velocity of the golf ball. From Newton's second law we have

$$\mathbf{F} = m\mathbf{a} = m\frac{\mathbf{v}_f - \mathbf{v}_0}{\Delta t}$$

Multiplying by Δt gives

$$\mathbf{F} \Delta t = m(\mathbf{v}_f - \mathbf{v}_0)$$

Impulse and Momentum

Figure 9-1 When a golf club strikes the ball, a force **F** acting through the time interval results in a change in its momentum.

*inter***NET**
CONNECTION

www.nhtsa.gov
buphy.pu.edu/~duffy/
mech/1M10_35
Car crashes are a demonstration of impulse and momentum. See these sites for more information.

or

$$\mathbf{F}\,\Delta t = m\mathbf{v}_f - m\mathbf{v}_0 \qquad\qquad 9\text{-}1$$

This equation is so useful in solving problems involving impact that the terms are given special names.

The impulse **F** Δ*t* **is a vector quantity equal in magnitude to the product of the force and the time interval in which it acts. Its direction is the same as that of the force.**

The momentum **p** **of a particle is a vector quantity equal in magnitude to the product of its mass** *m* **and its velocity** **v.**

$$\mathbf{p} = m\mathbf{v}$$

Thus, Eq. (9-1) can be stated verbally:

$$Impulse\ (\mathbf{F}\,\Delta t) = change\ in\ momentum\ (m\mathbf{v}_f - m\mathbf{v}_0)$$

The SI unit of impulse is the *newton-second* (N · s). The unit of momentum is the *kilogram-meter per second* (kg · m/s). It is convenient to distinguish these units, even though they are actually the same:

$$\mathrm{N}\cdot\mathrm{s} = \frac{\mathrm{kg}\cdot\mathrm{m}}{\mathrm{s}^2}\times\mathrm{s} = \mathrm{kg}\cdot\mathrm{m/s}$$

The corresponding units in the USCS are the *pound-second* (lb · s) and *slug-foot per second* (slug · ft/s).

EXAMPLE 9-1

A 3-kg sledgehammer is moving at a velocity of 14 m/s as it strikes a steel spike. It is brought to a stop in 0.02 s. Determine the average force on the spike.

Solution

Since $v_f = 0$, we have from Eq. (9-1),

$$F \Delta t = -mv_0$$

If we consider that the hammer is moving downward, we substitute $v_0 = -14$ m/s, giving

$$F = \frac{-mv_0}{\Delta t} = \frac{-(3 \text{ kg})(-14 \text{ m/s})}{0.02 \text{ s}}$$
$$= 2100 \text{ N}$$

This force, exerted on the hammer, is equal in magnitude but opposite in direction to the force exerted on the spike. It must be emphasized that the force found in this manner is an *average* force. At some instants the force will be much greater than 2100 N.

EXAMPLE 9-2

A 0.2-kg baseball moving toward the batter with a velocity of 30 m/s is struck with a bat that causes it to move in a reverse direction with a velocity of 50 m/s. (Refer to Fig. 9-2.) Find the impulse and the average force exerted on the ball if the bat is in contact with the ball for 0.008 s.

Before impact   Impulse   After impact

$v_0 = -44$ ft/s m mv_f

m $F\Delta t$ 88 ft/s

(a)   (b)   (c)

Figure 9-2   Impact of a bat and a baseball.

Solution

Consider the direction of the final velocity as positive. Applying Eq. (9-1), we solve for the impulse as follows:

$$F \Delta t = mv_f - mv_0 = m(v_f - v_0)$$
$$F \Delta t = (0.2 \text{ kg})[50 \text{ m/s} - (-30 \text{ m/s})]$$
$$= (0.2 \text{ kg})(80 \text{ m/s})$$
$$\text{Impulse} = F \Delta t = 16 \text{ kg} \cdot \text{m/s (or 16 N} \cdot \text{s)}$$

The average force is then found by substituting $t = 0.008$ s.

$$F(0.008 \text{ s}) = 16 \text{ N} \cdot \text{s}$$
$$F = 2000 \text{ N}$$

9-2 THE LAW OF CONSERVATION OF MOMENTUM

Let us consider the *head-on* collision of the masses m_1 and m_2, as shown in Fig. 9-3. Consider the surfaces to be frictionless. We denote their velocities before impact as $\mathbf{u}_1$ and $\mathbf{u}_2$ and after impact as $\mathbf{v}_1$ and $\mathbf{v}_2$. The impulse of the force $\mathbf{F}_1$ acting on the right mass is

$$F_1 \, \Delta t = m_1 v_1 - m_1 u_1$$

Similarly, the impulse of the force $\mathbf{F}_2$ on the left mass is

$$F_2 \, \Delta t = m_2 v_2 - m_2 u_2$$

During the time Δt, $\mathbf{F}_1 = -\mathbf{F}_2$, so that

$$F_1 \, \Delta t = -F_2 \, \Delta t$$

or

$$m_1 v_1 - m_1 u_1 = -(m_2 v_2 - m_2 u_2)$$

and finally, after rearranging,

$$m_1 u_1 + m_2 u_2 = m_1 v_1 + m_2 v_2 \qquad \text{9-2}$$

Total momentum before impact = total momentum after impact

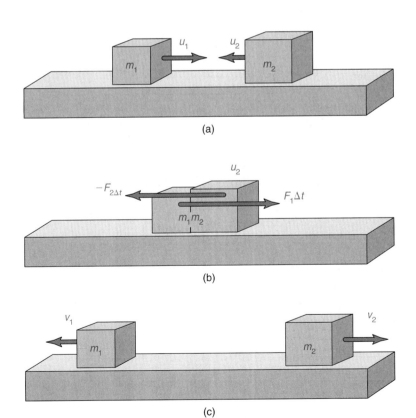

(a)

(b)

(c)

Figure 9-3 (a) Before impact: $m_1 u_1 + m_2 u_2$; (b) during impact: $F_1 \Delta t = -F_2 \, \Delta t$; (c) after impact: $m_1 v_1 + m_2 v_2$.

Thus, we have derived a statement of the law of conservation of momentum:

The total linear momentum of colliding bodies before impact is equal to their total momentum after impact.

EXAMPLE 9-3

Assume that m_1 and m_2 of Fig. 9-3 have masses of 8 and 6 kg, respectively. The mass m_1 moves initially to the right with a velocity of 4 m/s and collides with m_2, moving to the left at 5 m/s. What is the total momentum before and after the collision?

Solution

We choose the direction to the right as positive, taking care to affix the proper sign to the velocities.

$$p_0(\text{before impact}) = m_1 u_1 + m_2 u_2$$
$$p_0 = (8 \text{ kg})(4 \text{ m/s}) + (6 \text{ kg})(5 \text{ m/s})$$
$$= 32 \text{ kg} \cdot \text{m/s} - 30 \text{ kg} \cdot \text{m/s} = 2 \text{ kg m/s}$$

The same total momentum must exist after impact, and so we can write

$$p_f = m_1 v_1 + m_2 v_2 = 2 \text{ kg} \cdot \text{m/s}$$

If either v_1 or v_2 can be measured after the collision, the other can be determined from this relation.

EXAMPLE 9-4

A 1400-kg cannon mounted on wheels fires a 60-kg ball in a horizontal direction with a velocity of 50 m/s as shown in Fig. 9-4. Assuming that the cannon can move freely, what will be its recoil velocity?

Figure 9-4

Solution

First we recognize that momentum must be conserved; that is, the total momentum before the event must be the same as after the event:

$$m_1 u_1 + m_2 u_2 = m_1 v_1 + m_2 v_2$$

But the velocity of both the cannon and the ball is zero initially, so that

$$0 + 0 = m_1v_1 + m_2v_2 \quad \text{and} \quad m_1v_1 = -m_2v_2$$

Solving for the recoil velocity v_2 we obtain

$$v_1 = -\frac{m_2v_2}{m_1} = -\frac{(60 \text{ kg})(50 \text{ m/s})}{(1400 \text{ kg})}$$

Finally, the recoil velocity of the cannon is

$$v_1 = -2.14 \text{ m/s}$$

The negative sign tells us that the direction of the velocity of the cannon is opposite to that of the ball.

An interesting experiment that demonstrates the conservation of momentum can be performed with eight small marbles and a grooved track, as shown in Fig. 9-5. If one marble is released from the left, it will be stopped upon colliding with the others, and one on the right end will roll out with the same velocity. Similarly, when two, three, four, or five marbles are released from the left, the same number will roll out to the right with the same velocity, the others remaining at rest in the center.

You might reasonably ask why two marbles roll off in Fig. 9-5 instead of one with twice the velocity, since this condition would also conserve momentum. For example, if each marble has a mass of 50 g, and if two marbles approach from the left at a velocity of 20 cm/s, the total momentum before impact is 2000 g · cm/s. This same momentum can be achieved after impact if only one marble is left, assuming that it had a velocity of 40 cm/s. The answer lies in the fact that energy must be conserved. If one marble came off with twice the velocity, its kinetic energy would be much greater than that available from the other two. The kinetic energy put into the system would be

$$\begin{aligned} E_0 &= \frac{1}{2}mv^2 = \frac{1}{2}(0.1 \text{ kg})(0.2 \text{ m/s})^2 \\ &= 2 \times 10^{-3} \text{ J} \end{aligned}$$

The kinetic energy of one marble traveling at 40 cm/s is exactly twice this value.

$$\begin{aligned} E_f &= \frac{1}{2}mv^2 = \frac{1}{2}(0.05 \text{ kg})(0.4 \text{ m/s})^2 \\ &= 4 \times 10^{-3} \text{ J} \end{aligned}$$

Therefore, energy as well as momentum is important in describing impact phenomena.

Figure 9-5   Conservation of momentum.

From the experiment in the previous section, you might assume that kinetic energy as well as the momentum is unchanged by a collision. Although this assumption is approximately true for hard bodies like marbles and billiard balls, it is not true for soft bodies, which rebound much more slowly on impact. During impact, all bodies become slightly deformed, and small amounts of heat are liberated. The vigor with which a body restores itself to its original shape after deformation is a measure of its **elasticity,** or restitution.

If the kinetic energy remains constant in a collision (an ideal case), the collision is said to be **completely elastic.** In this instance no energy is lost through heat or deformation in a collision. A hardened steel ball dropped on a marble slab would approximate a completely elastic collision.

If the colliding bodies stick together and move off as a unit afterward, the collision is said to be **completely inelastic.** A bullet that becomes embedded in a wooden block is an example of this type of impact. The majority of collisions fall between these two extremes.

In a completely elastic collision between two masses m_1 and m_2, we can say that both energy and momentum will be unaltered. Hence, we can apply two equations:

$$\text{Energy:} \quad \frac{1}{2}m_1 u_1^2 + \frac{1}{2}m_2 u_2^2 = \frac{1}{2}m_1 v_1^2 + \frac{1}{2}m_2 v_2^2$$
$$\text{Momentum:} \quad m_1 u_1 + m_2 u_2 = m_1 v_1 + m_2 v_2$$

which can be simplified to give

$$m_1(u_1^2 - v_1^2) = m_2(v_2^2 - u_2^2)$$
$$m_1(u_1 - v_1) = m_2(v_2 - u_2)$$

Dividing the first equation by the second gives

$$\frac{u_1^2 - v_1^2}{u_1 - v_1} = \frac{v_2^2 - u_2^2}{v_2 - u_2}$$

Factoring the numerators and dividing out, we obtain

$$u_1 + v_1 = u_2 + v_2$$

or

$$v_1 - v_2 = u_2 - u_1 = -(u_1 - u_2) \qquad \text{9-3}$$

Thus, in the ideal case of a completely elastic collision, the relative velocity after collision, $v_1 - v_2$, is equal to the negative of the relative velocity before collision. The closer these quantities are to being equal, the more elastic the collision. The negative ratio of the relative velocity after collision to the relative velocity before collision provides a means of measuring the elasticity of a collision.

> **The coefficient of restitution e is the negative ratio of the relative velocity after impact to the relative velocity before impact.**

$$e = -\frac{v_1 - v_2}{u_1 - u_2}$$

Incorporating the minus sign into the numerator of this equation yields

$$e = \frac{v_2 - v_1}{u_1 - u_2} \qquad \text{9-4}$$

If the collision is completely elastic, $e = 1$. If the collision is completely inelastic, $e = 0$. In the inelastic case the two bodies move off with the same velocity, so that $v_2 = v_1$. In general, the coefficient of restitution has some value between 0 and 1.

A simple method of determining the coefficient of restitution is shown in Fig. 9-6. A sphere of the material being measured is dropped onto a fixed plate from a height h_1. Its rebound is measured to a height h_2. In this case, the mass of the plate is so large that v_2 is approximately 0. Therefore,

$$e = \frac{v_2 - v_1}{u_1 - u_2} = -\frac{v_1}{u_1}$$

The velocity u_1 is simply the velocity acquired in falling from a height h_1, which is found from

$$u_1^2 - u_0^2 = 2gh_1$$

But the initial velocity $u_0 = 0$, so that

$$u_1^2 = 2gh_1$$

or

$$u_1 = \sqrt{2gh_1}$$

We have considered the downward direction as positive. If the ball rebounds to a height h_2, its rebound velocity v_1 must be $-\sqrt{2gh_2}$. (The minus sign indicates the change in direction.) Hence, the coefficient of restitution is given by

$$e = -\frac{v_1}{u_1} = -\frac{-\sqrt{-2gh_2}}{\sqrt{2gh_1}}$$

or

$$e = \sqrt{\frac{h_2}{h_1}} \qquad \text{9-5}$$

The resulting coefficient is a joint property of the ball and the rebounding surface.

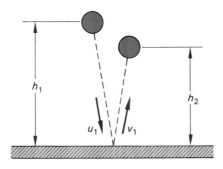

Figure 9-6

For an extremely elastic surface, e has a value of 0.95 or greater (steel or glass), whereas for less resilient substances e may be extremely small. It is interesting to note that the rebound height is a function of the vigor with which the impact deformation is restored. Contrary to popular conception, a steel ball or glass marble will bounce to a greater height than most rubber balls.

Problem-Solving Strategy

Conservation of Momentum: Collisions

1. Read the problem and then draw and label a rough sketch. Indicate the direction of motion for each mass by drawing vectors on the diagram.
2. Choose the x axis along the line of collision and indicate the positive direction. Velocities must be reckoned as positive or negative based on this choice.
3. Make a list of the given masses and velocities, taking care to use the appropriate sign and units for each velocity. The use of appropriate subscripts and letters will help you keep track of the different masses and velocities before and after a collision.
4. Write the conservation of momentum equation:

$$m_1 u_1 + m_2 u_2 = m_1 v_1 + m_2 v_2$$

5. Substitute all given quantities and simplify the resulting expression. When you substitute velocities, it is essential to include the appropriate sign for each velocity.
6. If the collision is completely *inelastic*, you proceed by solving the momentum equation for the unknown quantity.
7. If the collision is elastic, conservation of energy will yield a second independent equation:

$$v_2 - v_1 = e(u_1 - u_2)$$

where e is the *coefficient of restitution*. (For perfectly elastic collisions, $e = 1$.) Then proceed by solving this equation simultaneously with the momentum equation. Take care not to confuse the signs of substitution with the signs of operation.

CONCEPTUAL EXAMPLE 9-5

A 2-kg ball traveling to the left with a speed of 24 m/s collides head on with a 4-kg ball traveling to the right at 16 m/s. (a) Find the resulting velocity if the two balls stick together on impact. (b) Find their final velocities if the coefficient of restitution is 0.80.

Solution (a)

In this instance $v_2 = v_1$ and $e = 0$. Let us denote their final velocity by v. The law of conservation of momentum tells us that

$$m_1 u_1 + m_2 u_2 = m_1 v_1 + m_2 v_2 = (m_1 + m_2)v$$

since $v_1 = v_2 = v$. Choosing the right direction as positive, we have on substitution

$$(2 \text{ kg})(-24 \text{ m/s}) + (4 \text{ kg})(16 \text{ m/s}) = (2 \text{ kg} + 4 \text{ kg})v$$
$$-48 \text{ kg} \cdot \text{m/s} + 64 \text{ kg} \cdot \text{m/s} = (6 \text{ kg})v$$
$$16 \text{ kg} \cdot \text{m/s} = (6 \text{ kg})v$$

from which

$$v = 2.67 \text{ m/s}$$

The fact that this velocity is positive indicates that both bodies move together to the right after collision.

Solution (b)

In this case e is not zero, and the balls rebound after collision with different velocities. Therefore, more information is needed than that derived from the momentum equation alone. We resort to the given value of $e = 0.80$ and Eq. (9-4) to give us more information.

$$e = 0.80 = \frac{v_2 - v_1}{u_1 - u_2}$$

or

$$v_2 - v_1 = (0.80)(u_1 - u_2)$$

Substituting the known values for u_1 and u_2, we obtain

$$v_2 - v_1 = (0.80)(-24 \text{ m/s} - 16 \text{ m/s})$$
$$= (0.80)(-40 \text{ m/s})$$

or finally

$$v_2 - v_1 = -32 \text{ m/s}$$

We can now use the momentum equation to arrive at another relation between v_2 and v_1, allowing us to solve the two equations simultaneously.

$$m_1 u_1 + m_2 u_2 = m_1 v_1 + m_2 v_2$$

The left side of this equation has already been found in part (a) to equal 16 kg · m/s. Therefore, we have on substitution for m_1 and m_2 on the right side

$$16 \text{ kg} \cdot \text{m/s} = (2 \text{ kg})v_1 + (4 \text{ kg})v_2$$

from which

$$2v_1 + 4v_2 = 16 \text{ m/s}$$

or

$$v_1 + 2v_2 = 8 \text{ m/s}$$

Hence, we have two equations:

$$v_2 - v_1 = -32 \text{ m/s} \qquad v_1 + 2v_2 = 8 \text{ m/s}$$

Solving simultaneously, we obtain

$$v_1 = 24 \text{ m/s} \qquad v_2 = -8 \text{ m/s}$$

Thus, after colliding, the balls reverse their directions, m_1 moving to the right with a velocity of 24 m/s and m_2 moving to the left with a velocity of 8 m/s.

USING A CALCULATOR

Using the memory function on your calculator can simplify the evaluation of complex expressions. Note that since the second term is negative, the ± key is used to make the term negative before it is multiplied in.

$$m_1 u_1 + m_2 u_2 = m_1 v_1 + m_2 v_2 = (m_1 + m_2)v$$
$$m_1 u_1 + m_2 u_2 = (m_1 + m_2)v$$

Calculate the individual parts of the equation first, and add them into memory.

2 ⊠ +/- 2 4 = M_{in} − 48.
4 ⊠ 1 6 = 64.
+ M = 16.
◻ 6 = 2.67

EXAMPLE 9-6

A 12-g bullet is fired into a 2-kg block of wood suspended from a cord, as shown in Fig. 9-7. The impact of the bullet causes the block to swing 10 cm above its original level. Compute the velocity of the bullet as it strikes the block.

Solution

We can compute the combined velocity after impact from energy considerations. The kinetic energy of the block and bullet immediately after the collision is converted into potential energy as they rise to a height h. Hence, if v is the initial velocity of the block and bullet, we have

$$\frac{1}{2}(m_1 + m_2)v^2 = (m_1 + m_2)gh$$

Dividing by $m_1 + m_2$, we obtain

$$v^2 = 2gh$$

from which

$$v = \sqrt{2gh}$$

Therefore, the combined velocity after collision is

$$v = \sqrt{(2)(9.8 \text{ m/s}^2)(0.1 \text{ m})} = 1.4 \text{ m/s}$$

The momentum equation then becomes

$$m_1 u_1 + m_2 u_2 = (m_1 + m_2)v$$

or, since $u_2 = 0$.

$$(0.012 \text{ kg})u_1 = (0.012 \text{ kg} + 2 \text{ kg})(1.4 \text{ m/s})$$
$$= (2.012 \text{ kg})(1.4 \text{ m/s})$$
$$0.012u_1 = 2.82 \text{ m/s}$$

which gives an entrance velocity of

$$u_1 = 235 \text{ m/s}$$

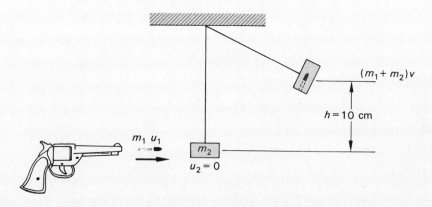

Figure 9-7 Computing the muzzle velocity u_1 from energy and momentum considerations.

Summary and Review

Summary

In this chapter you learned the relationship between impulse and momentum. Physical problems were then introduced to deal with elastic and inelastic collisions. The major concepts are summarized below:

- The *impulse* is the product of the average force **F** and the time interval Δt through which it acts.

$$\text{Impulse} = F\,\Delta t$$
$$\text{SI units: N} \cdot \text{s}$$
$$\text{USCS units: lb} \cdot \text{s}$$

- The *momentum* of a particle is its mass times its velocity.

$$\text{Momentum } p = mv$$
$$\text{SI units: kg} \cdot \text{m/s}$$
$$\text{USCS units: slug} \cdot \text{ft/s}$$

- The impulse is equal to the change in momentum:

$$F\,\Delta t = mv_f - mv_0$$

Note: $\text{N} \cdot \text{s} = \text{kg} \cdot \text{m/s}$ (equivalent units)

- *Conservation of momentum:* The total momentum before impact is equal to the total momentum after impact (see Fig. 9-3).

$$m_1 u_1 + m_2 u_2 = m_1 v_1 + m_2 v_2$$

- The *coefficient of restitution* is found from relative velocities before and after collision or from the rebound height:

$$e = \frac{v_2 - v_1}{u_1 - u_2} \qquad e = \sqrt{\frac{h_2}{h_1}}$$

- For a completely elastic collision, $e = 1$. For a completely inelastic collision, $e = 0$.

Key Terms

1. impulse
2. momentum
3. conservation of momentum
4. elastic impact
5. inelastic impact
6. coefficient of restitution

Review Questions

9-1. Show the equivalence of the units of impulse with the units of momentum in USCS units.

9-2. Discuss the vector nature of impulse and momentum.

9-3. How does the magnitude of the impulse 1 lb · s compare with the magnitude of the impulse 1 n · s?

9-4. Discuss the conservation of energy and momentum for (a) an elastic collision and (b) an inelastic collision.

9-5. If you hold a weapon loosely when firing, it appears to give a greater kick than when you hold it tight against your shoulder. Explain. What effect does the weight of the weapon have?

9-6. A mortar shell explodes in midair. How is momentum conserved? How is energy conserved?

9-7. Suppose you hit a tennis ball into the air with a racket. The ball first strikes the concrete court and then bounds over the fence, landing in the grass. How many impulses were involved, and which impulse was the greatest?

9-8. A father and his daughter stand facing each other on a frozen pond. If the girl pushes her father backward, describe their relative

Summary and Review

motion and velocities. Would they differ if the father pushed the daughter?

9-9. Two small cars of masses m_1 and m_2 are tied together with a compressed spring between them. The cord is burned with a match, releasing the spring and imparting an equal impulse to each car. Compare the ratio of their displacements to the ratio of their masses at some later instant.

Problems

Section 9-1—Impulse and Momentum

9-1. A 0.5-kg wrench is dropped from a height of 10 m. What is its momentum just before it strikes the floor? **Ans. 7 kg · m/s, down**

9-2. Compute the momentum and kinetic energy of a 2400-lb car moving north at 55 mi/h.

9-3. A 2500-kg truck traveling at 40 km/h strikes a brick wall and comes to a stop in 0.2 s. (a) What is the change in momentum? (b) What is the impulse? (c) What is the average force on the wall during the crash? **Ans. (a) -2.78×10^4 kg · m/s, (b) -2.78×10^4 N · s, (c) 1.39×10^5 N**

9-4. What is the momentum of a 3-g bullet moving at 600 m/s in a direction 30° above the horizontal? What are the horizontal and vertical components of this momentum?

9-5. A 0.2-kg baseball traveling to the left at 20 m/s is driven in the opposite direction at 35 m/s when it is hit by a bat. The average force on the ball is 6400 N. How long was it in contact with the bat? **Ans. 1.72 ms**

9-6. A bat exerts an average force of 248 lb on a 0.6-lb ball for 0.01 s. The incoming velocity of the ball was 44 ft/s. If it leaves in the opposite direction what is its velocity?

*9-7. A 500-g ball travels from left to right at 20 m/s. A bat drives the ball in the opposite direction with a velocity of 36 m/s. The time of contact was 0.003 s. What was the average force on the ball? **Ans. 9333 N**

*9-8. A 400-g rubber ball is dropped a vertical distance of 12 m onto the pavement. It is in contact with the pavement for 0.01 s and rebounds to a height of 10 m. What is the total change in momentum? What average force is exerted on the ball?

*9-9. A cue stick strikes an eight ball with an average force of 80 N over a time of 12 ms. If the mass of the ball is 200 g, what will be its velocity? **Ans. 4.80 m/s**

9-10. A golfer hits a 46-g golfball with an initial velocity of 50 m/s at 30°. What are the horizontal and vertical components of the momentum imparted to the ball?

*9-11. The face of the club in Prob. 9-10 is in contact with the ball for 1.5 ms. What are the horizontal and vertical components of the average force on the ball? (*Hint:* Work horizontal and vertical impulse and momentum independently.) **Ans. $F_x = 1330$ N, $F_y = 767$ N**

Section 9-2—Conservation of Momentum

9-12. A spring is tightly compressed between a 6-kg block and a 2-kg block and then tied with a string. When the string breaks, the 2-kg block moves to the right with a velocity of 9 m/s. What is the velocity of the 6-kg block?

9-13. Two masses, one three times that of the other, are compressed against a spring and then tied together on a frictionless surface as shown in Fig. 9-8. The connecting string breaks and sends the smaller mass to the left with a velocity of 10 m/s. What was the velocity of the larger mass? **Ans. 3.33 m/s**

9-14. A 70-kg person standing on a frictionless surface throws a football forward with a velocity of 12 m/s. If the person moves backward at 34 cm/s, what was the mass of the football?

Figure 9-8

9-15. A 20-kg child is at rest in a wagon. The child jumps forward at 2 m/s, sending the wagon backward at 12 m/s. What is the mass of the wagon?   **Ans. 3.33 kg**

9-16. Two children, one weighing 80 lb, and the other 50 lb, are at rest on roller skates. The larger child pushes the smaller one, who moves away at 6 mi/h. What is the velocity of the larger child?

9-17. A 60-g firecracker explodes, sending a 45-g piece to the left and another to the right with a velocity of 40 m/s. What is the velocity of the 45-g piece?   **Ans. −13.3 m/s**

*9-18. A 24-g bullet is fired with a muzzle velocity of 900 m/s from a 5-kg rifle. Find the recoil velocity of the rifle. What is the ratio of the kinetic energy of the bullet to that of the rifle?

*9-19. A 6-kg bowling ball collides head on with 1.8-kg pin. The pin moves forward at 3 m/s and the ball slows to 1.6 m/s. What was the initial velocity of the bowling ball?
   Ans. 2.50 m/s

*9-20. A 60-kg man on a lake of ice catches a 2-kg ball. The ball and man each move at 8 cm/s after the ball is caught. What was the velocity of the ball before it was caught? What energy was lost in the process?

*9-21. A 200-g rock traveling south at 10 m/s strikes a 3-kg block initially at rest. (a) If the two stick together on collision, what will be their common velocity? (b) What energy was lost in the collision?
   Ans. 62.5 cm/s, 9.38 J

Section 9-3—Elastic and Inelastic Impacts

9-22. A car traveling at 8 m/s crashes into a car of identical mass stopped at a traffic light. What is the velocity of the wreckage immediately after the crash, assuming that the cars stick together?

9-23. A 2000-kg truck traveling at 10 m/s crashes into a 1200-kg car initially at rest. What is the common velocity after the collision if they stick together? What is the loss in kinetic energy?   **Ans. 6.25 m/s, 37,500 J**

9-24. A 30-kg child stands on a frictionless surface. The father throws a 0.8-kg football with a velocity of 15 m/s. What velocity will the child have after catching the football?

9-25. A 20-g object traveling to the left at 8 m/s collides head on with a 10-g object traveling to the right at 5 m/s. What is their combined velocity after impact?
   Ans. 3.67 m/s to the left

*9-26. Find the percent loss of energy for the collision in Prob. 9-25.

*9-27. A 2-kg block of clay is tied to the end of a string as shown in Fig. 9-9. A 500-g steel ball embeds itself into the clay, causing both to rise to a height of 20 cm. Find the entrance velocity of the ball.
   Ans. 9.90 m/s

*9-28. In Prob. 9-27, suppose the 500-g ball passes entirely through the clay and emerges with a velocity of 10 m/s. What must be the new entrance velocity if the block is to rise to the same height of 20 cm?

*9-29. A 9-g bullet is embedded into a 2.0-kg ballistic pendulum (see Fig. 8-13). What was the initial velocity of the bullet if the combined masses rise to a height of 9 cm?
   Ans. 297 m/s

*9-30. A billiard ball moving to the left at 30 cm/s collides head on with another ball moving to the right at 20 cm/s. The masses of the balls are identical. If the collision is per-

Summary and Review

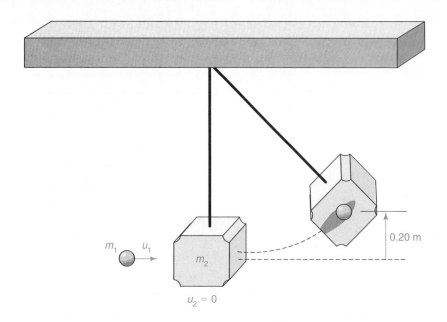

Figure 9-9

fectly elastic, what is the velocity of each ball after impact?

9-31. The coefficient of restitution for steel is 0.90. If a steel ball is dropped from a height of 7 m, how high will it rebound?

Ans. 5.67 m

*9-32. What is the time between the first contact with the surface and the second contact for Prob. 9-31?

*9-33. A ball dropped from rest onto a fixed horizontal plate rebounds to a height that is 81 percent of its original height. What is the coefficient of restitution? What is the required velocity on the first impact to cause the ball to rebound to a height of 8 m?

Ans. 0.9, 13.9 m/s

*9-34. A 300-g block moving north at 50 cm/s collides with a 200-g block moving south at 100 cm/s. If the collision is completely inelastic, what is the common velocity of the blocks after they stick together? What is the loss in energy?

*9-35. Suppose the collision in Prob. 9-34 is perfectly elastic. What are the velocities after impact? **Ans. +70 cm/s, −80 cm/s**

*9-36. A 5-lb and a 12-lb object approach each other with equal velocities of 25 ft/s. What will be their velocities after impact if the collision is (a) completely inelastic or (b) perfectly elastic?

Challenge Problems

9-37. An average force of 4000 N acts on a 400-g piece of metal, causing it to move from rest to a velocity of 30 m/s. What was the time of contact for this force? **Ans. 3.00 ms**

9-38. A 600-g object whose velocity is initially 12 m/s collides with a wall and rebounds with half of its original kinetic energy. What impulse was applied by the wall?

9-39. A 10-kg block at rest on a horizontal surface is struck by a 20-g projectile bullet moving at 200 m/s. The bullet passes entirely through the block and exits with a velocity of 10 m/s. What is the velocity of the block? **Ans. 38.0 cm/s**

9-40. In Prob. 9-39, how much kinetic energy was lost?

9-41. A 60-g body having an initial velocity of 100 cm/s to the right collides with a 150-g body moving to the left at 30 cm/s. The coefficient of restitution is 0.8. What are the velocities after impact? What percent of the energy is lost in collision?
Ans. −67.1 cm/s, +36.9 cm/s; 35.5 percent

9-42. The block in Fig. 9-6 weighs 6 lb. How high will it rise if a 0.4-lb projectile with an initial velocity of 90 ft/s embeds itself into the block? **Ans. 5.93 in**

9-43. A single railroad car moving north at 10 m/s strikes two identical, coupled cars initially moving south at 2 m/s. If all three cars couple after colliding, what is their common velocity? **Ans. 2.00 m/s, north**

*9-44. An atomic particle of mass 2.00×10^{-27} kg moves with a velocity of 4.00×10^6 and collides head on with a particle of mass 1.20×10^{-27} kg initially at rest. Assuming a perfectly elastic collision, what is the velocity of the incident particle after the collision?

*9-45. A bat strikes a 400-g softball moving horizontally to the left at 20 m/s. It leaves the bat with a velocity of 60 m/s at an angle of 30° with the horizontal. What are the horizontal and vertical components of the impulse imparted to the ball?
Ans. 28.8 N · s, 12 N · s

*9-46. If the bat in Prob. 9-45 was in contact with the ball for 5 ms, what was the magnitude of the average force on the softball?

*9-47. Cart A has a mass of 300 g and moves on a frictionless air track at 1.4 m/s when it hits Cart B at rest. The collision is perfectly elastic, and the 300-g cart's velocity is reduced to 0.620 m/s after the collision. What was the mass of the other cart, and what was its velocity after the collision?
Ans. 116 g, 2.02 m/s

*9-48. In Fig. 9-10, assume that the collision of the two masses is completely inelastic. What is the common velocity after the collision, and what is the ratio of the final kinetic energy to the initial kinetic energy?

*9-49. Assume that the collision in Prob. 9-48 is perfectly elastic. What is the velocity of each mass after the collision?
Ans. −5 m/s, 10 m/s

*9-50. A 2-kg mass moves to the right at 2 m/s and collides with a 6-kg mass moving to the left at 4 m/s. If the collision is completely inelastic, what is the common velocity of the two masses after they collide, and how much energy is lost in the collision?

*9-51. In Prob. 9-50, assume that the collision is perfectly elastic. What are the velocities after the collision? **Ans. −1 m/s, −7 m/s**

Figure 9-10

Summary and Review

Critical Thinking Questions

*9-52. An astronaut in orbit outside a capsule uses a revolver to control motion. The astronaut with gear weighs 200 lb on earth. If the revolver fires 0.05-lb bullets at 2700 ft/s, and 10 shots are fired, what is the final velocity of the astronaut? Compare the final kinetic energy of the 10 bullets with that of the astronaut. Why is the difference so great?   **Ans. −6.75 ft/s,**
bullets = 57,000 ft lb, astronaut = 142 ft lb

*9-53. In applying conservation of momentum for colliding objects to find final velocities, could we use the weight of the objects instead of the mass? Why, or why not? Verify your answer by applying it to one of the examples in the text.

*9-54. A 20-g bullet, moving at 200 m/s, strikes a 10-kg wooden block and passes entirely through it, emerging with a velocity of 10 m/s. What is the velocity of the block after impact? How much energy was lost?
Ans. 0.380 m/s, 399 J

*9-55. A 0.30-kg baseball moving horizontally at 40 m/s is struck by a bat. If the ball is in contact with the bat for a time of 5 ms, and it leaves with a velocity of 60 m/s at an angle of 30°, what are the horizontal and vertical components of the average force acting on the bat?

*9-56. When two masses collide, they produce equal but opposite impulses. The masses do not change in the collision, so the change in momentum of one should be the negative of the change for the other. Is this true whether the collision is elastic or inelastic? Verify your answer by using the data in Probs. 9-50 and 9-51.

*9-57. Two toy cars of masses m and $3m$ approach each other, each traveling at 5 m/s. If they couple together, what is their common speed afterward? What are the velocities of each car if the collision is perfectly elastic?

*9-58. An 8-g bullet is fired horizontally at two blocks resting on a frictionless surface. The first block has a mass of 1 kg, and the second has a mass of 2 kg. The bullet passes completely through the first block and lodges into the second. After their collisions, the 1-kg block moves with a velocity of 1 m/s and the 2-kg block moves with 2 m/s. What is the velocity of the bullet before and after emerging from the first block?   **Ans. 627 m/s, 502 m/s**

*9-59. A 1-kg mass A is attached to a support by a cord of length 80 cm and is held horizontally as in Fig. 9-11. After release it swings downward, striking the 2-kg mass B, which is at rest on a frictionless tabletop. Assuming that the collision is perfectly elastic, what are the velocities of each mass immediately after impact?

Figure 9-11

Uniform Circular Motion

Chapter 10

OBJECTIVES

After completing this chapter, you should be able to:

1. Demonstrate by definition and examples your understanding of the concepts of *centripetal acceleration* and *centripetal force.*
2. Apply your knowledge of centripetal force and centripetal acceleration to the solution of problems similar to those in this text.
3. Define and apply the concepts of frequency and period of rotation, and relate them to the linear speed of an object in uniform circular motion.
4. Apply your knowledge of centripetal force to problems involving *banking angles,* the *conical pendulum,* and motion in a *vertical circle.*
5. State and apply the universal law of gravitation.

In previous chapters we considered primarily motion in a straight line. This approach is sufficient to describe and apply most technical concepts. Bodies in the natural world, however, generally move along curved paths. Artillery shells travel along parabolic paths under the influence of the earth's gravitational field. Planets revolve around the sun in paths that are nearly circular. On the atomic level electrons circle about the nucleus of an atom. In fact, it is difficult to imagine any phenomenon in physics that does not involve motion in at least two dimensions.

10-1 MOTION IN A CIRCULAR PATH

Newton's first law tells us that all bodies moving in a straight line with constant speed will maintain their velocity unaltered unless acted on by an external force. The velocity of a body is a vector quantity consisting of both its speed and its direction. Just as a resultant force is required to change its speed, a resultant force must be applied to change its direction. Whenever this force acts in a direction other than the original direction of motion, the path of a moving particle is changed.

The simplest kind of two-dimensional motion occurs when a constant external force always acts at right angles to the path of a moving particle. In this case the resultant force will produce an acceleration that alters only the direction of motion, leaving the speed constant. This simple type of motion is referred to as **uniform circular motion.**

> Uniform circular motion **is motion in which there is no change in speed, only a change in direction.**

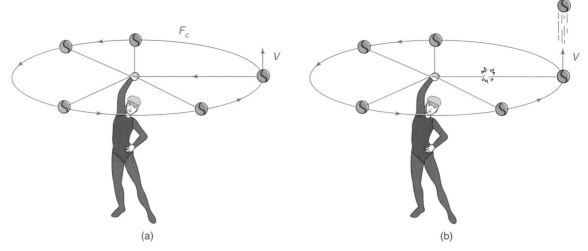

Figure 10-1 (a) The inward pull of the string on the rock causes it to move in a circular path. (b) If the string breaks, the rock will fly off at a tangent to the circle.

An example of uniform circular motion is afforded by swinging a rock in a circular path with a string, as shown in Fig. 10-1. As the rock revolves with constant speed, the inward force of the tension in the string constantly changes the direction of the rock, causing it to move in a circular path. If the string broke, the rock would fly off at a tangent perpendicular to the radius of its circular path.

10-2 CENTRIPETAL ACCELERATION

buphy.bu.edu
For demonstrations of the power of central forces, including a saw using a rotating paper disc that can cut paper, see this Web site (but do not type www.).

Newton's second law of motion states that a resultant force must produce an acceleration in the direction of the force. In uniform circular motion, the acceleration changes the velocity of a moving particle by altering its direction.

The position and velocity of a particle moving in a circular path of radius R are shown at two instants in Fig. 10-2. When the particle is at point A, its velocity is

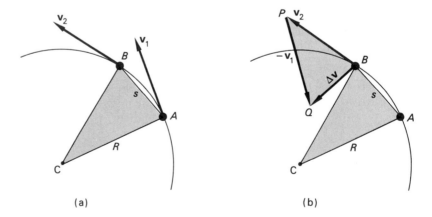

Figure 10-2 (a) A and B are the positions at two instants separated by a time interval Δt. (b) The change in velocity v is represented graphically. The vector will point directly toward the center if Δt is made small enough for the chord s to equal the arc joining points A and B.

represented by the vector $\mathbf{v}_1$. After a time interval Δt, its velocity is represented by the vector $\mathbf{v}_2$. The acceleration, by definition, is the change in velocity per unit time. Thus,

$$\mathbf{a} = \frac{\Delta\mathbf{v}}{\Delta t} = \frac{\mathbf{v}_2 - \mathbf{v}_1}{\Delta t} \qquad \text{10-1}$$

The change in velocity $\Delta\mathbf{v}$ is represented graphically in Fig. 10-2b. The difference between the two vectors $\mathbf{v}_2$ and $\mathbf{v}_1$ is constructed according to the methods introduced in Chap. 2. Since the velocities $\mathbf{v}_2$ and $\mathbf{v}_1$ have the same magnitude, they form the legs of an isosceles triangle BPQ whose base is $\Delta\mathbf{v}$. If we construct a similar triangle ABC, it can be noted that the magnitude of $\Delta\mathbf{v}$ has the same relationship to the magnitude of either velocity as the chord s has to the radius R. This proportionality is stated symbolically as

$$\frac{\Delta v}{v} = \frac{s}{R} \qquad \text{10-2}$$

where v represents the absolute magnitude of either $\mathbf{v}_1$ or $\mathbf{v}_2$.

The distance the particle actually covers in traveling from point A to point B is not the distance s but the length of the arc from A to B. The shorter the time interval Δt, the closer these points are until in the limit the chord length becomes equal to the arc length. In this case, the length s is given by

$$s = v\,\Delta t$$

which, when substituted into Eq. (10-2), yields

$$\frac{\Delta v}{v} = \frac{v\,\Delta t}{R}$$

Since the acceleration from Eq. (10-1) is $\Delta v/\Delta t$, we can rearrange terms and obtain

$$\frac{\Delta v}{\Delta t} = \frac{v^2}{R}$$

Therefore, the rate of change in velocity, or the ***centripetal acceleration,*** is given by

$$a_c = \frac{v^2}{R} \qquad \text{10-3}$$

where v is the linear speed of a particle moving in a circular path of radius R.

The term ***centripetal*** means that the acceleration is always directed toward the center. Notice in Fig. 10-2b that the vector $\Delta\mathbf{v}$ does not point toward the center. This is because we have considered a long time interval between the measurements at A and B. If we restrict the separation of these points to an infinitesimal distance, the vector $\Delta\mathbf{v}$ would point toward the center.

The units of centripetal acceleration are the same as those for linear acceleration. For example, in the SI, v^2/R would have the units

$$\frac{(\text{m/s})^2}{\text{m}} = \frac{\text{m}^2/\text{s}^2}{\text{m}} = \text{m/s}^2$$

EXAMPLE 10-1

A 2-kg body is tied to the end of a cord and whirled in a horizontal circle radius 1.5 m. If the body makes three complete revolutions every second, determine its linear speed and its centripetal acceleration.

Solution

If the body makes 3 rev/s, the time to travel one complete circle (a distance $2\pi R$) is $\frac{1}{3}$ s. Hence, the linear speed is

$$v = \frac{2\pi R}{0.333 \text{ s}} = \frac{2\pi(1.5 \text{ m})}{0.333 \text{ s}}$$

$$= 28.3 \text{ m/s}$$

Therefore, the centripetal acceleration, from Eq. (10-3), is

$$a_c = \frac{v^2}{R} = \frac{(28.3 \text{ m/s})^2}{1.5 \text{ m}} = 534 \text{ m/s}^2$$

The procedure used to compute linear speed in the above example is so useful that it is worth remembering. If we define the **period** as the time for one complete revolution and designate it by the letter T, the linear speed can be computed by dividing the period into the circumference. Thus,

$$v = \frac{2\pi R}{T} \qquad \text{10-4}$$

Another useful parameter in engineering problems is the rotational speed, expressed in *revolutions per minute* (rpm) or *revolutions per second* (rev/s). This quantity is called the *frequency of rotation* and is given by the reciprocal of the period.

$$f = \frac{1}{T} \qquad \text{10-5}$$

The validity of this relation is demonstrated by noting that the reciprocal of seconds per revolution (s/rev) is revolutions per second (rev/s). Substitution of this definition into Eq. (10-4) yields an alternative equation for determining the linear speed.

$$v = 2\pi f R \qquad \text{10-6}$$

For example, if the frequency is 1 rev/s and the radius 1 ft, the linear speed would be 2π ft/s.

10-3 CENTRIPETAL FORCE

The inward force necessary to maintain uniform circular motion is defined as the **centripetal force.** From Newton's second law of motion, the magnitude of this force must equal the product of mass and centripetal acceleration. Thus,

$$F_c = ma_c = \frac{mv^2}{R} \qquad \text{10-7}$$

where m is the mass of an object moving with a velocity v in a circular path of radius R. The units chosen for the quantities F_c, m, v, and R must be consistent for the system chosen. For example, the SI units for mv^2/R are

$$\frac{\text{kg} \cdot \text{m}^2/\text{s}^2}{\text{m}} = \text{kg} \cdot \text{m/s}^2 = \text{N}$$

An inspection of Eq. (10-7) reveals that the inward force $\mathbf{F}_c$ is directly proportional to the square of the velocity of the moving object. This means that increasing the linear speed to twice its original value will require four times the original force. Similar reasoning will show that doubling the mass or halving the radius will require twice the original centripetal force.

For problems in which the rotational speed is expressed in terms of the *frequency*, the centripetal force can be determined from

$$F_c = \frac{mv^2}{R} = 4\pi^2 f^2 mR \qquad \text{10-8}$$

This relation results from substitution of Eq. (10.6), which expresses the linear speed in terms of the frequency of revolution.

EXAMPLE 10-2

A 4-kg ball is swung in a horizontal circle by a cord 2 m long. What is the tension in the cord if the period is 0.5 s?

Solution
The tension in the cord will be equal to the centripetal force necessary to hold the 4-kg body in a circular path. The linear speed is obtained by dividing the period into the circumference.

$$v = \frac{2\pi R}{T} = \frac{2\pi(2 \text{ m})}{0.5 \text{ s}} = 25.1 \text{ m/s}$$

from which the centripetal force is

$$F_c = \frac{mv^2}{R} = \frac{(4 \text{ kg})(25.1 \text{ m/s})^2}{2 \text{ m}}$$

$$= 1260 \text{ N}$$

CONCEPTUAL EXAMPLE 10-3

Two 500-g masses rotate about the center axis at 12 rev/s, as shown in Fig. 10-3. (a) What is the constant force acting on each mass? (b) What is the tension in the supporting rod?

Math Tip

Does the orbital velocity of a satellite depend on its mass? For a satellite, the centripetal force $F_c = mv^2/R$ is equal to the weight mg. Thus,

$$F_c = W$$
$$\frac{mv^2}{R} = mg$$

Multiply both sides by R/m to solve for v^2.

$$\left(\frac{R}{m}\right)\left(\frac{mv^2}{R}\right) = \left(\frac{R}{m}\right)(mg)$$

This yields

$$v^2 = gR$$

Taking the square root of both sides, we obtain

$$v = \sqrt{gR}$$

Thus, a satellite's orbital velocity is determined from gravity g and the radius of its path. The velocity does not depend on the mass.

Theme Park Design Technician How much force keeps theme park goers in their seats on the tilt-o-whirl? Theme park design technicians use uniform circular motion to make rides safe, fun, and thrilling. ⚛

(a)

(b)

Figure 10-3 Objects traveling in a circular path. The resultant force of the rod on the objects provides the necessary centripetal force. According to Newton's third law, the objects exert an equal and opposite reaction force called the *centrifugal force.* These forces do not cancel each other because they act on different objects.

Solution (a)

The total downward force of the weights and rod is balanced up the upward force of the center support. Thus, the resultant force acting on each revolving mass is the pull toward the center equal to the centripetal force. First, we find the velocity of each mass.

$$v = 2\pi fR = 2\pi(12 \text{ rev/s})(0.30 \text{ m})$$
$$v = 22.6 \text{ m/s}$$

Now the centripetal force can be found from Eq. (10-7).

$$F_c = \frac{mv^2}{R} = \frac{(0.500 \text{ kg})(22.6 \text{ m/s})^2}{0.300 \text{ m}}$$

$$F_c = 853 \text{ N, } \textit{toward center}$$

The same calculation holds for the other mass.

Solution (b)

The resultant force on each mass is equal to 853 N directed toward the center. It is exerted *by* the center rod *on* on the mass. The outward force we often *think* acts on the mass is actually the reaction force exerted *by* the mass *on* the rod. The tension in the rod is due to this outward force, and it is equal in magnitude to the centripetal force of 853 N.

10-4 BANKING CURVES

When an automobile is driven around a sharp turn on a perfectly level road, friction between the tires and the road provides a centripetal force (see Fig. 10-4). If this centripetal force is not adequate, the car may slide off the road. The maximum value of the force of friction determines the maximum speed with which a car can negotiate a turn of a given radius.

EXAMPLE 10-4

What is the maximum speed at which an automobile can negotiate a curve of radius 100 m without sliding if the coefficient of static friction is 0.7?

Solution

As the car increases its speed, the force of static friction required to hold it increases. Finally, the car attains a speed so great that the centripetal force equals the maximum force of static friction. At that instant

$$\mathcal{F}_s = F_c = \frac{mv^2}{R}$$

and since $\mathcal{F}_s = \mu_s \mathcal{N}$, we can write

$$\frac{mv^2}{R} = \mu_s \mathcal{N}$$

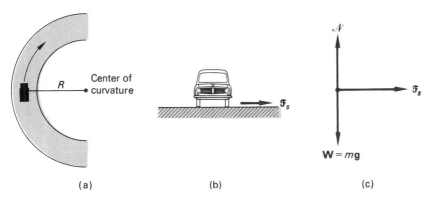

Figure 10-4 The centripetal force of friction. Note that there is no outward force on the car.

Applying the first condition for equilibrium to the vertical forces in Fig. 10-4 reveals that the normal force is equal to the weight of the car

$$\mathcal{N} = W = mg$$

Hence,

$$\frac{mv^2}{R} = \mu_s mg \quad \text{or} \quad v^2 = \mu_s gR$$

from which

$$v = \sqrt{\mu_s gR} \qquad\qquad \text{10-9}$$

Substituting known values for μ_s, g, and R, we can now compute the maximum speed.

$$v = \sqrt{(0.7)(9.8 \text{ m/s}^2)(100 \text{ m})} = 26.2 \text{ m/s}$$

or approximately 94.3 km/h (58.6 mi/h).

It is quite surprising that the weight of the automobile was not considered in determining the maximum speed. Our own experience tends to contradict the independence of weight. The answer to this paradox does not lie in the validity of the above equations but in the fact that the friction force acts on the wheels below the center of gravity of the car. There is also an uneven distribution of weight. Since the center of gravity of heavy cars is usually lower than for light cars, the difference in stability and distribution of weight will affect the turning speed. If we stipulate that the weight is uniformly distributed and neglect problems of vertical stability, however, the above equations are valid.

Now let us consider the effects of banking a turn to eliminate the friction force. As seen from Fig. 10-5, a road can be banked in such a manner that the normal force $\mathcal{N}$ has vertical and horizontal components:

$$\mathcal{N}_x = \mathcal{N} \sin \theta \qquad \mathcal{N}_y = \mathcal{N} \cos \theta$$

The horizontal component provides the necessary centripetal force. Therefore, if we represent the linear velocity by v and the radius of the turn by R, the banking

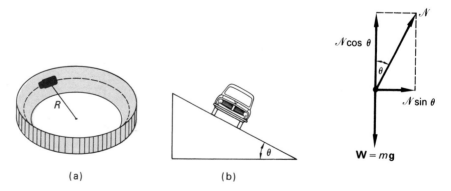

(a)   (b)

Figure 10-5   Effects of banking a turn. The horizontal component of the normal force $\mathcal{N} \sin \theta$ provides the necessary centripetal acceleration.

angle θ required to eliminate the need for friction is obtained from

$$\mathcal{N} \sin \theta = \frac{mv^2}{R}$$

Since the vertical forces are balanced,

$$\mathcal{N} \cos \theta = mg$$

Dividing the first equation by the second yields

$$\tan \theta = \frac{v^2}{Rg} \qquad\qquad \text{10-10}$$

EXAMPLE 10-5

Find the required banking angle for a curve of radius 300 m if the curve is to be negotiated at a speed of 80 km/h without the need of a friction force.

Solution

The **linear speed** is first converted to meters per second.

$$v = 80 \text{ km/h} = 22.2 \text{ m/s}$$

Substituting into Eq. (10-10) yields

$$\tan \theta = \frac{v^2}{Rg} = \frac{(22.2 \text{ m/s})^2}{(300 \text{ m})(9.8 \text{ m/s}^2)} = 0.168$$

from which the required banking angle is

$$\theta = 9.5°$$

10-5 THE CONICAL PENDULUM

A **conical pendulum** consists of a mass m revolving in a horizontal circle with constant speed v at the end of a cord of length L. A comparison of Fig. 10-6 with Fig. 10-5 shows that the formula derived for the banking angle also applies for the angle the cord makes with the vertical in a conical pendulum. In the latter problem, the necessary centripetal force is provided by the horizontal component of the tension in the cord. The vertical component is equal to the weight of the revolving mass. Hence,

$$T \sin \theta = \frac{mv^2}{R} \qquad\qquad T \cos \theta = mg$$

from which

$$\tan \theta = \frac{v^2}{Rg}$$

is obtained as in the previous section.

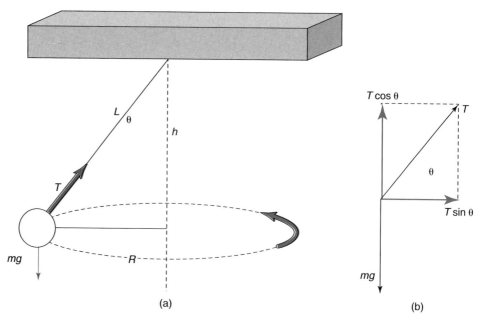

Figure 10-6

A careful study of Eq. (10-10) will show that, as the linear speed increases, the angle that the cord makes with the vertical also increases. Therefore, the vertical position of the mass (indicated in Fig. 10-6) is raised, causing a reduction in the distance h below the point of support. If we wish to express Eq. (10-10) in terms of the vertical position h, we should note that

$$\tan \theta = \frac{R}{h}$$

from which we obtain

$$\frac{R}{h} = \frac{v^2}{gR}$$

Thus, the distance of the weight below the support is a function of the linear speed and is given by

$$h = \frac{gR^2}{v^2}$$ 10-11

This principle includes governors used on some engines (Fig. 10-7). The position of the weight can be used to open or close fuel valves. Note that all the forces do not act at a single point in this case.

A more useful form for Eq. (10-11) can be obtained by expressing the linear speed in terms of the rotational frequency. Since $v = 2\pi fR$, we can write

$$h = \frac{gR^2}{4\pi^2 f^2 R^2} = \frac{g}{4\pi^2 f^2}$$

Solving for f, we obtain

$$f = \frac{1}{2\pi} \sqrt{\frac{g}{h}}$$ 10-12

Drive shaft

Weighted
arms

Link to
throttle valve

Spring

Adjusting
screw

Throttle valve closed by governor action

Figure 10-7

This form has the advantage of eliminating the radius of revolution R from consideration.

10-6 MOTION IN A VERTICAL CIRCLE

Motion in a vertical circle is different from the circular motion discussed in earlier sections. Since gravity always acts downward, the direction of the weight is the same at the top of the path as it is at the bottom. The forces that maintain the circular motion must always be directed, however, toward the center of the path. When more than one force acts on an object, it is the *resultant force* that produces the centripetal force.

Consider a mass m tied to the end of a rope and swung in a circle of radius R as shown by Fig. 10-8 on page 224. We assign v_1 as the velocity at the top of the circular path and v_2 as the velocity at the bottom. Let us first consider the resultant force on the object as it passes through its highest point. The weight mg and the tension T_1 in the rope are each directed downward. The resultant of these forces is the centripetal force. Hence,

$$T_1 + mg = \frac{mv_1^2}{R} \qquad \text{10-13}$$

On the other hand, when the mass passes through its lowest point, the weight mg is still directed downward, but the tension T_2 is directed upward. The resultant is still the necessary centripetal force, so that we have

$$T_2 - mg = \frac{mv_2^2}{R} \qquad \text{10-14}$$

It should be apparent from these equations that the tension in the cord at the bottom is greater than it is at the top. In one case, the weight adds to the tension, whereas in the other, the weight subtracts from the tension. The centripetal force (resultant) is a function of velocity, mass, and radius at any location.

Figure 10-8

Problem-Solving Strategy

Uniform Circular Motion

1. Read the problem and then draw and label a rough sketch.
2. Choose an axis perpendicular to the circular motion at the point where the centripetal force acts on a given mass.
3. Consider the direction of the centripetal force (toward the center) as positive.
4. The resultant force toward the center is the necessary centripetal force. If more than one force acts on the mass, the *net* force toward the center will be equal to mv^2/R. This is actually a statement of

Newton's second law for circulation motion.

$$\sum F = \frac{mv^2}{R} \qquad a_c = \frac{v^2}{R}$$

5. In calculating the resultant central force (ΣF), consider forces toward the center as positive and forces away from the center as negative. The right side of the equation, mv^2/R, will always be positive.
6. Substitute known quantities and solve the unknown factor. Be careful to distinguish between the weight and the mass of an object.

CONCEPTUAL EXAMPLE 10-6

In Fig. 10-8, assume that a 2-kg ball has a velocity of 5 m/s when it rounds the top of a circle whose radius is 80 cm. (a) What is the tension in the cord at that instant? (b) What is the minimum speed at the top necessary to maintain circular motion?

Solution (a)

Solving Eq. (10-13) for T_1, we obtain

$$T_1 = \frac{mv_1^2}{R} - mg$$

$$= \frac{(2\ \text{kg})(5\ \text{m/s})^2}{0.80\ \text{m}} - (2\ \text{kg})(9.8\ \text{m/s}^2)$$

$$= 42.9\ \text{N}$$

You should show that the centripetal force at this point is 62.5 N.

Solution (b)

The critical situation at the top of the path occurs when the tension drops to zero. At this instant, the centripetal force is provided only by the weight *mg*. Thus, Eq. (10-13) reduces to

$$mg = \frac{mv_1^2}{R}$$

Dividing out the mass *m* and solving for v_1, we obtain

$$v_1 = \sqrt{gR} \qquad\qquad \textbf{(10-15)}$$

$$= \sqrt{(9.8\ \text{m/s}^2)(0.8\ \text{m})}$$

$$= 2.8\ \text{m/s}$$

The velocity found from Eq. (10-15) is often called the *critical velocity*.

USING A CALCULATOR

If your answer is wrong, you might look for the source of your error in the units. Were you using similar units, and not a mixture of standard and metric units? Try canceling units on paper. Do the final units of your answer agree with the units you were expecting?

10-7 GRAVITATION

The earth and planets follow approximately circular orbits around the sun. Newton suggested that the inward force that maintains planetary motion is only one example of a universal force called *gravitation,* which acts between all masses in the universe. He stated his thesis in the following *universal law of gravitation:*

Each particle in the universe attracts each other particle with a force that is directly proportional to the product of their masses and inversely proportional to the square of the distance between them.

The proportionality is usually stated in the form of an equation:

$$F = G\frac{m_1 m_2}{r^2} \qquad\qquad \text{10-16}$$

where m_1 and m_2 are the masses of any two particles separated by a distance *r*, as shown in Fig. 10-9.

The proportionality constant *G* is a universal constant equal to

$$G = 6.67 \times 10^{-11}\ \text{N} \cdot \text{m}^2/\text{kg}^2$$

$$= 3.44 \times 10^{-8}\ \text{lb} \cdot \text{ft}^2/\text{slug}^2$$

interNET CONNECTION

www.physlink.com
For a physics resource for students, see this site.

Figure 10-9 The universal law of gravitation.

EXAMPLE 10-7

A 4-kg ball and a 2-kg ball are positioned so that their centers are 40 cm apart. With what force do they attract each other?

Solution

The force of attraction is found from Eq. (10-16):

$$F = G \frac{m_1 m_2}{r^2} = \frac{(6.67 \times 10^{-11} \text{ N} \cdot \text{m}^2/\text{kg}^2)(4 \text{ kg})(2 \text{ kg})}{(0.40 \text{ m})^2}$$

$$F = 3.34 \times 10^{-9} \text{ N}$$

The gravitational force is actually a small force. Because of the relatively large mass of the earth in comparison with objects on its surface, we often assume that gravitational forces are strong. If we consider two marbles closely separated on a horizontal surface, however, our experience certainly substantiates that the gravitational attraction is weak.

EXAMPLE 10-8

At the surface of the earth, the acceleration of gravity is 9.8 m/s². If the radius of the earth is 6.38×10^6 m, compute the mass of the earth.

Solution

Suppose we consider a mass m near the surface of the earth. The gravitational force that attracts this mass is equal to its weight given by

$$W = mg = G \frac{mm_e}{r^2}$$

Solving for m_e, we obtain

$$m_e = \frac{gr^2}{G}$$

from which the mass of the earth is found to be

$$m_e = \frac{(9.8 \text{ m/s}^2)(6.38 \times 10^6 \text{ m})^2}{6.67 \times 10^{-11} \text{ N} \cdot \text{m}^2/\text{kg}^2} = 5.98 \times 10^{24} \text{ kg}$$

10-8 THE GRAVITATIONAL FIELD AND WEIGHT

In earlier chapters, we defined **weight** as the attraction that the earth has on masses located near the surface of the earth. Perhaps it would be wise now to revisit this concept in light of Newton's law of gravitation. The attraction that any large spherical mass (such as the earth) has on another mass located outside the sphere can be calculated by assuming that the entire mass of the large sphere is concentrated at its center. Assume, as in the Example 10-8, that a mass m is located on the surface of the earth whose mass is m_e. Setting the weight mg equal to the gravitational force, we obtain

$$mg = \frac{Gmm_e}{R_e^2}$$

The radius of the earth is denoted by the symbol R_e. Now, if we divide out the mass m, we have the following value for the acceleration due to gravity.

$$g = \frac{Gm_e}{R_e^2} \qquad \text{10-17}$$

In the previous section, we used Eq. (10-16) to find the mass of the earth from the given radius. We can also see from this equation that gravity and, therefore, the weight of an object depend on its location above the earth's surface.

EXAMPLE 10-9

How far above the surface of the earth will a person's weight be reduced to one-half its value at the surface?

Solution
Since the weight mg is proportional to the mass, the weight will be reduced by one-half when $g = \frac{1}{2}(9.8 \text{ m/s}^2)$ or 4.90 m/s^2. Let r be the distance above the earth's surface. Then,

$$g = \frac{Gm_e}{(R_e + r)^2} = 4.90 \text{ m/s}^2$$

from which we obtain

$$(R_e + r) = \sqrt{\frac{Gm_e}{(4.90 \text{ m/s}^2)}}$$

Substituting known values for G and m_e, we find that

$$(R_e + r) = \sqrt{\frac{(6.67 \times 10^{-11} \text{ N} \cdot \text{m}^2/\text{kg}^2)(5.98 \times 10^{24} \text{ kg})}{(4.90 \text{ m/s}^2)}}$$

$$R_e + r = 9.02 \times 10^6 \text{ m}$$

Finally, after substituting $R_e = 6.38 \times 10^6 \text{ m}$, we find that

$$r = 2.64 \times 10^6 \text{ m} \quad \text{or} \quad 2640 \text{ km}$$

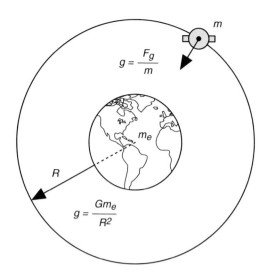

Figure 10-10 The gravitational field above the earth can be represented by the acceleration g that a small mass m would experience if it were placed at that point. The magnitude of the field can be found from the mass m_e of the earth and the distance R from the center of the earth.

If we know the acceleration due to gravity at any location above the earth's surface, we can determine the gravitational force (weight) acting on that object. The direction of this force will be toward the center of the earth. Refer to Fig. 10-10. We will find it useful to define a ***gravitational field*** as the *force per unit of mass* at a given location. The magnitude of this field is simply the acceleration due to gravity:

$$g = \frac{F_g}{m} = \frac{Gm_e}{R^2}$$

10-18

where R is the distance from the center of the earth to the point where gravity is to be determined. It should be noted that the gravitational field is a *property of space* and it exists at a certain point above the earth whether or not a mass is placed at that point. By knowing the gravitational field or the acceleration due to gravity at a point, we can immediately find the weight of a given mass placed at that point.

10-9 SATELLITES IN CIRCULAR ORBITS

An earth satellite is just a projectile that "falls" around the earth. In a thought experiment demonstrated by Fig. 10-11, imagine that you stand on the earth and throw baseballs at increasing speeds. The more velocity you give to the baseball, the longer is the curved path to the ground. Since the earth's surface is curved, one can at least imagine that if the velocity were great enough, the falling ball would simply follow the curved surface around the earth. Of course, there are at least two serious problems with this example. First, the surface of the earth is not uniform and there will definitely be obstructions. Second, because of the large acceleration near the earth's surface, the velocity would have to be exceptionally large. Calculations will show that velocities of the order of 29,000 km/h or 18,000 mi/h

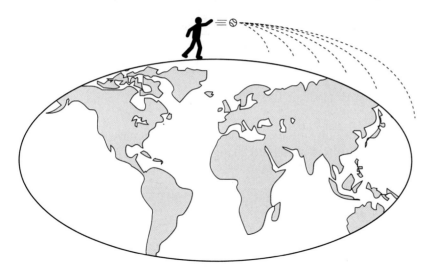

Figure 10-11 A baseball thrown with increasing horizontal velocity would eventually become a satellite as it "fell" around the earth.

Technology Today

Communications in Orbit

Satellites in geostationary earth orbit (GEO) enable fax, videoconferencing, Internet, long-distance fixed phone service, television, and broadband multimedia to be provided in developing areas of the world. Satellites in medium earth orbit (MEO) are used for mobile cell phones, fixed phones, and other personal communication. Satellites in low earth orbit (LEO) are used for handheld mobile phones, paging, fax, ship or truck tracking, fixed ordinary phones, and broadband multimedia.

are needed. The baseball would be burned to a cinder quickly at such speeds due to atmospheric friction. There are many satellites orbiting the earth today, however, at altitudes where resistance and excessive speeds are not a problem. Some are moving in orbits that are nearly circular as they "fall" around the earth. If a space station is in a circular orbit around the earth, neither the space vehicle nor the passengers are "weightless." On the contrary, the gravitational force (weight) provides the centripetal force necessary for circular motion.

Consider for a moment the satellite of mass m that moves around the earth in a circular orbit of radius R as shown in Fig. 10-12. The centripetal force mv^2/R is determined from Newton's law of gravitation:

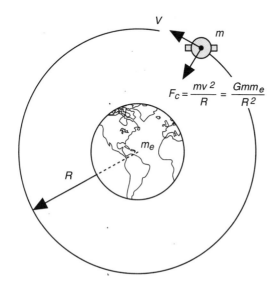

$$F_c = \frac{mv^2}{R} = \frac{Gmm_e}{R^2}$$

Figure 10-12 The centripetal force necessary for circular motion is provided by the gravitational force of attraction. Thus, there is only one speed v that a satellite may have if it is to remain in an orbit of fixed radius.

$$\frac{mv^2}{R} = \frac{Gmm_e}{R^2}$$

Solving for the speed, we have

$$v = \sqrt{\frac{Gm_e}{R}} \qquad \text{10-19}$$

Notice that there is only one speed v that a satellite can have if it is to remain in an orbit of fixed radius R. If the speed changes, then so does the radius of the orbit.

EXAMPLE 10-10

A person with a mass of 100 kg is riding in a space station that is moving in a circular orbit 900 km above the surface of the earth. (a) What is the speed of the space station? (b) What is the weight of the occupant?

Solution (a)

First, we must recognize that the radius R of the orbit is equal to the radius of the earth plus the height above the earth.

$$R = 6.38 \times 10^6 \text{ m} + 0.9 \times 10^6 \text{ m} = 7.28 \times 10^6 \text{ m}$$

Now the speed is found by substitution in Eq. (10-19):

$$v = \sqrt{\frac{Gm_e}{R}} = \sqrt{\frac{(6.67 \times 10^{-11} \text{ N} \cdot \text{m}^2/\text{kg}^2)(5.98 \times 10^{24} \text{ kg})}{7.28 \times 10^6 \text{ m}}}$$

$$= 7400 \text{ m/s } (16{,}600 \text{ mi/h})$$

Solution (b)

The weight of the 100 kg person can be found from Newton's law of gravitation.

$$W = \frac{Gmm_e}{R^2} = \frac{(6.67 \times 10^{-11} \text{ N} \cdot \text{m}^2/\text{kg}^2)(100 \text{ kg})(5.98 \times 10^{24} \text{ kg})}{(7.28 \times 10^6)^2}$$

$$= 753 \text{ N}$$

As an additional exercise, you should verify the same result from mv^2/R. Notice that the occupant is anything but "weightless." The person is merely in a free-fall situation, which gives the appearance of weightlessness since there is no upward or normal force acting to balance the weight.

For many satellites, the period T, or the time required for one orbital revolution of the satellite, is of significant interest. For example, communication satellites often need to circle the earth with a period equal to that of the earth, in other words, one day. Such orbits are said to be **geosynchronous** and the satellites are called *synchronous satellites*. As illustrated in Fig. 10-13, such satellites will remain accessible at an essentially constant latitude, permitting easy, direct-line communication between two points on the earth. Three such satellites are needed to permit direct-line communication between *all* points on the entire earth.

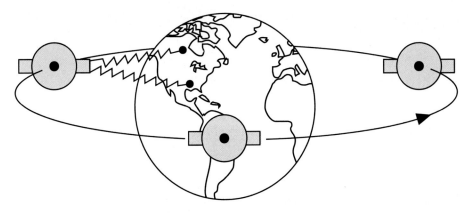

Figure 10-13 Geosynchronous satellites are positioned so that they move around the earth in equatorial orbits with a period that is equal to that of the earth (one day).

The derivation of a relationship between the period of a satellite (or of a planet) and the radius of its orbit can be accomplished by applying the concepts already discussed in this chapter. If we assume a circular orbit, the period T is related to the speed of a satellite by

$$v = \frac{2\pi R}{T}$$

Setting this equal to v in Eq. (10-19), we have

$$\sqrt{\frac{Gm_e}{R}} = \frac{2\pi R}{T}$$

From this we obtain an expression for calculating the period T.

$$T^2 = \left(\frac{4\pi^2}{Gm_e}\right)R^3 \qquad\qquad \textbf{10-20}$$

EXAMPLE 10-11

What must be the altitude of all synchronous satellites that are to be placed in orbit about the earth?

Solution

We recognize that R in Eq. (10-20) refers to the distance to the center of the earth. Thus, after finding the distance R, we will need to subtract the radius of the earth (6.38×10^6 m) if we are to obtain the height above the surface of the earth. The period of a synchronous satellite is equal to one day or 8.64×10^4 s. Solving for R^3 in Eq. (10-20), we obtain

$$R^3 = \frac{Gm_e T^2}{4\pi^2} = \frac{(6.67 \times 10^{-11}\ \text{N} \cdot \text{m}^2/\text{kg}^2)(5.98 \times 10^{24}\ \text{kg})(8.64 \times 10^4\ \text{s})^2}{4\pi^2}$$

$$= 7.54 \times 10^{22}\ \text{m}^3$$

Now we must take the cube root of each side, giving

$$R = 4.23 \times 10^7 \text{ m}$$

Finally, after subtracting the radius of the earth, we find that

$$h = 42.3 \times 10^6 \text{ m} - 6.38 \times 10^6 \text{ m} = 35.8 \times 10^6 \text{ m}$$

The geocentric orbit must be 35,800 km or more than 22,000 miles above the surface of the earth.

10-10 KEPLER'S LAWS

www3.adnc.com/ ~topquark/fun
To see the motion of two bodies affected only by their own gravity, see this site and click on Java.

The movements of the planets and the stars have been studied for thousands of years. As early as the second century A.D., the Greek astronomer Claudius Ptolemy established the theory that the earth was at the center of the universe. Many centuries later, Nicolaus Copernicus (1473–1543) was able to show that the earth and other planets actually moved in circular orbits about the sun.

The Danish astronomer Tycho Brahe (1546–1601) conducted extensive measurements of the motion of the planets over a period of twenty years. He was able to give fairly precise measurements of the planets and more than 700 stars visible to the naked eye. Since the telescope had not yet been invented, Brahe made his measurements using a large sextant and compass. The currently accepted model of the solar system has evolved from these early observations.

The German astronomer Johannes Kepler, who was Brahe's student, took the immense data compiled by Brahe and worked for many years trying to develop a mathematical model to agree with the observed data. Early in his investigation, it became obvious to Kepler that the orbits of the planets could not be circular. In fact, his studies demonstrated that the orbit of the planet Mars was really an ellipse, with the sun at one focus. This conclusion was later generalized to apply to all planets orbiting the sun, and Kepler was able to establish several mathematical statements relative to the solar system. Today these statements are known as Kepler's laws of planetary motion.

> **Kepler's First Law** **All planets move in elliptical orbits with the sun at one focus. This law is sometimes called the *law of orbits*.**

Figure 10-14 shows a planet of mass m_p moving in an elliptical orbit about the sun whose mass is M_s. The semimajor axis is labeled a and the semiminor axis is b. The smallest value of the distance r of the planet from the sun is called the *perihelion* and the largest value is called the *aphelion*. The distance c from the sun to the center of the ellipse must obey the equation: $a^2 = b^2 + c^2$. The ratio c/a is defined as the *eccentricity* of the orbit. With the exception of Mars, Mercury, and Pluto, most of the planetary orbits are nearly circular. They have an eccentricity that is approximately equal to 1 since c is about the same as a.

> **Kepler's Second Law** **A line that connects a planet to the sun sweeps out equal areas in equal times. This law can be called the *law of areas*.**

www.the-cosmos.com
Interested in finding out more about the motion of planets, stars, and galaxies? Visit this site.

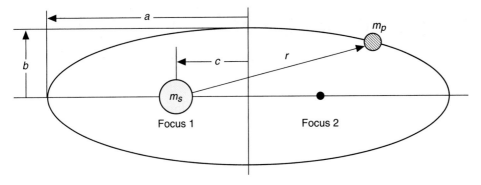

Figure 10-14 Kepler's first law states that all planets move in elliptical orbits with the sun at one focus. The semimajor axis a and the semiminor axis b are indicated.

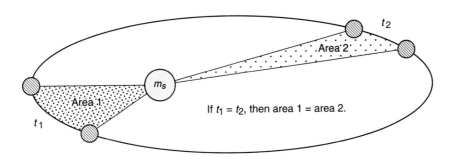

Figure 10-15

The second law is illustrated by Fig. 10-15. It means that the planet must move more slowly when it is at a great distance from the sun and more quickly when it is closer to the sun. Newton was able to show later that this observation as well as those from the other two laws were a consequence of his law of universal gravitation.

> **Kepler's Third Law** The square of the period of any planet is proportional to the cube of the average distance from the planet to the sun. This law is also known as the *law of periods.*

Kepler's third law is indicated clearly by Eq. (10-20), which was derived for a satellite in a circular orbit. It will also be true for ellipses if we replace R (the average distance from the planet to the sun) with a, the semimajor axis of the ellipse. Thus, a more general form for Eq. (10-20) might be written as:

$$T^2 = \frac{4\pi^2 a^3}{Gm_s} \qquad \textbf{10-21}$$

Note that when the path of the planet is circular, $a = R$, and Eq. (10-21) becomes the same as Eq. (10-20).

Math Tip

It's useful to review the algebra used in developing an expression for Kepler's third law. Two expressions determine the centripetal force acting on a satellite: Newton's law of gravitation and Newton's second law:

$$F_g = \frac{Gm_p m_s}{R^2}$$

and

$$F_c = \frac{m_s v^2}{R}$$

The subscripts p and s refer to the planet and its satellite, respectively. To determine the period T of the satellite, we also need to recall the relationship between velocity v and the period.

$$v = \frac{2\pi R}{T}$$

and

$$v^2 = \frac{4\pi^2 R^2}{T^2}$$

Next, substitute the expression for v^2 into the formula for centripetal force.

$$F_c = \frac{m_s v^2}{R} = \frac{m_s}{R}\left(\frac{4\pi^2 R^2}{T^2}\right)$$

or

$$F_c = \frac{4\pi^2 m_s R}{T^2}$$

This force F_c is also equal to F_g, so that we may write

$$\frac{Gm_p m_s}{R^2} = \frac{4\pi^2 m_s R}{T^2}$$

To solve for the period, we can multiply both sides by $(R^2 T^2 / Gm_s m_p)$ and simplify.

$$\left(\frac{\cancel{R^2}T^2}{\cancel{G}\cancel{m_p}\cancel{m_s}}\right)\left(\frac{\cancel{G}\cancel{m_p}\cancel{m_s}}{\cancel{R^2}}\right)$$

$$= \left(\frac{R^2 \cancel{T^2}}{Gm_p \cancel{m_s}}\right)\left(\frac{4\pi^2 \cancel{m_s} R}{\cancel{T^2}}\right)$$

$$T^2 = \frac{4\pi^2 R^3}{Gm_p}$$

This is Kepler's third law, in which a circular orbit is assumed.

Summary and Review

Summary

We have defined uniform circular motion as motion in a circle where the speed is constant and only the direction changes. The change in direction produced by a central force is referred to as *centripetal acceleration*. The major concepts presented in this chapter are as follows:

- The linear speed v of an object in uniform circular motion can be calculated from the period T or frequency f:

$$v = \frac{2\pi R}{T} \qquad v = 2\pi f R$$

- The centripetal acceleration a_c is found from the linear speed, the period, or the frequency as follows:

$$a_c = \frac{v^2}{R} \qquad a_c = \frac{4\pi^2 R}{T^2} \qquad a_c = 4\pi^2 f^2 R$$

- The centripetal force F_c is equal to the product of the mass m and the centripetal acceleration a_c. It is given by

$$F_c = \frac{mv^2}{R} \qquad F_c = 4\pi^2 f^2 mR$$

- Other useful formulas are as follows:

$$v = \sqrt{\mu_s g R} \qquad \text{\textit{Maximum Speed without Slipping}}$$

$$\tan \theta = \frac{v^2}{gR} \qquad \text{\textit{Banking Angle or Conical Pendulum}}$$

$$f = \frac{1}{2\pi} \sqrt{\frac{g}{h}} \qquad \text{\textit{Frequency of a Conical Pendulum}}$$

Key Terms

1. uniform circular motion
2. centripetal acceleration
3. centripetal force
4. gravitational constant
5. universal law of gravitation
6. linear speed
7. period
8. frequency
9. critical speed
10. conical pendulum

Review Questions

10-1. Explain with diagrams why the acceleration of a body moving in a circle at constant speed is directed toward the center.

10-2. A bicyclist leans to the side when negotiating a turn. Why? Describe with a free-body diagram the forces acting on the rider.

10-3. In negotiating a circular turn, a car hits a patch of ice and skids off the road. According to Newton's first law, the car will move forward in a direction tangent to the curve, not outward at right angles to it. Why?

10-4. If the force causing circular motion is directed toward the center of rotation, why is water thrown off clothes during the spin cycle of a washing machine?

10-5. When a ball tied at the end of a string is revolved in a circle at constant speed, the inward centripetal force is equal in magnitude to the outward centrifugal force. Does this represent a condition of equilibrium? Explain.

10-6. What factors contribute to the most desirable banking angles on roadways?

10-7. Does the centripetal force do work in uniform circular motion?

10-8. A motorcyclist negotiates a circular track at constant speed. What exerts the centripetal force, and on what does the force act? What exerts the centrifugal reaction force, and on what does it act?

10-9. A rock at the end of a string moves in a vertical circle. Under what conditions can its linear speed be constant? On what does the centripetal force act? On what does the centrifugal force act?

10-10. What is the value of the gravitational constant G on the moon?

10-11. Given the mass of the earth, its distance from the sun, and its orbital speed, explain how you would determine the mass of the sun.

10-12. What provides the force that keeps a satellite in orbit about the earth? Is the satellite actually weightless in orbit? What happens to the orbit if the satellite increases or decreases its velocity?

10-13. What happens to the gravitational force between two masses when the distance between them is doubled? What happens if each mass is doubled?

10-14. The earth moves more slowly in its orbit during the summer than it does during the winter. According to Kepler's laws, would you say that the earth is closer or farther away from the sun in the winter months?

Problems

Section 10-2—Centripetal Acceleration

10-1. A ball is attached to the end of a 1.5-m string, and it swings in a circle with a constant speed of 8 m/s. What is the centripetal acceleration? **Ans. 42.7 m/s^2**

10-2. What are the period and frequency of rotation for the ball in Prob. 10-1?

10-3. A drive pulley 6-cm in diameter is set to rotate at 9 rev/s. What is the centripetal acceleration of a point on the edge of the pulley? What would be the linear speed of a belt around the pulley?
Ans. 95.9 m/s^2, 1.70 m/s

10-4. An object revolves in a circle of diameter 3 m at a frequency of 6 rev/s. What is the period of revolution, the linear speed, and the centripetal acceleration?

10-5. A car moves around a curve 50 m in radius and undergoes a centripetal acceleration of 2 m/s^2. What is its constant speed?
Ans. 10.0 m/s

10-6. A 1500-kg car travels at a constant speed of 22 m/s along a circular track. If the centripetal acceleration is 6 m/s^2, what is the radius of the track? What is the centripetal force on the car?

10-7. An airplane dives along a curved path of radius R and velocity v. The centripetal acceleration is 20 m/s^2. If both the velocity and the radius are doubled, what will be the new acceleration? **Ans. 40 m/s^2**

Section 10-3—Centripetal Force

10-8. A 20-kg child riding a loop-the-loop at a fair moves at 16 m/s through a track of radius 16 m. What is the resultant force on the child?

10-9. A 3-kg rock, attached to a 2-m cord, swings in a horizontal circle so that it makes one revolution in 0.3 s. What is the centripetal force on the rock? Is there an outward force on the rock? **Ans. 2630 N, no**

10-10. An 8-lb object swings in a horizontal circle with a speed of 95 ft/s. What is the radius of the path if the centripetal force is 2000 lb?

10-11. Two 8-kg masses are attached to the end of a tin rod 400 mm long. The rod is supported in the middle and whirled in a circle. The rod can support a maximum tension of only 800 N. What is the maximum frequency of revolution?
Ans. 3.56 rev/s

10-12. A 500-g damp shirt rotates against the wall of a washing machine at 300 rpm. The diameter of the rotating drum is 70 cm. What are the magnitude and direction of the resultant force on the shirt?

10-13. A 70-kg runner rounds a track of radius 25 m at a speed of 8.8 m/s. What is the central force causing the runner to turn, and what exerts the force?   **Ans. 217 N, friction**

10-14. In Olympic bobsled competition, a team takes a turn of radius 24 ft at a speed of 60 mi/h. What is the acceleration? How many g's do passengers experience?

Section 10-4—Banking Curves

10-15. On a rainy day the coefficient of static friction between tires and the roadway is only 0.4. What is the maximum speed at which a car can negotiate a turn of radius 80 m?   **Ans. 63.8 km/h**

10-16. A bus negotiates a turn of radius 400 ft while traveling at a speed of 60 mi/h. If slipping just begins at this speed, what is the coefficient of static friction between the tires and the road?

10-17. Find the coefficient of static friction necessary to sustain motion at 20 m/s around a turn of radius 84 m.   **Ans. 0.486**

10-18. A 20-kg child sits 3 m from the center of a rotating platform. If $\mu_s = 0.4$, what is the maximum number of revolutions per minute that can be achieved before the child slips?

10-19. A platform rotates freely at 100 rev/m. If the coefficient of static friction is 0.5, how far from the center of the platform can a bolt be placed without slipping?   **Ans. 21.1 cm**

10-20. Find the required banking angle for the car to negotiate the curve described in Prob. 10-15 without slipping.

10-21. Find the required banking angle to prevent the bus in Prob. 10-16 from slipping.   **Ans. 31.2°**

10-22. The optimum banking angle for a curve of radius 20 m is found to be 28°. For what speed was this angle designed?

10-23. A curve in a road 9 m wide has a radius of 96 m. How much higher than the inside edge should the outside edge be so that an automobile can travel at the optimum speed of 40 km/h?   **Ans. 1.17 m**

Section 10-5—The Conical Pendulum

10-24. A conical pendulum swings in a horizontal circle of radius 30 cm. What angle does the supporting cord make with the vertical when the linear speed of the mass is 12 m/s?

10-25. What is the linear speed of the flyweights in Fig. 10-16 if $L = 20$ cm and $\theta = 60°$? What is the frequency of revolution?   **Ans. 1.71 m/s, 1.58 rev/s**

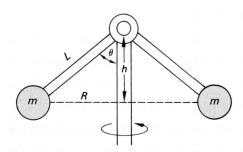

Figure 10-16

10-26. If the length of L in Fig. 10-16 is 60 cm, what velocity is required to cause the flyweights to move to an angle of 30° with the vertical?

10-27. Each of the flyweights in Fig. 10-16 has a mass of 2 kg. Length $L = 40$ cm, and the shaft rotates at 80 rev/min. What is the tension in each arm? What is the angle θ? What is the height h?   **Ans. 56.1 N, 69.6°, 14 cm**

10-28. In Fig. 10-16, assume that $L = 6$ in., each flyweight is 1.5 lb, and the shaft is rotating at 100 rev/min. What is the tension in each arm? What is the angle θ? What is the distance h?

10-29. Consider the rotating swings in Fig. 10-17. Length $L = 10$ m, and the distance $a = 3$ m. What must be the linear velocity of the seat if the rope is to make an angle of 30° with the vertical?   **Ans. 6.73 m/s**

Figure 10-17

10-30. What must be the frequency of revolution for the swing in Fig. 10-17 if the angle θ is to be equal to 25°?

Section 10-6—Motion in a Vertical Circle

10-31. A rock rests on the bottom of a bucket moving in a vertical circle of radius 70 cm. What is the least speed the bucket must have as it rounds the top of a circle if the rock is to remain in the bucket?

Ans. 6.86 m/s

10-32. A 1.2-kg rock is tied to the end of a 90-cm length of string. The rock is then whirled in a vertical circle at a constant speed. What is the critical velocity at the top of the path if the string is not to become slack?

10-33. Assume that the rock of Prob. 10-32 moves in a vertical circle at a constant speed of 8 m/s. What are the tensions in the rope at the top and bottom of the circle?

Ans. 73.6 N, 97.1 N

*10-34. A test pilot in Fig. 10-18 goes into a dive at 620 ft/s and pulls out in a curve of radius 2800 ft. If the pilot weighs 160 lb, what acceleration will be experienced at the lowest point? What is the force exerted by the seat on the pilot?

Figure 10-18 Forces exerted on an airplane at the upper and lower limits of a vertical loop.

Summary and Review

10-35. If the pilot in Prob. 10-34 is not to experience an acceleration greater than seven times gravity (7 g), what is the maximum velocity for pulling out of a dive of radius 1 km? **Ans. 943 km/h**

10-36. A 3-kg ball swings in a vertical circle at the end of an 8-m cord. When it reaches the top of its path, its velocity is 16 m/s. What is the tension in the cord? What is the critical speed at the top?

10-37. A 36-kg girl rides on the seat of a swing attached to two chains that are each 20 m long. If she is released from a position 8 m below the top of the swing, what force does the swing exert on the girl as she passes the lowest point? **Ans. 776 N**

Section 10-7—Gravitation

10-38. How far apart should a 2-ton weight be from a 3-ton weight if their mutual force of attraction is equal to 0.0004 lb?

10-39. A 4-kg mass is separated from a 2-kg mass by a distance of 8 cm. Compute the gravitational force of attraction between the two masses. **Ans. 8.34×10^{-8} N**

10-40. A 3-kg mass is located 10 cm away from a 6-kg mass. What is the resultant gravitational force on a 2-kg mass located at the midpoint of a line joining the first two masses?

10-41. On a distant planet, the acceleration due to gravity is 5.00 m/s² and the radius of the planet is roughly 4560 km. Use the law of gravitation to estimate the mass of this planet. **Ans. 1.56×10^{24} kg**

*10-42. The mass of the earth is about 81 times the mass of the moon. If the radius of the earth is four times that of the moon, what is the acceleration due to gravity on the moon?

*10-43. A 60-kg mass and a 20-kg mass are separated by 10 m. At what point on a line joining these charges will another mass experience zero resultant force? **Ans. 6.34 m from the 60-kg mass**

Section 10-10—Kepler's Laws

10-44. What speed must a satellite have if it is to move in a circular orbit of 800 km above the surface of the earth?

10-45. The mass of Jupiter is 1.90×10^{27} kg, and its radius is 7.15×10^{7} m. What speed must a spacecraft have in order to circle Jupiter at a height of 6.00×10^{7} m above the surface of Jupiter? **Ans. 31,400 m/s (about 69,800 mi/h)**

10-46. What is the orbital speed of a satellite that moves in an orbit 1200 km above the earth's surface?

10-47. The radius of the moon is 1.74×10^{6} m, and the acceleration due to its gravity is 1.63 m/s². Apply the law of universal gravitation to find the mass of the moon. **Ans. 7.40×10^{22} kg**

10-48. A satellite is located at a distance of 900 km above the earth's surface. What is the period of the satellite's motion?

10-49. How far above the earth's surface must a satellite be located if it is to circle the earth in a time of 28 h? **Ans. 4.04×10^{7} m**

Challenge Problems

10-50. At what frequency should a 6-lb ball be revolved in a radius of 3 ft to produce a centripetal acceleration of 12 ft/s²? What is the tension in the cord?

10-51. What centripetal acceleration is required to move a 2.6-kg mass in a horizontal circle

of radius 300 mm if its linear speed is 15 m/s? What is the centripetal force? **Ans. 750 m/s², 1950 N**

10-52. What must be the speed of a satellite located 1000 mi above the earth's surface if it is to travel in a circular path?

10-53. A 2-kg ball swings in a vertical circle at the end of a cord 2 m in length. What must the critical velocity be at the top if the orbit is to remain circular?   **Ans. 4.43 m/s**

10-54. A 4-kg rock swings at a constant speed of 10 m/s in a vertical circle at the end of a 1.4-m cord. What are the tensions in the cord at the top and bottom of the circular path?

10-55. What frequency of revolution is required to raise the flyweights in Fig. 10-16 a vertical distance of 25 mm above their lowest position? Assume that $L = 150$ mm.

Ans. 84.6 rev/min

10-56. The combined mass of a motorcycle and driver is 210 kg. If the driver is to negoti-

ate a loop-the-loop of radius 6 m, what is the critical speed at the top?

10-57. If the speed at the top of the loop in Prob. 10-56 is 12 m/s, what is the normal force at the top of the loop?   **Ans. 2980 N**

10-58. The speed limit at a certain turn of radius 200 ft is 45 mi/h. What is the optimum banking angle for this situation? Are roads actually constructed at the optimum angles?

10-59. For the conical pendulum shown in Fig. 10-17, assume that $a = 2$ m and $L = 4$ m. What linear speed is required to cause the swing to move out to an angle of 20°?

Ans. 3.47 m/s

Critical Thinking Questions

10-60. A coin rests on a rotating platform distance of 12 cm from the center of rotation. If the coefficient of static friction is 0.6, what is the maximum frequency of rotation so that the coin does not slip? Suppose the frequency is cut in half. Now, how far from the center can the coin be placed?

Ans. 1.11 rev/s, 48 cm

*10-61. The laboratory apparatus shown in Fig. 10-19 allows a rotating mass to stretch a spring so that the supporting cord is vertical at a particular frequency of rotation. Assume that the mass of the bob is 400 g and the radius of revolution is 14 cm. With a stopwatch, the time for 50 revolutions is found to be 35 s. What are the magnitude and direction of the force acting on the bob?

Figure 10-19

Summary and Review

10-62. In Prob. 10-59, assume that a 100-g mass is added to the 400-g bob. The force required to stretch the spring should be the same as before, but the rotating mass has increased. What changes when the experiment is performed again so that the centripetal force is the same as before? On what does the centripetal force act in this experiment?

*10-63. A 10-in. diameter platform turns at 78 rev/min. A bug rests on the platform 1 in. from the outside edge. If the bug weighs 0.02 lb, what force acts on it? What exerts this force? Where should the bug crawl in order to reduce this force by one-half?

10-64. The diameter of Jupiter is 11 times that of the earth, and its mass is about 320 times that of the earth. What is the acceleration due to gravity near the surface of Jupiter?
Ans. 25.9 m/s^2

10-65. Assume that $L = 50$ cm and $m = 2$ kg in Fig. 10-16. How many revolutions per second are needed to make the angle $\theta = 30°$? What is the tension in the supporting rod at that point? **Ans. 0.757 rev/s, 22.6 N**

*10-66. A 9-kg block rests on the bed of a truck as it turns a curve of radius 86 m. Assume that $\mu_k = 0.3$ and that $\mu_s = 0.4$. Does the friction force on the block act toward the center of the turn or away from it? What is the maximum speed with which the truck can make the turn without slipping? If the truck makes the turn at a much greater speed, what would be the resultant force on the block?

Rotation of Rigid Bodies

OBJECTIVES

After completing this chapter, you should be able to:

1. Define *angular displacement, angular velocity,* and *angular acceleration,* and apply these concepts to the solution of physical problems.
2. Draw analogies relating rotational-motion parameters (θ, ω, α) to linear-motion parameters, and solve angular acceleration problems in a manner similar to that learned in Chap. 6 for linear acceleration problems (refer to Table 11-1).
3. Write and apply the relationships between linear speed or acceleration and angular speed or acceleration.
4. Define the *moment of inertia* of a body and describe how this quantity and the angular speed can be used to calculate *rotational kinetic energy.*
5. Apply the concepts of *Newton's second law, rotational work, rotational power,* and *angular momentum* to the solution of physical problems.

We have been considering only ***translational motion,*** in which an object's position is changing along a straight line. But it is possible for an object to move in a curved path or to undergo rotational motion. For example, wheels, shafts, pulleys, gyroscopes, and many other mechanical devices rotate about their axes without translational motion. The generation and transmission of power are nearly always dependent on rotational motion of some kind. It is essential for you to be able to predict and control such motion. The concepts and formulas presented in this chapter are designed to provide you with these essential skills.

11-1 ANGULAR DISPLACEMENT

The ***angular displacement*** of a body describes the amount of rotation. If point A on the rotating disk in Fig. 11-1 rotates on its axis to point B, the angular displacement is denoted by the angle θ. There are several ways of measuring this angle. We are already familiar with the units of degrees and revolutions, which are related according to the definition

$$1 \text{ rev} = 360°$$

Neither of these units is useful in describing rotation of rigid bodies. A more applicable measure of angular displacement is the ***radian*** (rad). An angle of 1 rad is a central angle whose arc s is equal in length to the radius R. (See Fig. 11-2.) More generally, the radian is defined by the following equation:

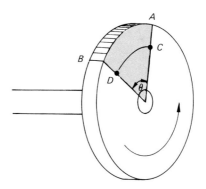

Figure 11-1 Angular displacement θ is indicated by the shaded portion of the disk. The angular displacement is the same from C to D as it is from A to B for a rigid body.

www.colorado.edu/ physics/2000/

If you want to learn more about the rotation of rigid bodies, the Physics 2000 Web site is very helpful.

$$\theta = \frac{s}{R} \qquad\qquad 11\text{-}1$$

where s is the arc of a circle described by the angle θ. Since the ratio of s to R is the ratio of two distances, the radian is a unitless quantity.

The conversion factor that relates radians to degrees is found by considering an arc length s equal to the circumference of a circle $2\pi R$. Such an angle in radians is given from Eq. (11-1) by

$$\theta = \frac{2\pi \cancel{R}}{\cancel{R}} = 2\pi \text{ rad}$$

Hence,

$$1 \text{ rev} = 360° = 2\pi \text{ rad}$$

from which we note that

$$1 \text{ rad} = \frac{360°}{2\pi} = 57.3°$$

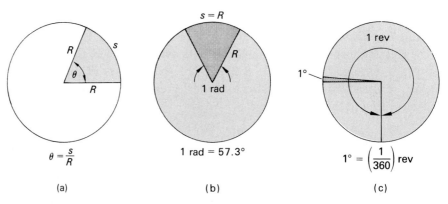

Figure 11-2 The measure of angular displacement and a comparison of units.

EXAMPLE 11-1

If the arc length s is 6 ft and the radius is 10 ft, compute the angular displacement θ in radians, degrees, and revolutions.

Solution

Substituting directly into Eq. (11-1), we obtain

$$\theta = \frac{s}{R} = \frac{6 \text{ ft}}{10 \text{ ft}} = 0.6 \text{ rad}$$

Converting to degrees yields

$$\theta = (0.6 \text{ rad}) \frac{57.3°}{1 \text{ rad}} = 34.4°$$

and since 1 rev = 360°

$$\theta = (34.4°) \frac{1 \text{ rev}}{360°} = 0.0956 \text{ rev}$$

EXAMPLE 11-2

A point on the edge of a rotating disk of radius 8 m moves through an angle of 37°. Compute the length of the arc described by the point.

Solution

Since Eq. (11-1) was defined for an angle measured in radians, we must first convert 37° to radian units.

$$\theta = (37°) \frac{1 \text{ rad}}{57.3°} = 0.646 \text{ rad}$$

The arc length is given by

$$s = R\theta = 8 \text{ m} (0.646 \text{ rad}) = 5.17 \text{ m}$$

The unit radian is dropped because it represents a ratio of length to length (m/m = 1).

11-2 ANGULAR VELOCITY

The time rate of change in angular displacement is called the ***angular velocity***. Thus, if an object rotates through an angle θ in a time t, its average angular velocity is given by

$$\overline{\omega} = \frac{\theta}{t} \qquad\qquad \text{11-2}$$

The symbol ω, the Greek letter *omega*, is used to denote rotational velocity. Although angular velocity may be expressed in *revolutions per minute* or *revolutions per second,* in most physical problems it is necessary to use *radians per second* to conform with convenient formulas. Since the rate of rotation in many technical problems is given in terms of the frequency of revolutions, the following relation will be useful:

$$\omega = 2\pi f \qquad\qquad 11\text{-}3$$

where ω is measured in *radians per second* and f is measured in *revolutions per second.*

EXAMPLE 11-3

A bicycle wheel with a diameter of 66 cm makes 40 revolutions in 1 min. (a) What is its angular velocity? (b) What linear distance will the wheel travel?

Solution (a)
The angular velocity depends only on the frequency of rotation. Since 1 rev $= 2\pi$ radians,

$$f = \left(\frac{40 \text{ rev}}{\text{min}}\right)\left(\frac{1 \text{ min}}{60 \text{ s}}\right) = 0.667 \text{ rev/s}$$

Substitution into Eq. (11-3) gives the angular velocity.

$$\omega = (2\pi \text{ rad})(0.667 \text{ rev/s}) = 4.19 \text{ rad/s}$$

Solution (b)
The linear displacement s can be found from the angular displacement θ in radians.

$$\theta = \left(\frac{2\pi \text{ rad}}{1 \text{ rev}}\right)(40 \text{ rev}) = 251 \text{ rad}$$

Solving Eq. (11-1) for s, we obtain

$$s = \theta R = (251 \text{ rad})(0.33 \text{ m}) = 82.8 \text{ m}$$

It is important to realize that the angular velocity discussed in this section represents an *average* velocity. The same distinction must be made between the average and the instantaneous angular velocity as that discussed in Chap. 6 for average and instantaneous linear velocities.

11-3 ANGULAR ACCELERATION

Like linear motion, rotational motion may be uniform or accelerated. The rate of rotation may increase or decrease under the influence of a resultant torque. For example, if the angular velocity changes from an initial value ω_0 to a final value ω_f in a time t, the angular acceleration is given by

$$\alpha = \frac{\omega_f - \omega_0}{t}$$

The Greek letter α (*alpha*) denotes angular acceleration. A more useful form for this equation is

$$\omega_f = \omega_0 + \alpha t \qquad \text{11-4}$$

A comparison of Eq. (11-4) with Eq. (5-4) for linear acceleration will show that their forms are identical if we draw analogies between angular and linear parameters.

Now that the concept of initial and final angular velocities has been introduced, we can express the average angular velocity in terms of its initial and final values:

$$\overline{\omega} = \frac{\omega_f + \omega_0}{2}$$

Substituting this equality for $\overline{\omega}$ in Eq. (11-2) yields a more useful expression for the angular displacement:

$$\theta = \overline{\omega}t = \frac{\omega_f + \omega_0}{2}\, t \qquad \text{11-5}$$

This equation is also similar to an equation derived for linear motion. In fact, the equations for angular acceleration have the same basic form as those derived in Chap. 6 for linear acceleration if we draw the following analogies:

$$s \text{ (m)} \leftrightarrow \theta \text{ (rad)}$$
$$v \text{ (m/s)} \leftrightarrow \omega \text{ (rad/s)}$$
$$a \text{ (m/s}^2) \leftrightarrow \alpha \text{ (rad/s}^2)$$

Time, of course, is the same for both types of motion and is measured in seconds. Table 11-1 illustrates the similarities between rotational and linear motion.

In applying these formulas, we must be careful to choose the appropriate units for each quantity. It is also important to choose a direction (clockwise or counterclockwise) as positive and to follow through consistently in affixing the appropriate sign to each quantity.

TABLE 11-1 Comparison of Linear Acceleration and Angular Acceleration Formulas

Constant Linear Acceleration	Constant Angular Acceleration
(1) $s = \dfrac{v_o + v_v}{2}t$	$\theta = \dfrac{\omega_f + \omega_o}{2}t$
(2) $v_f = v_o + at$	$\omega_f = \omega_o + \alpha t$
(3) $s = v_o t + \frac{1}{2}at^2$	$\theta = \omega_o t + \frac{1}{2}\alpha t^2$
(4) $s = v_f t - \frac{1}{2}at^2$	$\theta = \omega_f t - \frac{1}{2}\alpha t^2$
(5) $2as = v_f^2 - v_o^2$	$2\alpha\theta = \omega_f^2 - \omega_o^2$

EXAMPLE 11-4

A flywheel increases its rate of rotation from 6 to 12 rev/s in 8 s. What is its angular acceleration?

Solution

We will first compute the initial and final angular velocities:

$$\omega_0 = 2\pi f_0 = \left(\frac{2\pi \text{ rad}}{\text{rev}}\right)\left(\frac{6 \text{ rev}}{\text{s}}\right) = 12\pi \text{ rad/s}$$

$$\omega_f = 2\pi f_0 = \left(\frac{2\pi \text{ rad}}{\text{rev}}\right)\left(\frac{12 \text{ rev}}{\text{s}}\right) = 24\pi \text{ rad/s}$$

$$\alpha = \frac{\omega_f - \omega_0}{t} = \frac{(24\pi - 12\pi) \text{ rad/s}}{8 \text{ s}}$$

$$= 1.5\pi \text{ rad/s}^2 = 4.71 \text{ rad/s}^2$$

EXAMPLE 11-5

A grinding disk rotating initially with an angular velocity of 6 rad/s receives a constant acceleration of 2 rad/s^2. (a) What angular displacement will it describe in 3 s? (b) How many revolutions will it make? (c) What is its final angular velocity?

Solution (a)

The angular displacement is given by

$$\theta = \omega_0 t + \frac{1}{2}\alpha t^2$$

$$= (6 \text{ rad/s})(3 \text{ s}) + \frac{1}{2}(2 \text{ rad/s}^2)(3 \text{ s})^2$$

$$= 18 \text{ rad} + (1 \text{ rad/s}^2)(9 \text{ s}^2)$$
$$= 27 \text{ rad}$$

Solution (b)

Since 1 rev = 2π rad, we obtain

$$\theta = (27 \text{ rad})\frac{1 \text{ rev}}{2\pi \text{ rad}} = 4.30 \text{ rev}$$

Solution (c)

The final angular velocity is equal to the initial angular velocity plus the change in speed. Thus,

$$\omega_f = \omega_0 + \alpha t = 6 \text{ rad/s} + (2 \text{ rad/s}^2)(3 \text{ s})$$
$$= 12 \text{ rad/s}$$

11-4 RELATIONSHIP BETWEEN ROTATIONAL AND LINEAR MOTION

The **axis of rotation** of a rigid rotating body can be defined as that line of particles that remains stationary during rotation. This may be a line through the body, as with a spinning top, or it may be a line through space, as with a rolling hoop. In any case, our experience tells us that the farther a particle is from the axis of rotation, the greater its linear velocity. This fact was expressed in Chap. 10 by the formula

$$v = 2\pi f R$$

where f is the frequency of rotation. We now derive a similar relation in terms of angular velocity. The rotating particle in Fig. 11-3 turns through an arc s, which is given by

$$s = \theta R$$

from Eq. (11-1). If this distance is traversed in a time t, the linear velocity of the particle is given by

$$v = \frac{s}{t} = \frac{\theta R}{t}$$

Since $\theta / t = \omega$, the linear velocity can be expressed as a function of the angular velocity.

$$v = \omega R \qquad \textbf{11-6}$$

This result also follows from Eq. (11-3), in which the angular velocity is expressed as a function of the frequency of revolution.

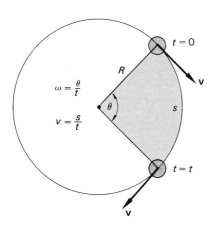

Figure 11-3 The relationship between angular velocity and linear velocity.

11-4   Relationship between Rotational and Linear Motion   **247**

Math Tip

When formulas have squared terms, answers often end up in radicals. Some reminders on adding and subtracting radicals will help you clean up your answers. If two radicals have the same index and the same radicand, they can be added algebraically. Sometimes, when we have a quantity inside a radical, that quantity (called the *radicand*) can be simplified or reduced. For example,

$$4\sqrt{3} + \sqrt{27}$$

can be simplified. First ask whether the radicands are prime numbers. Three is a prime number; that is, it can be divided only by itself and one. But 27 is not a prime number.

$$27 = 3^3 = 3 \times 3 \times 3$$
$$= 3 \times 3^2$$

Putting the number back in the radical, the 3^2 pops out as follows:

$$4\sqrt{3} + \sqrt{27} = 4\sqrt{3}$$
$$+ \sqrt{3 \times 3^2} = 4\sqrt{3} + 3\sqrt{3}$$

And since the index (2) and the radicand (3) are the same for both radical expressions, we can simply add the two terms algebraically to get:

$$4\sqrt{3} + 3\sqrt{3}$$
$$= (4 + 3)\sqrt{3} = 7\sqrt{3}$$

Reducing is a good idea because it is easier to determine the value of $7\sqrt{3}$ than it is to find the size of $\sqrt{3} + \sqrt{27}$.

EXAMPLE 11-6

A drive shaft has an angular velocity of 60 rad/s. At what distance should flyweights be positioned from the axis if they are to have a linear velocity of 12 m/s?

Solution

Solving for R in Eq. (11-6), we obtain

$$R = \frac{v}{\omega} = \frac{12 \text{ m/s}}{60 \text{ rad/s}} = 0.200 \text{ m}$$

Let us now return to a particle moving in a circle of radius R and assume that the linear velocity changes from some initial value v_0 to a final value v_f in a time t. The *tangential acceleration* a_T of such a particle is given by

$$a_T = \frac{v_f - v_0}{t}$$

Because of the close relationship between linear velocity and angular velocity, as represented by Eq. (11-6), we can also express the tangential acceleration in terms of a change in angular velocity.

$$a_T = \frac{\omega_f R - \omega_0 R}{t} = \frac{\omega_f - \omega_0}{t} R$$

or

$$a_T = \alpha R \qquad \text{11-7}$$

where α represents the **angular acceleration.**

We must be careful to distinguish between the tangential acceleration, as defined in Eq. (11-7), and the centripetal acceleration defined by

$$a_c = \frac{v^2}{R} \qquad \text{11-8}$$

The **tangential acceleration** represents a change in linear velocity, whereas the centripetal acceleration represents only a change in the direction of motion. The distinction is shown graphically in Fig. 11-4. The resultant acceleration can be found by computing the vector sum of the tangential and centripetal accelerations.

EXAMPLE 11-7

Find the resultant acceleration of a particle moving in a circle of radius 0.5 m at the instant when its angular velocity is 3 rad/s and its angular acceleration is 4 rad/s².

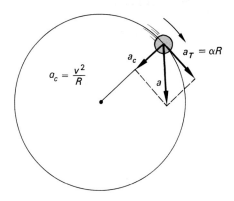

Figure 11-4   The relationship between tangential and centripetal acceleration.

Solution

The tangential acceleration from Eq. (11-7) is

$$a_T = \alpha R = (4 \text{ rad/s}^2)(0.5 \text{ m}) = 2 \text{ m/s}^2$$

Since $v = \omega R$, the centripetal acceleration is given by

$$a_c = \frac{v^2}{R} = \frac{\omega^2 R^2}{R} = \omega^2 R$$

from which we obtain

$$a_c = (3 \text{ rad/s})^2(0.5 \text{ m}) = 4.5 \text{ m/s}^2$$

Finally, the magnitude of the resultant acceleration is obtained from Pythagoras's theorem.

$$a = \sqrt{a_T^2 + a_c^2} = \sqrt{2^2 + 4.5^2}$$
$$= 4.92 \text{ m/s}^2$$

The direction of the resultant acceleration can be found from its components in the usual manner.

11-5 ROTATIONAL KINETIC ENERGY: MOMENT OF INERTIA

We have seen that a particle moving in a circle of radius R has a linear speed given by

$$v = \omega R$$

If the particle has a mass m, it will have a kinetic energy given by

$$E_k = \tfrac{1}{2}mv^2 = \tfrac{1}{2}m\omega^2 R^2$$

A rigid body like that in Fig. 11-5 can be considered as consisting of many particles of varying masses located at different distances from the axis of rotation O. The total kinetic energy of such an extended body will be the sum of the kinetic energies of each particle making up the body. Hence,

$$E_k = \sum \tfrac{1}{2}m\omega^2 r^2$$

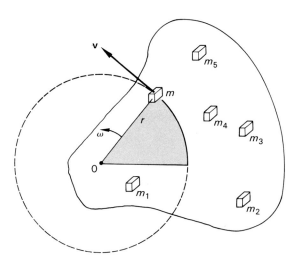

Figure 11-5 Rotation of an extended body. The body can be thought of as many individual masses all rotating with the same angular velocity.

Since the constant ½ and the angular velocity ω are the same for all particles, we can rearrange to obtain

$$E_k = \tfrac{1}{2}\left(\sum mr^2\right)\omega^2$$

The quantity in parentheses, $\sum mr^2$, has the same value for a given body regardless of its state of motion. We define this quantity as the **moment of inertia** and represent it by I:

$$I = m_1r_1^2 + m_2r_2^2 + m_3r_3^2 + \cdots$$

or

$$I = \sum mr^2 \qquad\qquad \textbf{11-9}$$

The SI unit for I is the *kilogram-square meter* and the USCS unit is the *slug-square foot.*

Using this definition, we can express the rotational kinetic energy of a body in terms of its moment of inertia and its angular velocity:

$$E_k = \tfrac{1}{2}I\omega^2 \qquad\qquad \textbf{11-10}$$

Note the similarity between the terms m for linear motion and I for rotational motion.

CONCEPTUAL EXAMPLE 11-8

Find the moment of inertia for the system illustrated in Fig. 11-6. The rods joining the masses are of negligible weight, and the system rotates with an angular velocity of 6 rad/s. What is the rotational kinetic energy? (Consider the masses to be point masses.)

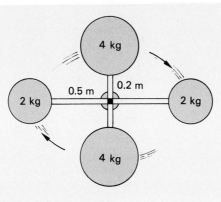

Figure 11-6 Computing the moment of inertia.

Solution
The moment of inertia from Eq. (11-9) is

$$I = m_1r_1^2 + m_2r_2^2 + m_3r_3^2 + m_4r_4^2$$
$$= (2 \text{ kg})(0.5 \text{ m})^2 + (4 \text{ kg})(0.2 \text{ m})^2 + (2 \text{ kg})(0.5 \text{ m})^2 + (4 \text{ kg})(0.2 \text{ m})^2$$
$$= (0.50 + 0.16 + 0.50 + 0.16)\text{kg} \cdot \text{m}^2$$
$$= 1.32 \text{ kg} \cdot \text{m}^2$$

The **rotational kinetic energy** is given by

$$E_k = \tfrac{1}{2}I\omega^2 = \tfrac{1}{2}(1.32 \text{ kg} \cdot \text{m}^2)(6 \text{ rad/s})^2$$
$$= 23.8 \text{ J}$$

For bodies that are not composed of separate masses but are actually continuous distributions of matter, the calculation of moments of inertia is more difficult and usually involves calculus. A few simple cases are given in Fig. 11-7, along with the formulas for computing their moments of inertia.

Sometimes it is desirable to express the rotational inertia of a body in terms of its **radius of gyration k.** This quantity is defined as the radial distance from the center of rotation to a circumference at which the total mass of the body might be concentrated without changing its moment of inertia. According to this definition, the moment of inertia can be found from the formula,

$$I = mk^2 \qquad\qquad \textbf{11-11}$$

where m represents the total mass of the rotating body and k is its radius of gyration.

11-6 THE SECOND LAW OF MOTION IN ROTATION

Suppose we analyze the motion of the rotating rigid body in Fig. 11-8. Consider a force **F** acting on the small mass m, indicated by a shaded portion of the object, at a distance r from the axis of rotation.

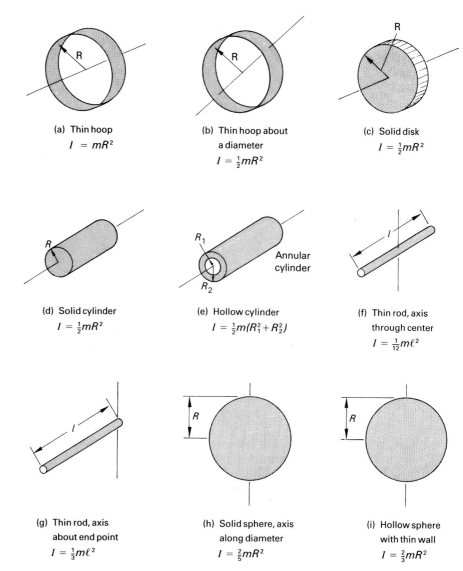

Figure 11-7   Moments of inertia for bodies about their indicated axes.

The force **F** applied perpendicular to r causes the body to rotate with a tangential acceleration:

$$a_T = \alpha r$$

where α is the angular acceleration. From Newton's second law of motion,

$$F = ma_T = m\alpha r$$

Multiplying both sides of this relation by r yields

$$Fr = (mr^2)\alpha$$

The quantity Fr is recognized as the torque produced by the force **F** about the axis. Thus, for the mass m we write

$$\tau = (mr^2)\alpha$$

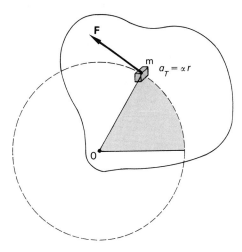

Figure 11-8 Newton's second law for rotational motion states the relationship between torque Fr and angular acceleration α.

A similar equation may be derived for all other portions of the rotating object. The angular acceleration will be constant for every portion, however, regardless of its mass or distance from the axis. Therefore, the resultant torque on the whole body is

$$\tau = \left(\sum mr^2\right)\alpha$$

or

$$\tau = I\alpha \qquad \qquad \text{11-12}$$

Torque = moment of inertia $\times$ angular acceleration

Note the similarity of Eq. (11-12) with the second law for linear motion, $F = ma$. Newton's *law for rotational motion* is as follows:

A resultant torque applied to a rigid body will always result in an angular acceleration that is directly proportional to the applied torque and inversely proportional to the body's moment of inertia.

In applying Eq. (11-12), it is important to recall that the torque produced by a force is equal to the product of its distance from the axis and the perpendicular component of the force. It must also be remembered that the angular acceleration is expressed in radians per second per second.

EXAMPLE 11-9

A circular grinding disk of radius 0.6 m and mass 90 kg is rotating at 460 rev/min. What frictional force, applied tangent to the edge, will cause the disk to stop in 20 s?

Solution

We first calculate the moment of inertia of the disk from the formula given in Fig. 11-7.

$$I = \tfrac{1}{2}mR^2 = \tfrac{1}{2}(90 \text{ kg})(0.6 \text{ m})^2$$
$$= 16.2 \text{ kg} \cdot \text{m}^2$$

Converting the rotational velocity to radians per second, we obtain

$$\omega = \left(\frac{2\pi \text{ rad}}{\text{rev}}\right)\left(\frac{460 \text{ rev}}{\text{min}}\right)\left(\frac{1 \text{ min}}{60 \text{ s}}\right)$$
$$= 48.2 \text{ rad/s}$$

Thus, the angular acceleration is

$$\alpha = \frac{\omega_f - \omega_0}{t}$$

$$= \frac{0 - (48.2 \text{ rad/s})}{20 \text{ s}} = -2.41 \text{ rad/s}^2$$

Applying Newton's second law gives

$$\tau = Fr = I\alpha$$

from which

$$F = \frac{I\alpha}{r} = \frac{(16.2 \text{ kg} \cdot \text{m}^2)(-2.41 \text{ rad/s}^2)}{0.6 \text{ m}}$$

$$= -65.0 \text{ N}$$

The negative sign appears because the force must be directed opposite to the direction of rotation of the disk.

11-7 ROTATIONAL WORK AND POWER

www.iop.org
For a Web site with links to interesting physics topics, see the home page of the Institute of Physics.

Work was defined in Chap. 8 as the product of a displacement and the component of the force in the direction of the displacement. We now consider the work done in rotational displacement under the influence of a resultant torque.

Consider a force **F** acting at the edge of a pulley of radius r, as shown in Fig. 11-9. The effect of such a force is to rotate the pulley through an angle θ while the point at which the force is applied moves through a distance s. The arc distance s is related to θ by

$$s = r\theta$$

Hence, the work of the force **F** is by definition

$$\text{Work} = Fs = Fr\theta$$

but Fr is the torque due to the force, so that we obtain

$$\text{Work} = \tau\theta \qquad \text{11-13}$$

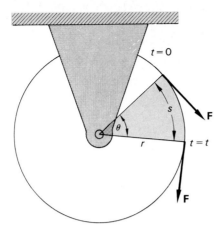

Figure 11-9 Work and power in rotation.

The angle θ must be expressed in radians for either system of units so that the work can be expressed in foot-pounds or joules.

Mechanical energy is usually transmitted in the form of ***rotational work.*** When we speak of the power output of engines, we are concerned with the rate at which rotational work is done. Thus, rotational power can be determined by dividing both sides of Eq. (11-13) by the time t required for the torque τ to effect a displacement θ:

$$\text{Power} = \frac{\text{work}}{t} = \frac{\tau\theta}{t} \qquad\qquad \textbf{11-14}$$

Since θ/t represents the average angular velocity $\bar{\omega}$, we write

$$\text{Power} = \tau\bar{\omega} \qquad\qquad \textbf{11-15}$$

Notice the similarity of this relation with its analog, $P = Fv$, derived earlier for linear motion. Both are measures of *average* power.

EXAMPLE 11-10

A wheel of radius 60 cm has a moment of inertia of 5 kg · m². A constant force of 60 N is applied at the rim. (a) Assuming it starts from rest, how much work is done in 4 s? (b) What power is developed?

Solution (a)

The work is a product of torque and angular displacement. We first calculate the applied torque:

$$\tau = Fr = (60 \text{ N})(0.6 \text{ m}) = 36 \text{ N} \cdot \text{m}$$

Next, we find α from Newton's second law.

$$\alpha = \frac{\tau}{I} = \frac{36 \text{ N} \cdot \text{m}}{5 \text{ kg} \cdot \text{m}^2} = 7.2 \text{ rad/s}^2$$

Now the angular displacement θ can be found.

$$\theta = \omega_0 t + \tfrac{1}{2}\alpha t^2$$
$$= (0) + \tfrac{1}{2}(7.2 \text{ m/s}^2)(4 \text{ s})^2 = 57.6 \text{ rad}$$

The work is, therefore,

$$\text{Work} = \tau\theta = (36 \text{ N} \cdot \text{m})(57.6 \text{ rad})$$
$$= 2070 \text{ J}$$

Solution (b)

The average power is

$$P = \frac{\text{work}}{t} = \frac{2070 \text{ J}}{4 \text{ s}} = 518 \text{ W}$$

The same result could be found by calculating the average angular velocity $\overline{\omega}$ and using Eq. (11-15). As an added example, you might show that the work done is equal to the change in rotational kinetic energy.

11-8 ANGULAR MOMENTUM

Consider a particle of mass m moving in a circle of radius r, as shown in Fig. 11-10a. If its linear velocity is v, it will have a linear momentum $p = mv$. With reference to the fixed axis of rotation, we define the *angular momentum L* of the particle as the product of its linear momentum and the perpendicular distance from the axis to the revolving particle.

$$L \text{ (particle)} = mvr \qquad\qquad \textbf{11-16}$$

Now let us consider the definition of angular momentum as it applies to an extended rigid body. Figure 11-10b describes such a body, which is rotating about its axis O. Each particle in the body has an angular momentum given by Eq. (11-16). Substituting $v = \omega r$, each particle has an angular momentum given by

$$mvr = m\omega r^2 = (mr^2)\omega$$

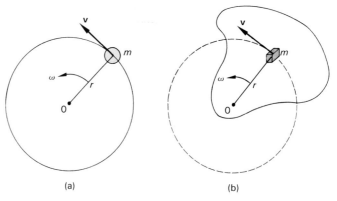

(a) (b)

Figure 11-10 Defining angular momentum.

Since the body is rigid, all particles within the body have the same angular velocity, and the total angular momentum of the body is

$$L = \left(\sum mr^2 \right) \omega$$

Thus, the total angular momentum is equal to the product of a body's angular velocity and its moment of inertia:

$$L = I\omega \qquad\qquad \textbf{11-17}$$

EXAMPLE 11-11

A thin uniform rod is 1 m long and has a mass of 6 kg. If the rod is pivoted at its center and set into rotation with an angular velocity of 16 rad/s, compute its angular momentum.

Solution

The moment of inertia of a thin rod is, from Fig. 11-7,

$$I = \frac{ml^2}{12} = \frac{(6 \text{ kg})(1 \text{ m})^2}{12} = 0.5 \text{ kg} \cdot \text{m}^2$$

Hence, its angular momentum is

$$L = I\omega = (0.5 \text{ kg} \cdot \text{m}^2)(16 \text{ rad/s})$$
$$= 8 \text{ kg} \cdot \text{m}^2/\text{s}$$

Notice that the SI unit of angular momentum is kg · m²/s. The USCS unit is slug · ft²/s.

Problem-Solving Strategy

Rotation of Rigid Bodies

1. It is helpful to realize that problems involving the rotation of a rigid body are similar to those you have worked before for constant linear acceleration. Review the analogies given in the summary at the end of this chapter.

2. When converting from linear to angular motion or vice versa, you should recall the following relationships:

$$s = \theta R \qquad v = \omega R \qquad a = \alpha R$$

When these relations are applied, the angular measures must be in radians (rad).

3. Uniform angular acceleration problems are approached in the same way as they were for linear acceleration in Chap. 6. (Refer to Table 11-1.) You need find only three of the given quantities and select the appropriate equation that contains one unknown factor and not the other. Be careful to use consistent units for displacement, velocity, and acceleration.

4. Applications of Newton's second law, work, energy, power, and momentum are also solved in the same manner as for previous chapters. Just recall that, for rotation, we use the rotational inertia instead of linear mass and we use the angular measures for displacement, velocity, and acceleration.

We can understand the definition of angular momentum better if we return to the fundamental equation for angular motion, $\tau = I\alpha$. Recalling the defining equation for angular acceleration,

$$\alpha = \frac{\omega_f - \omega_0}{t}$$

we can write Newton's second law as

$$\tau = I\frac{\omega_f - \omega_0}{t}$$

Multiplying by t, we obtain

$$\tau t = I\omega_f - I\omega_0 \qquad \text{11-18}$$
$$\textit{Angular impulse} = \textit{change in angular momentum}$$

The product τt is defined as the *angular impulse*. Note the similarity between this equation and the one derived in Chap. 9 for linear impulse.

If no external torque is applied to a rotating body, we can set $\tau = 0$ in Eq. (11-18), yielding

$$0 = I\omega_f - I\omega_0$$
$$I\omega_f = I\omega_0 \qquad \text{11-19}$$
$$\textit{Final angular momentum} = \textit{initial angular momentum}$$

Thus, we are led to a statement of the *conservation of angular momentum*:

If the sum of the external torques acting on a body or system of bodies is zero, the angular momentum remains unchanged.

This statement holds true even if the rotating body is not rigid but is altered so that its moment of inertia changes. In this case, the angular speed also changes so that the product $I\omega$ is always constant. Skaters, divers, and acrobats all control the rate at which their bodies rotate by extending or retracting their limbs to decrease or increase their angular speed.

An interesting experiment illustrating the conservation of angular momentum is shown in Fig. 11-11. A woman stands on a rotating platform with heavy weights in each hand. She is first set into rotation with her arms fully extended. By drawing her hands closer to her body, she decreases her moment of inertia. Since her angular momentum cannot change, she will notice a considerable increase in her angular speed. By extending her arms, she will be able to decrease her angular speed.

CONCEPTUAL EXAMPLE 11-12

Assume that the woman holding the extended weights in Fig. 11-11 has a rotational inertia of 6 kg · m^2 and that the rotational inertia is decreased to 2 kg · m^2 when she brings them close to her body. With the weights in their extended position, she is set into rotation at 1.4 rev/s. What will be her rate of rotation when she brings the weights close to her body?

Low angular velocity   High angular velocity

Figure 11-11 Experiment to demonstrate the conservation of angular momentum. The woman controls her rate of rotation by moving the weights inward to increase rotational speed or outward to decrease rotational speed.

Solution

In the absence of external torque, the system cannot change its rotational momentum. Therefore, the final angular momentum is equal to the initial angular momentum.

$$I_f\omega_f = I_o\omega_o$$
$$(2 \text{ kg} \cdot \text{m}^2)\omega_f = (6 \text{ kg} \cdot \text{m}^2)(1.4 \text{ rev/s})$$
$$\omega_f = 4.2 \text{ rev/s}$$

Note that if we are happy with rev/s as the unit for rotational speed, it is not necessary to convert to radians per second. Basically, we see that decreasing the rotational inertia to one-third causes the rate of rotation to triple in order to conserve momentum.

Summary and Review

Summary

In this chapter, we extended the concept of circular motion to include the rotation of a rigid body composed of many particles. We found that many problems can be solved by methods discussed earlier for linear motion. The essential concepts are now summarized.

- The similarities between rotational and linear motion:

Rotational	θ	ω	α	I	$I\omega$	τ	$I\alpha$	$\tau\theta$	$\frac{1}{2}I\omega^2$	$\tau\omega$
Linear	s	v	a	m	mv	F	ma	Fs	$\frac{1}{2}mv^2$	Fv

- The angle in radians is the ratio of the arc distance s to the radius R of the arc. Symbolically we write:

$$\theta = \frac{s}{R} \qquad s = \theta R$$

The radian is a unitless ratio of two lengths.

- Angular velocity, which is the rate of angular displacement, can be calculated from θ or from the frequency of rotation:

$$\overline{\omega} = \frac{\theta}{t} \qquad \overline{\omega} = 2\pi f \qquad \begin{array}{l}\textit{Average}\\ \textit{Angular Velocity}\end{array}$$

- Angular acceleration is the time rate of change in angular speed:

$$\alpha = \frac{\omega_f - \omega_0}{t} \qquad \textit{Angular Acceleration}$$

- By comparing θ to s, ω to v, and α to a, the following equations can be utilized for angular acceleration problems:

$$\theta = \frac{\omega_f + \omega_0}{2}t$$

$$\omega_f = \omega_0 + \alpha t$$

$$\theta = \omega_0 t + \frac{1}{2}\alpha t^2$$

$$2\alpha\theta = \omega_f^2 - \omega_0^2$$

When any three of the five parameters θ, α, t, ω_f, and ω_0 are given, the other two can be found from one of these equations. Choose a direction of rotation as being positive throughout your calculations.

- The following equations are useful when comparing linear motion with rotational motion:

$$v = \omega R \qquad a_T = \alpha R$$

- Other useful relationships:

$I = \sum mR^2$	*Moment of Inertia*
$I = mk^2$	*Radius of Gyration*
Work $= \tau\theta$	*Work*
$L = I\omega$	*Angular Momentum*
$E_k = \frac{1}{2}I\omega^2$	*Rotational Kinetic Energy*
$\tau = I\alpha$	*Newton's Law*
$P = \tau\omega$	*Power*
$I\omega_f = I\omega_0$	*Conservation of Angular Momentum*

Key Terms

1. radius of gyration
2. radian
3. angular velocity
4. angular momentum
5. moment of inertia
6. tangential acceleration
7. angular displacement
8. angular acceleration
9. rotational kinetic energy
10. rotational work

Review Questions

11-1. Make a list of the SI and USCS units for angular velocity, angular acceleration, moment of inertia, torque, and rotational kinetic energy.

11-2. State the angular analogies for the following translational equations:

a. $\omega_f = v_0 + at$
b. $s = v_0 t + \frac{1}{2}at^2$
c. $F = ma$
d. $E_k = \frac{1}{2}mv^2$
e. Work $= Fs$
f. Power $=$ work$/t = Fv$

11-3. A sphere, a cylinder, a disk and a hollow ring all have identical masses and are rotating with constant angular velocity about a common axis. Compare their individual rotational kinetic energies, assuming their outside diameters are equal.

11-4. Explain how a diver controls her motion so that she can strike the water feet first or head first.

11-5. A cat held feet up and dropped from any elevation above a few feet will always land feet first. How does it accomplish this?

11-6. When energy is supplied to a body so that both translation and rotation result, its total kinetic energy is given by

$$E_k = \frac{1}{2}mv^2 + \frac{1}{2}I\omega^2$$

How the energy is divided between rotational and translational effects is determined by the distribution of mass (moment of inertia). From these statements, which of the following objects will roll to the bottom of an inclined plane first?

a. A solid disk of mass M
b. A circular hoop of mass M

11-7. Refer to Question 11-6. If a solid sphere, a solid disk, a solid cylinder, and a hollow cylinder are released simultaneously from the top of an inclined plane, in what order will they reach the bottom?

11-8. A disk whose moment of inertia is I_1 and whose angular velocity is ω_1 is meshed with a disk whose moment of inertia is I_2 and whose angular velocity is ω_2. Write the conservation equation by denoting their combined angular velocity by ω.

11-9. Refer to Question 11-8. Suppose $\omega_1 = \omega_2$ and $I_1 = 2I_2$. How would their combined velocities compare with their initial velocity? Suppose $\omega_1 = 3\omega_2$ and $I_1 = I_2$?

Problems

Section 11-3—Angular Acceleration

11-1. A cable is wrapped around a drum 80 cm in diameter. How many revolutions of this drum will cause an object attached to the cable to move a linear distance of 2 m? What is the angular displacement?

Ans. 0.796 rev, 5 rad

11-2. A bicycle wheel is 26 in. in diameter. If the wheel makes 60 revolutions, what linear distance will it travel?

11-3. A point on the edge of a large wheel of radius 3 m moves through an angle of 37°. Find the length of the arc described by the point. **Ans. 1.94 m**

11-4. A person sitting on the edge of a 6-ft-diameter platform moves a linear distance of 2 ft. Express the angular displacement in radians, degrees, and revolutions.

11-5. An electric motor turns at 600 rpm. What is the angular velocity? What is the angular displacement after 6 s?

Ans. 62.8 rad/s, 377 rad

11-6. A rotating pulley completes 12 revolutions in 4 s. Determine the average angular velocity in revolutions per second, revolutions per minute, and radians per second.

11-7. A bucket is hung from a rope that is wrapped many times around a circular drum of radius 60 cm. The bucket starts from rest and is lifted to a height of 20 m in 5 s. (a) How many revolutions were made by the drum? (b) What was the average angular speed of the rotating drum? **Ans. (a) 5.31 rev; (b) 6.67 rad/s**

Summary and Review

11-8. A wheel of radius 15.0 cm starts from rest and makes 2.00 revolutions in 3.00 s. (a) What is its *average* angular velocity in radians per second? (b) What is the final *linear* velocity of a point on the rim of the wheel?

11-9. A cylindrical piece of stock 6 in. in diameter rotates in a lathe at 800 rev/min. What is the linear velocity at the surface of the cylinder? **Ans. 20.9 ft./s**

11-10. The proper tangential velocity for machining steel stock is about 70 cm/s. At what revolutions per minute should a steel cylinder 8 cm in diameter be turned in a lathe?

11-11. For the wheel in Prob. 11-8, what is the angular acceleration? What is the *linear* acceleration of a point on the edge of the wheel? **Ans. 2.79 rad/s², 0.419 m/s²**

11-12. A circular drum of radius 40 cm is initially rotating at 400 rev/min. It is then brought to a stop after making 50 revolutions. What were the angular acceleration and the stopping time?

11-13. A belt is wrapped around the edge of a pulley that is 40 cm in diameter. The pulley rotates with a constant angular acceleration of 3.50 rad/s². At $t = 0$, the rotational speed is 2 rad/s. What are the angular displacement and angular velocity of the pulley 2 s later? **Ans. (a) 11.0 rad, 9.00 rad/s**

11-14. In Prob. 11-13, what are the final linear speed and the final linear acceleration of the pulley belt as it moves around the edge of the pulley?

11-15. A wheel initially rotates at 6 rev/s and then undergoes a constant angular acceleration of 4 rad/s². What is the angular velocity after 5 s? How many revolutions will the wheel make? **Ans. 57.7 rad/s, 38.0 rev**

11-16. A grinding disk is brought to a stop in 40 revolutions. If the braking acceleration was −6 rad/s², what was the initial frequency or revolution in revolutions per second?

11-17. A pulley 320 mm in diameter and rotating initially at 4 rev/s receives a constant angular acceleration of 2 rad/s². What is the linear velocity of a belt around the pulley after 8 s? What is the tangential acceleration of the belt? **Ans. 6.58 m/s, 0.320 m/s²**

*11-18. A person initially at rest 4 m from the center of a rotating platform covers a distance of 100 m in 20 s. What is the angular acceleration of the platform? What is the angular velocity after 4 s?

Section 11-5—Rotational Kinetic Energy; Moment of Inertia

11-19. A 2-kg mass and a 6-kg mass are connected by a light 30-cm bar. The system is then rotated horizontally at 300 rpm about an axis 10 cm from the 6-kg mass. What is the moment of inertia about this axis? What is the rotational kinetic energy? **Ans. 0.140 kg m², 69.1 J**

11-20. A 1.2-kg bicycle wheel has a radius of 70 cm with spokes of negligible weight. If it starts from rest and receives an angular acceleration of 3 rad/s², what will be its rotational kinetic energy after 4 s?

11-21. A 16-lb grinding disk is rotating at 400 rev/min. What is the radius of the disk if its kinetic energy is 54.8 ft lb? What is the moment of inertia? **Ans. 6.00 in., 0.0625 slug ft²**

11-22. What must be the radius of a 4-kg circular disk if it is to have the same moment of inertia as a 1-kg rod 1 m long and pivoted at its midpoint?

*11-23. A wagon wheel 60 cm in diameter is mounted on a central axle where it spins at 200 rev/min. The wheel can be thought of as a circular hoop of mass 2 kg, and each of 12 wooden 500-g spokes can be thought of as thin rods rotating about their ends. Calculate the moment of inertia for the entire wheel. What is the rotational kinetic energy? **Ans. 0.360 kg m², 78.9 J**

11-24. Compare the rotational kinetic energies of three objects of equal radius and mass: a circular hoop, a circular disk, and a solid sphere.

Section 11-6—Newton's Second Law and Rotation

11-25. A rope wrapped around a 5-kg circular drum pulls with a tension of 400 N. If the radius of the drum is 20 cm and it is free to rotate about its central axis, what is the angular acceleration? **Ans. 800 rad/s^2**

11-26. The flywheel of an engine has a moment inertia of 24 slug ft^2. What torque is required to accelerate the wheel from rest to an angular velocity of 400 rpm in 10 s?

11-27. A 3-kg thin rod is 40 cm long and pivoted about its midpoint. What torque is required to cause it to make 20 revolutions while its rotational speed increases from 200 to 600 rev/min? **Ans. 0.558 N · m**

11-28. A large 120-kg turbine wheel has a radius of gyration of 1 m. A frictional torque of 80 N · m opposes the rotation of the shaft. What torque must be applied to accelerate the wheel from rest to 300 rev/min in 10 s?

11-29. A 2-kg mass swings in a circle of radius 50-cm at the end of a light rod. What resultant torque is required to give an angular acceleration of 2.5 rad/s^2? **Ans. 1.25 N · m**

11-30. A rope is wrapped several times around a cylinder of radius 0.2 m and mass 30 kg. What is the angular acceleration of the cylinder if the tension in the rope is 40 N and it turns without friction?

11-31. An 8-kg grinding disk has a diameter of 60 cm and is rotating at 600 rev/min. What braking force must be applied tangentially to the disk if it is to stop rotating in 5 s? **Ans. 15.1 N**

11-32. An unbalanced torque of 150 N · m imparts an angular acceleration of 12 rad/s^2 to the rotor of a generator. What is the moment of inertia?

Section 11-7—Rotational Work and Power

11-33. A rope wrapped around a 3-kg disk 20 cm in diameter is pulled for a linear distance of 5 m with a force of 40 N. What is the linear work done by the 40-N force? What is the rotational work done on the disk? **Ans. 200 J, 200 J**

11-34. Use the work-energy theorem to calculate the final angular velocity of the disk in Prob. 11-33 if it starts from rest.

11-35. A 1.2-kW motor acts for 8 s on a wheel having a moment of inertia of 2 kg · m^2. Assuming the wheel was initially at rest, what average angular speed was developed? **Ans. 98.0 rad/s**

11-36. A cord is wrapped around the rim of a cylinder that has a mass of 10 kg and a radius of 30 cm. If the rope is pulled with a force of 60 N, what is the angular acceleration of the cylinder? What is the linear acceleration of the rope?

11-37. A 600-W motor drives a pulley with an average angular velocity of 20 rad/s. What torque is developed? **Ans. 30 N · m**

11-38. The crankshaft on an automobile develops 350 lb · ft of torque at 1800 rpm. What is the output horsepower?

Section 11-8—Angular Momentum

11-39. A 500-g steel rod 30 cm in length is pivoted about its center and rotated at 300 rev/min. What is the angular momentum? **Ans. 0.118 kg m/s^2**

11-40. In Prob. 11-39, what average braking torque must be applied to stop the rotation in 2 s?

11-41. A sudden torque of 400 N · m is applied to the edge of a disk initially at rest. If the rotational inertia of the disk is 4 kg · m^2, and the torque acts for 0.02 s, what is the change in angular momentum? What is the final angular speed? **Ans. 8.00 kg · m^2/s, 2.00 rad/s**

11-42. In Fig. 11-12, a 6-kg disk A rotating clockwise at 400 rev/min engages with a 3-kg disk B initially at rest. The radius of disk A is 0.4 m, and the radius of disk B is 0.2 m. What is the combined angular speed after the two disks are meshed?

11-43. Assume that disk B in Prob. 11-42 is initially rotating clockwise at 200 rev/min in the same direction as disk A. What is the common angular speed as they mesh? **Ans. 378 rev/min**

11-44. Assume the same conditions as in Prob. 11-42 except that the disk B is rotating

Summary and Review

Figure 11-12

Figure 11-13

counterclockwise and *A* is rotating clockwise. What is the combined angular velocity after the disks are meshed?

11-45. The rod connecting the two weights in Fig. 11-13 has negligible weight but is configured to allow the weights to slip outward.

At the instant when the angular speed is 600 rev/min, the 2-kg masses are 10 cm apart. What is the rotational speed later when the masses are 34 cm apart?

Ans. 51.9 rpm

Challenge Problems

11-46. A 6-kg circular grinding disk is rotating initially at 500 rev/min. The radius of the disk is 40 cm. What is the angular acceleration of the disk if the axe exerts a tangential force of 120 N at the edge? How many revolutions will the disk make before stopping? What work is done, and what power is lost in the process?

11-47. A 3-kg wheel with spokes of negligible mass is free to rotate about its center without friction. The edge of the wheel of

radius 40 cm is struck suddenly with an average tangential force of 600 N lasting for 0.002 s. (a) What angular impulse is imparted to the wheel? (b) If the wheel was initially at rest, what was its angular speed at the end of the 0.002-s interval?

Ans. (a) 0.48 N · m · s; (b) 1 rad/s

11-48. Disk *A* has three times the rotational inertia of disk *B*. Disk *A* is rotating initially clockwise at 200 rev/min and disk *B* is rotating in the opposite direction at 800 rev/min. If the two are meshed together, what will be the common rate of rotation of the combined disks?

11-49. If the disks in Prob. 11-48 are initially rotating in the same direction, what would be the common angular speed after meshing?

Ans. 350 rev/min clockwise

11-50. The radius of gyration of an 8-kg wheel is 50 cm. Find its moment of inertia and its kinetic energy if it is rotating at 400 rev/min.

11-51. How much work is required to slow the wheel in Prob. 11-50 to 100 rev/min?

Ans. 1644 J

11-52. A wheel of radius 2 ft has a moment of inertia of 8.2 slug ft². A constant force of 12 lb acts tangentially at the edge of the wheel, which is initially at rest. What is the angular acceleration?

11-53. In Prob. 11-52, the wheel was brought to rest in 5 s. How much work was done? What horsepower was developed?

Ans. 878 ft · lb, 0.319 hp

11-54. An engine operating at 1800 rev/min develops 200 hp. What is the torque developed?

11-55. A constant force of 200 N acts at the edge of a wheel 36 cm in diameter causing it to make 20 revolutions in 5 s. What power is developed?

Ans. 905 W

*11-56. A 2-kg circular hoop rolls down an inclined plane from an initial height of 20 m. The kinetic energy that develops is shared between rotation and translation. What will be the speed when it reaches the bottom of the incline?

11-57. Suppose a circular disk rolls down the same incline as in Prob. 11-56. What is its speed when it reaches the bottom? **Ans. 16.2 m/s**

Critical Thinking Questions

11-58. A circular hoop of mass 2 kg and radius 60 cm spins freely about its center connected by light central spokes. A force of 50 N acts tangent to the edge of the wheel for a time of 0.02 s. (a) What is the angular impulse? (b) What is the change in angular momentum? (c) If the hoop was initially at rest, what was the final angular speed? (d) Use the work-energy theorem to calculate the angular displacement.

Ans. (a) 0.60 N · ms; (b) 0.60 kg · m²/s, (c) 0.833 rad/s, (d) 0.00833 rad

11-59. The spin cycle on a washing machine slows from 900 to 300 rev/min in 4 s. Determine the angular acceleration. Does a force act to throw off the water from the clothes or does the lack of a force do this? When the cycle operates at 900 rev/min, the output power is 4 kW. What is the torque developed? If the radius of the tub is 30 cm, what is the linear speed of the clothes near the inside edge?

11-60. A block is attached to a cord passing over a pulley through a hole in a horizontal tabletop as shown in Fig. 11-14. Initially the block is revolving at 4 rad/s at a distance r

Figure 11-14

from the center of the hole. If the cord is pulled from below until its radius is $r/4$, what is the new angular velocity?

Ans. 64 rad/s

11-61. Suppose in Fig. 11-14 the block has a mass of 2 kg and is rotating at 3 rad/s when $r = 1$ m. At what distance r will the tension in the cord be 25 N?

*11-62. Consider Fig. 11-15, in which $m = 2$ kg, $M = 8$ kg, $R = 60$ cm, and $h = 6$ m. Write Newton's second law for the disk in terms of the tension in the rope, the moment of inertia of the disk, and the angular acceleration. Next write Newton's second law for falling mass in terms of the tension in the

Figure 11-15

rope, the mass, and the linear acceleration. Eliminate T from these two equations. Find the linear acceleration of the 2-kg mass by recalling that $v = \omega R$, $a = \alpha R$, and $I = \frac{1}{2}mR^2$.   **Ans. 3.27 m/s²**.

11-63. Use conservation of energy to find the velocity of the 2-kg mass in Fig. 11-15 just before it strikes the floor 6 m below. Use data given in Prob. 11-62.

11-64. A student stands on a platform with arms outstretched, holding weights in each hand so that the rotational inertial is 6.0 kg · m². The platform is then set into constant, frictionless rotation at 90 rev/min. Now the student is able to reduce the rotational

inertia to 2 kg · m² by pulling the weights toward the body. (a) What will be the new rate of rotation in the absence of external torque? (b) What is the ratio of the final kinetic energy to the initial kinetic energy? (c) Explain the increase in energy.

Ans. (a) 270 rev/min; (b) 3;
(c) work done on masses

*11-65. Consider the apparatus shown in Fig. 11-16. Consider the large pulley as a 6-kg disk of radius 50 cm. The right mass is 4 kg, and the left mass is 2 kg. Consider both rotational and translational energies and find the velocity just before the 4-kg mass strikes the floor.

Figure 11-16

Simple Machines

OBJECTIVES

After completing this chapter, you should be able to:

1. Describe a simple machine and its operation in general terms to the extent that *efficiency* and *conservation of energy* are explained.
2. Write and apply formulas for computing the efficiency of a simple machine in terms of work or power.
3. Distinguish by definition and example between *ideal* mechanical advantage and *actual* mechanical advantage.
4. Draw a diagram of each of the following simple machines and beside each diagram write a formula for computing the ideal mechanical advantage: (a) lever, (b) inclined plane, (c) wedge, (d) gears, (e) pulley systems, (f) wheel and axle, (g) screw jack, (h) belt drive.
5. Compute the mechanical advantage and the efficiency of each of the simple machines listed in the previous objective.

A ***simple machine*** is any device that transmits the application of a force into useful work. With a chain hoist we can transmit a small downward force into a large upward force for lifting. In industry delicate samples of radioactive material are handled by machines that allow an applied force to be reduced significantly. Single pulleys may be used to change the direction of an applied force without affecting its magnitude. A study of machines and their efficiency is essential for the productive use of energy. In this chapter, you will become familiar with levers, gears, pulley systems, inclined planes, and other machines used routinely for many industrial applications.

12-1 SIMPLE MACHINES AND EFFICIENCY

In a simple machine, input work is done by the application of a single force, and the machine performs output work by means of a single force. During any such operation (Fig. 12-1), three processes occur:

1. Work is supplied to the machine.
2. Work is done against friction.
3. Output work is done by the machine.

According to the principle of conservation of energy, these processes are related as follows:

$$\text{Input work} = \text{work against friction} + \text{output work}$$

Input work

Output work

Heat lost
to friction

Figure 12-1 Three processes occur in the operation of a machine: (1) the input of a certain amount of work, (2) the loss of energy in doing work against friction, and (3) the output of useful work.

Math Tip

Estimation is an important skill for you to develop. Here are some helpful hints for estimation.

- Multiplication by 9 is the same as multiplying the number by 9 and then subtracting the number from the result.
- Using a rough sketch can often help you visualize problems.
- When estimating the area of an irregular shape, use a series of rectangles or equilateral triangles to approximate the total area.
- Use your rough estimate to check your final answer.

The amount of useful work performed by a machine can never be greater than the work supplied to it. There will always be some loss due to friction or other dissipative forces. For example, when pumping up a bicycle tire with a small hand pump, we exert a downward force on the plunger, causing air to be forced into the tire. That some of our input work is lost to friction can be verified easily by feeling how warm the wall of the hand pump becomes. The smaller we can make the friction loss in a machine, the greater the return for our effort. In other words, the effectiveness of a given machine can be measured by comparing its output work with the work supplied to it.

The efficiency E of a machine is defined as the ratio of the work output to the work input.

$$E = \frac{\text{work output}}{\text{work input}} \qquad \text{12-1}$$

The *efficiency* as defined in Eq. (12-1) will always be a number between 0 and 1. Common practice is to express this decimal as a percentage by multiplying by 100. For example, a machine that does 40 J of work when 80 J of work is supplied to it has an efficiency of 50 percent.

Another useful expression for efficiency can be noted from the definition of power as work per unit time. We can write

$$P = \frac{\text{work}}{t} \qquad \text{or} \qquad \text{Work} = Pt$$

The efficiency in terms of power input P_i and power output P_o is given by

$$E = \frac{\text{work output}}{\text{work input}} = \frac{P_o t}{P_i t}$$

or

$$E = \frac{\text{power output}}{\text{power input}} = \frac{P_o}{P_i} \qquad \text{12-2}$$

EXAMPLE 12-1

A 60-hp motor winds a cable around a drum. (a) If the cable lifts a 3-ton load of bricks to a height of 12 ft in 3 s, calculate the efficiency of the motor. (b) At what rate is work done against friction?

Solution (a)
First compute the output power.

$$P_o = \frac{Fs}{t} = \frac{(6000 \text{ lb})(12 \text{ ft})}{3 \text{ s}}$$

$$= (24,000 \text{ ft} \cdot \text{lb/s}) \frac{1 \text{ hp}}{550 \text{ ft} \cdot \text{lb/s}}$$

$$= 43.6 \text{ hp}$$

The efficiency is then found from Eq. (12-2):

$$E = \frac{P_o}{P_i} = \frac{43.6 \text{ hp}}{60 \text{ hp}} = 0.727$$

$$= 72.7\%$$

Solution (b)
The rate at which work is done against friction is the difference between the input power and the output power, or 16.4 hp.

12-2 MECHANICAL ADVANTAGE

Simple machines like the lever, block and tackle, chain hoist, gears, inclined plane, and the screw jack all play important roles in modern industry. We can illustrate the operation of any of these machines by the general diagram in Fig. 12-2.

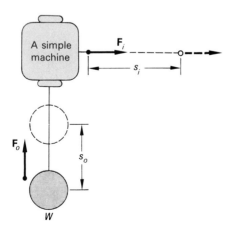

Figure 12-2 During the operation of any simple machine, an input force $\mathbf{F}_i$ acts through a distance s_i while an output force $\mathbf{F}_o$ acts through a distance s_o.

An input force $\mathbf{F}_i$ acts through a distance s_i, accomplishing the work $F_i s_i$. At the same time, an output force $\mathbf{F}_o$ acts through a distance s_o, performing the useful work $F_o s_o$.

The actual mechanical advantage M_A of a machine is defined as the ratio of the output force F_o to the input force F_i.

$$M_A = \frac{\text{output force}}{\text{input force}} = \frac{F_o}{F_i} \qquad \text{12-3}$$

An actual mechanical advantage greater than 1 indicates that the output force is greater than the input force. Although most machines have values of M_A greater than 1, this is not always the case. With small, fragile objects, it is sometimes desirable to make the output force smaller than the input force.

In the previous section, we noted that the efficiency of a machine increases as frictional effects become small. Applying the conservation-of-energy principle to the simple machine in Fig. 12-2 yields

$$\textit{Work input} = \textit{work against friction} + \textit{work output}$$
$$F_i s_i = (\text{work})_{\mathscr{F}} + F_o s_o$$

The most efficient engine possible would realize no losses due to friction. We can represent this ideal case by setting $(\text{work})_{\mathscr{F}} = 0$ in the above equation. Thus,

$$F_o s_o = F_i s_i$$

Since this equation represents an ideal case, we define the *ideal mechanical advantage M_I* as

$$M_I = \frac{F_o}{F_i} = \frac{s_i}{s_o} \qquad \text{12-4}$$

The ideal mechanical advantage of a simple machine is equal to the ratio of the distance the input force moves to the distance the output force moves.

The efficiency of a simple machine is the ratio of output work to input work. Therefore, for the general machine of Fig. 12-2, we have

$$E = \frac{F_o s_o}{F_i s_i} = \frac{F_o / F_i}{s_i / s_o}$$

Finally, utilizing Eqs. (12-3) and (12-4), we obtain

$$E = \frac{M_A}{M_I} \qquad \text{12-5}$$

All the above concepts have been treated as they apply to a general machine. In the following sections, we shall apply them to specific machines.

12-3 THE LEVER

Possibly the oldest and most generally useful machine is the simple lever. A *lever* consists of any rigid bar pivoted at a certain point called the *fulcrum.* Fig. 12-3 illustrates the use of a long rod to lift a weight **W.** We can calculate the ideal mechanical advantage of such a device in two ways. The first method involves the principle of equilibrium, and the second uses the principle of work, as discussed in the previous section. Since the equilibrium method is easier for the lever, we shall apply it first.

Because no translational motion is involved during the application of a lever, the condition for equilibrium is that the input torque equal the output torque:

$$F_i r_i = F_o r_o$$

The ideal mechanical advantage can be found from

$$M_I = \frac{F_o}{F_i} = \frac{r_i}{r_o} \qquad\qquad \textbf{12-6}$$

The ratio F_o/F_i is considered the *ideal* case because no friction forces are considered.

The same result is obtained from work considerations. Note from Fig. 12-3b that the force $\mathbf{F}_i$ moves through the arc distance s_i while the force $\mathbf{F}_o$ moves through the arc distance s_o. The two arcs are subtended by the same angle θ, however, and so we can write the proportion:

$$\frac{s_i}{s_o} = \frac{r_i}{r_o}$$

Substitution into Eq. (12-4) will verify the result obtained from equilibrium considerations, that is, $M_I = r_i/r_o$.

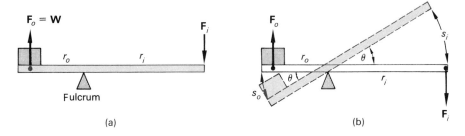

Figure 12-3   The lever.

EXAMPLE 12-2

An iron bar 3 m long is used to lift a 60-kg block. The bar is used as a lever, as shown in Fig. 12-3. The fulcrum is placed 80 cm from the block. What is the ideal mechanical advantage of the system, and what input force is required?

Solution

The distance $r_o = 0.8$ m, and the distance $r_i = 3$ m $- 0.8$ m $= 2.2$ m. Therefore, the ideal mechanical advantage is

$$M_I = \frac{r_i}{r_o} = \frac{2.2 \text{ m}}{0.8 \text{ m}} = 2.75$$

The output force in this case is equal to the weight of the 60-kg block ($W = mg$). Therefore, the required input force is given by

$$F_i = \frac{F_o}{M_I} = \frac{(60 \text{ kg})(9.8 \text{ m/s}^2)}{2.75}$$

$$= 214 \text{ N}$$

Before leaving the subject of the lever, it should be noted that little of the input work is lost to friction forces. For all practical purposes, the actual mechanical advantage for a simple lever is equal to the ideal mechanical advantage. Other examples of the lever are illustrated in Fig. 12-4.

Figure 12-4 The lever forms the operating principle of many simple machines.

12-4 APPLICATIONS OF THE LEVER PRINCIPLE

A serious limitation of the elementary *lever* is the fact that it operates through a small angle. There are many ways of overcoming this restriction by allowing for continuous rotation of the lever arm. For example, the *wheel and axle* (Fig. 12-5) allows for the continued action of the input force. Applying the reasoning described in Sec. 12-2 for a general machine, it can be shown that

$$M_I = \frac{F_o}{F_i} = \frac{R}{r} \qquad \text{12-7}$$

Thus, the ideal mechanical advantage of a wheel and axle is the ratio of the radius of the wheel to the radius of the axle.

Another application of the lever concept is through the use of pulleys. A single *pulley*, as shown in Fig. 12-6, is simply a lever whose input moment arm is equal to its output moment arm. From the principle of equilibrium, the input forces will equal the output force, and the ideal mechanical advantage will be

$$M_I = \frac{F_o}{F_i} = 1 \qquad \text{12-8}$$

The only advantage of such a device lies in its ability to change the direction of an input force.

A single movable pulley (Fig. 12-7), on the other hand, has an ideal mechanical advantage of 2. Note that the two supporting ropes must each be shortened by 1 ft to lift the load through a distance of 1 ft. Therefore, the input force moves through a distance of 2 ft while the output force moves a distance of only 1 ft. Applying the principle of work, we have

$$F_i(2 \text{ ft}) = F_o(1 \text{ ft})$$

from which the ideal mechanical advantage is

$$M_I = \frac{F_o}{F_i} = 2 \qquad \text{12-9}$$

*inter***NET**
CONNECTION

www.harken.com
A wide variety of pulleys are used for sailing. For pictures and descriptions of pulleys and block-and-tackle assemblies, see this Web site.

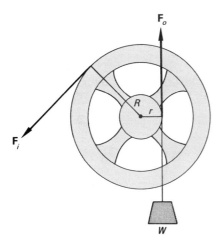

Figure 12-5   The wheel and axle.

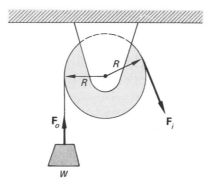

Figure 12-6   A single fixed pulley serves only to change the direction of the input force.

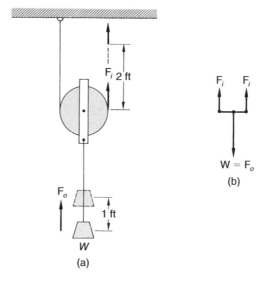

Figure 12-7 A single movable pulley. (a) The input force moves through twice the distance that the output force travels. (b) The free-body diagram shows that $2F_i = F_o$.

The same result can be shown by constructing a free-body diagram, as in Fig. 12-7b. From the figure it is evident that

$$2F_i = F_o$$

or

$$M_I = \frac{F_o}{F_i} = 2$$

The latter method is usually applied to problems involving movable pulleys since it allows one to associate M_I with the number of strands supporting the movable pulley.

EXAMPLE 12-3

Calculate the ideal mechanical advantage of the block-and-tackle arrangement shown in Fig. 12-8.

Solution

We first construct a free-body diagram, as shown in Fig. 12-8b. From the figure we note that

$$4F_i = F_o$$

from which

$$M_I = \frac{F_o}{F_i} = \frac{4F_i}{F_i} = 4$$

Note that the uppermost pulley serves only to change the direction of the input force. The same M_I would result if $\mathbf{F}_i$ were applied upward at point *a*.

Figure 12-8 The block and tackle. This arrangement has an ideal mechanical advantage of 4, since four strands support the movable block.

12-5 THE TRANSMISSION OF TORQUE

The simple machines discussed so far are used to transmit and apply forces to move loads. In most mechanical applications, work is done by transmitting torque from one drive to another. For example, the belt drive (Fig. 12-9) transmits the torque from a driving pulley to an output pulley. The mechanical advantage of such a system is the ratio of the torques between the output pulley and the driving pulley:

$$M_I = \frac{\text{output torque}}{\text{input torque}} = \frac{\tau_o}{\tau_i}$$

From the definition of torque, we can write this expression in terms of the radii of the pulleys:

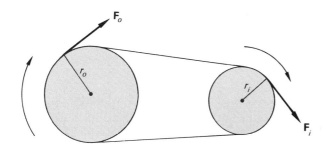

Figure 12-9 The belt drive.

$$M_I = \frac{\tau_o}{\tau_i} = \frac{F_o r_o}{F_i r_i}$$

If there is no slippage between the belt and pulleys, it is safe to say that the tangential input force $\mathbf{F}_i$ is equal to the tangential output force $\mathbf{F}_o$. Thus,

$$M_I = \frac{F_o r_o}{F_i r_i} = \frac{r_o}{r_i}$$

Since the diameters of pulleys are usually specified instead of their radii, a more convenient expression is

$$M_I = \frac{D_o}{D_i} \qquad\qquad\qquad \text{12-10}$$

where D_i is the diameter of the driving pulley and D_o is the diameter of the output pulley.

Suppose we now apply the principle of work to the **belt drive.** Remember that work is defined in rotary motion as the product of torque τ and angular displacement θ. For the belt drive, assuming ideal conditions, the input work equals the output work. Thus,

$$\tau_i \theta_i = \tau_o \theta_o$$

The power input must also be equal to the power output. Dividing the above equation by the time t required to rotate through the angles θ_i and θ_o, we obtain

$$\tau_i \frac{\theta_i}{t} = \tau_o \frac{\theta_o}{t} \qquad \text{or} \qquad \tau_i \omega_i = \tau_o \omega_o$$

where ω_i and ω_o are the angular speeds of the input and output pulleys. Note that the ratio τ_o/τ_i represents the ideal mechanical advantage. Therefore, we can add another expression to Eq. (12-10) to obtain

$$M_I = \frac{D_o}{D_i} = \frac{\omega_i}{\omega_o} \qquad\qquad\qquad \text{12-11}$$

This important result shows that the mechanical advantage is achieved at the expense of rotary motion. In other words, if the mechanical advantage is 2, the input shaft must rotate with twice the angular speed of the output shaft. The ratio ω_i/ω_o is sometimes referred to as the *speed ratio.*

If the speed ratio is greater than 1, the machine produces an output torque that is greater than the input torque. As we have seen, this feat is accomplished at the expense of rotation. On the other hand, many machines are designed to increase the rotational output speed. In these cases, the speed ratio is less than 1, and the increased rotational speed is accomplished with reduced torque output.

EXAMPLE 12-4

Consider the belt drive illustrated in Fig. 12-9, where the diameter of the small driving pulley is 6 in. and the diameter of the driven pulley is 18 in. A 6-hp motor drives the input pulley at 600 rpm. Calculate the revolutions per minute and torque delivered to the driven wheel if the system is 75 percent efficient.

Solution

We first calculate the ideal mechanical advantage (100 percent efficiency) of the system. From Eq. (12-11),

$$M_I = \frac{D_o}{D_i} = \frac{18 \text{ in.}}{6 \text{ in.}} = 3$$

Since the efficiency is 75 percent, the actual mechanical advantage is given from Eq. (12-5).

$$M_A = EM_I = (0.75)(3) = 2.25$$

Now the actual mechanical advantage is the simple ratio of output torque τ_o to input torque τ_i. Recalling that the power in rotational motion is equal to the product of torque and angular velocity, we can solve for τ_i as follows:

$$\tau_i = \frac{P_i}{\omega_i} = \frac{(6 \text{ hp})[(550 \text{ ft} \cdot \text{lb/s})/\text{hp}]}{(600 \text{ rev/min})(2\pi \text{ rad/rev})(1 \text{ min/60 } s)}$$

$$= \frac{(6)(550 \text{ ft} \cdot \text{lb/s})}{20\pi \text{ rad/s}} = 52.5 \text{ ft} \cdot \text{lb}$$

Since $M_A = \tau_o/\tau_i$, the output torque is given by

$$\tau_o = M_A\tau_i = (2.25)(52.5 \text{ ft} \cdot \text{lb})$$
$$= 118 \text{ ft} \cdot \text{lb}$$

Assuming the belt does not slip, it will move with the same linear velocity v around each pulley. Since $v = \omega r$, we can write the equality

$$\omega_i r_i = \omega_o r_o \qquad \text{or} \qquad \omega_i D_i = \omega_o D_o$$

from which

$$\omega_o = \frac{\omega_i D_i}{D_o} = \frac{(600 \text{ rpm})(6 \text{ in.})}{18 \text{ in.}} = 200 \text{ rpm}$$

Note that the ratio of ω_i to ω_o yields the ideal mechanical advantage and not the actual mechanical advantage. The difference between M_I and M_A is due to friction, both in the belt and in the shaft bearings. Since greater tension on the belt will result in greater friction forces, maximum efficiency is obtained by reducing the belt tension until it just prevents the belt from slipping on the pulleys.

Before leaving our discussion of the transmission of torque, we must consider the application of *gears*. A **gear** is simply a notched wheel that can transmit torque by meshing with another notched wheel, as shown in Fig. 12-10. A pair of meshing gears differs from a belt drive only in the sense that the gears rotate in opposite directions. The same relationships derived for the belt drive hold for gears:

$$M_I = \frac{D_o}{D_i} = \frac{\omega_i}{\omega_o} \qquad\qquad \text{12-12}$$

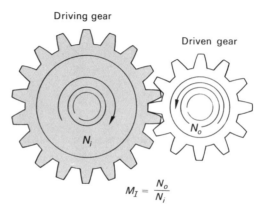

Driving gear

Driven gear

N_i

N_o

$$M_I = \frac{N_o}{N_i}$$

Figure 12-10   Spur gears. The ideal mechanical advantage is the ratio of the number of teeth on the output gear to the number of teeth on the input gear.

A more useful expression makes use of the fact that the number of teeth N on the rim of a gear is proportional to its diameter D. Because of this dependence, the ratio of the number of teeth on the driven gear N_o to the number of teeth on the driving gear N_i is the same as the ratio of their diameters. Hence, we can write

$$M_I = \frac{N_o}{N_i} = \frac{D_o}{D_i}$$ 12-13

The use of gears avoids the problem of slippage, which is common with belt drives. It also conserves space and allows for a greater torque to be transmitted.

In addition to the *spur* gears illustrated in Fig. 12-10, there are several other types of gears. Four common types are worm, helical, bevel, and planetary gears. Examples of each are shown in Fig. 12-11. The same general relationships apply for all these gears.

12-6 THE INCLINED PLANE

The only machines we have discussed so far involve application of the lever principle. A second fundamental machine is the ***inclined plane.*** Suppose you have to move a heavy load from the ground to a truck bed without hoisting equipment. You should probably select a few long boards and form a ramp from the ground to the bed of the truck. Experience has taught you that it takes less effort to push a load up a small elevation than it does to lift the load directly. Since a smaller input force results in the same output force, a mechanical advantage is realized. The smaller input force is accomplished, however, at the cost of greater distance.

Consider the movement of a weight **W** up the inclined plane in Fig. 12-12. The slope angle θ is such that the weight must be moved through a distance s to reach a height h at the top of the incline. If we neglect friction, the work required to push the weight up the plane is the same as the work required to lift it up vertically. We can express this equality as

Work input = work output
$$F_i s = Wh$$

(a) Helical gears

(b) Planetary gears

(c) Bevel gears

(d) The worm gear

Figure 12-11 Four common types of gears: (a) helical, (b) planetary, (c) bevel, (d) worm. (The spur gear, which is the most common type, is shown in Fig. 12-10.)

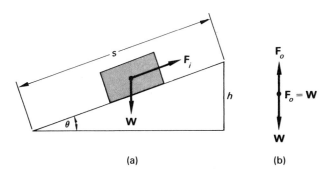

(a)

(b)

Figure 12-12 The inclined plane. The input force represents the effort required to push the block up the plane; the output force is equal to the weight of the block.

where F_i is the input force and W is the output force. The ideal mechanical advantage will be the ratio of the weight to the input force. Stating this symbolically, we have

$$M_I = \frac{W}{F_i} = \frac{s}{h} \qquad \text{12-14}$$

EXAMPLE 12-5

The 200-lb wooden crate in Fig. 12-13 is to be raised to a loading platform 6 ft high. A ramp 12 ft long is used to slide the crate from the ground to the platform. Assume that the coefficient of fraction is 0.3. (a) What is the ideal mechanical advantage of the ramp? (b) What is the actual mechanical advantage?

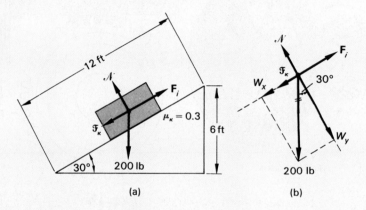

(a)   (b)

Figure 12-13

Solution (a)

The ideal mechanical advantage, from Eq. (12-14), is

$$M_I = \frac{s}{h} = \frac{12 \text{ ft}}{6 \text{ ft}} = 2$$

This value represents the mechanical advantage of the ramp if it were frictionless.

Solution (b)

The actual mechanical advantage is the ratio of the weight lifted to the required input force, considering friction. Applying the first condition for equilibrium to the free-body diagram (Fig. 12-13b), we find that the normal force $\mathcal{N}$ is given by

$$\mathcal{N} = W_y = (200 \text{ lb})(\cos 30°) = 173 \text{ lb}$$

from which the friction force must be

$$\mathcal{F} = \mu \mathcal{N} = (0.3)(173 \text{ lb}) = 51.9 \text{ lb}$$

Summing the forces along the plane, we obtain

$$F_i - \mathscr{F} - W_x = 0$$

But $W_x = (200\ \text{lb})(\sin 30°) = 100\ \text{lb}$, so that we have

$$F_i - 51.9\ \text{lb} - 100\ \text{lb} = 0$$
$$F_i = 51.9\ \text{lb} + 100\ \text{lb} = 152\ \text{lb}$$

We can now compute the actual mechanical advantage:

$$M_A = \frac{F_o}{F_i} = \frac{200\ \text{lb}}{152\ \text{lb}} = 1.32$$

It is left as an exercise for you to show that the efficiency of the ramp is 66 percent.

12-7 APPLICATIONS OF THE INCLINED PLANE

Many machines apply the principle of the inclined plane. The simplest is the *wedge* (Fig. 12-14), which is actually a double inclined plane. In the ideal case, the mechanical advantage of a wedge of length L and thickness t is given by

$$M_I = \frac{L}{t} \qquad \text{12-15}$$

This equation is a direct consequence of the general relation expressed by Eq. (12-14). The ideal mechanical advantage is always much greater than the actual mechanical advantage because of the large friction forces between the surfaces in contact. The wedge finds its application in axes, knives, chisels, planers, and all other cutting tools. A *cam* is a kind of rotary wedge that is used to lift valves in internal combustion engines.

(a)   (b)

Figure 12-14   The wedge is actually a double inclined plane.

Figure 12-15 The screw jack.

One of the most useful applications of the inclined plane is the ***screw.*** This principle can be explained by examining a common tool known as the *screw jack* (Fig. 12-15). The threads are essentially an inclined plane wrapped continuously around a cylindrical shaft. When the input force $\mathbf{F}_i$ turns through a complete revolution $(2\pi R)$, the output force $\mathbf{F}_o$ will advance through the distance p. This distance p is actually the distance between two adjacent threads, and it is called the ***pitch*** of the screw. The ideal mechanical advantage is the ratio of the input distance to the output distance.

$$M_I = \frac{s_i}{s_o} = \frac{2\pi R}{p}$$

12-16

The screw is an example of an inefficient machine, but in this case it is usually an advantage since friction forces are needed to hold in place while the input force is not being applied.

Summary and Review

Summary

A simple machine has been defined as a device that converts a single input force F_i into a single output force F_o. In general, the input force moves through a distance s_i, and the output force moves through a distance s_o. The purpose is to accomplish useful work in a manner suited to a particular application. The major concepts are given below.

- A simple machine is a device that converts a single input force F_i into a single output force F_o. The input force moves through a distance s_i, and the output force moves a distance s_o. There are two mechanical advantages:

$$M_A = \frac{F_o}{F_i} \quad \text{actual mechanical advantage (friction considered)}$$

$$M_I = \frac{s_i}{s_o} \quad \text{ideal mechanical advantage (assumes no friction)}$$

- The efficiency of a machine is a ratio of output to input work. It is normally expressed as a percentage and can be calculated from any of the following relations:

$$E = \frac{\text{work output}}{\text{work input}} \qquad E = \frac{\text{power output}}{\text{power input}}$$

$$E = \frac{M_A}{M_I}$$

- The ideal mechanical advantages for a number of simple machines are given below.

$$M_I = \left(\frac{F_o}{F_i}\right)_{\text{ideal}} = \frac{r_i}{r_o} \qquad \textit{Lever}$$

$$M_I = \left(\frac{F_o}{F_i}\right)_{\text{ideal}} = \frac{R}{r} \qquad \textit{Wheel and Axle}$$

$$M_I = \frac{D_o}{D_i} = \frac{\omega_i}{\omega_o} \qquad \textit{Belt Drive}$$

$$M_I = \frac{W}{F_i} = \frac{s}{h} \qquad \textit{Inclined Plane}$$

$$M_I = \frac{L}{t} \qquad \textit{Wedge}$$

$$M_I = \frac{N_o}{N_i} = \frac{D_o}{D_i} \qquad \textit{Gears}$$

$$M_I = \frac{s_i}{s_o} = \frac{2\pi R}{P} \qquad \textit{Screw Jack}$$

Key Terms

1. machine
2. efficiency
3. lever
4. pulley
5. gears
6. wedge
7. screw
8. pitch
9. inclined plane
10. wheel and axle
11. actual mechanical advantage
12. ideal mechanical advantage
13. belt drive

Summary and Review

Review Questions

12-1. What is meant by *useful work* or *output work*? What is meant by *input work*? Write the general relationship between input work and output work.

12-2. Two jacks are operated simultaneously to lift the front end of a car. Immediately afterward, the left jack feels warmer than the right one. Which jack is more efficient? Explain.

12-3. A machine may alter the magnitude and/or the direction of an input force. (a) Name several examples in which both changes occur. (b) Give examples in which only the magnitude of the input force is altered. (c) Give some examples in which only the direction is altered.

12-4. A machine lifts a load through a vertical distance of 4 ft while the input force moves through a distance of 2 ft. Would this machine be helpful in lifting large weights? Explain.

12-5. A bicycle can be operated in three gear ranges. In *low range* the pedals describe two complete revolutions while the rear wheel turns through one revolution. In *medium range* the pedals and the wheels turn at the same rate. In *high range* the rear wheel of the bicycle completes two revolutions for every complete pedal revolution. Discuss the advantages and disadvantages of each range.

12-6. What happens to the ideal mechanical advantage if a simple machine is operated in reverse? What happens to its efficiency?

12-7. Give several examples of machines that have an actual mechanical advantage less than 1.

12-8. Why do buses and trucks often use larger steering wheels than those found on automobiles? What principle is used?

12-9. Draw diagrams of pulley systems that have ideal mechanical advantages of 2, 3, and 5.

12-10. Usually the road to the top of a mountain winds around the mountain instead of going straight up the side. Why? If we neglect friction, is more work required to reach the top along the spiral road? Is more power required? If we consider friction, would it require less work to drive straight up the side of the mountain? Explain.

Problems

Section 12-1—Simple Machines, Efficiency, and Mechanical Advantage

12-1. A 25 percent efficient machine performs external work of 200 J. What input work is required?   **Ans. 800 J**

12-2. What is the input work of a 30 percent efficient gasoline engine if during each cycle it performs 400 J of useful work?

12-3. A 60-W motor lifts a 2-kg mass to a height of 4 m in 3 s. Compute the output power.   **Ans. 26.1 W**

12-4. What is the efficiency of the motor in Prob. 12-3? What is the rate at which work is done against friction?

12-5. A 60 percent efficient machine lifts a 10-kg mass at a constant speed of 3 m/s. What is the required input power?   **Ans. 490 W**

12-6. During the operation of a 300-hp engine, energy is lost to friction at the rate of 200 hp. What is the useful output power, and what is the efficiency of the engine?

12-7. A frictionless machine lifts a 200-lb load through a vertical distance of 10 ft. The input force moves through a distance of 300 ft. What is the ideal mechanical advantage of the machine? What is the magnitude of the input force?   **Ans. 30, 6.67 lb**

Section 12-3—Applications of the Lever Principle

12-8. One edge of a 50-kg safe is lifted with a 1.2-m steel rod. What input force is required at the end of the rod if a fulcrum is placed 12 cm from the safe? (*Hint:* To lift one edge, a force equal to one-half the weight of the safe is required.)

12-9. For the nutcracker in Fig. 12-4a, the nut is located 2 cm from the fulcrum and an input force of 20 N is applied at the handles, which are 10 cm from the fulcrum. What is the force applied to crack the nut?

Ans. 100 N

12-10. For the wheelbarrow in Fig. 12-4b, the center of gravity of a net load of 40 kg is located 50 cm from the wheel. What upward lift must be applied at a point on the handles that is 1.4 m from the wheel?

12-11. What is the ideal mechanical advantage of the wheelbarrow in Prob. 12-10?

Ans. 2.80

12-12. Find the ideal mechanical advantage of the crowbar in Fig. 12-4c if the input force is applied 30 cm from the nail and the fulcrum is located 2 cm from the nail.

12-13. The input force exerted by a muscle in the forearm (see Fig. 12-4d) is 120 N and acts at a distance of 4 cm from the elbow. The total length of the forearm is 25 cm. What weight is being lifted?

Ans. 19.2 N

12-14. A wheel 20 cm in diameter is attached to an axle that has a diameter of 6 cm. If a weight of 400 N is attached to the axle, what force must be applied to the rim of the wheel to lift the weight at constant speed? Neglect friction.

12-15. A 20-kg mass is to be lifted with a rod 2 m long. If you can exert a downward force of 40 N on one end of the rod, where should you place a block of wood to act as a fulcrum?

Ans. 33.9 cm from mass

12-16. Determine the force **F** required to lift a 200-N load W with the pulley shown in Fig. 12-16a.

12-17. What input force is needed to lift the 200-N load with the arrangement drawn in Fig. 12-16b?

Ans. 50.0 N

12-18. What are the input forces needed to lift the 200-N load for the arrangements in Fig. 12-16c and d?

12-19. What is the mechanical advantage of a screwdriver used as a wheel and axle if

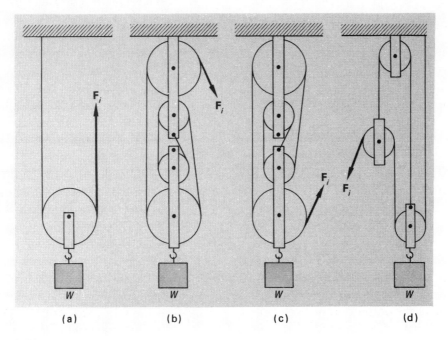

| (a) | (b) | (c) | (d) |

Figure 12-16

its blade is 0.3 in. wide and its handle is 0.8 in. long? **Ans. 2.67**

12-20. The chain hoist in Fig. 12-17 is a combination of the wheel and axle and the block and tackle. Show that the ideal mechanical advantage of such a device is given by

$$M_I = \frac{2R}{R - r}$$

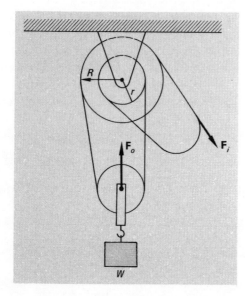

Figure 12-17

12-21. Assume that the larger radius in Fig. 12-17 is three times the smaller radius. What input force is required to lift a 10-kg load with no friction? **Ans. 32.7 N**

Section 12-5—The Transmission of Torque

12-22. A 1500-rev/min motor has a drive pulley 3 in. in diameter, and the driven pulley is 9 in. in diameter. What is the ideal mechanical advantage, and what are the revolutions per minute for the output pulley?

12-23. A 30-cm diameter input pulley turns at 200 rev/min on a belt drive connected to an output pulley 60 cm in diameter. What is the ratio of the output torque to the input torque? What are the output revolutions per minute? **Ans. 100 rev/min**

12-24. A V-belt pulley system has output and input drives of diameters 6 in. and 4 in., respectively. A torque of 200 lb · in. is applied to the input drive. What is the output torque?

12-25. The ratio of output speed to input speed for a gear drive is 2 : 1. What is the mechanical advantage? **Ans. ½**

12-26. A set of two spur gears has 40 teeth and 10 teeth. What are the possible ideal mechanical advantages?

12-27. For the spur gears in Prob. 12-26, what is the rotational speed of the smaller gear if the speed of the larger gear is 200 rev/min? **Ans. 800 rev/min**

Section 12-7—Applications of the Inclined Plane

12-28. What must be the thickness of the base if a wedge is 20 cm long and it is desired that the input force be one-tenth of the output force?

12-29. What should be the apex angle of a wedge if it is to have a mechanical advantage of 10? **Ans. 5.71°**

12-30. A 10-kg crate is moved from the ground to a loading platform by means of a ramp 6 m long and 2 m high. Assume that $\mu_k = 0.25$. What are the ideal and actual mechanical advantages of the ramp?

12-31. For the ramp in Prob. 12-30, what is the efficiency of the ramp? **Ans. 58.6 percent**

*12-32. An input force of 20 lb is applied to the 6-in. handle of a wrench used to tighten a ¼-in.-diameter bolt. An actual output force of 600 lb is produced. If the bolt has 10 threads per inch, what is the ideal mechanical advantage and what is the efficiency?

12-33. The lever of a screw jack is 24 in. long. If the screw has six threads per inch, what is the ideal mechanical advantage? **Ans. 904**

12-34. If the screw jack in Prob. 11-33 is 15 percent efficient, what force is needed to lift 2000 lb?

Challenge Problems

*12-35. An inclined plane is 6 m long and 1 m high. The coefficient of kinetic friction is 0.2. What force is required to pull a weight of 2400 N up the incline at constant speed? What is the efficiency of the inclined plane?
Ans. 873 N, 29.3 percent

*12-36. A wheel and axle are used to raise a mass of 700 kg. The radius of the wheel is 0.50 m, and the radius of the axle is 0.04 m. If the actual efficiency is 60 percent, what input force must be applied to the wheel?

*12-37. A shaft rotating at 800 rev/min delivers a torque of 240 N · m to an output shaft that is rotating at 200 rev/min. If the efficiency of the machine is 70 percent, compute the output torque. What is the output power?
Ans. 672 N · m, 14.1 kW

*12-38. A screw jack has a screw with a pitch of 0.25 in. Its handle is 16 in. long, and a load of 1.9 tons is being lifted. Neglecting friction, what force is required at the end of the handle? What is the mechanical advantage?

*12-39. A certain refrigeration compressor comes equipped with a 250-mm-diameter pulley

and is designed to operate at 600 rev/min. What should be the diameter of the motor pulley if the motor speed is 2000 rev/min?
Ans. 75.0 mm

*12-40. In a fan belt, the driving wheel is 20 cm in diameter and the driven wheel is 50 cm in diameter. The power input comes from a 4-kW motor that causes the driving wheel to rotate at 300 rev/min. If the efficiency is 80 percent, calculate the revolutions per minute and the torque delivered to the driven wheel.

12-41. A log-splitting wedge has a side length of 16 cm, and the angle at the apex is 10°. What is the ideal mechanical advantage?
Ans. 5.76

12-42. A machine has an efficiency of 72 percent. An input force of 500 N moves through a parallel distance of 40 cm. How much energy is lost in the process?

*12-43. A motor with an efficiency of 80 percent operates a winch with an efficiency of 50 percent. If the power supplied to the motor is 6 kW, how far will the winch lift a 400-kg mass in a time of 4 s?   **Ans. 2.45 m**

Critical Thinking Questions

12-44. A 60-N weight is lifted in the three different ways shown in Fig. 12-18. Compute the ideal mechanical advantage and the required input force for each application.
Ans. 2, 30N; 3, 20N; 0.33, 180N

12-45. A worm drive similar to that shown in Fig. 12-11 has n teeth in the gear wheel. (If $n = 80$, one complete turn of the worm will advance the wheel one-eightieth of a revolution.) Derive an expression for the ideal mechanical advantage of the worm gear in terms of the radius of the input pulley R, the radius of the output shaft r, and the number of teeth n in the gear wheel.

12-46. The worm drive of Prob. 12-45 has 80 teeth in the gear wheel. If the radius of the input wheel is 30 cm and the radius of the output shaft is 5 cm, what input force is

required to lift a 1200-kg load? Assume an efficiency of 80 percent.   **Ans. 30.6 N**

12-47. The oarlock on a 3.5 m oar is 1 m from the handle end. A person in a rowboat applies a force of 50 N to the handle end. What is the ideal mechanical advantage and output force? Does moving the oarlock closer to the handle end increase or decrease the mechanical advantage? How far from the handle end should the oarlock be placed to produce a 20 percent increase in the output force?

12-48. Sketch a block and tackle system with a mechanical advantage of 5.

12-49. A 60-W motor drives the input pulley of a belt drive at 150 rev/min. The diameters of the input and output pulleys are 60 cm and 20 cm, respectively. Assume an actual

Figure 12-18

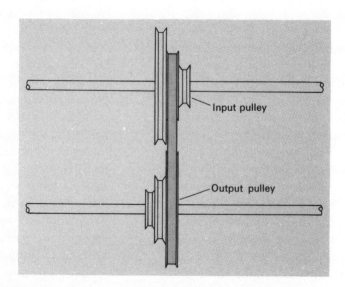

Figure 12-19

mechanical advantage of 0.25. (a) What is the output torque? (b) What is the output power? (c) What is the efficiency?

*12-50. A pair of step pulleys (Fig. 12-19) makes it possible to change output speeds merely by shifting the belt. If an electric motor turns the input pulley at 2000 rev/min, find the possible angular speeds of the output shaft. The pulley diameters are 4, 6, and 8 cm. **Ans.** (a) Small input pulley: 2000, 1330, 1000 rev/min; (b) middle input pulley: 3000, 2000, 1500 rev/min; (c) large input pulley: 4000, 2670, 2000 rev/min.

Chapter 13

Elasticity

Until now we discussed objects in motion or at rest. The objects have been assumed to be rigid and absolutely solid. But we know that wire can be stretched, that rubber tires will compress, and that bolts will sometimes break. A more complete understanding of nature requires a study of the mechanical properties of matter. The concepts of *elasticity, tension,* and *compression* are analyzed in this chapter. As the kinds of alloys increase and the demands on them become greater, our knowledge of such concepts becomes more important. For example, the stress placed on space ships or on cables in modern bridges is of a magnitude unheard of a few years ago.

13-1 ELASTIC PROPERTIES OF MATTER

We define an **elastic body** as one that returns to its original size and shape when a deforming force is removed. Rubber bands, golf balls, trampolines, diving boards, footballs, and springs are common examples of elastic bodies. Putty, dough, and clay are examples of inelastic bodies. For all elastic bodies, we shall find it convenient to establish a cause-and-effect relationship between a deformation and the deforming forces.

Consider the coiled spring of length l shown in Fig. 13-1. We can study its **elasticity** by adding successive weights and observing the increase in length. A 20 N weight lengthens the spring by 1 cm, a 40-N weight lengthens the spring by 2 cm, and a 60 N weight lengthens the spring by 3 cm. Evidently, there is a direct relationship between the elongation of a spring and the applied force.

Robert Hooke first stated this relationship in connection with the invention of a balance spring for a clock. In general, he found that a force **F** acting on a spring (Fig. 13-2) produces an elongation s that is directly proportional to the magnitude of the force. *Hooke's law* can be written

$$F = ks \qquad \text{13-1}$$

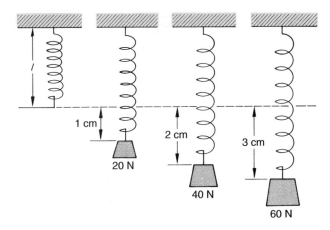

Figure 13-1 The uniform elongation of a spring.

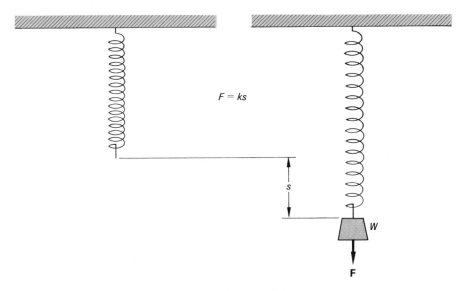

$F = ks$

Figure 13-2 The relation between a stretching force **F** and the elongation it produces.

The proportionality constant k varies extremely with the type of material and is called the **spring constant**. For the example illustrated in Fig. 13-1, the spring constant is

$$k = \frac{F}{s} = 20 \text{ N/cm}$$

Hooke's law is by no means restricted to coiled springs; it will apply to the deformation of all elastic bodies. To make the law more generally applicable, it will be convenient to define the terms *stress* and *strain*. **Stress** refers to the *cause* of an elastic deformation, whereas **strain** refers to the *effect*, in other words, the deformation itself.

Three common types of stresses and their corresponding deformations are shown in Fig. 13-3. A **tensile stress** occurs when equal and opposite forces are directed away from each other. A **compressive stress** occurs when equal and

www-sci.lib.uci.edu/SEP/ SEP.html
For science resources and a variety of answers to questions, see the Web site of Frank Potter's Science Gems.

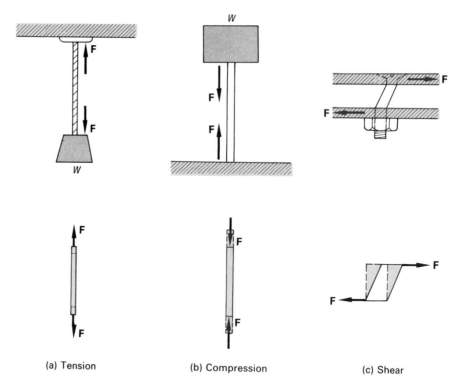

(a) Tension   (b) Compression   (c) Shear

Figure 13-3 Three common stresses shown with their corresponding deformations: (a) tension, (b) compression, and (c) shear.

opposite forces are directed toward each other. A *shearing stress* occurs when equal and opposite forces do not have the same line of action.

The effectiveness of any force producing a stress is highly dependent on the area over which the force is distributed. For this reason, a more complete definition of *stress* can be stated as follows:

> **Stress is the ratio of an applied force to the area over which it acts, for example, newtons per square meter or pounds per square foot.**

As mentioned earlier, the term *strain* must represent the effect of a given stress. The general definition of strain might be as follows:

> **Strain is the relative change in the dimensions or shape of a body as the result of an applied stress.**

In the case of a tensile or compressive stress, strain may be considered a change in length per unit length. A shearing stress, on the other hand, may alter only the shape of a body without changing its dimensions. Shearing strain is usually measured in terms of an angular displacement.

The *elastic limit* is the maximum stress a body can experience without becoming permanently deformed. For example, an aluminum rod whose cross-sectional area is 1 in.^{2} will become permanently deformed by the application of a tensile force greater than 19,000 lb. This does not mean that the aluminum rod will break at this point; it means only that the rod will not return to its original size. In fact, the tension can be increased to about 21,000 lb before the rod breaks. It is this property of metals that allows them to be drawn out into wires of smaller cross sections. The greatest stress a wire can withstand without breaking is known as its **ultimate strength.**

If the elastic limit of a material is not exceeded, we can apply Hooke's law to any elastic deformation. Within the limits of a given material, it has been experimentally verified that the ratio of a given stress to the strain it produces is a constant. In other words, the stress is directly proportional to the strain. **Hooke's law** states:

> **Provided that the elastic limit is not exceeded, an elastic deformation (strain) is directly proportional to the magnitude of the applied force per unit area (stress).**

If we call the proportionality constant the *modulus of elasticity*, we can write Hooke's law in its most general form:

$$\text{Modulus of elasticity} = \frac{stress}{strain} \qquad \textbf{13-2}$$

In the following sections we shall discuss the specific applications of this fundamental relation.

13-2 YOUNG'S MODULUS

In this section we consider longitudinal stresses and strains as they apply to wires, rods, or bars. For example, in Fig. 13-4 a force **F** is applied to the end of a wire of cross-sectional area A. The longitudinal stress is given by

$$\text{Longitudinal stress} = \frac{F}{A}$$

USING A CALCULATOR

Some problems may involve cube roots or higher-order roots. Calculators have a square root button, but many do not have a button for higher roots. To find a cube root or higher use the power button ⌃ in combination with the reciprocal button ⅟ₓ. For example, the fifth root of 9 is the same as 9 raised to the one-fifth power. To perform this function with the calculator, enter the following:

⑨ ⌃ ⑤ ⅟ₓ ⑤ 1.5518 . . .

The power button is y^x on some calculators.

Figure 13-4 Computing Young's modulus for a wire of cross section A. The elongation Δl is exaggerated for clarity.

The metric unit for stress is the *newton per square meter*, which is identical to the *pascal* (Pa).

$$1 \text{ Pa} = 1 \text{ N/m}^2$$

The USCS unit for stress is the *pound per square inch* (lb/in.^{2}). Since the pound per square inch remains in common use, it will be helpful to compare it with the SI unit:

$$1 \text{ lb/in.}^2 = 6895 \text{ Pa} = 6.895 \text{ kPa}$$

The effect of such a stress is to stretch the wire, in other words, to increase its length. Hence, the longitudinal strain can be represented by the change in length per unit length. We can write

$$\text{Longitudinal strain} = \frac{\Delta l}{l}$$

where l is the original length and Δl is the elongation. Experimentation has shown that a comparable decrease in length occurs for a compressive stress. The same equations will apply whether we are discussing an object under tension or an object under compression.

If we define the longitudinal modulus of elasticity of **Young's modulus** Y, we can write Eq. (13-2) as

$$\text{Young's modulus} = \frac{\text{longitudinal stress}}{\text{longitudinal strain}}$$

$$Y = \frac{F/A}{\Delta l/l} = \frac{Fl}{A \, \Delta l} \qquad \text{13-3}$$

The units of Young's modulus are the same as the units of stress: pounds per square inch or pascals. This makes sense because the longitudinal strain is a unitless quantity. Representative values for some of the most common materials are listed in Tables 13-1 and 13-2.

TABLE 13-1 Elastic Constants for Various Materials in SI Units

Material	Young's Modulus Y, MPa*	Shear Modulus S, MPa	Bulk Modulus B, MPa	Elastic Limit, MPa	Ultimate Strength, MPa
Aluminum	68,900	23,700	68,900	131	145
Brass	89,600	35,300	58,600	379	455
Copper	117,000	42,300	117,000	159	338
Iron	89,600	68,900	96,500	165	324
Steel	207,000	82,700	159,000	248	489

*(1 MPa = 10^6 Pa)

TABLE 13-2 Elastic Constants for Various Materials in USCS Units

Material	Young's Modulus Y, lb/in.^{2}	Shear Modulus S, lb/in.^{2}	Bulk Modulus B, lb/in.^{2}	Elastic Limit, lb/in.^{2}	Ultimate Strength, lb/in.^{2}
Aluminum	10×10^6	3.44×10^6	10×10^6	19,000	21,000
Brass	13×10^6	5.12×10^6	8.5×10^6	55,000	66,000
Copper	17×10^6	6.14×10^6	17×10^6	23,000	49,000
Iron	13×10^6	10×10^6	14×10^6	24,000	47,000
Steel	30×10^6	12×10^6	23×10^6	36,000	71,000

EXAMPLE 13-1

A telephone wire 120 m long and 2.2 mm in diameter is stretched by a force of 380 N. What is the longitudinal stress? If the length after stretching is 120.10 m, what is the longitudinal strain? Determine Young's modulus for the wire.

Solution

The cross-sectional area of the wire is

$$A = \frac{\pi D^2}{4} = \frac{\pi (2.2 \times 10^{-3} \text{ m})^2}{4} = 3.8 \times 10^{-6} \text{ m}^2$$

$$\text{Stress} = \frac{F}{A} = \frac{380 \text{ N}}{3.8 \times 10^{-6} \text{ m}^2}$$

$$= 100 \times 10^6 \text{ N/m}^2 = 100 \text{ MPa}$$

$$\text{Strain} = \frac{\Delta l}{l} = \frac{0.10 \text{ m}}{120 \text{ m}} = 8.3 \times 10^{-4}$$

$$Y = \frac{\text{stress}}{\text{strain}} = \frac{100 \text{ MPa}}{8.3 \times 10^{-4}} = 120,000 \text{ MPa}$$

CONCEPTUAL EXAMPLE 13-2

What is the maximum mass that can be hung from a steel wire 6 mm in diameter if its elastic limit is not to be exceeded? Determine the increase in length under this load if the original length is 2 m.

Solution

From Table 13-2, the elastic limit for steel is 248 MPa or 2.48×10^8 Pa. Since this value represents the limiting stress, we write

$$\frac{F}{A} = 2.48 \times 10^8 \text{ Pa}$$

where A is the area found from

$$A = \frac{\pi D^2}{4} = \frac{\pi (0.006 \text{ m})^2}{4} = 2.83 \times 10^{-5} \text{ m}^2$$

Thus, the limiting load F is the limiting stress times the area:

$$F = (2.48 \times 10^8 \text{ Pa})(2.83 \times 10^{-5} \text{ m}^2) = 7.01 \times 10^3 \text{ N}$$

The largest mass that can be supported is found from this weight:

$$m = \frac{W}{g} = \frac{7.01 \times 10^3 \text{ N}}{9.8 \text{ m/s}^2} = 716 \text{ kg}$$

The increase in length under such a load is found from Eq. (13-3), as follows:

$$\Delta L = \frac{l}{Y}\left(\frac{F}{A}\right) = \frac{2.00 \text{ m}}{2.07 \times 10^{11} \text{ Pa}}(2.48 \times 10^8 \text{ Pa})$$

$$\Delta L = 2.40 \times 10^{-3} \text{ m}$$

The length increases by 2.40 mm to a new length of 2.0024 m.

13-3 SHEAR MODULUS

Compressive and tensile stresses produce a slight change in linear dimensions. As mentioned earlier, a shearing stress alters only the shape of a body, leaving its volume unchanged. For example, consider the parallel noncurrent forces acting on the cube in Fig. 13-5. The applied force causes each successive layer of atoms to slip sideways, much like the pages of a book under similar stress. The interatomic forces restore the block to its original shape when the stress is relieved.

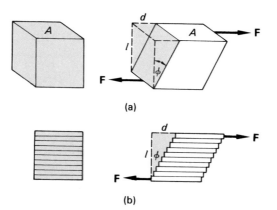

Figure 13-5 Shearing stress and shearing strain.

The **shearing stress** is defined as the ratio of the tangential force **F** to the area A over which it is applied. The **shearing strain** is defined as the angle ϕ (in radians), which is called the *shearing angle* (refer to Fig. 13-5b). Applying Hooke's law, we can now define the **shear modulus** S as follows:

$$S = \frac{\text{shearing stress}}{\text{shearing strain}} = \frac{F/A}{\phi} \qquad \text{13-4}$$

The angle ϕ is usually so small that it is approximately equal to tan ϕ. Making use of this fact, we can rewrite Eq. (13-4) in the form

$$S = \frac{F/A}{\tan \phi} = \frac{F/A}{d/l} \qquad \text{13-5}$$

Since the value of S is an indication of the rigidity of a body, it is sometimes referred to as the *modulus of rigidity*.

EXAMPLE 13-3

A steel stud (Fig. 13-6) 1 in. in diameter projects 1.5 in. out from the wall. If the end of the bolt is subjected to a shearing force of 8000 lb, compute its downward deflection.

Figure 13-6

Solution

The cross-sectional area is

$$A = \frac{\pi D^2}{4} = \frac{\pi (1 \text{ in.})^2}{4} = 0.785 \text{ in.}^2$$

If we represent the downward deflection by d, we can solve as follows:

$$S = \frac{F/A}{d/l} = \frac{Fl}{Ad}$$

$$d = \frac{Fl}{AS} = \frac{(8000 \text{ lb})(1.5 \text{ in.})}{(0.785 \text{ in.}^2)(12 \times 10^6 \text{ lb/in.}^2)}$$

$$= 1.27 \times 10^{-3} \text{ in.}$$

13-4 VOLUME ELASTICITY; BULK MODULUS

So far we have considered stresses that cause a change in the shape of an object or result in primarily one-dimensional strains. In this section we shall be concerned with changes in volume. For example, consider the cube in Fig. 13-7 on which forces are applied uniformly over the surface. The initial volume of the cube is denoted by V, and the area of each face is represented by A. The resultant force **F** applied normally to each face causes a change in volume $-\Delta V$. The minus sign indicates that the change represents a volume reduction. The *volume stress* F/A is the normal force per unit area, whereas the *volume strain* $-\Delta V/V$ is the change in volume per unit volume. Applying Hooke's law, we define the modulus of volume elasticity, or **bulk modulus,** as follows:

$$B = \frac{\text{volume stress}}{\text{volume strain}} = -\frac{F/A}{\Delta V/V} \qquad \textbf{13-6}$$

This type of strain applies to liquids as well as solids. Table 13-3 lists the bulk moduli for a few of the more common liquids. When working with liquids, it is sometimes more convenient to represent the stress as pressure P, which is defined as the force per unit area F/A. With this definition, we can rewrite Eq. (13-6):

$$B = \frac{-P}{\Delta V/V} \qquad \textbf{13-7}$$

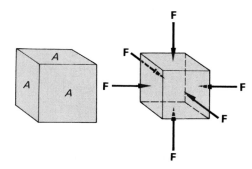

Figure 13-7 The bulk modulus. The original volume is reduced by the application of a uniform compressive force on each face.

The reciprocal of the bulk modulus is called the *compressibility k*. Often it is more convenient to study the elasticity of materials by measuring their compressibilities. By definition,

$$k = \frac{1}{B} = -\left(\frac{1}{P}\right)\frac{\Delta V}{V_0} \qquad \textbf{13-8}$$

Equation (13-8) indicates that the compressibility is the fractional change in volume per unit increase in pressure.

TABLE 13-3 Bulk Moduli for Liquids

| Liquid | Bulk Modulus B | |
	MPa	lb/in.^{2}
Benzene	1,050	1.5×10^5
Ethyl alcohol	1,100	1.6×10^5
Mercury	27,000	40×10^5
Oil	1,700	2.5×10^5
Water	2,100	3.1×10^5

EXAMPLE 13-4

A hydrostatic press contains 5 liters of water. Find the decrease in volume of the water when it is subjected to a pressure of 2000 kPa.

Solution

The decrease in volume is found by solving for ΔV from Eq. (13-7):

$$\Delta V = -\frac{PV}{B} = \frac{(2 \times 10^6 \text{ Pa})(5 \text{ L})}{2.1 \times 10^9 \text{ Pa}}$$
$$= -0.00476 \text{ L} = -4.76 \text{ mL}$$

13-5 OTHER PHYSICAL PROPERTIES OF METALS

In addition to the elasticity, tensile strength, and shearing strength of materials, there are other important properties of metals. A solid consists of molecules arranged so closely together that they attract each other strongly. This attraction, called *cohesion*, gives a solid a definite shape and size. It also affects its usefulness to industry as a working material. Properties such as hardness, ductility, malleability, and conductivity must be understood before metals are chosen for specific applications. Three of these properties are illustrated in Fig. 13-8.

USING A CALCULATOR

Try working problems with a partner or partners. Compare your answers. If there are differences, often the "team" can compare steps to spot where the error was introduced.

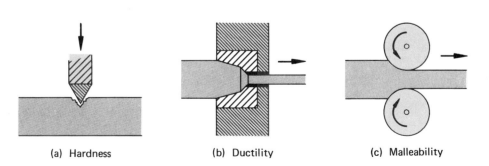

   (a) Hardness   (b) Ductility   (c) Malleability

Figure 13-8 Illustrating the working properties of metals: (a) a hard metal resists penetration, (b) a ductile metal can be drawn into a wire, and (c) a malleable metal can be rolled into sheets.

Hardness is an industrial term used to describe the ability of metals to resist forces that tend to penetrate them. Hard materials resist being scratched, worn away, penetrated, or otherwise damaged physically. Some metals such as sodium and potassium are soft, while iron and steel are two of the hardest materials. The hardness of metals is tested with machines that push a cone-shaped diamond point into test materials. The penetration is measured, and a hardness reading is taken directly from a dial.

Two other special properties of materials are *ductility* and *malleability*. The meaning of each of these terms is seen from Fig. 13-8. Ductility is defined as the ability of a metal to be drawn out into a wire. Tungsten and copper are extremely ductile. Malleability is the property that enables us to hammer or bend metals into any desired shape or to roll them into sheets. Most metals are malleable, gold being the most malleable.

Conductivity refers to the ability of metals to permit the flow of electricity. The best conductors are silver, copper, gold, and aluminum, in that order. More will be said about this property in later chapters.

Summary and Review

Summary

In industry, we must utilize materials effectively and for appropriate situations. Otherwise, metal failures will result in costly damage or serious injury to employees. In this chapter, we have discussed the elastic properties of matter and some of the formulas used to predict the effects of stress on certain solids. The following points will summarize this chapter:

- According to *Hooke's law,* an elastic body will deform or elongate an amount s under the application of a force $\mathbf{F}$. The constant of proportionality k is the *spring constant:*

$$F = ks \qquad k = \frac{F}{s} \qquad \textit{Hooke's Law}$$

- *Stress* is the ratio of an applied force to the area over which it acts. *Strain* is the relative change in dimensions that results from the stress.

  For example,

$$\text{Longitudinal stress} = \frac{F}{A}$$

$$\text{Longitudinal strain} = \frac{\Delta l}{l}$$

- The *modulus of elasticity* is the constant ratio of stress to strain:

$$\text{Modulus of elasticity} = \frac{stress}{strain}$$

- *Young's modulus Y* is for longitudinal deformations:

$$Y = \frac{F/A}{\Delta l/l}$$

$$\text{or} \qquad Y = \frac{Fl}{A \cdot \Delta l} \qquad \textit{Young's Modulus}$$

- A shearing strain occurs when an angular deformation ϕ is produced:

$$S = \frac{F/A}{\tan \phi}$$

$$\text{or} \qquad S = \frac{F/A}{d/l} \qquad \textit{Shear Modulus}$$

- Whenever an applied stress results in a change in volume ΔV, you will need the *bulk modulus B,* given by

$$B = -\frac{F/A}{\Delta V/V} \qquad \textit{Bulk Modulus}$$

The reciprocal of the bulk modulus is called the compressibility.

Key Terms

1. elasticity
2. Hooke's law
3. spring constant
4. tensile stress
5. compressive stress
6. shear stress
7. strain
8. elastic limit
9. ultimate strength
10. Young's modulus
11. shear modulus
12. bulk modulus

Review Questions

13-1. Explain clearly the relationship between stress and strain.

13-2. Two wires have the same length and cross-sectional area but are not of the same material. Each wire is hung from the ceiling with a 2000-lb weight attached. The wire on the left stretches twice as far as the one on the right. Which has the greater Young's modulus?

13-3. Does Young's modulus depend on the length and cross-sectional area? Explain.

13-4. Two wires, A and B, are made of the same material and are subjected to the same loads. Discuss their relative elongations when (a) wire A is twice as long as wire B and has twice the diameter of wire B and (b) wire A is twice as long as wire B and has one-half the diameter of wire B.

13-5. After studying the various elastic constants given in Tables 13-1 and 13-2, would you say that it was usually easier to stretch a material or to shear a material? Explain.

13-6. A 200-kg mass is supported evenly by three wires, one of copper, one of aluminum, and one of steel. If the wires have the same dimensions, which wire experiences the greatest stress, and which the least? Which has the greatest strain and the least strain?

13-7. Discuss the various stresses resulting when a machine screw is tightened?

13-8. Give several practical examples of longitudinal, shearing, and volume strains.

13-9. For a given metal, would you expect there to be any relationship between its modulus of elasticity and its coefficient of restitution? Discuss.

13-10. Which has the greater compressibility, steel or water?

Problems

Section 13-1—Elastic Properties of Matter

13-1. When a mass of 500 g is hung from a spring, the spring stretches 3 cm. What is the spring constant?　　**Ans. 163 N/m**

13-2. What will be the increase in stretch for the spring of Prob. 13-1 if an additional 500-g mass is hung below the first?

13-3. The spring constant for a certain spring is found to be 3000 N/m. What force is required to compress the spring for a distance of 5 cm?　　**Ans. 150 N**

13-4. A 6-in. spring has a 4-lb weight hung from one end, causing the new length to be 6.5 in. What is the spring constant? What is the strain?

13-5. A coil spring 12 cm long is used to support a 1.8-kg mass producing a strain of 0.10. How far did the spring stretch? What is the spring constant?　　**Ans. 1.2 cm, 1470 N/m**

13-6. For the coil of Prob. 13-5, what total mass should be hung if an elongation of 4 cm is desired?

Section 13-2—Young's Modulus

13-7. A 60-kg weight is suspended by means of a cable that has a diameter of 9 mm. What is the stress?　　**Ans. 9.24 Mpa**

13-8. A 50-cm length of wire is stretched to a new length of 50.01 cm. What is the strain?

13-9. A 12-m rod receives a compressional strain of -0.0004. What is the new length of the rod?　　**Ans. 11.995 m**

13-10. Young's modulus for a certain rod is 4×10^{11} Pa. What strain will be produced by a tensile stress of 420 Mpa?

13-11. A 500-kg mass is hung from the end of a 2-m length of metal wire 1 mm in diameter. If the wire stretches by 1.40 cm, what are the stress and strain? What is Young's modulus for this metal?
Ans. 6.24×10^9 Pa, 7.00×10^{-3}, 8.91×10^{11} Pa

13-12. A 16-ft steel girder with a cross-sectional area of 10 in.^{2} supports a compressional load of 20 tons. What is the decrease in length of the girder?

13-13. How much will a 60-cm length of brass wire, 1.2 mm in diameter, elongate when a 3-kg mass is hung from an end?

Ans. 0.174 mm

13-14. A wire of cross section 4 mm^2 is stretched 0.1 mm by a certain weight. How far will a wire of the same material and length stretch if its cross-sectional area is 8 mm^2 and the same weight is attached?

13-15. A wire 15 ft long and 0.1 in.^{2} in cross section is found to increase its length by 0.01 ft under a tension of 2000 lb. What is Young's modulus for this wire? Can you identify the material?

Ans. 30 × 10^6 lb/in.^{2}, steel

Section 13-3—Shear Modulus

13-16. A shearing force of 40,000 N is applied to the top of a cube that is 30 cm on a side. What is the shearing stress?

13-17. If the cube in Prob. 13-16 is made of copper, what will be the lateral displacement of the upper surface of the cube?

Ans. 3.15 μm

13-18. A shearing force of 26,000 N is distributed uniformly over the cross section of a pin 1.3 cm in diameter. What is the shearing stress?

13-19. An aluminum rod 20 mm in diameter projects 4.0 cm from the wall. The end of the bolt is subjected to a shearing force of 48,000 N. Compute the downward deflection.

Ans. 0.257 mm

13-20. A steel rod projects 1.0 in. above a floor and is 0.5 in. in diameter. The shearing force F is 6000 lb, and the shear modulus is 11.6 × 10^6 lb/in.^{2}. What is the shearing stress, and what is the horizontal deflection?

13-21. A 1500-kg load is supported at the end of a 5-m aluminum beam as shown in Fig. 13-9. The beam has a cross-sectional area of 26 cm^2, and the shear modulus is 23,700 MPa. What are the shearing stress and the downward deflection of the beam?

Ans. 5.65 × 10^6 Pa, 1.19 mm

5 m

1500 kg

Figure 13-9

13-22. A steel plate 0.5 in. thick has an ultimate shearing strength of 50,000 lb/in.^{2}. What force must be applied to punch a ¼-in. hole through the plate?

Section 13-4—Bulk Modulus

13-23. A pressure of 3 × 10^8 Pa is applied to a block of volume 0.500 m^3. If the volume decreases by 0.004 m^3, what is the bulk modulus? What is the compressibility?

Ans. 37,500 MPa, 2.67 × 10^{-11} Pa^{-1}

13-24. The bulk modulus for a certain grade of oil is 2.8 × 10^{10} Pa. What pressure is required to decrease its volume by a factor of 1.2 percent?

13-25. A solid brass sphere ($B = 35,000$ Mpa) of volume 0.8 m^3 is dropped into the ocean to a depth where the water pressure is 20 Mpa greater than it is at the surface. What is the change in volume of the sphere?

Ans. −4.57 × 10^{-4} m^3

13-26. A certain fluid compresses 0.40 percent under a pressure of 6 MPa. What is the compressibility of this fluid?

13-27. What is the fractional decrease in the volume of water when it is subjected to a pressure of 15 MPa?

Ans. −0.00714

Summary and Review

Challenge Problems

13-28. A 10-m steel wire, 2.5 mm in diameter, stretches a distance of 0.56 mm when a load is attached to its end. What was the mass of the load?

13-29. A shearing force of 3000 N is applied to the upper surface of a copper cube 40 mm on a side. Assume $S = 4.2 \times 10^{10}$ Pa. What is the shearing angle? **Ans. 4.46 × 10⁻⁵ rad**

13-30. A solid cylindrical steel column is 6 m long and 8 cm in diameter. What is the decrease in length if the column supports a 90,000-kg load?

13-31. A piston, 8 cm in diameter, exerts a force of 2000 N on 1 liter of benzene. What is the decrease in volume of the benzene? **Ans. −3.79 × 10⁻⁷ m³**

13-32. How much will a 600-mm length of brass wire, 1.2 mm in diameter, stretch when a 4-kg mass is hung from its end?

13-33. A solid cylindrical steel column is 12 ft tall and 6 in. in diameter. What load does it support if its decrease in length is −0.0255 in.? **Ans. 1.50 × 10⁵ lb**

13-34. Compute the volume contraction of mercury if its original volume of 1600 cm³ is subjected to a pressure of 400,000 Pa.

13-35. What is the minimum diameter of a brass rod if it is to undergo a 400-N tension without exceeding the elastic limit? **Ans. 1.16 mm**

13-36. A cubical metal block 40 cm on a side is given a shearing force of 400,000 N at the top edge. What is the shear modulus for this metal if the upper edge deflects a distance of 0.0143 mm?

13-37. A steel piano wire has an ultimate strength of about 35,000 lb/in.². How large a load can a 0.5-in.-diameter steel wire hold without breaking? **Ans. 6870 lb**

Critical Thinking Questions

13-38. A metal wire increases its length by 2 mm when subjected to tensile force. What elongation could be expected from this same force if the diameter of the wire was reduced to one-half of its initial value? Suppose the metal wire maintains its diameter, but doubles its length. What elongation would be expected for the same load? **Ans. 0.500 mm, 4 mm**

13-39. A cylinder 4 cm in diameter is filled with oil. What total force must be exerted on the oil to produce a 0.8 percent decrease in volume? Compare the forces necessary if the oil is replaced by water. By mercury.

*13-40. A 15-kg ball is connected to the end of a steel wire 6 m long and 1.0 mm in diameter. The other end of the wire is connected to a high ceiling, forming a pendulum. If

we ignore the small change in length, what is the maximum speed that the ball can have as it passes through its lowest point without exceeding the elastic limit? How much will the length of the wire increase under the limiting stress? What effect will this change have on the maximum velocity? **Ans. 4.38 m/s, 7.19 mm, slightly increases maximum velocity**

13-41. A cylinder 10 in. in diameter is filled to a height of 6 in. with glycerin. A piston of the same diameter pushes downward on the liquid with a force of 800 lb. The compressibility of glycerin is 1.50×10^{-6} in.²/lb. What is the stress on the glycerin? How far down does the piston move?

*13-42. The twisting of a cylindrical shaft (Fig. 13-10) through an angle θ is an example of a shearing strain. An analysis of the situation shows that the angle of twist in radians is given by

$$\theta = \frac{2\tau l}{\pi S R^4}$$

where τ = applied torque
l = length of cylinder
R = radius of cylinder
S = shear modulus

If a torque of 100 lb · ft is applied to the end of a cylindrical steel shaft 10 ft long and 2 in. in diameter, what will be the angle of twist in radians?   **Ans. 0.00764 rad**

*13-43. An aluminum shaft 1 cm in diameter and 16 cm high is subjected to a torsional shearing stress as explained in the previous problem. What applied torque will cause a twist of 1° as defined in Fig. 13-10?

*13-44. Two sheets of aluminum on an aircraft wing are to be held together by aluminum rivets of cross-sectional area 0.25 in.². The shearing stress on each rivet must not

Figure 13-10   A torque τ applied at one end of a solid cylinder causes it to twist through an angle θ.

exceed one-tenth of the elastic limit for aluminum. How many rivets are needed if each rivet supports the same fraction of the total shearing force of 25,000 lb?

Ans. 53 rivets

Simple Harmonic Motion

OBJECTIVES

After completing this chapter, you should be able to:

1. Describe and apply the relationship between force and displacement in simple harmonic motion.
2. Use the reference circle to describe the variation in magnitude and direction of displacement, velocity, and acceleration for simple harmonic motion.
3. Write and apply formulas for the determination of displacement x, velocity v, or acceleration a in terms of time, frequency, and amplitude.
4. Write and apply a relationship between the frequency of motion and the mass of a vibrating object when the spring constant is known.
5. Compute the frequency or period in simple harmonic motion when the position and acceleration are given.
6. Describe the motion of a simple pendulum and calculate the length required to produce a given frequency.

Until now we have discussed the motion of objects under the influence of a constant, unchanging force. Such motion was described by calculating the position and velocity as functions of time. The real world often consists, however, of varying forces. Common examples are swinging pendulums, balance wheels of watches, tuning forks, a mass vibrating at the end of a coiled spring, and vibrating air columns in musical instruments. In these cases and many others, we need a more complete description of motion caused by a resultant force that varies in a predictable manner.

14-1 PERIODIC MOTION

Whenever an object is deformed, an elastic restoring force appears that is proportional to the deformation. When released, such an object will vibrate back and forth about its equilibrium position. For example, after a swimmer springs from a diving board (Fig. 14-1), the board continues to vibrate about its normal position for a definite length of time.

This type of motion is said to be *periodic* because the position and velocity of the moving particles are repeated as a function of time. Since the restoring force is reduced after each vibration, the board eventually comes to rest.

Figure 14-1 The periodic vibration of a diving board.

Periodic motion is that motion in which a body moves back and forth over a fixed path, returning to each position and velocity after a definite interval of time.

An ice puck coupled to a spring is shown in Fig. 14-2. The ice puck is a laboratory device that rides on a cushion of carbon dioxide, approximating frictionless motion. Suppose we pull the ice puck to the side and release it so that it oscillates about its initial position O without friction. According to Hooke's law, the restoring force **F** is directly proportional to the displacement x of the ice puck from its equilibrium position O. Since the restoring force is always opposed to the displacement, a negative sign must be introduced. Thus, we write

$$F = -kx \qquad \text{14-1}$$

The maximum displacement of the ice puck from its equilibrium position is called the **amplitude A.** At this position, the ice puck experiences the maximum restoring force and consequently its maximum acceleration. The force decreases as the puck approaches its center of oscillation, becoming zero at that point. The momentum of the puck carries it on past the center, however, bringing the restoring force of the spring back into play again. This force increases until the puck comes to a stop at its maximum displacement, whereupon it begins its return trip. If there were no loss of energy due to friction, the back-and-fourth motion would continue indefinitely. This type of oscillatory motion in the absence of friction is known as **simple harmonic motion (SHM).**

www.mip.berkeley.edu/ physics/physics.html
If you need help picturing periodic motion, try this site.

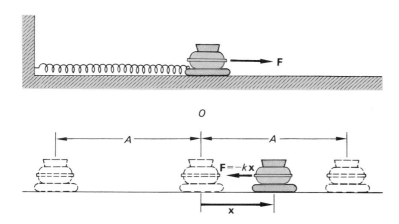

Figure 14-2 The frictionless ice puck can be used to illustrate simple harmonic motion. The restoring force **F** is always directed toward the center of oscillation O.

Simple harmonic motion is periodic motion in the absence of friction and produced by a restoring force that is directly proportional to the displacement and oppositely directed.

The *period T* is defined as the time for one complete trip, or oscillation. For example, if the ice puck is released from the right at a distance A from its equilibrium position, the time required for it to return to that position is its period of vibration. It must be pointed out, however, that every point in the vibratory path must be covered in measuring a complete oscillation. The time required to move from the center of oscillation to the distance A and back represents only one-half of a period.

The *frequency f* is the number of complete oscillations per unit time. Since the period is the time for one oscillation, the frequency must be the reciprocal of the period, or

$$f = \frac{1}{T}$$

The unit of frequency is often expressed as vibrations per second or as inverse seconds (s^{-1}); the SI unit for frequency is the *hertz* (Hz).

$$1 \text{ Hz} = \frac{1}{s}$$

Thus, a frequency of 500 vibrations per second becomes 500 Hz.

14-2 THE REFERENCE CIRCLE

It has been shown that an object vibrating with SHM is influenced by a restoring force that is directly proportional to its displacement. It we apply Newton's second law to Eq. (14-1), we shall see that the acceleration is also proportional to the displacement. Thus, $F = ma = -kx$, from which

$$a = -\frac{k}{m}x \qquad \text{14-2}$$

As long as the mass m remains constant, the ratio k/m will also be constant, indicating that the magnitude of the acceleration increases with the displacement. The minus sign appears because the acceleration is always directed opposite to the displacement.

Since the acceleration in SHM is not constant, the equations derived in earlier chapters will not apply. To determine new relationships directly requires the use of calculus. Fortunately, these equations can be deduced by geometrical methods. When a body moves in a circular path with uniform speed, its linear projection moves with SHM. This fact is illustrated in Fig. 14-3, where the shadow of a ball attached to a rotating disk oscillates back and forth with periodic motion. This experiment suggests that our knowledge of uniform circular motion may be helpful in describing SHM.

The *reference circle* in Fig. 14-4 is used to compare the motion of an object moving in a circle with its horizontal projection. Since it is the motion of the projec-

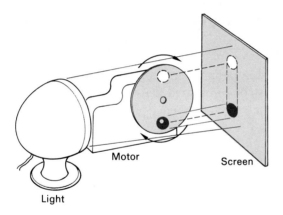

Motor   Screen

Light

Figure 14-3 The projection or shadow of a ball attached to a rotating disk moves with simple harmonic motion.

tion that we wish to study, we shall refer to the position P of the object moving in a circle as the *reference point*. The radius of the reference circle is equal to the amplitude of the horizontal oscillation. If the linear speed v_T and angular velocity ω of the reference point are constant, the projection Q will move back and forth with SHM. Time is assigned a zero value when the reference point is at B in Fig. 14-4. At any later time t, the reference point P will have moved through an angle θ. The displacement x of the projection Q is therefore

$$x = A \cos \theta$$

Recalling that the angle $\theta = \omega t$, we can now write the displacement as a function of the angular velocity of the reference point.

$$x = A \cos \theta = A \cos \omega t \qquad \textbf{14-3}$$

Although the angular velocity ω is useful for describing motion of the reference point P, it is not applied directly to the projection Q. We should recall, however,

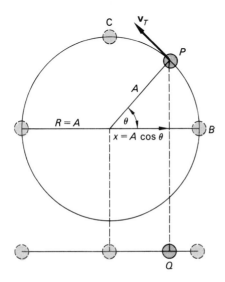

Figure 14-4 Displacement in simple harmonic motion.

that the angular velocity is related to the frequency of revolution by

$$\omega = 2\pi f$$

where ω is expressed in radians per second and f is the number of revolutions per second. It should also be recognized that the projection Q will describe one complete oscillation, whereas the reference point describes one complete revolution. Thus, the frequency f is the same for each point. Substituting $\omega = 2\pi f$ into Eq. (14-3), we obtain

$$x = A \cos 2\pi ft \qquad \text{14-4}$$

This equation can be applied to compute the displacement of a body moving with SHM of amplitude A and frequency f. Remember that the displacement x is always measured from the center of oscillation.

14-3 VELOCITY IN SIMPLE HARMONIC MOTION

Let us consider a body moving back and forth with SHM under the influence of a restoring force. Since the direction of the vibrating body is reversed at the end points of its motion, its velocity must be zero when its displacement is a maximum. It is then accelerated toward the center by the restoring force until it reaches its maximum speed at the center of oscillation, when its displacement is zero.

In Fig. 14-5 the velocity of a vibrating body is compared at three instants with corresponding points on a reference circle. It will be noticed that the velocity $\mathbf{v}$ of the vibrating body at any instant is the horizontal component of the tangential velocity $\mathbf{v}_T$ of the reference point. At point B the reference point is moving vertically upward and has no horizontal velocity. This point therefore corresponds to the zero velocity of the vibrating body when it reaches its amplitude A. At point C the horizontal component of $\mathbf{v}_T$ is equal to its entire magnitude. This point corresponds to a position of maximum velocity for the vibrating body, in other words, at its center of oscillation. In general, the velocity of the vibrating body at any point Q is determined from the reference circle to be

$$v = -v_T \sin \theta = -v_T \sin \omega t \qquad \text{14-5}$$

The minus sign appears since the direction of the velocity is toward the left. We can make this equation more useful by recalling the relationship between the tangential velocity v_T and the angular velocity:

$$v_T = \omega A = 2\pi fA$$

Substitution into Eq. (14-5) yields

$$v = -2\pi fA \sin 2\pi ft \qquad \text{14-6}$$

This equation will give the velocity of a vibrating body at any instant if it is remembered that $\sin \theta$ is negative when the reference point lies below the diameter of the reference circle.

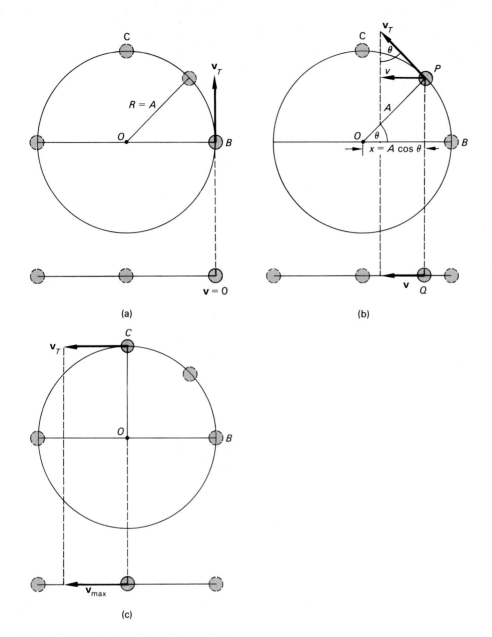

Figure 14-5 Velocity and the reference circle.

CONCEPTUAL EXAMPLE 14-1

An ice puck attached to a spring, as in Fig. 14-2, is pulled to the right a distance of 6 cm and released. (a) If it returns to the point of release in 2 s and continues to vibrate with SHM, compute its position and velocity after 5.2 s. (b) What is its maximum velocity?

Solution (a)

The time for one complete vibration is 2 s. Thus, the frequency is given by

$$f = \frac{1}{T} = \frac{1}{2 \text{ s}} = 0.5 \text{ Hz}$$

The position after 5.2 s is

$$x = A \cos 2\pi ft$$
$$= (6 \text{ cm}) \cos [(2\pi)(0.5 \text{ Hz})(5.2 \text{ s})]$$
$$= (6 \text{ cm}) \cos (16.34 \text{ rad})$$

Before evaluating cos (16.34 rad) make sure that the calculator is set to read angles in radians (rad). It is also advisable to avoid rounding of numbers until your final answer. A small error in the radian measure is significant. Once these precautions have been taken, the displacement x becomes

$$x = (6 \text{ cm})(-0.809) = -4.85$$

The minus sign indicates that the ice puck is 4.85 cm to the left of its equilibrium position.

The velocity of the puck after 5.2 s is found from

$$v = -2\pi fA \sin 2\pi ft$$
$$= -2\pi(0.5 \text{ Hz})(6 \text{ cm})(\sin 16.34 \text{ rad})$$
$$= (-18.8 \text{ cm/s})(-0.588)$$
$$= 11.1 \text{ cm/s}$$

The velocity is positive, indicating that it is moving to the right.

Solution (b)

The maximum velocity occurs when the displacement is zero or when the reference angle is 90 or 270°. Thus,

$$v_{\text{max}} = -2\pi fA \sin 90° = -2\pi fA$$
$$= -2\pi(0.5 \text{ Hz})(6 \text{ cm})$$
$$= -18.8 \text{ cm/s}$$

This represents the maximum velocity directed to the left, since we chose 90° as a reference angle. If 270° had been chosen, a positive value would have resulted.

14-4 ACCELERATION IN SIMPLE HARMONIC MOTION

The velocity of a vibrating body is never constant. Therefore, although it was not often mentioned, acceleration plays an important role in the equations derived in the previous section. We now attempt to obtain an expression that will allow us to determine the acceleration of objects under the influence of a restoring force.

At the position of maximum displacement, the velocity of a vibrating object is zero. It is at this instant that the body is acted on by the maximum restoring force. Thus, the acceleration of the body is a maximum when its velocity is zero. As the object approaches its equilibrium position, the restoring force (and therefore the acceleration) is reduced until it reaches zero at the center of oscillation. At the equilibrium position, the acceleration is zero and the velocity has its maximum value.

Figure 14-6 demonstrates that the acceleration a of a particle moving with SHM is equal to the horizontal component of the centripetal acceleration a_c of the reference point. From the figure,

$$a = -a_c \cos \theta = -a_c \cos \omega t \qquad \text{14-7}$$

The minus sign indicates that the acceleration is directed toward the left. Recalling that $a_c = \omega^2 R$ and that $R = A$, we can rewrite Eq. (14-7) as

$$a = -\omega^2 A \cos \omega t$$

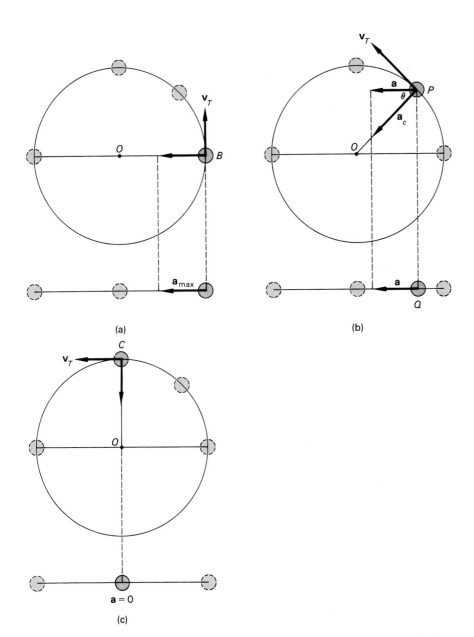

(a)

(b)

(c)

Figure 14-6 Acceleration and the reference circle.

Substituting $\omega = 2\pi f$, as in the previous section, we obtain

$$a = -4\pi^2 f^2 A \cos 2\pi ft \qquad \text{14-8}$$

Equation (14-8) can be simplified by noting from Eq. (14-4) that

$$\cos \theta = \cos 2\pi ft = \frac{x}{A}$$

Substituting this result into Eq. (14-8), we have

$$a = -4\pi^2 f^2 A \frac{x}{A}$$

or

$$a = -4\pi^2 f^2 x \qquad \text{14-9}$$

Therefore, the acceleration is directly proportional to the displacement and opposite in direction.

14-5 THE PERIOD AND FREQUENCY

From the information now established concerning displacement, velocity, and acceleration of vibrating bodies, we can derive some useful formulas for computing the period or frequency of vibration. For example, if we solve Eq. (14-9) for the frequency f, we obtain

$$f = \frac{1}{2\pi} \sqrt{-\frac{a}{x}} \qquad \text{14-10}$$

Since the **displacement** x and the acceleration are always opposite in signs, the term $-a/x$ is always positive.

The period T is the reciprocal of the frequency. Making use of this fact in Eq. (14-10), we define the period by

$$T = 2\pi \sqrt{-\frac{x}{a}} \qquad \text{14-11}$$

Thus, if the acceleration is known at a particular displacement, the period of vibration can be computed.

When the motion of bodies under the influence of an elastic **restoring force** is considered, it is more convenient to express the period as a function of the spring constant and mass of the vibrating body. This can be done by comparing Eqs. (14-2) and (14-9):

$$a = -\frac{k}{m}x \qquad a = -4\pi^2 f^2 x$$

Combining these relations, we obtain

$$4\pi^2 f^2 = \frac{k}{m}$$

from which the frequency is

$$f = \frac{1}{2\pi}\sqrt{\frac{k}{m}}$$ **14-12**

Finally, the period T is given by the reciprocal of the frequency. Thus,

$$T = 2\pi\sqrt{\frac{m}{k}}$$ **14-13**

Note that neither the period nor the frequency depends on the amplitude (maximum displacement) of the vibrating body. They depend only on the spring constant and the mass of the vibrating body.

**www3.adnc.com/
~topquark/fun**
For an applet demonstrating the concepts of velocity, position, energy, and damping coefficient, see this site.

EXAMPLE 14-2

A 2-kg steel ball is attached to the end of a flat strip of metal that is clamped at its base, as shown in Fig. 14-7. (a) If a force of 5 N is required to displace the ball by 16 cm, what will its period of vibration be upon release? (b) What is its maximum acceleration?

Figure 14-7

Solution (a)
A force of 5 N displaces the mass by 16 cm. Thus, the spring constant is

$$k = \frac{F}{x} = \frac{5\text{ N}}{0.16\text{ m}} = 31.2\text{ N/m}$$

The period is found from Eq. (14-13).

$$T = 2\pi\sqrt{\frac{m}{k}} = 2\pi\sqrt{\frac{2\text{ kg}}{31.2\text{ N/m}}}$$

$$= 2\pi(0.253) = 1.59\text{ s}$$

Solution (b)

The maximum acceleration occurs when the displacement is a maximum: when $x = 0.16$ m. Thus,

$$a = -4\pi^2 f^2 x = -\frac{4\pi^2}{T^2}x$$

$$= -\frac{4\pi^2(0.16 \text{ m})}{(1.59 \text{ s})^2} = -2.5 \text{ m/s}^2$$

14-6 THE SIMPLE PENDULUM

When a heavy pendulum bob is swinging at the end of a light cord or rod, as in Fig. 14-8, it approximates SHM. If we assume that all the mass is concentrated at the center of gravity of the bob and that the restoring force acts at a single point, we refer to this apparatus as a *simple* pendulum. Although this assumption is never strictly true, a close approximation is obtained by making the mass of the connecting rod or cord small in comparison to the pendulum bob.

Notice that the displacement x of the bob is not along a straight line but lies along the arc subtended by the angle θ. From methods discussed in Chap. 11, the length of the displacement is simply the product of the angle θ and the length of the cord. Thus,

$$x = l\theta$$

If the motion of the bob is SHM, the restoring force must be given by

$$F = -kx = -kl\theta \qquad \text{14-14}$$

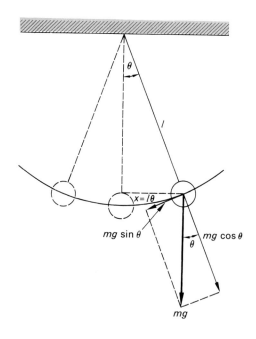

Figure 14-8

which means that the restoring force should be proportional to θ since the length l is constant. Let us examine the restoring force to see if this is the case. In the back-and-forth movement of the pendulum bob, the necessary restoring force is provided by the tangential component of the weight. From Fig. 14-8, we can write

$$F = -mg \sin \theta \qquad \textbf{14-15}$$

Thus, the restoring force is proportional to $\sin \theta$ instead of to θ. The conclusion must be that the bob does not oscillate with SHM. If we stipulate that the angle θ is small, however, $\sin \theta$ will be approximately equal to the angle θ in radians. You can verify this approximation by considering several small angles:

sin θ	θ (rad)
sin 6° = 0.1045	6° = 0.1047
sin 12° = 0.208	12° = 0.209
sin 27° = 0.454	27° = 0.471

When the approximation $\sin \theta \approx \theta$ is used, Eq. (14-15) becomes

$$F = -mg \sin \theta = -mg\theta$$

Comparing this relation with Eq. (14-14), we obtain

$$F = -kl\theta = -mg\theta$$

from which

$$\frac{m}{k} = \frac{l}{g}$$

Substitution of this proportion into Eq. (14-13) yields an expression for the period of a *simple pendulum:*

$$T = 2\pi \sqrt{\frac{l}{g}} \qquad \textbf{14-16}$$

Notice that for small amplitudes the period of a simple pendulum is a function of neither the mass of the bob nor the amplitude of vibration. In fact, since the acceleration of gravity is constant, the period is determined solely by the length of the connecting cord or rod.

EXAMPLE 14-3

In a laboratory experiment, a student is given a stopwatch, a wooden bob, and a piece of cord. He is then asked to determine the acceleration due to gravity g. If he constructs a simple pendulum of length 1 m and measures the period to be 2 s, what value will he obtain for g?

Squaring both sides of Eq. (14-16) gives

$$T^2 = 4\pi^2 \frac{l}{g}$$

from which

$$g = \frac{4\pi^2 l}{T^2} = \frac{4\pi^2 (1 \text{ m})}{(2 \text{ s})^2}$$

$$= 9.87 \text{ m/s}^2$$

14-7 THE TORSION PENDULUM

Another example of SHM is the torsion pendulum (Fig. 14-9), which consists of a solid disk or cylinder supported at the end of a thin rod. If the disk is twisted through an angle θ, the restoring torque τ is directly proportional to the angular displacement. Thus,

$$\tau = -k'\theta \qquad \qquad \textbf{14-17}$$

where k' is a constant that depends on the material from which the thin rod is made. (See Prob. 13-42.)

When the disk is released, the restoring torque produces an angular acceleration that is directly proportional to the angular displacement. The period of the simple angular harmonic motion thus produced is given by

$$T = 2\pi \sqrt{\frac{I}{k'}} \qquad \qquad \textbf{14-18}$$

where I is the moment of inertia of the vibrating system and k' is the ***torsion constant*** defined in Eq. (14-17).

Figure 14-9

EXAMPLE 14-4

A solid disk of mass 0.16 kg and radius of 0.12 m is twisted through an angle of 1 rad and released. (a) If the torsion constant of the wire supporting the disk is 0.025 N · m/rad, calculate the maximum angular acceleration. (b) What is the period of oscillation?

Solution (a)

The moment of inertia of the disk is

$$I = \tfrac{1}{2}mR^2 = \tfrac{1}{2}(0.16 \text{ kg})(0.12 \text{ m})^2$$
$$= 1.15 \times 10^{-3} \text{ kg} \cdot \text{m}^2$$

The angular acceleration is a maximum when the angular displacement is 1 rad. From Eq. (14-17) we have

$$\tau = I\alpha = -k'\theta$$

$$\alpha = -\frac{k'\theta}{I} = \frac{-(0.025 \text{ N} \cdot \text{m/rad})(1 \text{ rad})}{1.15 \times 10^{-3} \text{ kg} \cdot \text{m}^2}$$

$$= -21.7 \text{ rad/s}^2$$

Solution (b)

The period from Eq. (14-18) is

$$T = 2\pi \sqrt{\frac{I}{k'}} = 2\pi \sqrt{\frac{1.15 \times 10^{-3} \text{ kg} \cdot \text{m}^2}{0.025 \text{ N} \cdot \text{m/rad}}}$$

$$= 2\pi(0.214)\text{s} = 1.35 \text{ s}$$

Notice that the period is not a function of the angular displacement.

Summary and Review

Summary

Simple harmonic motion is periodic motion in which the restoring force is proportional to the displacement. Such vibratory motion without friction produces predictable variations in displacement and velocity. The major concepts discussed in this chapter are summarized below.

- Simple harmonic motion is produced by a *restoring force* **F** given by:

$$F = -kx \qquad \textit{Restoring Force}$$

- Since $F = ma = -kx$, the acceleration produced by a restoring force is

$$a = -\frac{k}{m}x \qquad \textit{Acceleration}$$

- A convenient way to study simple harmonic motion is to use the *reference circle*. The variations in displacement x, velocity v, and acceleration a can be seen by reference to Figs. 14-4, 14-5, and 14-6.
- For SHM the displacement, velocity, and acceleration may be expressed in terms of the amplitude A, the time t, and the frequency of vibration f:

$$x = A \cos 2\pi ft \qquad \textit{Displacement}$$
$$v = -2\pi fA \sin 2\pi ft \qquad \textit{Velocity}$$
$$a = -4\pi^2 f^2 x \qquad \textit{Acceleration}$$

- The period T and the frequency f in simple harmonic motion are found from

$$f = \frac{1}{2\pi} \sqrt{-\frac{a}{x}}$$

or

$$f = \frac{1}{2\pi} \sqrt{\frac{k}{m}} \qquad \textit{Frequency}$$

$$T = 2\pi \sqrt{-\frac{x}{a}}$$

or

$$T = 2\pi \sqrt{\frac{m}{k}} \qquad \textit{Period}$$

- For a simple pendulum of length l, the period is given by

$$T = 2\pi \sqrt{\frac{l}{g}} \qquad \begin{array}{l}\textit{Period of}\\ \textit{Simple Pendulum}\end{array}$$

- A torsion pendulum consists of a solid disk or cylinder of moment of inertia I suspended at the end of a thin rod. If the torsion constant k' is known, the period is given by

$$T = 2\pi \sqrt{\frac{I}{k'}} \qquad \begin{array}{l}\textit{Period of}\\ \textit{Torsion Pendulum}\end{array}$$

Key Terms

1. simple harmonic motion
2. spring constant
3. restoring force
4. simple pendulum
5. torsion constant
6. displacement
7. amplitude
8. frequency
9. period
10. hertz

Review Questions

14-1. Give several examples of motion that is SHM.

14-2. What effect will doubling the amplitude A of a body moving with SHM have on

(a) the period, (b) the maximum velocity, and (c) the maximum acceleration?

14-3. A 2-kg mass m_1 moves in SHM with a frequency f_1. What mass m_2 will cause the system to vibrate with twice the frequency?

14-4. Explain, with the use of diagrams, why the velocity in SHM is greatest when the acceleration is the least.

14-5. An ice puck is attached to a spring of force constant k and set into vibration of amplitude A, as shown in Fig. 14-2. The spring is then replaced with one with a force constant of $4k$, and the ice puck is set into vibration at the same amplitude. Compare their periods and frequencies of oscillation.

14-6. A pendulum clock runs too slow and loses time. What adjustment should be made?

14-7. Given a spring of known force constant, a meterstick, and a stopwatch, how can you determine the value of an unknown mass?

14-8. How can the principle of the pendulum be used to compute (a) length, (b) mass, and (c) time?

14-9. Explain clearly why the motion of a pendulum is not simple harmonic when the amplitude is large. Is the period larger or smaller than it should be if the motion were strictly simple harmonic?

Problems

Sections 14-1 and 14-2—Periodic Motion and the Reference Circle

14-1. A rock swings in a circle at constant speed on the end of a string, making 50 revolutions in 30 s. What are the frequency and the period for this motion?

Ans. 1.67 rev/s, 0.600 s

14-2. A child sits at the edge of a platform rotating at 30 rev/min. The platform is 10 m in diameter. What is the period of the motion, and what is the child's speed?

14-3. A rubber ball swings in a horizontal circle 2 m in diameter and makes 20 revolutions in 1 min. The shadow of the ball is projected on a wall by a distant light. What are the amplitude, frequency, and period for the motion of the shadow?

Ans. 1.00 m, 0.333 Hz, 3.00 s

14-4. Assume that a ball makes 300 rev/min while moving in a circle of radius 12 cm. What are the amplitude, frequency, and period for the motion of its shadow projected on a wall?

14-5. A mass oscillates at a frequency of 3 Hz and an amplitude of 6 cm. What are its positions at time $t = 0$ and $t = 2.5$ s?

Ans. 6 cm, 1.85 cm

14-6. A 50-g mass oscillates with an SHM of frequency 0.25 Hz. Assume that $t = 0$ when the mass is at its maximum displacement.

At what time will its displacement be zero? At what time will it be located at half of its amplitude?

14-7. When a mass of 200-g is hung from a spring, the spring is displaced downward a distance of 1.5 cm. What is the spring constant k?

Ans. 131 N/m

14-8. An additional mass of 400 kg is added to the initial 200-g mass in Prob. 14-7. What will be the increase in downward displacement?

*14-9. A mass at the end of a spring vibrates up and down with a frequency of 0.600 Hz and an amplitude of 5 cm. What is its displacement 2.56 s after it reaches a maximum?

Ans. −4.87 cm

*14-10. An object vibrates with an amplitude of 6 cm and a frequency of 0.490 Hz. Starting from maximum displacement in the positive direction, when will be the first time that its displacement is 2 cm?

Section 14-3—Velocity in Simple Harmonic Motion

14-11. A body vibrates with a frequency of 1.4 Hz and an amplitude of 4 cm. What is the maximum velocity? What is its position when the velocity is zero?

Ans. ±0.351 m/s, x = 4 cm

Summary and Review

14-12. An object oscillates at a frequency of 5 Hz and an amplitude of 6 cm. What is the maximum velocity?

14-13. A smooth block on a frictionless surface is attached to a spring, pulled to the right a distance of 4 cm, and then released. Three seconds later it returns to the point of release. What is the frequency, and what is the maximum speed?

Ans. 0.333 Hz, 8.38 cm/s

14-14. In Prob. 14-13, what are the position and velocity 2.55 s after release?

*14-15. Show that the velocity of an object in SHM can be written as a function of its amplitude and displacement:

$$v = \pm 2\pi f \sqrt{A^2 - x^2}$$

14-16. Use the relation derived in Prob. 14-15 to verify the answers obtained for position and velocity in Prob. 14-14.

14-17. A mass vibrating at a frequency of 0.5 Hz has a velocity of 5 cm/s as it passes the center of oscillation. What are the amplitude and the period of vibration?

Ans. 1.59 cm, 2 s

*14-18. A body vibrates with a frequency of 8 Hz and an amplitude of 5 cm. At what time after it is released from $x = +5$ cm will its velocity be equal to $+2.00$ m/s?

Section 14-4—Acceleration in Simple Harmonic Motion

14-19. A 400-g mass is attached to a spring, causing it to stretch a vertical distance of 2 cm. The mass is then pulled downward a distance of 4 cm and released to vibrate with SHM as shown in Fig. 14-10. What is the spring constant? What are the magnitude and direction of the acceleration when the mass is located 2 cm below its equilibrium position? **Ans. 196 N/m, 9.8 m/s² upward**

14-20. What is the maximum acceleration for the system described in Prob. 14-19, and what is its acceleration when it is located 3 cm above its equilibrium position?

Figure 14-10

14-21. A body makes one complete oscillation in 0.5 s. What is its acceleration when it is displaced a distance of $x = +2$ cm from its equilibrium position? **Ans. −3.16 m/s²**

14-22. Find the maximum velocity and the maximum acceleration of an object moving in SHM of amplitude 16 cm and frequency 2 Hz.

*14-23. An object vibrating with a period of 2 s is displaced a distance of $x = +6$ cm and released. What are the velocity and acceleration 3.20 s after release?

Ans. +11.1 cm/s, 0.479 m/s²

*14-24. A body vibrates with SHM of period 1.5 s and amplitude of 6 in. What are its maximum velocity and acceleration?

*14-25. For the body in Prob. 14-24, what are its velocity and acceleration after a time of 7 s?

Ans. 1.81 ft/s, 4.39 ft/s²

Section 14-5—The Period and Frequency

14-26. The prong of a tuning fork vibrates with a frequency of 330 Hz and an amplitude of 2 mm. What is the velocity when the displacement is 1.5 mm?

*14-27. A 400-g mass stretches a spring 20 cm. The 400-g mass is then removed and replaced with an unknown mass m. When the unknown mass is pulled down 5 cm and released, it vibrates with a period of 0.1 s. Compute the mass of the object.

Ans. 4.96 g

*14-28. A long thin piece of metal is clamped at its lower end and has a 2-kg ball fastened to its top end. When the ball is pulled to one side and released, it vibrates with a period of 1.5 s. What is the spring constant for this device?

*14-29. A car and its passengers have a total mass of 1600 kg. The frame of the car is supported by four springs, each having a force constant of 20,000 N/m. Find the frequency of vibration of the car when it drives over a bump in the road. **Ans. 1.13 Hz**

Section 14-6—The Simple Pendulum

14-30. What are the period and frequency of a simple pendulum 2 m in length?

*14-31. A simple pendulum clock beats seconds every time the bob reaches its maximum amplitude on either side. What is the period of this motion? What should be the length of the pendulum at the point at which $g = 9.80$ m/s^2? **Ans. 2.00 s, 0.993 m**

14-32. A 10-m length of cord is attached to a steel bob hung from the ceiling. What is the period of its natural oscillation?

*14-33. On the surface of the moon, the acceleration due to gravity is only 1.67 m/s^2. A pendulum clock adjusted for the earth is taken to the moon. What fraction of its length on earth must the new length be? **Ans. 0.17**

*14-34. A student constructs a pendulum of length 3 m and determines that it makes 50 complete vibrations in 2 min and 54 s. What is the acceleration due to gravity at this student's location?

Section 14-6—The Torsion Pendulum

*14-35. A torsion pendulum oscillates at a frequency of 0.55 Hz. What is the period of its vibration? What is the angular acceleration when its angular displacement is 60°? **Ans. −12.5 rad/s^2**

*14-36. The maximum angular acceleration of a torsional pendulum is 20 rad/s^2 when the angular displacement is 70°. What is the frequency of vibration?

*14-37. A disk 20 cm in diameter forms the base of a torsion pendulum. A force of 20 N applied to the rim causes it to twist an angle of 12°. If the period of the angular vibration after release is 0.5 s, what is the moment of inertia of the disk? **Ans. 0.0605 kg m^2**

*14-38. An irregular object is suspended by a wire as a torsion pendulum. A torque of 40 lb · ft causes it to twist through an angle of 15°. When released, the body oscillates with a frequency of 3 Hz. What is the moment of inertia of the irregular body?

Challenge Problems

14-39. The spring constant of a metal spring is 2000 N/m. What mass will cause this spring to stretch a distance of 4 cm? **Ans. 8.16 kg**

14-40. A 4-kg mass hangs from a spring whose constant k is 400 N/m. The mass is pulled downward a distance of 6 cm and released. What is the acceleration at the instant of release?

14-41. What is the natural frequency of vibration for the system described in Prob. 14-28? What is the maximum velocity? **Ans. 1.59 Hz, ±59.9 cm/s**

*14-42. A 50-g mass on the end of a spring ($k = 20$ N/m) is moving at a speed of 120 cm/s when located a distance of 10 cm from the equilibrium position. What is the amplitude of the vibration?

*14-43. A 40-g mass is attached to a spring ($k = 10$ N/m) and released with an amplitude of 20 cm. What is the velocity of the mass when it is halfway to the equilibrium position? **Ans. 2.74 m/s**

14-44. What is the frequency of the motion for the mass in Prob. 14-31?

14-45. A 2-kg mass is hung from a light spring. When displaced and released, the mass makes 20 oscillations in 25 s. Find the period and the spring constant. **Ans. 1.25 s, 50.5 N/m**

14-46. What length of pendulum is required if the period is to be 1.6 s at a point where $g = 9.80$ m/s^2?

*14-47. An object is moving with SHM of amplitude 20 cm and frequency 1.5 Hz. What are the maximum acceleration and velocity?

Ans. ± 17.8 m/s^2, ± 188 cm/s

*14-48. For the object in Prob. 14-47, what are the position, velocity, and acceleration 1.4 s after it reaches its maximum displacement?

Critical Thinking Questions

*14-49. A mass m connected to the end of a spring oscillates with a frequency $f = 2$ Hz and an amplitude A. If the mass m is doubled what is the new frequency for the same amplitude? If the mass is unchanged and the amplitude is doubled, what is the frequency?

Ans. 1.41 Hz, 2.00 Hz

*14-50. Consider a 2-kg mass connected to a spring whose constant is 400 N/m. What is the natural frequency of vibration? If the system is stretched by +8 cm and then released, at what points are its velocity and acceleration maximized? Will it reach half its maximum velocity at half the amplitude? Calculate the maximum velocity and the velocity at $x = 4$ cm to verify your answer.

*14-51. A 200-g mass is suspended from a long, spiral spring. When displaced downward 10 cm, the mass is found to vibrate with a period of 2 s. What is the spring constant? What are its velocity and acceleration as it passes *upward* through the point +5 cm above its equilibrium position?

Ans. 1.97 N/m, 27.2 cm/s, -49.3 cm/s^2

*14-52. A pendulum clock beats seconds every time the bob passes through its lowest point. What must be the length of the pendulum at a place where $g = 32.0$ ft/s? If the clock is moved to a point where $g = 31.0$ ft/s^2, how much time will it lose in 1 day?

*14-53. A 500-g mass is connected to a device having a spring constant of 6 N/m. The mass is displaced to the right a distance $x = +5$ cm from its equilibrium position and then released. What are its velocity and acceleration when $x = +3$ cm and when $x = -3$ cm?

Ans. ± 0.139 m/s, -0.360 m/s^2; ± 0.139 m/s, $+0.360$ m/s^2

Fluids

OBJECTIVES

After completing this chapter, you should be able to:

1. Define and apply the concepts of fluid pressure and buoyant force to the solution of physical problems similar to examples in the text.
2. Write and illustrate with drawings your understanding of the four basic principles of fluid pressure as summarized in Sec. 15-3.
3. Define *absolute pressure, gauge pressure,* and *atmospheric pressure,* and demonstrate by examples your understanding of the relationships between these terms.
4. Write and apply formulas for calculating the mechanical advantage of a hydraulic press in terms of input and output forces or areas.
5. Define the rate of flow of a fluid and solve problems that relate the rate of flow to the velocity and cross-sectional area.
6. Write Bernoulli's equation in its general form and describe the equation as it would apply to (a) a fluid at rest, (b) fluid flow at constant pressure, and (c) flow through a horizontal pipe.
7. Apply Bernoulli's equation to the solution of problems involving absolute pressure P, density ρ, fluid elevation h, and fluid velocity v.

Liquids and gases are called fluids because they flow freely and fill their containers. In this chapter you will learn that fluids may exert forces on the walls of their containers. Such forces acting on definite surface areas create a condition of pressure. A hydraulic press utilizes fluid pressure to lift heavy loads. The structure of water basins, dams, and large oil tanks is determined largely by pressure considerations. The design of boats, submarines, and weather balloons must take into account the pressure and density of the surrounding fluid. We will also study the fundamental aspects of fluid flow and Bernoulli's laws that govern such motion.

15-1 DENSITY

Before discussing the statics and dynamics of fluids, it is important to understand the relation of a body's weight to its volume. For example, we refer to lead or iron as *heavy,* whereas wood or cork is considered *light.* What we really mean is that a block of wood is lighter than a block of lead *of similar size.* The terms *light* and *heavy* are comparative terms. As illustrated in Fig. 15-1, it is possible for a block of lead to weigh the same as a block of wood if their relative size differs greatly. On the other hand, 1 ft³ of lead weighs more than six times as much as 1 ft³ of wood.

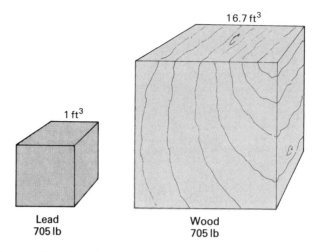

16.7 ft³

1 ft³

Lead
705 lb

Wood
705 lb

Figure 15-1 A comparison of weight and volume in lead and wood.

The quantity that relates a body's weight to its volume is known as its **weight density.**

> The **weight density D of a body is defined as the ratio of its weight W to its volume V. Units are the** *newton per cubic meter* **(N/m³) and the** *pound per cubic foot (lb/ft³).*

$$D = \frac{W}{V} \qquad W = DV \qquad \text{15-1}$$

Thus, if a 20-lb object occupies a volume of 4 ft³, it has a weight density of 5 lb/ft³.

As mentioned in Chap. 7, the weight of a body is not constant but varies according to location. A more useful relation for density takes advantage of the fact that *mass* is a universal constant, independent of gravity.

> The **mass density ρ of a body is defined as the ratio of its mass** *m* **to its volume V.**

$$\rho = \frac{m}{V} \qquad m = \rho V \qquad \text{15-2}$$

The units of mass density are the ratio of a mass unit to a volume unit, for example, grams per cubic centimeter, kilograms per cubic meter, or slugs per cubic foot.

The relationship between weight density and mass density is found by recalling that $W = mg$. Thus,

$$D = \frac{mg}{V} = \rho g \qquad \text{15-3}$$

In USCS units, matter is usually described in terms of its weight. For this reason, weight density is more often used when working with this system of units. In SI units, mass is the more convenient quantity, and the mass density is preferred. Table 15-1 lists the weight densities and mass densities of some common substances.

TABLE 15-1 Mass Density and Weight Density

Substance	D,lb/ft^3	ρ g/cm^3	ρ kg/m^3
Solids:			
Aluminum	169	2.7	2,700
Brass	540	8.7	8,700
Copper	555	8.89	8,890
Glass	162	2.6	2,600
Gold	1204	19.3	19,300
Ice	57	0.92	920
Iron	490	7.85	7,850
Lead	705	11.3	11,300
Oak	51	0.81	810
Silver	654	10.5	10,500
Steel	487	7.8	7,800
Liquids:			
Alcohol	49	0.79	790
Benzene	54.7	0.88	880
Gasoline	42	0.68	680
Mercury	850	13.6	13,600
Water	62.4	1.0	1,000
Gases (0°C):			
Air	0.0807	0.00129	1.29
Helium	0.0110	0.000178	0.178
Hydrogen	0.0058	0.000090	0.090
Nitrogen	0.0782	0.00126	1.25
Oxygen	0.0892	0.00143	1.43

EXAMPLE 15-1

A cylindrical tank for gasoline is 3 m long and 1.2 m in diameter. How many kilograms of gasoline will the tank hold?

Solution
First we find the volume:

$$V = \pi r^2 h = \pi (0.6 \text{ m})^2 (3 \text{ m}) = 3.39 \text{ m}^3$$

Substituting the volume and mass density into Eq. (15-2), we obtain

$$m = \rho V = (680 \text{ kg/m}^3)(3.39 \text{ m}^3) = 2310 \text{ kg}$$

Another method of indicating the densities of substances is by their *specific gravity,* which compares their densities to that for water. For example, a substance that is half as dense as water would have a specific gravity of 0.5.

The specific gravity of a substance is defined as the ratio of its density to the density of water at 4°C (1000 kg/m³).

A better name for this quantity is *relative density,* but the term *specific gravity* is much more widely used.

15-2 PRESSURE

The effectiveness of a given force often depends on the area over which it acts. For example, a woman wearing narrow heels will do much more damage to floors than she would with flat heels. Even though she exerts the same downward force in each case, with the narrow heels her weight is spread over a much smaller surface area. The *normal force per unit area* is called **pressure.** Symbolically, the pressure P is given by

$$P = \frac{F}{A}$$

15-4

where A is the area over which the perpendicular force F is applied. The unit of pressure is the ratio of any force unit to a unit of area. Examples are newtons per square meter and pounds per square inch. In SI units, the N/m^2 is renamed the *pascal* (Pa). The *kilopascal* (kPa) is the most appropriate measure for fluid pressure.

$$1 \text{ kPa} = 1000 \text{ N/m}^2 = 0.145 \text{ lb/in.}^2$$

EXAMPLE 15-2

A golf shoe has 10 cleats, each having an area of 0.01 in.² in contact with the floor. Assume that, in walking, there is one instant when all 10 cleats support the entire weight of a 180-lb person. What is the pressure exerted by the cleats on the floor? Express the answer in SI units.

Solution

The total area in contact with the floor is 0.1 in.² (10 × 0.01 in.²). Substitution into Eq. (15-4) yields

$$P = \frac{F}{A} = \frac{180 \text{ lb}}{0.1 \text{ in.}^2} = 1800 \text{ lb/in.}^2$$

Converting to SI units, we obtain

$$P = (1800 \text{ lb/in.}^2)\left(\frac{1 \text{ kPa}}{0.145 \text{ lb/in.}^2}\right) = 1.24 \times 10^4 \text{ kPa}$$

As the area of the shoe in contact with the floor decreases, the pressure becomes larger. It is easy to see why this factor must be considered in floor construction.

There is a significant difference between the way a force acts on a fluid and on a solid. Since a solid is a rigid body, it can withstand the application of a force without a significant change in shape. A liquid, on the other hand, can sustain a force only at an enclosed surface or boundary. If a fluid is not restrained, it will flow under a shearing stress instead of being deformed elastically.

The force exerted by a fluid on the walls of its container must always act perpendicular to the walls.

It is this characteristic property of fluids that makes the concept of pressure so useful. Holes bored in the bottom and sides of a barrel of water (Fig. 15-2) demonstrate that the force exerted by the water is everywhere perpendicular to the surface of the barrel.

A moment's reflection will show the student that a liquid also exerts an upward pressure. Anyone who has tried to keep a rubber float under the surface of water is immediately convinced of the existence of an upward pressure. In fact, we find that

Fluids exert pressure in all directions.

Figure 15-3 on page 330 shows a liquid under pressure. The forces acting on the face of the piston, the walls of the enclosure, and the surfaces of a suspended object are shown in the figure.

Just as larger volumes of solid objects exert greater forces against their supports, fluids exert greater pressure at increasing depths. The fluid at the bottom of a container is always under a greater pressure than that near the surface. This is due to the weight of the overlying liquid. We must point out, however, a distinct difference between the pressure exerted by solids and that exerted by liquids. A solid object can exert only a *downward* force due to its weight. At any particular depth in a fluid, the pressure is the same in all directions. If this were not true, the fluid would flow under the influence of a resultant pressure until a new condition of equilibrium was reached.

Pocket Lab

Juiced Up

Use a sharp pencil to make marks at 2.0 cm, 4.0 cm, and 6.0 cm on the inside of a juice can. Fill the can to the 2.0-cm mark and measure the mass of the can and water on a balance. Fill the can to the 6.0-cm mark and again measure the mass on a balance. Make a graph of mass (vertical) versus height (horizontal). Predict the mass when the can has 4.0 cm of water. The graph should approximate a straight line. Explain why the graph does not go through 0,0. What is the meaning of the slope of the graph?

Figure 15-2 The forces exerted by a fluid on the walls of its container are perpendicular at every point.

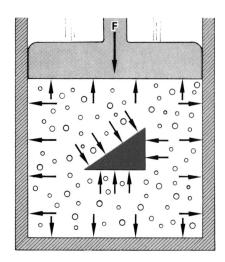

Figure 15-3 Fluids exert pressure in all directions.

Since the weight of the overlying fluid is proportional to its density, the pressure at any depth is also proportional to the density of the fluid. This can be seen by considering a rectangular column of water extending from the surface to a depth h, as shown in Fig. 15-4. The weight of the entire column acts on the surface area A at the bottom of the column.

From Eq. (15-1), we can write the weight of the column as

$$W = DV = DAh$$

where D is the weight density of the fluid. The pressure (weight per unit area) at the depth h is given by

$$P = \frac{W}{A} = Dh$$

or, in terms of mass density,

$$P = Dh = \rho p h \qquad \textbf{15-5}$$

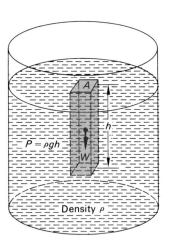

Figure 15-4 The relationship of pressure, density, and depth.

The fluid pressure at any point is directly proportional to the density of the fluid and to the depth below the surface of the fluid.

EXAMPLE 15-3

The water pressure in a certain house is 160 lb/in.^{2}. How high must the water level in a reservoir be above the point of release in the house?

Solution

The weight density of water is 62.4 lb/ft^3. The pressure is 160 lb/in.^{2}. To avoid inconsistency in units, we convert the pressure to units of pounds per square foot.

$$P = (160 \text{ lb/in.}^2) \, \frac{144 \text{ in.}^2}{1 \text{ ft}^2} = 23{,}040 \text{ lb/ft}^2$$

Solving for h in Eq. (15-5), we have

$$h = \frac{P}{D} = \frac{23{,}040 \text{ lb/ft}^2}{62.4 \text{ lb/ft}^2} = 369 \text{ ft}$$

In the above example, no mention was made of the size or shape of the reservoir containing the supply of water. Also, no information was given about the path of the water or size of the pipes connecting the reservoir to the home. Can we assume that our answer is correct when it is based only on the difference in water levels? Doesn't the shape or area of a container have any effect on liquid pressure? To answer these questions, we must recall some of the characteristics of fluids already discussed.

Consider a series of vessels of different areas and shapes interconnected, as shown in Fig. 15-5. It would seem at first glance that the greater volume of water in vessel A should develop a greater pressure at the bottom than vessel D. The effect of such a difference in pressure would then force the liquid to rise higher in vessel D. Filling the vessels with liquid shows the levels to be the same, however, for each vessel.

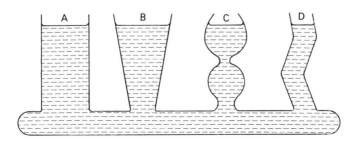

Figure 15-5 Water seeks its own level, indicating that the pressure is independent of the area or shape of the container.

Part of the problem in understanding this paradox results from confusing the terms *pressure* and *total force*. Since pressure is measured in terms of a unit area, we do not consider the *total area* when solving problems involving pressure. For example, in vessel A the area of the liquid at the bottom of the vessel is much greater than the area at the bottom of vessel D. This means that the liquid in vessel A will exert a greater **total force** on the bottom than the liquid in vessel D. But the greater force is applied over a larger area so that the pressure is the same in both vessels.

If the bottoms of vessels B, C, and D have the same area, we can say that the total forces are also equal at the bottoms of these containers. (Of course, the pressures are equal at any particular depth.) You may wonder how the total forces can be equal when vessels B and C contain a greater volume of water. The additional water in each case is supported by vertical components of forces exerted by the walls of the container on the fluid. (See Fig. 15-6.) When the walls of a container are vertical, the forces acting on the sides have no upward components. The total force at the bottom of a container is therefore equal to the weight of a straight column of water above the base area.

Figure 15-6 The pressure at the bottom of each vessel is a function only of the depth of the liquid and is the same in all directions. Since the area of the bottom is the same for both vessels, the total force exerted on the bottom of each container is also the same.

EXAMPLE 15-4

Assume that the vessels in Fig. 15-5 are filled with gasoline until the fluid level is 20 cm above the base of each vessel. The areas at the bases of vessels A and B are 20 cm^2 and 10 cm^2, respectively. Compare the pressure and the total force at the base of each container.

Solution

The pressure is the same at the base of either container and is given by

$$P = \rho g h = (680 \text{ kg/m}^3)(9.8 \text{ m/s}^2)(0.20 \text{ m}) = 1330 \text{ Pa}$$

The total force in each case is the product of the pressure and the base area ($F = PA$). Remember that 1 cm^2 = 1 × 10^{-4} m^2.

$$F = PA = (1330 \text{ Pa})(20 \times 10^{-4} \text{ m}^2) = 2.67 \text{ N}$$
$$F = PA = (1330 \text{ Pa})(10 \times 10^{-4} \text{ m}^2) = 1.33 \text{ N}$$

Problem-Solving Strategy

Before considering other applications of fluid pressure, let us summarize the principles discussed in this section for fluids at rest.

1. The forces exerted by a fluid on the walls of its container are always perpendicular to the walls.
2. The fluid pressure is directly proportional to the depth of the fluid and to its density.
3. At any particular depth, the fluid pressure is the same in all directions.
4. Fluid pressure is independent of the shape or area of its container.

15-4 MEASURING PRESSURE

The pressure discussed in the previous section is due only to the fluid itself and can be calculated from Eq. (15-5). Unfortunately, this is usually not the case. Any liquid in an open container, for example, is subjected to atmospheric pressure in addition to the pressure of its own weight. Since the liquid is relatively incompressible, the external pressure of the atmosphere is transmitted equally throughout the volume of the liquid. This fact, first stated by the French mathematician Blaise Pascal (1623–1662), is called **Pascal's law.** Generally, it can be stated as follows:

An external pressure applied to an enclosed fluid is transmitted uniformly throughout the volume of the liquid.

Most devices that measure pressure directly actually measure the difference between the **absolute pressure** and **atmospheric pressure.** The result is called the *gauge pressure.*

$$Absolute\ pressure = gauge\ pressure + atmospheric\ pressure$$

Atmospheric pressure at sea level is 101.3 kPa, or 14.7 lb/in.^{2}. Because atmospheric pressure enters into so many calculations, we often use a pressure unit of 1 **atmosphere** (atm), defined as the average pressure exerted by the atmosphere at sea level, that is, 101.3 kPa.

A common device for measuring gauge pressure is the open-tube **manometer** (muh-nom'-uh-ter), shown in Fig. 15-7. The manometer consists of a U-shaped tube containing a liquid, usually mercury. When both ends of the tube are open, the mercury seeks its own level because 1 atm of pressure is exerted at each of the open ends. When one end is connected to a pressurized chamber, the mercury will rise in the open tube until the pressures are equalized. The difference between the two levels of mercury is a measure of the gauge pressure: the difference between the absolute pressure in the chamber and the atmospheric pressure at the open end. The manometer is used so often in laboratory situations that atmospheric pressures and other pressures are often expressed in *centimeters of mercury* or *inches of mercury.*

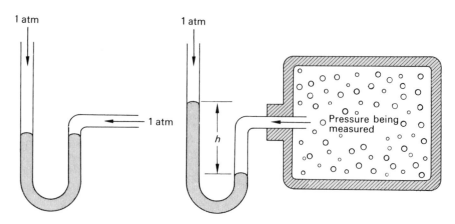

Figure 15-7 The open-tube manometer. Pressure is measured by the height h of the mercury column.

Atmospheric pressure is usually measured in the laboratory with a mercury barometer. The principle of its operation is shown in Fig. 15-8. A glass tube, closed at one end, is filled with mercury. The open end is covered, and the tube is inverted in a bowl of mercury. When the open end is uncovered, the mercury flows out of the tube until the pressure exerted by the column of mercury exactly balances atmospheric pressure acting on the mercury in the bowl. Since the pressure in the tube above the column of mercury is zero, the height of the column above the level of mercury in the bowl indicates the atmospheric pressure. At sea level an atmospheric pressure of 14.7 lb/in.^{2} will cause the level of the mercury in the tube to stabilize at a height of 76 cm, or 30 in.

In summary, we can write the following equivalent measures of atmospheric pressure:

$$1 \text{ atm} = 101.3 \text{ kPa} = 14.7 \text{ lb/in.}^2 = 76 \text{ cm of mercury}$$
$$= 30 \text{ in. of mercury} = 2116 \text{ lb/ft}^2$$

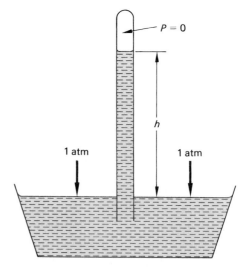

Figure 15-8 The barometer.

EXAMPLE 15-5

The mercury manometer is used to measure the pressure of a gas inside a tank. (Refer to Fig. 15-7.) If the difference between the two mercury levels is 36 cm, what is the absolute pressure inside the tank?

Solution

The gauge pressure is 36 cm of mercury, and the atmospheric pressure is 76 cm of mercury. Thus, the absolute pressure is found from Eq. (15-5).

$$\text{Absolute pressure} = 36 \text{ cm} + 76 \text{ cm} = 112 \text{ cm of mercury}$$

The pressure in the tank is equivalent to the pressure that would be exerted by a column of mercury 112 cm high.

$$\begin{aligned}
P = Dh &= \rho g h \\
&= (13{,}600 \text{ kg/m}^3)(9.8 \text{ m/s}^2)(1.12 \text{ m}) \\
&= 1.49 \times 10^5 \text{ N/m}^2 = 149 \text{ kPa}
\end{aligned}$$

You should verify that this absolute pressure is also 21.6 lb/in.^{2} or 1.47 atm.

15-5 THE HYDRAULIC PRESS

The most universal application of Pascal's law is found with the hydraulic press, shown in Fig. 15-9. According to Pascal's principle, a pressure applied to the liquid in the left column will be transmitted undiminished to the liquid in the column on the right. Thus, if an input force $\mathbf{F}_i$ acts upon a piston of area A_i, it will cause an output force $\mathbf{F}_o$ to act on a piston of area A_o so that

Input pressure = output pressure

$$\frac{F_i}{A_i} = \frac{F_o}{A_o} \qquad \textbf{15-6}$$

Figure 15-9 The hydraulic press.

The ideal mechanical advantage of such a device is equal to the ratio of the output force to the input force. Symbolically, we write

$$M_I = \frac{F_o}{F_i} = \frac{A_o}{A_i} \qquad \text{15-7}$$

A small input force can be multiplied to yield a much larger output force simply by having the output piston much larger in area than the input piston. The output force is given by

$$F_o = F_i \frac{A_o}{A_i} \qquad \text{15-8}$$

According to the methods developed in Chap. 12 for simple machines, the input work must equal the output work if we neglect friction. If the input force F_i travels through a distance s_i while the output force F_o travels through a distance s_o, we can write

$$Input\ work = output\ work$$
$$F_i s_i = F_o s_o$$

This relation leads to another useful expression for the ideal mechanical advantage of a hydraulic press.

$$M_I = \frac{F_o}{F_i} = \frac{s_i}{s_o} \qquad \text{15-9}$$

Notice that the mechanical advantage is gained at the expense of input distance. For this reason, most applications utilize a system of values to permit the output piston to be raised by a series of short input strokes.

EXAMPLE 15-6

The smaller and larger pistons of a hydraulic press have diameters of 2 and 24 in., respectively. (a) What input force is required to deliver a total output force of 2000 lb at the larger piston? (b) How far must the smaller piston travel to lift the larger piston 1 in.?

Solution (a)
The ideal mechanical advantage is

$$M_I = \frac{A_o}{A_i} = \frac{\pi d_o^2/4}{\pi d_i^2/4} = \left(\frac{d_o}{d_i}\right)^2$$

$$= \left(\frac{24\ \text{in.}}{2\ \text{in.}}\right)^2 = (12)^2 = 144$$

The required input force is given by

$$F_i = \frac{F_o}{M_I} = \frac{2000\ \text{lb}}{144} = 13.9\ \text{lb}$$

Solution (b)

Applying Eq. (15-9), we can compute the input distance.

$$s_i = M_I s_o = (144)(1 \text{ in.}) = 144 \text{ in.}$$

The principle of the hydraulic press is found in many engineering and mechanical devices. Power steering, the hydraulic jack, shock absorbers, and automobile braking systems are a few common examples.

Problem-Solving Strategy

Fluids at Rest

1. Draw a figure and label it with those quantities that are given and those that are to be found. Use consistent units for area, volume, density, and pressure.

2. Do not confuse *absolute* pressure with *gauge* pressure or *mass* density with *weight* density. You must use *absolute* pressure unless the problem involves a *difference* of pressure. Be careful with units if you attempt to use weight density, which is a *force* per unit volume.

3. The difference in pressure between two points is proportional to the density of the fluid and to the depth in the fluid:

$$P_2 - P_1 = \rho g h \qquad \rho = \frac{m}{V} \qquad P = \frac{F}{A}$$

4. Archimedes' principle states that an object completely or partially submerged in a fluid experiences an upward buoyant force equal to the weight of the displaced fluid:

$$F_B = mg = \rho g V \qquad \text{(buoyant force)}$$

5. Remember that the buoyant force depends on the density of the *displaced fluid* and the volume of the *displaced fluid*. It has nothing to do with the mass or density of the object submerged in the fluid. If the object happens to be *totally* submerged, it is true that the volume of the object and of the displaced fluid are equal. This fact can be used to find the buoyant force in those cases.

6. For an object that is *floating* in a fluid, the upward buoyant force must equal the weight of the object. This means that the weight of the object must be equal to the weight of the *fluid displaced*. Therefore, we may write

$$m_x g = m_f g \qquad \text{or} \qquad \rho_x V_x = \rho_f V_f$$

The subscript x refers to the floating object, and the subscript f refers to the displaced fluid. For example, if an object with a volume of 3 m³ is floating with two-thirds of its volume submerged, then $V_x = 3 \text{ m}^3$ and $V_f = 2 \text{ m}^3$.

15-6 ARCHIMEDES' PRINCIPLE

Anyone familiar with swimming and other water sports has observed that objects seem to lose weight when submerged in water. In fact, the object may even float on the surface because of the upward pressure exerted by the water. An ancient

Greek mathematician, Archimedes (287–212 B.C.), first studied the buoyant force exerted by fluids. *Archimedes' principle* can be stated as follows:

> **An object that is completely or partly submerged in a fluid experiences an upward force equal to the weight of the fluid displaced.**

Archimedes' principle can be demonstrated by studying the forces a fluid exerts on a suspended object. Consider a disk of area A and height H that is completely submerged in a fluid, as shown in Fig. 15-10. Recall that the pressure at any depth h in a fluid is given by

$$P = \rho g h$$

where ρ is the mass density of the fluid and g is the acceleration of gravity. Of course, if we wish to represent the absolute pressure within the fluid, we must add the external pressure exerted by the atmosphere. The total downward pressure P_1 on top of the disk in Fig. 15-10 is, therefore,

$$P_1 = P_a + \rho g h_1 \quad \text{(downward)}$$

where P_a is atmospheric pressure and h_1 is the depth at the top of the disk. Similarly, the upward pressure P_2 on the bottom of the disk is

$$P_2 = P_a + \rho g h_2 \quad \text{(upward)}$$

where h_2 is the depth at the bottom of the disk. Since h_2 is greater than h_1, the pressure on the bottom of the disk will exceed the pressure at the top, resulting in a net upward force. If we represent the downward force by F_1 and the upward force by F_2, we can write

$$F_1 = P_1 A \qquad F_2 = P_2 A$$

The net upward force exerted *by* the fluid *on* the disk is called the **buoyant force** and is given by

$$
\begin{aligned}
F_B &= F_2 - F_1 = A(P_2 - P_1) \\
&= A\,(P_a + \rho g h_2 - P_a - \rho g h_1) \\
&= A\rho g(h_2 - h_1) = A\rho g H
\end{aligned}
$$

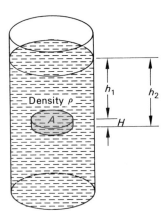

Figure 15-10   The buoyant force exerted on the disk is equal to the weight of the fluid it displaces.

where $H = h_2 - h_1$ is the height of the disk. Finally, if we recall that the volume of the disk is $V = AH$, we obtain this important result:

$$F_B = V\rho g = mg \qquad \text{15-10}$$

Buoyant force = weight of displaced fluid

which is Archimedes' principle.

In applying this result, it must be recalled that Eq. (15-10) allows us to compute only the *buoyant force* due to the difference in pressures. It does not represent the resultant force. A submerged body will sink if the weight of the fluid it displaces (the buoyant force) is less than the weight of the body. If the weight of the displaced fluid is exactly equal to the weight of the submerged body, it will neither sink nor rise. In this instance, the body will be in equilibrium. If the weight of displaced fluid exceeds the weight of a submerged body, the body will rise to the surface and float. When the floating body comes to equilibrium at the surface, it will displace its own weight of liquid. Figure 15-11 demonstrates this point with the use of an overflow can and a beaker to catch the fluid displaced by a wooden block.

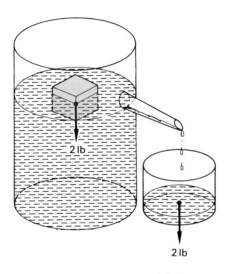

Figure 15-11 A floating body displaces its own weight of fluid.

CONCEPTUAL EXAMPLE 15-7

A cork float has a volume of 4 cm³ and a density of 207 kg/m³. (a) What volume of the cork is beneath the surface when the cork floats in water? (b) What downward force is needed to submerge the block completely?

Solution (a)

The floating cork will displace a volume of water equal to its own weight. Recalling that $1 \text{ cm}^3 = 1 \times 10^{-6} \text{ m}^3$, the weight of 4 cm³ of cork is

$$W = \rho g V = (207 \text{ kg/m}^3)(9.8 \text{ m/s}^2)(4 \times 10^{-6} \text{ m}^3)$$
$$= 8.11 \times 10^{-3} \text{ N}$$

Since $W = \rho g V$, the volume of water displaced is

$$V = \frac{W}{\rho g} = \frac{8.11 \times 10^{-3} \text{ N}}{(1000 \text{ kg/m}^3)(9.8 \text{ m/s}^2)}$$

$$= 8.28 \times 10^{-7} \text{ m}^3 \text{ or } 0.828 \text{ cm}^3$$

Thus, the volume of the cork under water is also 0.828 cm³.

If the area of the floating surface were known, you could calculate how deep the cork would sink into the water. Note that approximately 21 percent of the cork is under water. As an extra example, you should show that the fraction of volume submerged is equal to the specific gravity of an object.

Solution (b)

To submerge the cork, a downward force F must be exerted, in addition to the weight W of the cork. The sum of these forces equals the buoyant force F_B:

$$F + W = F_B$$

The required force F is therefore equal to the difference between the buoyant force and the weight of the cork.

$$F = F_B - W$$

The buoyant force is equal to the weight of 4 cm³ of water.

$$F_B = \rho g V = (1000 \text{ kg/m}^3)(9.8 \text{ m/s}^2)(4 \times 10^{-6} \text{ m}^3)$$

$$= 39.2 \times 10^{-3} \text{N}$$

The required force F to submerge the cork is

$$F = 39.2 \times 10^{-3} \text{ N} - 8.11 \times 10^{-3} \text{ N} = 31.1 \times 10^{-3} \text{ N}$$

EXAMPLE 15-8

A weather balloon is to operate at an altitude at which the density of air is 0.9 kg/m³. At this altitude, the balloon has a volume of 20 m³ and is filled with hydrogen ($\rho_H = 0.09$ kg/m³). If the balloon bag weighs 118 N, what load can it support at this level?

Solution

The buoyant force is equal to the weight of the displaced air. Thus,

$$F_B = \rho g V = (0.9 \text{ kg/m}^3)(9.8 \text{ m/s}^2)(20 \text{ m}^3) = 176 \text{ N}$$

The weight of 20 m³ of hydrogen is

$$W_H = \rho_H g V = (0.09 \text{ kg/m}^3)(9.8 \text{ m/s}^2)(20 \text{ m}^3) = 17.6 \text{ N}$$

The load supported is

$$W_L = F_B - W_H - W_B$$

$$= 176 \text{ N} - 17.6 \text{ N} - 118 \text{ N} = 40.4 \text{ N}$$

Large balloons can maintain a condition of equilibrium at any altitude by adjustment of their weight or buoyant force. The weight can be lightened by releasing the ballast provided for that purpose. The buoyant force can be decreased by releasing the gas from the balloon or increased by pumping more gas into the flexible balloon. Hot-air balloons use the lower density of heated air to provide their buoyancy.

15-7 FLUID FLOW

Until now our study of fluids has been restricted to conditions of rest, which are considerably simpler than dealing with fluids in motion. The mathematical difficulties encountered when treating the motion of a fluid are formidable. We will find it easier to make a few assumptions. First of all, we will consider that all fluids in motion exhibit *streamline flow.*

> **Streamline flow** is the motion of a fluid in which every particle in the fluid follows the same path (past a particular point) as that followed by previous particles.

Figure 15-12 illustrates the *streamlines* of air flowing past two stationary obstacles. Note that the streamlines break down as air passes over the second obstacle, setting up whirls and eddies. These little whirlpools represent **turbulent flow,** and they absorb much of the fluid energy, increasing the frictional drag through the fluid.

We shall further consider that fluids are incompressible and have essentially no internal friction. Under these conditions, certain predictions can be made about the rate of fluid flow through a pipe or other container.

> **The rate of flow** is defined as the volume of fluid that passes a certain cross-section per unit of time.

To express this rate quantitatively, we shall consider a liquid flowing through the pipe of Fig. 15-13 with an average velocity v. During a time interval t, each parti-

(a)   (b)

Figure 15-12 Streamline and turbulent flow of a fluid in its path.

Figure 15-13 Computing the rate of flow of a fluid through a pipe.

cle in the stream moves through a distance vt. The volume V flowing through a cross-section A is given by

$$V = Avt$$

Thus, the rate of flow (volume per unit time) can be calculated from

$$R = \frac{Avt}{t} = vA \qquad \text{15-11}$$

Rate of flow = velocity × cross section

The units of R express the ratio of a volume unit to a time unit. Common examples are cubic feet per second, cubic meters per second, liters per second, and gallons per minute.

If the fluid is incompressible and we ignore the effect of internal friction, the rate of flow R will remain constant. This means that a variation in the pipe cross section, as illustrated in Fig. 15-14, will result in a change in speed of the liquid so that the product vA remains constant. Symbolically, we write

$$R = v_1 A_1 = v_2 A_2 \qquad \text{15-12}$$

A liquid will flow faster through a narrow section of pipe and more slowly through the broad sections. This principle causes water to flow more rapidly when the banks of a small stream suddenly come closer together.

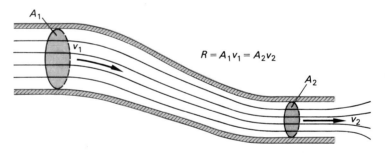

Figure 15-14   In streamline flow, the product of the fluid velocity and the cross-sectional area of the pipe is constant at any point.

Technology Today

Dimpled Baseball Bats?

You may have seen or even tested a new kind of baseball bat that has dimples along its length, similar to the dimples in a golf ball. These dimples actually help the bat swing faster through the air because of fluid dynamics. The dimples cause a microscopic turbulence, which, in turn, causes an overall more streamlined flow.

EXAMPLE 15-9

Water flows through a rubber hose 2 cm in diameter at a velocity of 4 m/s. (a) What must be the diameter of the nozzle if water is to emerge at 20 m/s? (b) What is the rate of flow in cubic meters per minute?

Solution (a)

The rate of flow is constant, so that $A_1 v_1 = A_2 v_2$. Since the area A is proportional to the square of the diameter, we have

$$d_1^2 v_1 = d_2^2 v_2 \qquad \text{or} \qquad d_2^2 = \frac{v_1}{v_2} d_1^2$$

from which

$$d_2^2 = \frac{4 \text{ m/s}}{20 \text{ m/s}}(2 \text{ cm})^2 = 0.8 \text{ cm}^2$$

Taking the square root of each side gives

$$d_2 = 0.894 \text{ cm}$$

Solution (b)

To find the rate of flow, we must first determine the area of the 2-cm hose.

$$A_1 = \frac{\pi d_1^2}{4} = \frac{\pi (2 \text{ cm})^2}{4} = 3.14 \text{ cm}^2$$

$$= 3.14 \text{ cm}^2 \left(\frac{1 \times 10^{-4} \text{ m}^2}{1 \text{ cm}^2} \right) = 3.14 \times 10^{-4} \text{ m}^2$$

The rate of flow is $R = A_1 v_1$, so that

$$R = (3.14 \times 10^{-4} \text{ m}^2)(4 \text{ m/s}) = 1.26 \times 10^{-3} \text{ m}^3/\text{s}$$
$$= (1.26 \times 10^{-3} \text{ m}^3/\text{s})(60 \text{ s/min}) = 0.0754 \text{ m}^3/\text{min}$$

The same rate would be found by considering the product $A_2 v_2$.

www.sciam.com/
1197issue/
1197amsci.html
For a build-at-home wind tunnel in which you can perform aerodynamic measurements, see this Web site.

Problem-Solving Strategy

Rate of Flow Problems

1. Read the problem carefully, and, after drawing a rough sketch, list the given information.
2. Remember that the rate of flow R represents the volume of fluid passing a given cross section in a unit of time.
3. When a volume of fluid passes from one cross section A_1 into another A_2, the rate of flow is unchanged.

$$R = v_1 A_1 = v_2 A_2$$

Be sure to use consistent units for volume and area.

4. Since the area A of a pipe is proportional to the square of its diameter d, a more useful form of the above equation might be:

$$v_1 d_1^2 = v_2 d_2^2$$

5. The units chosen for velocity or diameter in one section of the pipe must be the same as those in the second section of pipe.

15-8 PRESSURE AND VELOCITY

We have noted that a fluid's velocity increases when it flows through a constriction. An increase in velocity can result only through the presence of an accelerating force. To accelerate the liquid as it enters the constriction, the pushing force from the large cross section must be greater than the resisting force from the constriction. In other words, the pressure at points A and C in Fig. 15-15 must be greater than the pressure at B. The tubes inserted into the pipe above these points

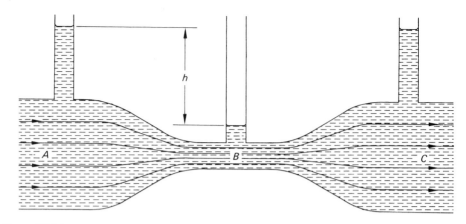

Figure 15-15 The increased velocity of a fluid flowing through a constriction causes a drop in pressure.

indicate clearly the difference in pressure. The fluid level in the tube above the restriction is lower than the level in the adjacent areas. If h is the difference in height, the pressure differential is given by

$$P_A - P_B = \rho g h \qquad \text{15-13}$$

This assumes that the pipe is horizontal and that no pressure changes are introduced because of a change in potential energy.

The preceding example, as illustrated in Fig. 15-15, shows the principle of the *venturi meter*. From a determination of the difference in pressure, this device makes it possible to calculate the velocity of water in a horizontal pipe.

The ***venturi effect*** has many other applications for both liquids and gases. The carburetor in an automobile utilizes the venturi principle to mix gasoline vapor and air. Air passing through a constriction on its way to the cylinders creates a low-pressure area as its velocity increases. The decrease in pressure is used to draw fuel into the air column, where it is readily vaporized.

Figure 15-16 shows two methods you can use to demonstrate the decrease in pressure due to an increase in velocity. The simplest example consists of blowing air past the top surface of a sheet of paper, as shown in Fig. 15-16a. The pressure

(a)   (b)

Figure 15-16 Demonstrations of the decrease in pressure resulting from an increase in air speeds.

in the air stream above the paper will be reduced. This allows the excess pressure on the bottom to force the paper upward.

A second demonstration requires a hollow spool, a cardboard disk, and a pin (Fig. 15-16b). The pin is driven through the cardboard disk and placed in one end of the hollow spool, as shown in the figure. If you blow through the open end, you will find that the disk becomes more tightly pressed to the other end. One would expect the cardboard disk to fly off immediately. The explanation is that air blown into the spool must escape through the narrow space between the disk and the end of the spool. This action creates a low-pressure area, allowing the external atmospheric pressure to push the disk tight against the spool.

15-9 BERNOULLI'S EQUATION

In our discussion of fluids, we have emphasized four quantities: the pressure P, the density ρ, the velocity v, and the height h above some reference level. The relationship between these quantities and their ability to describe fluids in motion was first established by Daniel Bernoulli (1700–1782), a Swiss mathematician. Steps leading to the development of this fundamental relationship can be understood by considering Fig. 15-17.

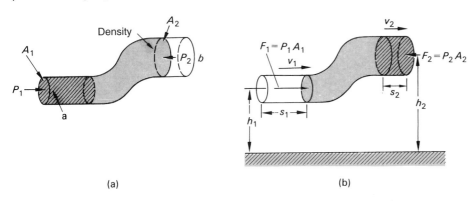

(a)   (b)

Figure 15-17 Derivation of Bernouilli's equation.

Since a fluid has mass, it must obey the same conservation laws established earlier for solids. Consequently, the work required to move a certain volume of fluid through a pipe must equal the total change in kinetic and potential energy. Let us consider the work required to move the fluid from point a to point b in Fig. 15-17a. The net work must be the sum of the work done by the input force $\mathbf{F}_1$ and the negative work done by the resisting force $\mathbf{F}_2$.

$$\text{Net work} = F_1 s_1 - F_2 s_2$$

But $F_1 = P_1 A_1$ and $F_2 = P_2 A_2$, so that

$$\text{Net work} = P_1 A_1 s_1 - P_2 A_2 s_2$$

The product of area and distance represents the volume V of the fluid moved through the pipe. Since this volume is the same at the bottom and at the top of the pipe, we can substitute

$$V = A_1 s_1 = A_2 s_2$$

*inter***NET**
CONNECTION

howthingswork. virginia.edu
For answers to such questions as, "Why does airspeed go up as it flows over an airplane wing?" see this Web site.

obtaining

$$\text{Net work} = P_1 V - P_2 V = (P_1 - P_2)V$$

The kinetic energy E_k of a fluid is defined as $\frac{1}{2}mv^2$, where m is the mass of the fluid and v is its velocity. Since the mass remains constant, a change in kinetic energy ΔE_k results only from the difference in fluid velocity. In our example, the change in kinetic energy is

$$\Delta E_k = \tfrac{1}{2}mv_2^2 - \tfrac{1}{2}mv_1^2$$

The potential energy of a fluid at a height h above some reference point is defined as mgh, where mg represents the weight of the fluid. The volume of fluid moved through the pipe is constant. Therefore, the change in potential energy ΔE_p results from the increases in height of the fluid from h_1 to h_2:

$$\Delta E_p = mgh_2 - mgh_1$$

We are now prepared to apply the principle of conservation of energy. The net work done on the system must equal the sum of the increases in kinetic and potential energy. Thus,

$$\text{Net work} = \Delta E_k + \Delta E_p$$

$$(P_1 - P_2)V = (\tfrac{1}{2}mv_2^2 - \tfrac{1}{2}mv_1^2) + (mgh_2 - mgh_1)$$

If the density of the fluid is ρ, we can substitute $V = m/\rho$, giving

$$(P_1 - P_2)\frac{m}{\rho} = \tfrac{1}{2}mv_2^2 - \tfrac{1}{2}mv_1^2 + mgh_2 - mgh_1$$

Multiplying through by ρ/m and rearranging, we obtain Bernoulli's equation:

$$P_1 + \rho gh_1 + \tfrac{1}{2}\rho v_1^2 = P_2 + \rho gh_2 + \tfrac{1}{2}\rho v_2^2 \qquad \text{15-14}$$

Since the subscripts 1 and 2 refer to any two points, Bernoulli's equation can be stated more simply as

$$P + \rho gh + \tfrac{1}{2}\rho v^2 = \text{constant} \qquad \text{15-15}$$

Bernoulli's equation finds application in almost every aspect of fluid flow. The pressure P must be recognized as the *absolute* pressure and not the *gauge* pressure. Remember that ρ is the mass density and not the weight density of the fluid. Notice that the units of each term in Bernoulli's equation are units of pressure.

15-10 APPLICATIONS OF BERNOULLI'S EQUATION

In many physical situations, the velocity, height, or pressure of a fluid is constant. In such cases Bernoulli's equation holds in simpler form. For example, when a liquid is stationary, both v_1 and v_2 are zero. Bernoulli's equation will then show

that the difference in pressure is

$$P_2 - P_1 = \rho g(h_1 - h_2) \qquad \text{15-16}$$

This equation is identical to the relationship discussed for fluids at rest.

Another important result occurs when there is no change in pressure ($P_1 = P_2$). In Fig. 15-18 a liquid emerges from a hole, or orifice, near the bottom of an open tank. Its velocity as it emerges from the orifice can be determined from Bernoulli's equation. We shall assume that the liquid level in the tank falls slowly in comparison with the emergent velocity, so the velocity v_2 at the top is assumed to be zero. In addition, it is noted that the liquid pressure both at the top and at the orifice is equal to atmospheric pressure. Thus, $P_1 = P_2$ and $v_2 = 0$, reducing Bernoulli's equation to

$$\rho g h_1 + \tfrac{1}{2}\rho v_1^2 = \rho g h_2$$

or

$$v_1^2 = 2g(h_2 - h_1) = 2gh$$

This relationship is known as **Torricelli's** (toh-ree-chel'-ees) **theorem:**

$$v = \sqrt{2gh} \qquad \text{15-17}$$

Note that the emergent velocity of a liquid at a depth h is the same as that of an object dropped from rest at a height h.

The rate at which a liquid flows from an orifice is given by vA from Eq. (15-10). Torricelli's relation allows us to express the rate of flow in terms of the height of a liquid above the orifice. Hence,

$$R = vA = A\sqrt{2gh} \qquad \text{15-18}$$

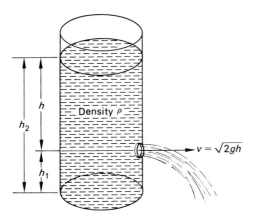

Figure 15-18 Torricelli's theorem.

EXAMPLE 15-10

A crack in a water tank has a cross-sectional area of 1 cm². At what rate is water lost from the tank if the water level in the tank is 4 m above the opening?

Solution

The area $A = 1 \text{ cm}^2 = 10^{-4} \text{ m}^2$ and the height $h = 4 \text{ m}$. Direct substitution into Eq. (15-18) gives

$$R = A\sqrt{2gh} = (10^{-4} \text{ m}^2)\sqrt{(2)(9.8 \text{ m/s}^2)(4 \text{ m})}$$
$$= (10^{-4} \text{ m}^2)(8.85 \text{ m/s}) = 8.85 \times 10^{-4} \text{ m}^3/\text{s}$$

An interesting example demonstrating Torricelli's principle is shown in Fig. 15-19. The discharge velocity increases with depth. Note that the maximum range occurs when the opening is at the middle of the water column. Although the discharge velocity increases below the midpoint, the water strikes the floor closer. This is because it strikes the floor sooner. Holes equidistant above and below the midpoint will have the same horizontal range.

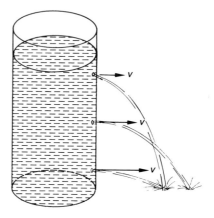

Figure 15-19   The discharge velocity increases with depth below the surface, but the range is a maximum at the midpoint.

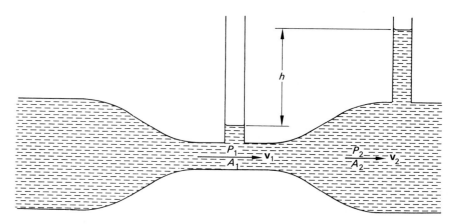

Figure 15-20   Fluid flow through a constriction in a horizontal pipe.

As a final application, let us return to the venturi effect, which describes the motion of a fluid through a constriction. If the pipe in Fig. 15-20 is horizontal, we can set $h_1 = h_2$ in Bernoulli's equation, giving

$$P_1 + \frac{1}{2}\rho v_1^2 = P_2 + \frac{1}{2}\rho v_2^2 \qquad \textbf{15-19}$$

Since v_1 is greater than v_2, it follows that the pressure P_1 must be less than the pressure P_2 for Eq. (15-19) to hold. This relationship between velocity and pressure has already been discussed.

Problem-Solving Strategy

Applications of Bernoulli's Equation

1. Read the problem carefully, and, after drawing a rough sketch, list the given information. Make sure that consistent units are used for pressure, height, and density.
2. The height h of a fluid is measured from a common reference point to the center of mass of the fluid. For example, a constriction in a horizontal pipe as in Fig. 15-20 does *not* represent a change in height ($h_1 = h_2$).
3. In Bernoulli's equation, the density ρ is *mass* density and the appropriate units are kg/m^3 and slugs/ft^3.
4. Write Bernoulli's equation for the problem and simplify by eliminating those factors that do not change:

$$P_1 + \rho g h_1 + \tfrac{1}{2}\rho v_1^2 = P_2 + \rho g h_2 + \tfrac{1}{2}\rho v_2^2$$

5. For a stationary fluid $v_1 = v_2$ and the third term on each side is eliminated; the middle terms drop out for a horizontal pipe ($h_1 = h_2$); and, if there is no change in pressure ($P_1 = P_2$), the first terms do not appear and the result is Torricelli's theorem (Eq. 15-17). Refer to the boxed equations in the summary.
6. Substitute the given quantities and solve for the unknown.

CONCEPTUAL EXAMPLE 15-11

Water flowing initially at 10 ft/s passes through a venturi tube like the one in Fig. 15-20. If $h = 4$ in., what is the velocity of the water in the constriction?

Solution

The difference in pressure, from Eq. (15-13), is

$$P_2 - P_1 = \rho g h$$

The pressure difference, from Bernoulli's equation, is

$$P_2 - P_1 = \tfrac{1}{2}\rho v_1^2 - \tfrac{1}{2}\rho v_2^2$$

Combining these two relations, we have

$$\rho g h = \tfrac{1}{2}\rho v_1^2 - \tfrac{1}{2}\rho v_2^2$$

Multiplying through by $2/\rho$ gives

$$2gh = v_1^2 - v_2^2$$

Note that this relation is similar to that for a freely falling body with an initial velocity v_2. Solving for v_1^2, we have

$$
\begin{aligned}
v_1^2 &= 2gh + v_2^2 \\
&= (2)(32 \text{ ft/s}^2)(0.333 \text{ ft}) + (10 \text{ ft/s})^2 \\
&= 21.3 \text{ ft}^2/\text{s}^2 + 100 \text{ ft}^2/\text{s}^2 = 121.3 \text{ ft}^2/\text{s}^2
\end{aligned}
$$

Therefore, the velocity through the constriction is

$$
v_1 = \sqrt{121.3} = 11 \text{ ft/s}
$$

In the preceding example, the density of the fluid did not enter into our calculations because the fluid in the tubes above the pipe was the same as that flowing through the pipe. When the density of the fluid is considered, it must be recognized that the symbol ρ is the mass density and not the weight density.

Summary and Review

Summary

We have presented the concepts of fluids at rest and in motion. Density, buoyant forces, and other quantities were defined and applied to many physical examples. The rate of flow of fluids was related to fluid velocity and to cross-sectional areas of pipes, and Bernoulli's equation was introduced to treat the more complete description of the dynamics of fluids. The essential concepts are summarized below:

- An important physical property of matter is its *density*. The weight density D and the mass density ρ are defined as follows:

$$Weight\ density = \frac{weight}{volume} \qquad D = \frac{w}{V}$$

$$\text{N/m}^3 \text{ or lb/ft}^3$$

$$Mass\ density = \frac{mass}{volume} \qquad \rho = \frac{m}{V}$$

$$\text{kg/m}^3 \text{ or slug/ft}^3$$

- Since $W = mg$, the relationship between D and ρ is

$$D = \rho g$$

Weight density = mass density × gravity

- Important points to remember about fluid pressure:

    a. The forces exerted by a fluid on the walls of its container are always perpendicular to the walls.
    b. The fluid pressure is directly proportional to the depth of the fluid and to its density.

$$P = \frac{F}{A} \qquad P = Dh \qquad P = \rho g h$$

    c. At any particular depth, the fluid pressure is the same in all directions.
    d. Fluid pressure is independent of the shape or area of the container.

- Pascal's law states that *an external pressure applied to an enclosed fluid is transmitted uniformly throughout the volume of the liquid.*
- When measuring fluid pressure, be sure to distinguish between *absolute* pressure and *gauge* pressure:

Absolute pressure
$$= gauge\ pressure + atmospheric\ pressure$$
Atmospheric pressure
$$= 1\ \text{atm} = 1.013 \times 10^5\ \text{N/m}^2$$
$$= 1.013 \times 10^5\ \text{Pa} = 14.7\ \text{lb/in.}^2$$
$$= 76\ \text{cm of mercury}$$

- Applying Pascal's law to the hydraulic press gives the following for the ideal mechanical advantage:

$$M_I = \frac{F_o}{F_i} = \frac{s_o}{s_i} \qquad \begin{array}{l} Ideal\ Mechanical\ Advantage \\ for\ Hydraulic\ Press \end{array}$$

- Archimedes' principle: *An object that is completely or partly submerged in a fluid experiences an upward force equal to the weight of the fluid displaced.*

$$F_B = mg \qquad \text{or} \qquad F_B = V\rho g \quad Buoyant\ Force$$

- The *rate of flow* is defined as the volume of fluid that passes a certain cross-section A per unit of time t. In terms of fluid velocity v, we write

$$R = \frac{V}{t} = vA \qquad \begin{array}{l} Rate\ of\ flow = velocity \\ \times\ cross\text{-}section \end{array}$$

- For an incompressible fluid flowing through pipes in which the cross-sections vary, the rate of flow is constant:

$$v_1 A_1 = v_2 A_2 \qquad d_1^2 v_1 = d_2^2 v_2$$

where v is the fluid velocity, A is the cross-sectional area of the pipe, and d is the diameter of the pipe.

Summary and Review

- The net work done on a fluid is equal to the changes in kinetic and potential energy of the fluid. Bernoulli's equation expresses this fact in terms of pressure P, the density ρ, the height of the fluid h, and its velocity v.

$$P + \rho g h + \tfrac{1}{2}\rho v^2 = \text{constant}$$

Bernoulli's Equation

If a volume of fluid changes from a state 1 to a state 2, as shown in Fig. 15-17, we can write

$$P_1 + \rho g h_1 + \tfrac{1}{2}\rho v_1^2 = P_2 + \rho g h_2 + \tfrac{1}{2}\rho v_2^2$$

- Special applications of Bernoulli's equation occur when one of the parameters does not change:

For a stationary liquid ($v_1 = v_2$)

$$P_2 - P_1 = \rho g(h_1 - h_2)$$

If the pressure is constant ($P_1 - P_2$)

$$v = \sqrt{2gh}$$

For a horizontal pipe ($h_1 = h_2$)

$$P_1 + \tfrac{1}{2}\rho v_1^2 = P_2 + \tfrac{1}{2}\rho v_2^2$$

Key Terms

1. weight density
2. mass density
3. pressure
4. total force
5. Pascal's law
6. absolute pressure
7. gauge pressure
8. manometer
9. Archimedes' principle
10. buoyant force
11. streamline flow
12. turbulent flow
13. rate of flow
14. venturi effect
15. Bernoulli's theorem
16. Torricelli's theorem

Review Questions

15-1. Make a list of the units for weight density and the similar units for mass density.

15-2. Which is numerically larger: the weight density of an object or its mass density?

15-3. The density of water is given in Table 15-1 as 62.4 lb/ft^3. In performing an experiment with water on the surface of the moon, would you trust this value? Explain.

15-4. Which is heavier, 870 kg of brass or 3.5 ft^3 of copper?

15-5. Why are dams so much thicker at the bottom than at the top? Does the pressure exerted on the dam depend on the length of the reservoir perpendicular to the dam?

15-6. A large block of ice floats in a bucket of water so that the level of the water is at the top of the bucket. Will the water overflow when the ice melts? Explain.

15-7. A tub of water rests on weighing scales that indicate 40 lb total weight. Will the total weight increase when a 5-lb fish is floating on the surface of the water? Discuss.

15-8. Suppose an iron block supported by a string is submerged completely in the tub of Question 15-7. How will the reading on the scales be affected?

15-9. A boy just learning to swim finds that he can float on the surface more easily after inhaling air. He also observes that he can hasten his descent to the bottom of the pool by exhaling air on the way down. Explain his observations.

15-10. A toy sailboat filled with pennies floats in a small tub of water. If the pennies are thrown into the water, what happens to the water level in the tub?

15-11. Is it more difficult to hold a cork float barely under the surface than it is at a depth of 5 ft? Explain.

15-12. Is it possible to construct a barometer using water instead of mercury? How high will the column of water be if the external pressure is 1 atm?

15-13. Discuss the operation of a submarine and a weather balloon. Why will a balloon rise to a definite height and stop? Will a submarine sink to a particular depth and stop if no changes are made after submerging?

15-14. What assumptions and generalizations are made concerning the study of fluid dynamics?

15-15. Why does the flow of water from a faucet decrease when someone turns on another faucet in the same building?

15-16. Two rowboats moving parallel to each other in the same direction are drawn together. Explain.

15-17. Explain what would happen in a modern jet airliner at high speed if a hijacker fired a bullet through the window or broke open an escape hatch.

15-18. During high-velocity windstorms or hurricanes, the roofs of houses are sometimes blown off; yet the houses are otherwise undamaged. Explain with the use of diagrams.

15-19. A small child knocks a balloon over the heating duct in his home and is surprised to find that the balloon remains suspended above the duct, bobbing from one side to the other. Explain.

15-20. What conditions would determine the maximum lift capacity of a streamlined aircraft wing? Draw figures to justify your answer.

15-21. Explain with diagrams how a baseball pitcher throws a rising fast ball, an outside curve ball, and a sinking fast ball. Would a pitcher prefer to throw into the wind or with the wind when delivering the three pitches discussed above?

15-22. Two identical reservoirs are placed on the floor side by side. One is filled with mercury, and the other is filled with water. A hole is bored in each reservoir at the same depth below the surface. Compare the ranges of the emergent fluids.

Problems

Section 15-1—Density

15-1. What volume does 0.4 kg of alcohol occupy? What is the weight of this volume?
Ans. 5.06×10^{-4} m^3, 3.92 N

15-2. An unknown substance has a volume of 20 ft^3 and weighs 3370 lb. What are the weight density and the mass density?

15-3. What volume of water has the same mass of 100 cm^3 of lead? What is the weight density of lead?
Ans. 1130 cm^3, 1.11×10^5 N/m^3

*15-4. A 200-mL flask (1 L = 1000 cm^3) is filled with an unknown liquid. An electronic balance indicates that the added liquid has a mass of 176 g. What is the specific gravity of the liquid? Can you guess the identity of the liquid?

Section 15-3—Fluid Pressure

15-5. Find the pressure in kilopascals due to a column of mercury 60 cm high. What is this pressure in lb/in.^{2} and in atmospheres?
Ans. 80.0 kPa, 11.6 lb/in.^{2}, 0.79 atm

15-6. A pipe contains water under a gauge pressure of 400 kPa. If you patch a 4-mm-diameter hole in the pipe with a piece of tape, what force must the tape be able to withstand?

*15-7. A submarine dives to a depth of 120 ft and levels off. The interior of the submarine is maintained at atmospheric pressure. What are the pressure and the total force applied to a hatch 2 ft wide and 3 ft long? The weight density of sea water is around 64 lb/ft^3. **Ans. 53.3 lb/in.^{2}, 46,100 lb**

15-8. If you constructed a barometer using water as the liquid instead of mercury, what

Summary and Review

height of water would indicate a pressure of one atmosphere?

15-9. A 20-kg piston rests on a sample of gas in a cylinder 8 cm in diameter. What is the gauge pressure on the gas? What is the absolute pressure? **Ans. 39.0 kPa, 140.3 kPa**

*15-10. An open U-shaped tube such as the one in Fig. 15-21 is 1 cm² in cross section. What volume of water must be poured into the right tube to cause the mercury in the left tube to rise 1 cm above its original position?

Figure 15-21

15-11. The gauge pressure in an automobile tire is 28 lb/in.². If the wheel supports 1000 lb, what area of the tire is in contact with the ground? **Ans. 35.7 in.²**

*15-12. Two liquids that do not react chemically are placed in a bent tube like the one in Fig. 15-21. Show that the heights of the liquids above their surface of separation are inversely proportional to their densities:

$$\frac{h_1}{h_2} = \frac{\rho_2}{\rho_1}$$

*15-13. Assume that the two liquids in the U-shaped tube in Fig. 15-21 are water and oil. Compute the density of the oil if the water

stands 19 cm above the interface and the oil stands 24 cm above the interface. Refer to Prob. 15-12. **Ans. 792 kg/m³**

15-14. A water-pressure gauge indicates a pressure of 50 lb/in.² at the foot of a building. What is the maximum height to which the water will rise in the building?

Section 15-5—The Hydraulic Press

15-15. The areas of the small and large pistons in a hydraulic press are 0.5 and 25 in.², respectively. What is the ideal mechanical advantage of the press? What force must be exerted to lift a 1-ton (2000-lb) load? Through what distance must the input force act to lift this load a distance of 1 in.? **Ans. 50, 40 lb, 50 in.**

15-16. A force of 400 N is applied to the small piston of a hydraulic press whose diameter is 4 cm. What must be the diameter of the large piston if it is to lift a 200-kg load?

15-17. The inlet pipe that supplies air pressure to operate a hydraulic lift is 2 cm in diameter. The output piston is 32 cm in diameter. What air pressure (gauge pressure) must be used to lift an 1800-kg automobile? **Ans. 219 kPa**

15-18. The area of a piston in a force pump is 10 in.². What force is required to raise water with the piston to a height of 100 ft?

Section 15-6—Archimedes' Principle

15-19. A 100-g cube 2 cm on each side is attached to a string and then totally submerged in water. What is the buoyant force and what is the tension in the rope? **Ans. 0.0784 N, 0.902 N**

*15-20. A solid object weighs 8 N in air. When this object is suspended from a spring scale and submerged in water, the apparent weight is only 6.5 N. What is the density of the object?

*15-21. A cube of wood 5.0 cm on each edge floats in water with three-fourths of its volume submerged. (a) What is the weight of the cube? (b) What is the mass of the cube? (c) What is the specific gravity of wood?

Ans. (a) 0.919 N, (b) 93.8 g, (c) 0.75

*15-22. A 20-g piece of metal has a density of 4000 kg/m^3. It is hung in a jar of oil (1500 kg/m^3) by a thin thread until it is completely submerged. What is the tension in the thread?

*15-23. The mass of a piece of rock is found to be 9.17 g in air. When the rock is submerged in a fluid of density 873 kg/m^3, its apparent mass is only 7.26 g. What is the density of this rock? **Ans. 4191 kg/m^3**

*15-24. A balloon 40 m in diameter is filled with helium. The mass of the balloon and attached basket is 18 kg. What additional mass can be lifted by this balloon?

Section 15-7—Fluid Flow

15-25. Gasoline flows through a 1-in.-diameter hose at an average velocity of 5 ft/s. What is the rate of flow in gallons per minute (1 ft^3 = 7.48 gal)? How much time is required to fill a 20-gal tank?

Ans. 12.2 gal/min, 1.63 min

15-26. Water flows from a terminal 3 cm in diameter and has an average velocity of 2 m/s. What is the rate of flow in liters per minute (1 L = 0.001 m^3)? How much time is required to fill a 40-L container?

15-27. What must be the diameter of a hose if it is to deliver 8 L of oil in 1 min with an exit velocity of 3 m/s? **Ans. 7.52 mm**

*15-28. Water flowing from a 2-in. pipe emerges horizontally at the rate of 8 gal/min. What is the emerging velocity? What is the horizontal range of the stream of water if the pipe is 4 ft from the ground?

15-29. Water flowing at 6 m/s through a 6-cm pipe is connected to a 3-cm pipe. What is the velocity in the small pipe? Is the rate of flow greater in the smaller pipe?

Ans. 24 m/s, no

Section 15-10—Applications of Bernoulli's Equation

15-30. Consider the situation described by Prob. 15-29. If the centers of each pipe are on the same horizontal line, what is the difference in pressure between the two connecting pipes?

15-31. What is the emergent velocity of water from a crack in its container 6 m below the surface? If the area of the crack is 1.3 cm^2, at what rate of flow does water leave the container? **Ans. 10.8 m/s, 1.41 × 10^{-3} m^3/s**

15-32. A 2-cm-diameter hole is in the side of a water tank, and it is located 5 m below the water level in the tank. What is the emergent velocity of the water from the hole? What volume of water will escape from this hole in 1 min?

*15-33. Water flows through a horizontal pipe at the rate of 82 ft^3/min. A pressure gauge placed on a 6-in. cross-section of this pipe reads 16 lb/in.^{2}. What is the gauge pressure in a section of pipe where the diameter is 3 in.? **Ans. 11.1 lb/in.^{2}**

*15-34. Water flows at the rate of 6 gal/min through an opening in the bottom of a cylindrical tank. The water in the tank is 16 ft deep. What is the rate of escape if an added pressure of 9 lb/in.^{2} is applied to the source of the water?

*15-35. Water moves through a pipe at 4 m/s under an absolute pressure of 200 kPa. The pipe narrows to one-half of its original diameter. What is the absolute pressure in the narrow part of the pipe? **Ans. 80.0 kPa**

*15-36. Water flows steadily through a horizontal pipe. At a point where the absolute pressure is 300 kPa, the velocity is 2 m/s. The pipe suddenly narrows, causing the absolute pressure to drop to 100 kPa. What will be the velocity of the water in this constriction?

Challenge Problems

*15-37. Human blood of density 1050 kg/m^3 is held a distance of 60 cm above an arm of a patient to whom it is being administered.

How much higher is the pressure at this position than it would be if it were held at the same level as the arm? **Ans. 6.17 kPa**

*15-38. A cylindrical tank 50 ft high and 20 ft in diameter is filled with water. (a) What is the water pressure on the bottom of the tank? (b) What is the total force on the bottom? (c) What is the pressure in a water pipe that is located 90 ft below the water level in the tank?

*15-39. A block of wood weighs 16 lb in air. A lead sinker, which has an apparent weight of 28 lb in water, is attached to the wood, and both are submerged in water. If their combined apparent weight in water is 18 lb, find the density of the wooden block.

Ans. 38.4 lb/ft³

*15-40. A 100-g block of wood has a volume of 120 cm³. Will it float in water? Gasoline?

*15-41. A vertical test tube has 3 cm of oil (0.8 g/cm³) floating on 9 cm of water. What is the pressure at the bottom of the tube?

Ans. 1.12 kPa

15-42. What percentage of an iceberg will remain below the surface of sea water (1030 kg/m³)?

*15-43. What is the smallest area of ice 30 cm thick that will support a 90-kg man? The ice is floating in fresh water. **Ans. 3.75 m²**

15-44. A spring balance indicates a weight of 40 N when an object is hung in air. When the same object is submerged in water, the indicated weight is reduced to only 30 N. What is the density of the object?

15-45. A thin-walled metal cup has a mass of 100 g and a total volume of 250 cm³. What is the maximum number of pennies that can be placed in the cup without sinking in water? The mass of a single penny is 3.11 g.

Ans. 48

15-46. What is the absolute pressure at the bottom of a lake that is 30 m deep?

15-47. A fluid is forced out of a 6-mm-diameter tube so that 200 mL emerges in 32 s. What is the average velocity of the fluid in the tube? **Ans. 0.221 m/s**

*15-48. A pump of 2 kW output power discharges water from a cellar into a street 6 m above. At what rate in liters per second is the cellar emptied?

15-49. A horizontal pipe of diameter 120 mm has a constriction of diameter 40 mm. The velocity of water in the pipe is 60 cm/s and the pressure is 150 kPa. (a) What is the velocity in the constriction? (b) What is the pressure in the constriction?

Ans. (a) 540 cm/s, (b) 136 kPa

*15-50. The water column in the container shown in Fig. 15-20 stands at a height H above the base of the container. Show that the depth h required to give the horizontal range x is given by

$$h = \frac{H}{2} \pm \frac{\sqrt{H^2 - x^2}}{2}$$

How does this equation show that the holes equidistant above and below the midpoint will have the same horizontal range?

*15-51. A column of water stands 16 ft above the base of its container. What are two hole depths at which the emergent water will have a horizontal range of 8 ft?

Ans. 1.07 ft, 14.9 ft

*15-52. Refer to Fig. 15-20 and Prob. 15-50. Show that the horizontal range is given by

$$x = 2\sqrt{h(H - h)}$$

Use this relation to show that the maximum range is equal to the height H of the water column.

*15-53. Water flows through a horizontal pipe at the rate of 60 gal/min (1 ft³ = 7.48 gal). What is the velocity in a narrow section of the pipe that is reduced from 6 to 1 in. in diameter? **Ans. 24.5 ft/s**

15-54. What must be the gauge pressure in a fire hose if the nozzle is to force water to a height of 20 m?

*15-55. Water flows through the pipe shown in Fig. 15-22 at the rate of 30 liters per second. The absolute pressure at point A is 200 kPa, and point B is 8 m higher than point A.

The lower section of pipe has a diameter of 16 cm and the upper section narrows to a diameter of 10 cm. (a) Find the velocities of the stream at points A and B. (b) What is the absolute pressure at point B?

Ans. (a) 1.49 m/s, 3.82 m/s; (b) 115 kPa

Figure 15-22

Critical Thinking Questions

15-56. A living room floor has floor dimensions of 4.50 m and 3.20 m and a height of 2.40 m. The density of air is 1.29 kg/m³. What does the air in the room weigh? What force does the atmosphere exert on the floor of the room? **Ans. 437 N, 1.46 x 10⁶ N**

15-57. A tin coffee can floating in water (1.00 g/cm³) has an internal volume of 180 cm³ and a mass of 112 g. How many grams of metal can be added to the can without causing it to sink in the water?

*15-58. A wooden block floats in water with two-thirds of its volume submerged. The same block floats in oil with nine-tenths of its volume submerged. What is the ratio of the density of the oil to the density of water (the specific gravity)? **Ans. 0.741**

*15-59. An aircraft wing 25 ft long and 5 ft wide experiences a lifting force of 800 lb. What is the difference in pressure between the upper and lower surfaces of the wing?

*15-60. Assume that air (ρ = 1.29 kg/m³) flows past the top surface of an aircraft wing at 36 m/s. The air moving past the lower surface of the wing has a velocity of 27 m/s. If the wing has a weight of 2700 N and an area of 3.5 m², what is the buoyant force on the wing? **Ans. 1280 N**

*15-61. Seawater has a weight density of 64 lb/ft³. This water is pumped through a system of pipes (see Fig. 15-23) at the rate of 4 ft³/min. Pipe diameters at the lower and upper ends are 4 in. and 2 in., respectively. The water is discharged into the atmosphere at the upper end, a distance of 6 ft higher than the lower section. What are the velocities of flow in the upper and lower pipes? What are the pressures in the lower and upper sections?

Figure 15-23

Thermodynamics, Mechanical Waves, and Sound

Chapter
16

Temperature and Expansion

We have discussed the behavior of systems at rest and in motion. The fundamental quantities of mass, length, and time were introduced to describe the state of a given mechanical system.

Consider, for example, a 10-kg block moving with a constant velocity of 20 m/s. The quantities of mass, length, and time are all present, and we find them sufficient to describe the motion. We can speak of the weight of the block, its kinetic energy, or its momentum, but a complete description of a system requires more than a simple statement of these quantities.

This becomes apparent when our 10-kg block encounters frictional forces. As the block slides to a stop, its energy seems to disappear, but the block and its supporting surface are slightly warmer. If energy is to be conserved, we must assume that the lost energy reappears in some form not yet considered. When energy disappears from the visible motion of objects and does not reappear in the form of visible potential energy, we frequently notice a rise in temperature. In this chapter we introduce the concept of temperature as a fourth fundamental quantity.

16-1 TEMPERATURE AND THERMAL ENERGY

Until now we have been concerned only with the causes and the effects of *external* motion. A block resting on a table is in translational and rotational equilibrium insofar as its surroundings are concerned. A much closer study of the block reveals, however, that it is active internally. Figure 16-1 shows a simple model of a solid. Individual molecules are held together by elastic forces analogous to the springs in the figure. These molecules oscillate about their equilibrium positions

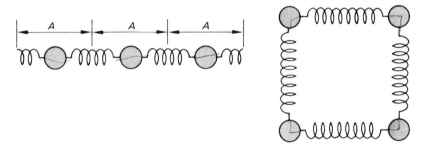

Figure 16-1   A simplified model of a solid in which the individual molecules are held together by elastic forces.

with a particular frequency and amplitude A. Thus, both potential energy and kinetic energy are associated with the molecular motion. Since this internal energy is related to the hotness or coldness of a body, it is often referred to as ***thermal energy.***

> **Thermal energy represents the total internal energy of an object: the sum of its molecular kinetic and potential energies.**

When two objects with different temperatures are placed in contact, energy is transferred from one to the other. For example, suppose hot coals are dropped into a container of water, as shown in Fig. 16-2. Thermal energy will be transferred from the coals to the water until the system reaches a stable condition, called ***thermal equilibrium.*** When touched, the coals and the water produce similar sensations and there is no more transfer of thermal energy.

Such changes in thermal energy states cannot be explained satisfactorily in terms of classical mechanics alone. Therefore, all objects must have a new fundamental property that determines whether they will be in thermal equilibrium with other objects. This property is called ***temperature.*** In our example, the coals and the water are said to have the same temperature when the transfer of energy is zero.

> **Two objects are said to be in thermal equilibrium if and only if they are at the same temperature.**

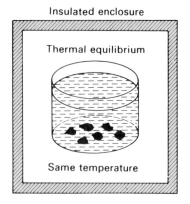

Figure 16-2 Thermal equilibrium.

Once we establish a means of measuring temperature, we have a necessary and sufficient condition for thermal equilibrium. The transfer of thermal energy that is due only to a difference in temperature is defined as *heat.*

> **Heat is defined as the transfer of thermal energy that is due to a difference of temperature.**

Before discussing the measurement of temperature, we should distinguish clearly between temperature and thermal energy. It is possible for two objects to be in thermal equilibrium (same temperature) with different thermal energies. For example, consider a pitcher of water and a small cup of water, each having a temperature of 90°C. If they are mixed together, there will be no transfer of energy, but the thermal energy is much greater in the pitcher because it contains many more molecules. Remember that thermal energy represents the *sum* of the kinetic and potential energies of all the molecules. If we pour the water from each container onto separate blocks of ice, as shown in Fig. 16-3, more ice will be melted by the larger volume, indicating that it had more thermal energy.

16-2 THE MEASUREMENT OF TEMPERATURE

Temperature is usually determined by measuring some mechanical, optical, or electrical quantity that varies with temperature. For example, most substances expand as their temperature increases. If a change in any dimension can be shown to have a one-to-one correspondence with changes in temperature, the variation can be calibrated to measure temperature. A device calibrated in this way is called a *thermometer.* The temperature of another object can then be measured by placing the thermometer in close contact with it and allowing the two to reach thermal equilibrium. The temperature indicated by a number on the graduated thermometer also corresponds to the temperature of the surrounding objects.

> **A thermometer is a device that, through marked scales, can give an indication of its own temperature.**

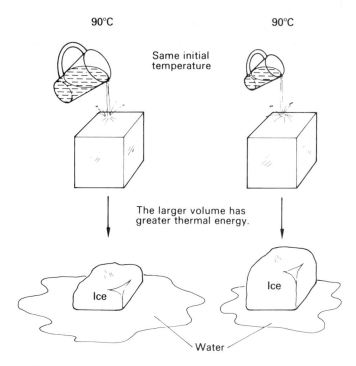

Figure 16-3 The distinction between thermal energy and temperature.

Two things are necessary when constructing a thermometer. First, we must have confirmation that some thermometer property X varies with temperature T. If the variation is linear, we can write

$$T = kX$$

where k is the proportionality constant. The thermometer property should be one that is easily measured, for example, the expansion of a liquid, the pressure in a gas, or the resistance in an electric circuit. Other quantities that vary with temperature are radiated energy, the color of emitted light, vapor pressure, and magnetic susceptibility. Thermometers have been constructed for each of these thermometric properties. The choice is dictated by the range of temperatures over which the thermometer is linear and by the mechanics of its use.

The second requirement in constructing a thermometer is the establishment of standard temperatures. Early temperature scales were based on the choice of upper and lower *fixed points,* which are temperatures readily available for laboratory measurements. Two convenient temperatures reproduced easily are the lower fixed point and the upper fixed point.

The lower fixed point (ice point) is the temperature at which water and ice coexist in thermal equilibrium under a pressure of 1 atm.

The upper fixed point (steam point) is the temperature at which water and steam coexist in equilibrium under a pressure of 1 atm.

A widely used temperature measurement for scientific work originated with a scale developed by Anders Celsius (1701–1744), a Swedish astronomer. The **Celsius scale** arbitrarily assigned the number 0 to the ice point and the number 100 to the steam point. Thus, at atmospheric pressure there are 100 divisions between the freezing point and the boiling point of water. Each division or unit on the scale is called a **degree** (°). For example, room temperature is often taken as 20°C, which is read as *twenty degrees Celsius.*

Another scale for measuring temperature was developed in 1714 by Gabriel Daniel Fahrenheit. The development of this scale was based on the choice of different fixed points. Fahrenheit chose the temperature of a freezing solution of salt water as his lower fixed point and assigned it the number and unit of 0°F. The upper fixed point was chosen to be the temperature of the human body. For some unexplained reason, he assigned the number and unit of 96°F for body temperature. The fact that body temperature is really 98.6°F indicates an experimental error in establishing the scale. Relating the **Fahrenheit scale** to the universally accepted fixed points on the Celsius scale, we note that 0 and 100°C correspond to 32 and 212°F.

We can compare the two scales by calibrating ordinary mercury-in-glass thermometers. This type of thermometer makes use of the fact that liquid mercury expands with increasing temperature. It consists of an evacuated glass capillary tube with a reservoir of mercury at the bottom and a closed top. Since mercury expands more than the glass tube, the mercury column rises in the tube until the mercury, glass, and its surroundings are in equilibrium.

Suppose we make two ungraduated thermometers and place them in a mixture of ice and water, as in Fig. 16-4. After allowing the mercury columns to stabilize, we mark 0°C on one thermometer and 32°F on the other. Next, we place the two thermometers directly above boiling water, allowing the mercury columns to sta-

Figure 16-4 Calibration of Celsius and Fahrenheit thermometers.

bilize at the steam point. Again we mark the two thermometers, inscribing 100°C and 212°F adjacent to the mercury level above the marks corresponding to the ice point. The level of mercury is the same in each thermometer. Thus, the only difference between the two thermometers will be in how they are graduated. There are 100 divisions, or Celsius degrees (C°), between the ice point and steam point on the Celsius thermometer, and there are 180 divisions, or Fahrenheit degrees (F°), on the Fahrenheit thermometer. Thus, 100 Celsius degrees represents the same temperature interval as 180 Fahrenheit degrees. Symbolically,

$$100 \text{ C}° = 180 \text{ F}° \qquad \text{or} \qquad 5 \text{ C}° = 9 \text{ F}° \qquad\qquad \textbf{16-1}$$

The degree mark (°) is placed after the C or F to emphasize that the numbers correspond to temperature intervals and not to specific temperatures. In other words, 20 F° is read "twenty Fahrenheit degrees" and corresponds to a *difference* between two temperatures on the Fahrenheit scale. The symbol 20°F, on the other hand, refers to a specific mark on the Fahrenheit thermometer. For example, suppose a pan of hot food cools from 98 to 76°F. These numbers correspond to specific temperatures, as indicated by the height of a mercury column. They represent a temperature interval, however, of

$$\Delta t = 98°\text{F} - 76°\text{F} = 22 \text{ F}°$$

where Δt is used to denote a change in temperature.

The physics that treats the transfer of thermal energy is nearly always concerned with changes in temperature. Thus, it often becomes necessary to convert a temperature interval from one scale to the corresponding interval on the other scale. This can best be accomplished by recalling from Eq. (16-1) that an interval of 5 C° is equivalent to an interval of 9 F°. The appropriate conversion factors can be written as

$$\frac{5 \text{ C}°}{9 \text{ F}°} = 1 = \frac{9 \text{ F}°}{5 \text{ C}°} \qquad\qquad \textbf{16-2}$$

When F° is converted to C°, the factor on the left should be used; when C° is converted to F°, the factor on the right should be used.

EXAMPLE 16-1

During a 24-h period, a steel rail varies in temperature from 20°F at night to 70°F in the middle of the day. Express this range of temperature in Celsius degrees.

Solution

The temperature interval is

$$\Delta t = 70°\text{F} - 20°\text{F} = 50 \text{ F}°$$

To convert the interval to Celsius degrees, we choose the conversion factor that will cancel the Fahrenheit units. Thus,

$$\Delta t = 50 \text{ F}° \times \frac{5 \text{ C}°}{9 \text{ F}°} = 27.8 \text{ C}°$$

Figure 16-5 Comparison of the Celsius and Fahrenheit scales.

It must be remembered that Eq. (16-2) applies for temperature intervals. It can be used only when working with *differences* in temperature. It is another matter entirely to find the temperature on the Fahrenheit scale that corresponds to the same temperature on the Celsius scale. Using ratio and proportion, we can develop an equation that will convert specific temperatures. Suppose, for example, we place two identical thermometers in a beaker of water, as shown in Fig. 16-5. One thermometer is graduated in Fahrenheit degrees, and the other in Celsius degrees. The symbols t_C and t_F represent the same temperature (the temperature of the water), but they are on different scales. It should be apparent from the figure that the difference between t_C and 0°C corresponds to the same interval as the difference between t_F and 32°F. The ratio of the former to 100 divisions should be the same as the ratio of the latter to 180 divisions. Hence,

$$\frac{t_C - 0}{100} = \frac{t_F - 32}{180}$$

Simplifying and solving for t_C, we obtain

$$t_C = \frac{5}{9}(t_F - 32) \qquad \text{16-3}$$

or, solving for t_F,

$$t_F = \frac{9}{5}t_C + 32 \qquad \text{16-4}$$

EXAMPLE 16-2

The melting point of lead is 330°C. What is the corresponding temperature on the Fahrenheit scale?

Solution

Substitution into Eq. (16-4) yields

$$t_F = \frac{9}{5}t_C + 32 = \frac{9}{5}(330) + 32$$

$$= 594 + 32 = 626°F$$

It is important to recognize that the t_F and t_C of Eqs. (16-3) and (16-4) represent identical temperatures. The numbers are different because the origin of each scale was at a different point and the degrees are of different size. What these equations tell us is the relationship between the *numbers* that are assigned to specific temperatures on two *different* scales.

16-3 THE GAS THERMOMETER

Although the mercury-in-glass thermometer is the best known and most widely used, it is not as accurate as many other thermometers. In addition, mercury freezes at around −40°C, restricting the range over which it can be used. An accurate thermometer with an extensive measuring range can be constructed by utilizing the properties of a gas. All gases subjected to heating expand in nearly the same manner. If their expansion is prevented by maintaining a constant volume, the pressure will increase in proportion to temperature.

In general, there are two kinds of gas thermometers. One kind maintains a constant pressure and utilizes the increases in volume as an indicator. This type is called a **constant-pressure thermometer.** The other kind, called a **constant-volume thermometer,** measures the increase in pressure as a function of temperature. The constant-volume thermometer is illustrated in Fig. 16-6. The gas is contained in

Figure 16-6   The constant-volume thermometer.

bulb *B*, and the pressure it exerts is measured by the mercury manometer. As the temperature of the gas increases, it expands, forcing the mercury down in the closed tube and up in the open tube. To maintain a constant volume of the gas, the open tube must be raised until the level of mercury in the closed tube is brought back to the reference mark *R*. The difference in the two mercury levels is then an indication of the gas pressure at constant volume. The instrument can be calibrated for temperature measurements through the use of fixed points, as mentioned in the previous section.

The same apparatus can be used as a constant-pressure thermometer, as illustrated in Fig. 16-7. In this instance the volume of gas in bulb *B* is allowed to increase at constant pressure. The pressure exerted on the gas is maintained constant at 1 atm by lowering or raising the open tube until the mercury levels are the same in both tubes. The change in volume with temperature can then be indicated by the mercury level in the closed tube. Calibration consists of marking the mercury level at the ice point and again at the steam point.

Gas thermometers are useful because of their almost unlimited range. For this reason and because they are so accurate, they are commonly used in research laboratories and in bureaus of standards. They are also large and bulky, however, rendering them useless for many minute technical measurements.

Bulb

B

Figure 16-7   The constant-pressure thermometer.

16-4 THE ABSOLUTE TEMPERATURE SCALE

It has probably occurred to you that Celsius and Fahrenheit scales have a serious limitation. Neither 0°C nor 0°F represents a true zero of temperature. Consequently, for temperatures much lower than the ice point, a negative temperature results. Even more serious is the fact that a formula involving temperature as a variable will not work with the existing scales. For example, we have discussed the expansion of a gas with an increase in temperature. We can state this proportionality as

$$V = kt$$

where k is the proportionality constant and t is the temperature. Certainly, the volume of a gas is not zero at 0°C or negative at negative temperatures, and yet these conclusions might be drawn from the above relationship.

This example provides a clue for establishing an *absolute* scale. If we can determine the temperature at which the volume of a gas under constant pressure becomes zero, we can establish a true zero of temperature. Suppose we use a constant-pressure gas thermometer, like the one in Fig. 16-7. The volume of the gas in the bulb can be measured carefully, first at the ice point and then at the steam point. These two points can be plotted on a graph, as in Fig. 16-8, with the volume as the ordinate and the temperature as the abscissa. The points A and B correspond to the gas volume at temperatures of 0 and 100°C, respectively. A straight line through these two points, extended both to the left and to the right, provides a mathematical description of the change in volume as a function of temperature. Note that the line can be extended indefinitely to the right, indicating that there is no upper limit to temperature. We cannot extend the line indefinitely to the left, however, because it will eventually intercept the temperature axis. At this theoretical point, the gas would have zero volume. Further extension of the line would indicate a negative volume, which is meaningless. Therefore, the point at which the line intercepts the temperature axis is called the **absolute zero** of temperature. (Actually, any real gas would liquefy before reaching this point.)

If the above experiment is performed for several different gases, the slope of the curves will vary slightly. But the temperature intercept will always be the same and near −273°C. Ingenious theoretical and experimental procedures have established that the absolute zero of temperature is −273.15°C. In this text we shall assume that it is −273°C without fear of significant error. Conversion to the Fahrenheit scale shows that absolute zero is −460°F on that scale.

An absolute temperature scale has as its zero point the absolute zero of temperature. One such scale was devised by Lord Kelvin (1824–1907). The standard interval on this scale, the *kelvin,* has been adopted by the international (SI) metric system as the base unit for temperature measurement. The interval on the **Kelvin scale** represents the same change in temperature as the Celsius degree. Thus, an interval of 5 K (reads "five kelvins") is exactly the same 5 C°.

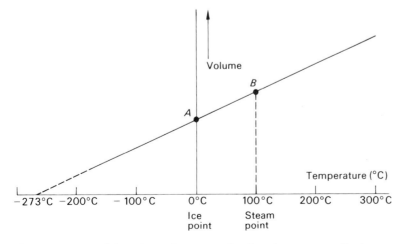

Figure 16-8 The variation of the volume of a gas as a function of temperature. Absolute zero can be defined by extrapolation to zero volume.

The Kelvin scale is related to the Celsius scale by the formula

$$T_K = t_C + 273 \qquad\qquad \text{16-5}$$

For example, 0°C would correspond to 273 K, and 100°C would correspond to 373 K. (See Fig. 16-9). Hereafter, we will reserve the symbol T for absolute temperature and the symbol t for other temperatures.

Problems with the reproducibility of accurate measurements for the ice and steam points of water caused the International Committee on Weights and Measures to establish a new standard in 1954. This standard is based on the **triple point of water,** which is the single temperature and pressure at which water, water vapor, and ice can coexist in thermal equilibrium. This convenient event occurs at a temperature of around 0.01°C and at a pressure of 4.58 mm of mercury. To remain somewhat consistent with earlier measurements, the temperature for the triple point of water was set at exactly 273.16 K. Thus, the **kelvin** is now defined as the fraction 1/273.16 of the temperature of the triple point of water. The SI temperature is now fixed by this definition, and all other scales must be redefined based on this single standard temperature.

A second absolute scale, called the *Rankine scale,* remains in limited use despite efforts by various organizations to eliminate its use entirely. The Rankine degree is included here only for the sake of completeness. Its absolute zero point is at −460°F, and the degree intervals are identical to the Fahrenheit degree interval. The relationship between the temperature in degrees Rankine (°R) and the corresponding temperature in degrees Fahrenheit is

$$T_R = t_F + 460 \qquad\qquad \text{16-6}$$

For example, 0°F corresponds to 460°R and 212°F corresponds to 672°R.

Figure 16-9 A comparison of the four common temperature scales.

Remember that Eqs. (16-5) and (16-6) apply for specific temperatures. If we are concerned with a change in temperature or a difference in temperature, the absolute change or difference is the same in kelvins as it is in Celsius degrees. It is helpful to recall that

$$1 \text{ K} = 1 \text{ C}° \qquad 1 \text{ R}° = 1 \text{ F}° \qquad\qquad 16\text{-}7$$

EXAMPLE 16-3

A mercury-in-glass thermometer may not be used at temperatures below $-40°C$. This is because mercury freezes at this temperature. (a) What is the freezing point of mercury on the Kelvin scale? (b) What is the difference between this temperature and the freezing point of water? Express the answer in kelvins.

Solution (a)
    Direct substitution of $-40°C$ into Eq. (16-5) yields

$$T_K = -40°C + 273 = 233 \text{ K}$$

Solution (b)
    The difference in the freezing points is

$$\Delta t = 0°C - (-40°C) = 40 \text{ C}°$$

Since the size of the kelvin is identical to that of the Celsius degree, the difference is also 40 kelvins.

At this point you may ask why we still retain the Celsius and Fahrenheit scales. When working with heat, one is nearly always concerned with differences in temperature. In fact, a difference in temperature is necessary for heat to be transferred. Otherwise, the system would be in thermal equilibrium. Since the Kelvin and Rankine scales are based on the same intervals as the Celsius and Fahrenheit scales, it makes no difference which scale is used for temperature intervals. On the other hand, if a formula calls for a specific temperature rather than a temperature difference, the absolute scale must be used.

16-5 LINEAR EXPANSION

The most common effect produced by temperature changes is a change in size. With a few exceptions, all substances increase in size with rising temperature. The atoms in a solid are held together in a regular pattern by electric forces. At any temperature the atoms vibrate with a certain frequency and amplitude. As the temperature is increased, the amplitude (maximum displacement) of the atomic vibrations increases. This results in an overall change in the dimensions of the solid.

A change in any *one* dimension of a solid is called ***linear expansion.*** It is found experimentally that an increase in a single dimension, for example, the length of

Figure 16-10 Linear expansion.

a rod, is dependent on the original dimension and the change in temperature. Consider, for example, the rod in Fig. 16-10. The initial length is L_0 and the initial temperature is t_0. When the rod is heated to a temperature t, its new length is denoted by L. Thus, a change in temperature, $\Delta t = t - t_0$, has resulted in a change in length, $\Delta L = L - L_0$. The proportional change in length is given by

$$\Delta L = \alpha L_0 \, \Delta t \qquad \text{16-8}$$

where α is the proportionality constant called the **coefficient of linear expansion.** Since an increase in temperature does not produce the same increase in length for all materials, the coefficient α is a property of the material. Solving Eq. (16-8) for α, we obtain

$$\alpha = \frac{\Delta L}{L_0 \, \Delta t} \qquad \text{16-9}$$

The coefficient of linear expansion of a substance can be defined as the change in length per unit length per degree change in temperature. Since the ratio $\Delta L/L_0$ has no dimensions, the units of α are in inverse degrees, that is, $1/C°$ or $1/F°$. The expansion coefficients for many common materials are given in Table 16-1.

TABLE 16-1 Linear Expansion Coefficients

Substance	α	
	$10^{-5}/C°$	$10^{-5}/F°$
Aluminum	2.4	1.3
Brass	1.8	1.0
Concrete	0.7–1.2	0.4–0.7
Copper	1.7	0.94
Glass, Pyrex	0.3	0.17
Iron	1.2	0.66
Lead	3.0	1.7
Silver	2.0	1.1
Steel	1.2	0.66
Zinc	2.6	1.44

EXAMPLE 16-4

An iron pipe is 300 m long at room temperature (20°C). If the pipe is to be used as a steam pipe, how much allowance must be made for expansion, and what will the new length of the pipe be?

Solution

The temperature of steam is 100°C and $\alpha_{iron} = 1.2 \times 10^{-5}/C°$. Thus, the increase in length is

$$\Delta L = \alpha L_0 \, \Delta t = (1.2 \times 10^{-5}/C°) \, (300 \text{ m})(100°C - 20°C)$$
$$= (1.2 \times 10^{-5}/C°)(300 \text{ m})(80 \text{ C}°) = 0.288 \text{ m}$$

Therefore, the length of the pipe at 100°C is

$$L = L_0 + \Delta L = 300.29 \text{ m}$$

We can see from this example that the new length may be calculated by the following relation:

$$L = L_0 + \alpha L_0 \, \Delta t \qquad\qquad \textbf{16-10}$$

Remember, when calculating ΔL, that the units of α must be consistent with the units of Δt.

Linear expansion has both useful and destructive properties when applied to physical situations. The destructive effects require engineers to use expansion joints or rollers to make allowances for expansion. The predictable expansion of some materials, on the other hand, can be used to open or close switches at certain temperatures. Such devices are called *thermostats*.

Probably the most common application of the principle of linear expansion is the bimetallic strip. This device, shown in Fig. 16-11, consists of two flat strips of different metals welded or riveted together. The strips are fused so that they are the same length at a chosen temperature t_0. If we heat the strip, causing a rise in temperature, the material with the larger expansion coefficient will expand more.

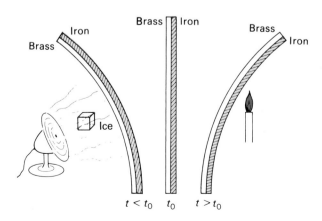

Figure 16-11 The bimetallic strip.

For example, a brass-iron strip will bend in an arc toward the iron side. When the source of heat is removed, the strip will gradually return to its original position. Cooling the strip below the initial temperature will cause the strip to bend in the other direction. The material with the higher coefficient of expansion also *decreases* in length at a faster rate. The bimetallic strip has many useful applications, from thermostatic control systems to blinking lights. Since the expansion is in direct proportion to an increase in temperature, the bimetallic strip can also be used as a thermometer.

16-6 AREA EXPANSION

Linear expansion is by no means restricted to the length of a solid. Any line drawn through the solid will increase in length per unit length at the rate given by its expansion coefficient α. For example, in a solid cylinder the length, diameter, and a diagonal drawn through the solid will all increase their dimensions in the same proportion. In fact, the expansion of a surface is exactly analogous to a photographic enlargement, as illustrated in Fig. 16-12. Notice also that if the material contains a hole, the area of the hole expands at the same rate it would if it were filled with material.

Let us consider the area expansion of the rectangular surface in Fig. 16-13. Both the length and the width of the material will expand at the rate given by Eq. (16-10). Thus, the new length and width are given, in factored form, by

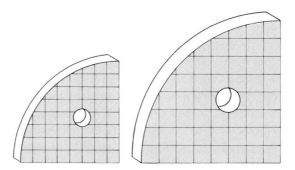

Figure 16-12   Thermal expansion is analogous to a photographic enlargement. Note that the hole increases in the same proportion as the material.

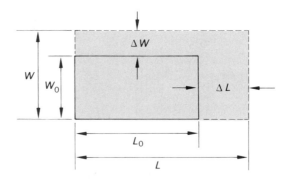

Figure 16-13   Area expansion.

$$L = L_0 (1 + \alpha \, \Delta t)$$
$$W = W_0(1 + \alpha \, \Delta t)$$

We can now derive an expression for area expansion by finding the product of these two equations.

$$LW = L_0 \, W_0(1 + \alpha \, \Delta t)^2$$
$$= L_0 \, W_0(1 + 2\alpha \, \Delta t + \alpha^2 \, \Delta t^2)$$

Since the magnitude of α is of the order of 10^{-5}, we may certainly neglect the term containing α^2. Hence, we can write

$$LW = L_0 W_0(1 + 2\alpha \, \Delta t)$$

or

$$A = A_0(1 + 2\alpha \, \Delta t)$$

where $A = LW$ represents the new area and $A_0 = L_0 \, W_0$ represents the original area. Rearranging terms, we obtain

$$A - A_0 = 2\alpha A_0 \, \Delta t$$

or

$$\Delta A = 2\alpha A_0 \, \Delta t \qquad \text{16-11}$$

The **coefficient of area expansion** γ (gamma) is approximately twice the coefficient of linear expansion. Symbolically,

$$\gamma = 2\alpha \qquad \text{16-12}$$

where γ is the change in area per unit initial area per degree change in temperature. Using this definition, we may write the following formulas for area expansion.

$$\Delta A = \gamma A_0 \, \Delta t \qquad \text{16-13}$$
$$A = A_0 + \gamma A_0 \, \Delta t \qquad \text{16-14}$$

EXAMPLE 16-5

A brass disk has a hole 80 mm in diameter punched in its center at 70°F. If the disk is placed in boiling water, what will be the new area of the hole?

Solution

We first compute the area of the hole at 70°F.

$$A_0 = \frac{\pi D^2}{4} = \frac{\pi(80 \text{ mm})^2}{4} = 5027 \text{ mm}^2$$

Now, the area expansion coefficient is

$$\gamma = 2\alpha = (2)(1.0 \times 10^{-5}/\text{F}°) = 2 \times 10^{-5}/\text{F}°$$

The increase in the area of the hole is found from Eq. (16-13) as follows:

$$\Delta A = \gamma A_0 \, \Delta t$$
$$= (2 \times 10^{-5}/\text{F}°)(5027 \text{ mm}^2)(212°\text{F} - 70°\text{F})$$
$$= 14.3 \text{ mm}^2$$

The new area is found by adding this increase to the original area. See Eq. (16-14).

$$A = A_0 + \Delta A$$
$$= 5027 \text{ mm}^2 + 14.3 \text{ mm}^2 = 5041.3 \text{ mm}^2$$

16-7 VOLUME EXPANSION

The expansion of heated material is the same in all directions. Therefore, the volume of a liquid, gas, or solid will have a predictable increase in volume with a rise in temperature. Reasoning similar to that of the previous sections will give us the following formulas for volume expansion.

$$\Delta V = \beta V_0 \Delta t \qquad \text{16-15}$$
$$V = V_0 + \beta V_0 \Delta t \qquad \text{16-16}$$

The symbol β (beta) is the ***volume expansion coefficient.*** It represents the change in volume per unit volume per degree change in temperature. For solid materials it is approximately three times the linear expansion coefficient.

$$\beta = 3\alpha \qquad \text{16-17}$$

When working with solids, we can compute β from the table of linear expansion coefficients (Table 16-1). For different liquids, the volume expansion coefficients are listed in Table 16-2. The molecular separation in gases is so great that they all expand at approximately the same rate. Volumetric expansion of gases will be discussed later.

TABLE 16-2 Volume Expansion Coefficients		
	β	
Liquid	**$10^{-4}/\text{C}°$**	**$10^{-4}/\text{F}°$**
Alcohol, ethyl	11	6.1
Benzene	12.4	6.9
Glycerin	5.1	2.8
Mercury	1.8	1.0
Water	2.1	1.2

CONCEPTUAL EXAMPLE 16-6

A Pyrex glass bulb is filled with 50 cm³ of mercury at 20°C. What volume will overflow if the system is heated uniformly to a temperature of 60°C? Refer to Fig. 16-14.

Figure 16-14 The volume overflow is found by subtracting the change in volume of the glass from the change in volume of the liquid.

Solution

The inside volume of the glass bulb is 50 cm³ initially and will increase according to Eq. (16-15). Remember that $\beta_g = 3\alpha_g$. At the same time, the mercury will increase in volume according to the value of β_m. The overflow will, therefore, be the difference between the two expansions.

$$\begin{pmatrix} \text{Volume} \\ \text{overflow} \end{pmatrix} = \begin{pmatrix} \text{volume increase} \\ \text{in mercury} \end{pmatrix} - \begin{pmatrix} \text{volume increase} \\ \text{in glass} \end{pmatrix}$$

$$\begin{aligned} V_{\text{overflow}} &= \Delta V_m - \Delta V_g \\ &= \beta_m V_m \Delta t - \beta_g V_g \Delta t \end{aligned}$$

We will compute the volume increases separately.

$$\Delta V_m = \beta_m V_m \Delta t = (1.8 \times 10^{-4}/\text{C}°)(50 \text{ cm}^3)(40 \text{ C}°) = 0.36 \text{ cm}^3$$
$$\Delta V_g = 3\alpha_g V_g \Delta t = 3(0.3 \times 10^{-5}/\text{C}°)(50 \text{ cm}^3)(40 \text{ C}°) = 0.018 \text{ cm}^3$$

Thus, the volume overflow is

$$\begin{aligned} V_{\text{overflow}} &= \Delta V_m - \Delta V_g = 0.36 \text{ cm}^3 - 0.018 \text{ cm}^3 \\ &= 0.342 \text{ cm}^3 \end{aligned}$$

The volume overflow is 0.342 cm³.

Problem-Solving Strategy

Temperature and Expansion

1. Read the problem carefully, and, after drawing a rough sketch, list the given information. Use zero subscripts to distinguish between initial and final values of length, area, volume, and temperature.

2. Do not confuse *specific* temperatures *t* with temperature *intervals* Δt. The practice of using the degree mark before and after the symbol is helpful, for example, $55°C - 22°C = 33 C°$.

3. For expansion problems, be sure to include the temperature unit with the constant to avoid multiplying by the incorrect temperature interval. If the coefficient is $1/C°$, the interval Δt must be in Celsius degrees.

4. The area and volume expansion coefficients for solids can be found by multiplying two or three times the linear values given in Table 16-1.

5. When you are asked to find an *initial* or *final* value of length, area, volume, or temperature, it is usually easier to calculate first the *change* in that parameter and then solve for the initial or final value. For example, you can find the final temperature t_f by first calculating Δt and then adding or subtracting to find t_f.

6. Simultaneous expansion of different materials must be adjusted to account for the differing rates of expansion. For a liquid in a solid container, the net increase or decrease in volume is equal to the difference in the changes experienced by each material. See Example 16-6.

16-8 THE UNUSUAL EXPANSION OF WATER

Suppose we fill the bulb of the tube in Fig. 16-15 with water at 0°C so that the narrow neck is partially filled. Expansion or contraction of the water can be measured easily by observing the water level in the narrow tube. As the temperature of the water increases, the water in the tube will gradually sink, indicating a contraction. The contraction continues until the temperatures of the bulb and the water are 4°C. As the temperature increases above 4°C, the water reverses its direction and rises continuously, indicating the normal expansion with an increase in temperature. This means that water has its minimum volume and its maximum density at 4°C.

The variation in the density of water with temperature is shown graphically in Fig. 16-16. If we study the graph from the high-temperature side, we note that the density gradually increases to a maximum of 1.0 g/cm³ at 4°C. The density then decreases gradually until the water reaches the ice point. Ice occupies a greater volume than water, and its formation sometimes results in cracked water pipes if proper precautions are not taken.

The greater volume in the ice results from the way groups of its molecules are bonded in its crystalline structure. As the ice melts, the water formed still contains groups of molecules bonded in this open crystal structure. As these structures begin to collapse, the molecules move closer together, increasing the density. This is the

Figure 16-15 The irregular expansion of water. As the temperature of water is increased from 0°C to 8°C, it first contracts and then expands.

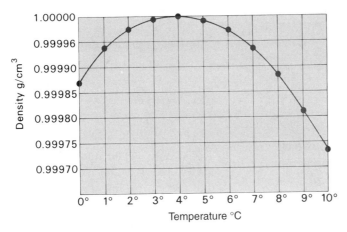

Figure 16-16 Variation in the density of water near 4°C.

dominant process until the water reaches a temperature of 4°C. From that point to higher temperatures, the increased amplitude of molecular vibrations takes over and the water expands.

Once again, the beginning student may be tempted to marvel that science can be so exact. The very thought that the density of water at 4°C "turns out to be exactly 1.00 g/cm^3" must certainly be a remarkable coincidence. Like the temperatures of the ice point and steam point, however, this result is also a consequence of a definition. The scientists who originally devised the metric system defined the **kilogram** as the mass of 1000 cm^3 of water at 4°C. Later the kilogram was redefined in terms of a platinum-iridium cylinder, which serves as a standard.

Summary and Review

Summary

We have seen that, because of the existence of four commonly used temperature scales, temperature conversions are important. You have also studied one important effect of changes in temperature of materials: a change in the physical dimensions. The major concepts are summarized below.

- There are four temperature scales with which you should be thoroughly familiar. These scales are compared in Fig. 16-9, giving values for the steam point, the ice point, and absolute zero on each scale. It is important for you to distinguish between a temperature interval Δt and a specific temperature t. For temperature intervals:

$$\frac{5 \text{ C}°}{9 \text{ F}°} = 1 = \frac{9 \text{ F}°}{5 \text{ C}°} \quad 1 \text{ K} = 1 \text{ C}° \quad 1 \text{ R}° = 1\text{F}°$$

Temperature Intervals

- For specific temperatures, you must correct for the interval difference, but you must also correct for the fact that different numbers are assigned for the same temperatures:

$$t_C = \frac{5}{9}(t_F - 32) \qquad t_F = \frac{9}{5}t_C + 32$$

Specific Temperatures

$$T_K = t_C + 273 \qquad T_R = t_F + 460$$

Absolute Temperatures

- The following relationships apply for thermal expansion of solids:

$$\Delta L = \alpha L_0 \Delta t \quad L = L_0 + \alpha L_0 \, \Delta t$$

Linear Expansion

$$\Delta A = \gamma A_0 \Delta t \quad A = A_0 + \gamma A_0 \, \Delta t \quad \gamma = 2\alpha$$

Area Expansion

$$\Delta V = \beta V_0 \, \Delta t \quad V = V_0 + \beta V_0 \Delta t \quad \beta = 3\alpha$$

Volume Expansion

- The volume expansion of a liquid uses the same relation as for a solid except, of course, that there is no linear expansion coefficient α for a liquid. Only β is needed.

Key Terms

1. thermal energy
2. temperature
3. thermal equilibrium
4. thermometer
5. ice point
6. steam point
7. Celsius scale
8. Fahrenheit scale
9. absolute zero
10. Kelvin scale
11. coefficient of linear expansion

Review Questions

16-1. Two lumps of hot iron ore are dropped into a container of water. The system is insulated and allowed to reach thermal equilibrium. Is it necessarily true that the iron ore and the water have the same thermal energy? Is it necessarily true that they have the same temperature? Discuss.

16-2. Distinguish clearly between thermal energy and temperature.

16-3. if a flame is placed underneath a mercury-in-glass thermometer, the mercury column first drops and then rises. Explain.

16-4. What factors must be considered in the design of a sensitive thermometer?

16-5. How good is our sense of touch as a means of judging temperature? Does the "hotter" object always have the higher temperature?

16-6. Given an unmarked thermometer, how would you proceed to graduate it in Celsius degrees?

16-7. A 6-in. ruler expands 0.0014 in. when the temperature is increased 1 C°. How much would a 6-cm ruler expand during the same temperature interval if it is made of the same material?

16-8. A brass rod connects the opposite sides of a brass ring. If the system is heated uniformly, will the ring remain circular?

16-9. A brass nut is used with a steel bolt. How is the closeness of fit affected when the bolt alone is heated? If the nut alone is heated? If they are both heated equally?

16-10. An aluminum cap is screwed tightly on the top of a pickle jar at room temperature. After the pickle jar has been stored in a refrigerator for a day or two, the cap cannot be removed easily. Explain. Suggest a way to remove the cap with little effort. How might this problem be solved by the manufacturer?

16-11. Describe the expansion of water near 4°C. Why does a lake freeze at the surface first? What temperature is likely to result at the bottom of the lake if its surface is frozen?

16-12. Follow reasoning similar to that for area expansion to derive Eqs. (16-15) and (16-17). In the text it was stated that γ is only approximately equal to twice α. Why is it not exactly twice α? Is the error larger in Eq. (16-13) or in Eq. (16-15)?

Problems

Section 16-2—The Measurement of Temperature

16-1. Body temperature is normal at 98.6°F. What is the corresponding temperature on the Celsius scale? **Ans. 37.0°C**

16-2. The boiling point of sulfur is 444.5°C. What is the corresponding temperature on the Fahrenheit scale?

16-3. A steel rail cools from 70 to 30°C in 1 h. What is the change in temperature in Fahrenheit degrees for the same time period? **Ans. 72 F°**

*16-4. At what temperature will the Celsius scale and Fahrenheit scale have the same numerical reading?

16-5. A piece of charcoal initially at 180°F experiences a decrease in temperature of 120 F°. Express this change of temperature in Celsius degrees. What is the final temperature on the Celsius scale?
Ans. 66.7 C°, 15.6°C

16-6. Acetone boils at 56.5°C, and liquid nitrogen boils at −196°C. Express these specific temperatures on the Kelvin scale. What is the difference in these temperatures on the Celsius scale?

16-7. The boiling point of oxygen is −297.35°F. Express this temperature in kelvins and in degrees Celsius. **Ans. 90.0 K, −183°C**

16-8. If oxygen cools from 120 to 70°F, what is the change of temperature in kelvins?

16-9. A wall of firebrick has an inside temperature of 313°F and an outside temperature of 73°F. Express the difference of temperature in kelvins. **Ans. 133 K**

16-10. Gold melts at 1336 K. What is the corresponding temperature in degrees Celsius and in degrees Fahrenheit?

16-11. A sample of gas cools from −120 to −180°C. Express the change of temperature in kelvins and in Fahrenheit degrees.
Ans. −60 K, −108 F°

Sections 16-5 to 16-7—Temperature and Expansion

16-12. A slab of concrete is 20 m long. What will be the increase in length if the temperature changes from 12°C to 30°C? Assume that $\alpha = 9 \times 10^{-6}/\text{C}°$.

16-13. A piece of copper tubing is 6 m long at 20°C. How much will it increase in length when heated to 80°C? **Ans. 6.12 mm**

Summary and Review

16-14. A silver bar is 1 ft long at 70°F. How much will it increase in length when it is placed into boiling water (212°F)?

16-15. The diameter of a hole in a steel plate is 9 cm when the temperature is 20°C. What will be the diameter of the hole at 200°C?
Ans. 9.02 cm

*16-16. A brass rod is 2.00 m long at 15°C. To what temperature must the rod be heated so that its new length is 2.01 m?

16-17. A square cooper plate 4 cm on a side at 20°C is heated to 120°C. What is the increase in the area of the copper plate?
Ans. 0.0544 cm²

*16-18. A circular hole in a steel plate has a diameter of 20.0 cm at 27°C. To what tempera-

ture must the plate be heated so that the area of the hole will be 314 cm²?

16-19. What is the increase in volume of 16.0 liters of ethyl alcohol when the temperature is increased by 30 C°?
Ans. 0.528 L

16-20. A Pyrex beaker has an inside volume of 600 mL at 20°C. At what temperature will the inside volume be 603 mL?

16-21. If 200 cm³ of benzene exactly fills an aluminum cup at 40°C, and the system cools to 18°C, how much benzene (at 18°C) can be added to the cup without overflowing?
Ans. 5.14 cm³

16-22. A Pyrex glass beaker is filled to the top with 200 cm³ of mercury at 20°C. How much mercury will overflow if the temperature of the system is increased to 68°C?

Challenge Problems

*16-23. The diameter of a hole in a copper plate at 20°C is 3.00 mm. To what temperature must the copper be cooled if its diameter is to be 2.99 mm?
Ans. −176°C

16-24. A rectangular sheet of aluminum measures 6 by 9 cm at 28°C. What is its area at 0°C?

*16-25. A steel tape measures the length of an aluminum rod as 60 cm when both are at 8°C. What will the tape read for the length of the rod if both are at 38°?
Ans. 60.022 cm

16-26. At 20°C, a copper cube measures 40 cm on a side. What is the volume of the cube when the temperature reaches 150°C?

16-27. A Pyrex beaker ($\alpha = 0.3 \times 10^{-5}$/C°) is filled to the top with 200 mL of glycerine ($\beta = 5.1 \times 10^{-4}$/C°). How much glycerine will overflow the top if the system is heated from 20°C to 100°C?
Ans. 8.02 mL

16-28. An oven is set at 450°F. If the temperature drops by 50 kelvins, what is the new temperature in degrees Celsius?

*16-29. A 100-ft steel tape correctly measures distance when the temperature is 20°C. What is the true measurement if this tape indi-

cates a distance of 94.62 ft on a day when the temperature is 36°C?
Ans. 94.64 ft

*16-30. The diameter of a steel rod is 3.000 mm when the temperature is 20°C. The diameter of a brass ring is 2.995 mm, also at 20°C. At what common temperature will the brass ring slip over the steel rod smoothly?

*16-31. The volume of a metal cube increases by 0.50 percent when the temperature of the cube increases by 100 C°. What is the linear expansion coefficient for this metal?
Ans. 1.67×10^{-5}/C°

16-32. By what percentage is the volume of a brass cube increased when heated from 20°C to 100°C?

16-33. A round brass plug has a diameter of 8.001 cm at 28°C. To what temperature must the plug be cooled if it is to fit snugly into an 8.000-cm hole?
Ans. 21.1°C

*16-34. Five hundred cubic centimeters of ethyl alcohol completely fill a Pyrex beaker. If the temperature of the system increases by 70 C°, what volume of alcohol overflows?

Critical Thinking Questions

16-35. The laboratory apparatus for measuring the coefficient of linear expansion is illustrated in Fig. 16-17. The temperature of a metal rod is increased by passing steam through an enclosed jacket. The resulting increase in length is measured with a micrometer screw at one end. Since the original length and temperature are known, the expansion coefficient can be calculated from Eq. (16-8). The following data were recorded during an experiment with a rod of unknown metal:

$$L_0 = 600 \text{ mm} \qquad t_0 = 23°C$$
$$\Delta L = 1.04 \text{ mm} \qquad t_f = 98°C$$

What is the coefficient of linear expansion for this metal? Can you identify the metal?

Ans. 2.3 × 10⁻⁵/C°, Al

*16-36. Assume that the end points of a rod are fixed rigidly between two walls to prevent expansion with increasing temperature. From the definitions of Young's modulus (Chap. 13) and your knowledge of linear expansion, show that the compressive force F exerted by the walls will be given by

$$F = \alpha AY \, \Delta t$$

where A = cross section of the rod
Y = Young's modulus
Δt = increase in temperature of rod.

*16-37. Prove that the density of a material changes with temperature so that the new density ρ is given by

$$\rho = \frac{\rho_0}{1 + \beta \Delta t}$$

where ρ_0 = original density
β = volume expansion coefficient
Δt = change in temperature

16-38. The density of mercury at 0°C is 13.6 g/cm³. Use the relationship in the previous example to find the density of mercury at 60°C?

16-39. A steel ring has an inside diameter of 4.000 cm at 20°C. The ring is to be slipped over a copper shaft that has a diameter of 4.003 cm at 20°C. To what temperature must the ring be heated? If the ring and the shaft are cooled uniformly, at what temperature will the ring just slip off the shaft?

Ans. 82.5°C, −130°C

Figure 16-17 Apparatus for measuring the coefficient of linear expansion.

Quantity of Heat

Thermal energy is the energy associated with random molecular motion, but it is not possible to measure the position and velocity of every molecule in a substance to determine its thermal energy. We can measure *change* in thermal energy, however, by relating it to a change in temperature.

For example, when two systems at different temperatures are placed together, they eventually reach a common intermediate temperature. From this observation, it is safe to say that the system at the higher temperature has lost thermal energy to the system at the lower temperature. The thermal energy lost or gained by objects is called **heat.** This chapter is concerned with the quantitative measurement of heat.

17-1 THE MEANING OF HEAT

It was originally believed that the two systems reach thermal equilibrium through the transfer of a substance called *caloric.* It was postulated that all bodies contain an amount of caloric in proportion to their temperature. Thus, when two objects were placed in contact, the object of higher temperature transferred caloric to the object of lower temperature until they reached the same temperature. The idea that a substance is transferred carries with it the implication that there is a limit to the amount of heat energy that can be withdrawn from a body. It was this point that eventually led to the downfall of the caloric theory.

Count Rumford of Bavaria was the first to shed doubt on the caloric theory. He made his discovery in 1798 while supervising the boring of a cannon. The bore of the cannon was kept full of water to prevent overheating. As the water boiled away, it was replenished. According to the existing theory, caloric had to be supplied to boil the water. The apparent production of caloric was explained by supposing that when matter is finely divided, it loses some of its ability to retain caloric. Rumford devised an experiment to show that even a dull boring tool that did not cut the gun metal at all produced enough caloric to boil the water. In fact, it seemed that as long as mechanical work was supplied, the tool was an inexhaustible source of caloric. Rumford ruled out the caloric theory on the basis of his experiments and suggested that the explanation must be related to motion. Hence, the idea that mechanical work is responsible for the creation of heat was introduced. The equivalence of heat and work as two forms of energy was established later by Sir James Prescott Joule.

Heat can produce electricity. With developments in space technology, heat can produce electricity. The technique, called *thermoelectric power generation*, works on the same technology that powers deep-space crafts, which cannot generate electricity from the sun. The system works using an array of semiconductor-thermocouples of a bismuth telluride alloy. A *thermocouple* is a device that bends when exposed to heat. This bending completes a circuit containing an n-type and a p-type transistor (see Chap. 40 for more on transistors), which generates electricity. This process is not very efficient—only 5 percent of the heat energy becomes electricity. But if the heat is waste heat, that 5 percent is akin to getting free energy. One company pursuing this technology hopes to use it to produce all the electricity needed to run the electric portions of a home heating system, which would then run on gas alone. If you are interested in this technology, you can find out more by seeing this company's Web site (www.hi-z.com) or the International Thermoelectric Society Web site (www.its.org).

17-2 THE QUANTITY OF HEAT

The idea of heat as a substance must be discarded. It is not something that an object *has* but something that it *gives up* or *absorbs*. Heat is simply another form of energy that can be measured only in terms of the effect it produces. The SI unit of energy, the *joule*, is also the preferred unit for heat since heat is a form of energy. There are three older units that remain in use today, however, and they will also be treated in this text. These earlier units were based on the thermal energy required to produce a standard change. They are the **calorie**, the *kilocalorie*, and the **British thermal unit**.

> One **calorie (cal)** is the quantity of heat required to change the temperature of one gram of water through one Celsius degree.

> One **kilocalorie (kcal)** is the quantity of heat required to change the temperature of one kilogram of water through one Celsius degree (kcal = 1000 cal).

> One **British thermal unit (Btu)** is the quantity of heat required to change the temperature of one standard pound (lb) through one Fahrenheit degree.

Aside from the fact that these older units imply that heat energy cannot be related to other forms of energy, there are other problems. The heat required to change the temperature of water from 92 to 93°C is not exactly the same as that needed to increase the temperature from 8 to 9°C. Thus, it is necessary to specify the interval of temperature in strict applications for the calorie and for the British thermal unit. The chosen intervals were 14.5 to 15.5°C and 63 to 64°F. Additionally, the pound unit that appears in the definition of the Btu must be recognized as the *mass of the standard pound*. This represents a departure from USCS units, in which the unit pound was reserved for weight. Therefore, in this chapter, when we refer to 1 lb of water, we shall be referring to a *mass* of water equivalent to about 1/32 slug. This distinction is necessary because the pound of water must represent a constant quantity of matter, independent of its location. By definition the pound mass is related to the gram and kilogram as follows:

$$1 \text{ lb} = 454 \text{ g} = 0.454 \text{ kg}$$

The difference between these older units for heat results from the difference in masses and the difference in temperature scales. It is left as an exercise for you to show that

$$1 \text{ Btu} = 252 \text{ cal} = 0.252 \text{ kcal} \qquad \textbf{17-1}$$

The first quantitative relationship between these older units and the traditional units for mechanical energy was established by Joule in 1843. Although Joule devised many different experiments to demonstrate the equivalence of heat units and energy units, the apparatus most often remembered is illustrated in Fig. 17-1. Falling weights supplied the mechanical energy, which rotated a set of paddles in a water can. The quantity of heat absorbed by the water was measured from the known mass and the measured increase in temperature of the water.

In modern times the mechanical equivalent of heat has been accurately established by a variety of techniques. The accepted results are:

$$1 \text{ cal} = 4.186 \text{ J}$$
$$1 \text{ kcal} = 4186 \text{ J}$$
$$1 \text{ Btu} = 778 \text{ ft} \cdot \text{lb}$$

Thus, 4.186 J of heat is needed to raise the temperature of 1 g of water from 14.5 to 15.5°C. Since each of the above units remains in common use, it will often become necessary to compare units or to convert from one unit to another.

Now that we have defined units for the quantitative measurement of heat, the distinction between quantity of heat and temperature should be clear. For example, suppose we pour 200 g of water into one beaker and 800 g of water into another

Figure 17-1 Joule's experiment for determining the mechanical equivalent of heat. The falling weights do work in stirring the water and raising its temperature.

Figure 17-2 The same quantity of heat is applied to different masses of water. The larger mass experiences a smaller rise in temperature.

beaker, as shown in Fig. 17-2. The initial temperature of the water in each beaker is measured to be 20°C. A flame is placed under each beaker for the same length of time, delivering 8000 J of heat energy to the water in each beaker. The temperature of the 800-g quantity of water increases by a little more than 2 C°, but the temperature of the 200-g quantity increases by almost 10 C°. Yet the same quantity of heat was applied to the water in each beaker.

17-3 SPECIFIC HEAT CAPACITY

We have defined a quantity of heat as the thermal energy required to raise the temperature of a given mass. The amount of thermal energy required to raise the temperature of a substance varies, however, for different materials. For example, suppose we apply heat to five balls, all of the same size but made of different materials, as shown in Fig. 17-3a. If we want to raise the temperature of each ball to 100°C, we shall find that some of the balls must be heated longer than the others. For illustration, let us assume that each ball has a volume of 1 cm³ and an initial temperature of 0°C. Each ball is heated with a burner capable of delivering thermal energy at the rate of 1 cal/s. The approximate times required to obtain a

Figure 17-3 A comparison of heat capacities for five balls of different materials.

temperature of 100°C for each ball are given in Fig. 17-3. Notice that the lead ball reaches the final temperature in only 37 s, whereas the iron ball requires 90 s of continuous heating. The glass, aluminum, and copper balls require intermediate times.

Since more heat was absorbed by the iron and copper balls, it might be expected that they would release more heat on cooling. This can be demonstrated by placing the five balls (at 100°C) simultaneously on a thin strip of paraffin, as in Fig. 17-3b. The iron and copper balls eventually melt through the paraffin and drop into the pan. The lead and glass balls never make it through. Clearly, there must be some property of a material that relates to the quantity of heat absorbed or released during a change in temperature. As a step in establishing this property, we first define *heat capacity.*

> The **heat capacity** of a body is the ratio of heat supplied to the corresponding rise in temperature of the body.

$$\text{Heat capacity} = \frac{Q}{\Delta t} \qquad \text{17-2}$$

The SI units for heat capacity are *joules per kelvin* (J/K), but since the Celsius interval is the same as the kelvin and more commonly used, the *joule per Celsius degree* (J/C°) is used in this text. Other units are *calories per Celsius degree* (cal/C°), *kilocalories per Celsius degree* (kcal/C°), and *Btu per Fahrenheit degree* (Btu/F°). In the preceding example, 89.4 cal of heat was required to raise the temperature of the iron ball by 100 C°. Thus, the heat capacity of this particular iron ball is 0.894 cal/C°.

The mass of an object is not included in the definition of heat capacity. Therefore, heat capacity is a property of the object. To make it a property of the material, the *heat capacity per unit mass* is defined. We call this property the *specific heat capacity*, denoted *c*.

> **The specific heat capacity of a material is the quantity of heat required to raise the temperature of a unit mass through one degree.**

$$c = \frac{Q}{m \, \Delta t} \qquad Q = mc \, \Delta t \qquad \qquad \text{17-3}$$

The SI unit of specific heat assigns the *joule* for heat, the *kilogram* for mass, and the *kelvin* for temperature. If we again replace the kelvin with the Celsius degree, the units of *c* are J/kg · C°. In industry most temperature measurements are made in C° or F°, and the calorie and the Btu are still used more than the SI units. We will therefore continue to speak of specific heat in units of cal/g · C° and Btu/lb · F°, but we will also use the preferred SI units in some cases. In the example of the iron ball, the mass is found to be 7.85 g. The specific heat of iron is therefore

$$c = \frac{Q}{m \, \Delta t} = \frac{89.4 \text{ cal}}{(7.85 \text{ g})(100 \text{ C°})} = 0.114 \text{ cal/g} \cdot \text{C°}$$

Notice that we speak of the heat capacity of the *ball* and the *specific* heat capacity of *iron*. The former relates to the object itself, whereas the latter relates to the material from which the object is made. In our experiment with the balls, we noted only the quantity of heat required to raise their temperature 100 C°. The density of the materials was not considered. If the sizes of the balls were adjusted so that each had the same mass, we would observe different results. Since the specific heat of aluminum is highest, more heat would be required for the aluminum ball than for the others. Similarly, the aluminum ball would release more heat in cooling.

For most practical applications, the specific heat of water can be taken as

$$\begin{array}{ccc} 4186 \text{ J/kg} \cdot \text{C°} & & 4.186 \text{ J/g} \cdot \text{C°} \\ 1 \text{ cal/g} \cdot \text{C°} & \text{or} & 1 \text{ Btu/lb} \cdot \text{F°} \end{array}$$

Notice that the numerical values are the same for the specific heat expressed in cal/g · C° and in Btu/lb · F°. This is a consequence of their definitions and can be demonstrated by unit conversion:

$$1 \frac{\text{Btu}}{\text{lb} \cdot \text{F°}} \times \frac{9 \text{ F°}}{5 \text{ C°}} \times \frac{1 \text{ lb}}{454 \text{ g}} \times \frac{252 \text{ cal}}{1 \text{ Btu}} = 1 \text{ cal/g} \cdot \text{C°}$$

The specific heats for many common substances are listed in Table 17-1.

Once the specific heats of a large number of materials have been established, the thermal energy released or absorbed in many experiments can be determined. For example, the quantity of heat *Q* required to raise the temperature of a mass *m* through an interval *t*, from Eq. (17-3), is

$$Q = mc \, \Delta t \qquad \qquad \text{17-4}$$

where *c* is the specific heat of the mass.

TABLE 17-1 Specific Heat Capacities

Substance	J/kg · C°	cal/g · C° or Btu/lb · F°
Aluminum	920	0.22
Brass	390	0.094
Copper	390	0.093
Ethyl alcohol	2500	0.60
Glass	840	0.20
Gold	130	0.03
Ice	2300	0.5
Iron	470	0.113
Lead	130	0.031
Mercury	140	0.033
Silver	230	0.056
Steam	2000	0.48
Steel	480	0.114
Turpentine	1800	0.42
Zinc	390	0.092

EXAMPLE 17-1

How much heat is required to raise the temperature of 200 g of mercury from 20 to 100°C?

Solution

Substitution into Eq. (17-4) yields

$$Q = mc \, \Delta t = (0.2 \text{ kg})(140 \text{ J/kg} \cdot \text{C}°)(80 \text{ C}°)$$
$$= 2200 \text{ J}$$

17-4 THE MEASUREMENT OF HEAT

We have often emphasized the distinction between thermal energy and temperature. The term *heat* has now been introduced as the thermal energy *absorbed* or *released* during a temperature change. The quantitative relationship between heat and temperature is best described by the concept of specific heat as it appears in Eq. (17-4). The physical relationships among all these terms are now beginning to fall into place.

The principle of thermal equilibrium tells us that whenever objects are placed together in an insulated enclosure, they will eventually reach the same temperature. This is the result of a transfer of thermal energy from the warmer bodies to the cooler bodies. If energy is to be conserved, we say that *the heat lost by the warm bodies must equal the heat gained by the cool bodies.* That is,

This equation expresses the net result of heat transfer within a system.

The heat lost or gained by an object is not related in a simple way to the molecular energies of the objects. Whenever thermal energy is supplied to an object, it can absorb the energy in many different ways. The concept of specific heat is needed to measure the abilities of different materials to utilize thermal energy to increase their temperatures. The same amount of applied thermal energy does not result in the same temperature increase for all materials. For this reason, we say that temperature is a *fundamental* quantity. Its measurement is necessary to determine the quantity of heat lost or gained in a given process.

In applying the general equation for the conservation of thermal energy, Eq. (17-5), the quantity of heat gained or lost by each item is calculated from the equation

$$Q = mc\,\Delta t$$

The term Δt represents the absolute change in temperature when applied to the conservation equation. The procedure is best demonstrated by an example.

USING A CALCULATOR

Remember to calculate expressions inside parentheses before performing other operations. To convert 56°F into degrees Celsius, use the following formula:

$$°C = \tfrac{5}{9}(°F - 32)$$

or

$$t_c = \tfrac{5}{9}(t_F - 32)$$

Start inside the parentheses.

⑤ ⑥ ⊟ ③ ② ⊜ 24
⊠ ⑤ ⊘ ⑨ ⊜ 13.33

Always ask: Does the answer make sense? 56°F is above freezing (water freezes at 32°F); and our answer, 13.33°C, is above freezing (0°C). The answer is reasonable.

EXAMPLE 17-2

A handful of copper shot is heated to 90°C and then dropped into 80 g of water at 10°C. The final temperature of the mixture is 18°C. What was the mass of the shot?

Solution

Applying Eq. (17-5), we write

$$\text{Heat lost by shot} = \text{heat gained by water}$$
$$m_s c_s\,\Delta t_s = m_w c_w\,\Delta t_w$$
$$m_s c_s (t_s - t_e) = m_w c_w (t_e - t_w)$$

The change in temperature of the shot is computed by subtracting the equilibrium temperature t_e from the initial temperature of the shot t_s. On the other hand, the change in temperature of the water is computed by subtracting the initial temperature of the water t_w from the equilibrium temperature. This does not represent an error in sign because the quantity on the left represents a heat *loss* and the quantity on the right represents a heat *gain*. Obtaining the required specific heat from Table 17-1 and substituting other known quantities, we have

$$m_s(0.093 \text{ cal/g} \cdot \text{C}°)(90°C - 18°C) = (80 \text{ g})(1 \text{ cal/g} \cdot \text{C}°)(18°C - 10°C)$$
$$m_s(0.093 \text{ cal/g} \cdot \text{C}°)(72 \text{ C}°) = (80 \text{ g})(1 \text{ cal/g} \cdot \text{C}°)(8 \text{ C}°)$$
$$m_s = 95.6 \text{ g}$$

In this simple example we have neglected two important facts: (1) the water must have a container, which will also absorb heat from the shot; (2) the entire system must be insulated from external temperatures. Otherwise, the equilibrium

Figure 17-4 The laboratory calorimeter. (Central Scientific Co.)

temperature will always be room temperature. A laboratory device called a *calorimeter* (Fig. 17-4) is used to control these difficulties. The calorimeter consists of a thin metallic vessel K, generally aluminum, held centrally within an outer jacket A by a nonconducting rubber support H. Loss of heat is minimized in three ways: (1) the rubber gasket prevents loss by conduction, (2) the dead air space between the container walls prevents heat loss by air currents, and (3) highly polished metal vessels reduce the loss of heat by radiation. These three methods of heat transfer will be discussed in Chap. 18. The wooden cover L has holes in the top for insertion of a thermometer and an aluminum stirrer.

CONCEPTUAL EXAMPLE 17-3

In a laboratory experiment, a calorimeter is used to find the specific heat of iron. Eighty grams of dry iron shot are placed in a cup and heated to a temperature of 95°C. The mass of the inner aluminum cup and of the aluminum stirrer is 60 g. The calorimeter is partially filled with 150 g of water at 18°C. The hot shot is poured quickly into the cup, and the calorimeter is sealed, as shown in Fig. 17-5. After the system has reached thermal equilibrium, the final temperature is 22°C. Compute the specific heat of iron.

Solution

The heat lost by the iron shot must equal the heat gained by the water plus the heat gained by the aluminum cup and stirrer. We can assume that the initial temperature of the cup is the same as that of the water and stirrer (18°C). We will calculate the heat gained by the water and by the aluminum separately.

Figure 17-5 A calorimeter can be used to compute the specific heat of a substance.

$$Q_{water} = mc \, \Delta t = (150 \text{ g})(1 \text{ cal/g} \cdot \text{C}°)(22°\text{C} - 18°\text{C})$$
$$= (150 \text{ g})(1 \text{ cal/g} \cdot \text{C}°)(4 \text{ C}°) = 600 \text{ cal}$$
$$Q_{Al} = mc \, \Delta t = (60 \text{ g})(0.22 \text{ cal/g} \cdot \text{C}°)(22°\text{C} - 18°\text{C})$$
$$= (60 \text{ g})(0.22 \text{ cal/g} \cdot \text{C}°)(4 \text{ C}°) = 52.8 \text{ cal}$$

The total heat gained is the sum of these values.

$$\text{Heat gained} = 600 \text{ cal} + 52.8 \text{ cal} = 652.8 \text{ cal}$$

This amount must equal the heat lost by the iron shot:

$$\text{Heat lost} = Q_s = mc_s \, \Delta t = (80 \text{ g})c_s(95°\text{C} - 22°\text{C})$$

Setting the heat lost equal to the heat gained gives

$$(80 \text{ g})c_s(73 \text{ C}°) = 652.8 \text{ cal}$$

Solving for c_s, we obtain

$$c_s = \frac{652.8 \text{ cal}}{(80 \text{ g})(73 \text{ C}°)} = 0.11 \text{ cal/g} \cdot \text{C}°$$

In this experiment the heat gained by the thermometer was neglected. In an actual experiment, the portion of the thermometer inside the calorimeter would absorb about the same amount of heat as an extra 0.5 g of water. This quantity, called the ***water equivalent*** of the thermometer, should be added to the mass of water in an accurate experiment.

17-5 CHANGE OF PHASE

When a substance absorbs a given amount of heat, the speed of its molecules usually increases and its temperature rises. Depending on the specific heat of the substance, the rise in temperature is directly proportional to the quantity of heat supplied and inversely proportional to the mass of the substance. A curious thing happens, however, when a solid melts or when a liquid boils. In these cases, the temperature remains constant until all the solid melts or until all the liquid boils.

To understand what happens to the applied energy, let us consider a simple model, as illustrated in Fig. 17-6. Under the proper conditions of temperature and pressure, all substances can exist in three *phases*, solid, liquid, or gas. In the solid phase, the molecules are held together in a rigid, crystalline structure so that the substance has a definite shape and volume. As heat is supplied, the energies of the particles in the solid gradually increase and its temperature rises. Eventually, the kinetic energy becomes so great that some of the particles overcome the elastic forces that hold them in fixed positions. The increased separation gives them the freedom of motion that we associate with the liquid phase. At this point, the energy absorbed by the substance is used in separating the molecules more than in the solid phase. The temperature does not increase during such a change of phase. The change of phase from a solid to a liquid is called **fusion,** and the temperature at which this change occurs is called the **melting point.**

The quantity of heat required to melt a unit mass of a substance at its melting point is called the **latent heat of fusion** for that substance.

The latent heat of fusion L_f of a substance is the heat per unit mass required to change the substance from the solid to the liquid phase at its melting temperature.

$$L_f = \frac{Q}{m} \qquad Q = mL_f \qquad \text{17-6}$$

The latent heat of fusion L_f is expressed in *joules per kilogram* (J/kg), *calories per gram* (cal/g), or *Btu per pound* (Btu/lb). At 0°C, 1 kg of ice will absorb about 334,000 J of heat in forming 1 kg of water at 0°C. Thus, the latent heat of fusion for water is 334,000 J/kg. The term *latent* arises from the fact that the temperature remains constant during the melting process. The heat of fusion for water is any of the following:

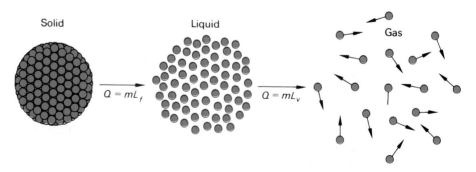

Figure 17-6 A simplified model showing the relative molecular separations in the solid, liquid, and gaseous phases. During a change of phase, the temperature remains constant.

$$3.34 \times 10^5 \text{ J/kg} \qquad 334 \text{ J/g}$$
$$80 \text{ cal/g} \qquad \text{or} \qquad 144 \text{ Btu/lb}$$

After all of the solid melts, the kinetic energy of the particles in the resulting liquid increases in accordance with its specific heat, and the temperature rises again. Eventually the temperature will level off as the thermal energy is used to change the molecular structure, forming a gas or vapor. The change of phase from a liquid to a vapor is called *vaporization,* and the temperature associated with this change is called the *boiling point* of the substance.

The quantity of heat required to vaporize a unit mass is called the *latent heat of vaporization.*

The latent heat of vaporization L_v of a substance is the heat per unit mass required to change the substance from a liquid to a vapor at its boiling temperature.

$$L_v = \frac{Q}{m} \qquad Q = mL_v \qquad\qquad \textbf{17-7}$$

The latent heat of vaporization L_v is expressed in units of *joules per kilogram, calories per gram,* or *Btu per pound.* It is found that 1 kg of water at 100°C absorbs 2,260,000 J of heat in forming 1 kg of steam at the same temperature. The heat of vaporization for water is

$$2.26 \times 10^6 \text{ J/kg} \qquad 2260 \text{ J/g}$$
$$540 \text{ cal/g} \qquad \text{or} \qquad 970 \text{ Btu/lb}$$

Values for the heat of fusion and the heat of vaporization of many substances are given in Table 17-2. They are given in SI units and in calories per gram. It should be noted that the Btu per pound (Btu/lb) equivalents can be obtained by multiplying the value in cal/g by (9/5). They differ only because of the difference in the temperature scales. The industrial use of the SI units of J/kg for both L_f and L_v is strongly encouraged, but few U.S. companies have made these conversions.

TABLE 17-2 Heats of Fusion and Heats of Vaporization for Various Substances

Substance	Melting point, °C	Heat of fusion		Boiling point, °C	Heat of vaporization	
		J/kg	cal/g		J/kg	cal/g
Alcohol, ethyl	−117.3	104×10^3	24.9	78.5	854×10^3	204
Ammonia	−75	452×10^3	108.1	−33.3	1370×10^3	327
Copper	1080	134×10^3	32	2870	4730×10^3	1130
Helium	−269.6	5.23×10^3	1.25	−268.9	20.9×10^3	5
Lead	327.3	24.5×10^3	5.86	1620	871×10^3	208
Mercury	−39	11.5×10^3	2.8	358	296×10^3	71
Oxygen	−218.8	13.9×10^3	3.3	−183	213×10^3	51
Silver	960.8	88.3×10^3	21	2193	2340×10^3	558
Water	0	334×10^3	80	100	2256×10^3	540
Zinc	420	100×10^3	24	918	1990×10^3	475

Figure 17-7 The variation of temperature with a change in thermal energy for water.

It is often helpful when studying the changes of phase of a substance to plot a graph showing how the temperature of the substance varies as thermal energy is applied. Such a graph is shown in Fig. 17-7 for water. If a quantity of ice is taken from a freezer at −20°C and heated, its temperature will increase gradually until the ice begins to melt at 0°C. For each degree rise in temperature, each gram of ice will absorb 0.5 cal of heat energy. During the melting process, the temperature remains constant, and each gram of ice will absorb 80 cal of heat energy in forming 1 g of water.

Once all the ice has melted, the temperature begins to rise again at a uniform rate until the water begins to boil at 100°C. For each degree increase in temperature each gram will absorb 1 cal of heat energy. During the vaporization process, the temperature remains constant. Each gram of water absorbs 540 cal of heat energy in forming 1 g of water vapor at 100°C. If the resulting water vapor is contained and the heating is continued until all the water is gone, the temperature will again start to rise. The specific heat of steam is 0.48 cal/g · C°.

EXAMPLE 17-4

What quantity of heat is required to change 20 lb of ice at 12°F to steam at 212°F?

Solution

The heat necessary to raise the temperature of the ice to its melting point is

$$Q_1 = mc\,\Delta t = (20\text{ lb})(0.5\text{ Btu/lb} \cdot \text{F°})(32°\text{F} - 12°\text{F})$$
$$= 200\text{ Btu}$$

The heat required to melt the ice is given from Eq. (17-6).

$$Q_2 = mL_f = (20\text{ lb})(144\text{ Btu/lb}) = 2800\text{ Btu}$$

The heat necessary to raise the temperature of the resulting water to 212°F is

$$Q_3 = mc\,\Delta t = (20\text{ lb})(1\text{ Btu/lb} \cdot \text{F}°)(212°\text{F} - 32°\text{F})$$
$$= 3600\text{ Btu}$$

The heat required to vaporize the water is, from Eq. (17-7),

$$Q_4 = mL_v = (20\text{ lb})(970\text{ Btu/lb}) = 19,400\text{ Btu}$$

The total heat required is

$$Q = Q_1 + Q_2 + Q_3 + Q_4$$
$$= (200 + 2880 + 3600 + 19,400)\text{ Btu}$$
$$= 26,080\text{ Btu}$$

When heat is removed from a gas, its temperature drops until it reaches the temperature at which it boiled. As more heat is removed, the vapor returns to the liquid phase. This process is referred to as **condensation.** In condensing, a vapor gives up an amount of heat equivalent to the heat required to vaporize it. Thus, the *heat of condensation* is equivalent to the heat of vaporization. The difference lies only in the direction of heat transfer.

Similarly, when heat is removed from a liquid, its temperature will drop until it reaches the temperature at which it melted. As more heat is removed, the liquid returns to its solid phase. This process is called **freezing** or **solidification.** The heat of solidification is exactly equal to the heat of fusion. Thus, the only distinction between freezing and melting lies in whether heat is being released or absorbed.

Under the proper conditions of temperature and pressure, it is possible for a substance to change from the solid phase directly to the gaseous phase without passing through the liquid phase. This process is referred to as **sublimation.** Solid carbon dioxide (dry ice), iodine, and camphor (mothballs) are examples of substances that are known to sublime at normal temperatures. The quantity of heat absorbed per unit mass in changing from a solid to a vapor is called the **heat of sublimation.**

Before we leave the subject of fusion and vaporization, it will be instructive to offer examples of their measurement. In any given mixture, the quantity of heat absorbed must equal the quantity of heat released. This principle holds even if a change of phase occurs. The procedure is demonstrated in the following examples.

EXAMPLE 17-5

After 12 g of crushed ice at $-10°\text{C}$ is dropped into a 50-g aluminum calorimeter cup containing 100 g of water at 50°C, the system is sealed and allowed to reach thermal equilibrium. What is the resulting temperature?

Solution

The heat lost by the calorimeter and water must equal the heat gained by the ice, including any changes of phase that take place. Let us assume that all the ice melts, leaving only water at the equilibrium temperature t_e.

$$\text{Total heat lost} = \text{heat lost by calorimeter} + \text{heat lost by water}$$
$$= m_c c_c (50°C - t_e) + m_w c_w (50°C - t_e)$$
$$= (50)(0.22)(50°C - t_e) + (100)(1)(50°C - t_e)$$
$$= 550°C - 11t_e + 5000°C - 100t_e$$
$$= 5550°C - 111t_e$$

$$\text{Total heat gained} = \begin{matrix} \text{heat gained} \\ \text{by ice} \end{matrix} + \begin{matrix} \text{heat to} \\ \text{melt ice} \end{matrix} + \begin{matrix} \text{heat to} \\ \text{bring water to } t_e \end{matrix}$$
$$= m_i c_i \, \Delta t_i + m_i L_f + m_i c_w (t_e - 0°C)$$
$$= (12)(0.5)(10) + (12)(80) + (12)(1)t_e$$
$$= 1020 + 12t_e$$

$$\text{Total heat lost} = \text{total heat gained}$$
$$5550 - 111t_e = 1020 + 12t_e$$
$$123t_e = 4530$$
$$t_e = 36.8°C$$

CONCEPTUAL EXAMPLE 17-6

If 10 g of steam at 100°C is introduced into a mixture of 200 g of water and 120 g of ice, find the final temperature and composition of the mixture.

Solution

The rather small amount of steam in comparison with ice and water makes us wonder whether enough heat can be released by the steam to melt all the ice. To check this suspicion, we shall compute the heat required to melt the 120 g of ice completely at 0°C.

$$Q_1 = m_i L_f = (120 \text{ g})(80 \text{ cal/g}) = 9600 \text{ cal}$$

The maximum heat we can expect the steam to give up is

$$Q_2 = m_s L_v + m_s c_w (100°C - 0°C)$$
$$= (10)(540) + (10)(1)(100) = 6400 \text{ cal}$$

Since 9600 cal was needed to melt all the ice and only 6400 cal can be delivered by the steam, the final mixture must consist of ice and water of 0°C.

To determine the final composition of the mixture, note that 3200 additional calories would be required to melt the remaining ice. Hence,

$$m_i L_f = 3200 \text{ cal}$$

$$m_i = \frac{3200 \text{ cal}}{80 \text{ cal/g}} = 40 \text{ g}$$

Therefore, there must be 40 g of ice in the final mixture. The amount of water remaining is

$$\text{Water remaining} = \text{initial water} + \text{melted ice} + \text{condensed steam}$$
$$= 200 \text{ g} + 80 \text{ g} + 10 \text{ g} = 290 \text{ g}$$

The final composition consists of a mixture of 40 g of ice in 290 g of water at 0°C.

Suppose we had assumed in the above example that all the ice melted and attempted to solve for t_e as in Example 17-5. In this case, we would have obtained a value for the equilibrium temperature that was below the freezing point (0°C). Clearly, this kind of answer could result only from a false assumption.

An alternative procedure for solving Example 17-6 would be to solve directly for the number of grams of ice that must have melted to balance the 6400 cal of heat energy released by the steam. It is left as an exercise for you to show that the same results are obtained.

Problem-Solving Strategy

Quantity of Heat and Calorimetry

1. Read the problem carefully and then draw a rough sketch, labeling it with the given information and stating what is to be found. Be careful to include units for all physical quantities.
2. If a heat loss or gain results in a change of temperature, you will need to decide which units are appropriate for the specific heat capacity. The need for the use of consistent units is demonstrated here:

$$Q = mc\,\Delta t \qquad J = (\text{kg})\left(\frac{\text{J}}{\text{kg}\cdot\text{C}°}\right)(\text{C}°)$$

It is wise to write the units of specific heat on substitution so that the choice of the correct units for other quantities is obvious.
3. If there is a change of phase, the latent heat of fusion or vaporization will be needed. Here again we must be careful to use those units consistent with *joules per kilogram, calories per gram,* or *Btu per pound.* The temperature remains constant during a phase change.
4. Conservation of energy demands that the total of the heat losses be equal to the total of the heat gains. Note that a decrease of temperature, condensation, or freezing of a liquid indicates a *loss* of heat. A rise in temperature, melting, or vaporizing occur when there is a *gain* of heat. We might add the absolute values of the losses on the left side and set them equal to the total gains on the right side. As an example, consider a mass m_s of steam at 100°C mixed with a mass m_i of ice at 0°C. The result is water at the equilibrium temperature t_e.

$$m_s L_v + m_s c_w(100 - t_e) = m_i L_f + m_i c_w(t_e - 0)$$

Note that the temperature differences are taken as *high minus low* in each case to yield the *absolute* values lost or gained.

17-6 HEAT OF COMBUSTION

Whenever a substance is burned, it releases a definite quantity of heat. The quantity of heat per unit mass, or per unit volume, when the substance is burned completely is called the **heat of combustion.** Commonly used units are Btu per pound mass, Btu per cubic foot, calories per gram, and kilocalories per cubic meter. For example, the heat of combustion of coal is approximately 13,000 Btu/lb$_m$. This means that each pound of coal, when completely burned, should release 13,000 Btu of heat energy.

Summary and Review

Summary

In this chapter you have studied the quantity of heat as a measurable quantity that is based on a standard change. The British thermal unit and the calorie are measures of the heat required to raise the temperature of a unit mass of water by a unit degree. By applying these standard units to experiments with a variety of materials, we have learned to predict heat losses or heat gains in a constructive fashion. The essential concepts presented in this chapter are as follows:

- The *British thermal unit* (Btu) is the heat required to change the temperature of one pound-mass of water one Fahrenheit degree.
- The *calorie* is the heat required to raise the temperature of one gram of water by one Celsius degree.
- Several conversion factors may be useful for problems involving thermal energy:

1 Btu = 252 cal = 0.252 kcal	1 cal = 4.186 J
1 Btu = 778 ft · lb	1 kcal = 4186 J

- The *specific heat capacity c* is used to determine the quantity of heat Q absorbed or released by a unit mass m as the temperature changes by an interval Δi.

$$c = \frac{Q}{m\,\Delta t} \qquad Q = mc\,\Delta t \qquad \text{Specific Heat Capacity}$$

- Conservation of thermal energy requires that in any exchange of thermal energy the heat lost must equal the heat gained.

$$Heat\ lost = heat\ gained$$

$$\sum (mc\,\Delta t)\text{loss} = \sum (mc\,\Delta t)\text{gain}$$

As an example, suppose body 1 transfers heat to bodies 2 and 3 as the system reaches an equilibrium temperature t_e:

$$m_1 c_1 (t_1 - t_e) = m_2 c_2 (t_e - t_2) + m_3 c_3 (t_e - t_3)$$

- The *latent heat of fusion* L_f and the *latent heat of vaporization* L_v are heat losses or gains by a unit mass m during a phase change. There is no change in temperature.

$$L_f = \frac{Q}{m} \qquad Q = mL_f \qquad \begin{array}{c}Latent\ Heat\\of\ Fusion\end{array}$$

$$L_v = \frac{Q}{m} \qquad Q = mL_v \qquad \begin{array}{c}Latent\ Heat\\of\ Vaporization\end{array}$$

If a change of phase occurs, the above relationships must be added to the calorimetry equation as appropriate.

Key Terms

1. heat
2. temperature
3. calorie
4. British thermal unit
5. mechanical equivalent of heat
6. heat capacity
7. specific heat capacity
8. calorimeter
9. water equivalent
10. fusion
11. melting point
12. latent heat of fusion
13. vaporization
14. boiling point
15. latent heat of vaporization
16. freezing
17. sublimation
18. heat of combustion

Review Questions

17-1. Discuss the caloric theory of heat. In what ways is it successful in explaining heat phenomena? Where does it fail?

17-2. Blocks of five different metals—aluminum, copper, zinc, iron, and lead—are constructed with the same mass and the same cross-sectional area. Each block is heated to a temperature of 100°C and placed on a block of ice. Which will melt the ice to the greatest depth? List the remaining four metals in the order of decreasing penetration depths.

17-3. On a winter day the snow is observed to melt from the concrete sidewalk before it melts from the road. Which has the higher heat capacity?

17-4. If two objects have the same heat capacity, are they necessarily constructed of the same material? What if they have the same specific heat capacities?

17-5. Why is temperature considered a *fundamental quantity*?

17-6. A mechanical analogy to the concept of thermal equilibrium is given in Fig. 17-8. When the valve is opened, the water will flow until it has the same level in each tube.

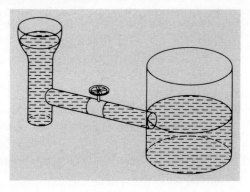

Figure 17-8 A mechanical analogy to the equalization of temperature.

What are the analogies to temperature and thermal energy?

17-7. In a mixture of ice and water, the temperature of both ice and water is 0°C. Why, then, does the ice feel colder to the touch?

17-8. Why does steam at 100°C produce a far worse burn than water at 100°C?

17-9. The temperature of 1 g of iron is raised by 1 C°. How much more heat would be required to raise the temperature of 1 lb of iron by 1 F°?

Problems

Note: Refer to Tables 17-1 and 17-2 for accepted values for specific heat, heat of vaporization, and heat of fusion for the substances in the problems below.

Sections 17-2 and 17-3—Quantity of Heat and Specific Heat Capacity

17-1. What quantity of heat is required to change the temperature of 200 g of lead from 20 to 100°C? **Ans. 496 cal**

17-2. A certain process requires 500 J of heat. Express this energy in calories and in Btu.

17-3. An oven applies 400 kJ of heat to a 4 kg of a substance, causing its temperature to increase by 80 C°. What is the specific heat capacity? **Ans. 1250 J/kg C°**

17-4. What quantity of heat will be released when 40 lb of copper cools from 78 to 32°F?

17-5. A lawn mower engine does work at the rate of 3 kW. What equivalent amount of heat will be given off in 1 hr? **Ans. 10.8 MJ**

17-6. An air conditioner is rated at 15,000 Btu/h. Express this power in kilowatts and in calories per second.

17-7. Hot coffee with a specific heat of 880 J/kg C° is poured into a 0.5-kg ceramic cup. How much heat is absorbed by the cup if its temperature increases from 20 to 80°C? **Ans. 26.4 kJ**

17-8. A 2-kW electric motor is 80 percent efficient. How much heat is lost in 1 hr?

17-9. An 8-kg copper sleeve must be heated from 25 to 140°C in order that it will expand to fit over a shaft. How much heat is required? **Ans. 359 kJ**

Summary and Review

17-10. How many grams of iron at 20°C must be heated to 100°C so that it releases 1800 cal of heat as it returns to its original temperature?

17-11. A 4-kg chunk of metal ($c = 320$ J/kg C°) is initially at 300°C. What will its final temperature be if it loses 50 kJ of heat energy?
Ans. 261°C

17-12. In a heat treatment, a hot copper part is quenched with water, cooling it from 400 to 30°C. What was the mass of the part if it loses 80 kcal of heat?

Section 17-4—The Measurement of Heat

17-13. A 400-g copper pipe initially at 200°C is dropped into a container filled with 3 kg of water at 20°C. Ignoring other heat exchanges, what is the equilibrium temperature of the mixture? **Ans. 22.2°C**

17-14. How much aluminum ($c = 0.22$ cal/g C°) at 20°C must be added to 400 g of hot water at 80°C in order that the equilibrium temperature be 30°C?

17-15. A 450-g chunk of metal is heated to 100°C and then dropped into a 50-g aluminum calorimeter cup containing 100 g of water. The initial temperature of the cup and water is 10°C and the equilibrium temperature is 21.1°C. Find the specific heat of the metal. **Ans. 0.0347 cal/g C°**

17-16. What mass of water initially at 20°C must be mixed with 2 kg of iron to bring the iron from 250°C to an equilibrium temperature of 25°C?

17-17. A worker removes a 2-kg piece of iron from an oven and places it into a 1-kg aluminum container partially filled with 2 kg of water. If the temperature of the water rises from 21 to 50°C, what was the initial temperature of the iron? **Ans. 337°C**

17-18. How much iron at 212°F must be mixed with 10 lb of water at 68°F in order to have an equilibrium temperature of 100°F?

17-19. A 1.3-kg copper block is heated to 200°C and then dropped into an insulated container partially filled with 2 kg of water at 20°C. What is the equilibrium temperature?
Ans. 30.3°C

17-20. Fifty grams of brass shot are heated to 200°C and then dropped into a 50-g aluminum cup containing 160 g of water. The temperature of the cup and water is initially 20°C. What is the equilibrium temperature?

Section 17-5—Change of Phase

17-21. A foundry has an electric furnace that can completely melt 540 kg of copper. If the temperature of the copper was initially 20°C, what total heat is needed to melt the copper? **Ans. 2.96×10^8 J**

17-22. How much heat is needed to completely melt 20 g of silver at its melting temperature?

17-23. What quantity of heat is needed to convert 2 kg of ice at −25°C to steam at 100°C?
Ans. 6.13×10^6 J

17-24. If 7.57×10^6 J of heat is absorbed in the process of completely melting a 1.60-kg chunk of an unknown metal, what is the latent heat of fusion and what is the metal?

17-25. How many grams of steam at 100°C must be mixed with 200 g of water at 20°C in order for the equilibrium temperature to be 50°C? **Ans. 10.2 g**

17-26. What total heat is released when 0.500 lb of steam at 212°F changes to ice at 10°F?

17-27. One hundred grams of ice at 0°C is mixed with 600 g of water at 25°C. What will be the equilibrium temperature for the mixture? **Ans. 10.0°C**

17-28. A certain grade of gasoline has a heat of combustion of 4.6×10^7 J/kg. Assuming 100 percent efficiency, how much gasoline must be burned to completely melt 2 kg of copper at its melting temperature?

Challenge Problems

17-29. If 1600 J of heat is applied to a brass ball, its temperature rises from 20 to 70°C. What is the mass of the ball? **Ans. 82.1 g**

17-30. How much heat does an electric freezer absorb in lowering the temperature of 2 kg of water from 80 to 20°C?

17-31. A heating element supplies an output power of 12 kW. How much time is needed to melt completely a 2-kg silver block? Assume that no power is wasted. **Ans. 14.7 s**

17-32. How much ice at $-10°C$ must be added to 200 g of water at 50°C to bring the equilibrium temperature to 40°C?

17-33. Assume that 5 g of steam at 100°C is mixed with 20 g of ice at 0°C. What will be the equilibrium temperature? **Ans. 64.0°C**

17-34. How much heat is developed by the brakes of a 4000-lb truck to bring it to a stop from a speed of 60 mi/h?

17-35. Two hundred grams of copper at 300°C is dropped into a 310-g copper calorimeter cup partially filled with 300 g of water. If the initial temperature of the cup and water was 15°C, what is the equilibrium temperature? **Ans. 30.3°C**

17-36. How many pounds of coal must be burned to melt completely 50 lb of ice in a heater that is 60 percent efficient?

17-37. If 80 g of molten lead at 327.3°C is poured into a 260-g iron casting initially at 20°C, what will be the equilibrium temperature neglecting other losses? **Ans. 58.9°C**

17-38. What equilibrium temperature is reached when 2 lb of ice at 0°F is dropped into a 3-lb aluminum cup containing 7.5 lb of water? The temperature of the cup and water is initially 200°F.

17-39. A solar collector has an area of 5 m², and the power of sunlight is delivered at 550 W/m². This power is used to increase the temperature of 200 g of water from 20 to 50°C. How much time is required? **Ans. 9.13 s**

17-40. If 10 g of milk at 12°C is added to 180 g of coffee at 95°C, what is the equilibrium temperature? Assume that milk and coffee are essentially water.

*17-41. How many grams of steam at 100°C must be added to 30 g of ice at 0°C in order to produce an equilibrium temperature of 40°C? **Ans. 6.00 g**

*17-42. A 5-g lead bullet moving at 200 m/s embeds itself into a wooden block. Half of its initial energy is absorbed by the bullet. What is the increase in temperature of the bullet?

*17-43. If 4 g of steam at 100°C is mixed with 20 g of ice at $-5°C$, find the final temperature of the mixture. **Ans. 37.9°C**

Critical Thinking Questions

17-44. A large, insulated container holds 120 g of coffee at 85°C. How much ice at 0°C must be added to cool the coffee to 50°C? Now, how much coffee at 100°C must be added to return the contents to 85°C? How many grams are finally in the container? **Ans. 32.3 g, 355 g, 508 g**

17-45. Four 200-g blocks are constructed of copper, aluminum, silver, and lead so that they have the same mass and the same base area (although of different heights). The temperature of each block is raised from 20 to 100°C by applying heat at the rate of 200 J/s. Find the time required for each block to reach 100°C.

17-46. Each of the blocks in Question 17-45 is placed on a large block of ice. Find out how much ice is melted by each block when all blocks reach equilibrium at 0°C. Which block sinks deepest, and which sinks the least? **Ans. Pb = 7.78 g, Ag = 13.8 g, Cu = 23.4 g, Al = 55.1 g, Al, Pb**

17-47. In an experiment to determine the latent heat of vaporization for water, a student measures the mass of an aluminum calorimeter cup to be 50 g. After a quantity of water is added, the combined mass of the water and cup is 120 g. The initial temperature of the cup and water is 18°C. A quantity of steam at 100°C is passed into the calorimeter, and the system is observed to reach equilibrium at 47.4°C. The total mass of the final mixture is 124 g. What value will the student obtain for the heat of vaporization?

*17-48. Equal masses of ice at 0°C, water at 50°C, and steam at 100°C are mixed and allowed to reach equilibrium. Will all the steam condense? What will be the temperature of the final mixture? What percentage of the final mixture will be water and what percentage will be steam? **Ans. no, 100°C, 19.1 percent steam and 80.9 percent water**

*17-49. If 100 g of water at 20°C is mixed with 100 g of ice at 0°C and 4 g of steam at 100°C, find the equilibrium temperature and the composition of the mixture.

*17-50. Ten grams of ice at −5°C is mixed with 6 g of steam at 100°C. Find the final temperature and the composition of the mixture.
Ans. 100°C, 13.4 g of water, 2.62 g of steam

Transfer of Heat

OBJECTIVES

After completing this chapter, you should be able to:

1. Demonstrate by definition and example your understanding of *thermal conductivity*, *convection*, and *radiation*.
2. Solve *thermal conductivity* problems involving parameters such as quantity of heat Q, surface area A, surface temperature t, time τ, and material thickness L.
3. Solve problems involving the transfer of heat by convection and discuss the meaning of the *convection coefficient*.
4. Define the *rate of radiation* and *emissivity* and use these concepts to solve problems involving thermal radiation.

We have referred to heat as a form of energy in transit. Whenever there is a difference of temperature between two bodies or between two portions of the same body, heat is said to *flow* in a direction from higher to lower temperature. There are three principal methods by which this heat exchange occurs: *conduction, convection,* and *radiation*. Examples of all three are shown in Fig. 18-1.

(a) Conduction   (b) Convection   (c) Radiation

Figure 18-1 The three methods of heat transfer: (a) conduction, (b) convection, and (c) radiation.

18-1 METHODS OF HEAT TRANSFER

Most of our discussion has involved heat transfer by conduction: by molecular collisions between neighboring molecules. For example, if we hold one end of an iron rod in a fire, the heat will eventually reach our hand through the process of conduction. The increased molecular activity at the heated end is passed on from molecule to molecule until it reaches the hand. The process will continue as long as there is a difference in temperature along the rod.

Conduction is the process in which heat energy is transferred by adjacent molecular collisions throughout a material medium. The medium itself does not move.

The most common application of the principle of conduction is probably cooking.

On the other hand, if we place our hand above a fire, as shown in Fig. 18-1b, we can feel the transfer of heat by the rising hot air. This process, called *convection*, differs from conduction because the material medium is moving. Heat is transferred by moving masses instead of being passed along by neighboring molecules.

Convection is the process by which heat is transferred by the actual mass motion of a fluid.

Convection currents form the basis for heating and cooling most houses.

When we hold our hand to the side of a fire, the primary source of heat is through ***thermal radiation.*** Radiation involves the emission or absorption of electromagnetic waves originating at the atomic level. These waves travel at the speed of light (3×10^8 m/s) and require no material medium for their passage.

Radiation is the process by which heat is transferred by electromagnetic waves.

The most obvious source of radiant energy is our own sun. Neither conduction nor convection can play a role in transferring heat energy through space to the earth. The enormous heat energy received on the earth is transferred by electromagnetic radiation. Where a material medium is involved, however, the transfer of heat due to radiation is usually small in comparison with that transferred by conduction and convection.

Unfortunately, there are many factors that affect the transfer of heat energy by all three methods. The computation of the quantity of thermal energy transferred in a given process is complicated. The relationships discussed in the following sections are based on empirical observations and depend on ideal conditions. The extent to which these conditions can be met usually determines the accuracy of their predictions.

18-2 CONDUCTION

When two parts of a material are maintained at different temperatures, energy is transferred by molecular collisions from higher to lower temperature. This process of conduction is also aided by the motion of free electrons within the substance. These electrons have become disassociated from their parent atoms and are free to move about when stimulated either thermally or electrically. Most metals are efficient conductors of heat because they have a number of free electrons that can distribute heat in addition to that propagated by molecular agitation. In general, a good conductor of electricity is also a good conductor of heat.

The fundamental law of heat conduction is a generalization of experimental results concerning the flow of heat through a slab of material. Consider the slab of thickness L and area A in Fig. 18-2. One face is maintained at a temperature t and the other at a temperature t'. The quantity of heat Q that flows perpendicular to the face during a time τ is measured. Repeating the experiment for many different materials of different thicknesses and face areas, we are able to make some general observations concerning the conduction of heat:

1. The quantity of heat transferred per unit of time is directly proportional to the temperature difference ($\Delta t = t' - t$) between the two faces.
2. The quantity of heat transferred per unit of time is directly proportional to the area A of the slab.
3. The quantity of heat transferred per unit of time is inversely proportional to the thickness L of the slab.

These results can be expressed in equation form by introducing the proportionality constant k. We write

$$H = \frac{Q}{\tau} = kA\frac{\Delta t}{L} \qquad \textbf{18-1}$$

where H represents the rate at which heat is transferred. Although the equation was introduced for a slab of material, it also holds for a rod of cross section A and length L.

The proportionality constant k is a property of the material called its **thermal conductivity.** From the defining equation, one can see that substances with a large thermal conductivity are good conductors of heat, whereas substances of low conductivity are poor conductors, or *insulators.*

The thermal conductivity of a substance is a measure of its ability to conduct heat and is defined by the relation

$$k = \frac{QL}{A\tau\,\Delta t} \qquad \textbf{18-2}$$

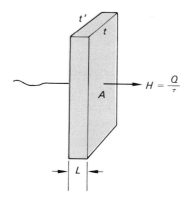

Figure 18-2   Measurement of thermal conductivity.

The numerical value for the thermal conductivity depends on the units chosen for heat, thickness, area, time, and temperature. A substitution of SI units for each of these quantities yields the following accepted units:

$$\text{SI units: J/s} \cdot \text{m} \cdot \text{C}° \qquad \text{or} \qquad \text{W/m} \cdot \text{K}$$

You will remember that the joule per second (J/s) is the power in watts (W) and that kelvin and Celsius temperature intervals are equal.

Unfortunately, the SI units are rarely used for conductivity in industry today. The choice of units is more often made on the basis of the convenience of measurement. For example, in the USCS, heat is measured in Btu, thickness in inches, area in square feet, time in hours, and the temperature interval in Fahrenheit degrees. Consequently, the units for thermal conductivity, from Eq. (18-2), are as follows:

$$\text{USCS: } k = \text{Btu} \cdot \text{in./ft}^2 \cdot \text{h} \cdot \text{F}°$$

In the metric system, heat transfer in calories is often encountered more than heat transfer expressed in joules. Therefore, the following units are frequently used:

$$\text{Metric units: } k = \text{kcal/m} \cdot \text{s} \cdot \text{C}°$$

The following conversion factors will be helpful:

$$1 \text{ kcal/s} \cdot \text{m} \cdot \text{C}° = 4186 \text{ W/m} \cdot \text{K}$$
$$1 \text{ Btu} \cdot \text{in./ft}^2 \cdot \text{h} \cdot \text{F}° = 3.445 \times 10^{-5} \text{ kcal/m} \cdot \text{s} \cdot \text{C}°$$

The thermal conductivities for various materials are listed in Table 18-1.

TABLE 18-1 Thermal Conductivities and R-values

Substance	Conductivity, k			R-values*
	W/m · K	kcal/m · s · C°	Btu · in./ft² · h · F°	ft² · h · F°/Btu
Aluminum	205	5.0×10^{-2}	1451	0.00069
Brass	109	2.6×10^{-2}	750	0.0013
Copper	385	9.2×10^{-2}	2660	0.00038
Silver	406	9.7×10^{-2}	2870	0.00035
Steel	50.2	1.2×10^{-2}	320	0.0031
Brick	0.7	1.7×10^{-4}	5.0	0.20
Concrete	0.8	1.9×10^{-4}	5.6	0.18
Corkboard	0.04	1.0×10^{-5}	0.3	3.3
Drywall	0.16	3.8×10^{-5}	1.1	0.9
Fiberglass	0.04	1.0×10^{-5}	0.3	3.3
Glass	0.8	1.9×10^{-4}	5.6	0.18
Polyurethane	0.024	5.7×10^{-6}	0.17	5.9
Sheathing	0.55	1.3×10^{-5}	0.38	2.64
Air	0.024	5.7×10^{-6}	0.17	5.9
Water	0.6	1.4×10^{-4}	4.2	0.24

*R-values based on 1-in. thickness.

EXAMPLE 18-1

The outside wall of a brick barbecue pit is 6 cm thick. The inside surface is at 150°C, and the outside surface is at 30°C. How much heat is lost in 1 h through an area of 1 m²?

Solution

Solving for Q in Eq. (18-1), we have

$$Q = kA\tau\frac{\Delta t}{L}$$

$$= (0.7 \text{ W/mK})(1 \text{ m}^2)(3600 \text{ s})\frac{150°C - 30°C}{0.06 \text{ m}}$$

$$= 5.04 \times 10^6 \text{ J}$$

It is always wise to carry the units of each quantity throughout the entire solution of a problem. This practice will save many needless errors. For example, it is sometimes easy to forget that in USCS units the thickness must be expressed in inches and the area in square feet. If the units of thermal conductivity are given with their numerical value in the equation, these errors will not be made.

When two materials of different thermal conductivities and similar cross sections are connected, the rate at which heat is conducted through each material must be constant. If there are no sources or sinks of heat energy within the materials and the end points are maintained at a constant temperature, a steady flow will eventually be reached. The heat cannot "pile up" or "speed up" at any point.

EXAMPLE 18-2

The wall of a freezing plant is composed of 10 cm of corkboard inside 14 cm of solid concrete (Fig. 18-3). The temperature of the inside cork surface is −20°C, and the outside surface of the concrete is at 24°C. (a) Find the temperature at the corkboard-concrete interface. (b) Calculate the rate of heat loss in watts per square meter.

Figure 18-3 Heat conduction through a compound wall.

Solution (a)

For steady flow, the rate of heat flow through the corkboard is equal to the rate of heat flow through the concrete. We will use the subscript 1 to refer to the corkboard and the subscript 2 to refer to the concrete. Thus, letting t_i be the temperature of the interface, we have

$$\frac{H_1}{A_1}(\text{corkboard}) = \frac{H_2}{A_2}(\text{concrete})$$

$$\frac{k_1[t_i - (-20°C)]}{0.10 \text{ m}} = \frac{k_2(24°C - t_i)}{0.14 \text{ m}}$$

$$\frac{(0.04 \text{ W/m} \cdot \text{K})(t_i + 20°C)}{0.10 \text{ m}} = \frac{(0.8 \text{ W/m} \cdot \text{K})(24°C - t_i)}{0.14 \text{ m}}$$

Now, we simplify by multiplying each term by 14 to obtain

$$5.6(t_i + 20°C) = 80(24°C - t_i)$$

$$t_i = 21.1°C$$

Solution (b)

The heat flow per unit area per unit time can now be found from Eq. (18-1) applied to either the corkboard or the concrete. For the concrete, we have

$$\frac{H_2}{A_2} = \frac{k_2(24°C - t_i)}{0.14 \text{ m}}$$

$$= \frac{(0.8 \text{ W/m} \cdot \text{K})(24°C - 21.1°C)}{0.14 \text{ m}}$$

$$= 16.4 \text{ W/m}^2$$

Notice that the kelvin interval and the Celsius interval cancel one another because they are equal. The same rate would be calculated through the corkboard. The difference in temperature between the end points of the corkboard is 41.1 C°, whereas the temperature difference in the concrete is only 2.9 C°. The different temperature intervals result primarily from the difference in the thermal conductivities of the walls.

18-3 INSULATION: THE *R*-VALUE

Heat losses in homes and in industry are often based on the insulating properties of various composite walls. For example, one might wish to know the effects of replacing dead air spaces in walls with fiberglass insulation. In such cases, the concept of *thermal resistance R* has been introduced into engineering applications. The *R*-value of a material of thickness L and of thermal conductivity k is defined as

$$R = \frac{L}{k}$$

If we recognize that the steady-state flow of heat through a composite wall is constant (see Example 18-2) and apply the thermal conductivity equation to a number of thicknesses of different materials, it can be shown that

$$\frac{Q}{\tau} = \frac{A \, \Delta t}{\Sigma_i \, (L_i/k_i)} = \frac{A \, \Delta t}{\Sigma_i \, R_i}$$ **18-3**

The quantity of heat flowing per unit of time (Q/τ) through a number of thicknesses of different materials is equal to the product of the area A and the temperature difference Δt divided by the sum of the R-values of the various materials. The R-values of common building materials are most often given in USCS units as shown in Table 18-1. For example, fiberglass ceiling insulation that is 6 in. thick has an R-value of 18.8 ft$^2 \cdot$ F$^\circ \cdot$ h/Btu. A 4-in. brick has an R-value of 4.00 ft$^2 \cdot$ F$^\circ \cdot$ h/Btu. These materials placed side by side would have a total R-value of 22.8 ft$^2 \cdot$ F$^\circ \cdot$ h/Btu.

18-4 CONVECTION

Convection has been defined as the process in which heat is transferred by the actual mass motion of a material medium. A current of liquid or gas that absorbs energy at one place and then moves to another place, where it releases heat to a cooler portion of the fluid, is called a *convection current*. A laboratory demonstration of a convection current is illustrated in Fig. 18-4. A rectangular section of

F.Y.I.

Forced convection is convection in which the fluid (or air) motion is maintained by something other than the heat content of the fluid.

Figure 18-4 An example of natural convection.

glass tubing is filled with water and heated at one of the lower corners. The water near the flame is heated and expands, becoming less dense than the cooler water above it. As the heated water rises, it is replaced by cooler water from the lower tube. This process continues until a counterclockwise convection current circulates throughout the tubing. The existence of such a current is demonstrated vividly by dropping ink into the opening at the top. The ink will be carried along by the convection current until it finally returns to the top from the right section of the tube.

If the motion of a fluid is caused by a difference in density that accompanies a change in temperature, the current produced is referred to as **natural convection.** The water flowing through the glass tubing in the above example represents a natural-convection current. When a fluid is called to move by the action of a pump or fan, the current produced is referred to as **forced convection.** Many homes are heated by using fans to force hot air from a furnace throughout the rooms.

Both forced- and natural-convection currents occur in the process of heating a room with a radiator. (Refer to Fig. 18-5.) A circulating water pump forces hot water through the pipes to the radiator and back to the heater or furnace. The heat from the water is conducted through walls of the radiator to the air in contact with it. The heated air rises and displaces the cooler air, thus establishing natural-convection currents throughout the room. Although some heating occurs by the process of radiation, conduction and convection play the large heating roles. The name *radiator* is a misnomer.

Figure 18-5 Forced convection currents circulate the heated water and return it to the furnace. The room is heated by natural convection currents in the air.

The calculation of heat transferred by convection is an enormously difficult task. So many physical properties of a fluid depend upon temperature and pressure that we can hope for only an estimate in most situations. For example, consider the conducting slab of material A and temperature t_s which is submerged completely in a cooler fluid at a temperature t_f as illustrated in Fig. 18-6. The fluid that comes in contact with the slab will rise and displace the cooler fluid. Experimental observation shows that the rate H at which heat is transferred by convection is proportional to the area A and to the difference in temperature Δt between the slab and the fluid.

Unlike thermal conductivity, convection is not a property of the solid or fluid but depends on many parameters of the system. It is known to vary with the geometry of the solid and its surface finish, the velocity of the fluid, the density of the fluid, and thermal conductivity. Differences in the pressure also affect the transfer of heat by convection. To understand how geometry affects convection, one only has to consider the obvious differences imposed by a floor facing upward or a ceiling facing downward. Several models have been developed to produce mathematical estimates of heat transfer by convection, but none are reliable enough for inclusion in this treatment.

18-5 RADIATION

The term *radiation* refers to the continuous emission of energy in the form of electromagnetic waves originating at the atomic level. Gamma rays, x rays, light waves, infrared rays, radio waves, and radar waves are all examples of electromagnetic radiation; they differ only in their wavelength. In this section, we shall be concerned with *thermal radiation*.

Thermal radiation consists of electromagnetic waves emitted by a solid, liquid, or gas by virtue of its temperature.

All objects are continuously emitting radiant energy. At low temperatures the rate of emission is small, and the radiation is predominantly of long wavelengths. As the temperature is increased, the rate of emission increases rapidly, and the pre-

Figure 18-6 When a heated slab is placed in a cool fluid, convection currents transfer heat away from the slab at a rate proportional to the difference in temperatures and to the area of the slab.

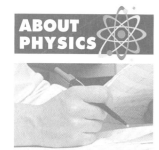

dominant radiation shifts to shorter wavelenghts. If an iron rod is heated continuously, it will eventually give off radiation in the visible region; hence, the terms *red hot* and *white hot.*

Experimental measurements have shown that the rate at which thermal energy is radiated from a surface *varies directly with the fourth power of the absolute temperature of the radiating body.* Thus, if the temperature of an object is doubled, the rate at which it emits thermal energy will be increased sixteenfold.

An additional factor that must be considered in computing the rate of heat transfer by radiation is the nature of the exposed surfaces. Objects that are efficient emitters of thermal radiation are also efficient absorbers of radiation. An object that absorbs all the radiation incident on its surface is called an **ideal absorber.** Such an object will also be an **ideal radiator.** There is no such thing as an *ideal* absorber; but, in general, the blacker a surface, the better it absorbs thermal energy. For example, a black shift absorbs more of the sun's radiant energy than a lighter shirt. Since the black shirt is also a good emitter, its external temperature will be higher than our body temperature, making us uncomfortable.

An ideal absorber or an ideal radiator is sometimes referred to as a **blackbody** for the reasons mentioned above. The radiation emitted from a blackbody is called *blackbody radiation.* Although such bodies do not actually exist, the concept is useful as a standard for comparing the abilities of various surfaces to absorb or emit thermal energy.

Emissivity e is a measure of a body's ability to absorb or emit thermal radiation.

The emissivity is a unitless quantity that has a numerical value between 0 and 1, depending on the nature of the surface. For a blackbody, the emissivity is equal to unity. For a highly polished silver surface, it is near zero.

The ***rate of radiation R*** of a body is formally defined as the radiant energy emitted per unit area per unit time, in other words, the power per unit area. Symbolically,

$$R = \frac{E}{\tau A} = \frac{P}{A}$$

18-4

If the radiant power P is expressed in watts and the surface area A in square meters, the rate of radiation will be in watts per square meter. As we have discussed earlier, this rate depends on two factors, the absolute temperature T and the emissivity e of the radiating body. The formal statement of this dependence, known as the **Stefan–Boltzmann law,** can be written

$$R = \frac{P}{A} = e\sigma T^4$$

18-5

The proportionality constant σ is a universal constant completely independent of the nature of the radiation. If the radiant power is expressed in watts and the surface in square meters, σ has the value of 5.67×10^{-8} W/m$^2 \cdot$ K^4. The emissivity e has values from 0 to 1, depending on the nature of the radiating surface. A summary of the symbols and their definitions is given in Table 18-2.

TABLE 18-2 Definition of Symbols in the Stefan–Boltzmann law ($R = e\sigma T^4$)

Symbol	Definition	Comment
R	Energy radiated per unit time per unit area	$\dfrac{E}{\tau A}$ or $\dfrac{P}{A}$
e	Emissivity of the surface	0–1
σ	Stefan's constant	5.67×10^{-8} W/m$^2 \cdot$ K^4
T^4	The fourth power of the absolute temperature	K^4

EXAMPLE 18-3

What power will be radiated from a spherical silver surface 10 cm in diameter if its temperature is 527°C? The emissivity of the surface is 0.04.

Solution

We must first compute the surface area from the known diameter of the sphere.

$$A = 4\pi R^2 = \pi D^2 = \pi(0.1 \text{ m})^2 = 0.0314 \text{ m}^2$$

The absolute temperature is

$$T = 527 + 273 = 800 \text{ K}$$

Solving for P in Eq. (18-6), we obtain

$$
\begin{aligned}
P &= e\sigma A T^4 \\
&= (0.04)(5.67 \times 10^{-8} \text{ W/m}^2 \cdot \text{K}^4)(0.0314 \text{ m}^2)(800 \text{ K})^4 \\
&= 29.2 \text{ W}
\end{aligned}
$$

We have said that all objects continuously emit radiation, regardless of their temperature. If this is true, why don't the objects eventually run out of fuel? The answer is that they *would* run down if no energy were supplied to them. The filament in an electric light bulb cools rather quickly to room temperature when the supply of electric energy is shut off. It does not cool further because, at this point, the filament is absorbing radiant energy at the same rate that it is emitting radiant energy. The law covering this phenomenon is known as *Prevost's law of heat exchange:*

A body at the same temperature as its surroundings radiates and absorbs heat at the same rates.

Figure 18-7 shows an isolated object in thermal equilibrium with the walls of its container.

The rate at which energy is absorbed by a body is also given by the Stefan–Boltzmann law [Eq. (18-5)]. Thus, we can figure the net transfer of radiant energy by an object surrounded by walls at a different temperature. For example, consider a thin wire filament in a lamp that is covered with an envelope, as shown in

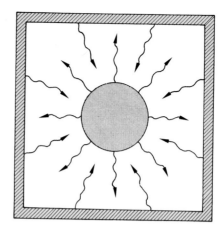

Figure 18-7 When an object and its surroundings are at the same temperature, the radiant energy emitted is the same as that absorbed.

Figure 18-8 The net energy emitted by a radiator in surroundings of a different temperature.

Fig. 18-8. Let the temperature of the filament be denoted by T_1 and the temperature of the surrounding envelope be denoted by T_2. The emissivity of the filament is e, and only radiative processes are considered. In this example, we note that

Net rate of radiation = rate of energy emission − rate of energy absorption
$$R = e\sigma T_1^4 - e\sigma T_2^4$$

$$R = e\sigma(T_1^4 - T_2^4) \tag{18-6}$$

Equation (18-6) can be applied to any system for computing the net energy emitted by a radiator of temperature T_1 and emissivity e in the presence of surroundings at a temperature T_2.

Summary and Review

Summary

Heat is the transfer of thermal energy from one place to another. We have seen that the rate of transfer by conduction, convection, and radiation can be predicted from experimental formulas. The effects of material, surface areas, and temperature differences must be understood for many industrial applications of heat transfer. The major concepts presented in this chapter are listed below.

- In the transfer of heat by conduction, the quantity of heat Q transferred per unit of time τ through a wall or rod of length L is given by

$$H = \frac{Q}{\tau} = kA\frac{\Delta t}{L} \qquad \textit{Conduction}$$

where A is the area and Δt is the difference in surface temperatures. The SI unit for H is the *watt* (W). Other commonly used units are kcal/s and Btu/h. From this relation, the thermal conductivity is

$$k = \frac{QL}{\tau A\,\Delta t} \qquad \textit{Thermal Conductivity}$$

The SI units for k are W/m · K. Useful conversions can be made from the following definitions:

1 kcal/m · s · C° = 4186 W/m · K

1 W/m · K = 6.94 Btu · in./ft² · h · F°

1 Btu · in./ft² · h · F° = 3.445 × 10⁻⁵ kcal/m · s · C°

- The *R*-value is an engineering term to measure the thermal resistance offered to the conduction of heat. It is defined as follows:

$$R = \frac{L}{K} \qquad \textit{R-Value}$$

This concept applied to a number of thicknesses of different materials yields the following useful equation:

$$\frac{Q}{\tau} = \frac{A\,\Delta t}{\Sigma_i\,(L_i/k_i)} = \frac{A\,\Delta t}{\Sigma_i\,R_i}$$

The quantity of heat flowing per unit of time (Q/τ) through a number of thicknesses of different materials is equal to the product of the area A and the temperature difference Δt divided by the sum of the R-values of the various materials. In keeping with current engineering practice, the units for the R-value are ft² · F° · h/Btu.

- For heat transfer by radiation, we define the rate of radiation as the energy emitted per unit area per unit time (or simply the power per unit area):

$$R = \frac{E}{\tau A} = \frac{P}{A} \qquad \textit{Rate of Radiation, W/m}^2$$

According to *Stefan–Boltzmann's law*, this rate is given by

$$R = \frac{P}{A} = e\sigma T^4 \qquad \sigma = 5.67 \times 10^{-8}\text{ W/m}^2 \cdot \text{K}^4$$

- Prevost's law of heat exchange states that *a body at the same temperature as its surroundings radiates and absorbs heat at the same rates.*

Key Terms

1. conduction
2. thermal conductivity
3. natural convection
4. forced convection
5. thermal radiation
6. blackbody
7. emissivity
8. Stefan–Boltzmann law
9. Prevost's law

Review Questions

18-1. Discuss the vacuum bottle and explain how it minimizes transfer of heat by conduction, convection, and radiation.

18-2. What determines the *direction* of heat transfer?

18-3. Heat flows both by conduction and by radiation. In what ways are they different? In what ways are they similar?

18-4. A hot chunk of iron is suspended centrally within an evacuated calorimeter. Can we determine the specific heat of iron by the techniques introduced in Chap. 18? Discuss.

18-5. Discuss the analogies that exist between steady-state heat flow and the flow of an incompressible fluid.

18-6. A pan of water is placed over a gas burner on a kitchen stove until the water boils vigorously. Discuss the heat transfers that take place. How would you explain the fact that the bubbles forming in the water are carried to the surface in the form of a pyramid instead of rising directly to the surface?

18-7. By placing a flame underneath a paper cup filled with water, it is possible to bring the water to a boil without burning the bottom of the cup. Explain.

18-8. When a piece of paper is wrapped around a stick of wood and the system is heated with a flame, the paper will begin to burn. But if the paper is wrapped tight around a copper rod and heated in the same manner, it does not burn. Why?

18-9. On a cold day, a piece of iron feels colder to the touch than a piece of wood. Explain.

18-10. Copper has about twice the thermal conductivity of aluminum, but its specific heat is a little less than half that of aluminum. A rectangular block is made from each material so that they have identical masses and the same surface area at their bases.

Each block is heated to 300°C and placed on the top of a large cube of ice. Which block will stop sinking first? Which will sink deeper?

18-11. Distinguish between thermal conductivity and specific heat as they relate to heat transfer.

18-12. The term *absorptivity* is sometimes used in place of the term *emissivity*. Can you justify this practice?

18-13. Why is more air conditioning required to cool the inside of a navy-blue car than a white car of the same size?

18-14. If a house is to be designed for maximum comfort in both summer and winter, would you prefer a light roof or a dark roof? Explain.

18-15. If you are interested in the number of kilocalories transferred by radiation in a unit of time, the Stefan–Boltzmann law can be written in the form

$$\frac{Q}{\tau} = e\sigma A T^4$$

For this form of the law, show that Stefan's constant σ is equal to

$$1.35 \times 10^{-11} \text{ kcal/m}^2 \cdot \text{s} \cdot \text{K}^4.$$

18-16. When a liquid is heated in a glass beaker, a wire gauze is usually placed between the flame and the bottom of the beaker. Why is this a wise practice?

18-17. Does the warm air over a burning fire rise, or is it forced upward by the flames?

18-18. Should a hot-water or steam radiator be painted with an efficient emitter or a poor one? If it is painted black, will it be more efficient? Why?

18-19. Which is a faster process, conduction or convection? Give an illustration to justify your conclusion.

Problems

Note: Refer to Tables 18-1 and 18-2 for thermal conductivities and other constants needed in the solution of the problems in this section.

Section 18-2—Thermal Conductivity

18-1. A block of copper has a cross section of 20 cm^2 and a length of 50 cm. The left end is maintained at 0°C and the right end is at 100°C. What is the rate of heat flow in watts? **Ans. 154 W**

18-2. In Prob. 18-1, what is the heat flow if the copper block is replaced with a block of glass having the same dimensions?

18-3. A 50-cm-long brass rod has a diameter of 3 mm. The temperature of one end is 76 C° higher than the other end. How much heat is conducted in 1 min? **Ans. 7.03 J**

18-4. A pane of window glass is 10 in. wide, 16 in. long, and ⅛ in. thick. The inside surface is at 60°F, and the outside surface is at 20°F. How many Btu are transferred to the outside in a time of 2 h?

18-5. One end of an ~~iron~~ Steel rod 30 cm long and 4 cm^2 in cross section is placed in a bath of ice and water. The other end is placed in a steam bath. How many minutes are needed to transfer 1.0 kcal of heat? In what direction does the heat flow?
Ans. 10.4 min, toward the ice

18-6. A steel plate of thickness 20 mm has a cross section of 600 cm^2. One side is at 170°C, and the other is at 120°C. What is the rate of heat transfer?

18-7. How much heat is lost in 12 h through a 3-in. brick firewall that has an area of 10 ft^2 if one side is at 330°F and the other is at 78°F? **Ans. 50,400 Btu**

*18-8. A composite wall 6 m long and 3 m high consists of 12 cm of concrete joined with 10 cm of corkboard. The inside temperature is 10°C, and the outside temperature is 40°C. Find the temperature at the interface of the two materials.

*18-9. What is the steady-state rate of heat flow through the composite wall of Prob. 18-8?
204. J/s ~~Ans. 4.83 J/s~~

Section 18-5—Thermal Radiation

18-10. What is the power radiated from a spherical blackbody of surface area 20 cm^2 if its temperature is 250°C?

18-11. What is the rate of radiation for a spherical blackbody at a temperature of 327°C? Will this rate change if the radius is doubled for the same temperature?
Ans. 7.35 kW/m^2, no.

18-12. The emissivity of a metallic sphere is 0.3, and at a temperature of 500 K it radiates a power of 800 W. What is the radius of the sphere?

*18-13. If a certain body absorbs 20 percent of the incident thermal radiation, what is its emissivity? What energy will be emitted by this body in 1 min if its surface area is 1 m^2 and its temperature is 727°C?
Ans. 680 kJ

*18-14. The operating temperature of the filament in a 25-W lamp is 2727°C. If the emissivity is 0.3, what is the surface area of the filament?

Challenge Problems

18-15. A glass window pane is 60 cm wide, 1.8 m high, and 3 mm thick. The inside temperature is 20°C, and the outside temperature is −10°C. How much heat leaves the house through this window in 1 hr? **Ans. 31 MJ**

18-16. A 20-cm thickness of fiberglass insulation covers a 20-m by 15-m attic floor. How many calories of heat are lost to the attic if the temperatures on the sides of the insulation are −10°C and +24°C?

18-17. The bottom of an aluminum pan is 3 mm thick and has a surface area of 120 cm^2. How many calories per minute are conducted through the bottom of the pan if the temperature of the outer surface is 114°C and the temperature of the inner surface is 117°C? **Ans. 36,000 cal**

18-18. A solid wall of concrete is 80 ft high, 100 ft wide, and 6 in. thick. The surface temperatures on the sides of the concrete are 30 and 100°F. What time is required for 400,000 Btu of heat to be transferred?

18-19. The bottom of a hot metal pan has an area of 86 cm² and a temperature of 98°C. The pan is placed on top of a corkboard 5 mm thick. The Formica tabletop under the corkboard is maintained at a constant temperature of 20°C. How much heat is conducted through the cork in 2 min?
Ans. 644 J

*18-20. What thickness of copper is required to have the same insulating value as 2 in. of corkboard?

*18-21. What thickness of concrete is required to have the same insulating value as 6 cm of fiberglass? **Ans. 1.20 m**

18-22. A plate-glass window in an office building measures 2 by 6 m and is 1.2 cm thick. Its outer surface is at 23°C, and its inner surface is at 25°C. How many joules of heat pass through the glass in an hour?

18-23. What must be the temperature of a black body if its rate of radiation is 860 W/m²?
Ans. 351 K

18-24. A gray steel ball has an emissivity of 0.75, and when its temperature is 570°C, the power radiated is 800 W. What is the surface area of the ball?

18-25. A 25-W lamp has a filament of area 0.212 cm². If the emissivity is 0.35, what is the operating temperature of the filament?
Ans. 2776 K

Critical Thinking Questions

*18-26. The wall of a freezing plant consists of 6 in. of concrete and 4 in. of corkboard. The temperature of the inside cork surface is −15°F, and the temperature of the outside surface is 70°F. What is the temperature at the interface between the cork and concrete? How much heat is conducted through each square foot of wall in one hour? **Ans. 63.7°F, 5.92 Btu/ft² h**

*18-27. A wooden icebox has walls that are 4 cm thick, and the overall effective surface area is 2 m². How many grams of ice will be melted in 1 min if the inside temperatures of the walls are 4°C an the outside temperatures are 20°C?

*18-28. The filament in a lamp operates at a temperature of 727°C and is surrounded by an envelope at 227°C. If the filament has an emissivity of 0.25 and a surface area of 0.30 cm², what is the operating power of the lamp? **Ans. 0.399 W.**

*18-29. The thermal conductivity of material A is twice that for material B. Which has the greater R value? When 6 cm of material A is between two walls, the rate of conduction is 400 W/m². What thickness of material B is needed to provide the same rate of heat transfer if the other factors remain constant?

Thermal Properties of Matter

OBJECTIVES

After completing this chapter, you should be able to:

1. Write and apply the relationship between the volume and the pressure of a gas at constant temperature *(Boyle's law).*
2. Write and apply the relationship between the volume and the temperature of a gas under conditions of constant pressure *(Charles's law).*
3. Write and apply the relationship between the temperature and pressure of a gas under conditions of constant volume *(Gay-Lussac's law).*
4. Apply the general gas law to the solution of problems involving changes in mass, volume, pressure, and temperature of gases.
5. Define *vapor pressure, dew point,* and *relative humidity,* and apply these concepts to the solution of problems.

Now that we have an understanding of the concepts of heat and temperature, we proceed to study the thermal behavior of matter. Four measurable quantities are of interest: the pressure, volume, temperature, and mass of a sample. Together, these variables determine the state of a given sample of matter. Depending on its state, matter may exist in the liquid, solid, or gaseous phase. Thus, it is important to distinguish between the terms *state* and *phase.* We begin by studying the thermal behavior of gases.

19-1 IDEAL GASES AND BOYLE'S LAW

In a gas the individual molecules are so far apart that the cohesive forces between them are usually small. Even though the molecular structure of different gases may vary considerably, their behavior is affected little by the size of the individual molecules. It is usually safe to say that when a large quantity of gas is confined in a rather small volume, the volume occupied by the molecules is still a tiny fraction of the total volume.

One of the most useful generalizations about gases is the concept of an ***ideal gas,*** whose behavior is completely unaffected by cohesive forces or molecular volumes. Of course, no real gas is *ideal,* but under ordinary conditions of temperature and pressure, the behavior of any gas conforms closely to the behavior of an ideal gas. Therefore, experimental observations of many real gases can lead to the derivation of general physical laws governing their thermal behavior. The degree to which any real gas obeys these relations is determined by how closely it approximates an ideal gas.

The first experimental measurements of the thermal behavior of gases were made by Robert Boyle (1627–1691). He made an exhaustive study of the changes in the volume of gases as a result of changes in pressure. All other variables, such as mass and temperature, were kept constant. In 1660, Boyle demonstrated that the volume of a gas is inversely proportional to its pressure. In other words, doubling the volume decreases the pressure to one-half its original value. This finding is now know as *Boyle's law.*

> **Boyle's law** **Provided that the mass and temperature of a sample of gas are held constant, the volume of the gas is inversely proportional to its absolute pressure.**

Another way of stating Boyle's law is to say that the product of the pressure P of a gas and its volume V will be constant as long as the temperature does not change. Consider, for example, a closed cylinder equipped with a movable piston, as shown in Fig. 19-1. In Fig. 19-1a, the initial state of the gas is described by its pressure P_1 and its volume V_1. If the piston is pressed downward until it reaches the new position shown in Fig. 19-1b, its pressure will increase to P_2 and its vol-

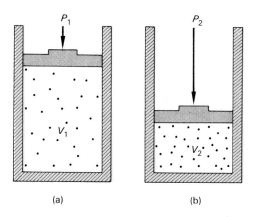

(a)   (b)

Figure 19-1   When a gas is compressed at constant temperature, the product of its pressure and its volume is always constant; that is $P_1 V_1 = P_2 V_2$.

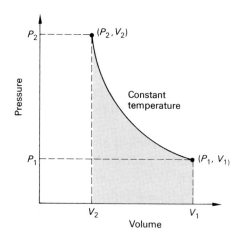

Figure 19-2   A *P-V* diagram illustrating that the pressure of an ideal gas varies inversely with its volume.

ume will decrease to V_2. This process is shown graphically in Fig. 19-2. If the process occurs without a change in temperature, Boyle's law reveals that

$$P_1V_1 = P_2V_2 \qquad \text{With Constant} \atop m \text{ and } T \qquad \text{19-1}$$

In other words, the product of pressure and volume in the initial state is equal to the product of pressure and volume in the final state. Equation (19-1) is a mathematical statement of Boyle's law. The pressure P must be the *absolute* pressure and not *gauge* pressure.

EXAMPLE 19-1

What volume of hydrogen gas at atmospheric pressure is required to fill a 5000-cm³ tank under a gauge pressure of 530 kPa?

Solution
Recall that atmospheric pressure is 101.3 kPa absolute. Thus, the absolute pressures initially and finally are

$$P_1 = 101.3 \text{ kPa} \qquad P_2 = 530 \text{ kPa} + 101.3 \text{ kPa} = 631 \text{ kPa}$$

The final volume V_2 is 5000 cm³. Applying Eq. (19-1), we have

$$P_1V_1 = P_2V_2$$
$$(101.3 \text{ kPa})V_1 = (631 \text{ kPa})(5000 \text{ cm}^3)$$
$$V_1 = 31{,}100 \text{ cm}^3$$

Notice that is was not necessary for the units for pressure to be consistent with the units for volume. Since P and V appear on each side of the equation, it is necessary only to choose the same units for pressure. The units for volume will then be the units substituted for V_2.

In Chap. 16 we used the fact that the volume of a gas increased directly with its temperature to help us define absolute zero. We found the result $(-273°C)$ by extending the line on the graph in Fig. 19-3. Of course, any real gas will become

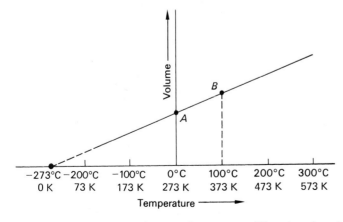

Figure 19-3 The variation of volume as a function of temperature. When the volume is extrapolated to zero, the temperature of a gas is at absolute zero (0 K).

a liquid before its volume reaches zero. But the direct relationship is a valid approximation for most gases that are not subjected to extreme conditions of temperature and pressure.

This direct proportionality between volume and temperature was first experimentally tested by Jacques Charles in 1787. *Charles's law* may be stated as follows:

Charles's law **Provided that the mass and pressure of a gas are held constant, the volume of the gas is directly proportional to its absolute temperature.**

If we use the subscript 1 to refer to an initial state of a gas and the subscript 2 to refer to the final state, a mathematical statement of Charles's law is obtained.

$$\frac{V_1}{T_1} = \frac{V_2}{T_2} \qquad \textit{With Constant} \atop \textit{m and P} \qquad \text{19-2}$$

In this equation V_1 refers to the volume of a gas at the *absolute* temperature T_1, and V_2 is the later volume of the same sample of gas when its absolute temperature is T_2.

EXAMPLE 19-2

A large balloon filled with air has a volume of 200 liters at 0°C. What will its volume be at 57°C if the pressure is unchanged?

Solution

Since Charles's law applies only for absolute temperatures, we must first convert the given temperatures to kelvins.

$$T_1 = 273 \text{ K} \qquad T_2 = 57 + 273 = 330 \text{ K}$$

Now we may substitute and solve Eq. (19-2) for V_2.

$$\frac{V_1}{T_1} = \frac{V_2}{T_2}$$

$$\frac{200 \text{ liters}}{273 \text{ K}} = \frac{V_2}{330 \text{ K}}$$

$$V_2 = \frac{(200 \text{ liters})(330 \text{ K})}{273 \text{ K}} = 242 \text{ liters}$$

19-2 GAY-LUSSAC'S LAW

The three quantities that determine the state of a given mass of gas are its pressure, volume, and temperature. Boyle's law deals with changes in pressure and volume under constant temperature, and Charles's law applies for volume and temperature under constant pressure. The variation in pressure as a function of temperature is described in a law attributed to Gay-Lussac.

Gay-Lussac's law **If the volume of a sample of gas remains constant, the absolute pressure of the gas is directly proportional to its absolute temperature.**

This means that doubling the pressure applied to a gas will cause its absolute temperature to double also. In equation form, Gay-Lussac's law may be written as

$$\frac{P_1}{T_1} = \frac{P_2}{T_2} \qquad \text{With Constant } m \text{ and } V \qquad \textbf{19-3}$$

EXAMPLE 19-3

An automobile tire is inflated to a gauge pressure of 30 lb/in.^{2} at a time when the surrounding pressure is 14.4 lb/in.^{2} and the temperature is 70°F. After the car is driven, the temperature of the air in the tire increases to 100°F. Assuming that the volume changes only slightly, what will be the new gauge pressure in the tire?

Solution

Gay-Lussac's law applies for constant volume, but we must first convert to absolute pressure and to absolute temperature.

$$P_1 = 30 \text{ lb/in.}^2 + 14.4 \text{ lb/in.}^2 = 44.4 \text{ lb/in.}^2$$
$$T_1 = 70 + 460 = 530°\text{R} \qquad T_2 = 100 + 460 = 560°\text{R}$$

Now we calculate the new pressure P_2 from Eq. (19-3).

$$\frac{P_1}{T_1} = \frac{P_2}{T_2}$$
$$\frac{44.4 \text{ lb/in.}^2}{530°\text{R}} = \frac{P_2}{560°\text{R}}$$
$$P_2 = \frac{(44.4 \text{ lb/in.}^2)(560°\text{R})}{530°\text{R}} = 46.9 \text{ lb/in.}^2$$

Note that 46.9 lb/in.^{2} will be the absolute pressure. The gauge pressure will be 14.4 lb/in.^{2} less.

$$\text{Gauge pressure} = 46.9 \text{ lb/in.}^2 - 14.4 \text{ lb/in.}^2 = 32.5 \text{ lb/in.}^2$$

19-3 GENERAL GAS LAWS

Thus far, we have discussed three laws that can be used to describe the thermal behavior of gases. Boyle's law, as given by Eq. (19-1) applies for a sample of gas whose temperature is unchanged. Charles's law, as given by Eq. (19-2), applies for a gas sample under a constant pressure. Gay-Lussac's law, in Eq. (19-3), is for a gas sample under constant volume. Unfortunately, none of these conditions is usually satisfied. Normally, a system undergoes changes in volume, temperature, and pressure as a result of a thermal process. A more general relation combines the three laws as follows:

$$\frac{P_1 V_1}{T_1} = \frac{P_2 V_2}{T_2} \qquad \text{With Constant } m \qquad \textbf{19-4}$$

where (P_1, V_1, T_1) may be considered coordinates of the initial state and (P_2, V_2, T_2) the coordinates of the final state. In other words, for a given mass the ratio PV/T is constant for any ideal gas. Equation (19-4) can be remembered by "a private (PV/T) is always a private."

EXAMPLE 19-4

An oxygen tank with an internal volume of 20 liters is filled with oxygen under an absolute pressure of 6×10^6 N/m^2 at 20°C. The oxygen is to be used in a high-flying aircraft, where the absolute pressure is 7×10^4 N/m^2 and the temperature is -20°C. What volume of oxygen can be supplied by the tank under these conditions?

Solution

After converting the temperatures to the absolute kelvin scale, we apply Eq. (19-4).

$$\frac{P_1 V_1}{T_1} = \frac{P_2 V_2}{T_2}$$

$$\frac{(6 \times 10^6 \text{ N/m}^2)(20 \text{ liters})}{293 \text{ K}} = \frac{(7 \times 10^4 \text{ N/m}^2) V_2}{253 \text{ K}}$$

$$V_2 = 1480 \text{ liters}$$

Let us now consider the effect of a change in mass on the behavior of gases. If the temperature and volume of an enclosed gas are held constant, the addition of more gas will result in a proportional increase in pressure. Similarly, if the pressure and temperature are fixed, an increase in the mass will result in a proportional increase in the volume of the container. We can combine these experimental observations with Eq. (19-4) to obtain the general relation:

$$\frac{P_1 V_1}{m_1 T_1} = \frac{P_2 V_2}{m_2 T_2} \qquad \text{19-5}$$

where m_1 is the initial mass and m_2 is the final mass. A study of this relation will reveal that Boyle's law, Charles's law, Gay-Lussac's law, and Eq. (19-4) are all special cases of the more general equation (19-5).

CONCEPTUAL EXAMPLE 19-5

The pressure gauge on a helium storage tank reads 2000 lb/in.^{2} when the temperature is 27°C. The container develops a leak overnight, and the gauge pressure the next morning is found to be 1500 lb/in.^{2} at a temperature of 17°C. What percentage of the original mass of helium remains inside the container?

Solution

Since the volume of the container remains constant, $V_1 = V_2$ and Eq. (19-5) gives

$$\frac{P_1}{m_1 T_1} = \frac{P_2}{m_2 T_2}$$

The ratio m_2/m_1 represents the fraction of the helium mass remaining. Hence,

$$\frac{m_2}{m_1} = \frac{P_2 T_1}{P_1 T_2}$$

The pressures and temperatures are adjusted to their absolute values as follows:

$$P_1 = 2000 \text{ lb/in.}^2 + 14.7 \text{ lb/in.}^2 = 2014.7 \text{ lb/in.}^2$$
$$P_2 = 1500 \text{ lb/in.}^2 + 14.7 \text{ lb/in.}^2 = 1514.7 \text{ lb/in.}^2$$
$$T_1 = 27 + 273 = 300 \text{ K}$$
$$T_2 = 17 + 273 = 290 \text{ K}$$

Substitution of these values yields

$$\frac{m_2}{m_1} = \frac{(1514.7)(300)}{(2014.7)(290)} = 0.778$$

Therefore, 77.8 percent of the helium still remains inside the container.

Equation (19-5) is general because it accounts for variance in the pressure, volume, temperature, and mass of a gas. The quantity that affects pressure and volume is not the mass of a gas, however, but the number of molecules in the gas. According to the kinetic theory of gases, the pressure is due to molecular collisions with the walls of the container. Increasing the number of molecules will increase the number of particles colliding per second, and the pressure of the gas will become greater. If we are considering a thermal process involving quantities of the same gas, it is safe to apply Eq. (19-5) because the mass is proportional to the number of molecules.

When dealing with different kinds of gas, such as hydrogen compared with oxygen, it is necessary to refer to equal numbers of molecules rather than equal masses. When they are placed in similar containers, 6 g of hydrogen will yield a much greater pressure than 6 g of oxygen. There are many more hydrogen molecules in 6 g of H_2 than there are oxygen molecules in 6 g of O_2. To be more general, we must revise Eq. (19-5) to account for differences in the number of gas molecules instead of the difference in mass. First, we must develop methods of relating the quantity of a gas to the number of molecules present.

19-4 MOLECULAR MASS AND THE MOLE

Although the mass of individual atoms is difficult to determine because of their size, experimental methods have been successful in measuring atomic mass. For example, we know that one atom of helium has a mass of 6.65×10^{-24} g. When working with macroscopic quantities, such as volume, pressure, and temperature, it is much more convenient to compare the *relative masses* of individual atoms.

The relative atomic masses are based on the mass of a reference atom known as *carbon 12*. By arbitrarily assigning exactly 12 *atomic mass units* (u) to this atom, we have a standard for comparison of other atomic masses.

The atomic mass of an element is the mass of an atom of that element compared with the mass of an atom of carbon taken as 12 atomic mass units.

On this basis, the atomic mass of hydrogen is approximately 1 u, and the atomic mass of oxygen is approximately 16 u.

A molecule consists of two or more atoms in chemical combination. The definition of molecular mass follows from the definition of relative atomic mass.

The molecular mass M is the sum of the atomic masses of all the atoms making up the molecule.

For example, a molecule of oxygen (O_2) contains two atoms of oxygen. Its molecular mass is 16 u $\times$ 2 = 32 u. A molecule of carbon dioxide (CO_2) contains one atom of carbon and two atoms of oxygen. Thus, the molecular mass of CO_2 is 44 u:

$$\begin{aligned} 1\ C &= 1 \times 12 = 12\ \text{u} \\ 2\ O &= 2 \times 16 = 32\ \text{u} \\ \hline CO_2 &= 44\ \text{u} \end{aligned}$$

In dealing with gases, we have noted that it is more meaningful to treat the amount of substance present in terms of the number of molecules present. This is accomplished by establishing a new unit of measure called the *mole* (mol).

The mole is that quantity of a substance that contains the same number of particles as there are atoms in 12 g of carbon 12.

On the basis of this definition, 1 mol of carbon must be equal to 12 g by definition. Since the molecular mass of any substance is based on carbon 12 as a standard, it follows that

One mole is the mass in grams equal numerically to the molecular mass of a substance.

For example, 1 mol of hydrogen (H_2) is 2 g, 1 mol of oxygen (O_2) is 32 g, and 1 mol of carbon dioxide (CO_2) is 44 g. In other words, 2 g of H_2, 32 g of O_2, and 44 g of CO_2 all have the same number of molecules. This number N_A is known as *Avogadro's number.*

The ratio of the number of molecules N to the number of moles n must equal Avogadro's number N_A. Symbolically,

$$N_A = \frac{N}{n} \qquad \textit{Molecules per Mole} \qquad \textbf{19-6}$$

There are several experimental methods of determining Avogadro's number. The accepted value for N_A is

$$N_A = 6.023 \times 10^{23} \text{ molecules per mole} \qquad \textbf{19-7}$$

The easiest way to determine the number of moles n contained in a gas is to divide its mass m in grams by its molecular mass M. Thus,

$$n = \frac{m}{M} \qquad \qquad \textit{Number of Moles} \qquad \textbf{19-8}$$

EXAMPLE 19-6

(a) How many moles of gas are present in 200 g of CO_2? (b) How many molecules are present?

Solution (a)

The molecular mass of CO_2 is 44 u or 44 g/mol. Therefore,

$$n = \frac{m}{M} = \frac{200 \text{ g}}{44 \text{ g/mol}} = 4.55 \text{ mol}$$

Solution (b)

From Eq. (19-6),

$$n = \frac{N}{N_A} = 4.55 \text{ mol}$$
$$N = (6.023 \times 10^{23} \text{ molecules/mol})(4.55 \text{ mol})$$
$$= 2.74 \times 10^{24} \text{ molecules}$$

19-5 THE IDEAL GAS LAW

Let us now return to our search for a more general gas law. If we substitute the number of moles n for the mass m in Eq. (19-5), we can write

$$\frac{P_1 V_1}{n_1 T_1} = \frac{P_2 V_2}{n_2 T_2} \qquad \qquad \textbf{19-9}$$

This equation represents the most useful form of a general gas law when all the parameters of an initial state and a final state are known except for a single quantity.

An alternative expression of Eq. (19-9) is

$$\frac{PV}{nT} = R \qquad \qquad \textbf{19-10}$$

where R is known as the *universal gas constant*. If we can evaluate R under certain known values of P, V, n, and T, Eq. (19-10) can be used directly without any information concerning initial and final states. The numerical value for R, of course, depends on the choice of units for P, V, n, and T. In SI units, the value has been determined as

$$R = 8.314 \text{ J/mol} \cdot \text{K}$$

Other choices of units lead to the following equivalent values:

$$R = 0.0821 \text{ L} \cdot \text{atm/mol} \cdot \text{K}$$
$$= 1.99 \text{ cal/mol} \cdot \text{K}$$

If the pressure is measured in pascals and the volume in cubic meters, one should use 8.314 J/mol · K for the constant R. Often, however, the pressure is expressed in atmospheres and the volume in liters. Rather than making appropriate conversions, it would probably be simpler to use $R = 0.0821$ L · atm/mol · K.

Equation (19-10) is known as the *ideal gas law* and is usually written in the form

$$PV = nRT \tag{19-11}$$

Another useful form of the ideal gas law makes use of the fact that $n = m/M$. Thus,

$$PV = \frac{m}{M}RT \tag{19-12}$$

EXAMPLE 19-7

Find the volume of 1 mole of any ideal gas at a condition of standard temperature (0°C = 273 K) and pressure (1 atm = 101.3 kPa).

Solution
Solving Eq. (19-11) for V, we obtain

$$V = \frac{nRT}{P} = \frac{(1 \text{ mol})(0.0821 \text{ L} \cdot \text{atm/mol} \cdot \text{K})(273 \text{ K})}{1 \text{ atm}}$$

$$= 22.4 \text{ L or } 0.0224 \text{ m}^3$$

EXAMPLE 19-8

How many grams of oxygen will occupy a volume of 1.6 m³ at a pressure of 200 kPa and a temperature of 27°C?

Solution
Solving for m in Eq. (19-12), we obtain

$$m = \frac{MPV}{RT} = \frac{(32 \text{ g/mol})(200,000 \text{ Pa})(1.6 \text{ m}^3)}{(8.314 \text{ J/mol} \cdot \text{K})(300 \text{ K})}$$

$$= 4110 \text{ g or } 4.11 \text{ kg}$$

19-6 LIQUEFACTION OF A GAS

We have defined an ideal gas as one whose thermal behavior is completely unaffected by cohesive forces or molecular volume. Such a gas compressed at a constant temperature will remain a gas no matter how great a pressure is supplied to it. In other words, it will obey Boyle's law at any temperature. The binding forces necessary for liquefaction are never present.

All real gases experience intermolecular forces. At rather low pressures and high temperatures, however, real gases behave much like an ideal gas. Boyle's law applies because the intermolecular forces under these conditions are practically negligible. A real gas at high temperatures can be compressed in a cylinder, as in Fig. 19-4, to relatively high pressures without liquefying. If the increase in pressure is plotted as a function of the volume, the curve A_1B_1 is obtained. Note the similarity between this curve and that for an ideal gas, as shown in Fig. 19-2.

If the same gas is compressed at a much lower temperature, it will begin to condense at a particular pressure and volume. Further compression will continue to liquefy the gas at essentially constant pressure until all the gas has condensed. At that point, a sharp rise in pressure occurs with a slight decrease in volume. The entire process is shown graphically as the curve A_2B_2 in Fig. 19-4.

Let us now begin with the high-temperature compression and perform the experiment at lower and lower constant temperatures. Eventually, a temperature will be reached at which the gas will just begin to liquefy under compression. The highest temperature at which this liquefaction occurs is called the **critical temperature.**

The **critical temperature** of a gas is that temperature above which the gas will not liquefy, regardless of the amount of pressure applied to it.

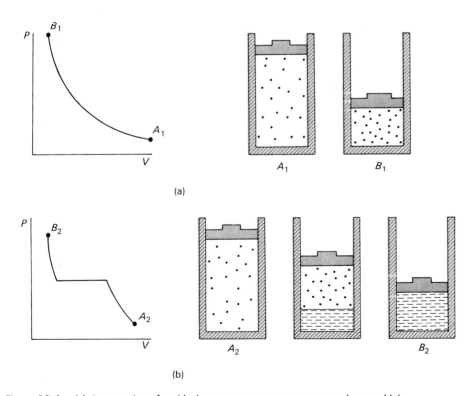

(a)

(b)

Figure 19-4 (a) Compression of an ideal gas at any temperature or a real gas at high temperature. (b) Liquefaction of a real gas when it is compressed at lower temperatures.

If any gas is to be liquefied, it must first be cooled below its critical temperature. Before this concept was understood, scientists attempted to liquefy oxygen by subjecting it to extreme pressures. Their attempts failed because the critical temperature of oxygen is −119°C. After cooling the gas below this temperature, it can be easily liquefied by compression.

19-7 VAPORIZATION

In Chap. 17 we discussed at length the process of vaporization in which a definite quantity of heat is required to change from the liquid phase to the vapor phase. There are three ways in which such a change may occur: (1) evaporation, (2) boiling, and (3) sublimation. During evaporation, vaporization occurs at the surface of a liquid as the more energetic molecules leave the surface. In the process of boiling, vaporization occurs within the body of the liquid. Sublimation occurs when a solid vaporizes without passing through the liquid phase. In each case, an amount of energy equal to the latent heat of vaporization or **sublimation** must be lost by the liquid or solid.

The molecular theory of matter assumes that a liquid consists of molecules crowded fairly close together. These molecules have an average kinetic energy that is related to the temperature of the liquid. Because of random collisions or vibratory motion, however, not all the molecules move at the same rate of speed; some move faster than others.

Because the molecules are so close together, the forces between them are relatively large. As a molecule approaches the surface of a liquid, as in Fig. 19-5, it experiences a resultant downward force. The net force results from the fact that there are no liquid molecules above the surface to offset the downward attraction of these below the surface. Only the *faster-moving* particles can approach the surface with sufficient energy to overcome the retarding forces. These molecules are said to *evaporate* because, on leaving the liquid, they become typical gas particles. They have not changed chemically; the only difference between a liquid and its vapor is the distance between molecules.

Since only the most energetic molecules are able to break away from the surface, the average kinetic energy of the molecules remaining in the liquid is reduced. Hence, *evaporation is a cooling process.* (If you place a few drops of alcohol on the back of your hand, you will feel a cooling sensation.) The rate of evaporation is affected by the temperature of the liquid, the number of molecules above the liquid (the pressure), the exposed surface area, and the extent of ventilation.

Figure 19-5 A molecule near the surface of a liquid experiences a net downward force. Only the more energetic molecules are able to overcome this force and leave the liquid.

19-8 VAPOR PRESSURE

A jar is partially filled with water, as shown in Fig. 19-6. The pressure exerted by the molecules above the surface of the water is measured by an open-tube mercury manometer. In Fig. 19-6a there are as many molecules of air inside the jar as are contained in an equal volume of air outside the jar. In other words, the pressure inside the jar is equal to 1 atm, as indicated by the equal levels of mercury in the manometer.

When a high-energy liquid molecule breaks away from the surface, it becomes a vapor molecule and mixes with the air molecules above the liquid. These vapor molecules collide with air molecules, other vapor molecules, and the walls of the jar. The additional vapor molecules cause a rise in pressure inside the jar, as indicated in Fig. 19-6b. The vapor molecules may also rebound back into the liquid, where they are held as liquid molecules. This process is called **condensation.** Eventually the rate of evaporation will equal the rate of condensation, and a condition of equilibrium will exist, as shown in Fig. 19-6c. Under these conditions, the space above the liquid is said to be *saturated.* The pressure exerted by the saturated vapor against the walls of the jar, over and above that exerted by the air molecules, is called the **saturated vapor pressure.** It is characteristic of the substance and the temperature but independent of the volume of the vapor.

> The saturated vapor pressure of a substance is the additional pressure exerted by vapor molecules on the substance and its surroundings under a condition of saturation.

Once a condition of saturation is obtained for a substance and its vapor at a particular temperature, the vapor pressure remains essentially constant. If the temperature is increased, the molecules in the liquid will acquire more energy and evaporation will occur more rapidly. The condition of equilibrium remains upset until once again the rate of condensation has caught up with the rate of evaporation. The saturated vapor pressure of a substance therefore increases with a rise in temperature.

The saturated-vapor-pressure curve for water is plotted in Fig. 19-7. Note that the vapor pressure increases rapidly with temperature. At room temperature

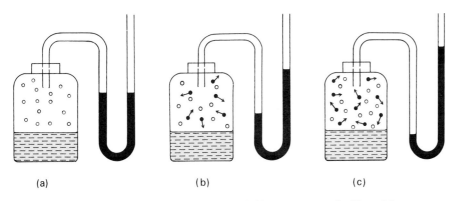

(a)   (b)   (c)

Figure 19-6 Measuring the vapor pressure of a liquid: (a) air pressure only, (b) partial vapor pressure, (c) saturated vapor pressure.

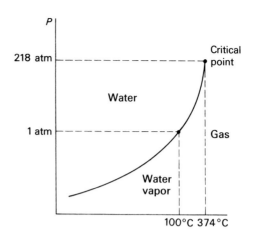

Figure 19-7 The vaporization curve for water. Any point on the curve represents a condition of pressure and temperature that allows water to boil. The curve ends abruptly at the critical temperature because water can exist only as a gas beyond that point.

(20°C), it is around 17.5 mm of mercury; at 50°C, it has increased to 92.5 mm; and at 100°C, it is equal to 760 mm, or 1 atm. The latter point is important in distinguishing between *evaporation* and *boiling.*

When a liquid boils, bubbles of its vapor can be seen rising toward the surface from within the liquid. The fact that these bubbles are stable and do not collapse indicates that the pressure inside the bubble is equal to the pressure outside the bubble. The pressure inside the bubble is the vapor pressure at that temperature; the pressure on the outside is the pressure at that depth in the liquid. Under this condition of equilibrium, vaporization occurs freely throughout the liquid, causing the liquid to become agitated.

> **Boiling is defined as vaporization within the body of a liquid when its vapor pressure equals the pressure in the liquid.**

If the pressure on the liquid surface is 1 atm, as it would be in an open container, the temperature at which boiling occurs is called the *normal boiling point* for that liquid. The normal boiling point for water is 100°C because that is the temperature at which the vapor pressure of water is 1 atm (760 mm of mercury). If the pressure on any liquid surface is lower than 1 atm, boiling will occur at a temperature lower than the normal boiling point. If the external pressure is greater than 1 atm, boiling will occur at a higher temperature.

19-9 TRIPLE POINT

We have discussed in detail the process of vaporization, and in Fig. 19-7 we constructed a vaporization curve for water. This curve is represented by the line *AB* in the general phase diagram of Fig. 19-8. Any point on this curve represents a temperature and pressure at which water and its vapor can coexist in equilibrium.

A similar curve can be plotted for the temperatures and pressures at which a substance in the solid phase can coexist with its liquid phase. Such a curve is called

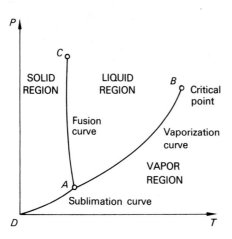

Figure 19-8 Triple-point phase diagram for water or other substance that expands on freezing.

a *fusion curve.* The fusion curve for water is represented by the line *AC* in the phase diagram. At any point on this curve, the rate at which ice is melting is equal to the rate at which water is freezing. Note that, as the pressure increases, the melting temperature (or freezing temperature) is lowered.

A third graph, called the *sublimation curve,* can be plotted to show the temperatures and pressures at which a solid may coexist with its vapor. The sublimation curve for water is represented by line *AD* of Fig. 19-8.

Let us now study the phase diagram for water more closely to illustrate the usefulness of such a graph for any substance. The coordinates of any point on the graph represent a particular pressure *P* and a particular temperature *T.* The volume must be considered constant for any thermal change indicated by the graph. For any point that falls in the fork between the vaporization and fusion curves, water will exist in its liquid phase. The vapor and the solid regions are also indicated on the diagram. The point *A,* at which all three curves intersect, is called the **triple point** for water. This is the temperature and pressure at which ice, liquid water, and water vapor coexist in equilibrium. Careful measurements have shown that the triple point for water is 0.01°C and 4.62 mm of mercury (Hg).

19-10 HUMIDITY

The air in our atmosphere consists largely of nitrogen and oxygen with small amounts of water vapor and other gases. It is often useful to describe the water-vapor content of the atmosphere.

> **The absolute humidity is defined as the mass of water per unit volume of air.**

For example, if 7 g of water vapor is contained in every cubic meter of air, the absolute humidity is 7 g/m³. Other units of absolute humidity are pounds per cubic foot and grains per cubic foot (7000 grains = 1 lb).

A more useful method of expressing the water-vapor content in air is to compare the actual vapor pressure at a particular temperature with the saturated vapor pressure at that temperature. When the atmosphere is holding all the water possible for a certain temperature, it is saturated. The addition of more vapor molecules will simply result in an equal amount of condensation.

The relative humidity is defined as the ratio of the actual vapor pressure in the air to the saturated vapor pressure at that temperature.

$$Relative\ humidity = \frac{actual\ vapor\ pressure}{saturated\ vapor\ pressure} \qquad \text{19-13}$$

The relative humidity is usually expressed as a percentage.

If the air in a room is not already saturated, it can be made so either by adding more water vapor to the air or by lowering the room temperature until the vapor already present is sufficient. The temperature to which the air must be cooled at constant pressure to produce saturation is called the **dew point.** Thus, if ice is placed in a glass of water, moisture will eventually collect on the outside walls of the glass when its temperature reaches the dew point. Given the temperature and the dew point, the relative humidity can be computed from saturated-vapor-pressure tables. Table 19-1 lists the saturated vapor pressure for water at various temperatures.

TABLE 19-1 Saturated Vapor Pressure for Water

Temperature °C	°F	Pressure mmHg	Temperature °C	°F	Pressure, mmHg
0	32	4.62	50	122	92.5
5	41	6.5	60	140	149.4
10	50	9.2	70	158	233.7
15	59	12.8	80	176	355.1
17	62.6	14.5	85	185	433.6
19	66.2	16.5	90	194	525.8
20	68	17.5	95	203	633.9
22	71.6	19.8	98	208.4	707.3
24	75.2	22.4	100	212	760.0
26	78.8	25.2	103	217.4	845.1
28	82.4	28.3	105	221	906.1
30	86	31.8	110	230	1074.6
35	95	42.2	120	248	1489.1
40	104	55.3	150	302	3570.5

EXAMPLE 19-9

On a clear day, the air temperature is 86°F, and the dew point is 50°F. What is the relative humidity?

Solution

The pressure of saturated vapor at 50°F is 9.2 mm, from Table 19-1. The pressure of saturated vapor at 86°F is 31.8 mm. From Eq. (19-13),

$$\text{Relative humidity} = \frac{9.2}{31.8} = 0.29$$

Thus, the relative humidity is 29 percent.

Summary and Review

Summary

The thermal properties of matter must be understood if we are to apply the many laws discussed in this chapter. The relationships among mass, temperature, volume, and pressure allow us to explain and predict the behavior of gases. The major concepts discussed in this chapter are summarized below.

- A useful form of the general gas law that does not involve the use of moles is written on the basis that PV/mT is constant. When a gas in state 1 changes to another state 2, we may write

$$\frac{P_1 V_1}{m_1 T_1} = \frac{P_2 V_2}{m_2 T_2}$$

P = pressure $\qquad$ V = volume
m = mass $\qquad$ T = absolute temperature

When one or more of the parameters m, P, T, or V is constant, that factor disappears from both sides of the above equation. *Boyle's law, Charles's law,* and *Gay-Lussac's law* are the following special cases:

$$P_1 V_1 = P_2 V_2 \qquad \frac{V_1}{T_1} = \frac{V_2}{T_2} \qquad \frac{P_1}{T_1} = \frac{P_2}{T_2}$$

- When applying the general gas law in any of its forms, it must be remembered that the pressure is *absolute pressure* and the temperature is *absolute temperature.*

$$\frac{\text{Absolute}}{\text{pressure}} = \frac{\text{gauge}}{\text{pressure}} + \frac{\text{atmospheric}}{\text{pressure}}$$

$$T_K = t_C + 273 \qquad T_R = t_F + 460$$

For example, a pressure measured in an auto tire is 30 lb/in.^{2} at 37°C. These values must be adjusted before substitution into the gas laws:

$P = 30$ lb/in.^{2} + 14.7 lb/in.^{2} $\qquad$ (absolute)
$T = 37 + 273 = 310$ K

- A more general form of the gas law is obtained by using the concepts of molecular mass M and the number of moles n for a gas. The number of molecules in 1 mol is Avogadro's number N_A.

$$N_A = \frac{N}{n}$$

$$N_A = 6.023 \times 10^{23} \text{ molecules/mol}$$

Avogadro's Number

The number of moles is found by dividing the mass of a gas (in grams) by its molecular mass M:

$$n = \frac{m}{M} \qquad \textit{Number of Moles}$$

One often wants to determine the mass, pressure, volume, or temperature of a gas in a single state. The ideal gas law uses the molar concept to arrive at a more specific equation:

$$PV = nRT \qquad R = 8.314 \text{ J/mol} \cdot \text{K}$$

It should be noted that use of the constant given above restricts the units of P, V, T, and n to those that are in the constant.

- The *relative humidity* can be computed from saturated-vapor-pressure tables according to the following definition:

$$\text{Relative humidity} = \frac{\textit{actual vapor pressure}}{\textit{saturated vapor pressure}}$$

Remember that the *actual* vapor pressure at a particular temperature is the same as the *saturated* vapor pressure for the dew-point temperature. Refer to Example 19-9.

Key Terms

1. ideal gas	7. boiling	13. vapor pressure
2. Boyle's law	8. sublimation	14. triple point
3. Charles's law	9. molecular mass	15. absolute humidity
4. atomic mass	10. Avogadro's number	16. dew point
5. mole	11. ideal gas law	17. Gay-Lussac's law
6. evaporation	12. critical temperature	

Review Questions

19-1. Distinguish between *state* and *phase*.

19-2. Explain Boyle's law in terms of the molecular theory of matter.

19-3. Explain Charles's law in terms of the molecular theory of matter.

19-4. Why must the absolute temperature be used in Charles's law?

19-5. A closed steel tank is filled with an ideal gas and heated. What happens to the (a) mass, (b) volume, (c) density, and (d) pressure of the enclosed gas?

19-6. Prove the accuracy of the following equations that involve the density ρ of an ideal gas.

$$\text{(a)} \quad \frac{P_1}{\rho_1 T_1} = \frac{P_2}{\rho_2 T_2}$$

$$\text{(b)} \quad \rho = \frac{PM}{RT}$$

19-7. Suppose we wish to express the pressure of an ideal gas in millimeters of mercury and the volume in cubic centimeters. Show that the universal gas constant will be equal to 6.23×10^4 mm · cm^3/mol · K.

19-8. A mole of any gas occupies 22.4 liters at STP. Can we also say that the same mass of any gas will occupy the same volume? Explain.

19-9. Distinguish between evaporation, boiling, and sublimation.

19-10. From your experience, would you expect alcohol to have a higher vapor pressure than water? Why?

19-11. Explain the principle of operation for the pressure cooker and the vacuum pan in cooking.

19-12. Can a solid have a vapor pressure? Explain.

19-13. If evaporation is a cooling process, is condensation a heating process? Explain.

19-14. Explain the cooling effects of evaporation in terms of the latent heat of vaporization.

19-15. Distinguish between a vapor and a gas by discussing critical temperature.

19-16. Will it take longer to cook an egg by boiling it in water on Mt. Everest or at the seashore? Why?

19-17. Water is brought to a vigorous boil in an open flask. When the flask is removed from the flame and tightly stoppered, the boiling stops. Why? The stoppered flask is then inverted and held under a stream of cold running water. The boiling begins again. Explain. As soon as the flask is removed from the water, boiling stops. If the flask is cooled, boiling begins again. How long can the process of making the water boil by cooling be continued?

19-18. On a cool day, the relative humidity inside a house is the same as the relative humidity outside the house. Are the dew points necessarily the same? Explain.

19-19. Explain what is meant by *critical pressure*.

19-20. Is it possible for ice to exist in equilibrium with boiling water? Explain.

19-21. The formation of moisture on the walls and windows inside a home can cause considerable damage. What causes this moisture? Discuss several ways of reducing or preventing the formation of moisture.

19-22. Discuss the formation of fog and clouds. Why are fog conditions usually worse in the fall and early spring?

Summary and Review

Problems

Section 19-3—General Gas Laws: Initial and Final States

19-1. An ideal gas occupies a volume of 4.00 m³ at 200 kPa absolute pressure. What will be the new pressure if the gas is slowly compressed to 2.00 m³ at constant temperature? **Ans. 400 kPa**

19-2. The absolute pressure of a sample of ideal gas is 300 kPa at a volume of 2.6 m³. If the pressure decreases to 101 kPa at constant temperature, what is the new volume?

19-3. Two hundred cubic centimeters of an ideal gas at 20°C expands to a volume of 212 cm³ at constant pressure. What is the final temperature? **Ans. 37.6°C**

19-4. The temperature of a gas sample decreases from 55 to 25°C at constant pressure. If the initial volume was 400 mL, what is the final volume?

19-5. A steel cylinder contains an ideal gas at 27°C. The *gauge* pressure is 140 kPa. If the temperature of the container increases to 79°C, what is the new *gauge* pressure? **Ans. 182 kPa**

19-6. The absolute pressure of a sample of gas initially at 300 K doubles as the volume remains constant. What is the new temperature?

19-7. A steel cylinder contains 2.00 kg of an ideal gas. Overnight the temperature and volume remain constant, but the absolute pressure decreases from 500 to 450 kPa. How many grams of the gas leaked out overnight? **Ans. 200 g**

19-8. Five liters of a gas at 25°C has an absolute pressure of 200 kPa. If the absolute pressure reduces to 120 kPa and the temperature increases to 60°C, what is the final volume?

19-9. An air compressor takes in 2 m³ of air at 20°C and one atmosphere (101.3 kPa) pressure. If the compressor discharges into a 0.3-m³ tank at an absolute pressure of 1500 kPa, what is the temperature of the discharged air? **Ans. 651 K**

19-10. A 6-L tank holds a sample of gas under an absolute pressure of 600 kPa and a temperature of 57°C. What will be the new pressure if the same sample of gas is placed into a 3-L container at 7°C?

19-11. If 0.8 L of a gas at 10°C is heated to 90°C at constant pressure, what will the new volume be? **Ans. 1.03 L**

19-12. The inside of an automobile tire is under a gauge pressure of 30 lb/in.² at 4°C. After several hours, the inside air temperature rises to 50°C. Assuming constant volume, what is the new gauge pressure?

19-13. A 2-L sample of gas has an absolute pressure of 300 kPa at 300 K. If both pressure and volume experience a two-fold increase, what is the final temperature? **Ans. 1200 K**

Section 19-4—Molecular Mass and the Mole

19-14. How many moles are contained in 600 g of air? (M = 29 g/mol.)

19-15. How many moles of gas are there in 400 g of nitrogen gas? (M = 28 g/mol.) How many molecules are in this sample? **Ans. 14.3 mol, 8.60 × 10²⁴ molecules**

19-16. What is the mass of a 4-mol sample of air? (M = 29 g/mol.)

19-17. How many grams of hydrogen gas (M = 2 g/mol) are there in 3.0 moles of hydrogen? How many grams of air (M = 29 g/mol) are there in 3.0 moles of air? **Ans. 6 g, 87 g**

*19-18. How many molecules of hydrogen gas (M = 2 g/mol) are needed to have the same mass as 4 g of oxygen (M = 32 g/mol)? How many moles are in each sample?

*19-19. What is the mass of one molecule of oxygen? (M = 32 g/mol). **Ans. 5.31 × 10⁻²⁶ kg**

*19-20. The molecular mass of CO_2 is 44 g/mol. What is the mass of a single molecule of CO_2?

Section 19-5—The Ideal Gas Law

19-21. Three moles of an ideal gas have a volume of 0.026 m^3 and a pressure of 300 kPa. What is the temperature of the gas in degrees Celsius? **Ans. 39.7°C**

19-22. A 16-L tank contains 200 g of air ($M = 29$ g/mol) at 27°C. What is the absolute pressure of this sample?

19-23. How many kilograms of nitrogen gas ($M = 28$ g/mol) will occupy a volume of 2000 L at an absolute pressure of 202 kPa and a temperature of 80°C? **Ans. 3.85 kg**

19-24. What volume is occupied by 8 g of nitrogen gas ($M = 28$ g/mol) at standard temperature and pressure (STP)?

19-25. A 2-L flask contains 2×10^{23} molecules of air ($M = 29$ g/mol) at 300 K. What is the absolute gas pressure? **Ans. 414 kPa**

19-26. A 2-m^3 tank holds nitrogen gas ($M = 28$ g/mol) under a gauge pressure of 500 kPa. If the temperature is 27°C, what is the mass of gas in the tank?

19-27. How many moles of gas are contained in a volume of 2000 cm^3 at conditions of standard temperature and pressure (STP)? **Ans. 0.0893 mol**

19-28. A 0.30-cm^3 cylinder contains 0.27 g of water vapor ($M = 18$ g/mol) at 340°C. What is its absolute pressure, assuming that the water vapor is an ideal gas?

Section 19-10—Humidity

19-29. If the air temperature is 20°C and the dew point is 12°C, what is the relative humidity? **Ans. 60.8 percent**

19-30. The dew point is 20°C. What is the relative humidity when the air temperature is 24°C?

19-31. The relative humidity is 77 percent when the air temperature is 28°C. What is the dew point? **Ans. 23.5°C**

19-32. What is the pressure of water vapor in the air on a day when the temperature is 86°F and the relative humidity is 80 percent?

19-33. The air temperature in a room during the winter is 28°C. What is the relative humidity if moisture first starts forming on a window when the temperature of its surface is 20°C? **Ans. 61.8 percent**

Challenge Problems

19-34. A sample of gas occupies 12 L at 7°C and at an absolute pressure of 102 kPa. Find its temperature when the volume reduces to 10 L and the pressure increases to 230 kPa.

19-35. A tractor tire contains 2.8 ft^3 of air at a gauge pressure of 70 lb/in.^{2}. What volume of air at 1 atm of pressure is required to fill this tire if there is no change in temperature or volume? **Ans. 16.1 ft^3**

19-36. A 3-L container is filled with 0.230 mol of an ideal gas at 300 K. What is the pressure of the gas? How many molecules are in this sample of gas?

19-37. How many moles of helium gas ($M = 4$ g/mol) are there in a 6-L tank when the pressure is 2×10^5 Pa and the temperature is 27°C? What is the mass of the helium? **Ans. 0.481 mol, 1.92 g**

19-38. How many grams of air ($M = 29$ g/mol) must be pumped into an automobile tire if it is to have a gauge pressure of 31 lb/in.^{2}? Assume that the volume of the tire is 5000 cm^3 and its temperature is 27°C.

19-39. The air temperature inside a car is 26°C. The dew point is 24°C. What is the relative humidity inside the car? **Ans. 88.9 percent**

19-40. The lens in a sensitive camera is clear when the room temperature is 71.6°F and the relative humidity is 88 percent. What is the lowest temperature of the lens if it is not to become foggy from moisture?

*19-41. What is the density of oxygen gas ($M = 32$ g/mol) at a temperature of 23°C and atmospheric pressure? **Ans. 1.32 kg/m^3**

*19-42. A 5000-cm^3 tank is filled with carbon dioxide ($M = 44$ g/mol) at 300 K and 1 atm of

Summary and Review

pressure. How many grams of CO_2 can be added to the tank if the maximum absolute pressure is 60 atm and there is no change in temperature?

*19-43. The density of an unknown gas at standard temperature and pressure (STP) is 1.25 kg/m³. What is the density of this gas at 18 atm and 60°C? **Ans. 18.4 kg/m³**

Critical Thinking Questions

*19-44. A tank with a capacity of 14 L contains helium gas at 24°C under a gauge pressure of 2700 kPa. (a) What will be the volume of a balloon filled with this gas if the helium expands to an internal absolute pressure of 1 atm and the temperature drops to −35°C? (b) Now suppose the system returns to its original temperature (24°C). What is the final volume of the balloon? **Ans. (a) 310 L, (b) 387 L**

*19-45. A steel tank is filled with oxygen. One evening, when the temperature is 27°C, the gauge at the top of the tank indicates a pressure of 400 kPa. During the night a leak develops in the tank. The next morning it is noticed that the gauge pressure is only 300 kPa and that the temperature is 15°C. What percentage of the original gas remains in the tank?

*19-46. A 2-L flask is filled with nitrogen ($M = 28$ g/mol) at 27°C and 1 atm of absolute pressure. A stopcock at the top of the flask is opened to the air, and the

system is heated to a temperature of 127°C. The stopcock is then closed, and the system is allowed to return to 27°C. What mass of nitrogen is in the flask? What is the final pressure? **Ans. 1.71 g, 0.750 atm**

*19-47. What is the volume of 8 g of sulfur dioxide ($M = 64$ g/mol) if it has an absolute pressure of 10 atm and a temperature of 300 K? If 10^{20} molecules leak from this volume every second, how long will it take to reduce the pressure by one-half?

*19-48. A flask contains 2 g of helium ($M = 4$ g/mol) at 57°C and 12 atm absolute pressure. The temperature then decreases to 17°C and the pressure falls to 7 atm. How many grams of helium have leaked out of the container? **Ans. 1.13 L, 0.672 g**

*19-49. What must be the temperature of the air in a hot-air balloon in order that the mass of the air is 0.97 times that of an equal volume of air at a temperature of 27°C?

Thermodynamics

OBJECTIVES

After completing this chapter, you should be able to:

1. Demonstrate by definition and examples your understanding of the *first* and *second laws of thermodynamics*.
2. Define and give illustrated examples of *adiabatic, isochoric,* and *isothermal* processes.
3. Write and apply a relationship for determining the *ideal efficiency* of a heat engine.
4. Define the *coefficient of performance* for a refrigerator and solve refrigeration problems similar to those discussed in the text.

Thermodynamics treats the transformation of heat energy into mechanical energy and the reverse process, the conversion of work into heat. Since almost all the energy available from raw materials is liberated in the form of heat, it is easy to see why thermodynamics plays such an important role in science and technology.

In this chapter, we shall study two basic laws that must be obeyed when heat energy is used to accomplish work. The first law is simply a restatement of the principle of conservation of energy. The second law places restrictions on the efficient use of the available energy.

20-1 HEAT AND WORK

The equivalence of heat and work as two forms of energy has been clearly established. Rumford destroyed the caloric theory of heat by showing that it is possible to remove heat indefinitely from a system as long as external work is supplied. Joule then sealed the case by demonstrating the mechanical equivalent of heat.

Work, like heat, involves a transfer of energy, but there is an important distinction between the two terms. In mechanics, we define **work** as a scalar quantity, equal in magnitude to the product of a force and a displacement. Temperature plays no role in this definition. Heat, on the other hand, is energy that flows from one body to another because of a difference in temperature. A temperature difference is a necessary condition for the transfer of heat. Displacement is the necessary condition for the performance of work.

The important point in this discussion is to recognize that both heat and work represent changes that occur in a given process. Usually these changes are accompanied by a change in internal energy. Consider the two situations illustrated in

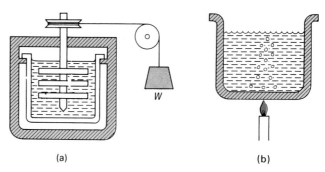

Figure 20-1 Increasing the internal energy of a system (a) by the performance of work and (b) by supplying heat to the system.

**homearts.com/helpers/
calculators/burncal.htm**
For an online calculator of
the number of calories you
burn doing exercise, see this
Web site (but do not type
www.).

Fig. 20-1. In Fig. 20-1a the internal energy of the water is increased by the performance of mechanical work. In Fig. 20-1b the internal energy of the water is increased through the flow of heat.

20-2 THE INTERNAL ENERGY FUNCTION

A system is said to be in *thermodynamic equilibrium* if there is no resultant force on the system and if the temperature of the system is the same as its surroundings. This condition requires that no work be done on or by the system and that there be no exchange of heat between the system and its surroundings. Under these conditions, the system has a definite internal energy U. Its *thermodynamic state* can be described by three coordinates: (1) its pressure P, (2) its volume V, and (3) its temperature T. Whenever energy is absorbed or released by such a system in the form of either heat or work, it will reach a new state of equilibrium in such a way that energy is conserved.

In Fig. 20-2, let us consider a general thermodynamic process in which a system is caused to change from an equilibrium state 1 to an equilibrium state 2. In Fig. 20-2a the system is in thermodynamic equilibrium, with an initial internal energy U_1 and thermodynamic coordinates (P_1, V_1, T_1). In Fig. 20-2b the system reacts with its surroundings. Heat Q may be absorbed by the system and/or released to its environment. The heat considered positive for heat input and negative for heat output. The net heat *absorbed* by the system is represented by ΔQ. Work W may be done *by* the system and/or *on* the system. Output work is considered positive, and input work is considered negative. Thus, ΔW represents the net work done *by* the system (output work). In Fig. 20-2c the system has reached its final state 2 and is again in equilibrium, with a final internal energy U_2. Its new thermodynamic coordinates are (P_2, V_2, T_2).

If energy is to be conserved, the change in internal energy

$$\Delta U = U_2 - U_1$$

must represent the difference between the net heat ΔQ absorbed by the system and the net work ΔW done by the system on its surroundings.

$$\Delta U = \Delta Q - \Delta W \qquad \text{20-1}$$

Hint for Success

When writing a résumé,
be consistent with punc-
tuation and capitalization
of headings and descrip-
tions. This makes your
résumé visually appealing
and shows that you have
an eye for detail.

(a) Initial state of system (P_1, V_1, T_1)

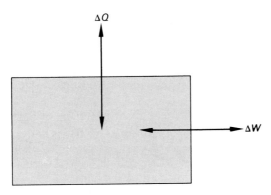

(b) System undergoing thermodynamic process

(c) Final state of system (P_2, V_2, T_2)

Figure 20-2 A schematic diagram of a thermodynamic process.

Thus, the change in internal energy is uniquely defined in terms of the measurable quantities heat and work. Equation (20-1) states the existence of an ***internal energy function*** that is determined by the thermodynamic coordinates of a system. Its value at the final state minus its value at the initial state represents the change in energy of the system.

20-3 THE FIRST LAW OF THERMODYNAMICS

The ***first law of thermodynamics*** is simply a restatement of the principle of *conservation of energy:*

Energy cannot be created or destroyed but can change from one form to another.

Applying this law to a thermodynamic process, we note from Eq. (20-1) that

$$\Delta Q = \Delta W + \Delta U$$

20-2

This equation represents a mathematical statement of the *first law of thermodynamics,* which can be stated as follows:

In any thermodynamic process, the net heat absorbed by a system is equal to the sum of the thermal equivalent of the work done by the system and the change in the internal energy of the system.

EXAMPLE 20-1

In a certain process, a system absorbs 400 cal of heat and at the same time does 80 J of work on its surroundings. What is the increase in the internal energy of the system?

Solution

Applying the first law, we have

$$\Delta U = \Delta Q - \Delta W$$

$$= 400 \text{ cal} - 80 \text{ J} \, \frac{1 \text{ cal}}{4.186 \text{ J}}$$

$$= 400 \text{ cal} - 19.1 \text{ cal} = 380.9 \text{ cal}$$

Thus, the 400 cal of input thermal energy is used to perform 19.1 cal of work while the internal energy of the system is increased by 380.9 cal. Energy is conserved.

20-4 THE *P-V* DIAGRAM

Many thermodynamic processes involve energy changes that occur to gases enclosed in cylinders. At this point, it will be useful to derive an expression for computing the work done by an expanding gas. We consider a system consisting of a gas in a cylinder equipped with a movable piston, as shown in Fig. 20-3a. The piston has a cross-sectional area A and rests on a column of gas under a pressure P. Heat can flow in or out of the gas through the cylinder walls. Work can be done on or by the gas by pushing the piston down or by allowing it to expand upward.

Let us first consider the work done by the gas when it expands at a constant pressure P. The force exerted by the gas on the piston will be equal to PA. If the piston moves upward through a distance Δx, the work ΔW of this force will be given by

$$\Delta W = F \, \Delta x = PA \, \Delta x$$

F = PA

Δx

P = F/A

(a)

Pressure

P

Work = P ΔV

Volume

V₁   V₂

(b)

Figure 20-3 (a) Calculating the work done by a gas expanding at constant pressure. (b) The work is equal to the area under the curve on a *P-V* diagram.

But $A \Delta x = \Delta V$, where ΔV represents the change in volume of the gas. Hence,

$$\Delta W = P \Delta V \qquad\qquad \textbf{20-3}$$

In other words, the work done by a gas expanding at constant pressure is equal to the product of the pressure and the change in volume of the gas.

The process can be shown graphically by plotting the increase in volume as a function of pressure (Fig. 20-3b). This representation, called a **P-V diagram,** is extremely useful in thermodynamics. In the above example, the pressure was constant, so that the graph is a straight line. Note that the area under the line, indicated by the shaded portion in the figure, is

$$\text{Area} = P(V_2 - V_1) = P \Delta V$$

which is equal to the work ΔW, from Eq. (20-3). This leads us to an important principle:

> **When a thermodynamic process involves changes in volume and/or pressure, the work done by the system is equal to the area under the curve on a P-V diagram.**

In general, the pressure will not be constant during a piston displacement. For example, in the power stroke of a gasoline engine, the fluid is ignited under high pressure, and the pressure decreases as the piston is displaced downward. The *P-V* diagram in this case is a sloping curve, as shown in Fig. 20-4a. The volume increases from V_1 to V_2 while the pressure decreases from P_1 to P_2. To compute the work in such a process, we must resort to a graphical analysis and measure the area under the *P-V* curve.

That the area under the curve is equal to the work when the pressure is not constant can be demonstrated in Fig. 20-4. The area of the narrow shaded rec-

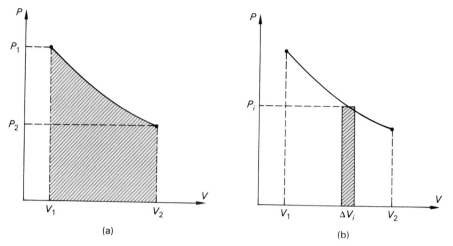

Figure 20-4 Calculating the work done by a gas expanding under a varying pressure.

tangle represents the work done by the gas in expanding by an increment ΔV_i under a constant pressure P_i. If the area under the entire curve is split up into many of these rectangles, we can sum all the products $P_i \Delta V_i$ to obtain the total work. Thus, the total work is simply the area under the P-V diagram between the points V_1 and V_2 on the volume axis.

20-5 THE GENERAL CASE FOR THE FIRST LAW

The first law of thermodynamics stipulates that energy must be conserved in any thermodynamic process. In the mathematical formulation

$$\Delta Q = \Delta W + \Delta U$$

there are three quantities that may undergo changes. The most general process is one in which all three quantities are involved. For example, the fluid in Fig. 20-5 is expanded while in contact with a hot flame. Considering the gas as a system, there is a net heat transfer ΔQ imparted to the gas. This energy is used in two ways: (1) The internal energy ΔU of the gas is increased by a portion of the input thermal energy, and (2) the gas does an amount of work ΔW on the piston that is equivalent to the remainder of the available energy.

Special cases of the first law arise when one or more of the three quantities—ΔQ, ΔW, or ΔU—do not undergo change. In these instances, the first law is considerably simplified. In the following sections we consider several of these special processes.

20-6 ADIABATIC PROCESSES

Suppose a system is completely isolated from its surroundings so that there can be no exchange of thermal energy Q. Any process that occurs completely inside such an isolated chamber is called an *adiabatic process,* and the system is said to be surrounded by *adiabatic* walls.

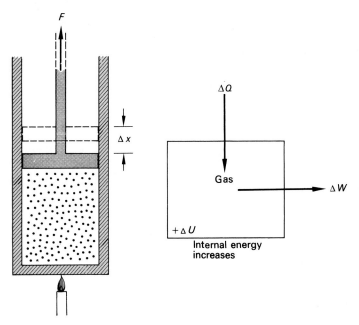

Figure 20-5 Part of the energy ΔQ supplied to the gas by the flame results in external work ΔW. The remainder increases the internal energy ΔU of the gas.

An adiabatic process is one in which there is no exchange of thermal energy ΔQ between a system and its surroundings.

Applying the first law to a process in which $\Delta Q = 0$, we obtain

$$\Delta W = -\Delta U \qquad\qquad \textit{Adiabatic} \quad \textbf{20-4}$$

Equation (20-4) tells us that, in an adiabatic process, the work is done at the *expense* of internal energy. The decrease in thermal energy is usually accompanied by a decrease in temperature.

As an example of an adiabatic process, consider Fig. 20-6, in which a piston is lifted by an expanding gas. If the walls of the cylinder are insulated and the expansion occurs rapidly, the process will be approximately adiabatic. As the gas expands, it does work on the piston but loses internal energy and experiences a drop in temperature. If the process is reversed by forcing the piston back down, work is done *on* the gas ($-\Delta W$), and there will be an increase in internal energy (ΔU) such that

$$-\Delta W = +\Delta U$$

In this instance the temperature will rise.

Another example of an adiabatic process that is useful in industrial refrigeration is referred to as a ***throttling process.***

A throttling process is one in which a fluid at high pressure seeps adiabatically through a porous wall or narrow opening into a region of low pressure.

Consider a gas forced by a pump to circulate through the apparatus in Fig. 20-7. Gas from the high-pressure side of the pump is forced through the narrow

Figure 20-6 In an adiabatic process, there is no transfer of heat, and work is done at the *expense* of internal energy.

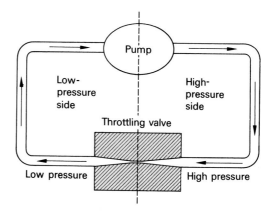

Figure 20-7 The throttling process.

constriction, called the *throttling valve,* to the low-pressure side. The valve is heavily insulated, so that the process is adiabatic and $\Delta Q = 0$. According to the first law, $\Delta W = -\Delta U$, and the net work done by the gas in passing through the valve is accomplished at the expense of internal energy. In refrigeration, a liquid coolant undergoes a drop in temperature and partial vaporization as a result of the throttling process.

20-7 ISOCHORIC PROCESSES

Another special case for the first law occurs when there is no work done, either *by* the system or *on* the system. This type of process is referred to as an **isochoric process.** It is also referred to as an **isovolumic process** since there can be no change in volume without the performance of work.

Figure 20-8 In an isochoric process, the volume of the system (water and vapor) remains constant.

An isochoric process is one in which the volume of the system remains constant.

Applying the first law to a process in which $\Delta W = 0$, we obtain

$$\Delta Q = \Delta U \qquad \text{Isochoric} \quad \textbf{20-5}$$

Therefore, in an isochoric process all the thermal energy absorbed by a system goes to increase its internal energy. In this instance there is usually a rise in the temperature of the system.

An isochoric process occurs when water is heated in a container of fixed volume, as shown in Fig. 20-8. As heat is supplied, the increase in internal energy results in a rise in the temperature of the water until it begins to boil. Further increases in internal energy go into the process of vaporization. The volume of the system, consisting of the water and its vapor, remains constant, however, and no external work is done.

When the flame is removed, the process is reversed as heat leaves the system through the bottom of the cylinder. The water vapor will condense, and the temperature of the resulting water will eventually drop to room temperature. This process represents a loss of heat and a corresponding decrease in internal energy, but, once again, no work is done.

20-8 ISOTHERMAL PROCESSES

It is possible for the pressure and volume of a gas to vary without a change in temperature. In Chap. 19 we introduced Boyle's law to describe volume and pressure changes during such a process. A gas can be compressed in a cylinder so slowly that it will essentially remain in thermal equilibrium with its surroundings. The pressure increases as the volume decreases, but the temperature is essentially constant.

An isothermal process is one in which the temperature of the system remains constant.

If there is no change of phase, a constant temperature indicates that there is no change in the internal energy of the system. Applying the first law to a process in which $\Delta U = 0$, we obtain

$$\Delta Q = \Delta W \qquad \qquad Isothermal \quad \textbf{20-6}$$

Thus, in an *isothermal process* all the energy absorbed by a system is converted into output work.

20-9 THE SECOND LAW OF THERMODYNAMICS

When we rub our hands together vigorously, the work done against friction increases the internal energy and causes a rise in temperature. The surrounding air forms a large reservoir at a lower temperature, and the heat energy is transferred to the air without changing its temperature appreciably. When we stop rubbing, our hands return to the same state as before. According to the first law of thermodynamics, mechanical energy has been transformed into heat with 100 percent efficiency.

$$\Delta W = \Delta Q$$

Such a transformation can be continued indefinitely as long as work is supplied.

Let us now consider the reverse process. Is it possible to convert heat energy into work with 100 percent efficiency? In the above example, is it possible to capture all the heat transferred to the air and return it to our hands, causing them to rub together indefinitely of their own accord? On a cold winter day, this process would be a boon to hunters with cold hands. Unfortunately, such a process cannot occur, even though it does not violate the first law. Neither is it possible to retrieve all the heat lost in braking a car to start the wheels rolling again.

We shall see that the conversion of heat energy into mechanical work is a losing process. The first law of thermodynamics tells us that we cannot win in such an experiment. In other words, it is impossible to get more work out of a system than the heat put into the system. It does not, however, preclude us from breaking even. Clearly, we need another rule stating that the 100 percent conversion of heat energy into useful work is not possible. This rule forms the basis for the *second law of thermodynamics.*

> **The second law of thermodynamics** **It is impossible to construct an engine that, operating continuously, produces no effect other than the extraction of heat from a reservoir and the performance of an equivalent amount of work.**

To give more insight and application to this principle, suppose we study the operation and efficiency of heat engines. A particular system might be a gasoline engine, a jet engine, a steam engine, or even the human body. The operation of a heat engine is best described by a diagram similar to that shown in Fig. 20-9. During the operating of such a general engine, three processes occur:

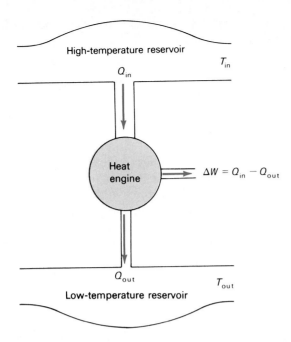

Figure 20-9 A schematic diagram for a heat engine.

1. A quantity of heat Q_{in} is supplied to the engine from a reservoir at a high temperature T_{in}.
2. Mechanical work W_{out} is done by the engine through the use of a portion of the heat input.
3. A quantity of heat Q_{out} is released to a reservoir at a low temperature T_{out}.

Since the system is returned periodically to its initial state, the net change in internal energy is zero. Thus, the first law tells us that

Work output = heat input − heat output

$$W_{out} = Q_{in} - Q_{out}$$ 20-7

The efficiency of a **heat engine** is defined as the ratio of the useful work done by the engine to the heat put into the engine, and it is usually expressed as a percentage.

$$\text{Efficiency} = \frac{\text{work output}}{\text{heat input}}$$

$$e = \frac{Q_{in} - Q_{out}}{Q_{in}}$$ 20-8

For example, an engine that is 25 percent efficient ($e = 0.25$) might absorb 1000 Btu, perform 250 Btu of work, and reject 750 Btu as wasted heat. A 100 percent efficient engine is one in which all the input heat is converted to useful work. In this case, no heat would be rejected to the environment ($Q_{out} = 0$). Although such a process would conserve energy, it violates the second law of thermodynamics. The most efficient engine is the one that rejects the *least* possible heat to the environment.

20-10 THE CARNOT CYCLE

All heat engines are subject to many practical difficulties. Friction and the loss of heat through conduction and radiation prevent actual engines from obtaining their maximum efficiency. An idealized engine free of such problems was suggested by Sadi Carnot in 1824. The **Carnot engine** has the maximum possible efficiency for an engine that absorbs heat from a high-temperature reservoir, performs external work, and deposits heat to a low-temperature reservoir. The effectiveness of a given engine can therefore be determined by comparing it with the Carnot engine operating between the same temperatures.

The **Carnot cycle** is illustrated in Fig. 20-10. A gas contained in a cylinder equipped with a movable piston is placed in contact with a reservoir at a high temperature T_{in}. A quantity of heat Q_{in} is absorbed by the gas, which expands isothermally as the pressure decreases. The first stage of a Carnot cycle is shown graphically as the curve AB in the P-V diagram (Fig. 20-11). The cylinder is next placed on an insulated stand, where it continues to expand adiabatically as the pressure drops to its lowest value. This stage is represented graphically by the curve BC. In the third stage the cylinder is removed from the insulated pad and placed on a reservoir at a low temperature T_{out}. A quantity of heat Q_{out} is exhausted from the gas as it is compressed isothermally from point C to point D in the P-V diagram. Finally, the cylinder is again placed on the insulated pad, where it is compressed adiabatically to its original state along the path DA. The engine does external work during the expansion processes and returns to its initial state during the compression processes.

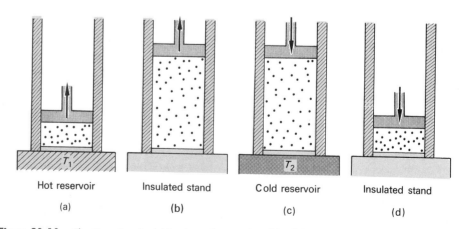

Hot reservoir   Insulated stand   Cold reservoir   Insulated stand

(a)   (b)   (c)   (d)

Figure 20-10 The Carnot cycle: (a) isothermal expansion, (b) adiabatic expansion, (c) isothermal compression, and (d) adiabatic compression.

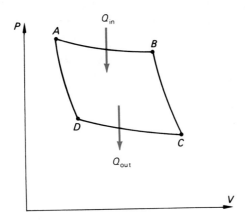

Figure 20-11 A *P-V* diagram of the ideal Carnot cycle.

20-11 THE EFFICIENCY OF AN IDEAL ENGINE

The efficiency of a real engine is difficult to predict from Eq. (20-8) because the quantities Q_{in} and Q_{out} are difficult to calculate. Frictional and heat losses through the cylinder walls and around the piston, incomplete burning of the fuel, and even the physical properties of different fuels all frustrate our attempts to measure the efficiency of such engines. We can image an *ideal engine*, however, one that is not restricted by these practical difficulties. The efficiency of such an engine depends only on the quantities of heat absorbed and rejected between two well-defined heat reservoirs. It does not depend on the thermal properties of the working fuel. In other words, regardless of the internal changes in pressure, volume, length, or other factors, all ideal engines have the same efficiency when they are operating between the same two temperatures (T_{in} and T_{out}).

An ideal engine is one that has the highest possible efficiency for the temperature limits within which it operates.

If we can define the efficiency of an engine in terms of input and output temperatures instead of in terms of the input and output of heat, we will have a more useful formula. For an ideal engine it can be shown that the ratio of Q_{in}/Q_{out} is the same as the ratio of T_{in}/T_{out}. The actual proof is beyond the scope of this text. The efficiency of an ideal engine can, therefore, be expressed as a function of the absolute temperatures of the input and output reservoirs. Equation (20-8), for an ideal engine, becomes

$$e = \frac{T_{in} - T_{out}}{T_{in}} \qquad \text{20-9}$$

It can be shown that no engine operating between the same two temperatures can be more efficient than would be indicated by Eq. (20-9). This ideal efficiency thus represents an upper limit to the efficiency of any practical engine. The greater the difference in temperature between two reservoirs, the greater the efficiency of any engine.

CONCEPTUAL EXAMPLE 20-2

(a) What is the efficiency of an ideal engine operating between two heat reservoirs at 400 and 300 K? (b) How much work is done by the engine in one complete cycle if 800 cal of heat is absorbed from the high-temperature reservoir? (c) How much heat is delivered to the low-temperature reservoir?

Solution (a)

The ideal efficiency is found from Eq. (20-9).

$$e = \frac{T_{in} - T_{out}}{T_{in}} = \frac{400 \text{ K} - 300 \text{ K}}{400 \text{ K}} = 0.25$$

Thus, the ideal efficiency is 25 percent.

Solution (b)

The efficiency is the ratio of W_{out}/Q_{in}, so that

$$\frac{W_{out}}{Q_{in}} = 0.25 \qquad \text{or} \qquad W_{out} = (0.25)Q_{in}$$

$$W_{out} = (0.25)(800 \text{ cal}) = 200 \text{ cal}$$

A 25 percent efficient engine delivers one-fourth of the input heat to useful work. The rest must be lost (Q_{out}).

Solution (c)

The first law of thermodynamics requires that

$$W_{out} = Q_{in} - Q_{out}$$

Solving for Q_{out} we obtain

$$Q_{out} = Q_{in} - W_{out} = 800 \text{ cal} - 200 \text{ cal} = 600 \text{ cal}$$

The work output is usually expressed in joules. Conversion to this unit gives

$$W_{out} = (200 \text{ cal})(4.186 \text{ J/cal}) = 837 \text{ J}$$

20-12 INTERNAL COMBUSTION ENGINES

An internal combustion engine generates the input heat within the engine itself. The most common engine of this variety is the four-stroke gasoline engine, in which a mixture of gasoline and air is ignited by a spark plug in each cylinder. The thermal energy released is converted into useful work by the pressure exerted on a piston by the expanding gases. The four-stroke process is illustrated in Fig. 20-12. During the *intake stroke* (Fig. 20-12a), a mixture of air and gasoline vapor enters the cylinder through the intake valve. Both valves are closed during the *compression stroke* (Fig. 20-12b), as the piston moves upward, causing a rise in pressure. Just before the piston reaches the top, the mixture is ignited, causing a sharp increase in temperature and pressure. In the *power stroke* (Fig. 20-12c) the expand-

(a) Intake stroke   (b) Compression stroke   (c) Power stroke   (d) Exhaust stroke

Figure 20-12 The four-stroke gasoline engine: (a) intake stroke, (b) compression stroke, (c) power stroke, and (d) exhaust stroke.

ing gases force the piston downward, performing external work. The fourth stroke (Fig. 20-12d) pushes the burned gases out of the cylinder through the exhaust valve. The entire cycle is then repeated for as long as the combustible fuel is supplied to the cylinder.

The ideal cycle used by the engineer to perfect the gasoline engine is shown in Fig. 20-13. It is named the *Otto cycle* after its inventor. The compression stroke is represented by the curve *ab*. The pressure increases adiabatically as the volume is reduced. At point *b* the mixture is ignited, supplying a quantity of heat Q_{in} to the system. This causes the sharp rise in pressure indicated by the line *bc*. In the power stroke (*cd*), the gases expand adiabatically, performing external work. The system then cools at constant volume to point *a*, releasing a quantity of heat Q_{out}. The

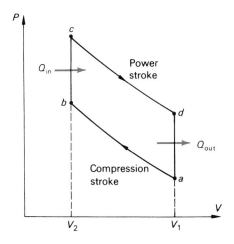

Figure 20-13 The Otto cycle for a four-stroke gasoline engine.

burned gases are exhausted on the next upward stroke, and more fuel is drawn in on the next downward stroke. Then the cycle begins all over again. The volume ratio V_1/V_2, as indicated in the P-V diagram, is called the *compression ratio* and is about 8 for most automobile engines.

The efficiency of the ideal Otto cycle can be shown in Eq. (20-10):

$$e = 1 - \frac{1}{(V_1/V_2)^{\gamma-1}}$$ 20-10

where γ is the adiabatic constant for the working substance. The adiabatic constant is defined by

$$\gamma = \frac{c_p}{c_v}$$

where c_p is the specific heat of the gas at constant pressure and c_v is the specific heat at constant volume. For monatomic gases $\gamma = 1.67$, and for diatomic gases $\gamma = 1.4$. In the gasoline engine the working substance is mostly air, for which $\gamma = 1.4$. In the ideal case, Eq. (20-10) shows that the higher compression ratios yield higher efficiencies since γ is always greater than 1.

EXAMPLE 20-3

Compute the efficiency of a gasoline engine for which the compression ratio is 8 and $\gamma = 1.4$.

Solution

From the given information, we note that

$$\frac{V_1}{V_2} = 8 \quad \text{and} \quad \gamma - 1 = 1.4 - 1 = 0.4$$

Thus, from Eq. (20-10),

$$e = 1 - \frac{1}{8^{0.4}} = 1 - \frac{1}{2.3} = 57\%$$

In the above example, 57 percent represents the maximum possible efficiency of a gasoline engine with the parameters given. Actually, the efficiency of such an engine is normally around 30 percent because of uncontrolled heat losses.

A second type of internal combustion engine is the diesel engine. In this engine the air is compressed to a high temperature and pressure near the top of the cylinder. Diesel fuel, which is injected into the cylinder at this point, ignites and forces the piston downward. The idealized diesel cycle is shown by the P-V diagram in Fig. 20-14. Starting at a, air is compressed adiabatically to point b, where the diesel fuel is injected. The diesel fuel, ignited by the hot air, delivers a quantity of heat Q_{in} at nearly constant pressure (line bc). The remainder of the power stroke consists of an adiabatic expansion to point d, performing external work. During the exhaust and intake strokes, the gas cools at constant volume to point a, losing a

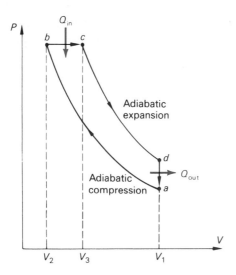

Figure 20-14 The ideal diesel cycle.

quantity of heat Q_{out}. The efficiency of a diesel engine is a function of the compression ratio (V_1/V_2) and the *expansion ratio* (V_1/V_3).

20-13 REFRIGERATION

A *refrigerator* can be thought of as a heat engine operated in reverse. A schematic diagram of a refrigerator is shown in Fig. 20-15. During every cycle, a compressor or similar device supplies mechanical work W to the system, extracting

If a refrigerator is plugged in and operates in a closed room with the refrigerator door open, the temperature of the room will increase.

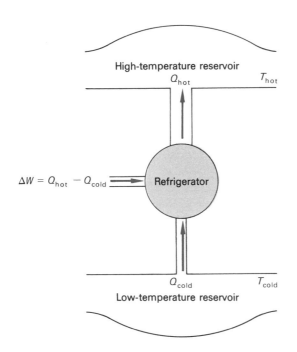

Figure 20-15 A schematic diagram for a refrigerator.

a quantity of heat Q_{cold} from a cold reservoir and depositing a quantity of heat Q_{hot} to a hot reservoir. According to the first law, the input work is given by

$$W = Q_{hot} - Q_{cold}$$

The effectiveness of any refrigerator is determined by the amount of heat Q_{cold} extracted for the least expenditure of mechanical work W. The ratio Q_{cold}/W is therefore a measure of the cooling efficiency of a refrigerator and is called its **coefficient of performance** K. Symbolically,

$$K = \frac{Q_{cold}}{W} = \frac{Q_{cold}}{Q_{hot} - Q_{cold}} \qquad \text{20-11}$$

The maximum efficiency can be expressed in terms of absolute temperatures:

$$K = \frac{T_{cold}}{T_{hot} - T_{cold}} \qquad \text{20-12}$$

To understand the refrigeration process better, let us consider the general schematic in Fig. 20-16. This diagram may refer to a number of refrigeration devices, from a commercial plant to a household refrigerator. The working substance, called the **refrigerant,** is a gas that is liquefied easily by an increase in pressure or a drop in temperature. In the liquid phase it can be vaporized readily by passing it through a throttling process (see Sec. 20-6) near room temperature. Common refrigerants are ammonia, Freon 12, methyl chloride, and sulfur dioxide. Ammonia, the most common industrial refrigerant, boils at $-28°F$ under a

Figure 20-16 The basic components of a refrigeration system.

pressure of 1 atm. Freon 12, the most common household refrigerant, boils at $-22°F$ at atmospheric pressure. Variation in pressure radically affects the condensation and evaporation temperatures of all refrigerants.

As shown in the schematic, a typical refrigeration system consists of a **compressor,** a **condenser,** a *liquid storage tank,* a *throttling valve,* and an **evaporator.** The compressor provides the necessary input work to move the refrigerant through the system. As the piston moves to the right, it sucks in the refrigerant through the intake valve at a little above atmospheric pressure and near room temperature. During the power stroke, the intake valve closes and the discharge valve opens. The emergent refrigerant, at high temperature and pressure, passes into the condenser, where it is cooled until it liquefies. The condenser may be cooled by running water or by an electric fan. It is during this phase that a quantity of heat Q_{hot} is rejected from the system. The condensed liquid refrigerant, still under a condition of high pressure and temperature, is collected in a liquid reservoir. Then the liquid refrigerant is drawn from the storage tank through a throttling valve, causing a sudden drop in temperature and pressure. As the cold liquid refrigerant flows through the evaporator coils, it absorbs a quantity of heat Q_{cold} from the space and products being cooled. This heat boils the liquid refrigerant and is carried away by the gaseous refrigerant as latent heat of vaporization. This phase is the "payoff" for the entire operation, and all components just contribute to the effective transfer of heat to the evaporator. Finally, the refrigerant vapor leaves the evaporator and is sucked into the compressor to begin another cycle.

Summary and Review

Summary

Thermodynamics is the science that treats the conversion of heat into work or the reverse process, the conversion of work into heat. We have seen that, not only must energy be conserved in such processes, but also there are limits on the efficiency. The major concepts presented in this chapter are summarized below.

- The *first law of thermodynamics* is a restatement of the conservation-of-energy principle. It says that the net heat ΔQ put into a system is equal to the net work ΔW done by the system plus the net change in internal energy ΔU of the system. Symbolically,

$$\Delta Q = \Delta W + \Delta U \qquad \text{First Law of Thermodynamics}$$

- In thermodynamics, work ΔW is often done on a gas. In such cases the work is often represented in terms of pressure and volume. A *P-V* diagram is also useful for measuring ΔW. If the pressure is constant,

$$\Delta W = P \, \Delta V$$

ΔW = area under *P-V* curve

- Special cases of the first law occur when one of the quantities doesn't undergo a change.

  (a) *Adiabatic process:*
  $\Delta Q = 0 \qquad \Delta W = -\Delta U$
  (b) *Isochoric process:*
  $\Delta V = 0 \qquad \Delta W = 0 \qquad \Delta Q = \Delta U$
  (c) *Isothermal process:*
  $\Delta T = 0 \qquad \Delta U = 0 \qquad \Delta Q = \Delta W$

- The *second law of thermodynamics* places restrictions on the possibility of satisfying the first. In short, it points out that in every process there is some loss of energy due to frictional forces or other dissipative forces. A 100 percent efficient engine, one that converts all input heat to useful output work, is not possible.

- A heat engine is represented generally by Fig. 20-9. The meaning of the symbols used in the following equations can be taken from that figure. The work done by the engine is the difference between input heat and output heat.

$$W = Q_{in} - Q_{out} \qquad \text{Work (kcal \emph{or} J)}$$

- The *efficiency e* of an engine is the ratio of the work output to the heat input. It can be calculated for an ideal engine from either of the following relations:

$$e = \frac{Q_{in} - Q_{out}}{Q_{in}}$$

$$e = \frac{T_{in} - T_{out}}{T_{in}} \qquad \text{Efficiency}$$

- A *refrigerator* is a heat engine operated in reverse. A measure of the performance of such a device is the amount of cooling you get for the work you must put into the system. Cooling occurs as a result of the extraction of heat Q_{cold} from the cold reservoir. The coefficient of performance K is given by either

$$K = \frac{Q_{cold}}{Q_{hot} - Q_{cold}} \qquad \text{or} \qquad K = \frac{T_{cold}}{T_{hot} - T_{cold}}$$

Key Terms

1. thermodynamics	8. compressor	15. second law of thermodynamics
2. *P-V* diagram	9. condenser	
3. adiabatic process	10. evaporator	16. thermal efficiency
4. isochoric process	11. internal energy function	17. Carnot cycle
5. heat engine	12. first law of thermodynamics	18. Carnot engine
6. refrigerator	13. throttling process	19. coefficient of performance
7. refrigerant	14. isothermal process	

Review Questions

20-1. The latent heat of vaporization of water is 540 cal/g. When 1 g of water is vaporized completely at constant pressure, however, the internal energy of the system increases by only 500 cal. What happened to the remaining 40 cal? Is this an isochoric process? Is it an isothermal process?

20-2. If both heat and work can be expressed in the same units, why is it necessary to distinguish between them?

20-3. Is it necessary to use the concept of molecular energy to describe and use the internal energy function? Explain.

20-4. A gas undergoes an adiabatic expansion. Does it perform external work? If so, what is the source of energy?

20-5. What happens to the internal energy of a gas undergoing (a) adiabatic compression, (b) isothermal expansion, and (c) a throttling process?

20-6. A gas performs external work during an isothermal expansion. What is the source of energy?

20-7. In the text, only one statement was given for the second law of thermodynamics. Discuss each of the following statements, showing them to be equivalent to that given in the text:

(a) It is impossible to construct a refrigerator that, working continuously, will extract heat from a cold body and exhaust it to a hot body without the performance of work on the system.

(b) The natural direction of heat flow is from a body at high temperature to a body at a low temperature, regardless of the size of each reservoir.

(c) All natural spontaneous processes are irreversible.

(d) Natural events always proceed in the direction from order to disorder.

20-8. It is energetically possible to extract the heat energy contained in the ocean and use it to power a steamship across the sea. What objections can you offer?

20-9. In an electric refrigerator, heat is transferred from the cool interior to warmer surroundings. Why isn't this in violation of the second law of thermodynamics?

20-10. Consider the performances of external work by the isothermal expansion of an ideal gas. Why isn't this process of converting heat into work in violation of the second law of thermodynamics?

20-11. If natural processes tend to decrease order in the universe, how can you explain the evolution of biological systems to a highly organized state? Does this violate the second law of thermodynamics?

20-12. Will keeping the door of an electric refrigerator open warm or cool a room? Explain.

20-13. What temperature must the cold reservoir have if a Carnot engine is to be 100 percent efficient? Can this ever happen? If it is impossible for a Carnot engine to have a 100 percent efficiency, why is it called the *ideal* engine?

20-14. What determines the efficiency of heat engines? Why is it generally so low?

Summary and Review

Problems

Section 20-3—The First Law of Thermodynamics

20-1. In an industrial chemical process, 600 J of heat is supplied to a system and 200 J of work is done *by* the system. What is the increase in the internal energy of the system? **Ans. 400 J**

20-2. Assume that the internal energy of a system decreases by 300 J while 200 J of work is done by a gas. What is the value of Q? Is heat lost or gained by the system?

20-3. In a thermodynamic process, the internal energy of the system increases by 500 J. How much work was done by the gas if 800 J of heat is absorbed? **Ans. 300 J**

20-4. A piston does 3000 ft · lb of work on a gas, which then expands, performing 2500 ft · lb of work on its surroundings. What is the change in internal energy of the system if the net heat exchange is zero?

20-5. In a chemical laboratory, a technician applies 340 J of energy to a gas while the system surrounding the gas does 140 J of work *on* the gas. What is the change in internal energy? **Ans. 480 J**

20-6. What is the change in internal energy for Prob. 20-5 if the 140 J of work is done *by* the gas instead of *on* the gas?

20-7. A system absorbs 200 J of heat as the internal energy increases by 150 J. What work is done by the gas? **Ans. 50 J**

*20-8. The specific heat of water is 4186 J/kg · C°. How much does the internal energy of 200 g of water change as it is heated from 20 to 30°C? Assume that the volume is constant.

*20-9. At a constant pressure of 101.3 kPa, 1 g of water (1 cm³) is vaporized completely and has a final volume of 1671 cm³ in its vapor form. What work is done by the system against its surroundings? What is the increase in internal energy? **Ans. 169 J, 2090 J**

*20-10. A 10-kg block slides down a plane from a height of 10 m and has a velocity of 10 m/s when it reaches the bottom. How many calories of heat were lost because of friction?

Sections 20-6 to 20-8—Thermodynamic Processes

20-11. An ideal gas expands isothermally while absorbing 4.80 J of heat. The piston has a mass of 3 kg. How high will the piston rise above its initial position? **Ans. 16.3 cm**

20-12. The work done on a gas during an adiabatic compression is 140 J. Calculate the increase in internal energy of the system in calories.

20-13. During an isobaric expansion, a steady pressure of 200 kPa causes the volume of a gas to change from 1 L to 3 L. What work is done by the gas? **Ans. 400 J**

20-14. A gas is confined to a copper can. How much heat must be supplied to increase the internal energy by 59 J? What type of thermodynamic process is involved?

20-15. A gas confined by a piston expands almost isobarically at 100 kPa. When 20,000 J of heat is absorbed by the system, its volume increases from 0.100 m³ to 0.250 m³. What work is done, and what is the change in internal energy? **Ans. 15.0 kJ, 5 kJ**

20-16. The specific heat of brass is 390 J/kg C°. A 4-kg piece of brass is heated isochorically, causing the temperature to rise by 10 C°. What is the increase in internal energy?

*20-17. Two liters of an ideal gas have a temperature of 300 K and a pressure of 2 atm. The gas undergoes an isobaric expansion while its temperature is increased to 500 K. What work is done by the gas? **Ans. 270 kJ**

*20-18. The diameter of a piston is 6.00 cm, and the length of its stroke is 12 cm. Assume that a constant force of 340 N moves the piston for a full stroke. First calculate the work based on force and distance. Then verify your answer by considering pressure and volume.

*20-19. For adiabatic processes, it can be shown that the pressure and volume are related by:

$$P_1 V_1^\gamma = P_2 V_2^\gamma$$

where γ is the adiabatic constant which is 1.40 for diatomic gases and also for the gasoline vapor/air mixture in combustion engines. Use the ideal gas law to prove the companion relationship:

$$T_1 V_1^{\gamma-1} = T_2 V_2^{\gamma-1}$$

*20-20. The compression ratio for a certain diesel engine is 15. The air-fuel mixture ($\gamma = 1.4$) is taken in at 300 K and 1 atm of pressure. Find the pressure and temperature of the gas after adiabatic compression. (Refer to Prob. 20-19).

Section 20-9—The Second Law of Thermodynamics

20-21. What is the efficiency of an engine that does 300 J of work in each cycle while discarding 600 J to the environment?
Ans. 33.3 percent

20-22. During a complete cycle, a system absorbs 600 cal of heat and rejects 200 cal to the environment. How much work is done? What is the efficiency?

20-23. A 37 percent efficient engine loses 400 J of heat during each cycle. What work is done, and how much heat is absorbed in each cycle?
Ans. 235 J, 635 J

20-24. What is the efficiency of an ideal engine that operates between the temperatures of 525 K and 300 K?

20-25. A steam engine takes superheated steam from a boiler at 200°C and rejects it directly into the air at 100°C. What is the ideal efficiency?
Ans. 21.1 percent

20-26. In a Carnot cycle, the isothermal expansion of a gas takes place at 400 K, and 500 cal of heat is absorbed by the gas. How much heat is lost if the system undergoes isothermal compression at 300 K? What is the heat loss, and what work is done?

20-27. A Carnot engine absorbs 1200 cal during each cycle as it operates between 500 K and 300 K. What is the efficiency? How much heat is rejected, and how much work in joules is done during each cycle?
Ans. 40 percent, 720 cal, 2010 J

20-28. The actual efficiency of an engine is 60 percent of its ideal efficiency. The engine operates between temperatures of 460 K and 290 K. How much work is done in each cycle if 1600 J of heat is absorbed?

20-29. A refrigerator extracts 400 J of heat from a box during each cycle and rejects 600 J to a high-temperature reservoir. What is the coefficient of performance?
Ans. 2.00

20-30. The coefficient of performance of a refrigerator is 5.0. How much heat is discarded if the compressor does 200 J of work during each cycle?

20-31. How much heat is extracted from the cold reservoir if the compressor of a refrigerator does 180 J of work during each cycle? The coefficient of performance is 4.0. What heat is rejected to the hot reservoir?
Ans. 720 J, 900 J

20-32. An ideal refrigerator extracts 400 J of heat from a reservoir at 200 K and rejects heat to a reservoir at 500 K. What is the ideal coefficient of performance, and how much work is done in each cycle?

*20-33. A Carnot refrigerator has a coefficient of performance of 2.33. If 600 J of work is done by the compressor in each cycle, how many joules of heat are extracted from the cold reservoir and how much is ejected to the environment?
Ans. 1400 J, 2000 J

Challenge Problems

20-34. In a thermodynamic process, 200 Btu is supplied to produce an isobaric expansion under a pressure of 100 lb/in.^{2}. The internal energy of the system does not change. What is the increase in volume of the gas?

20-35. A 100-cm^3 sample of gas at a pressure of 100 kPa is heated isochorically from point A to point B until its pressure reaches 300 kPa. Then it expands isobarically to point C, where its volume is 400 cm^3. The pressure then returns to 100 kPa at point D, with no change in volume. Finally, it returns to its original state at point A. Draw the P-V diagram for this cycle. What is the net work done for the entire cycle?

Ans. 60 J

20-36. Find the net work done by a gas as it is carried around the cycle shown in Fig. 20-17.

20-37. What is the net work done for the process ABCA as described by Fig. 20-18?

Ans. 85 J

Figure 20-17

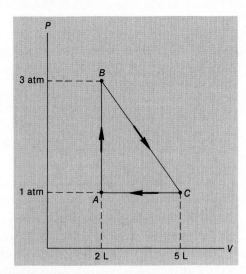

Figure 20-18

20-38. A real engine operates between 327 and 0°C and has an output power of 8 kW. What is the ideal efficiency for this engine? How much power is wasted if the actual efficiency is only 25 percent?

20-39. The Otto efficiency for a gasoline engine is 50 percent, and the adiabatic constant is 1.4. Compute the compression ratio.

Ans. 5.66

20-40. A heat pump takes heat from a water reservoir at 41°F and delivers it to a system of pipes in a house at 78°F. The energy required to operate the heat pump is about twice that required to operate a Carnot pump. How much mechanical work must be supplied by the pump so that it will deliver 1×10^6 Btu of heat energy to the house?

20-41. A Carnot engine has an efficiency of 48 percent. If the working substance enters the system at 400°C, what is the exhaust temperature?

Ans. 350 K

20-42. During the compression stroke of an automobile engine, the volume of the combustible mixture decreases from 18 to 2 in.^{3}. If the adiabatic constant is 1.4, what is the maximum possible efficiency for the engine?

20-43. How many joules of work must be done by the compressor in a refrigerator to change 1 kg of water at 20°C to ice at −10°C? The coefficient of performance is 3.5.

Ans. 126 kJ

20-44. In a mechanical refrigerator the low-temperature coils of the evaporator are at −30°C, and the condenser has a temperature of 60°C. What is the maximum possible coefficient of performance?

20-45. An engine has a thermal efficiency of 27 percent and an exhaust temperature of 230°C. What is the lowest possible input temperature?

Ans. 416°C

20-46. The coefficient of performance of a refrigerator is 5.0. If the temperature of the room is 28°C, what is the lowest possible temperature that can be obtained inside the refrigerator?

Critical Thinking Questions

20-47. A gas expands against a movable piston, lifting it through 2 in. at constant speed. How much work is done by the gas if the piston weighs 200 lb and has a cross-sectional area of 12 in.^{2}? If the expansion is adiabatic, what is the change in internal energy in Btu? Does ΔU represent an increase or decrease in internal energy?

Ans. 33.3 ft · lb, 0.0428 Btu, decrease

*20-48. Consider the P-V diagram shown in Fig. 20-19, where the pressure and volume are indicated for each of the points A, B, C, and D. Starting at point A, a 100 cm^3 sample of gas absorbs 200 J of heat, causing the pressure to increase from 100 to 200 kPa while the volume increases to 200 cm^3. Next the gas expands from B to C, absorbing an additional 400 J of heat while its volume increases to 400 cm^3. (a) Find the net work done and the change in internal energy for each of the processes AB and BC. (b) What are the net work and the total change in internal energy for the process ABC? (c) What kind of process is illustrated by BC?

*20-49. The cycle begun in the previous problem now continues from C to D while an additional 200 J of heat is absorbed. (a) Find the net work and the net change in internal energy for the process CD. (b) Suppose the system returns to its original state at point A. What is the net work for the entire cycle $ABCDA$, and what is the efficiency of the cycle?

*20-50. Consider a specific mass of gas that is forced through an adiabatic throttling process. Before entering the valve, it has internal energy U_1, pressure P_1, and volume V_1. After passing though the valve, it has internal energy U_2, pressure P_2, and volume V_2. The net work done is the work done *by* the gas minus the work done *on* the gas. Show that

$$U_1 + P_1V_1 = U_2 + P_2V_2$$

The quantity $U + PV$, called the *enthalpy*, is conserved during a throttling process.

*20-51. A gasoline engine takes in 2000 J of heat and delivers 400 J of work per cycle. The heat is obtained by burning gasoline, which has a heat of combustion of 50 kJ/g. What is the thermal efficiency? How much heat is lost per cycle? How much gasoline is burned in each cycle? If the engine goes through 90 cycles per second, what is the output power?

Ans. 20 percent, 1600 J, 0.040 g, 36 kW

*20-52. Consider a Carnot engine of efficiency e and a Carnot refrigerator whose coefficient of performance is K. If these devices operate between the same temperatures, derive the following relationship.

$$K = \frac{1-e}{e}$$

Figure 20-19

Chapter 21

Mechanical Waves

OBJECTIVES

After completing this chapter, you should be able to:

1. Demonstrate by definition and example your understanding of *transverse* and *longitudinal wave motion.*
2. Define, relate, and apply the meaning of the terms *frequency, wavelength,* and *speed* for wave motion.
3. Solve problems involving the *mass, length, tension,* and *wave velocity* for transverse waves in a string.
4. Write and apply an expression for determining the *characteristic frequencies* for a vibrating string with fixed end points.

Energy can be transferred from one place to another by several means. In driving a nail, the bulk kinetic energy of a hammer is converted into useful work on the nail. Wind, projectiles, and most simple machines also perform work at the expense of material motion. Even the conduction of heat and electricity involves the motion of elementary particles called *electrons.* In this chapter we study the transfer of energy from one point to another without the physical transfer of material between the points.

21-1 MECHANICAL WAVES

When a stone is dropped into a pool of water, it creates a disturbance that spreads out in concentric circles, eventually reaching all parts of the pool. A small stick, floating on the surface of the water, bobs up and down as the disturbance passes. Energy has been transferred from the point of impact of the stone into the water to the floating stick some distance away. This energy is passed along by the agitation of neighboring water particles. Only the disturbance moves through the water. The actual motion of any particular water particle is comparatively small. Energy propagation by means of a disturbance in a medium instead of the motion of the medium itself is called **wave motion.**

The example above is referred to as a **mechanical wave** because its existence depends on a mechanical source and a material medium.

> **A mechanical wave is a physical disturbance in an elastic medium.**

It is important to recognize that all disturbances are not necessarily mechanical. For example, light waves, radio waves, and heat radiation propagate their energy

Mechanical Waves

by means of electric and magnetic disturbances. No physical medium is necessary for the transmission of electromagnetic waves. Many of the basic ideas presented in this chapter for mechanical waves are also applicable, however, to electromagnetic waves.

21-2 TYPES OF WAVES

Waves are classified according to the motion of a local part of a medium with respect to the direction of wave propagation.

> **In a transverse wave the vibration of the individual particles of the medium is perpendicular to the direction of wave propagation.**

For example, suppose we fasten one end of a rope to a post and grasp the other end with the hand, as shown in Fig. 21-1. By moving the free end up and down quickly, we send a single disturbance called a *pulse* down the length of rope. Three equally spaced knots at points *a*, *b*, and *c* demonstrate that the individual particles move up and down while the disturbance moves to the right with velocity **v**.

Another type of wave, which may occur in a coiled spring, is illustrated in Fig. 21-2. The coils near the left end are pinched closely together to form a *condensation*. When the distorting force is removed, a condensation pulse is propagated throughout the length of the spring. No part of the spring moves very far from its equilibrium position, but the pulse continues to travel along the spring. Such a wave is called a *longitudinal wave* because the spring particles are displaced along the same direction in which the disturbance is traveling.

> **In a longitudinal wave the vibration of the individual particles is parallel to the direction of wave propagation.**

Figure 21-1 In a transverse wave, the individual particles move perpendicular to the direction of wave propagation.

You will need 10 dominoes and a ruler.

Place the dominoes on a table 1.0 cm apart and stand them on a short end. The dominoes represent a liquid's molecules. Start a chain reaction by pushing on the first domino. Notice the vibration that travels through each domino to the next one. Does each domino move far? How are the dominoes' movements similar to a pressure wave as it moves through a solid? How are the movements different? How can you decrease and increase the speed of the wave?

Explanation: Even though the wave goes the entire length of the domino chain, each domino moves only a small distance. To increase the speed of the wave, move the dominoes closer together. To decrease the wave's speed, move the dominoes farther apart.

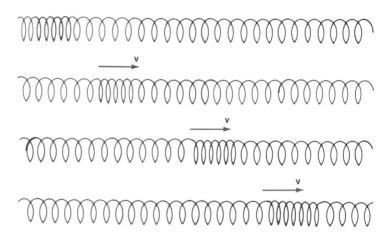

Figure 21-2   In a longitudinal wave, the motion of the individual particles is parallel to the direction of wave propagation. The illustration demonstrates the motion of a condensation pulse.

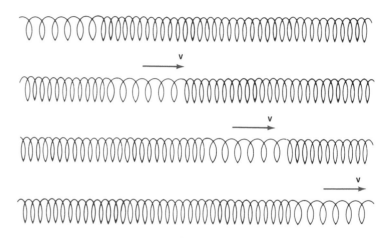

Figure 21-3   Longitudinal motion of a rarefaction pulse in a string.

If the coils of the spring in our example were forced apart at the left, a *rarefaction* would be formed, as shown in Fig. 21-3. Upon removal of the disturbing force, a longitudinal rarefaction pulse would be propagated along the spring. In general, a longitudinal wave consists of a series of condensations and rarefactions moving in a determined direction.

21-3 CALCULATING WAVE SPEED

The speed with which a pulse moves through a medium depends on the elasticity of the medium and the inertia of its particles. The more elastic materials yield greater restoring forces when distorted. The less dense materials offer less resistance to motion. In either case, the ability of particles to pass on a disturbance to neighboring particles is improved, and a pulse will travel at a greater speed.

Let us consider the motion of a transverse pulse down the string in Fig. 21-4. The string of mass m and length l is maintained under a constant tension F by the

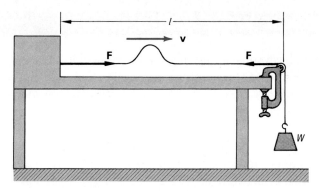

Figure 21-4 Computing the speed of a transverse pulse in a string.

suspended weight. When the string is plucked near the left end, a transverse pulse is propagated along the string. The elasticity of the string is measured by the tension F in the string. The inertia of the individual particles is determined by the *mass per unit length* μ of the string. It can be shown that the speed of a transverse pulse in a string is given by

$$v = \sqrt{\frac{F}{\mu}} = \sqrt{\frac{F}{m/l}} \qquad \textbf{21-1}$$

The mass per unit length μ is usually referred to as the **linear density** of the string. If F is expressed in newtons and μ in kilograms per meter, the speed is in meters per second.

EXAMPLE 21-1

The length l of the string in Fig. 21-4 is 2 m, and the string has a mass of 0.3 g. Calculate the speed of a transverse pulse in the string if it is under a tension of 20 N.

Solution
We first compute the linear density of the string.

$$\mu = \frac{m}{l} = \frac{0.3 \times 10^{-3} \text{ kg}}{2 \text{ m}}$$
$$= 1.5 \times 10^{-4} \text{ kg/m}$$

Then Eq. (21-1) yields

$$v = \sqrt{\frac{F}{\mu}} = \sqrt{\frac{20 \text{ N}}{1.5 \times 10^{-4} \text{ kg/m}}}$$
$$= 365 \text{ m/s}$$

Calculation of the speed of a longitudinal pulse will be reserved for the following chapter, in connection with sound waves.

So far, we have been considering single, nonrepeated disturbances called *pulses.* What happens if similar disturbances are repeated periodically? Suppose we attach the left end of a string to the end point of an electromagnetic vibrator, as shown in Fig. 21-5. The end of the metal vibrator is driven with harmonic motion by an oscillating magnetic field. Since the string is attached to the end of the vibrator, a series of periodic transverse pulses is sent down the string. The resulting waves consist of many crests and troughs, which move down the string at a constant rate of speed. The distance between any two adjacent crests or troughs in such a wave train is called the **wavelength,** denoted by λ.

As a wave travels along the string, each particle of the string vibrates about its equilibrium position with the same frequency and amplitude as the vibrating source. The particles of the string are not in corresponding positions, however, at the same times. Two particles are said to be **in phase** if they have the same displacement and if they are moving in the same direction. The particles A and B of Fig. 21-5b are in phase. Since particles at the crests of a given wave train are also in phase, we can provide a more general definition for the wavelength.

> The wavelength λ **of a periodic wave train is the distance between any two adjacent particles that are in phase.**

Each time the end point P of the vibrator makes a complete oscillation, the wave will move through a distance of one wavelength. The time required to cover this distance is therefore equal to the period T of the vibrating source. Hence, the wave speed v can be related to the wavelength λ and period T by the equation

$$v = \frac{\lambda}{T} \qquad \text{21-2}$$

The **frequency** f of a wave is the number of waves that pass a particular point in a unit of time. It is the same as the frequency of the vibrating source and is therefore equal to the reciprocal of the period ($f = 1/T$). The units of frequency may be expressed in waves per second, oscillations per second, or cycles per second. The SI unit for frequency is the **hertz** (Hz), which is defined as a cycle per second.

$$1 \text{ Hz} = 1 \text{ cycle/s} = \frac{1}{\text{s}}$$

Thus, if 40 waves pass a point every second, the frequency is 40 Hz.

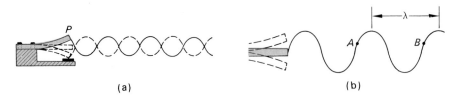

(a) (b)

Figure 21-5 (a) Production and propagation of a periodic transverse wave. (b) The wavelength λ is the distance between any two particles in phase, such as those at adjacent crests or at points A and B.

The speed of a wave is more often expressed in terms of its frequency rather than its period. Thus, Eq. (21-2) can be written

$$v = f\lambda \qquad \text{21-3}$$

Equation (21-3) represents an important physical relationship between the speed, frequency, and wavelength of *any* periodic wave. An illustration of each quantity is given in Fig. 21-6 for a periodic transverse wave.

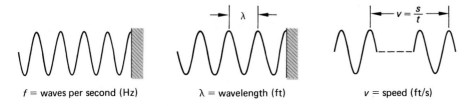

f = waves per second (Hz)   λ = wavelength (ft)   v = speed (ft/s)

Figure 21-6 The relationships among the frequency, wavelength, and speed of a transverse wave.

EXAMPLE 21-2

A man sits near the end of a fishing dock and counts the water waves as they strike a supporting post. In 1 min he counts 80 waves. If a particular crest travels 20 m in 8 s, what is the wavelength of the waves?

Solution

The frequency and velocity of the waves are calculated as follows:

$$f = \frac{80 \text{ waves}}{60 \text{ s}} = 1.33 \text{ Hz}$$

$$v = \frac{20 \text{ m}}{8 \text{ s}} = 2.5 \text{ m/s}$$

From Eq. (21-3), the wavelength is

$$\lambda = \frac{v}{f} = \frac{2.5 \text{ m/s}}{1.33 \text{ Hz}} = 1.88 \text{ m}$$

A longitudinal periodic wave can be generated by the apparatus shown in Fig. 21-7. The left end of a coiled spring is connected to a metal ball supported at the end of a clamped hacksaw blade. When the metal ball is displaced to the left and released, it vibrates with harmonic motion. The resulting condensations and rarefactions are passed along the spring, producing a periodic longitudinal wave. Each particle of the coiled spring oscillates back and forth horizontally with the same frequency and amplitude as the metal ball. The distance between any two adjacent particles that are in phase is the wavelength. As indicated in Fig. 21-7, the distance between adjacent condensations or adjacent rarefactions is a convenient measure of the wavelength. Equation (21-3) also applies for a periodic longitudinal wave.

Figure 21-7 Production and propagation of a periodic longitudinal wave.

21-5 ENERGY OF A PERIODIC WAVE

We have seen that each particle in a periodic wave oscillates with simple harmonic motion determined by the source of the wave. The energy content of a wave can be analyzed by considering the harmonic motion of the individual particles. For example, consider a periodic transverse wave in a string at the instant shown in Fig. 21-8. Particle a has reached its maximum *amplitude;* its velocity is zero, and it is experiencing its maximum restoring force. Particle b is passing through its equilibrium position, where the restoring force is zero. At this instance, particle b has its greatest speed and hence its maximum energy. Particle c is at its maximum displacement in the negative direction. As the periodic wave passes along the string, each particle oscillates back and forth through its own equilibrium position.

In Chap. 14 on harmonic motion, we found that the maximum velocity of a particle oscillating with frequency f and amplitude A is given by

$$v_{max} = 2\pi f A$$

When a particle has this speed, it is passing through its equilibrium position, where its potential energy is zero and its kinetic energy is a maximum. Thus, the total energy of the particle is

$$E = E_p + E_k = (E_k)_{max}$$
$$= \tfrac{1}{2}mv^2_{max} = \tfrac{1}{2}m(2\pi f A)^2$$
$$= 2\pi^2 f^2 A^2 m \qquad\qquad \text{21-4}$$

As a periodic wave passes through a medium, each element of the medium is continuously doing work on adjacent elements. Therefore, the energy transmitted

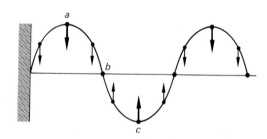

Figure 21-8 The restoring forces that act on particles of a vibrating string.

along the length of a vibrating string is not confined to one position. Let us apply the result obtained for a single particle to the entire length of a vibrating string. The energy content of the entire string is the sum of the individual energies of its constituent particles. If we let m refer to the entire mass of the string instead of the mass of an individual particle, Eq. (21-4) represents the total wave energy in the string. In a string of length l, the wave energy per unit length is given by

$$\frac{E}{l} = 2\pi^2 f^2 A^2 \frac{m}{l}$$

Substituting μ for the mass per unit length, we can write

$$\frac{E}{l} = 2\pi^2 f^2 A^2 \mu \qquad\qquad \textbf{21-5}$$

The wave energy is proportional to the square of the frequency f, to the square of the amplitude A, and to the linear density μ of the string. It must be recognized that the linear density is not a function of the length of string. This is true because the mass increases in proportion to the length l, so that μ is constant for any length.

Suppose that a wave travels down a length l of a given string with a speed v. The time t required for the wave to travel this length is

$$t = \frac{l}{v}$$

If the energy in this length of string is represented by E, the power P of the wave is given by

$$P = \frac{E}{t} = \frac{E}{l/v} = \frac{E}{l}v \qquad\qquad \textbf{21-6}$$

This represents the *rate* at which energy is propagated down the string. Substitution from Eq. (21-5) yields

$$P = 2\pi^2 f^2 A^2 \mu v \qquad\qquad \textbf{21-7}$$

The wave power is directly proportional to the energy per unit length and to the wave speed.

The dependence of wave energy and wave power on f^2 and A^2, as found in Eqs. (21-5) and (21-7), is a general conclusion for all kinds of waves. The same ideas will be applied in Chap. 22 when the energy of a sound wave is discussed.

21-6 THE SUPERPOSITION PRINCIPLE

Until now we have been considering the motion of a single train of pulses passing through a medium. We now consider what happens when two or more wave trains pass simultaneously through the same medium. Let us consider transverse waves in a vibrating string. The speed of a transverse wave is determined by the tension of the string and its linear density. Since these parameters are a function of the medium and not the source, any transverse wave will have the same speed for a given string under constant tension. However, the frequency and amplitude may vary considerably.

> **When two or more wave trains exist simultaneously in the same medium, each wave travels through the medium as though the other were not present.**

The resultant wave is a superposition of the component waves. In other words, the resultant displacement of a particular particle on the vibrating string is the algebraic sum of the displacements each wave would produce independently of the other. This is the *superposition principle:*

> **When two or more waves exist simultaneously in the same medium, the resultant displacement at any point and time is the algebraic sum of the displacements of each wave.**

It should be noted that the superposition principle, as stated here, applies only for *linear* media—those for which the response is directly proportional to the cause. Also, the sum of displacements is *algebraic* only if the waves have the same plane of polarization. For our purposes, a vibrating string will be assumed to satisfy both these conditions.

The application of this principle is shown graphically in Fig. 21-9. Two waves, indicated by the solid and dashed lines, superpose to form the resultant wave indicated by the heavy line. In Fig. 21-9a the superposition results in a wave of larger amplitude. These waves are said to interfere *constructively.* **Destructive interference** occurs when the resulting amplitude is smaller, as in Fig. 21-9b.

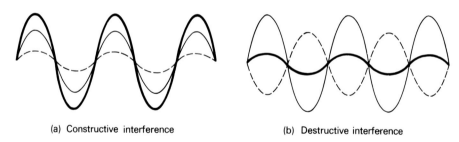

(a) Constructive interference   (b) Destructive interference

Figure 21-9   The superposition principle.

21-7 STANDING WAVES

Let us consider the reflection of a transverse pulse, as shown for the string in Fig. 21-10. When the end of the string is fixed rigidly to a support, the arriving pulse strikes the support, exerting an upward force on it. The reaction force exerted by the support kicks downward on the string, setting up a reflected pulse. Both the displacement and the velocity are reversed in the reflected pulse. In other words, a pulse that arrives as a crest is reflected as a trough with the same speed traveling in the opposite direction, and vice versa.

Suppose we consider the wave set up by a vibrating string whose end points are fixed, as in Fig. 21-11. We can use the superposition principle to analyze the resul-

Figure 21-10 The reflection of a transverse pulse at a fixed boundary.

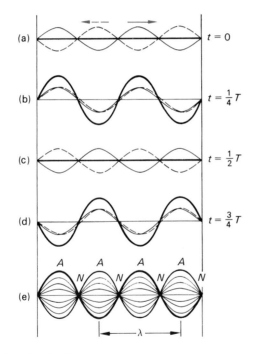

Figure 21-11 Production of a standing wave.

tant waveform at any instant. In Fig. 21-11a, we consider the incident and reflected waves at a particular time $t = 0$. The incident wave, traveling to the right, is indicated by a light solid line. The reflected wave, traveling to the left, is indicated by a dotted line. The two waves have the same speed and wavelength, but they are oppositely directed. At this instant all particles of the string lie in a straight horizontal line, as shown by the heavy line. Note that the heavy line is a superposition of the incident and reflected waves at a moment when their displacements add to zero. A snapshot of the string an instant later will show that, with a few exceptions, the particles have all changed positions. This is because the component waves have moved a finite distance.

Let us now consider the resultant wave at time t equal to one-fourth of a period later ($t = \frac{1}{2}T$), as in Fig. 21-11b. The component wave indicated by a solid line will have moved to the *right* a distance of one-fourth wavelength. The component wave indicated by a dashed line will have moved to the *left* a distance of one-fourth wavelength. The resultant wave and hence the shape of the string at this time are indicated on the heavy solid line. Constructive interference has resulted in a wave with twice the amplitude of either of the component waves. When the time t is one-half a period ($t = \frac{1}{2}T$), total destructive interference occurs, and once again the shape of the string is a straight line, as in Fig. 21-11c. At $t = \frac{3}{4}T$ the shape of the string reaches its maximum amplitude in the opposite direction. This constructive interference is shown by the heavy line in Fig. 21-11d.

A series of snapshots of the vibrating string at closely spaced time intervals would reveal a number of loops, as shown in Fig. 21-11e. Such a wave is called a **standing wave.** Notice that there are certain points along the string that remain at rest. These positions, called **nodes,** are labeled N in the figure. A flea resting on the vibrating string at these points would not be moved up and down by the wave motion.

Between the nodal points, the particles of the string move up and down with simple harmonic motion. The points of maximum amplitude occur midway between the nodes and are called **antinodes.** A flea resting on the string at any of these points, labeled A, would experience maximum speeds and displacements in the upward and downward oscillation of the string.

> **The distance between alternate nodes or alternate antinodes in a standing wave is a measure of the wavelength of the component waves.**

Longitudinal standing waves may also occur by the continuous reflection of condensation and rarefaction pulses. In this case the nodes exist where the particles of the medium are stationary, and the antinodes occur where the particles of the medium oscillate with maximum amplitude in the direction of propagation. Longitudinal standing waves will be discussed in Chap. 22 in connection with sound waves.

21-8 CHARACTERISTIC FREQUENCIES

Let us now consider the possible standing waves that can be set up in a string of length l whose ends are fixed, as in Fig. 21-12. When the string is set into vibration, the incident and reflected wave trains travel in opposite directions with the same wavelength. The fixed end points represent *boundary conditions* that restrict the possible wavelengths that will produce standing waves. These end points must be displacement nodes for any resulting wave patterns.

The simplest possible standing wave occurs when the wavelengths of incident and reflected waves are equal to twice the length of the string. The standing wave consists of a single loop with nodal points at each end, as shown in Fig. 21-12a. This pattern of vibration is referred to as the *fundamental mode of oscillation.* The higher modes of oscillation occur for shorter and shorter wavelengths. From the figure, it is noted that the allowable wavelengths are

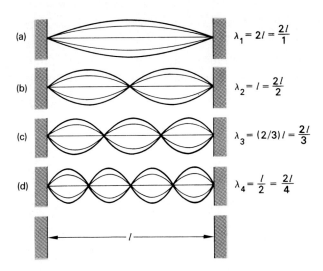

Figure 21-12 Possible standing-wave patterns in a vibrating string.

$$\frac{2l}{1}, \frac{2l}{2}, \frac{2l}{3}, \frac{2l}{4}, \ldots$$

or, in equation form,

$$\lambda_n = \frac{2l}{n} \qquad n = 1, 2, 3, \ldots \qquad \text{21-8}$$

The corresponding frequencies of vibration are, from $v = f\lambda$,

$$f_n = \frac{nv}{2l} = n\frac{v}{2l} \qquad n = 1, 2, 3, \ldots \qquad \text{21-9}$$

where v is the speed of the transverse waves. This speed is the same for all wavelengths because it depends only on the characteristics of the vibrating medium. The frequencies given by Eq. (21-9) are called the **characteristic frequencies of vibration.** In terms of string tension F and linear density μ, the characteristic frequencies are

$$f_n = \frac{n}{2l} \sqrt{\frac{F}{\mu}} \qquad n = 1, 2, 3, \ldots \qquad \text{21-10}$$

The lowest possible frequency ($v/2l$) is called the **fundamental frequency** f_1. The others, which are integral multiples of the fundamental, are known as the **overtones.** The entire series,

$$f_n = nf_1 \qquad n = 1, 2, 3, \ldots \qquad \text{21-11}$$

consisting of the fundamental and its overtones, is known as the **harmonic series.** The fundamental is the first harmonic, the first overtone ($f_2 = 2f_1$) is the second harmonic, the second overtone ($f_3 = 3f_1$) is the third harmonic, and so on.

EXAMPLE 21-3

A steel piano wire 50 cm long has a mass of 5 g and is under a tension of 400 N. What are the frequencies of its fundamental mode of vibration and the first two overtones?

Solution

The fundamental is found by setting $n = 1$ in Eq. (21-10).

$$f_1 = \frac{1}{2l}\sqrt{\frac{F}{\mu}} = \frac{1}{2l}\sqrt{\frac{F}{m/l}}$$

$$= \frac{1}{(2)(0.5\text{ m})}\sqrt{\frac{400\text{ N}}{(0.005\text{ kg})(0.5\text{ m})}}$$

$$= 200\text{ Hz}$$

The first and second overtones are

$$f_2 = 2f_1 = 400\text{ Hz}$$
$$f_3 = 3f_1 = 600\text{ Hz}$$

We shall see in the following chapter that, as a string vibrates in one or more of its possible modes, energy is transmitted to the surrounding air in the form of sound waves. These longitudinal waves consist of condensations and rarefactions of the same frequency as the vibrating strings. The human ear interprets these waves as sound.

Summary and Review

Summary

We have investigated mechanical wave motion in which energy is transferred by a physical disturbance in an elastic medium. The fundamental laws developed in this chapter are important because they also apply for many other types of waves that will be studied later. The essential concepts are summarized below.

- The velocity of a transverse wave in a string of mass m and length l is given by

$$v = \sqrt{\frac{F}{\mu}} \qquad \mu = \frac{m}{l} \qquad v = \sqrt{\frac{Fl}{m}}$$

Wave Speed

	Force F	Mass m	Length l	Speed v
SI units	N	kg	m	m/s
USCS units	lb	slug	ft	ft/s

- For any wave of period T or frequency f, the speed v can be expressed in terms of the wavelength λ as follows:

$$v = \frac{\lambda}{T} \qquad v = f\lambda$$

Frequency is in $Hz = 1/s$

- The *energy per unit length* and the *power* of wave propagation can be found from

$$\frac{E}{l} = 2\pi^2 f^2 A^2 \mu \qquad P = 2\pi^2 f^2 A^2 \mu v$$

- The characteristic frequencies for the possible modes of vibration in a stretched string are found from:

$$f_n = \frac{n}{2l}\sqrt{\frac{F}{\mu}} \qquad n = 1, 2, 3, \ldots$$

Characteristic Frequencies

- The series $f_n = nf_1$ is called the *harmonics*. They are integral multiples of the fundamental f_1. These are mathematical values, and all harmonics may not exist. The actual possibilities beyond the fundamental are called *overtones*. Since all harmonics are possible for the vibrating string, the first overtone is the second harmonic, the second overtone is the third harmonic, and so on.

Key Terms

1. wave motion
2. wave speed
3. wavelength
4. frequency
5. phase
6. nodes
7. antinodes
8. fundamental
9. overtone
10. harmonic
11. mechanical wave
12. transverse wave
13. longitudinal wave
14. linear density
15. amplitude
16. superposition principle
17. constructive interference
18. destructive interference
19. standing waves
20. characteristic frequencies

Review Questions

21-1. Explain how a water wave is both transverse and longitudinal.

21-2. Describe an experiment to demonstrate that energy is associated with wave motion.

21-3. In a *torsional wave* the individual particles of the medium vibrate with angular harmonic motion about the axis of propagation. Give a mechanical example of such a wave.

21-4. Discuss the interference of waves. Is there a loss of energy when waves interfere? Explain.

21-5. A transverse pulse is sent down a string of mass m and length l under a tension F. How will the speed of the pulse be affected if (a) the mass of the string is quadrupled, (b) the length of the string is quadrupled, and (c) the tension is reduced by one-fourth?

21-6. Draw graphs of a periodic transverse wave and a periodic longitudinal wave. Indicate on the figures the wavelength and amplitude of each wave.

21-7. Which harmonic is indicated by Fig. 21-12d? Which overtone is present?

21-8. We have seen that boundary conditions determine possible modes of vibration. Draw a diagram of the fundamental and of the first two overtones for a vibrating rod (a) clamped at one end and (b) clamped at its midpoint.

21-9. A vibrating string has a fundamental frequency of 200 Hz. If the length is reduced by one-fourth, what will the new fundamental frequency be? Has the wave speed been altered by shortening the string? Assume constant tension.

21-10. Show graphically the superposition of two waves traveling in the same direction. The second wave has three-fourths the amplitude and one-half the wavelength of the first wave.

21-11. In an experiment with the vibrating string, one end of the string is attached to the tip of a vibrator and the other end passes over a pulley. Suspended weights are used to produce the fundamental and the first three overtones. What effect will stretching the string have on frequency calculations?

Problems

Section 21-1—Mechanical Waves

21-1. A transverse wave has a wavelength of 30 cm and vibrates with a frequency of 420 Hz. What is the speed of this wave?
Ans. 126 m/s

21-2. A person on a pier counts the slaps of a wave as the crests hit a post. If 80 slaps are heard in 1 min and a particular crest travels a distance of 8 m in 4 s, what is the length of a single wave?

21-3. A transverse wave is pictured in Fig. 21-13. Find the amplitude, wavelength, period, and speed of the wave if it has a frequency of 12 Hz.
Ans. 12 cm, 28 cm, 83.3 ms, 3.36 m/s

21-4. For the longitudinal wave in Fig. 21-13, find the amplitude, wavelength, period, and speed of the wave if it has a frequency of 8 Hz. If the amplitude were doubled, would any of the other factors change?

21-5. A 500-g metal wire has a length of 50 cm and is under a tension of 80 N. What is the speed of a transverse wave in the wire?
Ans. 8.94 m/s.

21-6. If the wire in Prob. 21-5 is cut in half, what will be its new mass? Show that the speed of the wave is unchanged. Why?

21-7. A 3-m cord under a tension of 200 N sustains a transverse wave speed of 172 m/s. What is the mass of the rope? **Ans. 20.3 g**

21-8. A 200-g cord is stretched over a distance of 5.2 m and placed under a tension of 500 N. Compute the speed of a transverse wave in the cord.

21-9. What tension is required to produce a wave speed of 12.0 m/s in a 900-g string that is 2 m long? **Ans. 64.8 N**

21-10. A wooden float at the end of a fishing line makes eight complete oscillations in 10 s. If it takes 3.60 s for a single wave to travel 11 m, what is the wavelength of the water waves?

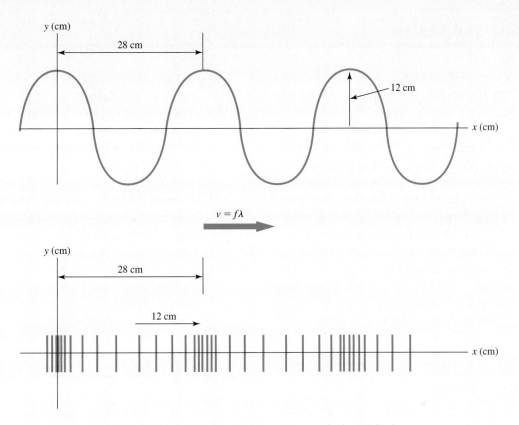

Figure 21-13 The wavelength, velocity, and amplitude for a transverse wave and a longitudinal wave ($A = 12$ cm and $\lambda = 28$ cm).

21-11. What frequency is required to cause a rope to vibrate with a wavelength of 20 cm when it is under a tension of 200 N? Assume the linear density of the rope to be 0.008 kg/m.
Ans. 791 Hz

21-12. A tension of 400 N causes a 300-g wire of length 1.6 m to vibrate with a frequency of 40 Hz. What is the wavelength of the transverse waves?

21-13. A horizontal spring is jiggled back and forth at one end by a device that makes 80 oscillations in 12 s. What is the speed of the longitudinal waves if condensations are separated by 15 cm as the wave progresses down the spring? **Ans. 1.00 m/s**

Section 21-5—Energy of a Periodic Wave

21-14. A 2-m length of string has a mass of 300 g and vibrates with a frequency of 2 Hz and an amplitude of 50 mm. If the tension in the rope is 48 N, how much power must be delivered to the string?

21-15. An 80-g string has a length of 40 m and vibrates with a frequency of 8 Hz and an amplitude of 4 cm. Find the energy per unit of length passing along the string.
Ans. 4.04 × 10⁻³ J/m

21-16. If the wavelength of the transverse wave in Prob. 21-11 is 1.6 m, what power is supplied by the source?

*21-17. A 300-g string has a length of 2.50 m and vibrates with an amplitude of 8.00 mm. The tension in the string is 46 N. What must be the frequency of the waves in order that the average power be 90.0 W?
Ans. 174 Hz

Sections 21-7 and 21-8—Standing Waves and Characteristic Frequencies

21-18. A string vibrates with a fundamental frequency of 200 Hz. What is the frequency of the second harmonic and of the third overtone?

Summary and Review

21-19. If the fundamental frequency of a wave is 330 Hz, what is the frequency of the fifth harmonic and the second overtone?

Ans. 1650 Hz, 990 Hz

21-20. The linear density of a string is 0.00086 kg/m. What should be the tension in the rope in order for a 2 m length of this string to vibrate at 600 Hz for its third harmonic?

21-21. A 10-g string, 4 m in length, has a tension of 64 N. What is the frequency of its fundamental mode of vibration? What are the frequencies of the first and second overtones?

Ans. 20, 40, and 60 Hz

21-22. The second harmonic of a vibrating string is 200 Hz. The length of the string is 3 m, and its tension is 200 N. Compute the linear density of the string.

21-23. A 0.500-g string is 4.3 m long and has a tension of 300 N. If it is fixed at each end and vibrates in three segments, what is the frequency of the standing waves?

Ans. 560 Hz

21-24. A string vibrates with standing waves in five loops when the frequency is 600 Hz. What frequency will cause the string to vibrate in only two loops?

21-25. A 120-g wire fixed at both ends is 8 m long and has a tension of 100 N. What is the longest possible wavelength for a standing wave? What is the frequency?

Ans. 16 m, 5.10 Hz

Challenge Problems

21-26. A longitudinal wave of frequency 400 Hz has a velocity of 60 m/s. What is the wavelength?

21-27. What is the speed of a transverse wave in a rope of length 2.00 m and mass 80 g under a tension of 400 N?

Ans. 100 m/s

21-28. A transverse wave travels at a speed of 8.00 m/s. A particular particle on the string moves from its highest point to its lowest point in a time of 0.03 s. What is the wavelength?

21-29. A bass guitar string 750 mm long is stretched with sufficient force to produce a fundamental vibration of 220 Hz. What is the velocity of the transverse waves in this string?

Ans. 330 m/s

21-30. A 5-kg mass is hung from the ceiling by a 30-g wire that is 1.8 m long. What is the fundamental frequency of vibration for this wire?

21-31. A steel guy wire supporting a pole is 18.9 m long and 9.5 mm in diameter. It has a linear density of 0.474 kg/m. When it is struck at one end by a hammer, the pulse returns in 0.3 s. What is the tension in the wire?

Ans. 7530 N

*21-32. A 30-m wire weighing 400 N is stretched with a tension of 1800 N. How much time is required for a pulse to make a round trip if it is struck at one end?

*21-33. Transverse waves have a speed of 20 m/s on a string that has a tension of 8 N. What tension is required to give a wave speed of 30 m/s for the same string?

Ans. 18.0 N

*21-34. The fundamental frequency for a given string is 80 Hz. If the mass of the string is doubled but other factors remain constant, what is the new fundamental frequency?

Critical Thinking Questions

21-35. In a laboratory experiment, an electromagnetic vibrator is used as a source of standing waves in a string. A 1-m length of the string is determined to have a mass of 0.6 g. One end of the string is connected to the tip of the vibrator, and the other passes over a pulley 1 m away and is attached to a weight hanger. A mass of 392 g hanging from the free end causes the string to vibrate in three segments. What is the frequency of the vibrator? What new mass attached to the free end will cause the string to vibrate in four loops? What is the fundamental frequency?

Ans. 120 Hz, 22.0 g

*21-36. To understand the parameters that affect wave velocity in a vibrating string, suppose that

$$v = \sqrt{\frac{F\ell}{m}} = 100 \text{ m/s}$$

What is the new wave speed v' for each of the following changes: (a) $F' = 2F$, (b) $m' = 2 \, m$, (c) $\ell' = 2\ell$?

*21-37. A 2-mW power source generates waves down rope A, and another power source generates waves down an identical rope B. The waves in each rope are of the same frequency f and velocity v. If the amplitude in rope B is twice that of rope A, what power is supplied to rope B? **Ans. 8.00 mW**

*21-38. The fundamental frequency of a steel piano wire is 253 Hz. By what fractional amount must the tension in the wire be increased in order that the frequency be the desired C note (256 Hz)?

*21-39. A variable oscillator allows a laboratory student to adjust the frequency of a source to produce standing waves in a vibrating string. A 1.20-m length of string ($\mu = 0.400$ g/m) is placed under a tension of 200 N. What frequency is necessary to produce three standing loops in the vibrating string? What is the fundamental frequency? What frequency will produce five loops?

Ans. 884 Hz, 295 Hz, 1470 Hz

Chapter

22

Sound

OBJECTIVES

After completing this chapter, you should be able to:

1. Define sound and solve problems involving its velocity in metal, in a liquid, and in a gas.
2. Use boundary conditions to derive and apply relationships for calculating the *characteristic frequencies* for an open pipe and for a closed pipe.
3. Compute the intensity level in decibels for a sound whose intensity is given in *watts per square meter*.
4. Use your understanding of the *Doppler effect* to predict the apparent change in sound frequency that occurs as a result of relative motion between a source and an observer.

When a periodic disturbance takes place in air, longitudinal *sound* waves travel out from it. For example, if a tuning fork is struck with a hammer, the vibrating prongs send out longitudinal waves, as shown in Fig. 22-1. An ear, acting as a receiver for these periodic waves, interprets them as sound.

Is the ear necessary for sound to exist? If the tuning fork were struck in the atmosphere of a distant planet, would there be sound, even though no ear could interpret the disturbance? The answer depends on the definition of sound.

The term **sound** is used in two different ways. Physiologists define *sound* in terms of the auditory sensations produced by longitudinal disturbances in air. For them,

Figure 22-1 A tuning fork acts as a source of longitudinal sound waves.

Sound

sound does not exist on a distant planet. In physics, on the other hand, we refer to the disturbances themselves rather than the sensations produced.

Sound is a longitudinal mechanical wave that travels through an elastic medium.

In this case, sound does exist on the planet. In this chapter *sound* will be used in its physical sense.

22-1 PRODUCTION OF A SOUND WAVE

Two things must exist to produce a sound wave. There must be a source of mechanical vibration, and there must be an elastic medium through which the disturbance can travel. The source may be a tuning fork, a vibrating string, or a vibrating air column in an organ pipe. *Sounds are produced by vibrating matter.* The requirement of an elastic medium can be demonstrated by placing an electric bell inside an evacuable flask, as shown in Fig. 22-2. When the bell is connected to a battery so that it rings continuously, the flask is slowly evacuated. As more and more of the air is pumped from the flask, the sound of the bell becomes fainter and fainter until finally it cannot be heard at all. When air is allowed to reenter the flask, the sound of the bell returns. Thus, air is necessary to transmit sound.

Let us now examine more closely the longitudinal sound waves in air as they proceed from a vibrating source. A thin strip of metal clamped tight at its base is pulled to one side and released. As the free end oscillates to and fro with simple harmonic motion, a series of periodic, longitudinal sound waves spread through the air away from the source. The air molecules in the vicinity of the metal strip are alternately compressed and expanded, sending out a wave like that illustrated in Fig. 22-3a. The dense regions where many molecules are packed tightly together are called **compressions.** They are exactly analogous to the *condensations* discussed for longitudinal waves in a coiled spring. The regions with relatively few molecules are referred to as **rarefactions.** The compressions and rarefactions alternate throughout the medium as the individual air particles oscillate to and fro in the

New microphone technology is getting smaller than ever before. One microphone is etched on a silicon wafer, and the vibration of a 400-nm-thick membrane converts the sound into an electrical signal. The entire device is only 2 mm². Another new microphone technology uses two tiny lanes of metal to measure the tiny ripples of air pushed aside by sound waves. The air causes more cooling on one lane than the other (the lanes are about as skinny as the wavelength of audible sound). This microphone captures sound using no moving parts and is said to be highly directional and to eliminate background noise. Another tiny microphone actually uses a laser shining on a gold-coated silicon nitride membrane and reflecting into an optical fiber. The reflected laser light is then converted into an electrical signal. This device is reported to be a thousand times more sensitive than the previous generation of fiber-optic pressure sensors.

Figure 22-2 A bell ringing in a vacuum cannot be heard. A material medium is necessary for the production of sound.

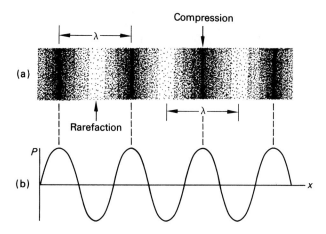

Figure 22-3 (a) Compressions and rarefactions in a sound wave in air at a particular instant. (b) The sinusoidal variation in pressure as a function of displacement.

direction of wave propagation. Since a compression corresponds to a high-pressure region and a rarefaction corresponds to a low-pressure region, a sound wave can also be represented by plotting the change in pressure P as a function of the distance x. (See Fig. 22-3b.) The distance between two successive compressions or rarefactions is the wavelength.

22-2 THE SPEED OF SOUND

Anyone who has seen a weapon being fired at a distance has observed the smoke from the weapon before hearing the report. Similarly, we observe the flash of lightning before hearing the thunder. Even though both light and sound travel with finite speeds, the speed of light is so much greater in comparison that it can be considered instantaneous. The speed of sound can be measured directly by observing the time required for the waves to move through a known distance. In air at 0°C, sound travels at a speed of 331 m/s or 1087 ft/s.

In Chap. 21 we established the idea that wave speed depends on the elasticity of the medium and the inertia of its particles. The more elastic materials sustain greater wave speeds, whereas the denser materials retard wave motion. The following empirical relationships are based on these proportionalities.

For longitudinal sound waves in a wire or rod, the wave speed is given by

$$v = \sqrt{\frac{Y}{\rho}} \qquad\qquad Rod \quad \textbf{22-1}$$

where Y is Young's modulus for the solid and ρ is its density. This relation is valid only for rods whose diameters are small in comparison with the longitudinal wavelengths of sound passing through them.

In an *extended solid*, the longitudinal wave speed is a function of the shear modulus S, the bulk modulus B, and the density p of the medium. The wave speed can be calculated from

$$v = \sqrt{\frac{B + \frac{4}{3}S}{\rho}} \qquad \text{Extended Solid} \quad \textbf{22-2}$$

For longitudinal waves in a fluid, the wave speed is found from

$$v = \sqrt{\frac{B}{\rho}} \qquad \text{Fluid} \quad \textbf{22-3}$$

where B is the bulk modulus for the fluid and ρ is its density.

In computing the speed of sound in a gas, the bulk modulus is given by

$$B = \gamma P$$

where γ is the adiabatic constant ($\gamma = 1.4$ for air and diatomic gases) and P is the pressure of the gas. Thus, the speed of longitudinal waves in a gas, from Eq. (22-3), is given by

$$v = \sqrt{\frac{B}{\rho}} = \sqrt{\frac{\gamma P}{\rho}} \qquad \textbf{22-4}$$

But for an ideal gas

$$\frac{P}{\rho} = \frac{RT}{M} \qquad \textbf{22-5}$$

where R = universal gas constant
T = absolute temperature of gas
M = molecular mass of gas

Substitution of Eq. (22-5) into Eq. (22-4) yields

$$v = \sqrt{\frac{\gamma P}{\rho}} = \sqrt{\frac{\gamma RT}{M}} \qquad \text{Gas} \quad \textbf{22-6}$$

EXAMPLE 22-1

Compute the speed of sound in an aluminum rod.

Solution

Young's modulus and the density for aluminum are

$$Y = 68{,}900 \text{ MPa} = 6.89 \times 10^{10} \text{ N/m}^2$$
$$\rho = 2.7 \text{ g/cm}^3 = 2.7 \times 10^3 \text{ kg/m}^3$$

From Eq. (22-1),

$$v = \sqrt{\frac{Y}{\rho}} = \sqrt{\frac{6.89 \times 10^{10} \text{ N/m}^2}{2.7 \times 10^3 \text{ kg/m}^3}}$$
$$= \sqrt{2.55 \times 10^7 \text{ m}^2/\text{s}^2} = 5050 \text{ m/s}$$

This speed is approximately 15 times the speed of sound in air.

EXAMPLE 22-2

Compute the speed of sound in air on a day when the temperature is 27°C. The molecular mass of air is 29.0, and the adiabatic constant is 1.4.

Solution

From Eq. (22-6),

$$v = \sqrt{\frac{\gamma RT}{M}} = \sqrt{\frac{(1.4)(8.31 \text{ J/mol} \cdot \text{K})(300 \text{ K})}{29 \times 10^{-3} \text{ kg/mol}}}$$
$$= 347 \text{ m/s}$$

The speed of sound is significantly greater at 27°C than at 0°C. At standard temperature and pressure (0°C, 1 atm) the speed of sound is 331 m/s. For each Celsius degree rise in temperature (above 0°C) the speed of sound in air increases by approximately 0.6 m/s. Hence, the speed v of sound can be approximated by

$$v = 331 \text{ m/s} + \left(0.6 \frac{\text{m/s}}{\text{C}°}\right)t \qquad \text{22-7}$$

where t is the Celsius temperature of the air.

EXAMPLE 22-3

What is the approximate speed of sound in air at room temperature (20°C)?

Solution

From Eq. (22-7),

$$v = 331 \text{ m/s} + \left(0.6 \frac{\text{m/s}}{\text{C}°}\right)(20°\text{C}) = 343 \text{ m/s}$$

22-3 VIBRATING AIR COLUMNS

In Chap. 21 we described the possible modes of vibration for a string fixed at both ends. The frequency of the sound waves set up in the air surrounding the string is identical with the frequency of the vibrating string. Thus, the possible frequencies, or the *harmonics,* of sound waves produced by a vibrating string are given by

$$f_n = \frac{nv}{2l} \qquad n = 1, 2, 3, \ldots \qquad \text{22-8}$$

where v is the velocity of transverse waves in the string.

Sound can also be produced by the longitudinal vibrations of an air column in an open or closed pipe. As in the vibrating string, the possible modes of vibrations are determined by the boundary conditions. The possible modes of vibration for the

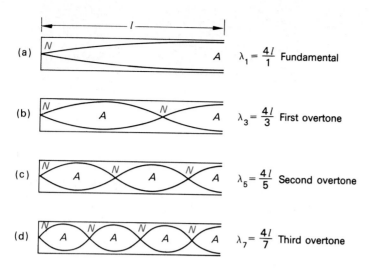

Figure 22-4 Possible standing waves in a closed pipe.

air in a closed pipe are illustrated in Fig. 22-4. When a compressional wave is set up in the pipe, the displacement of the air particles at the closed end must be zero.

The closed end of a pipe must be a displacement node.

The air at the open end of a pipe has the greatest freedom of motion, and so the displacement is free at an open end.

The open end of a pipe must be a displacement antinode.

The sinusoidal curves in Fig. 22-4 represent maximum displacements.

The fundamental mode of oscillation for an air column in a closed pipe has a node at the closed end and an antinode at the open end. Thus, the wavelength of the fundamental is four times the length l of the pipe (Fig. 22-4a). The next possible mode, which is the first overtone, occurs when there are two nodes and two antinodes, as shown in Fig. 22-4b. The wavelength of the first overtone is therefore equal to $4l/3$. Similar reasoning will show that the second and third overtones occur for wavelengths equal to $4l/5$ and $4l/7$, respectively. In summary, the possible wavelengths are

$$\lambda_n = \frac{4l}{n} \qquad n = 1, 3, 5, \ldots \qquad \text{22-9}$$

The speed of the sound waves is given by $v = f\lambda$, so that the possible frequencies for a *closed pipe* are

$$f_n = \frac{nv}{4l} \qquad n = 1, 3, 5, \ldots \qquad \textit{Closed Pipe} \quad \text{22-10}$$

Notice that only the *odd harmonics* are allowed for a closed pipe. The first overtone is the third harmonic, the second overtone is the fifth harmonic, and so on.

An air column vibrating in a pipe open at *both* ends must be bounded by displacement antinodes. Figure 22-5 shows the fundamental and first three overtones for an open pipe. Note that the fundamental wavelength is twice the length l of

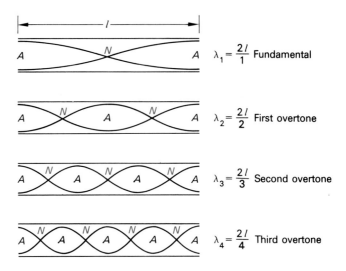

Figure 22-5 Possible standing waves in an open pipe.

the pipe. When the number of nodes is increased one at a time, the possible wavelengths in an open pipe are

$$\lambda_n = \frac{2l}{n} \qquad n = 1, 2, 3, \ldots \qquad \text{22-11}$$

The possible frequencies are, therefore,

$$f_n = \frac{nv}{2l} \qquad n = 1, 2, 3, \ldots \qquad \textit{Open Pipe} \quad \text{22-12}$$

where v is the velocity of the sound waves. Thus, all the harmonics are possible for a vibrating air column in an open pipe. Open pipes of varying lengths are used in many musical instruments, for example, organs, flutes, and trumpets.

EXAMPLE 22-4

What are the frequencies of the fundamental and first two overtones for a 12-cm closed pipe? Air temperature is 30°C.

Solution

The velocity of the sound is

$$v = 331 \text{ m/s} + \left(0.6 \, \frac{\text{m/s}}{\text{C}°}\right)(30°\text{C}) = 349 \text{ m/s}$$

Thus, from Eq. (22-10),

$$f_1 = \frac{1v}{4l} = \frac{349 \text{ m/s}}{(4)(0.12 \text{ m})} = 727 \text{ Hz}$$

The first and second overtones are the third and fifth harmonics. Hence,

$$\text{First overtone} = 3f_1 = 2181 \text{ Hz}$$
$$\text{Second overtone} = 5f_1 = 3635 \text{ Hz}$$

EXAMPLE 22-5

What length of open pipe will have a frequency of 1200 Hz as its first overtone? Take the speed of sound as 340 m/s.

Solution

The first overtone in an open pipe is equal to the second harmonic. Thus, we can set $n = 2$ in Eq. (22-12):

$$f_2 = \frac{2v}{2l} = \frac{v}{l}$$

$$l = \frac{v}{f_2} = \frac{340 \text{ m/s}}{1200 \text{ Hz}} = 0.283 \text{ m}$$

22-4 FORCED VIBRATION AND RESONANCE

When a vibrating body is placed in contact with another body, the second body is forced to vibrate with the same frequency as the original vibrator. For example, if a tuning fork is struck with a hammer and then placed with its base against a wooden tabletop, the intensity of the sound will suddenly be increased. When the tuning fork is removed from the table, the intensity decreases to its original level. The vibrations of the particles in the tabletop in contact with the tuning fork are called *forced vibrations.*

We have seen that elastic bodies have certain natural frequencies of vibration that are characteristic of the material and boundary conditions. A taut string of a particular length can produce sounds of characteristic frequencies. An open or closed pipe also has natural frequencies of vibration. Whenever a body is acted on by a series of periodic impulses having a frequency nearly equal to one of the natural frequencies of the body, the body is set into vibration with a relatively large amplitude. This phenomenon is referred to as *resonance* or *sympathetic vibration.*

An example of resonance is offered by a child sitting in a swing. Experience tells us that the swing can be set into vibration with large amplitude by a series of small pushes at just the right intervals. Such resonance occurs only when the pushes are in phase with the natural frequency of vibration for the swing. A slight variation of the input pulses would result in little or no vibration.

Reinforcement of sound by resonance has many useful applications as well as many unpleasant consequences. The resonance of an air column in an organ pipe amplifies the weak sound of a vibrating air jet. Many musical instruments are designed with resonant cavities to produce varying sounds. Electrical resonance in radio receivers enables the listener to hear weak signals clearly. When tuned to the frequency of a desired station, the signal is amplified by electrical resonance. In poorly designed auditoriums or long hallways, music and voices may have a hollow sound that is unpleasant to the ear. Bridges have been known to collapse because of sympathetic vibrations set up by gusts of wind.

22-5 AUDIBLE SOUND WAVES

We have defined **sound** as a *longitudinal mechanical wave traveling through an elastic medium.* This is a broad definition that makes no restriction whatsoever on the frequencies of sound. The physiologist is concerned primarily with sound waves that are capable of affecting the sense of hearing. Thus, it is useful to divide the spectrum of sound according to the following definitions.

> **Audible sound refers to sound waves in the frequency range from 20 to 20,000 Hz.**

> **Sound waves having frequencies below the audible range are termed infrasonic.**

> **Sound waves having frequencies above the audible range are termed ultrasonic.**

When studying audible sound, the physiologist uses the terms **loudness, pitch,** and **quality** to describe the sensations produced. Unfortunately, these terms represent sensory magnitudes and are therefore subjective. What is loud to one person is moderate to another. What one person perceives as quality, another considers inferior. As always, the physicist must deal with explicit measurable definitions. The physicist therefore attempts to correlate the sensory effects with the physical properties of waves. These correlations can be summarized as follows:

Sensory effects		Physical property
Loudness	$\longleftrightarrow$	Intensity
Pitch	$\longleftrightarrow$	Frequency
Quality	$\longleftrightarrow$	Waveform

The meaning of the terms on the left may vary considerably among individuals. The terms on the right are measurable and objective.

Sound waves constitute a flow of energy through matter. The **intensity** of a given sound wave is a measure of the rate at which energy is propagated through a given volume of space. A convenient method of specifying sound intensity is in terms of the rate at which energy is transferred through a unit area normal to the direction of wave propagation (see Fig. 22-6). Since the rate at which energy flows is the *power* of a wave, the intensity can be related to the power per unit area passing a given point.

> **Sound intensity is the power transferred by a sound wave through a unit of area normal to the direction of the propagation.**

$$I = \frac{P}{A}$$

22-13

The units for intensity are the ratio of a power unit to an area unit. In SI units, the intensity is in W/m², and that is the unit we will use in this text. The rate of

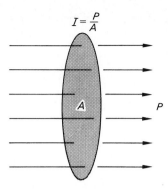

$$I = \frac{P}{A}$$

Figure 22-6 The intensity of a sound wave is a measure of the power transmitted per unit of area perpendicular to the direction of wave propagation.

energy flow in sound waves is small, however, and industry still uses the μW/cm^2 in many applications. The conversion factor is

$$1 \ \mu\text{W/cm}^2 = 1 \times 10^{-2} \ \text{W/m}^2$$

It can be shown by methods similar to those utilized for a vibrating string that the sound intensity varies directly with the square of the frequency f and with the square of the amplitude A of a given sound wave. Symbolically, the intensity I is given by

$$I = 2\pi^2 f^2 A^2 \rho v \qquad \text{22-14}$$

where v is the sound velocity in a medium of density ρ. The symbol A in Eq. (22-14) refers to the amplitude of the sound wave and not the unit area, as in Eq. (22-13).

The intensity I_0 of the faintest audible sound is of the order of 10^{-12} W/m^2. This intensity, which is referred to as the **hearing threshold,** has been adopted by acoustical experts as the minimum intensity for audible sound.

> **The hearing threshold represents the standard minimum of intensity for audible sound. Its value at a frequency of 1000 Hz is**

$$I_0 = 1 \times 10^{-12} \ \text{W/m}^2 = 1 \times 10^{-10} \ \mu\text{W/cm}^2 \qquad \text{22-15}$$

The range of intensities over which the human ear is sensitive is enormous. It extends from the hearing threshold I_0 to an intensity 10^{12} times as great. The upper extreme represents the point at which the intensity is intolerable for the human ear. The sensation becomes one of feeling or pain instead of simply hearing.

> **The pain threshold represents the maximum intensity that the average ear can record without feeling or pain. Its value is**

$$I_p = 1 \ \text{W/m}^2 = 100 \ \mu\text{W/cm}^2 \qquad \text{22-16}$$

In view of the wide range of intensities over which the ear is sensitive, it is more convenient to set up a logarithmic scale for the measurement of sound intensities. Such a scale is established by the following rule:

> **When the intensity I_1 of one sound is 10 times as great as the intensity I_2 of another, the ratio of intensities is said to be 1 bel (B).**

Thus, when comparing the intensities of two sounds, we refer to a difference in intensity levels given by

$$B = \log \frac{I_1}{I_2} \qquad \text{bels (B)} \qquad \text{22-17}$$

where I_1 is the intensity of one sound and I_2 is the intensity of the other.

EXAMPLE 22-6

Two sounds have intensities of 2.5×10^{-8} W/m^2 and 1.2 W/m^2. Compute the difference in intensity levels in bels.

Solution

$$B = \log \frac{I_1}{I_2} = \log \frac{1.2 \text{ W/m}^2}{2.5 \times 1.0^{-8} \text{ W/m}^2}$$
$$= \log 4.8 \times 10^7 = 7.68 \text{ B}$$

In practice, the unit of 1 B is too large. To obtain a more useful unit, we define a *decibel* (dB) as one-tenth of a bel. Thus, the answer to Example 22-6 can also be expressed as 76.8 dB.

By using the standard intensity I_0 as a comparison for all intensities, a general scale has been devised for rating any sound. The intensity level in decibels of any sound of intensity I can be found from the general relation

$$\beta = 10 \log \frac{I}{I_0} \qquad \text{decibels (dB)} \qquad \text{22-18}$$

where I_0 is the intensity at the hearing threshold (1×10^{-12} W/m^2). The intensity level for I_0 is zero decibels.

EXAMPLE 22-7

Compute the intensity level of a sound whose intensity is 1×10^{-4} W/m^2.

Solution

$$\beta = 10 \log \frac{I}{I_0} = 10 \log \left(\frac{10^{-4} \text{ W/m}^2}{10^{-12} \text{ W/m}^2} \right)$$
$$= 10 \log 10^8 = 10(8) = 80 \text{ dB}$$

Through the logarithmic decibel notation, we have reduced the wide range of intensities to intensity levels from 0 to 120 dB. We must remember, however, that the scale is not linear but logarithmic. A 40-dB sound is much more than twice as intense as a 20-dB sound. A sound that is 100 times as intense as another is only 20 dB larger. Several examples of the intensity levels for common sounds are given in Table 22-1.

TABLE 22-1 Intensity Levels for Common Sounds	
Sound	**Intensity level, dB**
Hearing threshold	0
Rustling leaves	10
Whisper	20
Quiet radio	40
Normal conversation	65
Busy street corner	80
Subway car	100
Pain threshold	120
Jet engine	140–160

The intensity of a sound decreases as a person moves away from a source. The change in intensity varies with the square of the distance from the source. For example, a person twice as far away from a source hears a sound that is one-fourth as intense; a person three times as far away hears one that is one-ninth as intense. To see why this occurs, consider that sound radiates outward in all directions from a point source as illustrated in Fig. 22-7. The sound wave is seen as a succession of spherical surfaces. Consider points A and B located at distances r_1 and r_2 from a source that produces a sound of power P. Recalling that $I = P/A$ and that the area of a sphere is $4\pi r^2$, we can write the intensities I_1 and I_2 as follows:

$$I_1 = \frac{P}{4\pi r_1^2} \quad \text{and} \quad I_2 = \frac{P}{4\pi r_2^2}$$

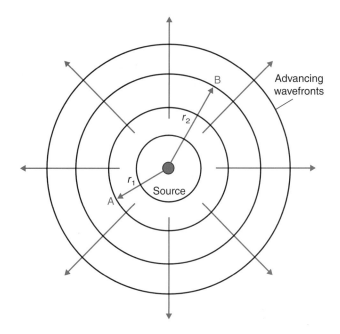

Figure 22-7 Spherical sound waves propagating radially outward from an isometric source. The intensity drops off inversely with the square of the distance from the source.

The power of the source does not change; eliminating P from the above equations gives a useful formula for relating intensities to distances from a source:

$$\frac{I_1}{I_2} = \frac{r_2^2}{r_1^2} \quad \text{or} \quad I_1 r_1^2 = I_2 r_2^2 \qquad \text{22-19}$$

CONCEPTUAL EXAMPLE 22-8

A point source emits sound with an average power of 40 W. What is the intensity at a distance r_1 of 3.5 m from the source? What will be the intensity at a distance r_2 of 5 m?

Solution

Consider that the sound emanates from the source in spherical waves. At a distance r_1 from the 40-W source of sound, the power is distributed over the entire surface of the spherical wave of area $4\pi r^2$. The intensity at r_1 is, therefore, the power per unit area.

$$I_1 = \frac{P}{4\pi r_1^2} = \frac{40 \text{ W}}{4\pi(3.5 \text{ m})^2} = 0.260 \text{ W/m}^2$$

Now, we can use Eq. (22-19) to find the intensity at the 5-m point. Solving for I_2, we have

$$I_2 = \frac{I_1 r_1^2}{r_2^2} = \frac{(0.260 \text{ W/m}^2)(3.5 \text{ m})^2}{(5 \text{ m})^2} = 0.127 \text{ W/m}^2$$

Notice that this inverse square relationship applies to *intensities* and not to **intensity levels.**

22-6 PITCH AND QUALITY

The effect of intensity on the human ear manifests itself as **loudness.** In general, sound waves that are more intense are also louder, but the ear is not equally sensitive to sounds of all frequencies. Therefore, a high-frequency sound may not seem as loud as one of lower frequency that has the same intensity.

The *frequency* of a sound determines what the ear judges as the **pitch** of the sound. Musicians designate pitch by letters corresponding to key notes on the piano. For example the C note, D note, and F note each refer to a specific pitch, or frequency. A siren disk, shown in Fig. 22-8, can be used to demonstrate how the pitch is determined by the frequency of a sound. A stream of air is directed against a row of evenly spaced holes. Varying the rate of rotation of the disk causes the pitch of the resulting sound to be increased or decreased.

Two sounds of the same pitch can be distinguished easily. For example, suppose we sound a C note (256 Hz) successively on a piano, a flute, a trumpet, and a violin. Even though each sound has the same pitch, there is a marked difference in the tones. This distinction is said to result from a difference in the *quality* of sound.

Figure 22-8 Demonstrating the relationship between pitch and frequency.

Regardless of the source of vibration in musical instruments, several modes of oscillation are usually excited simultaneously. Therefore, the sound produced consists not only of the fundamental but also of many of the overtones. *The quality of a sound is determined by the number and relative intensities of the overtones present.* The difference in quality between two sounds can be observed objectively by analyzing the complex *waveforms* resulting from each sound. In general, the more complex the wave, the greater the number of harmonics that contribute to it.

22-7 INTERFERENCE AND BEATS

In Chap. 21 we discussed the superposition principle as a method for studying interference in transverse waves. Interference also occurs in longitudinal sound waves, and the superposition principle can be applied for them also. A common example of the interference in sound waves occurs when two tuning forks (or other single-frequency sound sources) whose frequencies differ only slightly are struck simultaneously. The sound produced fluctuates in intensity, alternating between loud tones and virtual silence. These regular pulsations are referred to as **beats.** The *vibrato* effect obtained on some organs is an application of this principle. Every vibrato note is produced by two pipes tuned to slightly different frequencies.

To understand the origin of beats, let us examine the interference set up between sound waves proceeding from two tuning forks of slightly different frequency, as shown in Fig. 22-9. The superposition of waves *A* and *B* illustrates the origin of beats. The loud tones occur when the waves interfere constructively, and the quiet tones occur when the waves interfere destructively. Observation and calculation show that the two waves interfere constructively $f - f'$ times per second. Thus, we can write

$$\text{Number of beats per second} = |f - f'| \qquad \textbf{22-20}$$

For example, if tuning forks of frequencies 256 and 259 Hz are struck simultaneously, the resulting sound will pulsate three times every second.

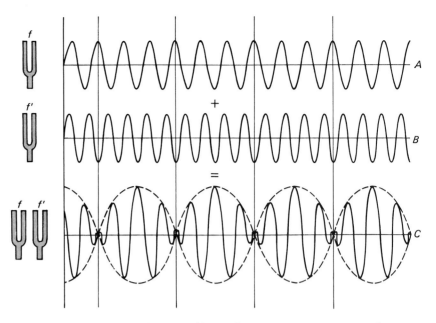

Figure 22-9 Diagram illustrating the origin of beats. The wave *C* is a superposition of waves *A* and *B*.

22-8 THE DOPPLER EFFECT

Whenever a source of sound is moving relative to an observer, the pitch of the sound, as heard by the observer, may not be the same as that perceived when the source is at rest. For example, if we stand near a railway track as a train blowing its whistle approaches, we notice that the pitch of the whistle is *higher* than the normal one when the train is stationary. As the train recedes, the pitch is observed to be *lower* than normal. Similarly, at race tracks the sound of cars driving toward the stands is considerably higher in pitch than the sound of cars driving away from the stands.

The phenomenon is not restricted to the motion of the source. If the source of sound is stationary, a listener moving toward the source will observe a similar increase in pitch. A listener leaving the source of sound will hear a lower-pitched sound. The change in frequency of sound resulting from relative motion between a source and an observer is called the ***Doppler effect.***

> **The Doppler effect refers to the apparent change in frequency of a source of sound when there is relative motion of the source and the listener.**

The origin of the Doppler effect can be demonstrated graphically by representing the periodic waves emitted by a source as concentric circles moving radially outward, as in Fig. 22-10. The distance between any two circles represents the wavelength λ of the sound traveling with a velocity *V.* The frequency with which these waves strike the ear determines the pitch of sound heard.

Let us first consider that the source is moving to the right toward a stationary observer *A,* as in Fig. 22-11. As the moving source emits sound waves, it tends to

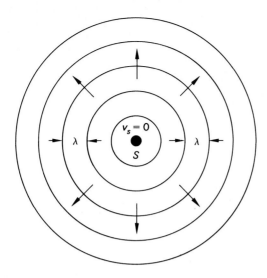

Figure 22-10 Graphic representation of sound waves emitted from a stationary source.

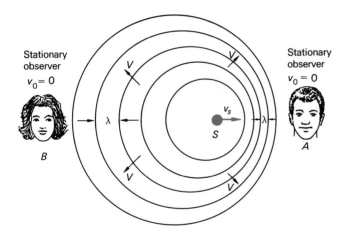

Figure 22-11 Illustration of the Doppler effect. The waves in front of a moving sound are closer together than the waves behind a moving source.

overtake waves traveling in the same direction as the source. Each successive wave is emitted from a point closer to the observer than its predecessor. The result is that the distance between successive waves, or the wavelength, is smaller than usual. A smaller wavelength results in a higher frequency of waves, which increases the pitch of the sound heard by observer A. Similar reasoning will show that an *increase* in the length of waves reaching observer B will cause B to hear a *lower*-frequency sound.

We can now derive a relationship for predicting the change in observed frequency. During one complete vibration of a stationary source (a time equal to the period T), each wave will move through a distance of one wavelength. This distance is represented by λ in Fig. 22-12a and is given by

$$\lambda = VT = \frac{V}{f_s} \qquad \textit{Stationary Source}$$

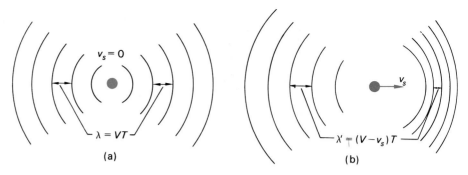

Figure 22-12 Computing the magnitude of sound emitted from a moving source. The source velocity v_s is considered positive for speeds of approach and negative for speeds of recession.

where V is the velocity of sound and f_s is the frequency of the source. If the source is moving to the right with a velocity v_s, as in Fig. 22-12b, the new wavelength λ' in front of the source will be given by

$$\lambda' = VT - v_sT = (V - v_s)T$$

But $T = 1/f_s$, so we write

$$\lambda' = \frac{V - v_s}{f_s} \qquad\qquad \text{Moving Source} \quad \textbf{22-21}$$

This equation will also apply for the wavelength on the left of the moving source if we follow the convention that speeds of approach are considered positive and speeds of recession are considered negative. Thus, if we were computing λ' on the left of our moving source, the negative value would be substituted for v_s, resulting in a larger wavelength.

The velocity of sound in a medium is a function of the properties of the medium and does not depend on the motion of the source. Thus, the frequency f_o heard by a stationary observer from a moving source of frequency f_s is given by

$$f_o = \frac{V}{\lambda'} = \frac{Vf_s}{V - v_s} \qquad\qquad \text{Moving Source} \quad \textbf{22-22}$$

where V is the speed of sound and v_s is the speed of the source. *The speed v_s is reckoned as positive for speeds of approach and negative for speeds of recession.*

EXAMPLE 22-9

A train whistle emits sound at a frequency of 400 Hz. (a) What is the pitch of the sound heard when the train is moving toward a stationary observer at a speed of 20 m/s? (b) What is the pitch heard when the train is moving away from the observer at this speed? Assume that the speed of sound is 340 m/s.

Solution (a)

Since the train is approaching the observer, its speed v_s is positive. Substitution into Eq. (22-22) yields

$$f_o = \frac{Vf_s}{V - v_s} = \frac{(340 \text{ m/s})(400 \text{ Hz})}{340 \text{ m/s} - 20 \text{ m/s}}$$

$$= \frac{1.36 \times 10^5}{320} = 425 \text{ Hz}$$

Solution (b)

Now v_s represents a speed of recession, so that -20 m/s should be substituted in Eq. (22-22).

$$f_o = \frac{Vf_s}{V - v_s} = \frac{(340 \text{ m/s})(400 \text{ Hz})}{340 \text{ m/s} - (-20 \text{ m/s})}$$

$$= \frac{1.36 \times 10^5}{360} = 378 \text{ Hz}$$

Now let us examine the situation in which a source is stationary and the observer moves toward the source with a velocity v_o. In this case, the wavelength of the sound received does not change, but the number of waves encountered per unit of time (the frequency) increases as a result of the observer's speed v_o. Hence, the observer will hear the frequency

$$f_o = \frac{f_s(V + v_o)}{V} \qquad \textit{Moving Observer} \quad \textbf{22-23}$$

Here, *the speed v_o of the observer should be reckoned as positive for speeds of approach and negative for speeds of recession.*

EXAMPLE 22-10

A stationary source of sound has a frequency of 800 Hz on a day when the speed of sound is 340 m/s. What pitch is heard by a person who is moving away from the source at a speed of 30 m/s?

Solution

Since v_o represents a speed of recession, $v_o = -30$ m/s must be used in Eq. (22-22). Therefore,

$$f_o = \frac{f_s(V + v_o)}{V} = \frac{(800 \text{ Hz})(340 \text{ m/s} - 30 \text{ m/s})}{340 \text{ m/s}}$$

$$= \frac{(800)(310)}{340} = 729 \text{ Hz}$$

If both the observer and the source are moving, Eq. (22-22) and (22-23) are combined to yield

$$f_o = f_s \frac{V + v_o}{V - v_s}$$
General Motion   **22-24**

The same sign convention must be used for this general equation as that developed earlier. Note that Eq. (22-24) will reduce to Eq. (22-22) or to Eq. (22-23) if the observer is stationary ($v_o = 0$) or if the source is stationary ($v_s = 0$).

Summary and Review

Summary

We have defined sound as a longitudinal mechanical wave in an elastic medium. Thus, the elasticity and density of a medium will affect the speed of sound as it travels in that medium. Under certain conditions we have seen that standing sound waves can produce characteristic frequencies that we observe as the pitch of the sound. The intensity of sound and the Doppler effect were also discussed in this chapter. The major concepts are summarized below.

- Sound is a longitudinal wave traveling through an elastic medium. Its speed in air at 0°C is 331 m/s or 1087 ft/s. At other temperatures the speed of sound is approximated by

$$v = 331 \text{ m/s} + \left(0.6 \frac{\text{m/s}}{\text{C}°}\right)t_C$$

Speed of Sound in Air

- The speed of sound in other media can be found from the following:

$$v = \sqrt{\frac{Y}{\rho}} \qquad \text{Rod}$$

$$v = \sqrt{\frac{\gamma P}{\rho}} = \sqrt{\frac{\gamma RT}{M}} \qquad \text{Gas}$$

$$v = \sqrt{\frac{B}{\rho}} \qquad \text{Fluid}$$

$$v = \sqrt{\frac{B + \frac{4}{3}S}{\rho}} \qquad \text{Extended Solid}$$

- Standing longitudinal sound waves may be set up in a vibrating air column for a pipe that is open at both ends or for one that is closed at one end. The characteristic frequencies are

$$f_n = \frac{nv}{2l} \qquad n = 1, 2, 3, \ldots$$

Open Pipe of Length l

$$f_n = \frac{nv}{4l} \qquad n = 1, 3, 5, \ldots$$

Closed Pipe of Length l

Note that *only the odd harmonics are possible for a closed pipe*. In this case, the first overtone is the third harmonic, the second overtone is the fifth harmonic, and so on.

- The intensity of a sound is the power P per unit area A perpendicular to the direction of propagation.

$$I = \frac{P}{A} = 2\pi^2 f^2 A^2 \rho v \qquad \text{Intensity, W/m}^2$$

- The intensity level in decibels is given by

$$\beta = 10 \log \frac{I}{I_0} \qquad I_0 = 1 \times 10^{-12} \text{ W/m}^2$$

Intensity Level

- Whenever two waves are nearly the same frequency and exist simultaneously in the same medium, beats are set up such that

$$\text{Number of beats per second} = f - f'$$

- The general equation for the Doppler effect is

$$f_o = f_s \frac{V + v_o}{V - v_s} \qquad \text{Doppler Effect}$$

where f_o = observed frequency
f_s = frequency of source
V = velocity of sound
v_o = velocity of observer
v_s = velocity of source

Note: Speeds are reckoned as positive for approach and negative for recession.

Summary and Review

Key Terms

1. sound
2. compression
3. condensation
4. resonance
5. infrasonic
6. ultrasonic
7. loudness
8. intensity
9. decibels
10. pitch
11. forced vibration
12. audible sound
13. hearing threshold
14. pain threshold
15. intensity level
16. frequency
17. quality
18. waveform
19. beats
20. Doppler effect

Review Questions

22-1. What is the physiological definition of sound? What is the meaning of sound in physics?

22-2. Why must astronauts on the surface of the moon communicate by radio? Can they hear another spacecraft as it lands nearby? Can they hear by touching helmets?

22-3. How is the sound of a person's voice affected by inhaling helium gas? Is the effect one of pitch, loudness, or quality?

22-4. Vocal sounds originate with the vibration of vocal cords. The mouth and nasal openings act as a resonant cavity to amplify and distinguish sounds. Suppose you hum at a constant pitch equal to the C note on a piano. By opening and closing your mouth, what physiological property of the sound is affected?

22-5. The distance in miles to a thunderstorm can be estimated by counting the number of seconds elapsing between the flash of lightning and the arrival of a clap of thunder and dividing the result by 5. Explain why this is a reasonable approximation.

22-6. A store window is broken by an explosion several miles away. A glass of thin crystal shatters when a high note is reached on a violin. Are the causes of damage similar? What physical property of sound was principally responsible in each case?

22-7. Compare the speed of sound in solids, liquids, and gases. Explain the reason for differences in speed.

22-8. Perform a unit analysis of Eq. (22-1) showing that $\sqrt{Y/\rho}$ will yield units of speed.

22-9. How will the speed of sound in a gas be affected if the temperature of the gas is quadrupled?

22-10. An electric bell operates inside an evacuated flask. No sound is heard because of the absence of a medium. Explain what happens if the flask is tilted until the bell touches the walls of the flask.

22-11. Draw diagrams to demonstrate the differences between a progressive longitudinal wave and a standing longitudinal wave.

22-12. A standing wave is set up in a vibrating string. How are the harmonics of the possible sounds related to the number of loops in the string? How are the harmonics related to the number of nodes?

22-13. What effect will closing one end of an open pipe have on the frequency of a vibrating air column?

22-14. Compare the quality of sound produced by a violin with that produced by a tuning fork.

22-15. If the average ear cannot hear sounds of frequencies much in excess of 15,000 Hz, what is the advantage of building stereo music systems that have frequency responses much higher than 15,000 Hz?

22-16. A vibrating tuning fork mounted on a resonating box is moved toward a wall and away from an observer. The resulting sound pulsates in intensity. Explain.

22-17. An instructor attempts to explain the Doppler effect by using baseballs and a bicycle. He proceeds as follows: "Suppose I am at rest, and I release one baseball in the same direction every second at constant speed. Consider me as the source of sound waves and the baseballs as advancing wavefronts. The spacing between the balls at any instant will be constant and analogous to the wavelength of sound waves. Now, suppose I ride a bicycle in the forward direction at a constant speed and continue to release balls in the forward direction and in the backward direction at the same rate and at the same speed. The spacing of the balls in front of me will be closer together because each time I release a ball in that direction I will have also moved in that direction. Similarly, the balls released in the backward direction will be spaced further apart than normal." Give a careful analysis of his explanation. In what ways is his analogy correct? In what important aspect does his analogy fail? Why would an equation similar to Eq. (22-21) fail as a means of predicting the spacing of the baseballs? Why does it work for sound waves?

Problems

Section 22-2—The Speed of Sound

22-1. Young's modulus for steel is 2.07×10^{11} Pa, and its density is 7800 kg/m³. Compute the speed of sound in a steel rod.
Ans. 5150 m/s

22-2. A 3-m length of copper rod has a density of 8800 kg/m³, and Young's modulus for copper is 1.17×10^{11} Pa. How much time will it take for sound to travel from one end of the rod to the other?

22-3. What is the speed of sound in air ($M = 29$ g/mol and $\gamma = 1.4$) on a day when the temperature is 30°C? Use the approximation formula to check this result. **Ans. 349 m/s**

22-4. The speed of longitudinal waves in a certain metal rod of density 7850 kg/m³ is measured to be 3380 m/s. What is the Young's modulus for the metal?

22-5. If the frequency of the waves in Prob. 22-4 is 312 Hz, what is the wavelength?
Ans. 10.8 m

22-6. Compare the theoretical speeds of sound in hydrogen ($M = 2.0$ g/mol, $\gamma = 1.4$) with helium ($M = 4.0$ g/mol, $\gamma = 1.66$) at 0°C.

*22-7. A sound wave is sent from a ship to the ocean floor, where it is reflected and returned. If the round trip takes 0.6 s, how deep is the ocean floor? Consider the bulk modulus for seawater to be 2.1×10^9 Pa and its density to be 1030 kg/m³.
Ans. 428 m

Section 22-3—Vibrating Air Columns

22-8. Find the fundamental frequency and the first three overtones for a 20-cm pipe at 20°C if the pipe is open at both ends.

22-9. Find the fundamental frequency and the first three overtones for a 20-cm pipe, at 20°C if the pipe is closed at one end.
Ans. 429, 1290, 2140, and 3000 Hz

22-10. What length of closed pipe will produce a fundamental frequency of 256 Hz at 20°C?

22-11. What length of open pipe will produce a fundamental frequency of 356 Hz at 20°C?
Ans. 48.2 cm

22-12. What length of open pipe will produce a frequency of 1200 Hz as its first overtone on a day when the speed of sound is 340 m/s?

22-13. The second overtone of a closed pipe is 1200 Hz at 20°C. What is the length of the pipe?
Ans. 35.7 cm

22-14. In a resonance experiment, the air in a closed tube of variable length is found to resonate with a tuning fork when the air column is first 6 cm and then 18 cm long. What is the frequency of the tuning fork if the temperature is 20°C?

*22-15. A closed pipe and an open pipe are each 3 m long. Compare the wavelength of the fourth overtone for each pipe at 20°C.
Ans. open = 1.20 m, closed = 1.33 m

Summary and Review

Section 22-5—Sound Intensity and Intensity level

22-16. What is the intensity level in decibels of a sound that has an intensity of 4×10^{-5} W/m^2?

22-17. The intensity of a sound is 6×10^{-8} W/m^2. What is the intensity level?   **Ans. 47.8 dB**

22-18. A 60-dB sound is measured at a particular distance from a whistle. What is the intensity of this sound in W/m^2?

22-19. What is the intensity of a 40-dB sound?

Ans. 1 $\times$ 10^{-8} W/m^2

22-20. Compute the intensities for sounds of 10 dB, 20 dB, and 30 dB.

22-21. Compute the intensity levels for sounds of 1×10^{-6} W/m^2, 2×10^{-6} W/m^2, and 3×10^{-6} W/m^2.

Ans. 60.0 dB, 63.0 dB, 64.8 dB

22-22. An isometric source of sound broadcasts a power of 60 W. What are the intensity and the intensity level of a sound heard at a distance of 4 m from this source?

22-23. A 3.0-W sound source is located 6.5 m from an observer. What are the intensity and the intensity level of the sound heard at that distance?

Ans. 5.65 $\times$ 10^{-3} W/m^2, 97.5 dB

22-24. A person located 6 m from a sound source hears an intensity of 2×10^{-4} W/m^2. What intensity is heard by a person located 2.5 m from the source?

*22-25. The intensity level 6 m from a source is 80 dB. What is the intensity level at a distance of 15.6 m from the same source?

Ans. 71.7 dB

Section 22-8—The Doppler Effect

Assume that the speed of sound in 343 m/s for all of these problems.

22-26. A stationary source of sound emits a signal at a frequency of 290 Hz. What are the frequencies heard by an observer (a) moving toward the source at 20 m/s and (b) moving away from the source at 20 m/s?

22-27. A car blowing a 560-Hz horn moves at a speed of 15 m/s as it first approaches a stationary listener and then moves away from a stationary listener at the same speed. What are the frequencies heard by the listener?   **Ans. 586 Hz, 537 Hz**

22-28. A person stranded in a car blows a 400-Hz horn. What frequencies are heard by the driver of a car passing at a speed of 60 km/h?

22-29. A train moving at 20 m/s blows a 300-Hz whistle as it passes a stationary observer. What are the frequencies heard by the observer as the train passes?

Ans. 319, 300, and 283 Hz

22-30. A child riding a bicycle north at 6 m/s hears a 600-Hz siren from a police car heading south at 15 m/s. What is the frequency heard by the child?

22-31. An ambulance moves northward at 15 m/s. Its siren has a frequency of 600 Hz at rest. A car heads south at 20 m/s toward the ambulance. What frequencies are heard by the car driver before and after the car and ambulance pass one another?

Ans. 664 Hz, 541 Hz

*22-32. A truck traveling at 24 m/s overtakes a car traveling at 10 m/s in the same direction. The trucker blows a 600-Hz horn. What frequency is heard by the car driver?

*22-33. A 500-Hz train whistle is heard by a stationary observer at a frequency of 475 Hz. What is the speed of the train? Is it moving toward the observer or away from the observer?   **Ans. 18.1 m/s, away**

Challenge Problems

22-34. The speed of sound in a certain metal rod is 4600 m/s, and the density of the metal is 5230 kg/m^3. What is Young's modulus for this metal?

22-35. A sonar beam travels in a fluid for a distance of 200 m in 0.12 s. The bulk modulus of elasticity for the fluid is 2600 MPa. What is the density of the fluid?
Ans. 936 kg/m^3

22-36. What is the frequency of the third overtone for a closed pipe of length 60 cm?

22-37. A 40-g string 2 m in length vibrates in three loops. The tension in the string is 270 N. What is the wavelength? What is the frequency?
Ans. 1.33 m, 87.1 Hz

22-38. How many beats per second are heard when two tuning forks of 256 and 259 Hz are sounded together?

22-39. What is the length of a closed pipe if the frequency of its second overtone is 900 Hz on a day when the temperature is 20°C?
Ans. 47.6 cm

22-40. The fundamental frequency for an open pipe is 360 Hz. If one end of this pipe is closed, what will be the new fundamental frequency?

*22-41. A 60-cm steel rod is clamped at one end as shown in Fig. 22-13a. Sketch the fundamental and the first overtone for these boundary conditions. What are the wavelengths in each case? **Ans. 2.40 m, 80.0 cm**

*22-42. The 60-cm rod in Fig. 22-13b is now clamped at its midpoint. What are the wavelengths for the fundamental and first overtone?

Figure 22-13

22-43. The velocity of sound in a steel rod is 5060 m/s. What is the length of a steel rod mounted as shown in Fig. 22-13a if the fundamental frequency of vibration for the rod is 3000 Hz? **Ans. 42.2 cm**

*22-44. Find the ratio of the intensities of two sounds if one is 12 dB higher than the other.

22-45. A certain loudspeaker has a circular opening of area 6 cm^2. The power radiated by this speaker is 6×10^{-7} W. What is the intensity of the sound at the opening? What is the intensity level?
Ans. 1 mW/m^2, 90 dB

22-46. The noon whistle at a textile mill has a frequency of 360 Hz. What are the frequencies heard by the driver of a car passing the mill at 25 m/s on a day when sound travels at 343 m/s?

*22-47. What is the difference in intensity levels (dB) for two sounds whose intensities are 2×10^{-5} W/m^2 and 0.90 W/m^2?
Ans. 46.5 dB

Critical Thinking Questions

*22-48. By inhaling helium gas, one can raise the frequency of the voice considerably. For air $M = 29$ g/mol and $\gamma = 1.4$; for helium $M = 4.0$ g/mol and $\gamma = 1.66$. At a temperature of 27°C, you sing a C note at 256 Hz. What is the frequency that will be heard if you inhale helium gas and other parameters are unchanged? Notice that both v and f were increased. How do you explain this in view of the fact that $v = f\lambda$? Discuss.
Ans. 751 Hz

*22-49. A toy whistle is made out of a piece of sugar cane that is 8 cm long. It is essentially an open pipe from the air inlet to the far end. Now suppose that a hole is bored at the midpoint so that a finger can alternately close and open the hole. If the velocity of sound is 340 m/s, what are the two possible fundamental frequencies obtained by closing and opening the hole at the center of the cane? What is the fundamental frequency if the center hole is covered and the far end is plugged?

*22-50. A tuning fork of frequency 512 Hz is moved away from an observer and toward a flat wall with a speed of 3 m/s. The speed of sound in the air is 340 m/s. What is the apparent frequency of the unreflected sound? What is the apparent frequency of the reflected sound? How many beats are heard each second?

Ans. 508 Hz, 517 Hz, 9 beats/s

*22-51. Using the logarithmic definition of decibels, derive the following expression, which relates the ratio of intensities of two sounds to the difference in decibels for the sounds:

$$\beta_2 - \beta_1 = 10 \log \frac{I_2}{I_1}$$

Use this relationship to work Probs. 22-44 and 22-47.

*22-52. The laboratory apparatus shown in Fig. 22-14 is used to measure the speed of sound in air by the resonance method. A vibrating tuning fork of frequency f is held over the open end of a tube, which is partly filled with water. The length of the air column can be varied by changing the water level. As the water level is gradually lowered from the top of the tube, the sound intensity reaches a maximum at the three levels shown in the figure. The maxima occur whenever the air column resonates with the tuning fork. Thus, the distance between successive resonance positions is the distance between adjacent notes for the standing waves in the air column. The frequency of the fork is 512 Hz, and the resonance positions occur at 17, 51, and 85 cm from the top of the tube. What is the velocity of sound in the air? What is the approximate temperature in the room?

Ans. 348 m/s, 28.3°C

22-53. What is the difference in intensity levels of two sounds, one being twice the intensity of the other?

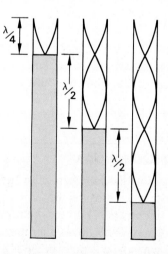

Figure 22-14 Laboratory apparatus for computing the velocity of sound by resonance methods.

Electricity, Magnetism, and Optics

Part 3

23

The Electric Force

OBJECTIVES

After completing this chapter, you should be able to:
1. Demonstrate the existence of two kinds of electric charge and verify the *first law of electrostatics* using laboratory materials.
2. Explain and demonstrate the processes of charging by *contact* and by *induction*, and use an *electroscope* to determine the nature of an unknown charge.
3. State *Coulomb's law* and apply it to the solution of problems involving electric forces.
4. Define the *electron*, the *coulomb*, and the *microcoulomb* as units of electric charge.

A hard-rubber comb or a plastic rod acquires a strange ability to attract other objects after it is rubbed on a coat sleeve. An annoying *shock* is sometimes experienced when you touch the handle of a car door after sliding across the seat. Stacked sheets of paper tend to resist separation. All these occurrences are examples of *electrification*, a phenomenon that frequently occurs as a result of rubbing objects together. Long ago, the name **charging** was given to the rubbing process, and the electrified object was said to be *charged*. This chapter begins our study of **electrostatics,** the science that treats electric charges at rest.

23-1 THE ELECTRIC CHARGE

The best way to begin a study of electrostatics is to experiment with objects that become electrified by rubbing. The materials illustrated in Fig. 23-1 are commonly found in the physics laboratory. In the order of their appearance in the figure, they are a hard-rubber rod resting on a piece of cat's fur, a glass rod resting on a piece of silk, the pith-ball electroscope, suspended pith balls, and the gold-leaf electroscope. A **pith ball** is a light sphere of wood pith painted with metallic paint and usually suspended from a silk thread. An **electroscope** is a sensitive laboratory instrument used to detect the presence of an electric charge.

The pith-ball electroscope can be used to study the effects of electrification. Two metallic-coated pith balls are suspended by silk threads from a common point. We begin by vigorously rubbing the rubber rod with cat's fur (or wool). Then, if the rubber rod is brought near the electroscope, the suspended pith balls will be attracted to the rod, as shown in Fig. 23-2a. After remaining in contact with the rod for an instant, the balls will be repelled from the rod and from each other.

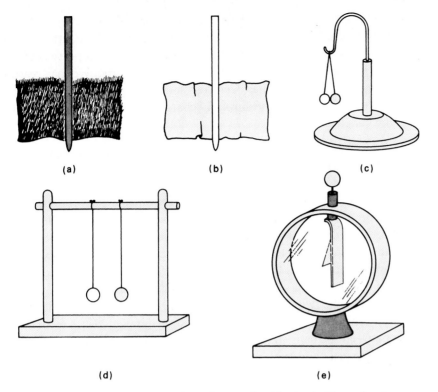

Figure 23-1 Laboratory materials for studying electrostatics: (a) a hard-rubber rod resting on a piece of cat's fur, (b) a glass rod resting on a piece of silk, (c) the pith-ball electroscope, (d) suspended pith balls, and (e) the gold-leaf electroscope.

www.colorado.edu
For more explanations of the electric charge and related topics, see the Physics 2000 Web site, and go to electro-magnetic radiation.

Figure 23-2 (a) Charging the pith-ball electroscope with a rubber rod. (b) Charging the pith balls with a glass rod.

When the rod is removed, the pith balls remain separated, as shown in the figure. The initial attraction (due to a redistribution of charge on the neutral pith balls) will be explained later. The repulsion must be due to some property acquired by the pith balls as a result of their contact with the charged rod. We may reasonably assume that some of the *charge* has been transferred from the rod to the pith balls and that all three objects become similarly charged. From these observations, we can state the conclusion that:

A force of repulsion exists between two substances that are electrified in the same way.

Let us continue our experimentation by picking up the glass rod and rubbing it vigorously on a silk cloth. When the charged rod is brought near the pith balls, the same sequence of events occurs as with the rubber rod. (See Fig. 23-2b.) Does this mean that the nature of the charge is the same on both rods? Our experiment neither proves nor disproves this assumption. In each case, the rod and balls were electrified in the same way, and so repulsion occurs in each case.

To test whether the two processes are identical, let us charge one pith ball with a glass rod and the other with a rubber rod. As shown in Fig. 23-3, a force of *attraction* exists between the balls charged in this manner. We conclude that the charges produced on the glass and rubber rods are opposite.

Similar experimentation with many different materials demonstrates that all electrified objects can be divided into two groups: (1) those that have a charge like that produced on glass and (2) those that have a charge like that produced on rubber. According to a convention established by Benjamin Franklin, objects in the former group are said to have a *positive* (+) charge, and objects belonging to the latter group are said to have a *negative* (−) charge. These terms have no mathematical significance; they simply denote the two opposite kinds of electric charge.

We are now in a position to state the *first law of electrostatics*, which is based on our experimentation:

Like charges repel and unlike charges attract.

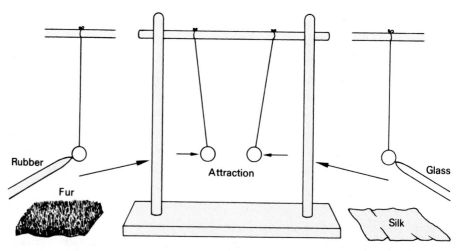

Figure 23-3 A force of attraction exists between two substances that are oppositely charged.

Two negatively charged objects or two positively charged objects repel each other, as demonstrated by Fig. 23-2a and b, respectively. Figure 23-3 demonstrates that a positively charged object attracts a negatively charged object.

23-2 THE ELECTRON

What actually occurs during a rubbing process that causes the phenomenon of electrification? Benjamin Franklin thought that all bodies contained a specified amount of electric fluid that served to keep them in an uncharged state. When two different substances were rubbed together, he postulated that one accumulated an excess of fluid and became positively charged, whereas the other lost fluid and became negatively charged. It is now known that the substance transferred is not a fluid but small amounts of negative electricity called *electrons.*

The modern atomic theory of matter holds that all substances are made up of atoms and molecules. Each atom has a positively charged central core, called the *nucleus*, which is surrounded by a cloud of negatively charged electrons. The nucleus consists of a number of *protons*, each with a single unit of positive charge, and (except for hydrogen) one or more *neutrons*. As the name suggests, a **neutron** is a neutral particle. Normally, an atom of matter is in a *neutral* or *uncharged* state because it contains the same number of protons in its nucleus as there are electrons surrounding the nucleus. A schematic diagram of the neon atom is shown in Fig. 23-4. If, for some reason, a neutral atom loses one or more of its outer electrons, the atom has a net positive charge and is referred to as a positive **ion.** A negative ion is an atom that has gained one or more additional charges.

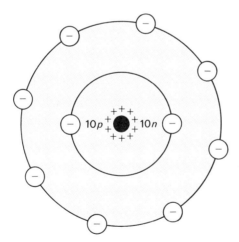

Figure 23-4   The neon atom consists of a tightly packed nucleus containing 10 protons (p) and 10 neutrons (n). The atom is electrically neutral because it is surrounded by 10 electrons.

When two particular materials are brought in close contact, some of the loosely held electrons may be transferred from one material to the other. For example, when a hard-rubber rod is rubbed against fur, electrons are transferred from the fur to the rod, leaving an *excess* of electrons on the rod and a *deficiency* of electrons on the fur. Similarly, when a glass rod is rubbed on a silk cloth, electrons are transferred from the glass to the silk. We can now state:

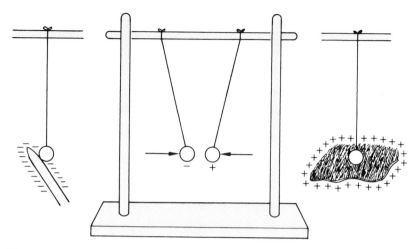

Figure 23-5 Rubbing a hard-rubber rod on a piece of fur transfers electrons from the fur to the rod.

An object that has an excess of electrons is negatively charged, and an object that has a deficiency of electrons is positively charged.

A laboratory demonstration of the transfer of charge is illustrated in Fig. 23-5. A hard-rubber rod is rubbed vigorously on a piece of fur. One pith ball is charged negatively with the rod, and the other is touched with the fur. The resulting attraction shows that the fur is oppositely charged. The process of rubbing has left a deficiency of electrons on the fur.

23-3 INSULATORS AND CONDUCTORS

A solid piece of matter is composed of many atoms arranged in a manner peculiar to that material. Some materials, primarily metals, have a large number of *free electrons*, which can move about through the material. These materials have the ability to transfer charge from one object to another, and they are called ***conductors***.

A conductor is a material through which charge can be easily transferred.

Most metals are good conductors. In Fig. 23-6 a copper rod is supported by a glass stand. The pith balls can be charged by touching the right end of the copper with a charged rubber rod. The electrons are transferred or *conducted* through the rod to the pith balls. Note that none of the charge is transferred to the glass support or to the silk thread. These materials are poor conductors and are referred to as ***insulators***.

An insulator is a material that resists the flow of charge.

Other examples of good insulators are rubber, plastic, mica, Bakelite, sulfur, and air.

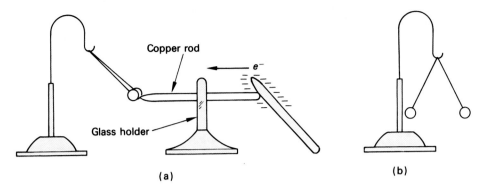

(a)   (b)

Figure 23-6   Electrons are conducted by the copper rod to charge the pith balls.

A semiconductor is a material intermediate in its ability to carry charge.

Examples are silicon, germanium, and gallium arsenide. The ease with which a semiconductor carries charge can be varied greatly by the addition of impurities or by a change in temperature.

23-4 THE GOLD-LEAF ELECTROSCOPE

The gold-leaf electroscope shown in Fig. 23-7 consists of a strip of gold foil attached to a conducting rod. The rod and the foil are protected from air currents by a cylindrical metal case with glass windows. The rod is fitted with a spherical knob at the top and is insulated from the case by a block of hard rubber or amber. Whenever the knob is given a charge, the repulsion of like charges on the rod and the gold leaf causes the leaf to diverge away from the rod.

Figure 23-8 illustrates charging the electroscope by *contact*. When the knob is touched with the negatively charged rod, electrons flow from the rod to the leaves,

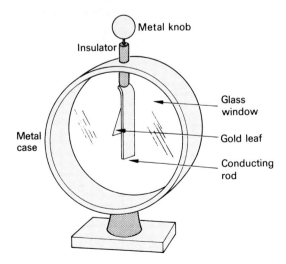

Figure 23-7   The gold-leaf electroscope.

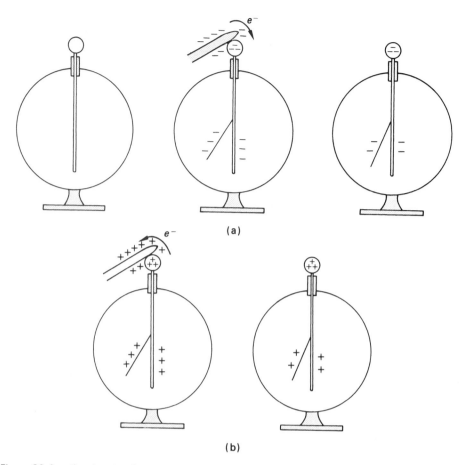

Figure 23-8 Charging the electroscope by contact with (a) a negatively charged rod and (b) a positively charged rod.

leaving an excess of electrons on the electroscope. When the knob is touched with a positively charged rod, electrons are transferred from the knob to the rod, leaving the electroscope deficient in electrons. Note that the residual charge on the electroscope is of the same sign as that of the charging rod.

Once the electroscope is charged, either negatively or positively, it can be used to detect the presence and nature of other charged objects. (See Fig. 23-9.) For example, consider what happens to the leaf of a negatively charged electroscope as a positively charged rod is brought near the knob. Some electrons are drawn from the leaf and rod up into the knob. As a result, the leaf converges. Bringing the rod closer produces a proportionately greater convergence of the leaf as more and more electrons are attracted to the knob. There appears to be a direct proportion between the number of charges on the leaf and rod and the force of repulsion between them. Moreover, there must be some *inverse* relation between the separation of the charged rod and the knob and the force attracting the electrons from the leaf and rod of the electroscope. This force becomes stronger as the separation decreases. These observations will help us understand Coulomb's law, developed in a later section.

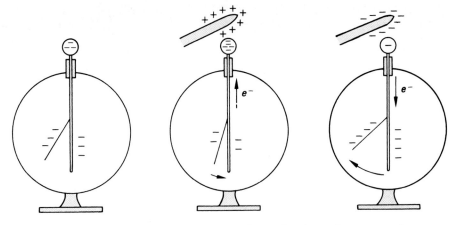

Figure 23-9 The negatively charged electroscope can be used to detect the presence of other charge.

Similar reasoning will show that the leaf of a negatively charged electroscope will be repelled further from the rod when the knob is placed near a negatively charged object. Thus, a charged electroscope can be used to indicate both the polarity and the presence of nearby charges.

23-5 REDISTRIBUTION OF CHARGE

When a negatively charged rod is brought close to an uncharged pith ball, there is an initial attraction, as shown in Fig. 23-10. The attraction of the uncharged object is due to the separation of positive and negative electricity within the neutral body. The proximity of the negatively charged rod repels loosely held electrons to the opposite side of the uncharged object, leaving a deficiency (positive charge) on the near side and an excess (negative charge) on the far side. Since the unlike charge is nearer to the rod, the force of attraction will exceed the force of repulsion and the electrically neutral object will be attracted to the rod. No charge is gained or lost during this process; the charge on the neutral body is simply redistributed.

F.Y.I.

Scientists are trying to reduce diesel emissions by zapping exhaust with a 50-ns burst of high-energy electrons. This breaks down harmful nitrogen oxides and other sooty particles before they are released into the atmosphere.

Figure 23-10 Attraction of a neutral body due to a redistribution of charge.

23-6 CHARGING BY INDUCTION

The redistribution of charge due to the presence of a nearby charged object can be useful in charging objects without contact. This process, called **charging by induction,** can be accomplished without any loss of charge from the charging body. For example, consider the two neutral metal spheres placed in contact as shown in Fig. 23-11. When a negatively charged rod is brought near the left sphere (without touching it), a redistribution of charge occurs. Electrons are forced from the left sphere to the right sphere through the point of contact. Now, if the spheres are separated in the presence of the charging rod, the electrons cannot return to the left sphere. Thus, the left sphere will have a deficiency of electrons (a **positive charge**), and the right sphere will have an excess of electrons (a **negative charge**).

A charge can also be induced on a single sphere. This process is illustrated with the electroscope in Fig. 23-12. A negatively charged rod is placed near the metal knob, causing a redistribution of charge. The repelled electrons cause the leaf to diverge, leaving a deficiency of electrons on the knob. By touching the knob with a finger or by connecting a wire from the knob to the earth, a path is provided for the repelled electrons to leave the electroscope. The body or the ground will acquire

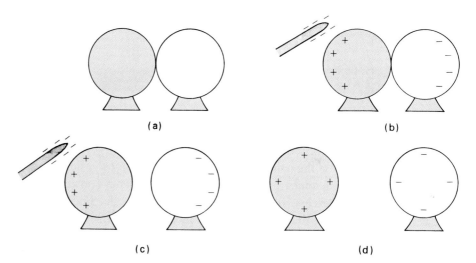

Figure 23-11   Charging two metal spheres by induction.

Figure 23-12   Charging an electroscope by induction. Note that the residual charge is opposite that of the charging body.

a negative charge equal to the positive charge (deficiency) left on the electroscope. When the charging rod is removed, the leaf of the electroscope will again diverge, as shown in the figure. Charging by induction always leaves a residual charge that is opposite that of the charging body.

23-7 COULOMB'S LAW

As usual, the task of the physicist is to measure the interactions between charged objects in some quantitative fashion. It is not sufficient to state that an electric force exists; we must be able to predict its magnitude.

The first theoretical investigation of the electric forces between charged bodies was accomplished by Charles Augustin de Coulomb in 1784. His studies were made with a torsion balance to measure the variation in force with separation and quantity of charge. The separation r of two charged objects is reckoned as the straight-line distance between their centers. The quantity of charge q can be thought of as the excess number of electrons or protons in the body.

Coulomb found that the force of attraction or repulsion between two charged objects is inversely proportional to the square of their separation distance. In other words, if the distance between two charged objects is reduced by one-half, the force of attraction or repulsion between them will be increased fourfold.

The concept of a quantity of charge was not clearly understood in Coulomb's time. There was no established unit of charge and no means for measuring it, but his experiments clearly showed that the electric force between two charged objects is directly proportional to the product of the quantity of charge on each object. Today, his conclusions are stated in **Coulomb's law:**

> **The force of attraction or repulsion between two point charges is directly proportional to the product of the two charges and inversely proportional to the square of the distance between them.**

To arrive at a mathematical statement of Coulomb's law, let us consider the charges in Fig. 23-13. In Fig. 23-13 the force **F** of attraction between two unlike charges is

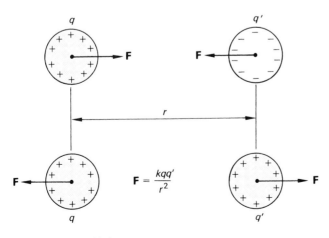

Figure 23-13 Illustrating Coulomb's law.

indicated, and a force of repulsion is shown for like charges. In either case, the magnitude of the force is determined by the magnitudes of the charges q and q' and by their separation r. From Coulomb's law, we can write

$$F \propto \frac{qq'}{r^2}$$

or

$$F = \frac{kqq'}{r^2} \qquad \text{23-1}$$

The proportionality constant k takes into account the properties of the medium separating the charged bodies and has the dimensions dictated by Coulomb's law.

In SI units, the practical system for the study of electricity, the unit of charge is expressed in **coulombs** (C). In this case, the quantity of charge is not defined by Coulomb's law but is related to the flow of charge through a conductor. We shall find later that this rate of flow is measured in *amperes*. A formal definition of the coulomb is as follows:

> **One coulomb is the charge transferred through any cross section of a conductor in one second by a constant current of one ampere.**

Since current theory is not a part of this chapter, it will suffice to compare the coulomb with the charge of an electron.

$$1 \text{ C} = 6.25 \times 10^{18} \text{ electrons}$$

Obviously the coulomb is an enormously large unit from the standpoint of most problems in electrostatics. The charge of one electron expressed in coulombs is

$$e^- = -1.6 \times 10^{-19} \text{ C} \qquad \text{23-2}$$

where $e-$ is the symbol for the electron and the minus sign denotes the nature of the charge.

A more convenient unit for electrostatics is the **microcoulomb** (μC), defined by

$$1 \ \mu\text{C} = 10^{-6} \text{ C} \qquad \text{23-3}$$

Since the SI units of force, charge, and distance do not depend on Coulomb's law, the proportionality constant k must be determined by experiment. A large number of experiments have shown that when the force is in newtons, the distance is in meters, and the charge is in coulombs, the proportionality constant is approximately

$$k = 9 \times 10^9 \text{ N} \cdot \text{m}^2/\text{C}^2 \qquad \text{23-4}$$

When applying Coulomb's law in SI units, one must substitute this value for k in Eq. (23-1):

$$F = \frac{(9 \times 10^9 \text{ N} \cdot \text{m}^2/\text{C}^2)qq'}{r^2} \qquad \text{23-5}$$

It must be remembered that **F** represents the force on a charged particle and is, therefore, a vector quantity. The direction of the force is determined solely by the nature of the charges q and q'. Therefore, it is easier to use the absolute values of q and q' in Coulomb's law, Eq. (23-5). Remember that like charges repel and opposite charges attract to get the direction of the force. When more than one force acts on a charge, the resultant force is the vector sum of the separate forces.

EXAMPLE 23-1

A -3-μC charge is placed 100 mm from a $+3$-μC charge. Calculate the force between the two charges.

Solution

First, we convert to appropriate units.

$$3 \ \mu C = 3 \times 10^{-6} \ C \qquad 100 \ mm = 100 \times 10^{-3} \ m = 0.1 \ m$$

Then, we use the absolute values, so that both q and q' are equal to 3×10^{-6} C. Applying Coulomb's law, we obtain

$$F = \frac{kqq'}{r^2} = \frac{(9 \times 10^9 \ N \cdot m^2/C^2)(3 \times 10^{-6} \ C)(3 \times 10^{-6} \ C)}{(0.1 \ m)^2}$$

$$F = 8.1 \ N \hspace{4cm} \textit{Attraction}$$

This is a force of *attraction* because the charges had opposite signs.

Problem-Solving Strategy

Electric Forces and Coulomb's Law

1. Read the problem and then draw and label a figure. Indicate positive and negative charges along with given distances. The charges must be in *coulombs* and the distances must be in *meters*. Remember that $1 \ \mu C = 1 \times 10^{-6}$ C and $1 \ nC = 1 \times 10^{-9}$ C.

2. Be careful not to confuse the *nature* of the charge (+ or −) with the sign given to forces and their components. Attraction and/or repulsion determines the direction of electric forces.

3. The *resultant force* on a given charge due to one or more nearby charges is found by vector addition of the force that each charge would exert acting alone. The magnitude of each force is found from Coulomb's law; the direction is determined from the fact that like charges repel and unlike charges attract. Construct a free-body diagram and proceed with vector addition as shown in text examples. You may wish to review vector addition by the component method as discussed in Chap. 3.

4. For charges in equilibrium, recall that the first condition for equilibrium provides that the sum of the *x*-components is zero and the sum of the *y*-components is zero.

CONCEPTUAL EXAMPLE 23-2

Two charges, $q_1 = -8 \ \mu C$ and $q_2 = +12 \ \mu C$, are placed 120 mm apart in the air. What is the resultant force on a third charge $q_3 = -4 \ \mu C$ placed midway between the other two charges?

Figure 23-14 Computing the resultant force on a charge placed midway between two other charges.

Solution

We convert the charges to coulombs ($1 \ \mu C = 1 \times 10^{-6}$ C), use their absolute values, and convert the distance to meters (120 mm = 0.12 m). One-half of 0.12 m is 0.06 m. A sketch is drawn in Fig. 23-14 to help visualize the forces. The force on q_3 due to q_1 is directed to the right and is calculated from Coulomb's law.

$$F_1 = \frac{kq_1q_3}{r^2} = \frac{(9 \times 10^9 \ N \cdot m^2/C^2)(8 \times 10^{-6} \ C)(4 \times 10^{-6} \ C)}{(0.06 \ m)^2}$$

$$F_1 = \frac{0.288 \ N \cdot m^2}{0.0036 \ m^2} = 80 \ N \qquad \textit{Repulsion, to the Right}$$

Similarly, the force F_2 on q_3 is equal to

$$F_2 = \frac{kq_2q_3}{r^2} = \frac{(9 \times 10^9 \ N \cdot m^2/C^2)(12 \times 10^{-6} \ C)(4 \times 10^{-6} \ C)}{(0.06 \ m)^2}$$

$$F_2 = 120 \ N \qquad \textit{Attraction, also to the Right}$$

The resultant force **F** is the vector sum of $\mathbf{F}_1$ and $\mathbf{F}_2$. Thus,

$$\mathbf{F} = 80 \ N + 120 \ N$$
$$= 200 \ N \qquad \textit{Directed to the Right}$$

Note that the signs of the charges were used only to determine the direction of the forces; they were not substituted into the calculations.

EXAMPLE 23-3

Three charges, $q_1 = +4 \times 10^{-9}$ C, $q_2 = -6 \times 10^{-9}$ C, and $q_3 = -8 \times 10^{-9}$ C, are arranged as shown in Fig. 23-15. What is the resultant force on q_3 due to the other two charges?

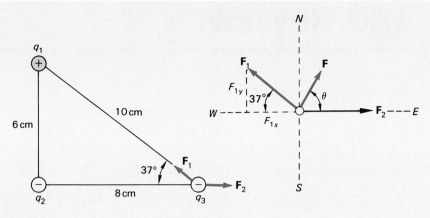

Figure 23-15

Solution

Let $\mathbf{F}_1$ be the force on q_3 due to q_1, and let $\mathbf{F}_2$ be the force on q_3 due to q_2. $\mathbf{F}_1$ is a force of attraction, and $\mathbf{F}_2$ is a force of repulsion, as shown in the figure. Their magnitudes are found from Coulomb's law.

$$F_1 = \frac{(9 \times 10^9 \text{ N} \cdot \text{m}^2/\text{C}^2)(4 \times 10^{-9} \text{ C})(8 \times 10^{-9} \text{ C})}{(0.1 \text{ m})^2}$$

$$= 2.88 \times 10^{-5} \text{ N} \qquad (37° \text{ north of west})$$

$$F_2 = \frac{(9 \times 10^9 \text{ N} \cdot \text{m}^2/\text{C}^2)(6 \times 10^{-9} \text{ C})(8 \times 10^{-9} \text{ C})}{(8 \times 10^{-2} \text{ m})^2}$$

$$= 6.75 \times 10^{-5} \text{ N} \qquad (\text{east})$$

We must now find the resultant $\mathbf{F}$ of the forces $\mathbf{F}_1$ and $\mathbf{F}_2$. From the free-body diagram, we note that

$$\begin{aligned} F_x &= F_2 - F_{1x} = F_2 - F_1 \cos 37° \\ &= (6.75 \times 10^{-5} \text{ N}) - (2.88 \times 10^{-5} \text{ N})(0.8) \\ &= 4.45 \times 10^{-5} \text{ N} \\ F_y &= F_1 \sin 37° = 1.73 \times 10^{-5} \text{ N} \end{aligned}$$

From the components we find

$$\tan \theta = \frac{F_y}{F_x} = \frac{1.73 \times 10^{-5} \text{ N}}{4.45 \times 10^{-5} \text{ N}} = 0.389$$

$$\theta = 21°$$

$$F = \frac{F_y}{\sin \theta} = \frac{1.73 \times 10^{-5} \text{ N}}{\sin 21°} = 4.80 \times 10^{-5} \text{ N}$$

Therefore, the resultant force on q_3 is 4.8×10^{-5} N directed 21° north of east.

Summary and Review

Summary

Electrostatics is the science that treats charges at rest. We have seen that there are two kinds of charge that exist in nature. If an object has an excess of electrons, it is said to be *negatively* charged; if it has a deficiency of electrons, it is *positively* charged. Coulomb's law was introduced to provide a quantitative measure of electrical forces between such charges. The major concepts are listed below.

- The first law of electrostatics states that *like charges repel each other and unlike charges attract each other.*
- Coulomb's law states that *the force of attraction or repulsion between two point charges is directly proportional to the product of the two charges and inversely proportional to the separation of the two charges.*

$$F = \frac{kqq'}{r^2}$$

$$k = 9 \times 10^9 \text{ N} \cdot \text{m}^2/\text{C}^2 \qquad \textit{Coulomb's Law}$$

The force F is in newtons (N) when the separation r is in meters (m) and the charge q is measured in coulombs (C).

- When solving the problems in this chapter, it is important to use the sign of the charges to determine the *direction* of forces and Coulomb's law to determine their *magnitudes*. The resultant force on a particular charge is then found by the methods of vector mechanics.

Key Terms

1. electrostatics
2. charging
3. electron
4. induced charge
5. conductor
6. insulator
7. semiconductor
8. negative charge
9. positive charge
10. ion
11. Coulomb's law
12. coulomb
13. microcoulomb
14. electroscope

Review Questions

23-1. Discuss several examples of static electricity in addition to those mentioned in the text.

23-2. In the process of rubbing a glass rod with a silk cloth, is charge *created*? Explain.

23-3. What is the nature of the charge on the silk cloth in Question 23-2?

23-4. An insulated stand supports a charged metal ball in the laboratory. Describe several procedures for determining the nature of the charge on the ball.

23-5. During an experiment in the laboratory, two bodies are seen to attract each other. Is this conclusive proof that they are both charged? Explain.

23-6. Two bodies are found to repel each other with an electric force. Is this conclusive proof that they are both charged? Explain.

23-7. One of the fundamental principles of physics is the principle of the conservation of charge, which states that *the total quantity of electric charge in the universe does not change.* Can you offer reasons for accepting this law?

23-8. Describe what happens to the leaf of a positively charged electroscope as (a) a negatively charged rod is brought closer and closer to the knob without touching it, (b) a positively charged rod is brought closer and closer to the knob.

23-9. When charging the leaf electroscope by induction, should the finger be removed before the charging rod is taken away? Explain.

23-10. List the units for each parameter in Coulomb's law for SI units.

23-11. Coulomb's law is valid only when the separation r is large in comparison with the radii of the charge. What accounts for this limitation?

23-12. How many electrons would be required to give a metal sphere a negative charge of (a) 1 C, (b) 1 μC?

Problems

Section 23-7—Coulomb's Law

23-1. Two balls, each having a charge of 3 μC, are separated by 20 mm. What is the force of repulsion between them? **Ans. 202 N**

23-2. Two point charges of -3 and $+4$ μC are 12 mm apart in a vacuum. What is the electrostatic force between them?

23-3. An alpha particle consists of two protons ($q_e = 1.6 \times 10^{-19}$ C) and two neutrons (no charge). What is the repulsive force between two alpha particles separated by 2 mm? **Ans. 2.30 $\times$ 10^{-10} N**

23-4. Assume that the radius of the electron's orbit around the proton in a hydrogen atom is approximately 5.2×10^{-11} m. What is the electrostatic force of attraction?

23-5. What is the separation of two -4-μC charges if the force of repulsion between them is 200 N? **Ans. 26.8 mm**

23-6. Two identical charges separated by 30 mm experience a repulsive force of 980 N. What is the magnitude of each charge?

*23-7. A 10-μC charge and a -6-μC charge are separated by 40 mm. What is the force between them? The spheres are placed in contact for a few moments and then separated again by 40 mm. What is the new force? Is it attractive or repulsive?
Ans. 338 N, attraction; 5.62 N, repulsion

*23-8. Two point charges initially attract each other with a force of 600 N. If their separation is reduced to one-third of its original value, what is the new force of attraction?

23-9. A $+60$-μC charge is placed 60 mm to the left of a $+20$=μC charge. What is the resultant force on a -35-μC charge placed midway between the two charges?
Ans. 1.40 $\times$ 10^4 N, left

23-10. A point charge of $+36$ μC is placed 80 mm to the left of a second point charge of -22 μC. What force is exerted on a third charge of $+10$ μC placed at the midpoint?

23-11. For Prob. 23-10, what is the resultant force on a third charge of $+12$ μC placed between the other charges and located 60 mm from the $+36$-μC charge?
Ans. 7020 N, right

23-12. A $+6$-μC charge is 44 mm to the right of a -8-μC charge. What is the resultant force on a -2-μC charge that is 20 mm to the right of the -8-μC charge?

*23-13. A 64-μC charge is located 30 cm to the left of a 16-μC charge. What is the resultant force on a -12-μC charge positioned exactly 50 mm below the 16-μC charge?
Ans. 2650 N, 113.3°

*23-14. A charge of $+60$ nC is located 80 mm above a -40-nC charge. What is the resultant force on a -50-nC charge located 45 mm horizontally to the right of the -40-nC charge?

*23-15. Three point charges, $q = +8$ μC, $q = -4$ μC, and $q = +2$ μC, are placed at the corners of an equilateral triangle, 80 mm on each side as shown in Fig. 23-15. What are the magnitude and direction of the resultant force on the $+8$-μC charge?
Ans. 39 N, 330°

Summary and Review

Challenge Problems

23-16. What should be the separation of two $+5$-μC charges so that the force of repulsion is 4 N?

23-17. The repulsive force between two pith balls is found to be 60 μN. If each pith ball carries a charge of 8 nC, what is their separation? **Ans. 98 mm**

23-18. Two identical unknown charges experience a mutual repulsive force of 48 N when separated by 60 mm. What is the magnitude of each charge?

23-19. One object contains an excess of 5×10^{14} electrons, and another has a deficiency of 4×10^{14} electrons. What is the force each exerts on the other if the objects are 30 mm apart? Is it attraction or repulsion? **Ans. 5.12×10^4 N attraction**

23-20. If it were possible to put 1 C of charge on each of two spheres separated by a distance of 1 m, what would be the repulsive force in newtons?

23-21. How many electrons must be placed on each of two spheres separated by 4 mm in order to produce a repulsive force of one 400 N? **Ans. 5.27×10^{12} electrons**

23-22. A -40-nC charge is placed 40 mm to the left of a $+6$-nC charge. What is the resultant force on a -12-nC charge placed 8 mm to the right of the $+6$-nC charge?

23-23. A 5-μC charge is placed 6 cm to the right of a 2-μC charge. What is the resultant force on a -9-nC charge placed 2 m to the left of the 2-μC charge? **Ans. 468 mN, right**

23-24. An equal number of electrons is placed on two metal spheres 3.0 cm apart in air. How many electrons are on each sphere if the resultant force is 4500 N?

23-25. A 4-nC charge is placed on a 4-g sphere that is free to move. A fixed 10-μC point charge is 4 cm away. What is the initial acceleration of the 4-μC charge? **Ans. 56.2 m/s²**

*23-26. Find the resultant force on a $+2$-μC charge that is 60 mm from each of two -4-μC charges that are 80 mm apart in air.

*23-27. Two charges of $+25$ and $+16$ μC are 80 mm apart? A third charge of $+60$ μC is placed between the other charges 30 mm from the $+25$-μC charge. Find the resultant force on the third charge. **Ans. 1.15×10^4 N**

*23-28. A 0.02-g pith ball is suspended freely. The ball is given a charge of $+20$ μC and placed 0.6 m from a charge of $+50$ μC. What will be the initial acceleration of the pith ball?

*23-29. A 4-μC charge is located 6 cm from an 8-μC charge. At what point on a line joining the two charges will the resultant force be zero? **Ans. 2.49 cm from 4-μC charge**

*23-30. A charge of $+8$ nC is placed 40 mm to the left of a -14-nC charge. Where should a third charge be placed if it is to experience a zero resultant force?

*23-31. A $+16$-μC charge is 80 mm to the right of a $+9$-μC charge. Where should a third charge be placed so that the resultant force is zero? **Ans. 34.3 mm right of the 9-μC charge**

*23-32. Two 3-g spheres are suspended from a common point with two 80-mm light silk threads of negligible mass. What charge must be placed on each sphere if their final positions are 50 mm apart?

Critical Thinking Questions

*23-33. A small metal sphere is given a charge of $+40$ μC, and a second sphere 8 cm away is given a charge of -12 μC. What is the force of attraction between them? If the two spheres are allowed to touch and are then again placed 8 cm apart, what new electric force exists between them? Is it attraction or repulsion? **Ans. 675 N, attraction 276 N, repulsion**

*23-34. The total charge on two metal spheres 50 mm apart is 80 μC. If they repel each other with a force of 800 N, what is the charge on each sphere?

*23-35. Four small spheres are each given charges of $q = +20$ μC and placed at the corners of a square with sides of length 6 cm. Show that the resultant force on each charge has a magnitude equal to 1914 N. What is the direction of the force? What will change if the charges are each $q = -20$ μC?

Ans. 1914 N, 45° away from center

*23-36. Two charges q_1 and q_2 are separated by a distance r. They experience a force F at this distance. If the initial separation is decreased by only 40 mm, the force between the two charges is doubled. What was the initial separation?

*23-37. Two 8-g pith balls are suspended from silk threads 60 cm long and attached to a common point. When the spheres are given equal amounts of negative charge, the balls come to rest 30 cm apart. Calculate the magnitude of the charge on each pith ball.

Ans. −450 nC

The Electric Field

OBJECTIVES
After completing this chapter, you should be able to:
1. Define the *electric field* and explain what determines its magnitude and direction.
2. Write and apply an expression that relates the electric field intensity at a point to the distance(s) from the known charge(s).
3. Explain and illustrate the concept of electric field lines, and discuss the two rules that must be followed in the construction of such lines.
4. Explain the concept of the *permittivity* of a *medium*, and how it affects the field intensity and the construction of field lines.
5. Write and apply *Gauss's law* as it relates to the electric fields surrounding surfaces of known charge density.

In our study of mechanics we discussed force and motion at great length. Newton's laws of motion were normally used to describe the application and consequences of *contact* forces. A moment's reflection on the universe as a whole convinces us of the enormous number of objects that are *not* in contact.

A projectile experiences a downward force that cannot be explained in terms of its interaction with air particles, planets revolve continuously through the void surrounding the sun, and the sun is pulled along an elliptical path by forces that do not touch it. Even at the atomic level, there are no "strings" to hold the electrons in their orbits about the nucleus.

If we are really to understand our universe, we must develop laws to predict the magnitude and direction of forces that are not transmitted by contact. Two such laws have already been discussed:

1. Newton's law of universal gravitation:

$$F_g = G\frac{m_1 m_2}{r^2}$$

24-1

2. Coulomb's law for electrostatic forces:

$$F_e = k\frac{q_1 q_2}{r^2}$$

24-2

Newton's law predicts the force that exists between two masses separated by a distance r; Coulomb's law deals with the electrostatic force, as discussed in Chap. 23. In applying such laws, we find it useful to develop certain properties of the space surrounding masses or charges.

24-1 THE CONCEPT OF A FIELD

Both the electric force and the gravitational force are examples of *action-at-a-distance forces*, which are extremely difficult to visualize. To overcome this fact, early physicists postulated the existence of an invisible material, called *ether*, that was thought to pervade all space. The gravitational force of attraction could then be due to strains in the ether caused by the presence of various masses. Certain optical experiments have now shown the ether theory to be untenable, and we are forced to consider whether space itself possesses properties of interest to the physicist.

It may be postulated that the mere presence of a mass alters the space surrounding it to produce a gravitational force on another nearby mass. We describe this alteration in space by introducing the concept of a *gravitational field* that surrounds all masses. Such a field may be said to exist in any region of space where a test mass will experience a gravitational force. The strength of the field at any point would be proportional to the force a given mass experiences at that point. For example, at every point in the vicinity of the earth, the gravitational field could be represented quantitatively by

$$\mathbf{g} = \frac{\mathbf{F}}{m}$$
24-3

where $\mathbf{g}$ = acceleration due to gravity
$\mathbf{F}$ = gravitational force
m = test mass (see Fig. 24-1)

If g is known at every point above the earth, the force $\mathbf{F}$ that will act on a given mass m placed at that point can be determined from Eq. (24-3).

The concept of a field can also be applied to electrically charged objects. The space surrounding a charged object is altered by the presence of the charge. We may postulate the existence of an **electric field** in this space.

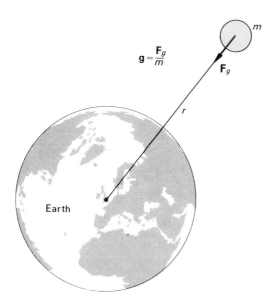

Figure 24-1 The gravitational field at any point above the earth can be represented by the acceleration **g** that a small mass *m* would experience if it were placed at that point.

An electric field is said to exist in a region of space in which an electric charge will experience an electric force.

This definition provides a test for the existence of an electric field. Simply place a charge at the point in question. If an electric force is observed, an electric field exists at that point.

Just as the force per unit mass provides a quantitative definition of a gravitational field, the strength of an electric field can be represented by the force per unit charge. We define the electric field intensity **E** at a point in terms of the force **F** experienced by a small positive charge $+q$ when it is placed at that point (see Fig. 24-2). The magnitude of the electric field intensity is given by

$$E = \frac{F}{q} \tag{24-4}$$

In the metric system, a unit of *electric field intensity* is the newton per coulomb (N/C). The usefulness of this definition rests with the fact that, if the field is known at a given point, we can predict the force that will act on any charge placed at that point.

Since the electric field intensity is defined in terms of a *positive* charge, its direction at any point is the same as the electrostatic force on a positive charge at that point.

The direction of the electric field intensity E at a point in space is the same as the direction in which a positive charge would move if it were placed at that point.

On this basis, the electric field in the vicinity of a positive charge $+Q$ would be outward, or away from the charge, as indicated by Fig. 24-3a. In the vicinity of a

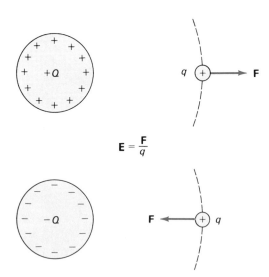

$$E = \frac{F}{q}$$

Figure 24-2 The direction of the electric field intensity at a point is the same as the direction in which a positive charge $+q$ would move when placed at that point. Its magnitude is the force per unit charge (F/q).

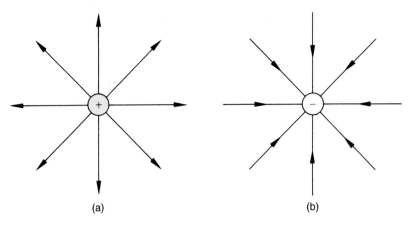

(a)   (b)

Figure 24-3 (a) The field in the vicinity of a positive charge is directed radially outward at every point. (b) The field is directed inward or toward a negative charge.

negative charge $-Q$, the direction of the field would be inward, or toward the charge (Fig. 24-3b).

It must be remembered that the electric field intensity is a property assigned to the *space* that surrounds a charged body. A gravitational field exists above the earth, whether or not a mass is positioned above the earth. Similarly, an electric field exists in the neighborhood of a charged body, whether or not a second charge is positioned in the field. If a charge *is* placed in the field, it will experience a force $\mathbf{F}$ give by

$$\mathbf{F} = q\mathbf{E} \qquad\qquad \text{24-5}$$

where $\mathbf{E}$ = field intensity
q = magnitude of charge placed in field

If q is positive, $\mathbf{E}$ and $\mathbf{F}$ will have the same direction; if q is negative, the force $\mathbf{F}$ will be directed opposite to the field $\mathbf{E}$.

EXAMPLE 24-1

The electric field intensity between the two plates in Fig. 24-4 is constant and directed downward. The magnitude of the electric field intensity is 6×10^4 N/C. What are the magnitude and the direction of the electric force exerted on an electron projected horizontally between the two plates?

Figure 24-4 An electron projected into an electric field of constant intensity.

Solution

Since the direction of the field intensity **E** is defined in terms of a positive charge, the force on an electron will be upward, or opposite to the field direction. The charge on an electron is -1.6×10^{-19} C. Thus, the electric force is given by Eq. (24-5).

$$F = qE = (1.6 \times 10^{-19} \text{ C})(6 \times 10^4 \text{ N/C})$$
$$= 9.6 \times 10^{-15} \text{ N} \qquad \text{(upward)}$$

Remember that the absolute value of the charge is used. The directions of **F** and **E** are the same for positive charges and opposite for negative charges.

EXAMPLE 24-2

Show that the gravitational force on the electron in Example 24-1 may be neglected. (The mass of an electron is 9.1×10^{-31} kg.)

Solution

The downward force due to the weight of the electron is

$$F_g = mg = (9.1 \times 10^{-31} \text{ kg})(9.8 \text{ m/s}^2)$$
$$= 8.92 \times 10^{-30} \text{ N}$$

The electric force is larger than the gravitational force by a factor of 1.08×10^{15}.

24-2 COMPUTING THE ELECTRIC INTENSITY

We have discussed one method of measuring the magnitude of the electric field intensity at a point in space. A known charge is placed at the point, and the resultant force is measured. The force per unit charge is then a measure of the electric intensity at that point. The disadvantage of this method is that it bears no obvious relationship to the charge Q that creates the field. Experimentation will quickly show that the magnitude of the electric field surrounding a charged body is directly proportional to the quantity of charge on the body. It can also be demonstrated that at points farther and farther away from a charge Q a test charge q will experience smaller and smaller forces. The exact relationship is derived from Coulomb's law.

Suppose we wish to calculate the field intensity E at a distance r from a single charge Q, as shown in Fig. 24-5. The force F that Q exerts on a test charge q at the point in question is, from Coulomb's law,

$$F = \frac{kQq}{r^2} \qquad \text{24-6}$$

Substituting this value for F into Eq. (24-4), we obtain

$$E = \frac{F}{q} = \frac{kQq/r^2}{q}$$

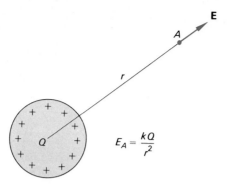

$$E_A = \frac{kQ}{r^2}$$

Figure 24-5 Calculating the electric field intensity at a distance *r* from the center of a single charge *Q*.

$$E = \frac{kQ}{r^2} \qquad\qquad 24\text{-}7$$

where *k* is equal to 9×10^9 N · m²/C². The direction of the field is away from *Q* if *Q* is positive and toward *Q* if *Q* is negative. We now have a relation that allows us to compute the field intensity at a point without having to place a second charge at the point.

EXAMPLE 24-3

What is the electric field intensity at a distance of 2 m from a charge of $-12\ \mu C$?

Solution

Since the charge *Q* is negative, the field intensity will be directed toward *Q*. Its magnitude, from Eq. (24-7), is

$$E = \frac{kQ}{r^2} = \frac{(9 \times 10^9\ \text{N} \cdot \text{m}^2/\text{C}^2)(12 \times 10^{-6}\ \text{C})}{(2\ \text{m})^2}$$

$$= 27 \times 10^3\ \text{N/C} \qquad (\text{toward } Q)$$

When more than one charge contributes to the field, as in Fig. 24-6, the resultant field is the vector sum of the contributions from each charge.

$$\mathbf{E} = \mathbf{E}_1 + \mathbf{E}_2 + \mathbf{E}_3 + \cdots$$

$$E = \frac{kQ_1}{r_1^2} + \frac{kQ_2}{r_2^2} + \frac{kQ_3}{r_3^2} + \cdots \qquad\qquad \textit{Vector Sum}$$

In an abbreviated form, the resultant field at a point in the vicinity of a number of charges is given by

$$E = \sum \frac{kQ}{r^2} \qquad\qquad \textit{Vector Sum} \qquad 24\text{-}8$$

It must be remembered that this is a *vector sum* and not an algebraic sum.

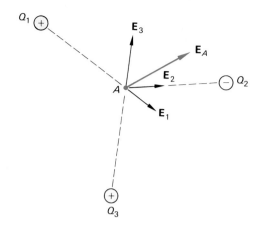

Figure 24-6 The field in the vicinity of a number of charges is equal to the vector sum of the fields due to the individual charges.

CONCEPTUAL EXAMPLE 24-4

Two point charges, $q_1 = -6$ nC and $q_2 = +6$ nC, are 12 cm apart, as shown in Fig. 24-7. Determine the electric field (a) at point A and (b) at point B.

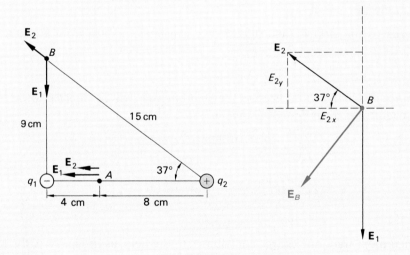

Figure 24-7

Solution (a)

By convention, let us consider any vector directed to the right or up as positive and any vector directed to the left or down as negative. The field at A due to q_1 is directed to the left since q_1 is negative, and its magnitude is

$$E_1 = \frac{kq_1}{r_1^2} = \frac{-(9 \times 10^9 \text{ N} \cdot \text{m}^2/\text{C}^2)(6 \times 10^{-9} \text{ C})}{(4 \times 10^{-2} \text{ m})^2}$$

$$= -3.38 \times 10^4 \text{ N/C} \qquad \text{(left)}$$

The electric intensity at A due to q_2 is also directed to the left since q_2 is positive, and it is given by

$$E_2 = \frac{kq_2}{r_2^2} = \frac{-(9 \times 10^9 \text{ N} \cdot \text{m}^2/\text{C}^2)(6 \times 10^{-9} \text{ C})}{(8 \times 10^{-2} \text{ m})^2}$$

$$= -0.844 \times 10^4 \text{ N/C}$$

Since the two vectors $\mathbf{E}_1$ and $\mathbf{E}_2$ have the same direction, the resultant intensity at A is simply

$$E_A = E_1 + E_2$$
$$= -3.38 \times 10^4 \text{ N/C} - 0.844 \times 10^4 \text{ N/C}$$
$$= -4.22 \times 10^4 \text{ N/C} \quad \text{(left)}$$

Solution (b)

The field intensity at B due to q_1 is directed downward and is equal to

$$E_1 = \frac{kq_1}{r_1^2} = \frac{-(9 \times 10^9 \text{ N} \cdot \text{m}^2/\text{C}^2)(6 \times 10^{-9} \text{ C})}{(9 \times 10^{-2} \text{ m})^2}$$

$$= -0.667 \times 10^4 \text{ N/C}$$

Similarly, the field due to q_2 is

$$E_2 = \frac{kq_2}{r_2^2} = \frac{(9 \times 10^9 \text{ N} \cdot \text{m}^2/\text{C}^2)(6 \times 10^{-9} \text{ C})}{(15 \times 10^{-2} \text{ m})^2}$$

$$= 0.240 \times 10^4 \text{ N/C} \quad \text{(at 37°)}$$

The resultant intensity at point B is the vector sum of $\mathbf{E}_1$ and $\mathbf{E}_2$. Referring to the vector diagram in Fig. 24-7, we can compute the x and y components of the resultant.

$$E_x = -E_{2x} = -(0.240 \times 10^4 \text{ N/C}) \cos 37°$$
$$= -0.192 \times 10^4 \text{ N/C}$$
$$E_y = E_{2y} - E_1$$
$$= (0.240 \times 10^4 \text{ N/C}) \sin 37° - 0.667 \times 10^4 \text{ N/C}$$
$$= 0.144 \times 10^4 \text{ N/C} - 0.667 \times 10^4 \text{ N/C}$$
$$= -0.523 \times 10^4 \text{ N/C}$$

The resultant intensity can now be calculated from its components:

$$\tan \phi = \frac{E_y}{E_x} = \frac{-0.523 \times 10^4 \text{ N/C}}{-0.192 \times 10^4 \text{ N/C}}$$

$$\phi = 69.8°$$

$$E = \frac{E_y}{\sin \theta} = \frac{0.523 \times 10^4 \text{ N/C}}{\sin 69.8°}$$

$$= 0.557 \times 10^4 \text{ N/C}$$

Therefore, the resultant field intensity at B is 0.557×10^4 N/C directed 69.8° downward and to the left.

Problem-Solving Strategy

Electric Fields

1. Read the problem and then draw and label a figure. Indicate positive and negative charges along with given distances. The charges must be in *coulombs* and the distances must be in *meters*. Remember that $1 \ \mu C = 1 \times 10^{-6} \ C$ and $1 \ nC = 1 \times 10^{-9} \ C$.

2. Remember that the electric field **E** is a property of *space* that allows us to determine the force **F** a unit positive charge *q* would experience *if* it were placed at a certain point in space. The field exists at a *point* in space independent of whether or not a charge is placed at that point.

3. The magnitude of the electric field due to a single charge is given by:

$$E = \frac{kQ}{r^2} \qquad k = 9 \times 10^9 \ N \cdot m^2/C^2$$

4. As discussed for forces, care must be taken not to confuse the nature of a charge (+ or −) with the sign given to electric fields or their components. The direction of the field **E** at a given point is consistent with the direction a *positive* test charge would move if placed at that point.

5. The *resultant electric field* due to a number of charges is found by vector addition of the electric fields due to each charge considered independently. Construct a free-body diagram and proceed with vector addition by the component method.

24-3 ELECTRIC FIELD LINES

An ingenious aid to the visualization of electric fields was introduced by Michael Faraday (1791–1867) in his early work in electromagnetism. The method consists of representing both the strength and the direction of an electric field by imaginary lines called *electric field lines.*

> **Electric field lines are imaginary lines drawn in such a manner that their direction at any point is the same as the direction of the electric field at that point.**

For example, the lines drawn radially outward from the positive charge in Fig. 24-3a represent the direction of the field at any point on the line. The electric lines in the vicinity of a negative charge would be radially inward and directed toward the charge, as in Fig. 24-3b. We shall see later that the density of these lines in any region of space is a measure of the *magnitude* of the field intensity in that region.

In general, the direction of the electric field in a region of space varies from place to place. Thus, the electric lines are normally curved. For instance, let us consider the construction of an electric field line in the region between a positive charge and a negative charge, as illustrated in Fig. 24-8.

The direction of the electric field line at any point is the same as the direction of the resultant electric field vector at that point. Two rules must be followed when constructing electric field lines:

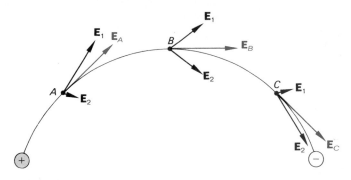

Figure 24-8   The direction of an electric field line at any point is the same as the direction of the resultant electric field intensity at that point.

1. The direction of the field line at any point is the same as the direction in which a positive charge would move if placed at that point.
2. The spacing of the field lines must be such that they are close together where the field is strong and far apart where the field is weak.

Following these general rules, one can construct the electric field lines for the two common cases shown in Fig. 24-9. As a consequence of how electric lines are drawn, *they will always leave positive charges and enter negative charges.* No lines can originate or terminate in space, although one end of an electric line may proceed to infinity.

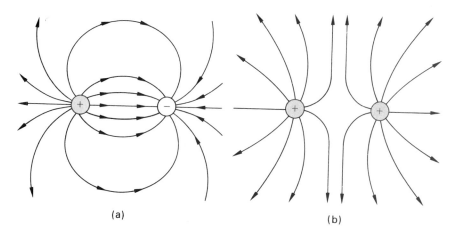

Figure 24-9   (a) A graphical illustration of the electric field lines in the region surrounding two opposite charges. (b) The field lines between two positive charges.

24-4 GAUSS'S LAW

For any given charge distribution, we can draw an infinite number of electric lines. Clearly, if the spacing of the lines is to be a standardized indication of field strength, we must set a limit on the number of lines drawn in any situation. For example, let us consider the field lines directed radially outward from a positive

books.nap.edu
www.nap.edu
For articles about the challenges of U.S. industries and global concerns, as well as book reviews of current technology publications, visit the Web site of the magazine *Issues in Science and Technology*, which can be found using either of the above two addresses.

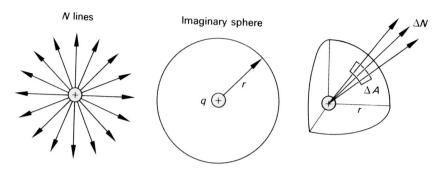

N lines   Imaginary sphere ΔN

Figure 24-10 Electric field intensity at a distance r from a point charge is directly proportional to the number of lines ΔN penetrating a unit area ΔA of an imaginary spherical surface constructed at that distance.

point charge. (Refer to Fig. 24-10). We shall use the letter N to represent the number of lines drawn. Now let us imagine a spherical surface surrounding the point charge at a distance r from the charge. The field intensity at every point on such a sphere would be given by

$$E = \frac{kq}{r^2}$$

24-9

From the way the field lines are drawn, we might also say that the field at a tiny element of surface area ΔA is proportional to the number of lines ΔN penetrating that area. In other words, the density of lines (lines per unit area) is directly proportional to the field strength. Symbolically,

$$\frac{\Delta N}{\Delta A} \propto E_n$$

24-10

The subscript n indicates that the field is everywhere normal to the surface area. This proportionality is true, regardless of the total number of lines N that may be drawn. Once we choose a proportionality constant for Eq. (24-10), however, we automatically set a limit to the number of lines drawn for any situation. It has been found that the most convenient choice for this spacing constant is ϵ_0. It is called the **permittivity of free space** and is defined by

$$\epsilon_0 = \frac{1}{4\pi k} = 8.85 \times 10^{-12} \ C^2/N \cdot m^2$$

24-11

where $k = 9 \times 10^9 \ N \cdot m^2/C^2$ from Coulomb's law. Hence, Eq. (24-10) can be written

$$\frac{\Delta N}{\Delta A} = \epsilon_0 E_n$$

24-12

or

$$\Delta N = \epsilon_0 E_n \Delta A$$

24-13

When E_n is constant over the entire surface, the total number of lines radiating outward from the enclosed charge is

$$N = \epsilon_0 E_n A \qquad \text{24-14}$$

It can be seen that the choice of ϵ_0 is a convenient one by substituting Eq. (24-11) into Eq. (24-9):

$$E_n = \frac{1}{4\pi\epsilon_0} \frac{q}{r^2}$$

Substituting this expression into Eq. (24-14) and recalling that the area of a spherical surface is $A = 4\pi r^2$, we obtain

$$N = \epsilon_0 E_n A$$

$$= \frac{\epsilon_0}{4\pi\epsilon_0} \frac{q}{r^2} (4\pi r^2) = q$$

The choice of ϵ_0 as the proportionality constant has resulted in the fact that *the total number of lines passing normally through a surface is numerically equal to the charge contained within the surface.* Although this result was obtained using a spherical surface, it will apply to any other surface. The more general statement of the result is known as *Gauss's law:*

The net number of electric lines of force crossing any closed surface in an outward direction is numerically equal to the net total charge within that surface.

$$N = \sum \epsilon_0 E_n A = \sum q \qquad \text{24-15}$$

Gauss's law can be used to compute the field intensity near surfaces of charge. This represents a distinct advantage over methods developed so far because the previous equations apply only to point charges. The best way to understand the application of Gauss's law is through examples.

24-5 APPLICATIONS OF GAUSS'S LAW

Since most charged conductors have large quantities of charge on them, it is not practical to treat the charges individually. Generally, we speak of the charge density σ, defined as the charge per unit area of surface.

$$\sigma = \frac{q}{A} \qquad q = \sigma A \qquad \text{24-16}$$

EXAMPLE 24-5

Calculate the electric field intensity at a distance r from an infinite sheet of positive charge, as shown in Fig. 24-11.

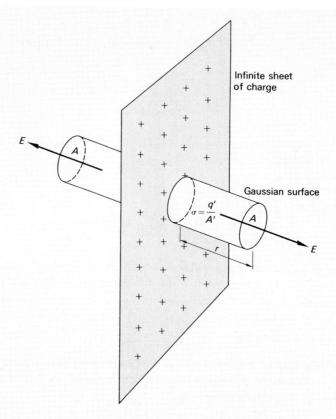

Figure 24-11 Computing the field outside an infinite sheet of positive charge.

Solution

The solution of problems using Gauss's law usually necessitates the construction of an imaginary surface of simple geometric form, for example, a sphere or a cylinder. These are referred to as **gaussian surfaces.** In this example, a cylindrical surface is imagined to penetrate the sheet of positive charge so that it projects a distance r on either side of the thin sheet. The area A of each end of the cylinder is the same as the area cut on the sheet of charge. Thus, the total charge contained inside the cylinder is

$$\sum q = \sigma A$$

where σ represents the **charge density.** Because of symmetry, the resultant field intensity **E** must be directed perpendicular to the sheet of charge at any point near the sheet. This means that no field lines will penetrate the curved sides of the cylinder, and the two ends of area A will represent the total area penetrated by the field lines. From Gauss's law,

$$\sum \epsilon_0 E_n A = \sum q$$
$$\epsilon_0 EA + \epsilon_0 EA = \sigma A$$
$$2\epsilon_0 EA = \sigma A$$
$$E = \frac{\sigma}{2\epsilon_0}$$

24-17

Note that the field intensity E is independent of the distance r from the sheet.

Before you assume that the example of an infinite sheet of charge is impractical, it should be pointed out that "infinity" in a practical sense implies only that the dimensions of the sheet are beyond the point of electrical interaction. In other words, Eq. (24-17) applies when the length and width of the sheet are very large in comparison with the distance r from the sheet.

CONCEPTUAL EXAMPLE 24-6

Show by using Gauss's law that all the excess charge resides on the surface of a charged conductor.

Solution

Let us first consider the charged solid conductor in Fig. 24-12. Within such a conductor, charges are free to move if a resultant force is applied to them. After a time, we could safely assume that all charges in the conductor are at rest. Under this condition, the electric field intensity within the conductor must be zero. Otherwise, the charges would be moving. Now, if we construct a gaussian surface just inside the surface of the conductor, as shown in the figure, we can write (for this surface)

$$\sum \epsilon_0 EA = \sum q$$

Substituting $E = 0$, we also find that $\Sigma q = 0$ or that no charge is enclosed by the surface. Since the gaussian surface can be drawn as close to the outside of the conductor as we wish, we can conclude that all the charge resides on the surface of the conductor. This conclusion holds, even if the conductor is hollow on the inside.

Figure 24-12   Gauss's law demonstrated that all the charge resides on the surface of a conductor.

An interesting experiment was devised by Michael Faraday to demonstrate that charge resides on the surface of a hollow conductor. In this experiment, known as the *ice-pail experiment,* a positively charged ball supported by a silk thread is lowered into a hollow metallic conductor. As shown in Fig. 24-13, a redistribution of

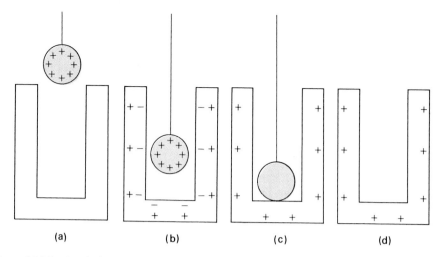

Figure 24-13 Faraday's ice-pail experiment.

charge occurs on the walls of the conductor, drawing the electrons to the inner surface. When the ball makes contact with the bottom of the conductor, the induced charge is neutralized, leaving a net positive charge on the outer surface. Probes with an electroscope will show that no charge resides inside the conductor and that a net positive charge remains on the outer surface.

EXAMPLE 24-7

A capacitor is an electrostatic device consisting of two metallic conductors of area A separated by a distance d. If equal and opposite charge densities σ are placed on the conductors, an electric field E will exist between them. Determine the electric field between the plates. (Refer to Fig. 24-14.)

$$E = \frac{\sigma}{\epsilon_0}$$

Figure 24-14 Electric field in the region between two oppositely charged plates is equal to the ratio of the charge density σ to the permittivity ϵ_0.

Solution

A gaussian cylinder is constructed as shown in Fig. 24-14 for the inner surface of either plate. There is no field inside the conducting plate, and the only area penetrated by the field lines is the surface A' projecting into the space between the plates. Applying Gauss's law to the left plate, we have

$$\sum \epsilon_0 EA = \sum q$$
$$\epsilon_0 EA' = \sigma A'$$
$$E = \frac{\sigma}{\epsilon_0} \qquad \textbf{24-18}$$

The same result would be obtained if the gaussian cylinder on the right were used. The lines would be directed inward, however, indicating an enclosed negative charge.

Notice that the field between the two plates in Example 24-7 is exactly twice the field due to a thin sheet of charge, as given by Eq. (24-17). One can understand this relationship by considering the field **E** between the plates as a superposition of the fields due to two oppositely charged sheets.

$$E = E_1 + E_2 = \frac{\sigma}{2\epsilon_0} + \frac{\sigma}{2\epsilon_0} = \frac{\sigma}{\epsilon_0}$$

The field $\mathbf{E}_1$ due to the positive sheet of charge is in the same direction as the field $\mathbf{E}_2$ due to the negative sheet of charge. To the left and right of the two plates, $\mathbf{E}_1$ and $\mathbf{E}_2$ are oppositely directed and cancel out.

Summary and Review

Summary

The concept of an electric field has been introduced to describe the region surrounding an electric charge. Its magnitude is determined by the force a unit charge will experience at a given location, and its direction is the same as the force on a positive charge at that point. Electric field lines were postulated to give a visual picture of electric fields, and the density of such field lines is an indication of the intensity of the electric field. The major concepts to be remembered are summarized below.

- An *electric field* is said to exist in a region of space in which an electric charge will experience an electric force. The *magnitude* of the electric field intensity E is given by the force F per unit of charge q.

$$E = \frac{F}{q} \qquad E = \frac{(9 \times 10^9 \text{ N} \cdot \text{m}^2/\text{C}^2)Q}{r^2}$$

The metric unit for the electric field intensity is the newton per coulomb (N/C). In the above equation, r is the distance from the charge Q to the point in question.

- The resultant field intensity at a point in the vicinity of a number of charges is the *vector* sum of the contributions from each charge.

$$\mathbf{E} = \mathbf{E}_1 + \mathbf{E}_2 + \mathbf{E}_3 + \cdots \qquad E = \sum \frac{kQ}{r^2}$$

Vector Sum

It must be emphasized that this is a vector sum and not an algebraic sum. Once the magnitude and direction of each vector is determined, the resultant can be found from vector mechanics.

- The permittivity of free space ϵ_0 is a fundamental constant defined as

$$\epsilon_0 = \frac{1}{4\pi k} = 8.85 \times 10^{-12} \text{ C}^2/\text{N} \cdot \text{m}^2$$

Permittivity

- Gauss's law states that the net number of electric field lines crossing any closed surface in an outward direction is numerically equal to the net total charge within that surface.

$$N = \sum \epsilon_0 E_n A = \sum q \qquad \textit{Gauss's Law}$$

- In applications of Gauss's law the concept of charge density σ as the charge q per unit area A of surface is often utilized:

$$\sigma = \frac{q}{A} \qquad q = \sigma A \qquad \textit{Charge Density}$$

Key Terms

1. electric field
2. electric field intensity
3. electric field lines
4. permittivity ϵ_0
5. charge density
6. Gauss's law
7. gaussian surface
8. Faraday's ice pail

Review Questions

24-1. Some texts refer to electric field lines as "lines of force." Discuss the advisability of this description.

24-2. Can an electric field exist in a region of space in which an electric charge would not experience a force? Explain.

24-3. Is it necessary that a charge be placed at a point to have an electric field at that point? Explain.

24-4. Using a procedure similar to that for electric fields, show that the gravitational acceleration can be calculated from

$$g = \frac{GM}{r^2}$$

where M = mass of earth
r = distance from center of earth

24-5. Discuss the similarities between electric fields and gravitational fields. In what ways do they differ?

24-6. In Gauss's law the constant ϵ_0 was chosen as the proportionality factor between line density and field intensity. In a theoretical sense, this is a wise choice because it leads to the conclusion that the total number of lines is equal to the enclosed charge. Is such a choice practical for graphically illustrating field lines? According to Gauss's relation, how many field lines would emanate from a charge of 1 C?

24-7. Justify the following statement: the electric field intensity on the surface of any charged conductor must be directed perpendicular to the surface.

24-8. Electric field lines will never intersect. Explain.

24-9. Suppose you connect an electroscope to the outside surface of Faraday's ice pail. Show graphically what will happen to the gold leaf during each of the steps illustrated in Fig. 24-13.

24-10. Can an electric field line begin and end on the same conductor? Discuss.

24-11. What form would Gauss's law take if we had chosen k for the proportionality constant instead of the permittivity ϵ_0?

24-12. In Gauss's law, demonstrate that the units of $\epsilon_0 EA$ are dimensionally equivalent to the units of charge.

24-13. Show that the field in the region outside the two parallel plates in Fig. 24-14 is equal to zero.

24-14. Why is the field intensity constant in the region between two oppositely charged plates? Draw a vector diagram of the field due to each plate at various points between the plates.

Problems

Section 24-1—The Concept of a Field

24-1. A charge of $+2\ \mu C$ placed at a point P in an electric field experiences a downward force of 8×10^{-4} N. What is the electric field intensity at that point?

Ans. 400 N/C, downward

24-2. A -5-nC charge is placed at point P in Prob. 24-1. What are the magnitude and direction of the force on the -5-nC charge?

24-3. A charge of $-3\ \mu C$ placed at point A experiences a downward force of 6×10^{-5} N. What is the electric field intensity at point A?

Ans. 20 N/C, upward

24-4. At a certain point, the electric field intensity is 40 N/C, due east. An unknown charge receives a westward force of 5×10^{-5} N. What is the nature and magnitude of the charge?

24-5. What are the magnitude and direction of the force that would act on an electron ($q = -1.6 \times 10^{-19}$ C) if it were placed at (a) point P in Prob. 24-1? or (b) point A in Prob. 24-3? **Ans. (a) 6.40×10^{-17} N, up, (b) 3.20×10^{-18} N, down**

24-6. What must be the magnitude and direction of the electric field intensity between two horizontal plates to produce an upward force of 6×10^{-4} N on a $+60$-μC charge?

Summary and Review

24-7. The uniform electric field between two horizontal plates is 8×10^4 C. The top plate is positively charged, and the lower plate has an equal negative charge. What are the magnitude and direction of the electric force acting on an electron as it passes horizontally through the plates?

Ans. 1.28×10^{-14} N, upward

24-8. Find the electric field intensity at a point P, located 6 mm to the left of an 8-μC charge. What are the magnitude and direction of the force on a -2-nC charge placed at point P?

24-9. Determine the electric field intensity at a point P, located 4 cm above a -12-μC charge. What are the magnitude and direction of the force on a $+3$-nC charge placed at point P?

Ans. 6.75×10^7 N/c, down, 0.202 N, down

Sections 24-2 and 24-3—Calculating the Resultant Electric Field Intensity

24-10. Determine the electric field intensity at the midpoint of a 70-mm line joining a -60-μC charge with a $+40$-μC charge.

24-11. An 8-nC charge is located 80 mm to the right of a $+4$-nC charge. Determine the field intensity at the midpoint of a line joining the two charges.

Ans. 2.25×10^4 N/C, left

24-12. Find the electric field intensity at a point 30 mm to the right of a 16-nC charge and 40 mm to the left of a 9-nC charge.

24-13. Two equal charges of opposite signs are separated by a horizontal distance of 60 mm. The resultant electric field at the midpoint of the line is 4×10^4 N/C. What is the magnitude of each charge?

Ans. 2 nC

*24-14. A 20-μC charge is 4 cm above an unknown charge q. The electric intensity at a point 1 cm above the 20-μC charge is 2.20×10^9 N/C and is directed upward. What are the magnitude and sign of the unknown charge?

*24-15. A charge of -20 μC is placed 50 mm to the right of a 49 μC charge. What is the resultant field intensity at a point located 24 mm directly above the -20-μC charge?

Ans. 2.82×10^8 N/C, 297.3°

*24-16. Two charges of $+12$ nC and $+18$ nC are separated horizontally by 28 mm. What is the resultant field intensity at a point 20 mm from each charge and above a line joining the two charges?

*24-17. A $+4$-nC charge is placed at $x = 0$, and a $+6$-nC charge is placed at $x = 4$ cm on an x-axis. Find the point at which the resultant electric field intensity will be zero.

Ans. $x = 1.80$ cm

Sections 24-4 and 24-5—Gauss's Law

24-18. Use Gauss's law to show that the field outside a solid charged sphere at a distance r from its center is given by

$$E = \frac{1}{4\pi\epsilon_0} \frac{Q}{r^2}$$

where Q is the total charge on the sphere.

*24-19. A charge of $+5$ nC is placed on the surface of a hollow metal sphere whose radius is 3 cm. Use Gauss's law to find the electric field intensity at a distance of 1 cm from the surface of the sphere. What is the electric field at a point 1 cm inside the surface?

Ans. 2.81×10^4 N/C, zero

*24-20. Two parallel plates, each 2 cm wide and 4 cm long, are stacked vertically so that the field intensity between the two plates is 10,000 N/C directed upward. What is the charge on each plate?

*24-21. A sphere 8 cm in diameter has a charge of 4 μC placed on its surface. What is the electric field intensity at the surface, 2 cm outside the surface, and 2 cm inside the surface?

Ans. 2.25×10^7 N/C, 9.99×10^6 N/C, zero

Challenge Problems

24-22. How far from a point charge of 90 nC will the field intensity be 500 N/C?

24-23. The electric field intensity at a point in space is found to be 5×10^5 N/C, directed due west. What are the magnitude and direction of the force on a $-4\text{-}\mu\text{C}$ charge placed at that point? **Ans. 2 N, east**

24-24. What are the magnitude and direction of the force on an alpha particle ($q = +3.2 \times 10^{-19}$ C) as it passes into an upward electric field of intensity 8×10^4 N/C?

24-25. What is the acceleration of an electron ($e = -1.6 \times 10^{-19}$ C) placed in a constant downward electric field of 4×10^5 N/C? What is the gravitational force on this charge if $m_e = 9.11 \times 10^{-31}$ kg? **Ans. 6.40×10^{-14} N, 8.93×10^{-30} N**

24-26. What is the electric field intensity at the midpoint of a 40-mm line between a 6-nC charge and a -9-nC charge? What force will act on a -2-nC charge placed at the midpoint?

*24-27. The charge density on each of two parallel plates is 4 $\mu\text{C/m}^2$. What is the electric field intensity between the plates? **Ans. 4.52×10^5 N/C**

*24-28. A -2 nC charge is placed at $x = 0$ on the x-axis. A $+8$-nC charge is placed at $x = 4$ cm. At what point will the electric field intensity be equal to zero?

*24-29. Charges of -2 and $+4$ μC are placed at the base corners of an equilateral triangle that has 10-cm sides. What are the magnitude and direction of the electric field intensity at the top corner? **Ans. 3.12×10^6 N/C, $150°$**

24-30. What are the magnitude and direction of the force that would act on a $-2\text{-}\mu\text{C}$ charge placed at the apex of the triangle described by Prob. 24-29?

*24-31. A 20-mg particle is placed in a uniform downward field of 2000 N/C. How many excess electrons must be placed on the particle for the electric and gravitational forces to balance one another? **Ans. 6.12×10^{11} electrons**

*24-32. Use Gauss's law to show that the electric field intensity at a distance R from an infinite line of charge is given by

$$E = \frac{\lambda}{2\pi\epsilon_0 R}$$

where λ is the charge per unit length. Construct a gaussian surface as in Fig. 24-15.

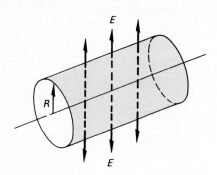

Figure 24-15

*24-33. Use Gauss's law to show that the field just outside any solid conductor is given by

$$E = \frac{\sigma}{\epsilon_0}$$

*24-34. What is the electric field intensity 2 cm from the surface of a sphere 20 cm in diameter having a surface charge density of $+8$ nC/m^2?

*24-35. A uniformly charged conducting sphere has a radius of 24 cm and a surface charge density of $+16$ $\mu\text{C/m}^2$. What is the total number of electric field lines leaving the sphere? **Ans. 1.16×10^{-5} lines**

*24-36. Two charges of $+16$ μC and $+8$ μC are 200 mm apart in air. At what point on a line joining the two charges will the electric field be zero?

*24-37. Two charges of $+8$ nC and -5 nC are 40 mm apart in air. At what point on a line joining the two charges will the electric field intensity be zero? **Ans. 151 mm outside of -5 nC charge**

Summary and Review

Critical Thinking Questions

*24-38. Two equal but opposite charges $+q$ and $-q$ are placed at the base corners of an equilateral triangle whose sides are of length a. Show that the magnitude of the electric field intensity at the apex is the same whether one of the charges is removed or not. What is the angle between the two fields so produced?

*24-39. What are the magnitude and direction of the electric field intensity at the center of the square of Fig. 24-16? Assume that $q = 1 \ \mu C$ and that $d = 4$ cm.

Ans. 3.56 × 10⁷ N, 153.4°

Figure 24-16

*24-40. The electric field intensity between the plates in Fig. 24-17 is 4000 N/C. What is the magnitude of the charge on the suspended pith ball whose mass is 3 mg?

Ans. 4.24 nC

Figure 24-17

*24-41. Two concentric spheres have radii of 20 and 50 cm. The inner sphere has a negative charge of $-4 \ \mu C$, and the outer sphere has a positive charge of $+6 \ \mu C$. Use Gauss's law to find the electric field intensity at distances of 40 and 60 cm from the center of the spheres.

*24-42. The electric field intensity between the two plates in Fig. 24-4 is 2000 N/C. The length of the plates is 4 cm, and their separation is 1 cm. An electron is projected into the field from the left with horizontal velocity of 2×10^7 m/s. What is the upward deflection of the electron at the instant it leaves the plates?

Ans. 0.700 mm

Electric Potential

In our study of mechanics, many problems were simplified by introducing the concepts of energy. The conservation of mechanical energy allowed us to say certain things about the initial and final states of systems without having to analyze the motion between states. The concept of a change of potential energy into kinetic energy avoids the problem of varying forces.

In electricity, many practical problems can be solved by considering the changes in energy experienced by a moving charge. For example, if a certain quantity of work is required to move a charge against electric forces, the charge should have a ***potential*** for giving up an equivalent amount of energy when it is released. In this chapter we develop the idea of ***electric potential energy.***

25-1 ELECTRIC POTENTIAL ENERGY

One of the best ways to understand the concept of electric potential energy is by comparing it with gravitational potential energy. In the gravitational case, consider that a mass *m* is moved from level *A* in Fig. 25-1 to level *B*. An external force **F** equal to the weight *mg* must be applied to move the mass against gravity. The work done by this force is the product of *mg* and *h*. When the mass *m* reaches level *B*, it has a potential for doing work relative to level *A*. The system has a *potential energy* (P.E.) that is equal to the work done against gravity.

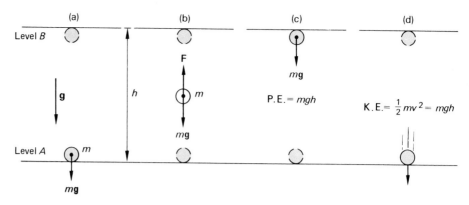

Figure 25-1 A mass m lifted against a gravitational field **g** results in a potential energy of mgh at level B. When released, this energy will be transformed entirely into kinetic energy as it falls to level A.

$$\text{P.E.} = mgh$$

This expression represents the potential for doing work after the mass m is released at level B and falls the distance h. Therefore, the magnitude of the potential energy at B does not depend on the path taken to reach that level.

Now, let us consider a positive charge $+q$ resting at point A in a uniform electric field **E** between two oppositely charged plates. (See Fig. 25-2.) An electric force $q\text{E}$ acts downward on the charge. The work done against the electric field in moving the charge from A to B is equal to the product of the force $q\text{E}$ and the distance d. Hence, the electric potential energy at point B relative to point A is

$$\text{P.E.} = qEd \qquad\qquad \text{25-1}$$

When the charge is released, the electric field will perform this amount of work and the charge q will have a kinetic energy.

$$\text{K.E.} = \frac{1}{2}mv^2 = qEd$$

when it returns to point A.

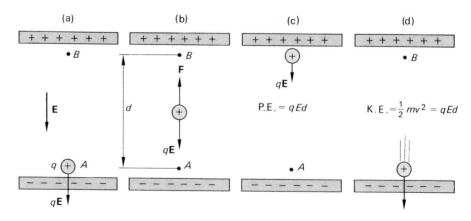

Figure 25-2 A positive charge $+q$ is moved against a constant electric field E through a distance d. At point B the potential energy will be qEd relative to point A. When released, the charge will gain an equivalent amount of kinetic energy.

The above statements and equations are valid regardless of the path through which the charge is moved. The same work against the gravitational field is required to slide a mass up an inclined plane as would be necessary to lift it vertically. Similarly, the potential energy due to the charge $+q$ at point B is independent of the path. As shown in Fig. 25-3, the potential energy would be the same whether $+q$ were moved along the path l or the path d. The only work which contributes to the potential energy is the work done against the electric field force qE. The effective distance moved against this downward electric force is

$$l \sin \theta = d$$

Therefore, the work is the same for either path.

Before we proceed, we should point out an important difference between gravitational potential energy and electric potential energy. In the case of gravity, there is only one kind of mass, and the forces involved are always forces of attraction. Therefore, a mass at higher elevations always has greater potential energy relative to the earth. This is not true in the electrical case because of the existence of negative charge. For example, in Fig. 25-4, a positive charge has a greater potential energy at point B than at point A. This is true regardless of the reference point chosen for measuring the energy because work has been done *against* the electric field. On the other hand, if a negative charge were moved from point A to point B, work would be done *by* the field. A negative charge would have a *lower* potential energy at B, which is exactly opposite to the situation for a positive charge.

Whenever a positive charge is moved against an electric field, the potential energy increases; whenever a negative charge moves against an electric field, the potential energy decreases.

The above rule is a direct consequence of the fact that the direction of the electric field is defined in terms of a positive charge.

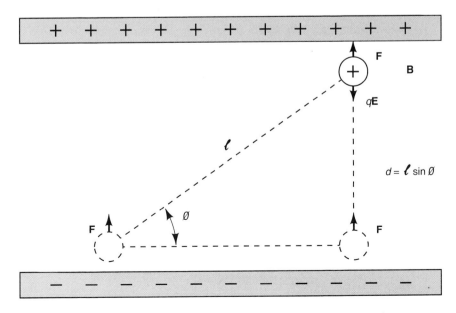

Figure 25-3 Electric potential energy at B is independent of the path taken to reach B.

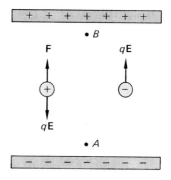

Figure 25-4 A positive charge increases its potential energy when moved from *A* to *B*; a negative charge *loses* potential energy when it moves from *A* to *B*.

25-2 CALCULATING POTENTIAL ENERGY

In considering the space between two oppositely charged plates, work computations are fairly simple because the electric field is uniform. The electric force that a charge experiences is constant as long as it remains between the plates. In general, however, the field will not be constant, and we must make allowances for a varying force. For example, consider the electric field in the vicinity of a positive charge *Q*, as illustrated in Fig. 25-5. The field is directed radially outward, and its intensity falls off inversely with the square of the distance from the center of the charge. The field at points *A* and *B* is

$$E_A = \frac{kQ}{r_A^2} \qquad E_B = \frac{kQ}{r_B^2}$$

where r_A and r_B are the respective distances to points *A* and *B*.

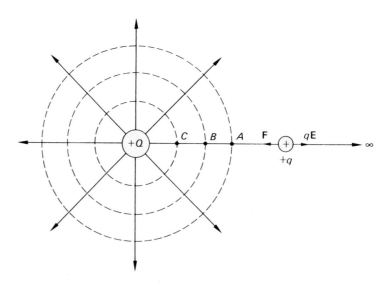

Figure 25-5 The potential energy due to a charge placed in an electric field is equal to the work done *against* electric forces in bringing the charge from infinity to the point in question.

The average electric force experienced by a charge $+q$ when it is moved from point A to point B is

$$F = \frac{kQq}{r_A r_B} \qquad \text{25-2}$$

Thus, the work done against the electric field in moving through the distance $r_A - r_B$ is equal to

$$\text{Work}_{A \to B} = \frac{kQq}{r_A r_B}(r_A - r_B)$$
$$= kQq\left(\frac{1}{r_B} - \frac{1}{r_A}\right) \qquad \text{25-3}$$

Note that the work is a function of the distances r_A and r_B. The path traveled is unimportant. The same work would be done against the field in moving a charge from any point on the dashed circle passing through A to any point on the circle passing through B.

Suppose now we compute the work done against electric forces in moving a positive charge from infinity to a point a distance r from the charge Q. From Eq. (25-3), the work is given by

$$\text{Work}_{\infty \to r} = kQq\left(\frac{1}{r} - \frac{1}{\infty}\right)$$
$$= \frac{kQq}{r} \qquad \text{25-4}$$

Since we have shown that the work done against the electric field equals the increase in potential energy, Eq. (25-4) is the potential energy at r relative to infinity. Often potential energy is taken as zero at infinity, so that the potential energy of a system composed of a charge q and another charge Q separated by a distance r is

$$\text{P.E.} = \frac{kQq}{r} \qquad \text{25-5}$$

The potential energy of the system is equal to the work done against the electric forces in moving the charge $+q$ from infinity to that point.

EXAMPLE 25-1

A charge of $+2$ nC is 20 cm away from another charge of $+4$ μC. (a) What is the potential energy of the system? (b) What is the change in potential energy if the 2-μC charge is moved to a distance of 8 cm from the $+4$-nC charge?

Solution (a)
The potential energy at 20 cm is given by Eq. (25-5).

$$P.E. = \frac{kQq}{r}$$

$$= \frac{(9 \times 10^9 \text{ N} \cdot \text{m}^2/\text{C}^2)(4 \times 10^{-6} \text{ C})(2 \times 10^{-9} \text{ C})}{0.2 \text{ m}}$$

$$= 36 \times 10^{-5} \text{ J}$$

Solution (b)

The potential energy at a distance of 8 cm is

$$P.E. = \frac{kQq}{r}$$

$$= \frac{(9 \times 10^9 \text{ N} \cdot \text{m}^2/\text{C}^2)(4 \times 10^{-6} \text{ C})(2 \times 10^{-9} \text{ C})}{0.08 \text{ m}}$$

$$= 90 \times 10^{-5} \text{ J}$$

The change in potential energy is

$$90 \times 10^{-5} \text{ J} - 36 \times 10^{-5} \text{ J} = 54 \times 10^{-5} \text{ J}$$

Notice that the difference is positive, indicating an increase in potential energy. If the charge Q were negative and all other parameters were unchanged, the potential energy would have decreased by this same amount.

25-3 POTENTIAL

When we first introduced the concept of an electric field as force per unit charge, we pointed out that the primary advantage of such a concept was that it allowed us to assign an electrical property to space. If the field intensity is known at some point, the force on a charge placed at that point can be predicted. It is equally convenient to assign another property to the space surrounding a charge that would allow us to predict the potential energy due to another charge placed at any point. This property of space is called *potential* and is defined as follows:

> **The potential V at a point a distance r from a charge Q is equal to the work per unit charge done against electric forces in bringing a positive charge +q from infinity to that point.**

In other words, the potential at some point A, as shown in Fig. 25-6, is equal to *the potential energy per unit charge*. The units of potential are expressed in *joules per coulomb*, defined as a **volt** (V).

$$V_A(\text{V}) = \frac{\text{P.E. (J)}}{q \text{ (C)}} \qquad \text{25-6}$$

Thus, a potential of *one volt* at point A means that *if a charge of one coulomb were to be placed at A, the potential energy would be one joule*. In general when the potential is known at a point A, the potential energy due to a charge q at that point can be found from

$$\text{P.E.} = qV_A \qquad \textbf{25-7}$$

Substitution of Eq. (25-5) into Eq. (25-6) yields an expression for directly computing the potential:

$$V_A = \frac{\text{P.E.}}{q} = \frac{kQq/r}{q}$$

$$V_A = \frac{kQ}{r} \qquad \textbf{25-8}$$

The symbol V_A refers to the potential at point A located at a distance r from the charge Q.

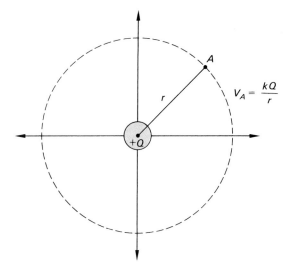

Figure 25-6 Calculating the potential at a distance r from a charge $+Q$.

It is noted now that the potential is the same at equal distances from a spherical charge. For this reason, the dashed lines in Figs. 25-5 and 25-6 are called *equipotential lines.* Note that the lines of equal potential are always perpendicular to the electric field lines. If this were not true, work would be done by a resultant force when a charge moved along an equipotential line. Such work would increase or decrease the potential.

Equipotential lines are always perpendicular to electric field lines.

Before offering an example, it must be pointed out that the potential at a point is defined in terms of a positive charge. This means that the potential will be negative at a point in space surrounding a negative charge. As a rule, we must remember that:

The potential due to a positive charge is positive, and the potential due to a negative charge is negative.

Using a negative sign for a negative charge Q in Eq. (25-8) results in a negative value for the potential.

EXAMPLE 25-2

(a) Calculate the potential at a point A that is 30 cm distant from a charge of $-2\ \mu C$. (b) What is the potential energy if a $+4$-nC charge is placed at A?

Solution (a)

From Eq. (25-8)

$$V_A = \frac{kQ}{r} = \frac{(9 \times 10^9\ \text{N} \cdot \text{m}^2/\text{C}^2)(-2 \times 10^{-6}\ \text{C})}{0.3\ \text{m}}$$
$$= -6 \times 10^4\ \text{V}$$

Solution (b)

The potential energy due to the $+4$-nC charge is given by direct substitution into Eq. (25-7).

$$\text{P.E.} = qV_A = (4 \times 10^{-9}\ \text{C})(-6 \times 10^4\ \text{V})$$
$$= -24 \times 10^{-5}\ \text{J}$$

A negative potential energy means that work must be done *against* the electric field in moving the charges apart. In Example 25-2, 24×10^{-5} J of work would have to be supplied by an external force to remove the charge to infinity.

Now let us consider the more general case, which deals with the potential in the neighborhood of a number of charges. As illustrated in Fig. 25-7:

The potential in the vicinity of a number of charges is equal to the algebraic sum of the potentials due to each charge.

$$V_A = V_1 + V_2 + V_3$$
$$= \frac{kQ_1}{r_1} + \frac{kQ_2}{r_2} + \frac{kQ_3}{r_3}$$

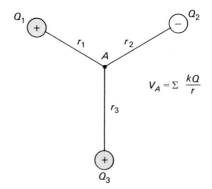

Figure 25-7 The potential in the vicinity of a number of charges.

Whenever the charge is negative, as with Q_2 in Fig. 25-7, the sign of the charge is inserted into the calculations. In general,

$$V = \sum \frac{kQ}{r} \qquad \text{25-9}$$

This equation is an algebraic sum since potential is a scalar quantity and not a vector quantity.

EXAMPLE 25-3

Two charges, $Q_1 = +6\ \mu C$ and $Q_2 = -6\ \mu C$, are separated by 12 cm, as shown in Fig. 25-8. Calculate the potential (a) at point A and (b) at point B.

Figure 25-8

Solution (a)

The potential at A is found from Eq. (25-9).

$$V_A = \frac{kQ_1}{r_1} + \frac{kQ_2}{r_2}$$
$$= \frac{(9 \times 10^9 \text{ N} \cdot \text{m}^2/\text{C}^2)(6 \times 10^{-6} \text{ C})}{4 \times 10^{-2} \text{ m}} + \frac{(9 \times 10^9 \text{ N} \cdot \text{m}^2/\text{C}^2)(-6 \times 10^{-6} \text{ C})}{8 \times 10^{-2} \text{ m}}$$
$$= 13.5 \times 10^5 \text{ V} - 6.75 \times 10^5 \text{ V}$$
$$= 6.75 \times 10^5 \text{ V}$$

This means that the electric field will do 6.75×10^5 J of work on each coulomb of positive charge that it moves from A to infinity.

Solution (b)

The potential at B is

$$V_B = \frac{kQ_1}{r_1} + \frac{kQ_2}{r_2}$$
$$= \frac{(9 \times 10^9 \text{ N} \cdot \text{m}^2/\text{C}^2)(6 \times 10^{-6} \text{ C})}{16 \times 10^{-2} \text{ m}} + \frac{(9 \times 10^9 \text{ N} \cdot \text{m}^2/\text{C}^2)(-6 \times 10^{-6} \text{ C})}{4 \times 10^{-2} \text{ m}}$$
$$= 3.38 \times 10^5 \text{ V} - 13.5 \times 10^5 \text{ V}$$
$$= -10.1 \times 10^5 \text{ V}$$

The negative value indicates that the field will hold onto a positive charge. To move 1 C of positive charge from A to infinity, another source of energy must perform 10.1×10^5 J of work. The field would perform negative work in this amount.

USING A CALCULATOR

Parentheses can be helpful when solving problems involving multiplication, division, and addition.

$$V_B = \frac{kQ_1}{r_1} + \frac{kQ_2}{r_2}$$

-10.1×10^5 V

25-4 POTENTIAL DIFFERENCE

In practical electricity, we are seldom interested in the work per unit charge to remove a charge to infinity. More often, we want to know the work requirements for moving charges between two points. This leads to the concept of *potential difference.*

> **The potential difference between two points is the work per unit positive charge done by electric forces in moving a small test charge from the point of higher potential to the point of lower potential.**

Another way of stating this would be to say that the potential difference between two points is the difference in the potentials at those points. For example, if the potential at some point A is 100 V and the potential at another point B is 40 V, the potential difference is

$$V_A - V_B = 100 \text{ V} - 40 \text{ V} = 60 \text{ V}$$

This means that 60 J of work will be done by the field on each coulomb of positive charge moved from A to B. In general, the **work done by the electric field** in moving a charge q from point A to point B can be found from

$$\text{Work}_{A \to B} = q(V_A - V_B) \qquad\qquad \textbf{25-10}$$

CONCEPTUAL EXAMPLE 25-4

What is the potential difference between points A and B in Fig. 25-8. How much work is done by the electric field in moving a -2-nC charge from point A to point B?

Solution

The potentials at points A and B were calculated in the previous example. They are

$$V_A = 6.75 \times 10^5 \text{ V} \qquad V_B = -10.1 \times 10^5 \text{ V}$$

Therefore, the potential difference between points A and B is

$$V_A - V_B = 6.75 \times 10^5 \text{ V} - (-10.1 \times 10^5 \text{ V})$$
$$= 16.9 \times 10^5 \text{ V}$$

Since A is at a higher potential than B, positive work would be done by the field when a *positive* charge is moved from A to B. If a *negative* charge is moved, the work done by the field in moving it from A to B will be negative. In this example, the work is

$$\text{Work}_{A \to B} = q(V_A - V_B)$$
$$= (-2 \times 10^{-9} \text{ C})(16.9 \times 10^5 \text{ V})$$
$$= -3.37 \times 10^{-3} \text{ J}$$

Since the work done by this field is negative, another source of energy must supply the work to move the charge.

Problem-Solving Strategy

Electric Potential and Potential Energy

1. Read the problem and then draw and label a figure. Indicate positive and negative charges along with given distances. The charges must be in coulombs and the distances must be in meters. Remember that $1 \ \mu C = 1 \times 10^{-6} \ C$ and $1 \ nC = 1 \times 10^{-9} \ C$.

2. Remember that the electric potential V is a property of *space* that allows us to determine the potential energy U when a charge q is placed at that point. The potential exists at a *point* in space independent of whether a charge is placed at that point.

3. The absolute potential at a point in the vicinity of a number of charges is the algebraic sum of the potentials due to each charge:

$$V = \sum \frac{kQ}{r} \qquad k = 9 \times 10^9 \mathrm{N} \cdot \mathrm{m}^2/\mathrm{C}^2$$

4. Only *changes* in potential are significant, and the reference point for zero potential can be at infinity or at any other point. Often the point of lowest absolute potential can be chosen as zero.

5. The *work* done *by* the electric field in moving a charge q from a point A to another point B is simply the product of the charge and the difference of potential:

$$\mathrm{Work}_{AB} = q(V_A - V_B)$$
Work by the Electric Field

6. Because the potential in the vicinity of a positive charge is positive and the potential near a negative charge is negative, the signs of the charge and potential can be used algebraically.

Let us now return to the example of a uniform electric field **E** between two oppositely charged plates, as in Fig. 25-9. We shall assume that the plates are separated by a distant d. A charge q placed in the region between the plates A and B will experience a force given by

$$\mathbf{F} = q\mathbf{E}$$

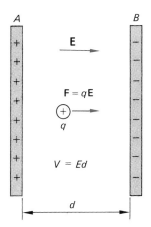

Figure 25-9 The potential between two oppositely charged plates.

The work done by this force in moving the charge q from plate A to plate B is given by

$$Fd = (qE)d$$

But this work is also equal to the product of the charge q and the potential difference $V_A - V_B$ between the two plates, so that we can write

$$q(V_A - V_B) = qEd$$

If we divide through by q and represent the potential difference by the single symbol V, we obtain

$$V = Ed \qquad \text{25-11}$$

The potential difference between two oppositely charged plates is equal to the product of the field intensity and the plate separation.

EXAMPLE 25-5

The potential difference between two plates 5 mm apart is 10 kV. Determine the electric field intensity between the plates.

Solution

Solving for E in Eq. (25-11) gives

$$E = \frac{V}{d} = \frac{10 \times 10^3 \text{ V}}{5 \times 10^{-3} \text{ m}} = 2 \times 10^6 \text{ V/m}$$

As an exercise, the student should show that the *volt per meter* is equivalent to the *newton per coulomb*. The electric field expressed in volts per meter is sometimes referred to as the **potential gradient**.

25-5 MILLIKAN'S OIL-DROP EXPERIMENT

Now that we have developed the concepts of the electric field and potential difference, we are ready to describe a classic experiment designed to determine the smallest unit of charge. Robert A. Millikan, an American physicist, devised a series of experiments in the early 1900s. A schematic diagram of his apparatus is shown in Fig. 25-10. Tiny oil droplets are sprayed into the region between the two metallic plates. Electrons are freed from air molecules by passing ionizing X rays through the medium. These electrons attach themselves to the oil droplets, giving them a net negative charge.

The downward motion of the oil droplets can be observed by a microscope as they fall slowly under the influence of their weight and the upward viscous force of air resistance. (Refer to Fig. 25-10.) The laws of hydrostatics can be used to calculate the mass m of a particular oil drop from its measured rate of fall.

(a)

(b)

Figure 25-10 Millikan's oil-drop experiment: (a) The mass m of the droplet is determined from its rate of fall against the viscous force of air resistance. (b) The magnitude of the charge is computed from the equilibrium conditions that suspend the charge between oppositely charged plates.

After the necessary data for determining the mass m have been recorded, an external battery is connected to establish a uniform electric field **E** between the oppositely charged plates. (See. Fig. 25-10.) The magnitude of the field intensity can be controlled by a variable resistor in the electric circuit. The field is adjusted until the upward electric force on the drop is equal to the downward gravitational force, so that the oil drop stops moving. Under these conditions

$$qE = mg \qquad\qquad \textbf{25-12}$$

where q = net charge of oil drop
m = mass of oil drop
g = acceleration of gravity

The field intensity **E,** as determined from Eq. (25-11), is a function of the applied voltage V and the plate separation d. Therefore, Eq. (25-12) becomes

$$q\frac{V}{d} = mg$$

and the magnitude of the charge on the oil drop is found from

$$q = \frac{mgd}{V} \qquad\qquad \textbf{25-13}$$

The potential difference V can be read directly from an indicating device called a *voltmeter* attached to the circuit. The other parameters are known.

The charges observed by Millikan were not always the same, but he observed that the magnitude of the charge was always an integral multiple of a basic quantity of charge. It was assumed that this *least* charge must be the charge of a single electron and that the other magnitudes resulted from two or more electrons. Computation of the electronic charge by this method yields

$$e = 1.6065 \times 10^{-19} \text{ C}$$

which agrees extremely well with the values obtained by other methods.

25-6 THE ELECTRONVOLT

Let us now consider the energy of a charged particle moving through a potential difference. There are several units in which to measure this energy, but most of the familiar units are inconveniently large. Consider, for example, a charge of 1 C accelerated through a potential difference of 1 V. Its kinetic energy will be

$$\text{K.E.} = qEd = qV$$
$$= (1 \text{ C})(1 \text{ V}) = 1 \text{ C} \cdot \text{V}$$

The coulomb-volt, of course, is a joule. But 1 C of charge is inconveniently large when applied to single particles, and the corresponding unit of energy (the joule) is also large. The most convenient unit of energy in atomic and nuclear physics is the *electronvolt* (eV).

> **The electronvolt is a unit of energy equivalent to the energy acquired by an electron that is accelerated through a potential difference of one volt.**

The electronvolt differs from the coulomb-volt by the same degree as the difference in the charge of an electron and the charge of 1 C. To compare the two units, suppose we compute the energy in joules acquired by an electron that has been accelerated through a potential difference of 1 V:

$$\text{K.E.} = qV$$
$$= (1.6 \times 10^{-19} \text{ C})(1 \text{ V})$$
$$= 1.6 \times 10^{-19} \text{ J}$$

Thus, 1 eV is equivalent to an energy of 1.6×10^{-19} J.

Summary and Review

Summary

The concepts of potential energy, potential, and potential difference have been extended to electrical phenomena. The many problems dealing with electrostatic potential have been designed to provide a base for direct current electricity that will be discussed later. The essential elements of this chapter are summarized below.

- When a charge q is moved against a constant electric force for a distance d, the potential energy of the system is

$$\text{P.E.} = qEd$$

where E is the constant electric field intensity. If the charge is released, it will acquire a kinetic energy

$$\text{K.E.} = \frac{1}{2}mv^2 = qEd$$

as it returns for the same distance.

- Because of the existence of positive and negative charges and the opposite effects produced by the same field, we must remember that: *The potential energy increases as a positive charge is moved against the electric field, and the potential energy decreases as a negative charge is moved against the same field.*

- In general, the potential energy due to a charge q placed at a distance r from another charge Q is equal to the work done against electric forces in moving the charge $+q$ from infinity.

$$\text{P.E.} = \frac{kQq}{r} \qquad \textit{Electric Potential Energy}$$

Note that the distance r is not squared, as it was for the electric field intensity.

- The electric *potential* V at a point a distance r from a charge Q is equal to the work per unit charge done against electric forces in bringing a positive charge $+q$ from infinity.

$$V = \frac{kQ}{r} \qquad \textit{Electric Potential}$$

- The unit of electric potential is the joule per coulomb (J/C), which is renamed the volt (V).

$$1\text{ V} = \frac{1\text{ J}}{1\text{ C}}$$

- The potential at a point in the vicinity of a number of charges is equal to the algebraic sum of the potentials due to each charge:

$$V = \sum \frac{kQ}{r} = \frac{kQ_1}{r_1} + \frac{kQ_2}{r_2} + \frac{kQ_3}{r_3} + \dots$$

$$\textit{Algebraic Sum}$$

- The potential difference between two points A and B is the difference in the potentials at those points.

$$V_{AB} = V_A - V_B \qquad \textit{Potential Difference}$$

- The work done by an electric field in moving a charge q from point A to point B can be found from

$$\text{Work}_{AB} = q(V_A - V_B)$$

$$\textit{Work and Potential Difference}$$

- The potential difference between two oppositely charged plates is equal to the product of the field intensity and the plate separation.

$$V = Ed \qquad E = \frac{V}{d}$$

Key Terms

1. electric potential energy
2. electric work
3. potential
4. volt
5. equipotential lines
6. potential difference
7. potential gradient
8. electronvolt

Review Questions

25-1. Distinguish clearly between positive and negative work. Distinguish between positive and negative potential energy.

25-2. It is possible for a mass m to increase the potential energy by moving it to a lower elevation? Is it possible for a charged object to increase the potential energy as it is moved to a position of lower potential? Explain.

25-3. Give an example in which the electric potential is zero at some point where the electric field intensity is not zero.

25-4. The electric field inside an electrostatic conductor is zero. Is the electric potential inside the conductor zero also? Explain.

25-5. If the electric field intensity is known at some point, can one determine the electric potential at that point? What information is needed?

25-6. The surface of any conductor is an equipotential surface. Justify this statement.

25-7. Is the direction of the electric field intensity from higher to lower potential? Illustrate.

25-8. Apply the potential concept to the gravitational field to obtain an expression similar to Eq. (25-9) for computing the potential energy per unit mass. Discuss the applications of such a formula.

25-9. Show that the volt per meter is dimensionally equivalent to the newton per coulomb.

25-10. Distinguish between potential difference and a difference in potential energy.

25-11. A potential difference of 220 V is maintained between the ends of a long high-resistance wire. If the center of the wire is grounded ($V = 0$), what is the potential difference between the center and the end points?

25-12. The potential due to a negative charge is negative, and the potential due to a positive charge is positive. Why? Is it also true that the potential energy due to a negative charge is negative? Explain.

25-13. Is potential a property assigned to space or to a charge? What is potential energy assigned to?

Problems

Section 25-1—Electric Potential Energy

25-1. A positively charged plate is 30 mm above a negatively charged plate, and the electric field intensity has a magnitude of 6×10^4 N/C. How much work is done *by* the electric field when a $+4\text{-}\mu\text{C}$ charge is moved from the negative plate to the positive plate? **Ans. −7.20 mJ**

25-2. In Prob. 25-1, how much work is done *on* or against the electric field? What is the electric potential energy at the positive plate?

25-3. The electric field intensity between two parallel plates, separated by 25 mm in 8000 N/C. How much work is done *by* the electric field in moving a $-2\text{-}\mu\text{C}$ charge from the negative plate to the positive plate?

What is the work done *by* the field in moving the same charge back to the positive plate?   **Ans. +4.00 × 10⁻⁴ J, −4.00 × 10⁻⁴ J**

25-4. In Prob. 25-3, what is the potential energy when the charge is at (a) the positive plate and (b) the negative plate?

25-5. What is the potential energy of a +6-nC charge located 50 mm away from a +80-μC charge? What is the potential energy if the same charge is 50 mm from a −80-μC charge?   **Ans. +86.4 mJ, −86.4 mJ**

25-6. At what distance from a −7-μC charge will a −3-nC charge have a potential energy of 60 mJ? What initial force will the −3-nC charge experience?

25-7. A +8-nC charge is placed at a point *P*, 40 mm from a +12-μC charge. What is the potential energy per unit charge at point *P* in joules per coulomb? Will this change if the 8-nC charge is removed?
    Ans. 2.70 × 10⁶ J/C, no

25-8. A charge of +6 μC is 30 mm away from another charge of 16 μC. What is the potential energy of the system?

25-9. In Prob. 25-8, what is the change in potential energy if the 6-μC charge is moved to a distance of only 5 mm? Is this an increase or decrease in potential energy?
    Ans. 144 J, increase

25-10. A −3-μC charge is placed 6 mm away from a −9-μC charge. What is the potential energy? Is it negative or positive?

25-11. What is the change in potential energy when a 3-nC charge is moved from a point 8 cm away from a −6-μC charge to a point that is 20 cm away? Is this an increase or decrease of potential energy?
    Ans. +1.22 J, increase

25-12. At what distance from a −7-μC charge must a charge of −12 nC be placed if the potential energy is to be 9 × 10⁻⁵ J?

25-13. The potential energy of a system consisting of two identical charges is 4.50 mJ when their separation is 38 mm. What is the magnitude of each charge?   **Ans. 139 nC**

Sections 25-3 and 25-4—Electric Potential and Potential Difference

25-14. What is the electric potential at a point that is 6 cm from a 8.40-μC charge? What is the potential energy of a 2-nC charge placed at that point?

25-15. Calculate the potential at point *A* that is 50 mm from a −40-μC charge. What is the potential energy if a +3-μC charge is placed at point *A*?   **Ans. −7.20 MV, −21.6 J**

25-16. What is the potential at the midpoint of a line joining a −12-μC charge with a +3-μC charge located 80 mm away from the first charge?

25-17. A +45-nC charge is 68 mm to the left of a −9-nC charge. What is the potential at a point located 40 mm to the left of the −9-nC charge?   **Ans. 12.4 kV**

***25-18.** Points *A* and *B* are 68 mm and 26 mm away from a 90-μC charge. Calculate the potential difference between points *A* and *B*. How much work is done *by* the electric field as a −5-μC charge moves from *A* to *G*?

***25-19.** Points *A* and *B* are 40 and 25 mm away from a +6-μC charge. How much work must be done against the electric field (by external forces) in moving a +5-μC charge from point *A* to point *B*?   **Ans. +4.05 J**

***25-20.** A +6-μC charge is located at $x = 0$ on the *x*-axis, and a −2-μC charge is located at $x = 8$ cm. How much work is done *by* the electric field in moving a −3-μC charge from the point $x = 10$ cm to the point $x = 3$ cm?

Challenge Problems

25-21. Point *A* is 40 mm above a −9-μC charge, and point *B* is located 60 mm below the same charge. A −3-nC charge is moved

from point *B* to point *A*. What is the change in potential energy?   **Ans. +2.02 mJ**

25-22. Two parallel plates are separated by 50 mm in air. If the electric field intensity between

Summary and Review

the plates is 2×10^4 N/C, what is the potential difference between the plates?

25-23. The potential difference between two parallel plates 60 mm apart is 4000 V. What is the electric field intensity between the plates? **Ans. 66.7 kV/m**

25-24. If an electron is located at the plate of lower potential in Prob. 25-23, what will be its kinetic energy when it reaches the plate of higher potential? What is the energy expressed in electronvolts?

25-25. Show that the potential gradient V/m is equivalent to the unit N/C for electric field.

25-26. What is the difference in potential between two points 30 and 60 cm away from a $-50\text{-}\mu C$ charge?

25-27. The potential gradient between two parallel plates 4 mm apart is 6000 V/m. What is the potential difference between the plates? **Ans. 24.0 V**

25-28. The electric field between two plates separated by 50 mm is 6×10^5 V/m. What is the potential difference between the plates?

25-29. What must be the separation of two parallel plates if the field intensity is 5×10^4 V/m and the potential difference is 400 V? **Ans. 8.00 mm**

25-30. The potential difference between two parallel plates is 600 V. A 6-μC charge is accelerated through the entire potential difference. What is the kinetic energy given to the charge?

25-31. Determine the kinetic energy of an alpha particle $(+2e)$ that is accelerated through a potential difference of 800 kV. Give the answer in both electronvolts and joules. **Ans. 1.60 MeV, 2.56 × 10⁻¹³ J**

25-32. A linear accelerator accelerates an electron through a potential difference of 4 MV. What is the energy of an emergent electron in electronvolts and in joules?

25-33. An electron acquires an energy of 2.8×10^{-15} J as it passes from point A to point

B. What is the potential difference between these points in volts? **Ans. 17.5 kV**

*25-34. Show that the total potential energy of the three charges placed at the corners of the equilateral triangle shown in Fig. 25-11 is given by:

$$-\frac{3kq^2}{d}$$

*25-35. Assume that $q = 1$ μC and $d = 20$ mm. What is the potential energy of the system of charges in Fig. 25-11? **Ans. 1.50 × 10⁻¹⁰ J**

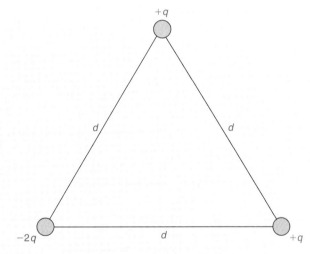

Figure 25-11

*25-36. The potential at a certain distance from a point charge is 1200 V, and the electric field intensity at that point is 400 N/C. What is the distance to the charge, and what is the magnitude of the charge? **Ans. 3 m, 400 NC**

*25-37. Two large plates are 80 mm apart and have a potential difference of 800 kV. What is the magnitude of the force that would act on an electron placed at the midpoint between these plates? What would be the kinetic energy of the electron moving from low potential plate to the high potential plate? **Ans. 1.60 × 10⁻¹² N, 1.28 × 10⁻¹³ J**

Critical Thinking Questions

25-38. Plate A has a potential that is 600 V higher than plate B, which is 50 mm below plate A. A $+2$-μC charge moves from plate A to plate B. What is the electric field intensity between the plates? What are the sign and magnitude of the work done by the electric field? Does the potential energy increase or decrease? Now answer the same questions if a -2-μC charge is moved from A to B. **Ans. 12 kV/m, +1.20 mJ, decreases, −1.2 mJ, increases**

25-39. Point A is a distance $x = +a$ to the right of a $+4$-μC charge. The rightward electric field at point A is 4000 N/C. What is the distance a? What is the potential at point A? What are the electric field and the potential at the point $x = -a$? Find the electric force and the electric potential energy when a -2-nC charge is placed at each point.

*25-40. Points A, B, and C are at the corners of an equilateral triangle that is 100 mm on each side. At the base of the triangle, a $+8$-μC charge is 100 mm to the left of a -8-μC charge. What is the potential at the apex C? What is the potential at a point D that is 20 mm to the left of the -8-μC charge? How much work is done by the electric field in moving a $+2$-μC charge from point C to point D? **Ans. 0, −2.70 MV, +5.40 J**

*25-41. Two charges, of $+12$ and -6 μC, are separated by 160 mm. What is the potential at the midpoint A of a line joining the two charges? At what point B is the electric potential equal to zero?

*25-42. For the charges and distances shown in Fig. 25-12, find the potential at points A, B, and C? How much work is done by the electric field in moving a $+2$-μC charge from C to A? At what point B is the electric potential equal to zero? **Ans. $V_A = -600$ V, $V_B = +600$ V, $V_C = -300$ V, work = +0.6 mJ, work = −2.4 mJ**

*25-43. The horizontal plates in Millikan's oil-drop experiment are 20 mm apart. The diameter of a particular drop of oil is 4 μm, and the density of oil is 900 kg/m^3. Assuming that two electrons attach themselves to the droplet, what potential difference must exist between the plates to establish equilibrium?

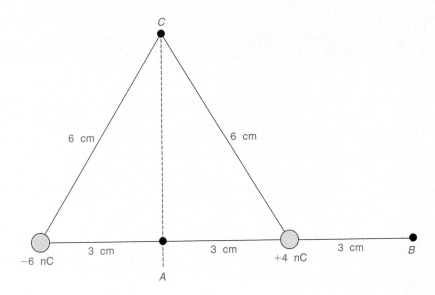

Figure 25-12

Capacitance

Chapter

OBJECTIVES

After completing this chapter, you should be able to:

1. Define *capacitance* and apply a relationship between *capacitance*, applied *voltage* and total *charge*.
2. Compute the capacitance of a *parallel-plate capacitor* when the area of the plates and their separation in a medium of known dielectric constant are given.
3. Write and apply expressions for calculating the *dielectric constant* as a function of the voltage, the electric field, or the capacitance before and after insertion of a dielectric.
4. Calculate the equivalent capacitance of a number of capacitors connected in *series* and in *parallel*.
5. Determine the energy of a charged capacitor, given the appropriate information.

Any charged conductor may be viewed as a reservoir, or source, of electric charge. If a conducting wire is connected to such a reservoir, electric charge can be transferred to perform useful work. In many applications of electricity, large quantities of charge are stored upon a conductor or group of conductors. Any device designed to store electric charge is called a **capacitor.** In this chapter we discuss the nature and application of these devices.

26-1 LIMITATIONS ON CHARGING A CONDUCTOR

How much electric charge can be placed on a conductor? Are there practical limits to the number of electrons that can be transferred to or from a conductor? Suppose we connect a large reservoir of positive and negative charges, such as the earth, to a conducting object, as illustrated in Fig. 26-1a. The energy necessary to transfer electrons from the earth to the conductor can be provided by an electrical device called a *battery*. Charging the conductor is analogous to pumping air into a hollow steel tank. (Refer to Fig. 26-1b). As more air is pumped into the tank, the pressure opposing the flow of additional air becomes greater. Similarly, as more charge Q is transferred to the conductor, the potential V of the conductor becomes higher, making it increasingly difficult to transfer more charge. We can say that the increase in potential V is directly proportional to the charge Q placed on the conductor. Symbolically,

$$V \propto Q$$

Figure 26-1 Charging a conductor is analogous to pumping air into a hollow steel tank.

Therefore, the ratio of the quantity of charge Q to the potential V produced will be a constant for a given conductor. This ratio reflects the ability of a conductor to store charge and is called its **capacitance** C.

$$C = \frac{Q}{V} \qquad \text{26-1}$$

The unit of capacitance is the *coulomb per volt,* which is redefined as a **farad** (F). Thus, *if a conductor has a capacitance of one farad, a transfer of one coulomb of charge to the conductor will raise its potential by one volt.*

Let us return to our original question about the limitations placed on charging a conductor. We have said that every conductor has a capacitance C for storing charge. The value of C for a given conductor is not a function of either the charge placed on a conductor or the potential produced. In principle, the ratio Q/V will remain constant as charge is added indefinitely, but the capacitance depends on the *size* and *shape* of a conductor as well as on the nature of the *surrounding medium.*

Suppose we try to place an indefinite quantity of charge Q on a spherical conductor of radius r, as illustrated in Fig. 26-2. The air surrounding the conductor is an insulator, sometimes called a **dielectric,** which contains few charges free to move. The electric field intensity E and the potential V at the surface of the sphere are given by

$$E = \frac{kQ}{r^2} \qquad \text{and} \qquad V = \frac{kQ}{r}$$

Since the radius r is constant, both the field intensity and the potential at the surface of the sphere increase in direct proportion to the charge Q. There is a limit, however, to the field intensity that can exist on a conductor without ionizing the surrounding air. When this occurs, the air essentially becomes a conductor, and any additional charge placed on the sphere will "leak off" to the air. This limiting value of electric field intensity for which a material loses its insulation properties is called the **dielectric strength** of that material.

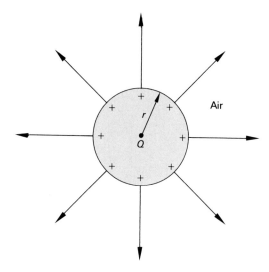

Figure 26-2 The amount of charge that can be placed on a conductor is limited by the dielectric strength of the surrounding medium.

> **The dielectric strength for a given material is that electric field intensity for which the material ceases to be an insulator and becomes a conductor.**

The dielectric strength for dry air at 1 atm pressure is around 3 MN/C. Since the dielectric strength of a material varies considerably with environmental conditions, such as pressure and humidity, it is difficult to compute accurate values.

EXAMPLE 26-1

What is the maximum charge that may be placed on a spherical conductor of radius 50 cm?

Solution

The field intensity at the surface of the sphere, at the point of breakdown in the air, is given by

$$E = \frac{kQ}{r^2} = 3 \text{ MN/C}$$

where 3 MN/C is assumed to be the dielectric strength of air. Solving for Q gives

$$
\begin{aligned}
Q &= \frac{(3 \times 10^6 \text{ N/C})r^2}{k} \\
&= \frac{(3 \times 10^6 \text{ N/C})(0.5 \text{ m})^2}{9 \times 10^9 \text{ N} \cdot \text{m}^2/\text{C}^2} \\
&= 8.33 \times 10^{-5} \text{ C} = 83.3 \ \mu\text{C}
\end{aligned}
$$

This example illustrates the enormous magnitude of the coulomb when it is applied as a unit of electrostatic charge.

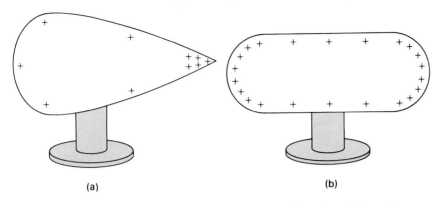

Figure 26-3 The charge density on a conductor is greatest at regions of greatest curvature.

Note that the amount of charge that can be placed on a spherical conductor decreases with the radius of the sphere. Thus, smaller conductors can usually hold less charge. But the shape of a conductor also influences its ability to retain charge. Consider the charged conductors illustrated in Fig. 26-3. If these conductors are tested with an electroscope, it will be discovered that the charge on the surface of a conductor is concentrated at points of greatest curvature. Because of the greater charge density in these regions, the electric field intensity is also greater in regions of smaller curvature. If the surface is reshaped to a sharp point, the field intensity may become great enough to ionize the surrounding air. A slow leakage of charge sometimes occurs at these locations, producing a *corona discharge,* which is often observed as a faint violet glow in the vicinity of the sharply pointed conductor. It is important to remove all sharp edges from electrical equipment to minimize this leakage of charge.

26-2 THE CAPACITOR

When a number of conductors are placed near one another, the potential of each is affected by the presence of the other. For example, suppose we connect a negatively charged plate *A* to an electroscope, as in Fig. 26-4. The divergence of the gold leaf in the electroscope provides a measure of the potential of the conductor. Now let us suppose that another conductor *B* is placed parallel to *A* a short distance away. When the second conductor is grounded, a positive charge will be induced on it as the electrons are forced into the ground. Immediately, the gold leaf will collapse slightly, indicating a drop in the potential of conductor *A*. Because of the presence of the induced charge on *B*, less work is required to bring additional units of charge to conductor *A*. In other words, the capacitance of the system for holding charge has been increased by the proximity of the two conductors. Two such conductors in close proximity, carrying equal and opposite charges, constitute a *capacitor*.

> **A capacitor consists of two closely spaced conductors carrying equal and opposite charges.**

The simplest capacitor is the *parallel-plate capacitor,* illustrated in Fig. 26-4. A potential difference between two such plates can be realized by connecting a bat-

Pocket Lab

Modeling a Capacitor

Materials: Two strips of aluminum foil ¾ by 9 in. Two strips of plastic kitchen film, 1 by 8 in.

Purpose: To model the production of a capacitor. Hold the aluminum strips together and cut out a rectangle from the bottom corner that is ½ by 1½ in. The remaining ¼- by 1½-in strip will be the capacitor's leads.

What to do: Lay the strips on top of each other in the order plastic, aluminum, plastic, aluminum. The aluminum will reach beyond the end of the plastic at the bottom. Lay the top aluminum piece with its lead coming out the opposite side from the middle aluminum piece (though both leads should come out of the same short end).

Starting with the end away from the leads, roll the four layers in a cylinder toward the end where the leads are. Notice that the aluminum never touches either itself or the other strip of aluminum.

Explanation: Your capacitor model is similar to the actual production method of capacitors. The dielectric material in this case is the plastic between the aluminum. The dielectric keeps the capacitor plates (the aluminum in this case) at exactly the same distance apart. The extra piece of plastic film is used to keep the plates from shorting out when the capacitor is rolled up.

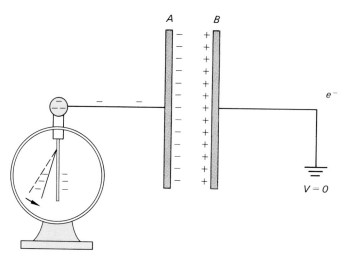

Figure 26-4 A capacitor consists of two closely spaced conductors.

tery to them, as shown in Fig. 26-5. Electrons are transferred from plate A to plate B, producing an equal and opposite charge on the plates. The capacitance of this arrangement is defined as follows:

> **The capacitance between two conductors having equal and opposite charges is the ratio of the magnitude of the charge on either conductor to the resulting potential difference between the two conductors.**

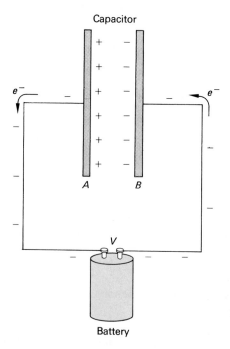

Figure 26-5 Charging a capacitor by transferring charge from one plate to another.

The equation for the capacitance of a capacitor is the same as Eq. (26-1) for a single conductor, except that the symbol V now applies to the *potential difference* and the symbol Q refers to the charge on *either* conductor.

$$C = \frac{Q}{V}$$

26-2

$$1 \text{ F} = \frac{1 \text{ C}}{1 \text{ V}}$$

Because of the enormous size of the coulomb as a unit of charge, the farad as a unit of capacitance is usually too large for practical application. Consequently, the following submultiples are commonly used:

$$1 \text{ microfarad } (\mu\text{F}) = 10^{-6} \text{ F}$$
$$1 \text{ picofarad } (\text{pF}) = 10^{-12} \text{ F}$$

Capacitances as low as a few picofarads are not uncommon in some electrical communication applications.

EXAMPLE 26-2

A capacitor having a capacitance of 4 μF is connected to a 60-V battery. What is the charge on the capacitor?

Solution

The charge *on* a capacitor refers to the magnitude of the charge on either plate of the capacitor. From Eq. (26-2),

$$Q = CV = (4 \ \mu\text{F})(60 \text{ V}) = 240 \ \mu\text{C}$$

26-3 COMPUTING THE CAPACITANCE

In general, a larger conductor can hold a greater quantity of charge, and a capacitor can store more charge than a single conductor because of the inductive effect of two closely spaced conductors. The closer the spacing of these conductors, the greater the inductive effect and hence the easier it becomes to transfer additional charge from one conductor to the other. On the basis of these observations, one might suspect that *the capacitance of a given capacitor will be directly proportional to the area of the plates and inversely proportional to their separation*. The exact relationship can be determined by considering the electric field intensity between the capacitor plates.

The electric field intensity between the plates of the charged capacitor in Fig. 26-6 can be found from

$$E = \frac{V}{d}$$

26-3

where V = potential difference between plates, V
d = separation of plates, m

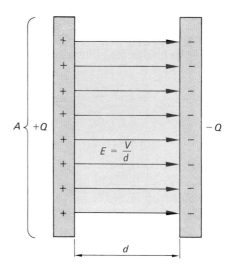

Figure 26-6 The capacitance is directly proportional to the area of either plate and inversely proportional to the plate separation.

An alternative equation for computing the electric field intensity was derived in Chap. 24, using Gauss's law. It relates the field intensity E to the charge density σ as follows:

$$E = \frac{\sigma}{\epsilon_0} = \frac{Q}{A\epsilon_0}$$

26-4

where Q = charge on either plate
A = area of either plate
ϵ_0 = permittivity of vacuum $(8.85 \times 10^{-2} \text{ C}^2/\text{N} \cdot \text{m}^2)$

For a capacitor with a vacuum between its plates, we combine Eqs. (26-3) and (26-4) to get

$$\frac{V}{d} = \frac{Q}{A\epsilon_0}$$

Realizing that the capacitance C is the ratio of charge to voltage, we can rearrange terms and obtain

$$C_0 = \frac{Q}{V} = \epsilon_0 \frac{A}{d}$$

26-5

The subscript 0 is used to indicate that a vacuum exists between the plates of the capacitor. To a close approximation, Eq. (26-5) can also be used when air is between the capacitor plates.

EXAMPLE 26-3

The plates of a parallel-plate capacitor are 3 mm apart in air. If the air of each plate is 0.2 m², what is the capacitance?

Solution

Direct substitution into Eq. (26-5) gives

$$C_0 = \epsilon_0 \frac{A}{d} = \frac{(8.85 \times 10^{-12} \text{ C}^2/\text{N} \cdot \text{m}^2)(0.2 \text{ m}^2)}{3 \times 10^{-3} \text{ m}}$$

$$= 590 \times 10^{-12} \text{ F} = 590 \text{ pF}$$

Parallel-plate capacitors are frequently made with a stack of several plates by connecting alternate plates as shown in Fig. 26-7. By making one of the sets of plates movable, a ***variable capacitor*** can be constructed. Rotating one set of plates relative to the other set varies the effective area of the capacitor plates, causing a variation in the capacitance. Variable capacitors are often used in the tuning circuits of radios.

26-4 DIELECTRIC CONSTANT; PERMITTIVITY

The amount of charge that can be put on a conductor is determined to a large degree by the dielectric strength of the surrounding medium. Similarly, the dielectric strength of the material between the plates of a capacitor limits its ability to store charge. Most capacitors have a nonconducting material, called a *dielectric*, between the plates to provide a ***dielectric strength*** greater than that for air. The following advantages are realized:

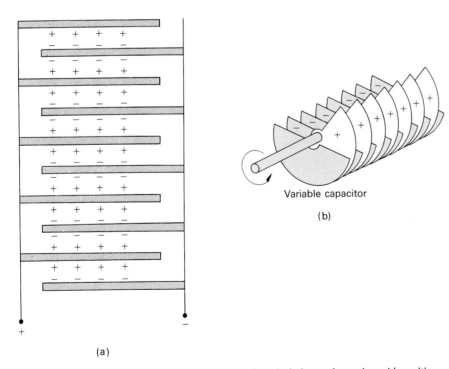

Variable capacitor

(b)

(a)

Figure 26-7 (a) A capacitor consisting of a number of stacked plates, alternating with positive and negative charges. (b) A variable capacitor allows one set of plates to be rotated relative to the other, causing a variation in the effective area.

1. A dielectric material provides for a small plate separation without contact.
2. A dielectric increases the capacitance of a capacitor.
3. Higher voltages can be used without danger of dielectric breakdown.
4. A dielectric often provides greater mechanical strength.

Common dielectric materials are mica, paraffined paper, ceramics, and plastics. Alternating sheets of metal foil and paraffin-coated paper can be rolled up to provide a compact capacitor with a capacitance of several microfarads.

To understand the effect of a dielectric, let us consider the insulating material of Fig. 26-8 placed between capacitor plates having a potential difference V. The electrons in the dielectric are not free to leave their parent atoms, but they do shift toward the positive plate. The protons and electrons of each atom align themselves as shown in the figure. The material is said to become polarized, and the atoms form *dipoles*. All the positive and negative charges inside the dotted ellipse in Fig. 26-8a neutralize each other. However, a layer of negative charge on one surface and a layer of positive charge on the other are not neutralized. An electric field $\mathbf{E}_D$ is set up in the dielectric *opposing* the field $\mathbf{E}_0$, which would exist without the dielectric. The resulting electric field intensity is

$$\mathbf{E} = \mathbf{E}_0 - \mathbf{E}_D \qquad\qquad 26\text{-}6$$

Therefore, insertion of a dielectric results in a reduction of the field intensity between the capacitor plates.

Since the potential difference V between the plates is proportional to the electric field intensity, $V = Ed$, a reduction in the intensity will cause a drop in potential difference. This fact is illustrated by the example shown in Fig. 26-9. The insertion of a dielectric causes a divergence of the gold leaf on the electroscope.

It can be seen from the definition of capacitance, $C = Q/V$, that a drop in voltage will result in an increased capacitance. If we represent the capacitance before

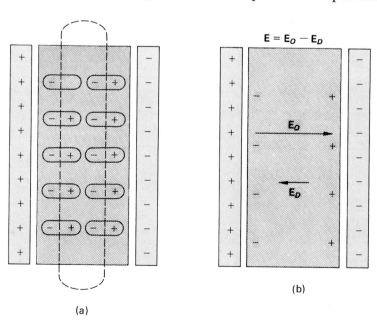

(a)

(b)

Figure 26-8 (a) The polarization of a dielectric when it is inserted between the plates of a capacitor. (b) The polarization results in an overall reduction in the electric field intensity.

Figure 26-9 The insertion of a dielectric between the plates of a capacitor causes a drop in potential difference, resulting in an increased capacitance.

insertion of a dielectric by C_0 and the capacitance after insertion by C, the ratio C/C_0 will denote the relative increase in capacitance. Although this ratio varies from material to material, it is constant for a particular dielectric.

> **The dielectric constant K for a particular material is defined as the ratio of the capacitance C of a capacitor with the material between its plates to the capacitance C_0 for a vacuum.**

$$K = \frac{C}{C_0} \qquad \text{26-7}$$

The *dielectric constant* for various dielectric materials is given in Table 26-1 along with the dielectric strength for the materials. Note that, for air, K is approximately 1.0.

On the basis of proportionalities, it can be shown that the dielectric constant is also given by

$$K = \frac{V_0}{V} = \frac{E_0}{E} \qquad \text{26-8}$$

> where V_0, E_0 = voltage and electric field with vacuum between
> capacitor plates
> V, E = respective values after insertion of dielectric material

The capacitance C of a capacitor having a dielectric between its plates is, from Eq. (26-7),

$$C = KC_0$$

TABLE 26-1 Dielectric Constant and Dielectric Strength

Material	Average Dielectric Constant	Average Dielectric Strength, MV/m
Air, dry at 1 atm	1.006	3
Bakelite	7.0	16
Glass	7.5	118
Mica	5.0	200
Nitrocellulose plastics	9.0	250
Paper, paraffined	2.0	51
Rubber	3.0	28
Teflon	2.0	59
Transformer oil	4.0	16

Substituting from Eq. (26-5), we have a relation for computing C directly:

$$C = K\epsilon_0 \frac{A}{d} \qquad \text{26-9}$$

where A is the area of the plates and d is their separation.

The constant ϵ_0 has been defined earlier as the **permittivity** of a vacuum. Recall from our discussions of Gauss's law that ϵ_0 is actually the proportionality constant that relates the density of electric field lines to the electric field intensity in a vacuum. The permittivity ϵ of a dielectric is greater than ϵ_0 by a factor equal to the dielectric constant K. Thus,

$$\epsilon = K\epsilon_0 \qquad \text{26-10}$$

On the basis of this relation, we can understand why the dielectric constant, $K = \epsilon/\epsilon_0$, is sometimes referred to as the *relative permittivity*. When we substitute Eq. (26-10) into Eq. (26-9), the capacitance for a capacitor containing a dielectric is simply

$$C = \epsilon \frac{A}{d} \qquad \text{26-11}$$

This relation is the most general equation for computing capacitance. When a vacuum or air is between the capacitor plates, $\epsilon = \epsilon_0$ and Eq. (26-11) reduces to Eq. (26-5).

CONCEPTUAL EXAMPLE 26-4

A certain capacitor has a capacitance of 4 μF when its plates are separated by 0.2 mm of vacant space. A battery is used to charge the plates to a potential difference of 500 V and is then disconnected from the system. (a) What will be the potential difference across the plates if a sheet of mica 0.2 mm thick is inserted between the plates? (b) What will the capacitance be after the dielectric is inserted? (c) What is the permittivity of mica?

Solution (a)

The dielectric constant of mica is 5. Thus, Eq. (26-8) gives

$$V = \frac{V_0}{K} = \frac{500 \text{ V}}{5} = 100 \text{ V}$$

Solution (b)

From Eq. (26-7),

$$C = KC_0 = (5)(4 \text{ } \mu\text{F}) = 20 \text{ } \mu\text{F}$$

Solution (c)

The permittivity is found from Eq. (26-10):

$$\epsilon = K\epsilon_0 = (5)(8.85 \times 10^{-12} \text{ C}^2/\text{N} \cdot \text{m}^2)$$
$$= 44.2 \times 10^{-12} \text{ C}^2/\text{N} \cdot \text{m}^2$$

It should be noted that the charge on the capacitor is the same before and after insertion since the voltage source did not stay connected to the capacitor.

EXAMPLE 26-5

Assume that the source of voltage stays connected to the capacitor in Example 26-4. What will be the increase in charge as a result of insertion of the mica sheet?

Solution

In this instance, the voltage stays at 500 V when the dielectric is inserted. Since the capacitance is increased by the dielectric, an increase in charge must result. The charge across the capacitor initially was

$$Q_0 = C_0 V_0 = (4 \text{ } \mu\text{F})(500 \text{ V}) = 2000 \text{ } \mu\text{C}$$

The change after insertion of the mica is determined by the new capacitance of 20 μF:

$$Q = CV = (20 \text{ } \mu\text{F})(500 \text{ V}) = 10,000 \text{ } \mu\text{C}$$

Thus, the increase in charge is

$$\Delta Q = Q - Q_0 = 10,000 \text{ } \mu\text{C} - 2000 \text{ } \mu\text{C} = 8000 \text{ } \mu\text{C}$$

This 8000 μC was provided by the voltage source.

26-5 CAPACITORS IN PARALLEL AND IN SERIES

Electric circuits often contain two or more capacitors grouped together. In considering the effect of such a grouping, it is convenient to resort to the *circuit diagram*, in which electrical devices are represented by symbols. Four symbols com-

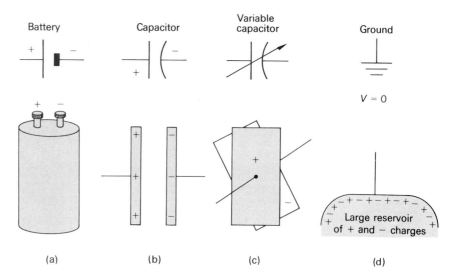

Figure 26-10 Definition of symbols frequently used with capacitors.

monly used with capacitors are defined in Fig. 26-10. The high-potential side of a battery is denoted by the longer line. The high-potential side of a capacitor may be represented as a straight line, with a curved line representing the low-potential side. An arrow indicates a variable capacitor. A *ground* is an electrical connection between the wiring of an apparatus and its metal framework or any other large reservoir of positive and negative charges.

First, we consider the effect of a group of capacitors connected along a single path, as shown in Fig. 26-11. Such a connection, in which the positive plate of one capacitor is connected to the negative plate of another, is called a ***series connection.*** The battery maintains a potential difference V between the positive plate of C_1 and the negative plate of C_3, transferring electrons from one to the other. The charge cannot pass between the plates of a capacitor. Therefore, all the charge inside the dotted circle in Fig. 26-11a is induced charge. For this reason, the charge on each capacitor is identical. We write

$$Q = Q_1 = Q_2 = Q_3$$

where Q is the effective charge transferred by the battery.

All three capacitors can be replaced by an equivalent capacitance C_e without changing the external effect. We now drive an expression for calculating this equiv-

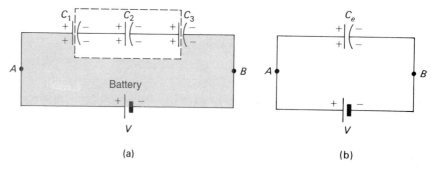

Figure 26-11 Computing the equivalent capacitance of a group of capacitors connected in series.

alent capacitance for the series connection. Since the potential difference between A and B is independent of the path, the battery voltage must equal the sum of the potential drops across each capacitor.

$$V = V_1 + V_2 + V_3 \qquad \textbf{26-12}$$

If we recall that the capacitance C is defined by the ratio Q/V, Eq. (26-12) becomes

$$\frac{Q}{C_e} = \frac{Q_1}{C_1} + \frac{Q_2}{C_2} + \frac{Q_3}{C_3}$$

For a series connection, $Q = Q_1 = Q_2 = Q_3$, so that we can divide out the charge, yielding

$$\frac{1}{C_e} = \frac{1}{C_1} + \frac{1}{C_2} + \frac{1}{C_3} \qquad \textit{Series Connection} \quad \textbf{26-13}$$

The total effective capacitance for *two* capacitors in series is

$$C_e = \frac{C_1 C_2}{C_1 + C_2} \qquad \textbf{26-14}$$

The derivation of Eq. (26-14) is left as an exercise.

Now, let us consider a group of capacitors connected so that charge can be shared between two or more conductors. When several capacitors are all connected directly to the same source of potential, as in Fig. 26-12, they are said to be connected in *parallel*. From the definition of capacitance, the charge on each parallel capacitor is

$$Q_1 = C_1 V_1 \qquad Q_2 = C_2 V_2 \qquad Q_3 = C_3 V_3$$

The total charge Q is equal to the sum of the individual charges.

$$Q = Q_1 + Q_2 + Q_3 \qquad \textbf{26-15}$$

The equivalent capacitance of the entire circuit is $Q = CV$, so that Eq. (26-15) becomes

$$CV = C_1 V_1 + C_2 V_2 + C_3 V_3 \qquad \textbf{26-16}$$

For a **parallel connection**,

$$V = V_1 = V_2 = V_3$$

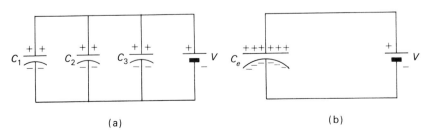

(a)   (b)

Figure 26-12 Equivalent capacitance of a group of capacitors connected in parallel.

since all capacitors are connected to the same potential difference. Hence, the voltages divide out of Eq. (26-16), giving

$$C = C_1 + C_2 + C_3 \qquad \textit{Parallel Connection} \quad \textbf{26-17}$$

EXAMPLE 26-3

(a) Find the equivalent capacitance of the circuit illustrated in Fig. 26-13a. (b) Determine the charge on each capacitor. (c) What is the voltage across the 4-μF capacitor?

Figure 26-13   Simplifying a problem by substituting equivalent values for capacitance.

Solution (a)

The 4- and 2-μF capacitors are in series. Their combined capacitance is found from Eq. (26-14).

$$C_{2,4} = \frac{C_2 C_4}{C_2 + C_4} = \frac{(2\ \mu\text{F})(4\ \mu\text{F})}{2\ \mu\text{F} + 4\ \mu\text{F}}$$

$$= 1.33\ \mu\text{F}$$

These two capacitors can be replaced by their equivalent capacitance, as shown in Fig. 26-13b. The two remaining capacitors are in parallel. Thus, the equivalent capacitance is

$$C_e = C_3 + C_{2,4} = 3\ \mu\text{F} + 1.33\ \mu\text{F}$$
$$= 4.33\ \mu\text{F}$$

Solution (b)

The total charge in the network is

$$Q = C_e V = (4.33\ \mu\text{F})(120\ \text{V}) = 520\ \mu\text{C}$$

The charge Q_3 on the 3-μF capacitor is

$$Q_3 = C_3 V = (3\ \mu\text{F})(120\ \text{V}) = 360\ \mu\text{C}$$

The remainder of the charge,

$$Q - Q_3 = 520 \ \mu C - 360 \ \mu C = 160 \ \mu C$$

must be deposited on the series capacitors. Hence,

$$Q_2 = Q_4 = 160 \ \mu C$$

As a check on these values for Q_2 and Q_4, the equivalent capacitance of the two series capacitors can be multiplied by the voltage drop across it:

$$Q_{2,4} = C_{2,4} \ V = (1.33 \ \mu F)(120 \ V) = 160 \ \mu C$$

Solution (c)

The voltage across the 4-μF capacitor is

$$V_4 = \frac{Q_4}{C_4} = \frac{160 \ \mu C}{4 \mu F} = 40 \ V$$

The remaining 80 V is across the 2-μF capacitor.

26-6 ENERGY OF A CHARGED CAPACITOR

Consider a capacitor that is initially uncharged. When a source of potential difference is connected to the capacitor, the potential difference between the plates increases as charge is transferred. As more and more charge builds up on the capacitor, it becomes increasingly difficult to transfer additional charge. Suppose we represent the total charge transferred by Q and the final potential difference by V. The *average* potential difference through which the charge is moved is given by

$$V_{av} = \frac{V_{final} + V_{initial}}{2} = \frac{V + 0}{2} = \frac{1}{2}V$$

Since the total charge transferred is Q, the total work done against electric forces is equal to the product of Q and the average potential difference V_{av}. Thus,

$$\text{Work} = Q\left(\tfrac{1}{2}V\right) = \tfrac{1}{2}QV$$

This work is equivalent to the electrostatic potential energy of a charged capacitor. From the definition of capacitance ($Q = CV$), this potential energy can be written in alternative forms:

$$\text{P.E.} = \tfrac{1}{2}QV$$

$$= \tfrac{1}{2}CV^2$$

$$= \frac{Q^2}{2C} \qquad \text{26-18}$$

When C is in farads, V is in volts, and Q is in coulombs, the potential energy will be in joules. These equations apply equally to all capacitors, regardless of their construction.

Summary and Review

Summary

Stored electric charge is a necessity if large quantities of electrical energy are to be delivered on demand to a modern industrial world. We have studied in this chapter the basic principles that determine the amount of charge that can be stored on capacitors. We have discussed the insertion of capacitors into electric circuits and the factors affecting the distribution of charge in such circuits. The fundamental concepts are summarized below.

- Capacitance is the ratio of charge Q to the potential V for a given conductor. For two oppositely charged plates, the Q refers to the charge on either plate and the V refers to the potential difference between the plates.

$$C = \frac{Q}{V} \qquad 1 \text{ farad (F)} = \frac{1 \text{ coulomb (C)}}{1 \text{ volt (V)}}$$

Capacitance

- The dielectric strength is that value for E for which a given material ceases to be an insulator and becomes a conductor. For air, this value is

$$E = \frac{kQ}{r^2} = 3 \times 10^6 \text{ N/C}$$

Dielectric Strength, Air

- For a parallel-plate capacitor, the material between the plates is called the dielectric. The insertion of such a material has an effect on the electric field and the potential between the plates. Consequently, it changes the capacitance. The dielectric constant K for a particular material is the ratio of the capacitance with the dielectric C to the capacitance for a vacuum C_0.

$$K = \frac{C}{C_0} \qquad K = \frac{V_0}{V} \qquad K = \frac{E}{E_0}$$

Dielectric Constant

- The permittivity of a dielectric is greater than the permittivity of a vacuum by a factor equal to the dielectric constant. For this reason, K is sometimes referred to as the *relative permittivity*.

$$K = \frac{\epsilon}{\epsilon_0} \qquad\qquad \epsilon = K\epsilon_0$$

$$\epsilon_0 = 8.85 \times 10^{-12} \text{ C}^2/\text{N} \cdot \text{m}^2$$

- The capacitance for a parallel-plate capacitor depends on the surface area A of each plate, the plate separation d, and the permittivity or dielectric constant. The general equation is

$$C = \epsilon \frac{A}{d} \qquad\qquad C = K\epsilon_0 \frac{A}{d} \qquad \textit{Capacitance}$$

For a vacuum, $K = 1$ in the above relationship.
- Capacitors may be connected in series, as shown in Fig. 26-11, or in parallel, as shown in Fig. 26-12.

(a) For *series connections,* the charge on each capacitor is the same as the total charge, the potential difference across the battery is equal to the sum of the drops across each capacitor, and the net capacitance is found from

$$Q_T = Q_1 = Q_2 = Q_3$$

$$V_T = V_1 + V_2 + V_3$$

$$\frac{1}{C_e} = \frac{1}{C_1} + \frac{1}{C_2} + \frac{1}{C_3} \qquad \begin{array}{l}\textit{Series}\\\textit{Connections}\end{array}$$

(b) For *parallel connections,* the total charge is equal to the sum of the charges across each capacitor, the voltage drop across each capacitor is the same as the drop across the battery, and the effective capacitance is equal to the sum of the individual capacitances.

$$Q_T = Q_1 + Q_2 + Q_3$$

$$V_B = V_1 = V_2 = V_3$$

$$C_e = C_1 + C_2 + C_3$$

Parallel Connections

- The potential energy stored in a charged capacitor can be found from any of the following relationships:

$$\text{P.E.} = \tfrac{1}{2}QV \qquad \text{P.E.} = \tfrac{1}{2}CV^2$$

$$\text{P.E.} = \frac{Q^2}{2C}$$

When *C* is in *farads*, *V* is in *volts*, and *Q* is in *coulombs*, the potential energy will be in *joules*.

Key Terms

1. capacitor
2. capacitance
3. dielectric
4. permittivity
5. farad
6. corona discharge
7. variable capacitor
8. dielectric strength
9. dielectric constant
10. parallel connection
11. series connection

Review Questions

26-1. Discuss several factors that limit the ability of a conductor to store charge.

26-2. Air is pumped from one metal tank to another, creating a partial vacuum in one tank and high pressure in the other. When the pump is removed, potential energy is stored. The energy is released if the two tanks are reconnected and the pressure in each tank becomes equal. In what ways is this mechanical example analogous to charging and discharging a capacitor?

26-3. Large sparks are often seen jumping from the leather belts driving machinery. Explain.

26-4. The Leyden jar is a capacitor that consists of a glass jar coated inside and out with tinfoil, as shown in Fig. 26-14. Contact with the inside coating is made with a metal chain connected to the central metal rod. From the figure, explain how the capacitor becomes charged. What is the function of the ground wire? What purpose does the glass serve?

26-5. Can lightning be considered a capacitor discharge? Explain.

26-6. Discuss the following statement: the permittivity is a measure of how easily a dielec-

Figure 26-14 The Leyden jar.

tric will permit the establishment of electric field lines within the dielectric.

26-7. A dielectric with a larger permittivity allows

for the storage of greater quantities of charge. Explain.

26-8. Distinguish the dielectric strength of a material from its dielectric constant. What part does each play in the design of a capacitor?

26-9. The term *breakdown voltage* is often used in electronics for capacitors. How would you define such a term? In what ways does it differ from the dielectric strength?

26-10. If two point charges are surrounded by a dielectric, will the force each exerts on the other be reduced or increased?

26-11. The unit of permittivity is the $C^2/N \cdot m^2$. Show that the permittivity can be expressed as farads per meter.

26-12. Prove that each of the expressions for potential energy, as given in Eq. (26-18), will yield a proper unit of energy (the joule).

Problems

Section 26-2—The Capacitor

26-1. What is the maximum charge that can be placed on a metal sphere 30 mm in diameter and surrounded by air? **Ans. 75 nC**

26-2. How much charge can be placed on a metal sphere of radius 40 mm if it is immersed in transformer oil that has a dielectric strength of 16 MV/m?

26-3. What would be the radius of a metal sphere in air if it could theoretically hold a charge of 1 C? **Ans. 54.8 m**

26-4. A 28-μF parallel-plate capacitor is connected to a 120-V source of potential difference. How much charge will be stored on this capacitor?

26-5. A potential difference of 110 V is applied across the plates of a parallel-plate capacitor. If the total charge on each plate is 1200 μC, what is the capacitance? **Ans. 10.9 μF**

26-6. Find the capacitance of a parallel-plate capacitor if 1600 μC of charge is on each plate when the potential difference is 80 V.

26-7. What potential difference is required to store a charge of 800 μC on a 40-μF capacitor? **Ans. 20 V**

26-8. Write an equation for the potential at the surface of a sphere of radius r in terms of the permittivity of the surrounding medium. Show that the capacitance of such a sphere is given by $C = 4\pi\epsilon r$.

*26-9. A spherical capacitor has a radius of 50 mm and is surrounded by a medium that has a permittivity of 3×10^{-11} $C^2/N \cdot m^2$. How

much charge can be transferred to this sphere by a potential difference of 400 V? **Ans. 4.71×10^{-14} C**

Section 26-3—Computing the Capacitance

26-10. A 5-μF capacitor has a plate separation of 0.3 mm of air. What will be the charge on each plate for a potential difference of 400 V? What is the area of each plate?

26-11. The plates of a certain capacitor are 3 mm apart and have an area of 0.04 m². What is the capacitance if air is the dielectric? **Ans. 118 pF**

26-12. A capacitor has plates of area 0.034 m² and a separation of 2 mm in air. The potential difference between the plates is 200 V. What is the capacitance, and what is the electric field intensity between the plates? How much charge is on each plate?

26-13. A capacitor of plate area 0.06 m² and plate separation 4 mm has a potential difference of 300 V when air is the dielectric. What is the capacitance for dielectrics of air ($K = 1$) and mica ($K = 5$)? **Ans. 133 pF, 664 pF**

26-14. What is the electric field intensity for mica and for air in Prob. 26-13?

26-15. Find the capacitance of a parallel-plate capacitor if the area of each plate is 0.08 m², the separation of the plates is 4 mm, and the dielectric is (a) air or (b) paraffined paper ($K = 2$)? **Ans. (a) 177 pF, (b) 354 pF**

26-16. Two parallel plates of a capacitor are 4.0 mm apart and the plate area is 0.03 m². Glass

($K = 7.5$) is the dielectric, and the plate voltage is 800 V. What is the charge on each plate, and what is the electric field intensity between the plates?

*26-17. A parallel-plate capacitor with a capacitance of 2.0 nF is to be constructed with mica ($K = 5$) as the dielectric, and it must be able to withstand a maximum potential difference of 3000 V. The dielectric strength of mica is 200 MV/m. What is the minimum area the plates of the capacitor can have? **Ans. 6.78 × 10⁻⁴ m²**

Figure 26-15

Section 26–5—Capacitors in Parallel and in Series

26-18. Find the equivalent capacitance of a 6-μF capacitor and a 12-μF capacitor connected (a) in series and (b) in parallel.

26-19. Find the effective capacitance of a 6-μF capacitor and a 15-μF capacitor connected (a) in series and (b) in parallel.
Ans. 4.29 μF, 21.0 μF

26-20. What is the equivalent capacitance for capacitors of 4, 7, and 12 μF connected (a) in series and (b) in parallel?

26-21. Find the equivalent capacitance for capacitors of 2, 6, and 8 μF connected (a) in series and (b) in parallel.
Ans. 16 μF, 1.26 μF

26-22. A 20- and a 60-μF capacitor are connected in parallel. Then the pair are connected in series with a 40-μF capacitor. What is the equivalent capacitance?

*26-23. If a potential difference of 80 V is placed across the group of capacitors in Prob. 26-22, what is the charge on the 40-μF capacitor? What is the charge on the 20-μF capacitor? **Ans. 2133 μC, 533 μC**

*26-24. Find the equivalent capacitance of a circuit in which a 6-μF capacitor is connected in series with two parallel capacitors whose capacitances are 5 and 4 μF.

*26-25. What is the equivalent capacitance for the circuit drawn in Fig. 26-15? **Ans. 6.00 μF**

*26-26. What is the charge on the 4-μF capacitor in Fig. 26-15? What is the voltage across the 6-μF capacitor?

*26-27. A 6- and a 3-μF capacitor are connected in series with a 24-V battery. What are the charge and voltage across each capacitor?
Ans. V_3 = 16.0 V, Q = 48.0 μC,
V_6 = 8.00 V, Q_6 = 48.0 μC

*26-28. If the 6- and 3-μF capacitors of Prob. 26-27 are reconnected in parallel with a 24-V battery, what are the charge and voltage across each capacitor?

Figure 26-16

*26-29. Compute the equivalent capacitance for the entire circuit shown in Fig. 26-16. What is the total charge on the equivalent capacitance? **Ans. 1.74 μF, 20.9 μF**

*26-30. What are the charge and voltage across each capacitor of Fig. 26-16?

Section 26-6—Energy of a Charged Capacitor

26-31. What is the potential energy stored in the electric field of a 200-μF capacitor when it is charged to a voltage of 2400 V?
Ans. 576 J

26-32. What is the energy stored on a 25-μF capacitor when the charge on each plate is 2400 μF? What is the voltage across the capacitor?

Summary and Review

26-33. How much work is required to charge a capacitor to a potential difference of 30 kV if 800 μC is on each plate? **Ans. 12.0 J**

*26-34. A parallel plate capacitor has a plate area of 4 cm^2 and a separation of 2 mm. A dielectric of constant $K = 4.3$ is placed between the plates, and the capacitor is connected to a 100-V battery. How much energy is stored in the capacitor?

Challenge Problems

26-35. What is the breakdown voltage for a capacitor with a dielectric of glass ($K = 7.5$) if the plate separation is 4 mm? The average dielectric strength is 118 MV/m.

Ans. 472 kV

26-36. A capacitor has a potential difference of 240 V, a plate area of 5 cm^2, and a plate separation of 3 mm. What are the capacitance and the electric field between the plates? What is the charge on each plate?

26-37. Suppose the capacitor of Prob. 26-36 is disconnected from the 240-V battery and then mica ($K = 5$) is inserted between the plates? What are the new voltage and electric field? If the 240-battery is reconnected, what charge will be on the plates?

Ans. 48.0 V, 1.60 × 10^4 V/m, 1.79 nC

26-38. A 6-μF capacitor is charged with a 24-V battery and then disconnected. When a dielectric is inserted, the voltage drops to 6 V. What is the total charge on the capacitor after the battery has been reconnected?

26-39. A capacitor is formed from 30 parallel plates, each 20 × 20 cm. If each plate is separated by 2 mm of dry air, what is the total capacitance? **Ans. 5.13 nF**

*26-40. Four capacitors, A, B, C, and D, have capacitances of 12, 16, 20, and 26 μF, respectively. Capacitors A and B are connected in parallel. The combination is then connected in series with C and D. What is the effective capacitance?

*26-41. Consider the circuit drawn in Fig. 26-17. What is the equivalent capacitance of the circuit? What are the charge and voltage across the 2-μF capacitor?

Ans. 6.00 μF, 18 μC, 9.00 V

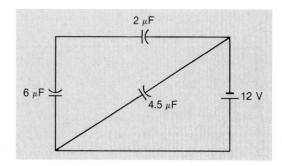

Figure 26-17

*26-42. Two identical 20-μF capacitors, A and B, are connected in parallel with a 12-V battery. What is the charge on each capacitor if a sheet of porcelain ($K = 6$) is inserted between the plates of capacitor B and the battery remains connected?

*26-43. Three capacitors, A, B, and C, have respective capacitances of 2, 4, and 6 μF. Compute the equivalent capacitance if they are connected in series with an 800-V source. What are the charge and voltage across the 4-μF capacitor? **Ans. 1.09 μF, 873 μC, 218 V**

*26-44. Suppose that the capacitors of Prob. 26-43 are reconnected in parallel with the 800-V source. What is the equivalent capacitance? What are the charge and voltage across the 4-μF capacitor?

*26-45. Show that the total capacitance of a multiple-plate capacitor containing N plates separated by air is given by:

$$C_0 = \frac{(N-1)\epsilon_0 A}{d}$$

where A is the area of each plate and d is the separation of each plate.

*26-46. The energy density u of a capacitor is defined as the potential energy (P.E.) per unit volume (Ad) of the space between the plates. Using this definition and several formulas from this chapter, derive the following relationship for finding the energy density u:

$$u = \frac{1}{2}\epsilon_0 E^2$$

where E is the electric field intensity between the plates.

*26-47. A capacitor with a plate separation of 3.4 mm is connected to a 500-V battery. Use the relation derived in Prob. 26-46 to calculate the energy density between the plates.

Ans. 95.7 mJ/m^3

Critical Thinking Questions

26-48. A certain capacitor has a capacitance of 12 μF when its plates are separated by 0.3 mm of vacant space. A 400-V battery charges the plates and is then disconnected from the capacitor. (a) What is the potential difference across the plates if a sheet of Bakelite $(K = 7)$ is inserted between the plates? (b) What is the total charge on the plates? (c) What is the capacitance with the dielectric inserted? (d) What is the permittivity of Bakelite? (e) How much additional charge can be placed on the capacitor if the 400-V battery is reconnected?

Ans. (a) 57.1 V, (b) 4800 μC, (c) 84.0 μF, (d) 6.20 × 10^{-11} C^2/N m^2, (e) 28.8 mC

*26-49. A medical defibrillator uses a capacitor to revive heart attack victims. Assume that a 65-μF capacitor in such a device is charged to 5000 V. What is the total energy stored? If 25 percent of this energy passes through a victim in 3 ms, what power is delivered?

*26-50. Consider three capacitors of 10, 20, and 30 μF. Show how these might be connected to produce the maximum and minimum equivalent capacitances and list the values. Draw a diagram of a connection that would result in an equivalent capacitance of 27.5 μF. Show a connection that will result in a combined capacitance of 22.0 μF.

Ans. 60 μF, 5.45 μF

*26-51. A 4-μF air capacitor is connected to a 500-V source of potential difference. The capacitor is then disconnected from the source and a sheet of mica $(K = 5)$ is inserted between the plates? What is the new voltage on the capacitor? Now reconnect the 500-V battery and calculate the final charge on the capacitor. By what percentage did the total energy on the capacitor increase due to the dielectric?

*26-52. A 3-μF capacitor and a 6-μF capacitor are connected in series with a 12-V battery. What is the total stored energy of the system? What is the total energy if they are connected in parallel? What is the total energy for each of these connections if mica $(K = 5)$ is used as a dielectric for each capacitor?

Ans. 0.144 mJ, 0.648 mJ, 0.720 mJ, 3.24 mJ

Chapter
27

Current and Resistance

OBJECTIVES

After completing this chapter, you should be able to:

1. Demonstrate by definition and example your understanding of *electric current* and *electromotive force.*
2. Write and apply *Ohm's law* to the solution of problems involving electric resistance.
3. Compute *power losses* as a function of voltage, current, and resistance.
4. Define the *resistivity* of a material and solve problems similar to those in the text.
5. Define the *temperature coefficient of resistance* and calculate the change in resistance that occurs with a change in temperature.

We now leave electrostatics and enter a discussion of charges in motion. We have been concerned with forces, electric fields, and potential energies as they relate to charged conductors. For example, excess electrons, evenly distributed over an insulated spherical surface, will remain at rest. If a wire is connected from the sphere to ground, however, the electrons will flow through the wire to the ground. The flow of charge constitutes an **electric current.** In this chapter the foundation is laid for a study of direct currents and electric resistance.

27-1 THE MOTION OF ELECTRIC CHARGE

Let us begin our discussion of moving charges by considering the discharge of a capacitor. The potential difference V between the two capacitor plates in Fig. 27-1a is indicated by the electroscope. The total charge Q on either plate is given by

$$Q = CV$$

where C is the capacitance. If a path is provided, electrons on one plate will travel to the other, decreasing the net charge and causing a drop in the potential difference. Thus, a drop in potential, as indicated by the collapsing leaf of the electroscope, means that charge has been transferred. Any conductor used to connect the plates of a capacitor will cause it to discharge. The rate of discharge varies considerably, however, with the size, shape, material, and temperature of the conductor.

If a short, thick wire is connected between the plates of the capacitor, as shown in Fig. 27-1b, the electroscope leaf collapses instantly, indicating a rapid transfer

Figure 27-1 (a) A charged capacitor is a source of current. (b) If the capacitor plates are joined by a short, thick wire, the capacitor will discharge instantly. (c) A long, thin wire allows for a gradual discharge.

of charge. This current, which exists for a short time, is called a ***transient current.*** If we replace the short, thick wire with a long, thin wire of the same material, we shall observe a gradual collapse of the electroscope leaf (Fig. 27-1c). Such opposition to the flow of electricity is called ***electric resistance.*** A quantitative description of electric resistance will be presented in a later section. It is introduced here to illustrate that the rate at which charge flows through a conductor varies. This rate is referred to as the ***electric current.***

> **The electric current I is the rate of the flow of charge Q past a given point P on an electric conductor.**

$$I = \frac{Q}{t} \qquad \qquad \text{27-1}$$

The unit of electric current is the *ampere.* One ***ampere*** (A) represents a flow of charge at the rate of *one coulomb per second* past any point.

$$1 \text{ A} = \frac{1 \text{ C}}{1 \text{ s}}$$

In the example of a discharging capacitor, the current arises from the motion of electrons, as illustrated in Fig. 27-2. The positive charges in a wire are tightly bound and cannot move. The electric field created in the wire because of the potential difference between the plates causes the free electrons in the wire to experience a drift toward the positive plate. The electrons are repeatedly deflected or stopped by processes relating to impurities and thermal motions of the atoms. Consequently, the motion of the electrons is not an accelerated one but a drifting or diffusion process. The average drift velocity of electrons is typically of the order

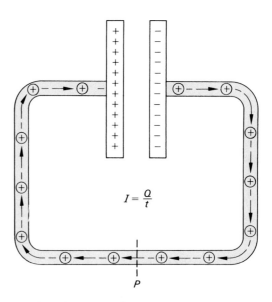

$$I = \frac{Q}{t}$$

P

Figure 27-2 Current arises from the motion of electrons and is a measure of the quantity of charge passing a given point in a unit of time.

of 4 m/h. This velocity of charge, which is a *distance* per unit of time, should not be confused with current, which is a *quantity* of charge per unit of time.

An analogy to water flowing through a pipe is useful in understanding current flow. The rate of flow of the water in gallons per minute is analogous to the rate of flow of charge in coulombs per second. For a current of 1 A, 6.25×10^{18} electrons (1 C) flow past a given point every second. Just as the size and length of a pipe affect the flow of water, the size and length of a conductor affect the flow of electrons.

EXAMPLE 27-1

How many electrons pass a point in 5 s if a constant current of 8 A is maintained in a conductor?

Solution

From Eq. (27-1),

$$\begin{aligned}
Q &= It = (8 \text{ A})(5 \text{ s}) \\
&= (8 \text{ C/s})(5 \text{ s}) = 40 \text{ C} \\
&= (40 \text{ C})(6.25 \times 10^{18} \text{ electrons/C}) \\
&= 2.50 \times 10^{20} \text{ electrons}
\end{aligned}$$

27-2 THE DIRECTION OF ELECTRIC CURRENT

Thus far, we have discussed only the magnitude of electric current. The choice of direction is purely arbitrary as long as we apply our definition consistently. The flow of charge caused by an electric field in a gas or a liquid consists of a flow of positive ions in the direction of the field or a flow of electrons opposite to the field direction. As we have seen, the current in a metallic material consists of electrons flowing against the field direction. However, a current that consists of negative particles moving in one direction is electrically the same as a current consisting of positive charges moving in the opposite direction.

There are a number of reasons for preferring the motion of positive charge as an indicator of direction. In the first place, all the concepts introduced in electrostatics, for example the electric field, potential energy, and potential difference, were defined in terms of positive charges. An electron flows contrary to the electric field and "up a potential hill" from the negative plate to the positive plate. If we define current as a flow of *positive* charge, the loss in energy as charge encounters resistance will be from + to − or "down a potential hill." By convention, we consider all currents as consisting of a flow of positive charge.

The direction of conventional current is always the same as the direction in which positive charges would move, even if the actual current consists of a flow of electrons.

For a metallic conducting wire, both electron flow and conventional current are indicated in Fig. 27-3. The zigzag line is used to indicate the electric resistance R. Note that the conventional current flows from the positive plate of the capacitor, neutralizing negative charge on the other plate. Conventional current is in the same direction as the electric field **E** producing the current.

27-3 ELECTROMOTIVE FORCE

The currents discussed in the preceding sections were called *transient currents* because they exist only for a short time. Once the capacitor has been completely discharged, there will no longer be a potential difference to promote the flow of additional charge. If some means were available to keep the capacitor continually charged, a continuous current could be maintained. This would require that electrons be supplied continuously to the negative plate to replace those leaving. In

Electron flow

R

E

Conventional current

Figure 27-3 In a metallic conductor, the conventional current is in a direction opposite to the actual flow of electrons.

other words, energy must be supplied to replace the energy lost by the charge in the external circuit. In this manner, the potential difference between the plates could be maintained, allowing for a continuous flow of charge. A device with the ability to maintain potential difference between two points is called a *source of electromotive force* (emf).

The most familiar sources of emf are batteries and generators. Batteries convert chemical energy into electric energy, and generators transform mechanical energy into electric energy. The detailed nature and operation of these devices will be discussed in a later chapter.

> **A source of electromotive force (emf) is a device that converts chemical, mechanical, or other forms of energy into the electric energy necessary to maintain a continuous flow of electric charge.**

In an electric circuit, the source of emf is usually represented by the symbol $\mathcal{E}$.

The function of a source of emf in an electric circuit is similar to the function of a water pump in maintaining the continuous flow of water through a system of pipes. In Fig. 27-4a the water pump must perform the work on each unit volume of water necessary to replace the energy lost by each unit volume flowing through the pipes. In Fig. 27-4b the source of emf must do work on each unit of charge that passes through it to raise it to a higher potential. This work must be supplied at a rate equal to the rate at which energy is lost in flowing through the circuit.

By convention, we have assumed that the current consists of a flow of positive charge, even though in most cases it is negative electrons. Therefore, the charge loses energy in passing through the resistor from a high potential to a low potential. In the hydraulic analogy, water passes from high pressure to low pressure. When the shut-off valve is closed, pressure exists but there is no water flow. Similarly, when the electric switch is *open*, there is voltage but no current.

Since emf is work per unit charge, it is expressed in the same unit as potential difference: the *joule per coulomb*, or *volt*.

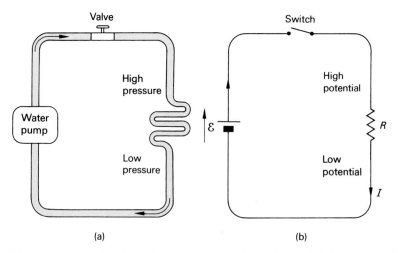

Figure 27-4 The mechanical analogy of a water pump can be used to explain the function of a source of emf in an electric circuit.

A source of emf of one volt will perform one joule of work on each coulomb of charge that passes through it.

For example, a 12-V battery performs 12 J of work on each coulomb of charge transferred from the low-potential side ($-$ terminal) to the high-potential side ($+$ terminal). An arrow ($\uparrow$) is usually drawn next to the symbol $\mathscr{E}$ for an emf to indicate the direction in which the source, acting alone, would cause a positive charge to move through the external circuit. The conventional current is directed away from the $+$ terminal of a battery, and the hypothetical positive charge flows "downhill" through external resistance of the $-$ terminal of the battery.

In the following sections, circuit diagrams, like that in Fig. 27-4b, will frequently be used to describe electric systems. Many of the symbols we shall use are defined in Fig. 27-5.

27-4 OHM'S LAW; RESISTANCE

Resistance (R) is defined as the opposition to the flow of electric charge. Although most metals are good conductors of electricity, all offer some opposition to the surge of electric charge through them. This electric resistance is fixed for many specific materials of known size, shape, and temperature. It is independent of the applied emf and the current passing through it.

jersey.uoregon.edu
You can study Ohm's law on the virtual light bulb simulator at this Web site.

(a) Resistor   (b) Rheostat   (c) Switch

(d) Battery   (e) Voltmeter   (f) Ammeter

Figure 27-5   Conventional symbols used in electric circuit diagrams.

The effects of resistance in limiting the flow of charge were first studied quantitatively by Georg Simon Ohm in 1826. He discovered that, *for a given resistor at a particular temperature, the current is directly proportional to the applied voltage.* Just as the rate of flow of water between two points depends on the difference of height between them, the rate of flow of electric charge between two points depends on the difference in potential between them. This proportionality is usually stated as ***Ohm's law:***

The current produced in a given conductor is directly proportional to the difference of potential between its end points.

The current I that is observed for a given voltage V is therefore an indication of resistance. Mathematically, the resistance R of a given conductor can be calculated from

$$R = \frac{V}{I} \qquad V = IR \qquad \qquad \text{27-2}$$

The greater the resistance R, the smaller the current I for a given voltage V. The unit of measurement of resistance is the **ohm,** for which the symbol is the Greek capital letter *omega* (Ω). From Eq. (27-2),

$$1\,\Omega = \frac{1\,\text{V}}{1\,\text{A}}$$

A resistance of *one ohm* will allow a current of *one ampere* when a potential difference of *one volt* is impressed across its terminals.

EXAMPLE 27-2

The difference of potential between the terminals of an electric heater is 80 V when there is a current of 6 A in the heater. What will the current be if the voltage is increased to 120 V?

Solution

According to Ohm's law, the resistance of the coils in the heater is

$$R = \frac{V}{I} = \frac{80\,\text{V}}{6\,\text{A}} = 13.3\,\Omega$$

Therefore, if the voltage is increased to 120 V, the new current will be

$$I = \frac{V}{R} = \frac{120\,\text{V}}{13.3\,\Omega} = 9\,\text{A}$$

Here we have neglected any change in resistance due to a rise in temperature of the heating coils.

Four devices commonly used in the laboratory to study Ohm's law are the battery, the voltmeter, the ammeter, and the rheostat. As their names imply, the **voltmeter** and **ammeter** are devices to measure voltage and current. The **rheostat** is simply a variable resistor. A sliding contact changes the number of resistance coils

(a)

Rheostat

Battery   Voltmeter   Ammeter

(b)

Figure 27-6 (a) A circuit diagram for studying Ohm's law. (b) A pictorial diagram showing how the various elements are connected in the laboratory.

through which charge can flow. A laboratory collection of these electrical devices is illustrated in Fig. 27-6. You should study the circuit diagram in Fig. 27-6a and justify the electrical connections shown pictorially in Fig. 27-6b. Note that the voltmeter is connected in parallel with the battery, whereas the ammeter is connected in series. In general, the positive terminals are color-coded red, and the negative terminals are black.

CONCEPTUAL EXAMPLE 27-3

The apparatus shown in Fig. 27-6 is used to study Ohm's law in the laboratory. The voltage V is determined by the source of emf and remains at 6 V. (a) What is the resistance when the rheostat is varied to indicate a current of 0.4 A? (b) If this resistance is doubled, what will the new current be?

Solution (a)
From Ohm's law,

$$R = \frac{V}{I} = \frac{6 \text{ V}}{0.4 \text{ A}} = 15 \text{ } \Omega$$

Solution (b)
Doubling the resistance to 30 Ω would result in a current given by

$$I = \frac{6 \text{ V}}{30 \text{ } \Omega} = 0.2 \text{ A}$$

Note that doubling the resistance in a circuit reduces the current by one-half.

We have seen that electric charge gains energy within a generating source of emf and loses energy in passing through external resistance. Inside the source of emf, work is done *by the source* in raising the potential energy of charge. As the charge passes through the external circuit, work is done *by* the charge on the components of the circuit. In the case of a pure resistor, the energy is dissipated in the form of heat. If a motor is attached to the circuit, the energy loss is divided between heat and useful work. In any case, the energy gained in the source of emf must equal the energy lost in the entire circuit.

Let us examine the work accomplished inside a source of emf more closely. By definition, *one joule* of work is accomplished for each *coulomb* of charge moved through a potential difference of *one volt*. Thus,

$$\text{Work} = Vq \qquad\qquad \textbf{27-3}$$

where q is the quantity of charge transferred during a time t. But $q = It$, so that Eq. (27-3) becomes

$$\text{Work} = VIt \qquad\qquad \textbf{27-4}$$

where I is the current in *coulombs per second*. This work represents the energy gained by a charge in passing through the source of emf during the time t. An equivalent amount of energy will be dissipated in the form of heat as the charge moves through an external resistance.

The rate at which heat is dissipated in an electric circuit is referred to as the *power loss*. When charge is flowing continuously through a circuit, this power loss is given by

$$P = \frac{\text{work}}{t} = \frac{VIt}{t} = VI \qquad\qquad \textbf{27-5}$$

When V is in volts and I is in amperes, the power loss is measured in watts. That the product of voltage and current will give a unit of power is shown as follows:

$$(\text{V})(\text{A}) = \frac{\text{J}}{\text{C}}\frac{\text{C}}{\text{s}} = \frac{\text{J}}{\text{s}} = \text{W}$$

Equation (27-5) can be expressed in alternative forms by using Ohm's law ($V = IR$). Substituting for V, we can write

$$P = VI = I^2R \qquad\qquad \textbf{27-6}$$

Substitution for I in Eq. (27-6) gives another variation:

$$P = VI = \frac{V^2}{R} \qquad\qquad \textbf{27-7}$$

The relation expressed by Eq. (27-6) is so often used in electrical work that heat loss in electrical wiring is often referred to as an "I-squared-R" loss.

EXAMPLE 27-4

A current of 6 A flows through a resistance of 300 Ω for 1 h. What is the power loss? How much heat is generated in joules?

Solution
From Eq. (27-6),

$$P = I^2R = (6 \text{ A})^2(300 \text{ Ω}) = 10,800 \text{ W}$$

Since the power represents the heat dissipated per unit of time, we obtain

$$\text{Work} = Pt = (10,800 \text{ W})(3600 \text{ s})$$
$$\text{Heat loss} = 3.89 \times 10^7 \text{ J}$$

This represents about 36,900 Btu.

27-6 RESISTIVITY

Just as capacitance is independent of the voltage and quantity of charge, the resistance of a conductor is independent of current and voltage. Both capacitance and resistance are inherent properties of a conductor. The resistance of a wire of uniform cross-sectional area, like the one shown in Fig. 27-7, is determined by the following four factors:

1. The kind of material
2. The length
3. The cross-sectional area
4. The temperature

Ohm, the German physicist who discovered the law that now bears his name, also reported that *the resistance of a conductor at a given temperature is directly proportional to its length, inversely proportional to its cross-sectional area, and dependent upon the material from which it is made.* For a given conductor at a given temperature, the resistance can be computed from

$$R = \rho\frac{l}{A} \qquad\qquad 27\text{-}8$$

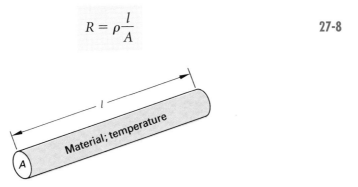

Figure 27-7 The resistance of a wire depends on the kind of material, the length, the cross-sectional area, and the temperature of the wire.

Technology Today

Bill Reader Advances

Have you ever wondered how vending machines recognize a dollar bill? The earliest machines used a magnetic head, similar to the magnetic head in an audiocassette player, to "read" the change in a magnetic field as the dollar was rolled through. This was possible because the ink on dollar bills had a high iron content. Later machines used a photoelectric cell to read light transmitted through the dollar bill from a diode on the other side. The latest technology measures the capacitance of the bill. This new technology can detect phony watermarks or phony identifying threads in counterfeit bills.

where R = resistance
l = length
A = area

The proportionality constant ρ is a property of the material called its **resistivity**, given by

$$\rho = \frac{RA}{l}$$

27-9

It varies considerably with different materials and is also affected by changes in temperature. When R is in ohms, A is in square meters, and l is in meters, the unit of resistivity is the ohm-meter ($\Omega \cdot$ m):

$$\frac{(\Omega)(m^2)}{m} = \Omega \cdot m$$

Table 27-1 lists the resistivities of several common metals.

TABLE 27-1 Resistivities of Various Materials at 20°C

	Resistivity	
	$\Omega \cdot$ m	$\Omega \cdot$ cmil/ft
Aluminum	2.8×10^{-8}	17.0
Constantan	49×10^{-8}	295
Copper	1.72×10^{-8}	10.4
Gold	2.2×10^{-8}	13.0
Iron	9.5×10^{-8}	57.0
Nichrome	100×10^{-8}	600
Silver	1.63×10^{-8}	9.6
Tungsten	5.5×10^{-8}	33.2

EXAMPLE 27-5

What is the resistance of a 20-m length of copper wire with a diameter of 0.8 mm?

Solution

We first compute the cross-sectional area of the wire in square meters.

$$A = \frac{\pi D^2}{4} = \frac{\pi (8 \times 10^{-4}\ m)^2}{4} = 5.03 \times 10^{-7}\ m^2$$

The resistivity of copper is $1.72 \times 10^{-8}\ \Omega \cdot$ m. Substitution into Eq. (27-8) gives

$$R = \rho \frac{l}{A} = \frac{(1.72 \times 10^{-8}\ \Omega \cdot m)(20\ m)}{5.03 \times 10^{-7}\ m^2}$$
$$= 0.684\ \Omega$$

In many engineering applications, the resistivity is given in mixed units. The length is measured in feet, and the area is measured in *circular mils* (cmil).

One circular mil (cmil) is defined as the cross-sectional area of a wire 1 mil (0.001 in.) in diameter.

To calculate the area of a wire in circular mils, we first convert its diameter to mils. Since 1 mil is 0.001 in., a diameter can be converted from inches to mils simply by moving the decimal three places to the right. For example,

$$0.128 \text{ in.} = 128.0 \text{ mils}$$

The area of a wire in *square mils* is found from

$$A = \frac{\pi D^2}{4} \qquad \text{(square mils)} \qquad \text{27-10}$$

By definition, however, a wire having a diameter of 1 mil has an area of 1 cmil. Thus,

$$1 \text{ cmil} = \frac{\pi}{4} \qquad \text{(square mils)} \qquad \text{27-11}$$

Comparing Eq. (27-10) and (27-11), we see that the *area in circular mils equals the square of the diameter in mils.* Symbolically,

$$A_{\text{cmils}} = (D_{\text{mils}})^2 \qquad \text{27-12}$$

If the area of a wire is expressed in circular mils and its length in feet, the unit for resistivity from Eq. (27-9) is

$$\rho = \frac{RA}{l} \rightarrow \frac{\Omega \cdot \text{cmil}}{ft}$$

Table 27-1 also lists the resistivity of materials in the *mil-foot system* of units.

EXAMPLE 27-6

What length of aluminum wire 0.025 in. in diameter is required to construct a 12-Ω resistor?

Solution
The diameter is 25 mils, and the area is found from Eq. (27-12).

$$A_{\text{cmils}} = (D_{\text{mils}})^2 = (25 \text{ mils})^2 = 625 \text{ cmils}$$

From Eq. (27-8), the length required is

$$l = \frac{RA}{\rho} = \frac{(12 \ \Omega)(625 \text{ cmils})}{17 \ \Omega \cdot \text{cmils/ft}} = 441 \text{ ft}$$

The resistivity ρ was taken from Table 27-1.

27-7 TEMPERATURE COEFFICIENT OF RESISTANCE

For most metallic conductors, the resistance tends to increase as the temperature increases. The increased atomic and molecular movement in the conductor hinders the flow of charge. The increase in resistance for most metals is approximately linear when compared with temperature changes. Experiments have shown that the increase in resistance ΔR is proportional to the initial resistance R_0 and the change in temperature Δt. We can write

$$\Delta R = \alpha R_0 \, \Delta t \qquad \text{27-13}$$

The constant α is a characteristic of the material known as the **temperature coefficient of resistance.** The defining equation for α can be found by solving Eq. (27-13):

$$\alpha = \frac{\Delta R}{R_0 \Delta t} \qquad \text{27-14}$$

The temperature coefficient of resistance **is the change in resistance per unit resistance per degree change in temperature.**

Since the units of ΔR and R_0 are the same, the unit of the coefficient α is inverse degrees (1/C°).

EXAMPLE 27-7

An iron wire has a resistance of 200 Ω at 20°C. What will its resistance be at 80°C if the temperature coefficient of resistance is 0.006/C°?

Solution

We first compute the change in resistance from Eq. (27-13):

$$\begin{aligned}
\Delta R &= \alpha R_0 \Delta t \\
&= (0.006/\text{C}°)(200 \ \Omega)(80°\text{C} - 20°\text{C}) \\
&= 72 \ \Omega
\end{aligned}$$

Therefore, the resistance at 80°C is

$$R = R_0 + \Delta R = 200 \ \Omega + 72 \ \Omega = 272 \ \Omega$$

The increase in resistance of a conductor with temperature is large enough to be measured easily. This fact is used in resistance thermometers to measure temperatures accurately. Because of the high melting point of some metals, resistance thermometers can be used to measure extremely high temperatures.

27-8 SUPERCONDUCTIVITY

In 1911 the Dutch physicist Heike Kamerlingh-Onnes was experimenting with the resistivity of metals at low temperatures. Using liquid helium as the coolant, he was able to cool the metals below 4.2 K ($^-$269°C). While some metals such as

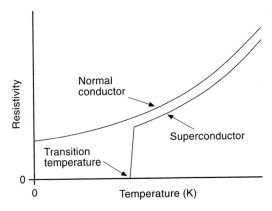

Figure 27-8 The change in resistivity as a function of temperature is shown for normal metallic conductors and for superconductors. Note that the resistivity of a superconductor drops rather abruptly to zero at its critical transition temperature.

platinum and gold reached a constant resistivity at low temperatures, other metals like mercury exhibited no resistance below a *critical* or *transition temperature,* as shown in Fig. 27-8. This phenomenon of zero resistance or infinite conductivity is called **superconductivity.** After many years of study, over 26 elements and a variety of alloys, compounds, and semiconductors have been classified as superconductors, each having a characteristic transition temperature (T_c).

In addition to their electrical properties, superconductors also exhibit extraordinary magnetic properties below their transition temperature. Superconductors placed in a magnetic field will expel all magnetic flux from its interior, in other words, they have perfect diamagnetism. A consequence of this phenomenon, known as the *Meissner-Ochsenfeld effect,* is shown in Fig. 27-9. As the magnet is brought close to the superconductor, the perfect diamagnetism (see Chap. 29) repels the magnet, levitating the magnet above the superconductor.

Until 1986, the highest recorded transition temperature was 23 K ($-250°C$) for a niobium-germanium alloy. Experiments done that year show that ceramic materials, usually having insulator properties, exhibit transition temperatures of about 90 K ($-183°C$). These new high-temperature superconductors have immense technological importance, primarily because liquid nitrogen, with a boiling temperature of 77 K ($-196°C$), can be used as the coolant. Liquid nitrogen is readily available, much less expensive, and easier to handle than liquid helium or liquid

Figure 27-9 The levitation of a small piece of metal is accomplished by a superconductor cooled to the temperature of liquid nitrogen. (Courtesy of Central Scientific Company.)

hydrogen. An experimental group from Switzerland was able to achieve transition temperatures near 135 K ($-138°C$). Andreas Schilling, Marco Cantoni, J. D. Guo, and Hans Ott accomplished this feat in 1993 using a compound containing mercury, calcium, barium, copper, and oxygen. Superconductors with room-temperature transitions remain a possibility. The illustration in Fig. 27-10 shows the rapid advance of research since 1986, when J. Georg Bednorz and Karl Alex Müller reported superconductivity at a transitional temperature of 30 K.

Practical applications of superconductors are already realized and under development. Magnets utilizing superconductor coils can achieve higher fields and have cheaper operating costs than conventional magnets because the zero resistivity of superconductors means no loss of energy due to resistive heating. In fact, a persistent current can exist in a superconductor without an applied potential difference. These magnets are now part of diverse instruments, such as supercolliders, magnetic energy storage, and medical diagnostic systems, for example, magnetic resonance imaging (MRI). Superconductor coils are used for windings in motors and generators, and potential applications include underground power transmission lines.

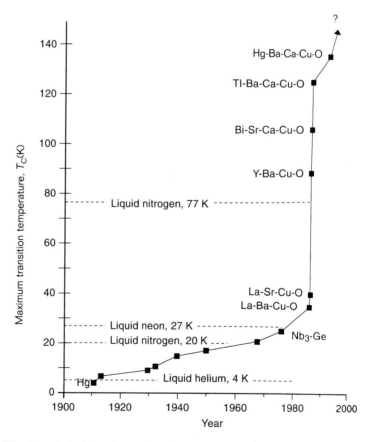

Figure 27-10 A graph depicting the rapid progress with experimentation on superconductors since the work of Bednorz and Müller in 1986. The most recent point indicates the transition temperature of 135 K, which was achieved in 1993 with a compound of mercury, barium, calcium, copper, and oxygen.

Another application is the magnetic levitated (MAGLEV) vehicle. These vehicles utilize the principle of levitation described earlier in Fig. 27-9. Superconducting magnets on the moving vehicle are located above normal sheet metal. The moving magnets create induced currents (eddy currents) in the metal sheet. These small, induced currents create in turn a magnetic field, which repeals the moving magnets. A prototype train has already been built in Japan using liquid helium as the coolant.

It should be noted that most of the applications described above do not use high-temperature (Type II) superconductors. The older (Type I) superconductors are more flexible and are able to carry more current then the new ceramic superconductors. Small devices called *SQUIDs* (*Superconducting Quantum Interference Devices*), made from high-temperature superconductors, are used, however, in sensitive, electronic measuring devices and computer components.

Summary and Review

Summary

In this chapter, we introduced the *ampere* as a unit of electric current, and we discussed the various quantities that affect its magnitude. Ohm's law described mathematically the relationship among current, resistance, and applied voltage. We also learned the factors that affect electric resistance, and applied these concepts to the solution of basic problems in elementary electricity. The major points are summarized below.

- Electric current I is the rate of flow of charge Q past a given point on a conductor:

$$I = \frac{Q}{t} \qquad 1 \text{ ampere (A)} = \frac{1 \text{ coulomb (C)}}{1 \text{ second (s)}}$$

- By convention, the *direction* of electric current is the same as the direction in which *positive* charges would move, even if the actual current consists of a flow of negatively charged electrons.

- Ohm's law states that *the current produced in a given conductor is directly proportional to the difference of potential between its end points:*

$$R = \frac{V}{I} \qquad V = IR \qquad \textit{Ohm's Law}$$

The symbol R represents the resistance in ohms (Ω) defined as

$$1 \text{ ohm } (\Omega) = \frac{1 \text{ volt (V)}}{1 \text{ ampere (A)}}$$

- The electric power in watts is given by any of the following:

$$P = VI \qquad P = I^2 R \qquad P = \frac{V^2}{R} \qquad \textit{Power}$$

- The resistance of a wire depends on four factors: (a) the kind of *material,* (b) the *length,* (c) the cross-sectional *area,* and (d) the *temperature.* By introducing a property for the material called its *resistivity* ρ, we can write

$$R = \rho \frac{l}{A} \qquad \rho = \frac{RA}{l} \qquad \textit{SI unit for } \rho: \ \Omega \cdot \text{m}$$

- The *temperature coefficient of resistance* α is the change in resistance per unit resistance per degree change in temperature.

$$\alpha = \frac{\Delta R}{R_0 \Delta t} \qquad \Delta R = \alpha R_0 \Delta t$$

Key Terms

1. current
2. resistance
3. ampere
4. emf
5. ohm
6. rheostat
7. ammeter
8. voltmeter
9. transient current
10. source of emf
11. electric power
12. Ohm's law
13. resistivity
14. circular mil
15. temperature coefficient of resistance

Review Questions

27-1. Distinguish clearly between electron flow and conventional current. What are some reasons for preferring the conventional current?

27-2. Use the mechanical analogy of water flowing through pipes to describe the flow of charge through conductors of various lengths and cross-sectional areas.

27-3. A rheostat is connected across the terminals of a battery. What determines the positive and negative terminals on the rheostat?

27-4. Is the electromotive force really a *force?* What is the function of a source of emf?

27-5. What is wrong with the following statement? The resistivity of a material is directly proportional to its length.

27-6. How many circular mils are equivalent to an area of 1 mil^2?

27-7. Use Ohm's law to verify Eqs. (27-6) and (27-7).

Problems

Sections 27-1 to 27-4—Electric Current and Ohm's Law

27-1. How many electrons pass a point every second in a wire carrying a current of 20 A? How much time is needed to transport 40 C of charge past this point?
Ans. 1.25 × 10^{20} electrons, 2 s

27-2. If 600 C of charge passes a given point in 3 s, what is the electric current in amperes?

27-3. Find the current in amperes when 690 C of charge passes a given point in 2 min.
Ans. 5.75 A

27-4. If a current of 24 A exists for 50 s, how many coulombs of charge have passed through the wire?

27-5. What is the potential drop across a 4-Ω resistor with a current of 8 A passing through it? **Ans. 32.0 V**

27-6. Find the resistance of a rheostat if the drop in potential is 48 V and the current is 4 A.

27-7. Determine the current through a 5-Ω resistor that has a 40-V drop in potential across it. **Ans. 8.00 A**

27-8. A 2-A fuse is placed in a circuit with a battery having a terminal voltage of 12 V. What is the minimum resistance for a circuit containing this fuse?

27-9. What emf is required to pass 60 mA through a resistance of 20 kΩ? If this same emf is applied to a resistance of 300 Ω, what will be the new current? **Ans. 1200 V, 4 A**

Section 27-5—Electric Power and Heat Loss

27-10. A soldering iron draws 0.75 A at 120 V. How much energy will it use in 15 min?

27-11. An electric lamp has an 80-Ω filament connected to a 100-V dc line. What is the current through the filament? What is the power loss in watts? **Ans. 1.38 A, 151 W**

27-12. Assume that the cost of energy in a home is 8 cents per kilowatt-hour. A family goes on a 2-week vacation leaving a single 80-W light bulb burning. What is the cost?

27-13. A 120-V dc generator delivers 2.4 kW to an electric furnace. What current is supplied? What is the resistance? **Ans. 20 A, 6 Ω**

27-14. A resistor develops heat at the rate of 250 W when the potential difference across its ends is 120 V. What is its resistance?

27-15. A 120-V motor draws a current of 4.0 A. How many joules of electrical energy are used in 1 h? How many kilowatt-hours? **Ans. 1.73 MJ, 0.48 kWh**

27-16. A household hair dryer is rated at 2000 W and is designed to operate on a 120-V outlet. What is the resistance of the device?

Section 27-6—Resistivity

27-17. What length of copper ($\rho = 1.78 \times 10^{-8}$ Ω m) wire 1/16 in. in diameter is needed to make a 20-Ω resistor at 20°C? What length of nichrome wire is needed? ($\rho = 100 \times 10^{-8}$ Ω · m) **Ans. 7510 ft, 130 ft**

27-18. A 3.0-m length of copper wire ($\rho = 1.78 \times 10^{-8}$ Ω · m) at 20°C has a cross section of 4 mm^2. What is the electrical resistance of this wire?

27-19. Find the resistance of 40 m of tungsten wire having a diameter of 0.8 mm at 20°C. ($\rho = 5.5 \times 10^{-8}$ Ω · m) **Ans. 4.37 Ω**

27-20. A certain wire has a diameter of 3 mm and a length of 150 m. It has a resistance of 3.00 Ω at 20°C. What is the resistivity?

27-21. What is the resistance of 200 ft of iron ($\rho = 9.5 \times 10^{-8}$ Ω · m) wire with a diameter of 0.002 in. at 20°C? **Ans. 2850 Ω**

Summary and Review

*27-22. A nichrome wire has a length of 40 m at 20°C. What is the diameter if the total resistance is 5 Ω? ($\rho = 100 \times 10^{-8}$ Ω · m)

*27-23. A 115-V source of emf is attached to a heating element that is a coil of nichrome wire ($\rho = 100 \times 10^{-8}$ Ω · m) of cross section 1.20 mm². What must be the length of the wire if the resistive power loss is to be 800 W? **Ans. 19.8 m**

Section 27-7—Temperature Coefficient of Resistance

27-24. The resistance of a length of wire ($\alpha = 0.0065/$C°) is 4.00 Ω at 20°C. What is the resistance at 80°C?

27-25. If the resistance of a conductor is 100 Ω at 20°C, and 116 Ω at 60°C, what is its temperature coefficient of resistivity? **Ans. 4.00 x 10⁻³/C°**

27-26. A length of copper ($\alpha = 0.0043/$C°) wire has a resistance of 8 Ω at 20°C. What is the resistance at 90°C? At −30°C?

*27-27. The copper windings of a motor experience a 20 percent increase in resistance over their value at 20°C. What is the operating temperature? ($\alpha = 0.0043/$C°). **Ans. 66.5°C**

*27-28. The resistivity of copper at 20°C is 1.78×10^{-8} Ω · m. What change in temperature will produce a 25 percent increase in resistivity?

Challenge Problems

27-29. A water turbine delivers 2000 kW to an electric generator that is 80 percent efficient and has an output terminal voltage of 1200 V. What current is delivered, and what is the electrical resistance? **Ans. 1.33 kA, 0.900 Ω**

27-30. A 110-V radiant heater draws a current of 6.0 A. How much heat energy in joules is delivered in 1 h?

27-31. A power line has a total resistance of 4 kΩ. What is the power loss through the wire if the current is reduced to 6.0 mA? **Ans. 0.144 W**

27-32. A certain wire has a resistivity of 2×10^{-8} Ω · m at 20°C. If its length is 200 m and its cross section is 4 mm², what will be its electrical resistance at 100°C? Assume that $\alpha = 0.005/$C° for this material.

27-33. Determine the resistivity of a wire made of an unknown alloy if its diameter is 0.007 in. and 100 ft of the wire is found to have a resistance of 4.0 Ω. **Ans. 1.96 Ω · cmil/ft**

27-34. The resistivity of a certain wire is 1.72×10^{-8} Ω m at 20°C. A 6-V battery is connected to a 20-m coil of this wire, which has a diameter of 0.8 mm. What is the current in the wire?

27-35. A certain resistor is used as a thermometer. Its resistance at 20°C is 26.00 Ω, and its resistance at 40°C is 26.20 Ω. What is the temperature coefficient of resistance for this material? **Ans. 3.85 × 10⁻⁴/C°**

*27-36. What length of copper wire at 20°C has the same resistance as 200 m of iron wire at 20°C? Assume the same cross section for each wire.

*27-37. The power loss in a certain wire at 20°C is 400 W. If $\alpha = 0.0036/$C°, by what percentage will the power loss increase when the operating temperature is 68°C? **Ans. 17.3 percent**

Critical Thinking Questions

27-38. A 150-Ω resistor at 20°C is rated at 2.0 W maximum power. What is the maximum voltage that can be applied across the resistor without exceeding the maximum allowable power? What is the current at this voltage?

27-39. The current in a home is alternating current, but the same formulas apply. Suppose a fan motor operating a home cooling system is rated at 10 A for a 120-V line. How much energy is required to operate the fan for a 24-h period? At a cost of 9 cents per kilowatt-hour, what is the cost of operating this fan continuously for 30 days?

Ans. 95.0 MJ, $53.36

*27-40. The power consumed in an electrical wire ($\alpha = 0.004/\text{C}°$) is 40 W at 20°C. If all other factors are held constant, what is the power consumption when (a) the length is doubled, (b) the diameter is doubled, (c) the resistivity is doubled, and (d) the absolute temperature is doubled?

Ans. (a) 80 W, (b) 10 W, (c) 80 W, (d) 86.9 W

*27-41. What must be the diameter of an aluminum wire if it is to have the same resistance as an equal length of copper wire of diameter 2.0 mm? What length of nichrome wire is needed to have the same resistance as 2 m of iron wire of the same cross section?

*27-42. An iron wire ($\alpha = 0.0065/\text{C}°$) has a resistance of 6.00 Ω at 20°C and a copper wire ($\alpha = 0.0043/\text{C}°$) has a resistance of 5.40 Ω at 20°C. At what temperature will the two wires have the same resistance?

Ans. −18.0°C

Direct-Current Circuits

Two types of current are in use. **Direct current** (dc) is the continuous flow of charge in only one direction. Alternating current (ac) is a flow of charge continually changing in both magnitude and direction. In this chapter we analyze current, voltage, and resistance for dc circuits. Many of the same methods and procedures can also be applied to ac circuits. The variations required for alternating currents build logically from a strong foundation in dc analysis.

28-1 SIMPLE CIRCUITS; RESISTORS IN SERIES

An electric circuit consists of any number of branches joined together so that at least one closed path is provided for current. The simplest circuit consists of a single source of emf joined to a single external resistance, as shown in Fig. 28-1. If $\mathcal{E}$ represents the emf and R indicates the total resistance, Ohm's law yields

$$\mathcal{E} = IR \qquad\qquad 28\text{-}1$$

where I is the current through the circuit. All the energy gained by a charge in passing through the source of emf is lost in flowing through the resistance.

Let us consider the addition of a number of elements to a circuit. Two or more elements are said to be in **series** if they have only *one* point in common that is not connected to some third element. Current can follow only a single path through elements in series. Resistors R_1 and R_2 of Fig. 28-2a are in series because point A is common to both resistors. The resistors in Fig. 28-2b, however, are not in series, because point B is common to three current branches. Electric current entering such a junction may follow two separate paths.

Suppose that three resistors (R_1, R_2, and R_3) are connected in series and enclosed in a box, indicated by the shaded portion of Fig. 28-3. The effective resistance R

Figure 28-1 A simple electric circuit.

of the three resistors can be determined from the external voltage V and current I, as recorded by the meters. From Ohm's law

$$R = \frac{V}{I}$$

28-2

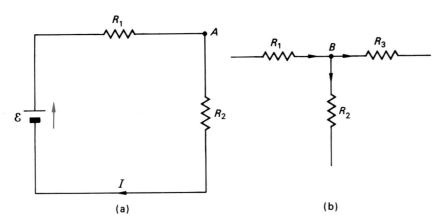

(a) (b)

Figure 28-2 (a) Resistors connected in series. (b) Resistors not connected in series.

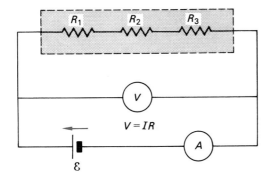

$$V = IR$$

Figure 28-3 The voltmeter-ammeter method of measuring the effective resistance of a number of resistors connected in series.

But what is the relationship of R to the three internal resistances? The current through each resistor must be identical since a single path is provided. Thus,

$$I = I_1 = I_2 = I_3 \qquad\qquad \textbf{28-3}$$

Utilizing this fact and noting that Ohm's law applies equally well to any part of a circuit, we write

$$V = IR \qquad V_1 = IR_1 \qquad V_2 = IR_2 \qquad V_3 = IR_3 \qquad \textbf{28-4}$$

The external voltage V represents the sum of the energies lost per unit of charge in passing through each resistance. Therefore,

$$V = V_1 + V_2 + V_3$$

Finally, if we substitute from Eq. (28-4) and divide out the current, we obtain

$$IR = IR_1 + IR_2 + IR_3$$

$$R = R_1 + R_2 + R_3 \qquad\qquad Series \quad \textbf{28-5}$$

To summarize what has been learned about resistors connected in *series:*

1. The current in all parts of a series circuit is the same.
2. The voltage across a number of resistances in series is equal to the sum of the voltage across the individual resistors.
3. The effective resistance of a number of resistors in series is equivalent to the sum of the individual resistances.

EXAMPLE 28-1

The resistances R_1 and R_2 in Fig. 28-2a are 2 and 4 Ω, respectively. If the source of emf maintains a constant potential difference of 12 V, what is the current delivered to the external circuit? What is the potential drop across each resistor?

Solution

The effective resistance is

$$R = R_1 + R_2 = 2\ \Omega + 4\ \Omega = 6\ \Omega$$

The current is then found from Ohm's law:

$$I = \frac{V}{R} = \frac{12\ \text{V}}{6\ \Omega} = 2\ \text{A}$$

The voltage drops are therefore

$$V_1 = IR_1 = (2\ \text{A})(2\ \Omega) = 4\ \text{V}$$
$$V_2 = IR_2 = (2\ \text{A})(4\ \Omega) = 8\ \text{V}$$

Note that the sum of the voltage drops $(V_1 + V_2)$ is equal to the applied 12 V.

28-2 RESISTORS IN PARALLEL

There are several limitations to the operation of series circuits. If a single element in a series circuit fails to provide a conducting path, the entire circuit is opened and current ceases. It would be quite annoying if all electrical devices in a home were to cease functioning whenever one lamp burned out. Moreover, each element in a series circuit adds to the total resistance of the circuit, thereby limiting the total current that can be supplied. These objections can be overcome by providing alternative paths for electric current. Such a connection, in which current can be divided between two or more elements, is called a ***parallel connection.***

A ***parallel circuit*** is one in which two or more components are connected to two common points in the circuit. For example, in Fig. 28-4, the resistors R_2 and R_3 are in parallel because they both have points A and B in common. Note that the current I, provided by the source of emf, is divided between resistors R_2 and R_3.

To arrive at an expression for the equivalent resistance R of a number of resistances connected in parallel, we follow a procedure similar to that discussed for a series connection. Assume that three resistors (R_1, R_2, and R_3) are placed inside a box, as shown in Fig. 28-5. The total current I delivered to the box is determined by its effective resistance and the applied voltage:

$$I = \frac{V}{R} \qquad\qquad \text{28-6}$$

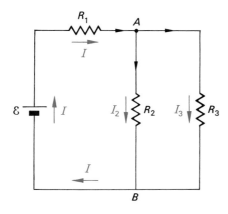

Figure 28-4 The resistors R_2 and R_3 are connected in parallel.

Figure 28-5 Computing the equivalent resistance of a number of resistors connected in parallel.

> **Math Tip**
>
> When using the following equation for the equivalent resistance for a set of resistors connected in parallel,
>
> $$\frac{1}{R} = \frac{1}{R_1} + \frac{1}{R_2} + \frac{1}{R_3} + \frac{1}{R_4}$$
>
> it is sometimes common to write it as
>
> $$R = \frac{R_1 R_2}{R_1 + R_2}$$
>
> when only two resistors are involved.

In a parallel connection, the voltage drop across each resistor is the same and equivalent to the total drop in voltage.

$$V = V_1 = V_2 = V_3 \qquad \text{28-7}$$

We see the truth of this statement when we consider that the same energy must be lost by a unit of charge, regardless of the path it travels in the circuit. In this example, charge may flow through any one of the three resistors. Thus, the total current delivered is divided among the resistors.

$$I = I_1 + I_2 + I_3 \qquad \text{28-8}$$

Applying Ohm's law to Eq. (28-8) yields

$$\frac{V}{R} = \frac{V_1}{R_1} + \frac{V_2}{R_2} + \frac{V_3}{R_3}$$

But the voltages are equal, and we can divide them out.

$$\frac{1}{R} = \frac{1}{R_1} + \frac{1}{R_2} + \frac{1}{R_3} \qquad \textit{Parallel} \quad \text{28-9}$$

In summary, for parallel resistors:

1. The total current in a parallel circuit is equal to the sum of the currents in the individual branches.
2. The voltage drops across all branches in a parallel circuit must be of equal magnitude.
3. The reciprocal of the equivalent resistance is equal to the sum of the reciprocals of the individual resistances connected in parallel.

In the case of only two resistors in parallel,

$$\frac{1}{R} = \frac{1}{R_1} + \frac{1}{R_2}$$

Solving this equation algebraically for R, we obtain a simplified formula for computing the equivalent resistance.

$$R = \frac{R_1 R_2}{R_1 + R_2} \qquad \text{28-10}$$

The equivalent resistance of two resistors connected in parallel is equal to their product divided by their sum.

EXAMPLE 28-2

The total applied voltage to the circuit in Fig. 28-6 is 12 V, and the resistances R_1, R_2, and R_3 are 4, 3, and 6 Ω, respectively. (a) Determine the equivalent resistance of the circuit. (b) What is the current through each resistor?

Solution (a)

The best approach to a problem that contains both series and parallel resistors is to reduce the circuit by steps to its simplest form, as shown in

Figure 28-6 Reducing a complex circuit to a simple equivalent circuit.

Fig. 28-6. We first find the equivalent resistance R' of the pair of resistors R_2 and R_3.

$$R' = \frac{R_2 R_3}{R_2 + R_3} = \frac{(3 \ \Omega)(6 \ \Omega)}{(3 + 6) \ \Omega} = 2 \ \Omega$$

Since the equivalent resistance R' is in series with R_1, the total equivalent resistance is

$$R = R' + R_1 = 2 \ \Omega + 4 \ \Omega = 6 \ \Omega$$

Solution (b)

The total current can be found from Ohm's law:

$$I = \frac{V}{R} = \frac{12 \ \text{V}}{6 \ \Omega} = 2 \ \text{A}$$

The current through R_1 and R' is therefore 2 A since the resistances are in series. To find the currents I_2 and I_3 we must know the voltage drop V' across the equivalent resistance R'.

$$V' = IR' = (2 \ \text{A})(2 \ \Omega) = 4 \ \text{V}$$

Thus, the potential must drop by 4 V through each of the resistors R_2 and R_3. The currents are found from Ohm's law:

$$I_2 = \frac{V'}{R_2} = \frac{4 \ \text{V}}{3 \ \Omega} = 1.33 \ \text{A}$$

$$I_3 = \frac{V'}{R_3} = \frac{4 \ \text{V}}{6 \ \Omega} = 0.67 \ \text{A}$$

Note that $I_2 + I_3 = 2$ A, which is the total current delivered to the circuit.

28-3 EMF AND TERMINAL POTENTIAL DIFFERENCE

In all the preceding problems, we have assumed that all resistance to current flow is due to elements of a circuit that are external to the source of emf. This is not strictly true, however, because there is an inherent resistance within every source of emf. This **internal resistance** is represented by the symbol r and is shown schematically as a small resistance in series with the source of emf. (See Fig. 28-7). When a current I is flowing through the circuit, there is a loss of energy through the external load R_L and also there is a heat loss due to the internal resistance. Thus, the actual terminal voltage V_T across a source of emf $\mathscr{E}$ with an internal resistance r is given by

$$V_T = \mathscr{E} - Ir \qquad \text{28-11}$$

The voltage applied to the external load is therefore less than the emf by an amount equal to the internal potential drop. Since $V_T = IR_L$, Eq. (28-11) can be rewritten

$$V_T = IR_L = \mathscr{E} - Ir \qquad \text{28-12}$$

Solving Eq. (28-12) for the current I, we have

$$I = \frac{\mathscr{E}}{R_L + r} \qquad \text{28-13}$$

The current in a simple circuit containing a single source of emf is equal to the emf $\mathscr{E}$ divided by the total resistance in the circuit (including internal resistance).

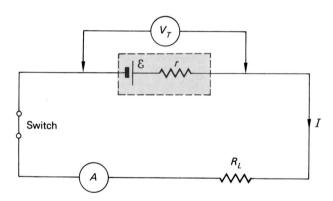

Figure 28-7 Internal resistance.

EXAMPLE 28-3

A load resistance of 8 Ω is connected to a battery whose internal resistance is 0.2 Ω. (a) If the emf of the battery is 12 V, what current is delivered to the load? (b) What is the terminal voltage of the battery?

Solution (a)
The current delivered is found from Eq. (28-13).

$$I = \frac{\mathscr{E}}{R_L + r} = \frac{12 \text{ V}}{8 \text{ } \Omega + 0.2 \text{ } \Omega} = 1.46 \text{ A}$$

Solution (b)

The terminal voltage is

$$V_T = \mathscr{E} - Ir = 12 \text{ V} - (1.46 \text{ A})(0.2 \text{ } \Omega)$$
$$= 12 \text{ V} - 0.292 \text{ V} = 11.7 \text{ V}$$

As a check, we can find the voltage drop across the load R_L:

$$V_T = IR_L = (1.46 \text{ A})(8 \text{ } \Omega) = 11.7 \text{ V}$$

28-4 MEASURING INTERNAL RESISTANCE

The internal resistance of a battery can be measured in the laboratory by using a voltmeter, an ammeter, and a known resistance. A voltmeter is an instrument that has an extremely high resistance. When a voltmeter is attached directly to the terminals of a battery, negligible current is drawn from the battery. We can see from Eq. (28-11) that, for zero current, this terminal voltage is equal to the emf ($V_T = \mathscr{E}$). In fact, the emf of a battery is sometimes referred to as its "open-circuit" potential difference. Thus, the emf can be measured with a voltmeter. By connecting a known resistance to the circuit, we can determine the internal resistance by measuring the current delivered to the circuit.

EXAMPLE 28-4

A dry cell gives an open-circuit reading of 1.5 V when a voltmeter is connected to its terminals. When the voltmeter is removed and a load of 3.5 Ω is placed across the terminals of the battery, a current of 0.4 A is measured. What is the internal resistance of the battery?

Solution

Solving for r in Eq. (28-12), we obtain

$$r = \frac{\mathscr{E} - IR_L}{I} = \frac{1.5 \text{ V} - (0.4 \text{ A})(3.5 \text{ } \Omega)}{0.4 \text{ A}}$$

$$= \frac{0.10 \text{ V}}{0.4 \text{ A}} = 0.25 \text{ } \Omega$$

As a dry cell ages, its internal resistance increases whereas its emf remains relatively constant. The increased internal resistance causes a reduction in the current delivered. This fact accounts for the difference in intensity of light between a flashlight using old batteries and one using fresh batteries.

28-5 REVERSING THE CURRENT THROUGH A SOURCE OF EMF

In a battery, chemical energy is converted into electric energy to maintain current flow in an electric circuit. A generator performs a similar function by converting mechanical energy into electric energy. In either case, the process is reversible. If a source of higher emf is connected in direct opposition to a source of lower emf, the current will pass through the latter from its positive terminal to its negative terminal. Reversing the flow of charge in this manner results in a loss of energy as electric energy is converted into chemical or mechanical energy.

Let us consider the process of charging a battery, as illustrated in Fig. 28-8. As charge flows through the higher source of emf $\mathscr{E}_1$, it gains energy. The terminal voltage for $\mathscr{E}_1$ is given by

$$V_1 = \mathscr{E}_1 - Ir_1$$

in accordance with Eq. (28-12). The output voltage is reduced because of the internal resistance r_1.

Energy is lost in two ways as charge is forced through the battery against its normal output direction:

1. Electric energy in the amount equal to $\mathscr{E}_2$ is stored as chemical energy in the battery.
2. Energy is lost to the internal resistance of the battery.

Therefore, the terminal voltage V_2, which represents the total drop in potential across the battery, is given by

$$V_2 = \mathscr{E}_2 + Ir_2 \qquad \text{28-14}$$

where r_2 is the internal resistance. Note that in this case the terminal voltage is *greater* than the emf or the battery. The remainder of the potential supplied by the higher source of emf is lost through the external resistance R.

Throughout the entire circuit, the energy lost must equal the energy gained. Thus, we can write

Energy gained per unit charge = energy lost per unit charge
$$\mathscr{E}_1 = \mathscr{E}_2 + Ir_1 + Ir_2 + IR$$

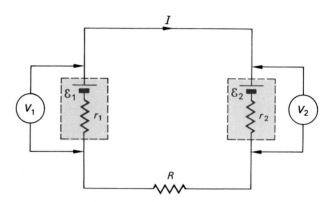

Figure 28-8   Reversing the current through a source of emf.

Solving for the current I yields

$$I = \frac{\mathscr{E}_1 - \mathscr{E}_2}{r_1 + r_2 + R}$$

The current supplied to a continuous electric circuit is equal to the net emf divided by the total resistance of the circuit, including internal resistance.

$$I = \frac{\Sigma \mathscr{E}}{\Sigma R} \qquad \text{28-15}$$

For the purposes of applying Eq. (28-15), an emf is considered negative when the current flows against its normal output direction.

EXAMPLE 28-5

Assume the following values for the parameters of the circuit in Fig. 28-8: $\mathscr{E}_1 = 12$ V, $\mathscr{E}_2 = 6$ V, $r_1 = 0.2$ Ω, $r_2 = 0.1$ Ω, and $R = 4$ Ω. (a) What is the current in the circuit? (b) What is the terminal voltage across the 6-V battery?

Solution (a)
From Eq. (28-15), the current is

$$I = \frac{\mathscr{E}_1 - \mathscr{E}_2}{r_1 + r_2 + R} = \frac{12 \text{ V} - 6 \text{ V}}{0.2 \text{ } \Omega + 0.1 \text{ } \Omega + 4 \text{ } \Omega}$$

$$= \frac{6 \text{ V}}{4.3 \text{ } \Omega} = 1.40 \text{ A}$$

Solution (b)
The terminal voltage of the battery being charged is, from Eq. (28-14),

$$V_2 = \mathscr{E}_2 + Ir_2$$
$$= 6 \text{ V} + (1.4 \text{ A})(0.1 \text{ } \Omega)$$
$$= 6.14 \text{ V}$$

28-6 KIRCHHOFF'S LAWS

An electrical network is a complex circuit consisting of a number of current loops or meshes. For example networks containing several meshes and a number of sources of emf, the application of Ohm's law becomes difficult. A more straightforward procedure for analyzing such circuits was developed in the nineteenth century by Gustav Kirchhoff, a German scientist. His method involves the use of two laws.

Kirchhoff's first law **The sum of the currents entering a junction is equal to the sum of the currents leaving that junction.**

$$\Sigma \, I_{\text{entering}} = \Sigma \, I_{\text{leaving}} \qquad \text{28-16}$$

Did You Know?

Microchip designers are faced with the challenge of making smaller transistors and thinner and better conductors in order to produce faster microchips. Utilizing copper, a better conductor than aluminum, and refining the methods of lithography in the process of etching patterns on silicon wafers are two keys to faster microchips. Currently, ultraviolet light is used to etch patterns, but shorter wavelengths produced by a new light source could make etching more precise.

Kirchhoff's second law **The sum of the emfs around any closed current loop is equal to the sum of all the *IR* drops around that loop.**

$$\Sigma \, \mathscr{E} = \Sigma \, IR \qquad\qquad 28\text{-}17$$

A *junction* refers to any point in a circuit at which three or more wires come together. The first law simply states that charge must flow continuously; it cannot pile up at a junction. In Fig. 28-9, if 12 C of charge enters the junction every second, then 12 C must leave it every second. The current delivered to each branch is inversely proportional to the resistance of that branch.

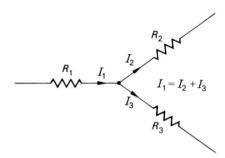

Figure 28-9 The sum of the currents entering a junction must equal the sum of the currents leaving that junction.

The second law is a restatement of the conservation of energy. If we begin at any point in a circuit and travel around any closed current loop, the energy gained by a unit of charge must equal the energy lost by that charge. Energy is gained through the conversion of chemical or mechanical energy into electric energy by a source of emf. Energy may be lost either in the form of *IR* potential drops or in the process of reversing the current through a source of emf. In the latter case, electric energy is converted into the chemical energy necessary to charge a battery or electric energy is converted to mechanical energy for the operation of a motor.

In applying Kirchhoff's rules, definite procedures must be followed. The steps in the general procedure will be presented by considering the example offered by Fig. 28-10a.

1. Assume a current direction for each loop in the network.

The three loops that may be considered are those illustrated in Fig. 28-10b, c, and d. Considering the entire circuit shown in Fig. 28-10a, the current I_1 is assumed to flow counterclockwise in the top loop, I_2 is assumed to travel to the left in the middle branch, and I_3 is assumed to flow counterclockwise in the lower loop. If we have guessed correctly, the solution to the problem will give a positive value for the current; if we have guessed incorrectly, a negative value will indicate that the current is actually in the other direction.

2. Apply Kirchhoff's first law to write a current equation for all but one of the junction points.

Writing the current equation for *every* junction would result in a duplicate equation. In our example, there are two junction points that are labeled *m* and *n*. The current equation for *m* is

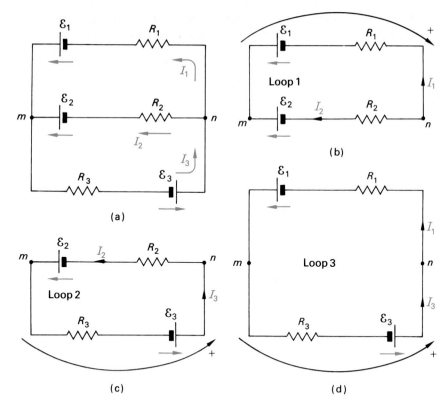

Figure 28-10 Applying Kirchhoff's laws to a complex circuit.

$$\Sigma\, I_{\text{entering}} = \Sigma\, I_{\text{leaving}}$$
$$I_1 + I_2 = I_3 \qquad\qquad \textbf{28-18}$$

The same equation would result if we considered junction n, and no new information would be given.

3. Indicate by a small arrow, drawn next to the symbol for each emf, the direction in which the source, acting alone, would cause a positive charge to move through the circuit.

In our example, $\mathscr{E}_1$ and $\mathscr{E}_2$ are directed to the left, and $\mathscr{E}_3$ is directed to the right.

4. Apply Kirchhoff's second law ($\Sigma\, \mathscr{E} = \Sigma\, IR$) for one loop at a time. There will be one equation for each loop.

In applying Kirchhoff's second rule, one must begin at a specific point on a loop and *trace* around the loop in a consistent direction back to the starting point. The choice of the *tracing direction* is arbitrary, but, once established, it becomes the positive (+) direction for sign conventions. (Tracing directions for the three loops in our example are labeled in Fig. 28-10.) The following sign conventions apply:

1. When the emfs are summed around a loop, the value assigned to the emf is positive if its output (see step 3) is with the tracing direction; it is considered negative if the output is against the tracing direction.
2. An *IR* drop is considered positive when the assumed current is with the tracing direction and negative when the assumed current opposes the tracing direction.

Let us now apply Kirchhoff's second law to each loop in our example.

Loop 1 Starting at point m and tracing clockwise, we have

$$-\mathcal{E}_1 + \mathcal{E}_2 = -I_1 R_1 + I_2 R_2 \qquad \text{28-19}$$

Loop 2 Starting at point m and tracing counterclockwise, we have

$$\mathcal{E}_3 + \mathcal{E}_2 = I_3 R_3 + I_2 R_2 \qquad \text{28-20}$$

Loop 3 Starting at m and tracing counterclockwise we have

$$\mathcal{E}_3 + \mathcal{E}_1 = I_3 R_3 + I_1 R_1 \qquad \text{28-21}$$

If the equation for loop 1 is subtracted from the equation for loop 2, the equation for loop 3 is obtained, showing that the last loop equation gives no new information.

We now have three independent equations involving only three unknowns. They can be solved simultaneously to find the unknown, and the third loop equation can be used to check the results.

CONCEPTUAL EXAMPLE 28-6

Solve for the unknown currents in Fig. 28-11, using Kirchhoff's laws.

Figure 28-11

Solution

Current directions are assumed, as indicated in the figure, for I_1, I_2, and I_3. If we apply Kirchhoff's first law to junction m, we obtain

$$\Sigma I_{\text{entering}} = \Sigma I_{\text{leaving}}$$
$$I_2 = I_1 + I_3 \qquad \text{28-22}$$

Next, the direction of positive output is indicated in the figure, adjacent to each source of emf. Since there are three unknowns, we need at least two more equations from the application of Kirchhoff's second law. Starting at m and tracing counterclockwise around the left loop, we write the voltage equation

$$\Sigma \mathcal{E} = \Sigma IR$$
$$6\text{ V} + 2\text{ V} = I_1(1\ \Omega) + I_2(3\ \Omega)$$
$$8\text{ V} = (1\ \Omega)I_1 + (3\ \Omega)I_2$$

Dividing through by 1 Ω and transposing, we obtain

$$I_1 + 3I_2 = 8 \text{ A} \qquad \text{28-23}$$

The unit *ampere* arises from the fact that

$$1 \text{ V}/\Omega = 1 \text{ A}$$

Another voltage equation can be written by starting at *m* and tracing clockwise around the right loop:

$$-3 \text{ V} = I_3(2 \ \Omega) + I_3(4 \ \Omega) + I_2(3 \ \Omega)$$

The negative sign arises from the fact that the output of the source opposes the tracing direction. Simplifying, we have

$$\begin{aligned} 2I_3 + 4I_3 + 3I_2 &= -3 \text{ A} \\ 6I_3 + 3I_2 &= -3 \text{ A} \\ I_2 + 2I_3 &= -1 \text{ A} \qquad \text{28-24} \end{aligned}$$

The three equations that must be solved simultaneously for I_1, I_2, and I_3 are

$$\begin{aligned} I_1 - I_2 + I_3 &= 0 && \text{(Eq. 28-22)} \\ I_1 + 3I_2 &= 8 \text{ A} && \text{(Eq. 28-23)} \\ I_2 + 2I_3 &= -1 \text{ A} && \text{(Eq. 28-24)} \end{aligned}$$

From Eq. (28-22), we note that

$$I_1 = I_2 - I_3$$

which, substituted into Eq. (28-23), yields

$$\begin{aligned} (I_2 - I_3) + 3I_2 &= 8 \text{ A} \\ 4I_2 - I_3 &= 8 \text{ A} \qquad \text{28-25} \end{aligned}$$

Now we can solve Eqs. (28-25) and (28-24) simultaneously by eliminating I_3 from the two equations by addition:

$$\begin{array}{lrcr} (28\text{-}24)\text{:} & I_2 + 2I_3 &=& -1 \text{ A} \\ 2 \times (28\text{-}25)\text{:} & \underline{8I_2 - 2I_3} &=& \underline{16 \text{ A}} \\ & 9I_2 &=& 15 \text{ A} \\ & I_2 &=& 1.67 \text{ A} \end{array}$$

Substituting $I_2 = 1.67$ A into Eqs. (28-23) and (28-24) gives values for the other currents:

$$I_1 = 3 \text{ A} \qquad I_3 = -1.33 \text{ A}$$

The negative value obtained for I_3 indicates that our assumed current direction was incorrect. Actually, the current flows opposite the assumed direction. In working problems, however, the minus sign should be retained until all unknowns have been determined.

As a check on the above results, we can write one more voltage equation by applying Kirchhoff's second law to the outside loop. Starting at *m* and tracing counterclockwise, we obtain

$$(6 + 2 + 3) \text{ V} = I_1(1 \text{ } \Omega) - I_3(4 \text{ } \Omega) - I_3(2 \text{ } \Omega)$$
$$I_1 - 6I_3 = 11 \text{ A}$$

Substituting for I_1 and I_3, we obtain

$$3 \text{ A} - (6)(-1.33 \text{ A}) = 11 \text{ A}$$
$$11 \text{ A} = 11 \text{ A} \qquad\qquad Check$$

Note again that the negative value for I_3 was used in the mathematics, even though it indicates an incorrect assumption.

28-7 THE WHEATSTONE BRIDGE

A precise laboratory method for measuring an unknown resistance uses a **Wheatstone bridge.** The circuit diagram for the apparatus is shown in Fig. 28-12. The bridge consists of a battery, a galvanometer, and four resistors. The galvanometer is a highly sensitive instrument that indicates the flow of electric charge. In the diagram, an unknown resistance R_x is balanced with the three known resistances (R_1, R_2, and R_3). Normally, when the switch is closed, the galvanometer will indicate a current between points B and D. If the contact point D is moved to the right or left, the resistances R_1 and R_2 can be varied until the galvanometer deflection is zero, indicating zero current between points B and D. Under these conditions, the bridge is said to be balanced, and points B and D must be at the same potential.

$$V_B = V_D \qquad\qquad Balanced$$

Thus, the potential drop from A to B must equal the potential drop from A to D.

$$V_{AB} = V_{AD} \qquad\qquad \textbf{28-26}$$

Similarly,

$$V_{BC} = V_{DC} \qquad\qquad \textbf{28-27}$$

The current I coming from the source of emf splits at point A, sending the current I_1 through the branch ABC and the current I_2 through the branch ADC. Applying Ohm's law to Eqs. (28-26) and (28-27) gives the relations

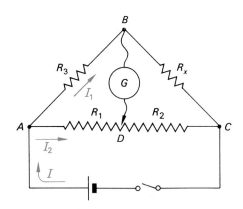

Figure 28-12 A circuit diagram for the Wheatstone bridge.

$$I_1 R_3 = I_2 R_1 \qquad \text{and} \qquad I_1 R_x = I_2 R_2$$

Dividing the second equation by the first and solving for R_x, we have

$$R_x = R_3 \frac{R_2}{R_1} \qquad\qquad \textbf{28-28}$$

The laboratory apparatus for the Wheatstone bridge is illustrated in Fig. 28-13. The resistance R_3 is usually a plug-type resistance box, which gives a positive indication of its resistance. Removing plugs will increase the known resistance. The resistances R_1 and R_2 normally consist of a wire of uniform resistance attached to a meterstick. Since the resistance of such a wire is directly proportional to its length, the ratio R_2/R_1 of Eq. (28-28) is equivalent to the ratio of the corresponding wire lengths. If l_1 and l_2 are used to represent the wire segments AD and DC, the unknown resistance can be found from

$$R_x = R_3 \frac{l_2}{l_1} \qquad\qquad \textbf{28-29}$$

Thus, it is not necessary to know the actual resistance of R_1 and R_2.

Figure 28-13 A laboratory apparatus for measuring unknown resistance with the Wheatstone bridge.

EXAMPLE 28-7

A Wheatstone bridge like the one shown in Fig. 28-13 is used to measure the resistance R_x of a spool of copper wire. The galvanometer gives a small deflection when 4 Ω is placed on the resistance box. When the contact key is positioned at the 40-cm mark (measured from A), the bridge is balanced. Determine the unknown resistance.

Solution

In this example $l_1 = 40$ cm, $l_2 = 60$ cm, and $R_3 = 4\ \Omega$. Substitution into Eq. (28-29) yields

$$R_x = R_3 \frac{l_2}{l_1} = (4\ \Omega)\frac{60\ \text{cm}}{40\ \text{cm}} = 6\ \Omega$$

During preliminary adjustments, when the bridge may be far from balanced, a resistor should be placed in series with the galvanometer to protect this extremely sensitive instrument from large currents. The resistor is removed from the circuit when final adjustments are made.

28-8 ELECTRIC CONDUCTION IN LIQUIDS

Direct electric currents can also be maintained in liquids under certain conditions. Two important applications that involve the motion of charges through a liquid are with batteries and with electroplating processes. In these operations, chemical energy and electrical energy interact in ways to accomplish useful work. The chemical energy available in lead storage batteries, for example, is converted into direct electric current. The reverse process uses the electric current to produce chemical changes in materials as with the depositing of silver or other metals onto electrodes inserted into a conducting solution.

In a metallic conductor the electric current consists largely of a movement of electrons in a direction opposite to the electric field. The mechanism for electric conduction is different in liquids and gases because the nature of fluids permits the movement of entire atoms or molecules that are electrically charged. Particles with an excess or deficiency of electrons are called *ions.* If a potential difference is placed across a fluid containing ions, a current is set up that consists of positive ions moving with the electric field and negative ions moving against the electric field.

Water is normally a poor conductor of electricity, but many substances when dissolved in water break up into charged particles (ions) that react to an electric field. Such substances are called *electrolytes.*

> **An electrolyte is a substance that conducts an electric current when it is dissolved in water or melted.**

Common electrolytes are hydrochloric acid (HCl), sulfuric acid (H_2SO_4) sodium chloride (NaCl), sodium hydroxide (NaOH), and barium hydroxide [$Ba(OH)_2$].

A simple apparatus for testing electrolytes is shown in Fig. 28-14. A potential difference is placed across two graphite *electrodes.* The positive electrode is referred to as the *anode,* and the negative electrode is called the *cathode.* The conductivity of a particular solution is indicated by the ammeter reading. Weak electrolytes indicate only a slight current.

28-9 ELECTROLYSIS

What are the effects of passing an electric current through a liquid? Consider ordinary table salt, sodium chloride (NaCl), as an electrolyte. When dissolved in water or melted, NaCl disassociates into sodium (Na^+) ions and chlorine (Cl^-) ions. Remember that the charge on an ion results from the loss or gain of elec-

Figure 28-14   Measuring the conductivity of an electrolyte.

trons; atoms are neutral in their normal, or ground, state. As illustrated in Fig. 28-15, a neutral sodium atom (Na^0) becomes a positive ion (Na^+) when it loses one electron.

$$Na^0 - 1e^- \rightarrow Na^+$$

The atom contains one more proton in the nucleus than there are electrons surrounding the nucleus. The process of losing electrons is referred to as *oxidation.*

Oxidation is a process by which particles lose electrons.

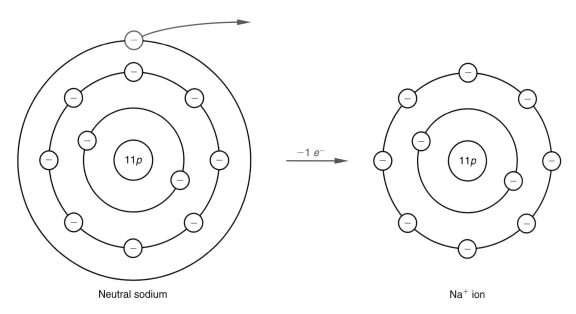

Neutral sodium Na^+ ion

Figure 28-15   A neutral sodium atom becomes a positive ion by losing an electron.

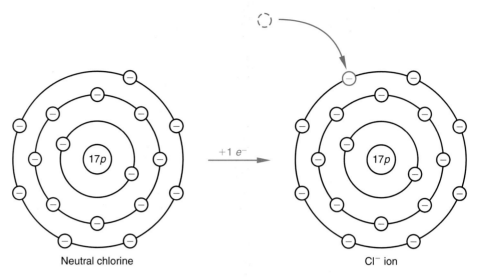

Neutral chlorine Cl^- ion

Figure 28-16 A neutral chlorine atom becomes a negative ion by gaining one electron.

A neutral chlorine atom, as shown in Fig. 28-16, becomes a negative ion by gaining an electron.

$$Cl^0 + 1e^- \rightarrow Cl^-$$

The resulting atom is a negative ion because it contains one more electron than it has protons in its nucleus. This reaction is called a **reduction.**

> **Reduction is a process by which particles gain electrons.**

The disassociation of the neutral salt molecule into sodium and chlorine ions illustrates a very important chemical principle:

> **When oxidation occurs in a given reaction, it must be accompanied by a corresponding reduction.**

In other words, charge is conserved. When one particle loses an electron, another particle gains an electron.

Now, let us consider how a current is conducted through a medium of molten sodium chloride (NaCl), chosen because it is simpler to discuss than a solution of NaCl. The process and the results are the same in either case. Two iron or graphite electrodes are inserted into the molten NaCl, and a battery or generator supplies the necessary potential difference. (See Fig. 28-17). The cathode (negatively charged electrode) attracts the positively charged particles, and the Na^+ ions drift to the right. The anode (positively charged electrode) causes the negatively charged Cl^- ions to drift to the left. At the cathode, a sodium ion (Na^+) picks up an electron and becomes a neutral sodium atom. The reaction at the cathode can be represented by

$$Na^+ + e^- \rightarrow Na$$

The neutral sodium atom is not soluble, and it is deposited on the cathode as ordinary metallic sodium. At the same time, a chemical reaction occurs at the anode when a chlorine ion (Cl^-) gives up an electron:

Figure 28-17   Electrolysis.

$$Cl^- \rightarrow Cl + e^-$$

The insoluble chlorine atoms thus formed come off the anode as bubbles of chlorine gas. Electrons released at the anode move through the wire under the influence of the emf provided by the battery until they reach the cathode. Thus a current is maintained. Positive and negative ions move in opposite directions through the molten NaCl, and electrons flow in one direction through the wire. The total charge transferred through a portion of the circuit is the sum of the positive charges moving to the right and the negative charges moving to the left. The entire phenomenon is an example of *electrolysis.*

> **Electrolysis is the process by which chemical changes are brought about by passing an electric current through a liquid.**

In the above example, the chemical changes involved the liberation of chlorine gas at the anode and the deposition of metallic sodium on the electrode.

In the electrolysis of water, a little sulfuric acid (H_2SO_4) is added to water to make it a conductor. The chemical reaction is

$$H_2SO_4 + 2H_2O \rightarrow 2H_3O^+ + SO_4^-$$

The hydronium ions (H_3O^+) and the sulfate ions (SO_4^-) become the charge carriers of the solution. A hydronium ion is simply a hydrogen ion (H^+) attached to a water molecule. Oxygen collects at the anode, and hydrogen collects at the cathode. Two volumes of hydrogen are released for each volume of oxygen.

Another useful product of electrolysis is called *electroplating.* Metallic sodium was deposited on the cathode in the above example. By choosing the proper electrodes and a suitable electrolyte, a thin coating of one metal can be applied to the surface of another. The object to be plated is thoroughly cleaned and attached to the cathode. The plating material (silver, nickel, etc.) is attached to the anode. The electrolyte should be a salt of the plating metal. For example, the plating material was sodium in our earlier example, and the electrolyte was sodium chloride.

28-10 DC VOLTAGE SOURCES; THE LEAD STORAGE BATTERY

A source of potential difference is needed to produce a direct electric current. One of the most common sources of dc voltage is the battery, which is a device that converts chemical energy into electrical energy. The most common batteries are the dry batteries used in flashlights, calculators, and other applications. Unfortunately, most of these batteries use up their chemicals and cannot be effectively recharged. The example we will present is that of a lead storage battery in which the process can be reversed; that is, the chemicals can be restored so that the battery is rechargeable.

The lead storage cells used in automobiles are the most common application of rechargeable chemical cells. In this type of cell, the cathode consists of a plate of spongy lead (Pb), the anode is lead peroxide (PbO_2), and the electrolyte is dilute sulfuric acid (H_2SO_4). Let us first consider what occurs when the cell is discharging, as illustrated in Fig. 28-18a.

On discharge, the lead atoms of the spongy lead cathode ionize and go into solution, leaving two electrons behind. The oxidation reaction is

$$Pb \rightarrow Pb^{2+} + 2e^- \qquad \textbf{28-30}$$

These lead ions combine with the surfate ions in solution to form a deposit of lead sulfate ($PbSO_4$):

$$Pb^{2+} + SO_4^{2-} \rightarrow PbSO_4 \qquad \textbf{28-31}$$

Thus the cathode is left with a negative charge. At the anode, the hydronium ions H_3O^+ remove oxygen from the lead peroxide, forming water and a deposit of lead sulfate. The anode reaction is

$$4H_3O^+ + 2e^- + PbO_2 + SO_4^{2-} \rightarrow PbSO_4 + 6H_2O \qquad \textbf{28-32}$$

This reaction gives the anode a positive charge. Because of the difference of potential, a current is maintained as electrons flow through the external circuit from the cathode to the anode. Note that when the cell is discharging, sulfuric acid is being removed from the solution, and water is being added. Also, the electrodes are becoming coated with lead sulfate.

Now suppose that the entire reactions discussed above are reversed by forcing a current through the cell in the opposite direction. This requires a source of higher emf to be connected opposite to the output of the cell, as shown in Fig. 28-18b. The reversed chemical action for the cathode and anode can be written

Cathode: $PbSO_4 + 2e^- \rightarrow Pb + SO_4^{2-}$ **28-33**
Anode: $PbSO_4 + 6H_2O - 2e^- \rightarrow PbO_2 + SO_4^{2-} + 4H_3O^+$ **28-34**

Under these conditions, sulfuric acid is restored to the solution, and water is taken from the solution. The electrodes return to their previous composition, the anode becoming lead peroxide and the cathode spongy lead. In this manner, the input electric energy is stored as chemical energy.

The output of a lead-acid storage cell over most of its discharge time is about 2 V. Most automobile batteries consist of six storage cells connected in series to give a 12-V output.

Figure 28-18 A lead-acid storage battery (a) on discharge, and (b) being charged by a dc generator.

28-11 CAPACITY RATING OF A BATTERY

The capacity rating of a battery is usually given in terms of *ampere-hours* (Ah). A battery having a rating of 80 Ah will theoretically provide a current of 1 A for 80 h, a current of 2 A for 40 h, or a current of 4 A for 20 h. These determinations are made on the basis of the equation

$$Life \ (hours) = \frac{ampere\text{-}hour \ rating}{amperes \ delivered} \qquad \textbf{28-35}$$

The ampere-hour rating is strongly affected by temperature and the rate of discharge. On very cold mornings when large amperage is required to start an automobile engine, the ampere-hour rating of a battery is reduced sharply.

Summary and Review

Summary

An understanding of direct-current circuits is essential as an introduction to electrical technology. Most of the advanced study builds on the ideas presented in this chapter. You should review in detail any sections that are not clear to you. A summary of the major relationships is given below.

- In dc circuits, resistors may be connected in series (refer to Fig. 28-3) or in parallel (refer to Fig. 28-5).

  (a) For **series connections,** the current in all parts of the circuit is the same, the total voltage drop is the sum of the individual drops across each resistor, and the effective resistance is equal to the sum of the individual resistances:

$$I_T = I_1 = I_2 = I_3$$

$$V_T = V_1 + V_2 + V_3$$

$$R_e = R_1 + R_2 + R_3 \quad \textit{Series Connections}$$

  (b) For *parallel connections,* the total current is the sum of the individual currents, the voltage drops are all equal, and the effective resistance is given by

$$I_T = I_1 + I_2 + I_3$$

$$V_T = V_1 = V_2 = V_3$$

$$\frac{1}{R_e} = \frac{1}{R_1} + \frac{1}{R_2} + \frac{1}{R_3}$$

Parallel Connections

For two resistors connected in parallel, a simpler form is

$$R_e = \frac{R_1 R_2}{R_1 + R_2} \quad \textit{Two Resistors in Parallel}$$

- The current supplied to an electric circuit is equal to the *net* emf divided by the total resistance of the circuit, including internal resistances.

$$I = \frac{\Sigma \mathscr{E}}{\Sigma R}$$

for example, $\quad I = \dfrac{\mathscr{E}_1 - \mathscr{E}_2}{r_1 + r_2 + R_L}$

The example is for two opposing batteries of internal resistances r_1 and r_2 when the circuit load resistance is R_L.

- According to Kirchhoff's laws, the current entering a junction must equal the current leaving the junction, and the net emf around any loop must equal the sum of the *IR* drops. Symbolically,

$$\Sigma I_{\text{entering}} = \Sigma I_{\text{leaving}}$$
$$\Sigma \mathscr{E} = \Sigma IR \qquad \textit{Kirchhoff's Laws}$$

- The following steps should be applied to solving circuits with Kirchhoff's law (see Fig. 28-10):

  Step 1 *Assume a current direction for each loop in the network.*

  Step 2 *Apply Kirchhoff's first law to write a current equation for all but one of the junction points* ($\Sigma I_{\text{in}} = \Sigma I_{\text{out}}$).

  Step 3 *Indicate by a small arrow the direction in which each emf acting alone would cause a positive charge to move.*

  Step 4 *Apply Kirchhoff's second law* ($\Sigma \mathscr{E} = \Sigma IR$) *to write an equation for all possible current loops. Choose an arbitrary positive tracing direction. An emf is considered positive if its output direction is the same as your tracing direction. An IR drop is considered positive when the assumed current direction is the same as your tracing direction.*

Summary and Review

Step 5 *Solve the equations simultaneously to determine the unknown quantities.*

- A Wheatstone bridge is a device that allows one to determine an unknown resistance R_x by balancing the voltage drops in the circuit.

If R_3 is known and the ratio R_2/R_1 can be determined, we have

$$R_x = R_3 \frac{R_2}{R_1} \qquad \text{Wheatstone Bridge}$$

Key Terms

1. DC circuit
2. series connection
3. parallel connection
4. terminal potential difference
5. internal resistance
6. Kirchhoff's first law
7. Kirchhoff's second law
8. Wheatstone bridge
9. electrolysis
10. cathode
11. anode
12. oxidation
13. reduction
14. ionization

Review Questions

28-1. Defend the following statement: the effective resistance of a group of resistors connected in parallel will be less than any of the individual resistances.

28-2. Discuss the advantages and disadvantages of connecting Christmas-tree lights (a) in series, (b) in parallel.

28-3. What is meant by the "open-circuit" potential difference of a battery?

28-4. Distinguish clearly between terminal potential difference and emf.

28-5. Many electrical devices and appliances are designed to operate at the same voltage. How should such devices be connected in an electric circuit?

28-6. Should elements connected in series be designed to function at a constant current or at a constant voltage?

28-7. In an electric circuit, it is desired to decrease the effective resistance by adding resistors. Should these resistors be connected in parallel or in series?

28-8. Describe a method for measuring the resistance of a spool of wire by using a voltmeter, an ammeter, a rheostat, and a source of emf. Draw the circuit diagram. (The rheostat is used to adjust the current to the range required for the ammeter.)

28-9. In a laboratory experiment with the Wheatstone bridge, the contact key is adjusted for zero galvanometer current. When the key has remained in contact with the wire for a while, a student notes that the bridge becomes unbalanced and there is a slight galvanometer deflection. Suggest a possible explanation.

28-10. Why is it acceptable to use lengths of wire rather than actual resistances when the Wheatstone bridge is balanced?

28-11. Given the emf of a battery, describe a laboratory procedure for determining its internal resistance.

28-12. Compare the formulas for computing equivalent capacitance with the formulas developed in this chapter for resistances in series and in parallel.

28-13. Can the terminal voltage of a battery ever be greater than its emf? Explain.

28-14. Solve Eq. (28-9) explicitly for the equivalent resistance R.

28-15. Distinguish between emf (open-circuit potential) and terminal voltage of a battery. What is meant by electrode potential? How is the electrode potential affected by an external discharge current?

28-16. The state of charge of a lead-acid storage

battery can be determined by a hydrometer, a device which indicates relative density (in comparison with pure water) of the electrolytic solution. Explain how this information is indicative of the state of charge of a storage battery.

28-17. The emf of a lead-acid storage battery varies with the state of charge. At one instant, the density of the electrolytic solution is found to be 1.29 g/cm³; at another time, the density is only 1.11 g/cm³. Compare the electromotive forces of the battery for these two situations.

28-18. Explain why a car battery "goes dead" faster on a cold morning than on a normal day.

28-19. Lead-acid batteries should never be left in a discharged condition for long periods of time. Why is this a bad practice?

28-20. As a part of routine maintenance, distilled water is sometimes added to a lead-acid storage battery. Why is this necessary? Why should the water be distilled? If some acid is spilled from a battery, what precautions should be taken in replacing the acid?

Problems

Section 28-2—Resistors in Series and Parallel

Ignore internal resistances for batteries in this section.

28-1. A 5-Ω resistor is connected in series with a 3-Ω resistor and a 16-V battery. What is the effective resistance, and what is the current in the circuit?   **Ans. 8.00 Ω, 2.00 A**

28-2. A 15-Ω resistor is connected in parallel with a 30-Ω resistor and a 30-V source of emf. What is the effective resistance, and what total current is delivered?

28-3. In Prob. 28-2, what is the current in 15- and 30-Ω resistors?   **Ans. 2.00 A, 1.00 A**

28-4. What is the equivalent resistance of 2-, 4-, and 6-Ω resistors connected in parallel?

28-5. An 18-Ω resistor and a 9-Ω resistor are first connected in parallel and then in series with a 24-V battery. What is the effective resistance for each connection? Neglecting internal resistance, what is the total current delivered by the battery in each case?
    Ans. 6.00 Ω, 27.0 Ω; 4.00 A, 0.899 A

28-6. A 12-Ω resistor and an 8-Ω resistor are first connected in parallel and then in series with a 28-V source of emf. What is the effective resistance and total current in each case?

28-7. An 8-Ω resistor and a 3-Ω resistor are first connected in parallel and then in series with a 12-V source. Find the effective resistance and total current for each connection.
    Ans. 2.18 Ω, 5.50 A; 11.0 Ω, 1.09 A

28-8. Given three resistors of 80, 60, and 40 Ω, find their effective resistance when connected in series and when connected in parallel.

28-9. Three resistances of 4, 9, and 11 Ω are connected first in series and then in parallel. Find the effective resistance for each connection.   **Ans. 24.0 Ω, 2.21 Ω**

*****28-10.** A 9-Ω resistor is connected in series with two parallel resistors of 6 and 12 Ω. What is the terminal potential difference if the total current from the battery is 4 A?

*****28-11.** For the circuit described in Prob. 28-10, what is the voltage across the 9-Ω resistor and what is the current through the 6-Ω resistor?   **Ans. 36.0 V, 2.67 A**

*****28-12.** Find the equivalent resistance of the circuit drawn in Fig. 28-19.

Figure 28-19

*28-13. Find the equivalent resistance of the circuit shown in Fig. 28-20.　　Ans. 2.22 Ω

Figure 28-20

*28-14. If a potential difference of 24 V is applied to the circuit in Fig. 28-19, what are the current and voltage across the 1-Ω resistor?

*28-15. If a potential difference of 12-V is applied to the free ends in Fig. 28-20, what are the current and voltage across the 2-Ω resistor?
　　　　　　　　Ans. 1.60 A, 3.20 V

Section 28-3—EMF and Terminal Potential Difference

28-16. A load resistance of 8 Ω is connected in series with a 18-V battery with an internal resistance of 1.0 Ω. What current is delivered, and what is the terminal voltage?

28-17. A resistance of 6 Ω is placed across a 12-V battery with an internal resistance of 0.3 Ω. What is the current delivered to the circuit? What is the terminal potential difference?
　　　　　　　　Ans. 1.90 A, 11.4 V

28-18. Two resistors, of 7 and 14 Ω, are connected in parallel with a 16-V battery with an internal resistance of 0.25 Ω. What is the terminal potential difference, and what is the current delivered to the circuit?

28-19. The open-circuit potential difference of a battery is 6 V. The current delivered to a 4-Ω resistor is 1.40 A. What is the internal resistance?　　Ans. 0.286 Ω

28-20. A dc motor draws 20 A from a 120-V dc line. If the internal resistance is 0.2 Ω, what is the emf of the motor?

28-21. For the motor in Prob. 28-20, what is the electric power drawn from the line? What portion of this power is dissipated because of heat losses? What power is delivered by the motor?　　Ans. 2400 W, 2320 W, 80 W

28-22. A 2- and a 6-Ω resistor are connected in series with a 24-V battery of internal resistance 0.5 Ω. What is the terminal voltage and the power lost to internal resistance?

*28-23. Determine the total current and the current through each resistor for Fig. 28-21 when $\mathscr{E} = 24$ V, $R_1 = 6$ Ω, $R_2 = 3$ Ω, $R_3 = 1$ Ω, $R_4 = 2$ Ω, and $r = 0.4$ Ω.
　　　　Ans. $I = 15$ A, $I_1 = 2$ A, $I_2 = 4$ Ω, $I_3 = 6$ A, $I_4 = 9$ A

Figure 28-21

*28-24. Find the total current and the current through each resistor for Fig. 28-21 when $\mathscr{E} = 50$ V, $R_1 = 12$ Ω, $R_2 = 6$ Ω, $R_3 = 6$ Ω, $R_4 = 8$ Ω, and $r = 0.4$ Ω.

Section 28-6—Kirchhoff's Laws

28-25. Apply Kirchhoff's second rule to the current loop in Fig. 28-22. What is the net voltage around the loop? What is the net IR drop? What is the current in the loop?
　　　　　　　　Ans. 16 V, 16 V, 2.00 A

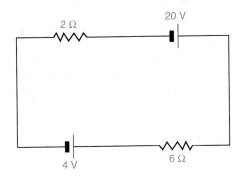

Figure 28-22

28-26. Answer the same questions for Prob. 28-25 when the polarity of the 20-V battery is changed, that is, its output direction is now to the left.

*28-27. Use Kirchhoff's laws to solve for the currents through the circuit shown in Fig. 28-23.

Ans. 190 mA, 23.8 mA, 214 mA

Figure 28-23

*28-28. Use Kirchhoff's laws to solve for the currents in Fig. 28-24.

Figure 28-24

*28-29. Apply Kirchhoff's laws to the circuit of Fig. 28-25. Find the currents in each branch.

Ans. 536 mA, 732 mA, 439 mA, 634 mA

Figure 28-25

Section 28-7—The Wheatstone Bridge

28-30. A Wheatstone bridge is used to measure the resistance R_x of a coil of wire. The resistance box is adjusted for 6 Ω, and the contact key is positioned at the 45-cm mark when measured from point A of Fig. 28-13. Find R_x.

28-31. Commercially available Wheatstone bridges are portable and have a self-contained galvanometer. The ratio R_2/R_1 can be set at any integral power of 10 between 0.001 and 1000 by a single dual switch. If this ratio is set to 100 and the known resistance R is adjusted to 46.7 Ω, the galvanometer current drops to zero. What is the unknown resistance?

Ans. 4670 Ω

28-32. In a commercial Wheatstone bridge, R_1 and R_2 have the resistances of 20 and 40 Ω, respectively. If the resistance R_x is 14 Ω, what must be the known resistance R_3 for zero galvanometer deflection?

Challenge Problems

28-33. Resistances of 3, 6, and 9 Ω are first connected in series and then in parallel with a 36-V source of potential difference. Neglecting internal resistance, what is the current leaving the positive terminal of the battery? **Ans. 2.00 A, 22.0 A**

28-34. Three 3-Ω resistors are connected in parallel. This combination is then placed in series with another 3-Ω resistor. What is the equivalent resistance?

*28-35. Three resistors of 4, 8, and 12 Ω are connected in series with a battery. A switch allows the battery to be connected or disconnected from the circuit. When the switch is open, a voltmeter across the terminals of the battery reads 50 V. When the switch is closed, the voltmeter reads 48 V. What is the internal resistance in the battery? **Ans. 1.00 Ω**

*28-36. The generator in Fig. 28-26 develops an emf of $\mathcal{E}_1 = 24$ V and has an internal resistance of 0.2 Ω. The generator is used to charge a battery $\mathcal{E}_2 = 12$ V that has an internal resistance of 0.3 Ω. Assume that $R_1 = 4$ Ω and $R_2 = 6$ Ω. What is the terminal voltage across the generator? What is the terminal voltage across the battery?

Figure 28-26

*28-37. What is the power consumed in charging the battery for Prob. 28-36? Show that the power delivered by the generator is equal to the power loss due to resistance and the power consumed in charging the battery.
Ans. 27.4 W = 13.7 W + 13.7 W

*28-38. Assume the following values for the parameters of the circuit illustrated in Fig. 28-8: $\mathcal{E}_1 = 100$ V, $\mathcal{E}_2 = 20$ V, $r_1 = 0.3$ Ω, $r_2 = 0.4$ Ω, and $R = 4$ Ω. What are the terminal voltages V_1 and V_2? What is the power lost through the 4-Ω resistor?

*28-39. Solve for the currents in each branch for Fig. 28-27. **Ans. 1.20 A, 0.600, 0.600 A**

Figure 28-27

*28-40. If the current in the 6-Ω resistor of Fig. 28-28 is 2 A, what is the emf of the battery? Neglect internal resistance. What is the power loss through the 1-Ω resistor?

Figure 28-28

Summary and Review

Critical Thinking Questions

*28-41. A three-way light bulb uses two resistors, a 50-W filament, and a 100-W filament. A three-way switch allows each to be connected in series and provides a third possibility by connecting the two filaments in parallel. Draw a possible arrangement of switches that will accomplish these tasks. Assume that the household voltage is 120 V. What are the resistances of each filament? What is the power of the parallel combination?   **Ans. 288 Ω, 144 Ω, 150 W**

*28-42. The circuit illustrated in Fig. 28-7 consists of a 12-V battery, a 4-Ω resistor, and a switch. When the battery is new, its internal resistance is 0.4 Ω, and a voltmeter is placed across the terminals of the battery. What will be the reading of the voltmeter when the switch is open and when it is closed? After a long period of time, the experiment is repeated and it is noted that open circuit reading is unchanged, but the terminal voltage has reduced by 10 percent. How do you explain the lower terminal voltage? What is the internal resistance of the old battery?

*28-43. Given three resistors of 3, 9, and 18 Ω, list all the possible equivalent resistances that can be obtained through various connections?

Ans. 2.00 Ω, 2.70 Ω, 6.30 Ω, 9.00 Ω, 11.6 Ω, 20.2 Ω, and 30.0 Ω

*28-44. Referring to Fig. 28-21, assume that $\mathscr{E} = 24$ V, $R_1 = 8$ Ω, $R_2 = 3$ Ω, $R_3 = 2$ Ω, $R_4 = 4$ Ω, and $r = 0.5$ Ω. What current is delivered to the circuit by the 24-V battery? What are the voltage and current for the 8-Ω resistor?

*28-45. What is the effective resistance of the external circuit for Fig. 28-29 if internal resistance is neglected? What is the current through the 1-Ω resistor?

Ans. 6.08 Ω, 2.58 A

Figure 28-29

Magnetism and the Magnetic Field

OBJECTIVES

After completing this chapter, you should be able to:

1. Demonstrate by definition and example your understanding of *magnetism, induction, retentivity, saturation,* and *permeability.*
2. Write and apply an equation relating the magnetic force on a moving charge to its velocity, its charge, and its direction in a field of known magnetic flux density.
3. Determine the magnetic force on a current-carrying wire placed in a known *B* field.
4. Calculate the magnetic flux density (a) at a known distance from a current-carrying wire, (b) at the center of a current loop or coil, and (c) at the interior of a solenoid.

In previous chapters, we saw that electrical charges exert forces on one another. In this chapter, we will study magnetic forces. A magnetic force may be generated by electric charges in motion, and an electric force may be generated by a magnetic field in motion. The operation of electric motors, generators, transformers, circuit breakers, televisions, radios, and most electric meters depends on the relationship between electric and magnetic forces. We shall begin this chapter by studying the magnetic effects associated with materials and conclude with a discussion of the magnetic effects of charges in motion.

29-1 MAGNETISM

The first magnetic phenomena to be observed were associated with rough fragments of lodestone (an oxide of iron) found near the ancient city of Magnesia some 2000 years ago. These *natural magnets* were observed to attract bits and pieces of unmagnetized iron. This force of attraction is referred to as **magnetism,** and the device that exerts a magnetic force is called a **magnet.**

If a bar magnet is dipped into a pan of iron filings and removed, the tiny pieces of iron are observed to cling most strongly to small areas near the ends (see Fig. 29-1). These regions where the magnet's strength appears to be concentrated are called **magnetic poles.**

When any magnetic material is suspended from a string, it turns about a vertical axis. As illustrated in Fig. 29-2, the magnet aligns itself in a north–south direction. The end pointing toward the north is called the *north-seeking pole* or the north (N) pole of the magnet. The opposite, *south-seeking*, end is referred to as

Figure 29-1   The strength of a magnet is concentrated near its ends.

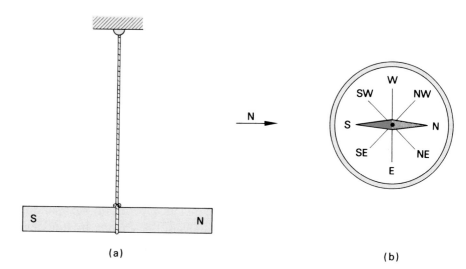

(a)   (b)

Figure 29-2   (a) A suspended bar magnet will come to rest in a north-south direction. (b) The top view of a magnetic compass.

the south (S) pole of the magnet. It is the polarization of magnetic material that accounts for its usefulness as a compass for navigation. The compass consists of a light magnetized needle pivoted on a low-friction support.

That the north and south poles of a magnet are different can be demonstrated easily. When another bar magnet is brought near a suspended magnet, as in Fig. 29-3, two north poles or two south poles repel each other, whereas the north pole of one and the south pole of another attract each other. The **law of magnetic force** states:

> **Like magnetic poles repel each other; unlike magnetic poles attract each other.**

Isolated poles do not exist. No matter how many times a magnet is broken in half, each piece will become a magnet, having both a north and a south pole. We know

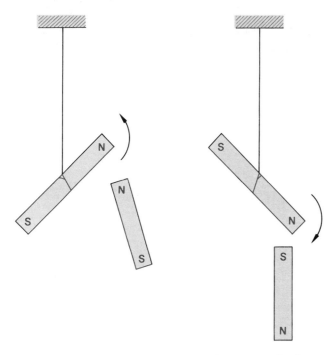

Figure 29-3 Like magnetic poles repel each other; unlike poles attract each other.

of no single particle that could create a magnetic field in the same way that a proton or electron can create an electric field.

The attraction of a magnet for unmagnetized iron and the interacting forces between magnetic poles act through all substances. In industry, ferrous materials in trash are separated by magnets for recycling.

29-2 MAGNETIC FIELDS

Every magnet is surrounded by a space in which its magnetic effects are present. Such regions are called *magnetic fields.* Just as electric field lines were useful in describing electric fields, magnetic field lines, called *flux lines,* are useful for visualizing magnetic fields. The direction of a flux line at any point is the same as the direction of the magnetic force on an *imaginary* isolated north pole positioned at that point (see Fig. 29-4a on page 644). Accordingly, lines of magnetic flux leave the north pole of a magnet and enter the south pole. Unlike electric field lines, magnetic flux lines do not have beginning or ending points. They form continuous loops, passing through the metallic bar, as shown in Fig. 29-4b. The flux lines in the region between two like or unlike poles are illustrated in Fig. 29-5 on page 644.

29-3 THE MODERN THEORY OF MAGNETISM

Magnetism in matter is currently believed to result from the movement of electrons in the atoms of substances. If this is true, magnetism is a property of *charge in motion* and is closely related to electric phenomena. According to classical

F.Y.I.

According to data from the Mars Global Surveyor, the planet Mars has a magnetic "fingerprint" on its surface, suggesting that at one time Mars may have had plate tectonic activity just as Earth does.

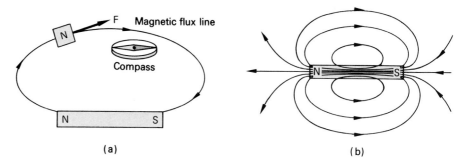

(a)   (b)

Figure 29-4 (a) Magnetic flux lines are in the direction of the force exerted on an independent north pole. (b) The flux lines in the vicinity of a bar magnet.

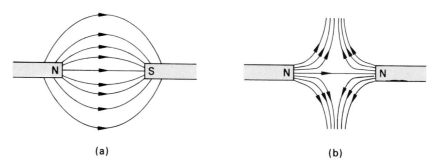

(a)   (b)

Figure 29-5 (a) Magnetic flux lines between two unlike magnetic poles. (b) Flux lines in the space between two like poles.

There's no such thing as a magnetic monopole — a magnet with only a north or only a south pole.

theory, individual atoms of a magnetic substance are, in effect, tiny magnets with north and south poles. The magnetic polarity of atoms is thought to arise primarily from the spin of electrons and is due only partially to their orbital motions around the nucleus. Figure 29-6 illustrates the two kinds of electron motion. Diagrams such as this should not be taken too seriously because there is still a lot we do not know about the motion of electrons. But we do firmly believe that magnetic fields of all particles must be caused by charge in motion, and such models help us to describe the phenomena.

The atoms in a magnetic material are grouped into microscopic magnetic regions called **domains.** All the atoms within a domain are believed to be magnetically polarized along a crystal axis. In an unmagnetized material, these domains are oriented in random directions, as indicated by the arrows in Fig. 29-7a. A dot is used to indicate an arrow directed out of the paper, and a cross is used to denote a direction into the paper. If a large number of the domains become oriented in

Figure 29-6 Two kinds of electron motion responsible for magnetic properties.

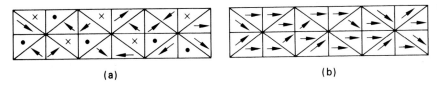

Figure 29-7 (a) Magnetic domains are randomly oriented in an unmagnetized material. (b) The preferred orientation of domains in a magnetized material.

the same direction, as in Fig. 29-7b, the material will exhibit strong magnetic properties.

This theory of magnetism is highly plausible because it offers an explanation for many of the observed magnetic effects of matter. For example, an unmagnetized iron bar can be made into a magnet simply by holding another magnet near it or in contact with it. This process, called *magnetic induction,* is illustrated by Fig. 29-8. The tacks become temporary magnets by induction. Note that the tacks on the right become magnetized, even though they do not actually touch the magnet. Magnetic induction is explained by the domain theory. The introduction of a magnetic field aligns the domains, resulting in magnetization.

Induced magnetism is often only temporary, and when the field is removed, the domains gradually become disoriented. If the domains remain aligned to some degree after the field has been removed, the material is said to be *permanently* magnetized. The ability to retain magnetism is referred to as *retentivity.*

Another property of magnetic materials that is explained easily by the domain theory is *magnetic saturation.* There appears to be a limit to the degree of magnetization experienced by a material. Once this limit has been reached, no greater strength of an external field can increase the magnetization. It is believed that all its domains have been aligned.

29-4 FLUX DENSITY AND PERMEABILITY

In Chap. 24 we stated that electric field lines are drawn so that their spacing at any point will determine the strength of the electric field at that point (refer to

Figure 29-8 Magnetic induction.

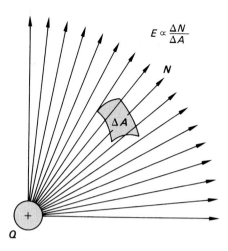

$$E \propto \frac{\Delta N}{\Delta A}$$

Figure 29-9 Electric field intensity is proportional to the electric line density.

Fig. 29-9). The number of lines ΔN drawn through a unit of area ΔA is directly proportional to the electric field intensity E.

$$\frac{\Delta N}{\Delta A} = \epsilon E \qquad\qquad 29\text{-}1$$

The proportionality constant ϵ, which determines the number of lines drawn, is the permittivity of the medium through which the lines pass.

An analogous description of a magnetic field can be presented by considering the magnetic flux ϕ passing through a unit of perpendicular area $A_\perp$. This ratio B is called the ***magnetic flux density.***

> **The mangetic flux density in a region of a magnetic field is the number of flux lines that pass through a unit of perpendicular area in that region.**

$$B = \frac{\phi \ (\text{flux})}{A_\perp \ (\text{area})} \qquad\qquad 29\text{-}2$$

The SI unit of magnetic flux is the ***weber*** (Wb). The unit of flux density would then be webers per square meter, which is redefined as the *tesla* (T). An older unit that remains in use is the *gauss* (G). In summary,

$$1 \ \text{T} = 1 \ \text{Wb/m}^2 = 10^4 \ \text{G} \qquad\qquad 29\text{-}3$$

EXAMPLE 29-1

A rectangular loop 10 cm wide and 20 cm long makes an angle of 30° with respect to the magnetic flux in Fig. 29-10. If the flux density is 0.3 T, compute the magnetic flux ϕ penetrating the loop.

Solution

The effective area penetrated by the flux is that component of the area that is perpendicular to the flux. Thus, Eq. (29-2) becomes

Figure 29-10 Computing the magnetic flux through a rectangular conductor.

$$B = \frac{\phi}{A \sin \theta} \quad \text{or} \quad \phi = BA \sin \theta$$

The magnetic flux in webers is found by substitution into this relation.

$$\phi = (0.3 \text{ T})(0.1 \text{ m} \times 0.2 \text{ m})(\sin 30°)$$
$$= (0.3 \text{ T})(0.02 \text{ m}^2)(0.5)$$
$$= 3 \times 10^{-3} \text{ Wb}$$

The flux density at any point in a magnetic field is affected strongly by the nature of the medium or by the nature of a material placed in the medium. For this reason it is convenient to define a new magnetic field vector, the *magnetic field intensity* **H,** which does not depend on the nature of a medium. In any case, the number of lines established per unit area is directly proportional to the magnetic field intensity **H.** We can write

$$B = \frac{\phi}{A} = \mu H \qquad\qquad \textbf{29-4}$$

where the proportionality constant μ is the **permeability** of the medium through which the flux lines pass. Equation (29-4) is exactly analogous to Eq. (29-1), which was developed for electric fields. The permeability of a medium can thus be thought of as a measure of its ability to establish magnetic flux lines. The greater the permeability of a medium, the more flux lines will pass through a unit of area.

The permeability of free space (vacuum) is denoted by μ_0 and has the following magnitude for SI units:

$$\mu_0 = 4\pi \times 10^{-7} \text{ Wb/A} \cdot \text{m} = 4\pi \times 10^{-7} \text{ T} \cdot \text{m/A}$$

The full meaning of the unit webers per ampere-meter will come later. It is determined by the units of ϕ, A, and H of Eq. (29-4), which, for a vacuum, can be written

$$B = \mu_0 H \qquad\qquad \textit{Vacuum} \quad \textbf{29-5}$$

If a nonmagnetic material, such as glass, is placed in a magnetic field like that in Fig. 29-11, the flux distribution will not vary appreciably from that established for a vacuum. However, when a highly permeable material, such as soft iron, is placed in the field, the flux distribution will be altered considerably. The

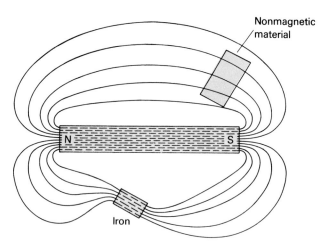

Figure 29-11 A permeable material becomes magnetized by induction, resulting in a greater flux density in that region.

permeable material becomes magnetized by induction, resulting in a greater field strength for that region. For this reason, the flux density B is also referred to as the *magnetic induction*.

Magnetic materials are classified according to their permeability compared with that of free space. The ratio of the permeability of a material to that of a vacuum is called its **relative permeability** and is given by

$$\mu_\tau = \frac{\mu}{\mu_0} \qquad \text{29-6}$$

Consideration of Eqs. (29-5) and (29-6) shows that the relative permeability of a material is a measure of its ability to change the flux density of a field from its value in a vacuum.

Materials with a relative permeability slightly less than unity have the property of being repelled by a strong magnet. Such materials are said to be **diamagnetic,** and the property is referred to as *diamagnetism.* On the other hand, materials with slightly greater permeability than that of a vacuum are said to be **paramagnetic.** These materials are feebly attracted by a strong magnet.

A few materials, for example, iron, cobalt, nickel, steel, and alloys of these metals, have extremely high permeabilities, ranging from a few hundred to thousands of times that for free space. Such materials are strongly attracted by a magnet and are said to be **ferromagnetic.**

29-5 MAGNETIC FIELD AND ELECTRIC CURRENT

Although the modern theory of magnetism holds that a magnetic field results from the motion of charges, science has not always accepted this proposition. It is fairly easy to show that a powerful magnet exerts no force on a static charge. In the course of a lecture demonstration in 1820, Hans Oersted set up an experiment to show his students that *moving* charges and magnets also do not interact. He placed the magnetic needle of a compass near a conductor, as illustrated in

Figure 29-12 Oersted's experiment.

Fig. 29-12. To his surprise, when a current was sent through the wire, a twisting force was exerted on the compass needle until it pointed almost perpendicular to the wire. Further, the magnitude of the force depended on the relative orientation of the compass needle and the current direction. The maximum twisting force occurred when the wire and compass needle were parallel before the current was established. If they were initially perpendicular, no force was experienced. Evidently, a magnetic field is set up by the charge in motion through the conductor.

In the same year that Oersted made his discovery, Ampère found that forces exist between two current-carrying conductors. Two wires with current in the same direction were found to attract each other, whereas oppositely directed currents caused a force of repulsion. A few years later, Faraday found that the motion of a magnet toward or away from an electric circuit produces a current in the circuit. The relationship between magnetic and electric phenomena could no longer be doubted. Today, all magnetic phenomena can be explained in terms of electric charges in motion.

29-6 THE FORCE ON A MOVING CHARGE

Let us investigate the effects of a magnetic field by observing the magnetic force exerted on a charge that passes through the field. In studying these effects, it is useful to imagine a positive-ion tube like that in Fig. 29-13. Such a tube allows us

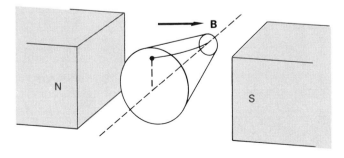

Figure 29-13 The magnetic force **F** on a moving charge is perpendicular both to the flux density **B** and to the charge velocity **v**.

to inject a positive ion of constant charge and velocity into a field of magnetic flux density **B.** By pointing the tube in various directions, we can observe the force exerted on the moving charge. The most striking observation is that the charge experiences a force that is perpendicular to both the magnetic flux density **B** and to the velocity **v** of the moving charge. Note that when the magnetic flux is directed from the left to the right and the charge is moving toward the reader, the charge is deflected upward. Reversing the polarity of the magnets causes the charge to be deflected downward.

The direction of the magnetic force **F** on a positive charge moving with a velocity **v** in a field of flux density **B** can be reckoned by the *right-hand-screw rule* (see Fig. 29-14):

> **The direction of the magnetic force F on a moving positive charge is the same as the direction of advance of a right-hand screw if rotated from v to B.**

If the moving charge is negative, as for an electron, the magnetic force will be directed *opposite* to the advance of a right-hand screw.

Let us now consider the magnitude of the force on a moving charge. Experimentation has shown that the magnitude of the magnetic force is directly proportional to the magnitude of the charge q and to its velocity **v.** Greater deflections will be indicated by our positive-ion tube if either of these parameters is increased.

An unexpected variation in the magnetic force will be observed if the ion tube is rotated slowly with respect to the magnetic flux density **B.** As indicated by Fig. 29-15, for a given charge of constant velocity **v,** the magnitude of the force varies with the angle the tube makes with the field. Particle deflection is a maximum when the charge velocity is perpendicular to the field. As the tube is rotated slowly toward **B,** the particle deflection becomes less and less. Finally, when the charge velocity is directed parallel to **B,** no deflection occurs, indicating that the magnetic

Figure 29-14 The right-hand-screw rule.

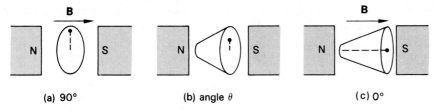

(a) 90° (b) angle θ (c) 0°

Figure 29-15 The magnitude of the magnetic force varies with the angle the moving charge makes with the direction of the magnetic field.

force has dropped to zero. Evidently, the magnitude of the force is a function not only of the magnitude of the charge and its velocity but also varies with the angle θ between **v** and **B**. This variation is accounted for by stating that the magnetic force is proportional to the component of the velocity, **v** sin θ, perpendicular to the field direction. (Refer to Fig. 29-16.)

The above observations are summarized by the proportionality

$$F \propto qv \sin \theta \qquad\qquad \textbf{29-7}$$

If the proper units are chosen, the proportionality constant can be equated to the magnetic flux density B of the field causing the force. In fact, this proportionality is often used to *define* magnetic flux density as the constant ratio:

$$B = \frac{F}{qv \sin \theta} \qquad\qquad \textbf{29-8}$$

A magnetic field having a flux density of one tesla (one weber per square meter) will exert a force of one newton on a charge of one coulomb moving perpendicular to the field with a velocity of one meter per second.

As a consequence of Eq. (29-8), it can be noted that

$$1 \text{ T} = 1 \text{ N}/(\text{C} \cdot \text{m/s}) = 1 \text{ N/A} \cdot \text{m} \qquad\qquad \textbf{29-9}$$

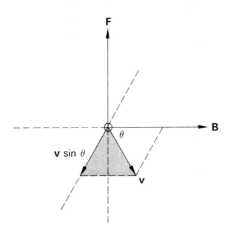

Figure 29-16

These unit relationships are useful in solving problems involving magnetic forces. Solving for the force F in Eq. (29-8), we obtain

$$F = qvB \sin \theta$$

which is the more useful form for direct calculation of magnetic forces. The force F is in newtons when the charge q is expressed in coulombs, the velocity v is in meters per second, and the flux density is in teslas. The angle θ indicates the direction of **v** with respect to **B**. The force **F** is *always* perpendicular to both **v** and **B**.

EXAMPLE 29-2

An electron is projected from left to right into a magnetic field directed vertically downward. The velocity of the electron is 2×10^6 m/s, and the magnetic flux density of the field is 0.3 T. Find the magnitude and direction of the magnetic force on the electron.

Solution

The electron is moving in a direction perpendicular to **B**. Thus, $\sin \theta = 1$ in Eq. (29-8), and we solve for the force F as follows:

$$
\begin{aligned}
F &= qvB \sin \theta \\
&= (1.6 \times 10^{-19} \text{ C})(2 \times 10^6 \text{ m/s})(0.3 \text{ T})(1) \\
&= 9.6 \times 10^{-14} \text{ N}
\end{aligned}
$$

Application of the right-hand-screw rule will show that the direction of the force is *out of the page*, or toward the reader. (It would be into the page for a positive charge like a proton.)

29-7 FORCE ON A CURRENT-CARRYING WIRE

The Hall effect is demonstrated by a current flowing through a thin film, which is deflected to one side by a magnet.

When an electric current passes through a conductor lying in a magnetic field, each charge q flowing through the conductor experiences a magnetic force. These forces are transmitted to the conductor as a whole, causing each unit of length to

Figure 29-17 Magnetic force on a current-carrying conductor.

experience a force. If a total quantity of charge Q passes through the length l of the wire (Fig. 29-17) with an average velocity $\bar{v}$, perpendicular to a magnetic field B, the net force on that segment of wire is

$$F = Q\bar{v}B$$

The average velocity for each charge passing through the length l in the time t is l/t. Thus, the net force on the entire length becomes

$$F = Q\frac{l}{t}B$$

Rearranging and simplifying, we obtain

$$F = \frac{Q}{t}\,lB = IlB$$

where I represents the current in the wire.

Just as the magnitude of the force on a moving charge varies with velocity direction, the force on a current-carrying conductor depends on the angle the current makes with the flux density. In general, if a wire of length l makes an angle θ with the **B** field, as illustrated in Fig. 29-18, it will experience a force given by

$$F = BIl \sin \theta \qquad\qquad \text{29-10}$$

where I is the current through the wire. When B is in **teslas**, l is in *meters*, and I is in *amperes*, the force will be in *newtons*.

The direction of the magnetic force on a current-carrying conductor can be determined by the right-hand-screw rule in the same way as for a moving charge. When **I** is rotated into **B**, the direction of the force **F** is in the same direction as the advance of a right-hand screw. The force **F** is *always* perpendicular to both **I** and **B**.

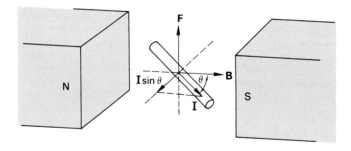

Figure 29-18

EXAMPLE 29-3

The wire in Fig. 29-18 makes an angle of 30° with respect to a B field of 0.2 T. If the length of the wire is 8 cm and a current of 4 A passes through it, determine the magnitude and direction of the resultant force on the wire.

Solution

Direct substitution into Eq. (29-10) yields

$$F = BIl \sin \theta$$
$$= (0.2 \text{ T})(4 \text{ A})(0.08 \text{ m})(\sin 30°)$$
$$= 0.032 \text{ N}$$

Application of the right-hand-screw rule shows the direction of the force to be upward, as indicated in Fig. 29-18. If the current direction were reversed, the force would be directed downward.

29-8 MAGNETIC FIELD OF A LONG, STRAIGHT WIRE

Oersted's experiment demonstrated that electric charge in motion, or a current, sets up a magnetic field in the space surrounding it. Up to this point, we have been discussing the force that such a field will exert on a second current-carrying conductor or on a charge moving in the field. We shall now begin to calculate the magnetic fields produced by electric currents.

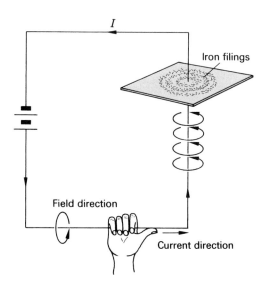

Figure 29-19 The magnetic field surrounding a straight, current-carrying conductor.

Let us first examine the flux density surrounding a long, straight wire carrying a constant current. If iron filings are sprinkled on the paper surrounding the wire in Fig. 29-19, they will become aligned in concentric circles around the wire. Similar investigation of the area surrounding the wire with a magnetic compass will confirm that the magnetic field is circular and directed in a clockwise fashion, as viewed along the direction of conventional (positive) current. A convenient method devised by Ampère to determine the direction of the field surrounding a straight wire is called the *right-hand-thumb rule* (refer to Fig. 29-19).

> **If the wire is grasped with the right hand so that the thumb points in the direction of the conventional current, the curled fingers of that hand will point in the direction of the magnetic field.**

The magnetic induction, or flux density, at a perpendicular distance d from a long, straight wire carrying a current I, as shown in Fig. 29-20, can be calculated from

$$B = \frac{\mu I}{2\pi d} \qquad \textit{Long Wire} \qquad \textbf{29-11}$$

where μ is the permeability of the medium surrounding the wire. In the special cases of a vacuum, air, and nonmagnetic media, the permeability μ_0 is

$$\mu_0 = 4\pi \times 10^{-7} \text{ T} \cdot \text{m/A} \qquad \textbf{29-12}$$

The units are determined from Eq. (29-11).

Figure 29-20 The magnetic field **B** at a perpendicular distance d from a long, current-carrying conductor.

EXAMPLE 29-4

Determine the magnetic induction in the air 5 cm from a long wire carrying a current of 10 A.

Solution
From Eq. (29-11),

$$B = \frac{\mu_0 I}{2\pi d} = \frac{(4\pi \times 10^{-7} \text{ T} \cdot \text{m/A})(10 \text{ A})}{(2\pi)(0.05 \text{ m})}$$
$$= 4 \times 10^{-5} \text{ T}$$

The direction of the magnetic induction is determined from the right-hand-thumb rule. For the case illustrated in Fig. 29-20, it would be out of the paper.

For a derivation of Eq. (29-11) and other relations that follow, refer to the *Biot-Savart law* or to *Ampère's law*. Many conventional physics texts provide a complete analysis, which usually includes the methods of calculus.

29-9 OTHER MAGNETIC FIELDS

If a wire is bent into a circular loop and connected to a source of current, as shown in Fig. 29-21a on page 656, a magnetic field similar to that for a bar magnet will be set up. The right-hand-thumb rule will still serve to give the field direction in a rough manner, but now the flux lines are no longer circular. The magnetic flux density varies considerably from point to point.

The magnetic induction at the center of a circular loop of radius r carrying a current I is given by Eq. (29-13).

Did You Know?

The earth's magnetic North Pole is called that because it attracts the north end of the magnetic dipoles we use as compasses. But since the north end of the compass is attracted to it, the North Pole is actually a magnetic South Pole.

$$B = \frac{\mu I}{2r} \qquad \qquad \textit{Center of Loop} \qquad \textbf{29-13}$$

The direction of B is perpendicular to the plane of the loop. If the wire consists of a coil having N turns, Eq. (29-13) becomes

$$B = \frac{\mu N I}{2r} \qquad \qquad \textit{Center of Coil} \qquad \textbf{29-14}$$

A *solenoid* consists of many circular turns of wire wound in the form of a helix, as illustrated in Fig. 29-22. The magnetic field produced is similar to that of a bar magnet. The magnetic induction in the interior of a solenoid is given by

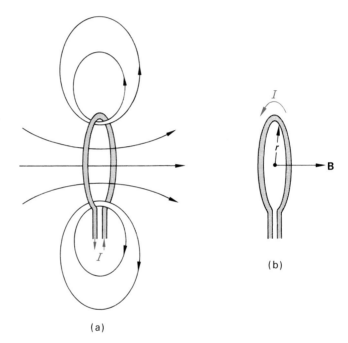

(a)

(b)

Figure 29-21 The magnetic field at the center of a circular loop.

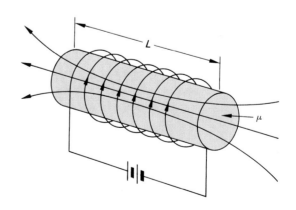

Figure 29-22 The solenoid.

$$B = \frac{\mu N I}{L} \qquad \text{\textit{Solenoid}} \qquad \textbf{29-15}$$

where N = number of turns
$\quad\quad I$ = current, A
$\quad\quad L$ = length of the solenoid, m

EXAMPLE 29-5

A solenoid is constructed by winding 400 turns of wire on a 20-cm iron core. The relative permeability of the iron is 13,000. What current is required to produce a magnetic induction of 0.5 T in the center of the solenoid?

Solution

The permeability of the core is

$$\mu = \mu_r \mu_0 = (13,000)(4\pi \times 10^{-7} \text{ T} \cdot \text{m/A})$$
$$= 1.63 \times 10^{-2} \text{ T} \cdot \text{m/A}$$

Solving for I in Eq. (29-15) and substituting known values, we obtain

$$I = \frac{BL}{\mu N} = \frac{(0.5 \text{ T})(0.2 \text{ m})}{(1.63 \times 10^{-2} \text{ T} \cdot \text{m/A})(400 \text{ turns})}$$
$$= 0.015 \text{ A}$$

The diameter of the solenoid is not significant provided that it is small relative to the length L.

One type of solenoid, called a *toroid*, is often used in studying magnetic effects. As will be seen in the following section, the toroid consists of a tightly wound coil of wire in the shape of a doughnut. The magnetic flux density in the core of a toroid is also given by Eq. (29-15).

29-10 HYSTERESIS

We have seen that the lines of magnetic flux are more numerous for a solenoid with an iron core than for a solenoid in air. The flux density is related to the permeability μ of the material serving as a core for the solenoid. Recall that the field intensity H and flux density B are related to each other according to the equation

$$B = \mu H$$

Comparison of this relationship with Eq. (29-15) shows that, for a solenoid,

$$H = \frac{NI}{L} \qquad\qquad \textbf{29-16}$$

Note that the magnetic intensity is independent of the permeability of the core. It is a function only of the number of turns N, the current I, and the solenoid length L. The magnetic intensity is expressed in *amperes per meter*.

We can study the magnetic properties of matter by observing the flux density B produced as a function of the magnetizing current or as a function of the magnetic intensity H. This can be done more easily when a substance is fashioned into the form of a toroid, as in Fig. 29-23. The magnetic field set up by a current in the magnetizing windings is confined wholly to the toroid. Such a device is often called a *Rowland ring* after J. H. Rowland, who used it to study the magnetic properties of many materials.

Suppose we begin by studying the magnetic properties of a material with an unmagnetized Rowland ring fashioned out of the substance. Initially, $B = 0$ and $H = 0$. The switch is closed, and the magnetizing current I is gradually increased, producing a magnetic intensity given by

$$H = \frac{NI}{L}$$

where L is the circumference of the ring. As the material is subjected to an increasing magnetic intensity H, *the flux density B increases until the material is saturated.* Refer to the curve AB in Fig. 29-24. Now, if the current is gradually reduced to

Figure 29-23   Rowland's ring.

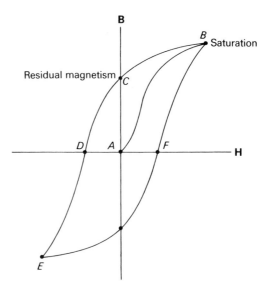

Figure 29-24   The hysteresis loop.

zero, the flux density B throughout the core does not return to zero but lags behind the magnetic intensity, as illustrated by the curve BC. (This refers essentially to residual magnetism.) The lack of retraceability is known as *hysteresis*.

Hysteresis is the lagging of the magnetization behind the magnetic intensity.

The only way to bring the flux density B within the ring back to zero is to reverse the direction of the current through the windings. This procedure builds up the magnetic intensity H in the opposite direction, as shown by curve CD. If the magnetization continues to increase in the negative direction, the material will eventually become saturated again with a reversed polarity. Refer to curve DE. Reducing the current to zero again and then increasing it in the positive direction will yield the curve EFB. The entire curve is called the *hysteresis loop.*

The area enclosed by a hysteresis loop is an indication of the quantity of energy that is lost (in the form of heat) by carrying a given material through a complete magnetization cycle. The efficiency of many electromagnetic devices depends on the selection of magnetic materials with low hysteresis. On the other hand, materials that are to remain well magnetized should have large hysteresis.

Summary and Review

Summary

We have seen that magnetic fields are created by charge in motion. This basic principle will underlie much of what follows in your study of electromagnetism. The operation of electric motors, generators, transformers, and an endless variety of industrial instruments requires an understanding of the magnetic field. The major concepts are summarized below.

- The magnetic flux density **B** in a region of a magnetic field is the number of flux lines that pass through a unit of area perpendicular to the flux.

$$B = \frac{\phi}{A_\perp} = \frac{\phi}{A \sin \theta} \qquad \textit{Magnetic Flux Density}$$

where ϕ = flux, Wb
  A = unit area, m^2
  θ = angle that plane of area makes with flux
  B = magnetic flux density, T (1 T = 1 Wb/m^2)

- The magnetic flux density **B** is proportional to the magnetic field intensity **H.** The constant of proportionality is the permeability of the medium in which the field exists.

$$B = \frac{\phi}{A_\perp} = \mu H \qquad \begin{array}{l}\textit{For a Vacuum} \\ \mu_0 = 4\pi \times 10^{-7} \text{ T} \cdot \text{m/A}\end{array}$$

- The *relative permeability* μ_r is the ratio of μ/μ_0. We can write

$$B = \mu_0 \mu_r H \quad \text{where} \quad \mu_r = \frac{\mu}{\mu_0} \quad \begin{array}{l}\textit{Relative} \\ \textit{Permeability}\end{array}$$

- A magnetic field of flux density equal to 1 T will exert a force of 1 N on a charge of 1 C moving perpendicular to the field with a velocity of 1 m/s. The general case is described by Fig. 29-25, in which the charge moves at an angle θ with the field.

$$F = qvB \sin \theta \qquad B = \frac{F}{qv \sin \theta}$$

Magnetic Force on a Moving Charge

The direction of the magnetic force is given by the right-hand-screw rules, as illustrated in Fig. 29-25.

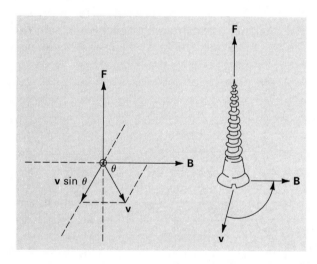

Figure 29-25

- The force **F** on a wire carrying a current I at an angle θ with a flux density B is given by

$$F = BIl \sin \theta \qquad \begin{array}{l}\textit{Magnetic Force} \\ \textit{on a Conductor}\end{array}$$

where l is the length of the conductor.

- Equations for many common magnetic fields are given below:

$$B = \frac{\mu I}{2\pi d} \quad \begin{array}{l}\textit{Long} \\ \textit{Wire}\end{array} \qquad B = \frac{\mu I}{2r} \quad \begin{array}{l}\textit{Center} \\ \textit{of Loop}\end{array}$$

$$B = \frac{\mu N I}{2r} \quad \begin{array}{l}\textit{Center} \\ \textit{of Coil}\end{array} \qquad B = \frac{\mu N I}{L} \quad \textit{Solenoid}$$

Key Terms

1. magnetism
2. magnet
3. domains
4. retentivity
5. permeability
6. weber
7. tesla
8. diamagnetic
9. paramagnetic
10. ferromagnetic
11. magnetic poles
12. law of magnetic force
13. Coulomb's law for magnetic forces
14. magnetic field
15. magnetic flux lines
16. magnetic induction
17. magnetic saturation
18. flux density
19. relative permeability
20. solenoid
21. hysteresis

Review Questions

29-1. How can you positively determine whether a piece of steel is magnetized? How would you determine its polarity if magnetized?

29-2. In general, magnetic materials with high permeability have low retentivity. Why do you think this is true?

29-3. The earth acts like a huge magnet with one pole in the Arctic circle and the other in the Antarctic region. Can you justify the following statement: The geographic North Pole is actually near a magnetic south pole? Explain.

29-4. If an iron bar is placed parallel to a north–south direction and is hammered on one end, the bar becomes a temporary magnet. Explain.

29-5. When a bar magnet is broken into several pieces, each part becomes a magnet with a north and a south pole. Apparently, an isolated pole cannot exist. Explain this, using the domain theory of magnetism.

29-6. Heating magnets or passing electric currents through them will cause a reduction in field strength. Explain.

29-7. The strength of a U magnet will be preserved much longer if an iron plate, called a *keeper*, is placed across the north and south poles. Explain.

29-8. A wire lying along a north–south direction supports an electric current from south to north. What happens to the needle of a compass if the compass is placed (a) above the wire, (b) below the wire, and (c) on the right side of the wire?

29-9. Use the right-hand-thumb rule and the right-hand-screw rule to explain why two adjacent wires experience a force of attraction when the currents are in the same direction. Illustrate your point by drawings.

29-10. Explain with the use of diagrams the repulsion of two adjacent wires carrying oppositely directed currents.

29-11. A circular coil in the plane of the paper supports an electric current. Determine the direction of the magnetic flux near the center of the coil when the current is counterclockwise.

29-12. When an electron beam is projected from left to right into a **B** field directed into the paper, the beam is deflected into a circular path. Do the electrons travel clockwise or counterclockwise? Why is the path circular? What if the beam consisted of protons?

29-13. A proton passes through a region of space without being deflected. Can we say positively that no magnetic field exists in that region? Discuss.

29-14. Many sensitive electrical instruments are shielded from magnetic effects by surrounding the device with ferromagnetic material. Explain.

29-15. If B is expressed in teslas and μ is in tesla-meters per ampere, what is the SI unit of H?

29-16. Hardened steel has a thick hysteresis loop whereas soft iron has a thin loop. Which should be used to produce a permanent magnet? Which should be used if strong temporary magnetization is desired?

Summary and Review

Problems

Section 29-1—Magnetic Fields

29-1. The area of a rectangular loop is 200 cm² and the plane of the loop makes an angle of 41° with a 0.28-T magnetic field. What is the magnetic flux penetrating the loop?
Ans. 3.67 × 10⁻³ Wb

29-2. A coil of wire 30 cm in diameter is perpendicular to a 0.6-T magnetic field. If the coil turns so that it makes an angle of 60° with the field, what is the change in flux?

29-3. A constant horizontal field of 0.5 T pierces a rectangular loop 120 mm long and 70 mm wide. Determine the magnetic flux through the loop when its plane makes the following angles with the **B** field: 0°, 30°, 60°, and 90°.
Ans. 0, 2.10 mWb, 3.64 mWb, 4.20 mWb

29-4. A flux of 13.6 mWb penetrates a coil of wire 240 mm in diameter. Find the magnitude of the magnetic flux density if the plane of the coil is perpendicular to the field.

29-5. A magnetic flux of 50 μWb passes through a perpendicular loop of wire having an area of 0.78 m². What is the magnetic flux density?
Ans. 64.1 μT

29-6. A rectangular loop 25 × 15 cm is oriented so that its plane makes an angle θ with a 0.6-T **B** field. What is the angle θ if the magnetic flux linking the loop is 0.015 Wb?

Section 29-6—The Force on a Moving Charge

29-7. A proton ($q = +1.6 \times 10^{-19}$ C) is injected to the right into a **B** field of 0.4 T directed upward. If the velocity of the proton is 2×10^6 m/s, what are the magnitude and direction of the magnetic force on the proton?
Ans. 1.28 × 10⁻¹³ N, into paper

29-8. An alpha particle ($+2e$) is projected with a velocity of 3.6×10^6 m/s into a 0.12-T magnetic field. What is the magnetic force on the charge at the instant its velocity is directed at an angle of 35° with the magnetic flux?

29-9. An electron moves with a velocity of 5×10^5 m/s at an angle of 60° north of an eastward **B** field. The electron experiences a force of 3.2×10^{-18} N directed into the paper. What are the magnitude of **B** and direction of the velocity?
Ans. 46.2 μT, v is 60° S of E

29-10. A proton ($+1e$) is moving vertically upward with a velocity of 4×10^6 m/s. It passes through a 0.4-T magnetic field directed to the right. What are the magnitude and direction of the magnetic force?

29-11. If an electron replaces the proton in Prob. 29-10, what will be the magnitude and direction of the magnetic force?
Ans. 2.56 × 10⁻¹³ N, out of paper

*29-12. A particle having a charge q and a mass m is projected into a **B** field directed into the paper. If the particle has a velocity v, show that it will be deflected into a circular path of radius:

$$R = \frac{mv}{qB}$$

Draw a diagram of the motion, assuming a positive charge entering the **B** field from left to right. *Hint*: The magnetic force provides the necessary centripetal force for the circular motion.

*29-13. A deuteron is a nuclear particle consisting of a proton and a neutron bound together by nuclear forces. The mass of a deuteron is 3.347×10^{-27} kg, and its charge is $+1e$. A deuteron projected into a magnetic field of flux density 1.2 T is observed to travel in a circular path of radius 300 mm. What is the velocity of the deuteron? See Prob. 29-12.
Ans. 1.72 × 10⁷ m/s

Section 29-7—Force on a Current-Carrying Wire

29-14. A wire 1 mm in length supports a current of 5.00 A and is perpendicular to a **B** field of 0.034 T. What is the magnetic force on the wire?

29-15. A long wire carries a current of 6 A in a direction 35° north of an easterly 0.04-T magnetic field. What are the magnitude and direction of the force on each centimeter of wire? **Ans. 1.72 × 10⁷ m/s**

29-16. A 12-cm segment of wire carries a current of 4.0 A directed at an angle of 41° north of an easterly **B** field. What must be the magnitude of the **B** field if it is to produce a 5-N force on this segment of wire? What is the direction of the force?

29-17. An 80-mm segment of wire is at an angle of 53° south of a westward, 2.3-T **B** field. What are the magnitude and direction of the current in this wire if it experiences a force of 2 N directed out of the paper? **Ans. 13.6 A**

*29-18. The linear density of a certain wire is 50.0 g/m. A segment of this wire carries a current of 30 A in a direction perpendicular to the **B** field. What must be the magnitude of the magnetic field required to suspend the wire by balancing its weight?

Section 29-9—Other Magnetic Fields

29-19. What is the magnetic induction **B** in air at a point 4 cm from a long wire carrying a current of 6 A? **Ans. 30 μT**

29-20. Find the magnetic induction in air 8 mm from a long wire carrying a current of 14.0 A.

29-21. A circular coil having 40 turns of wire in air has a radius of 6 cm and is in the plane of the paper. What current must exist in the coil to produce a flux density of 2 mT at its center? **Ans. 4.77 A**

29-22. If the direction of the current in the coil of Prob. 29-21 is clockwise, what is the direction of the magnetic field at the center of the loop?

29-23. A solenoid of length 30 cm and diameter 4 cm is closely wound with 400 turns of wire around a nonmagnetic material. If the current in the wire is 6 A, determine the magnetic induction along the center of the solenoid. **Ans. 10.1 mT**

29-24. A circular coil having 60 turns has a radius of 75 mm. What current must exist in the coil to produce a flux density of 300 μT at the center of the coil?

*29-25. A circular loop 240 mm in diameter supports a current of 7.8 A. If it is submerged in a medium of relative permeability 2.0, what is the magnetic induction at the center? **Ans. 81.7 μT**

*29-26. A circular loop of radius 50 mm in the plane of the paper carries a counterclockwise current of 15 A. It is submerged in a medium that has a relative permeability of 3.0. What are the magnitude and direction of the magnetic induction at the center of the loop?

Challenge Problems

29-27. A $+3$-μC charge is projected with a velocity of 5×10^5 m/s along the positive x-axis perpendicular to a **B** field. If the charge experiences an upward force of 6.0×10^{-3} N, what must be the magnitude and direction of the **B** field?
Ans. 4.00 mT, into paper

29-28. An unknown charge is projected with a velocity of 4×10^5 m/s from right to left into a 0.4-T **B** field directed out of the paper. The perpendicular force of 5×10^{-3} N causes the particle to move in a clockwise circle. What are the magnitude and sign of the charge?

29-29. A -8-nC charge is projected upward at 4×10^5 m/s into a 0.60-T **B** field directed into the paper. The field produces a force ($F = qvB$) that is also a centripetal force (mv^2/R). This force causes the negative charge to move in a circle of radius 20 cm. What is the mass of the charge, and does it move clockwise or counterclockwise?
Ans. 2.40 × 10⁻¹⁵, clockwise

29-30. What are the magnitude and direction of the **B** field 6 cm above a long segment of wire carrying a 9-A current directed out of the paper? What are the magnitude and direction of the **B** field 6 cm *below* the segment?

29-31. A 24-cm length of wire makes an angle of 32° above a horizontal **B** field of 0.44 T along the positive x-axis. What are the magnitude and direction of the current required to produce a force of 4 mN directed out of the paper?

Ans. 71.5 mA, 212°

*29-32. A velocity selector is a device (Fig. 29-26) that utilizes crossed **E** and **B** fields to select ions of only one velocity v. Positive ions of charge q are projected into the perpendicular fields at varying speeds. Ions with velocities sufficient to make the magnetic force equal and opposite to the electric force pass through the bottom of the slit undeflected. Show that the speed of these ions can be found from

$$v = \frac{E}{B}$$

Source of positive ions
of various velocities

$$v = \frac{E}{B}$$

Figure 29-26 The velocity selector.

29-33. What is the velocity of protons $(+1e)$ injected through a velocity selector (see

Prob. 29-32) if $E = 3 \times 10^5$ V/m and $B = 0.25$ T? **Ans. 1.20×10^6 m/s**

*29-34. A singly charged Li^7 ion $(+1e)$ is accelerated through a potential difference of 500 V and then enters at right angles to a magnetic field of 0.4 T. The radius of the resulting circular path is 2.13 cm. What is the mass of the lithium ion?

*29-35. A singly charged sodium ion $(+1e)$ moves through a **B** field with a velocity of 4×10^4 m/s. What must be the magnitude of the **B** field if the ion is to follow a circular path of radius 200 mm? (The mass of the ion is 3.818×10^{-27} kg.) **Ans. 4.77 mT**

*29-36. The cross sections of two parallel wires are shown in Fig. 29-27 located 8 cm apart in air. The left wire carries a current of 6 A out of the paper, and the right wire carries a current of 4 A into the paper. What is the resultant magnetic induction at the midpoint A due to both wires? (*Hint:* Superimpose the fields due to each wire to find the resultant field.)

*29-37. What is the resultant magnetic field at point B located 2 cm to the right of the 4-A wire? **Ans. 28.0 μT, downward**

*29-38. Two parallel wires (see Fig. 29-28) carrying currents I_1 and I_2 are separated by a distance d. Show that the force per unit length F/l each wire exerts on the other is given by

$$\frac{F}{\ell} = \frac{\mu I_1 I_2}{2\pi d}$$

*29-39. Two wires lying in a horizontal plane carry parallel currents of 15 A each and are 200 mm apart in air. If both currents are directed to the right, what are the magnitude and direction of the flux density at a point midway between the wires? **Ans. 0**

*29-40. What is the force per unit length that each wire in Prob. 29-39 exerts on the other? Is it attraction or repulsion?

*29-41. A solenoid of length 20 cm and 220 turns carries a coil current of 5 A. What should be the relative permeability of the core to

Figure 29-27

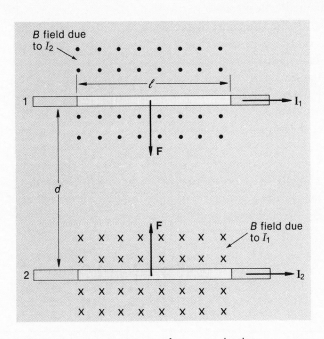

Figure 29-28 Two parallel wires carrying a steady current exert a force on each other.

produce a magnetic induction of 0.2 T at the center of the coil? **Ans. 28.9**

*29-42. A 1-m segment of wire is fixed so that it cannot move, and it carries a current of 6 A directed north. Another 1-m wire segment is located 2 cm above the fixed wire. If the upper wire has a mass of 40 g, what must be the magnitude and direction of the current in the upper wire if its weight is to be balanced by the magnetic force due to the field of the fixed wire?

*29-43. What is the resultant magnetic field at point C in Fig. 29-27?

Ans. 21.8 μT, 96.6°

Summary and Review

Critical Thinking Questions

*29-44. A magnetic field of 0.4 T is directed into the paper. Three particles are injected into the field in an upward direction, each with a velocity of 5×10^5 m/s. Particle 1 is observed to move in a clockwise circle of radius 30 cm, particle 2 continues to travel in a straight line, and particle 3 is observed to move counterclockwise in a circle of radius 40 cm. What are the magnitude and sign of the charge per unit mass (q/m) for each of the particles?

Ans. −4.17 MC/kg, zero, +3.12 MC/kg

*29-45. A 4.0-A current flows through the circular coils of a solenoid in a counterclockwise direction as viewed along the positive x-axis, which is aligned with the air core of the solenoid. What is the direction of the **B** field along the central axis? How many turns per meter of length are required to produce a **B** field of 0.28 T? If the air core is replaced by a material that has a relative permeability of 150, what current would be needed to produce the same 0.28-T field as before?

*29-46. The plane of a current loop 50 cm long and 25 cm wide is parallel to a 0.3 T **B** field directed along the positive x-axis. The 50-cm segments are parallel with the field,

and the 25 cm segments are perpendicular to the field. When viewed from the top, the 6-A current is clockwise around the loop. Draw a sketch to show the directions of the **B** field and the directions of the currents in each wire segment. (a) What are the magnitude and direction of the magnetic force acting on each wire segment? (b) What is the resultant torque on the current loop?

Ans. (a) 0.450 N (up), −0.450 N (down); (b) 0.225 N m., counterclockwise about the +z-axis.

*29-47. Consider the two wires in Fig. 29-29, in which the dot indicates current out of the page and the cross indicates current into the page. What is the resultant flux density at points A, B, and C?

*29-48. Two long, fixed, parallel wires A and B are 10 cm apart in air and carry currents of 6 A and 4 A, respectively, in opposite directions. (a) Determine the net flux density at a point midway between the wires. (b) What is the magnetic force per unit length on a third wire placed midway between A and B and carrying a current of 2 A in the same direction as A?

Ans. (a) 40 μT; (b) 80 μN/m, toward A

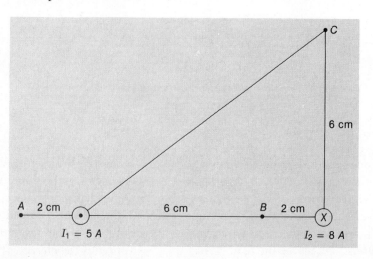

Figure 29-29

Forces and Torques in a Magnetic Field

OBJECTIVES

After completing this chapter, you should be able to:

1. Determine the direction of the magnetic force on a current-carrying conductor in a known B field.

2. Write and apply equations for calculating the magnetic torque on a coil or a solenoid of known area, number of turns, and current when located in a magnetic field of known flux density.

3. Explain with drawings the function of each part of a laboratory *galvanometer*; describe how it may be converted to an ammeter and to a voltmeter.

4. Calculate the *multiplier resistance* necessary to increase the range of a dc voltmeter that contains a galvanometer of fixed sensitivity.

5. Calculate the *shunt resistance* necessary to increase the range of a galvanometer or ammeter of constant sensitivity.

6. Explain the operation of a simple *dc motor,* discussing the function of each of its parts, with particular emphasis on the *split-ring commutator.*

We have seen that a current-carrying conductor placed in a magnetic field will experience a force that is perpendicular both to the current I and to the magnetic induction B. A coil suspended in a magnetic field will experience a *torque* due to equal and opposite magnetic forces on the sides of the coil. Such forces and torques provide the operating principle of many useful devices. In this chapter we discuss the galvanometer, the voltmeter, the ammeter, and the dc motor as applications of electromagnetic forces.

30-1 FORCE AND TORQUE ON A LOOP

A current-carrying conductor suspended in a magnetic field, as illustrated in Fig. 30-1, will experience a magnetic force given by

$$F = BIl \sin \theta = BI_\perp l \qquad \text{30-1}$$

where $I_\perp$ refers to the current perpendicular to the **B** field and l is the length of the conductor. The direction of the force is determined from the right-hand-screw rule.

Now let us examine the forces acting on a rectangular current-carrying loop suspended in a magnetic field, shown in Fig. 30-2. The lengths of the sides are a and b, and a current I passes around the loop, as indicated. (The seat of emf and

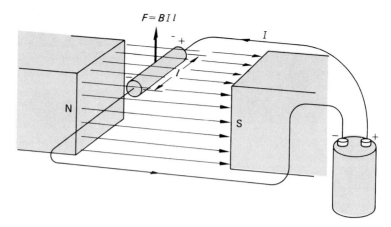

Figure 30-1 The force on a current-carrying conductor is directed perpendicular to the magnetic field.

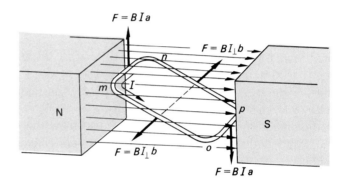

Figure 30-2 Magnetic forces on a current-carrying loop.

method of leading current into the loop are not shown for simplicity.) Sides *mn* and *op* of the loop are each of length *a* and perpendicular to the magnetic induction **B**. Thus, there are exerted on the sides equal and opposite forces of magnitude.

$$F = BIa \qquad\qquad \textbf{30-2}$$

The force is directed upward for the segment *mn* and downward for the segment *op*.

Similar reasoning will show that equal and opposite forces are also exerted on the other two sides. These forces have a magnitude of

$$F = BIb \sin \alpha$$

where α is the angle that the sides *np* and *mo* make with the magnetic field.

Evidently, the loop is in translation equilibrium since the resultant force on the loop is zero. However, the nonconcurrent forces on the sides of length *a* produce a resultant torque that tends to rotate the coil clockwise. As can be seen from Fig. 30-3, each force produces a torque equal to

$$\tau = BIa \, \frac{b}{2} \cos \alpha$$

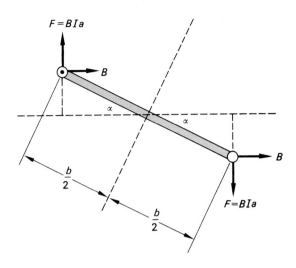

F = BIa

B

α

α

b/2

b/2

F = BIa

B

Figure 30-3 Calculating the torque on a current loop.

Because the total torque is equal to twice this value, the resultant torque can be found from

$$\tau = BI(a \times b) \cos \alpha \qquad \qquad \text{30-3}$$

Since $a \times b$ is the area A of the loop, Eq. (30-3) can be written

$$\tau = BIA \cos \alpha \qquad \qquad \text{30-4}$$

Note that the torque is a maximum when $\alpha = 0°$, that is, when the plane of the loop is parallel with the magnetic field. As the coil turns about its axis, the angle α increases, reducing the rotational effect of the magnetic forces. When the plane of the loop is perpendicular to the field, the angle $\alpha = 90°$ and the resultant torque is zero. The momentum of the coil will cause it to pass this point slightly, but the direction of the magnetic forces ensures that it will oscillate until it reaches equilibrium with the plane of the loop perpendicular to the field.

If the loop is replaced with a closely wound coil having N turns of wire, the general equation for computing the resultant torque is

$$\tau = NIBA \cos \alpha \qquad \qquad \text{30-5}$$

This equation applies to any complete circuit of area A, and its use need not be restricted to rectangular loops. Any plane loop obeys the same relationships.

EXAMPLE 30-1

A rectangular coil consisting of 100 turns of wire has a width of 16 cm and a length of 20 cm. The coil is mounted in a uniform magnetic field of flux density 8 mT, and a current of 20 A is sent through the windings. When the coil makes an angle of 30° with the magnetic field, what is the torque tending to rotate the coil?

Solution

Substituting in Eq. (30-5), we obtain

$$\tau = NBIA \cos \alpha$$
$$= (100 \text{ turns})(8 \times 10^{-3} \text{ T})(20 \text{ A})(0.16 \text{ m} \times 0.20 \text{ m})(\cos 30°)$$
$$= 0.443 \text{ N} \cdot \text{m}$$

30-2 MAGNETIC TORQUE ON A SOLENOID

howthingswork. virginia.edu

For more information on how electric motors and electric power generators work, go to this Web site, and refer to earlier questions in Chap. 12.

The relationship expressed by Eq. (30-5) applies for computing the torque on a solenoid of area *A* having *N* turns of wire. In applying the relation, however, we must remember that the angle α is the angle that each turn of wire makes with the field. It is the *complement* of the angle θ between the solenoid axis and the magnetic field (refer to Fig. 30-4). An alternative equation for computing the torque on a solenoid is therefore

$$\tau = NBIA \sin \theta \qquad \qquad \textit{Solenoid} \quad \textbf{30-6}$$

You should verify that sin θ is equal to cos α by looking at the figure.

The action of the solenoid in Fig. 30-4 can also be explained in terms of magnetic poles. Applying the right-hand-thumb rule to each turn of wire shows that the solenoid will act as an electromagnet, with north and south poles as indicated.

30-3 THE GALVANOMETER

Any device used to detect an electric current is called a **galvanometer**. The operating principle of the majority of such instruments is based on the torque exerted on a coil in a magnetic field. The essential parts are shown in Fig. 30-5. A coil, wrapped around a soft iron core, is pivoted on jeweled bearings between the poles of a permanent magnet. Its rotational motion is restrained by a pair of spiral springs, which also serve as current leads to the coil. Depending on the direction of the current being measured, the coil and pointer will rotate either in a clockwise or counterclockwise direction.

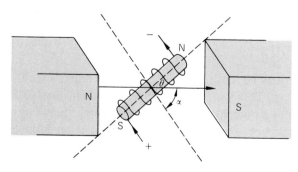

Figure 30-4 Magnetic torque on a solenoid.

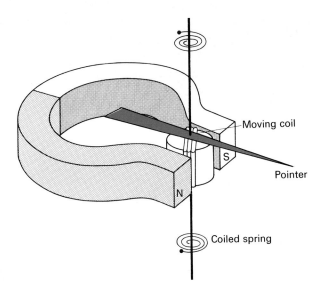

Figure 30-5 The essential components of a galvanometer.

Figure 30-6 The laboratory galvanometer.

For the laboratory galvanometer illustrated in Fig. 30-6 the coil and pointer are shown in the equilibrium position. Note that the permanent magnets are shaped to provide a uniform *radial magnetic field*. This ensures that the point deflection will be directly proportional to the current in the coil. If the current through the coil passes into the page on the right and out of the page on the left, magnetic forces in the radial field will produce a clockwise torque. The pointer and coil will move clockwise until the resisting counterclockwise spring torque equals the ***magnetic torque*** produced by the current. Thus, the position of the pointer on the marked scale is a measure of the current magnitude. Reversing the current direction will cause an equal pointer deflection in a counterclockwise direction.

The sensitivity of a galvanometer of the type indicated in Fig. 30-6 is determined by the spring torque, the friction in the bearings, and the strength of the magnetic field.

EXAMPLE 30-2

The galvanometer in Fig. 30-6 has a sensitivity of 50 μA per scale division. What current is required to give full deflection of the pointer through 25 scale divisions to the right or left of the equilibrium position?

Solution

The current required is equal to the product of the sensitivity and the number of divisions:

$$I = (50\ \mu\text{A/div})(25\ \text{div}) = 1250\ \mu\text{A}$$

Galvanometers of this type may be designed to read currents as low as 1 μA (10^{-6} A). Greater sensitivities require a modified design. One such design consists of suspending a coil in a magnetic field by thin thread. The motion of the coil is then observed by mounting a mirror on the thread, utilizing the principle of an optical lever.

30-4 THE DC VOLTMETER

Many important dc measuring instruments utilize a galvanometer as an indicating element. Two of the most common are the **voltmeter** and the **ammeter.** The first of these will be discussed in this section as an example of how the galvanometer can be used to measure voltages.

The potential difference across a galvanometer is extremely small, even when a large-scale deflection occurs. Thus, if a galvanometer is to be used to measure voltage, it must be converted to a high-resistance instrument. Suppose, for example, one wants to measure the drop in voltage across the battery in Fig. 30-7. This voltage must be measured without appreciably disturbing the current through the circuit. In other words, the voltmeter must draw negligible current. To accomplish this, a large **multiplier resistance** R_m is placed in series with the galvanometer as an integral part of a dc voltmeter.

Note that the galvanometer used in a voltmeter is adjusted so that its equilibrium position is to the extreme left on the scale. This allows for a greater range of measurement but unfortunately requires that the current pass through the coils in one direction only. The **sensitivity** of the galvanometer is determined by the current I_g required for **full-scale deflection** (maximum pointer deflection) as indicated in Fig. 30-7.

Suppose that the galvanometer coil has a resistance R_g and that the meter is designed to yield full-scale deflection for the current I_g. Such a galvanometer, acting alone, could be calibrated to record voltages from zero up to a maximum value given by

$$V_g = I_g R_g \qquad \textbf{30-7}$$

By properly choosing the multiplier resistance R_m, we can calibrate the meter to read any desired voltage.

Suppose, for example, we want full-scale deflection of the voltmeter for the voltage V_B in Fig. 30-7. The multiplier resistance R_m must be chosen so that only the

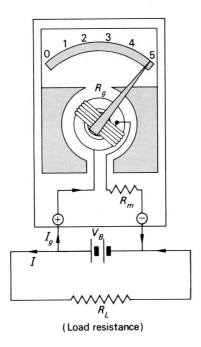

Figure 30-7 The dc voltmeter.

small current I_g passes through the galvanometer. Under these conditions,

$$V_B = I_g R_g + I_g R_m$$

Solving for R_m, we obtain

$$R_m = \frac{V_B}{I_g} - R_g \qquad \textbf{30-8}$$

Thus, we see that the multiplier resistance R_m is equal to the total resistance V_B/I_g less the galvanometer resistance R_g.

CONCEPTUAL EXAMPLE 30-3

A certain galvanometer has an internal resistance of 30 Ω and gives a full-scale deflection for a current of 1 mA. Calculate the multiplier resistance necessary to convert this galvanometer into a voltmeter that has a maximum range of 50 V.

Solution
The multiplier resistance R_m must be such that the total drop in voltage through R_g and R_m is 50 V. From Eq. (30-8),

$$R_m = \frac{50 \text{ V}}{1 \times 10^{-3} \text{ A}} - 30 \text{ } \Omega = 50{,}000 \text{ } \Omega - 30 \text{ } \Omega$$

$$= 49{,}970 \text{ } \Omega$$

Note that the total resistance of the voltmeter $(R_m + R_g)$ is 50 kΩ.

A voltmeter must be connected in *parallel* with the part of the circuit whose potential difference is to be measured. This is necessary so that the large resistance of the voltmeter will not greatly alter the circuit.

30-5 THE DC AMMETER

An *ammeter* is a device that, through calibrated scales, gives an indication of the electric current without appreciably altering it. A *galvanometer* is an ammeter, but its range is limited by the extreme sensitivity of the moving coil. The range of the galvanometer can be extended simply by adding a low resistance, called a *shunt*, in parallel with the galvanometer coil (refer to Fig. 30-8). Placing the shunt in parallel ensures that the ammeter as a whole will have a low resistance, which is necessary if the current is to be essentially unaltered. The major portion of the current will pass through the shunt. Only the small current I_g required for galvanometer deflection will be drawn from the circuit. For example, if 10 A goes through an ammeter, 9.99 A may go through the shunt and only 0.01 A through the coil itself.

Figure 30-8 The dc ammeter.

Suppose the range of a galvanometer is to be extended to measure a maximum current I of the circuit in Fig. 30-8. A shunt resistance R_s must be chosen such that only the current I_g, required for full-scale deflection, passes through the galvanometer coil. The remainder of the current I_s must pass through the shunt. Since R_g and R_s are in parallel, the IR drop across each resistance must be identical:

$$I_s R_s = I_g R_g \qquad \text{30-9}$$

The shunt current I_s is the difference between the circuit current I and the galvanometer current I_g. Thus, Eq. (30-9) becomes

$$(I - I_g)R_s = I_gR_g$$

Solving for the shunt resistance R_s, we obtain the following useful relation:

$$R_s = \frac{I_gR_g}{I - I_g} \qquad \text{30-10}$$

EXAMPLE 30-4

A certain galvanometer has an internal coil resistance of 46 Ω, and a current of 200 mA is required for full-scale deflection. What shunt resistance must be used to convert the galvanometer into an ammeter whose maximum range is 10 A?

Solution

Equation (30-10) gives

$$R_s = \frac{(0.2\text{ A})(46\text{ }\Omega)}{10\text{ A} - 0.2\text{ A}} = \frac{9.2\text{ V}}{9.8\text{ A}} = 0.939\text{ }\Omega$$

It is important to remember that an ammeter must be connected in *series* with the portion of a circuit through which the current is to be measured. The circuit must be opened at some convenient point and the ammeter inserted. If by mistake the ammeter were placed in parallel, the circuit would be shorted across the ammeter because of its extremely low resistance.

30-6 THE DC MOTOR

An **electric motor** is a device that transforms electric energy into mechanical energy. The *dc motor*, like the moving coil of a galvanometer, consists of a current-carrying coil in a magnetic field. However, the motion of the coil in a motor is unrestrained by springs. The design is that the coil will rotate continuously under the influence of magnetic torque.

A simple dc motor, consisting of a single current-carrying loop suspended between two magnetic poles, is illustrated in Fig. 30-9. Normally, the torque exerted on a current-carrying loop would diminish to zero when its plane becomes perpendicular to the field. To provide for continuous rotation of the loop, the current in the loop must be reversed automatically each time it turns through 180°.

The current reversal is accomplished by using a **split-ring commutator,** as shown in Fig. 30-9. The commutator consists of two metal half rings fused to each end of the conducting loop and insulated from each other. As the loop rotates, each brush touches first one half ring and then the other. Thus, the electrical connections are reversed every half revolution at times when the loop is perpendicular to the magnetic field. In this manner, the torque acting on the loop is always in the same direction, and the loop will rotate continuously.

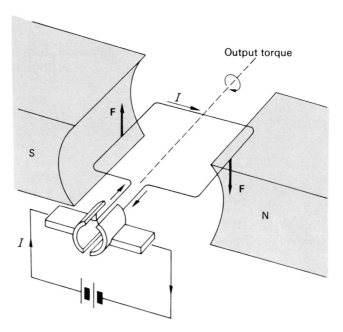

Figure 30-9 The dc motor.

Figure 30-10 Greater, more uniform torque is possible in commercial motors with many armature coils.

Although actual dc motors operate on the principle described for Fig. 30-9, there are a number of designs that increase the available torque and make it more uniform. One such design is shown in Fig. 30-10. A greater magnetic field is established by replacing the permanent magnets with electromagnets. Additionally, the torque can be increased and made more uniform by adding a number of different coils, each having a large number of turns around a slotted iron core called the *armature.* The commutator is an automatic switching arrangement that maintains the currents in the directions shown in the figure, regardless of the orientation of the armature. More will be said about the dc motor in the following chapter.

Summary and Review

Summary

Magnetic torques on current loops form the basis for so many applications that a strong foundation is essential. The operation of generators, motors, ammeters, voltmeters, and many industrial instruments is directly affected by magnetic forces and torques. The major concepts to be remembered are summarized below.

- The magnetic torque on a current-carrying coil of wire having N turns of wire is given by

$$\tau = NBIA \cos \alpha \qquad \textit{Magnetic Torque}$$

where n = number of turns of wire
B = flux density, T
I = current, A
A = area of the coil of wire, m^2
α = angle plane of coil makes with field

- The same equation applies for a solenoid, except that the angle α is generally replaced with θ, the angle the solenoid axis makes with the field.

$$\tau = NBIA \sin \theta \qquad \textit{Torque on a Solenoid}$$

- The multiplier resistance R_m that must be placed in series with a voltmeter to give full-scale deflection for V_B is found from

$$R_m = \frac{V_B - I_g R_g}{I_g} \qquad \textit{Multiplier Resistance}$$

I_g is the galvanometer current, and R_g is its resistance.

- The **shunt resistance** R_s that must be placed in parallel with an ammeter to give full-scale deflection for a current I is

$$R_s = \frac{I_g R_g}{I - I_g} \qquad \textit{Shunt Resistance}$$

Key Terms

1. magnetic torque
2. galvanometer
3. voltmeter
4. ammeter
5. full-scale deflection
6. shunt resistance
7. multiplier resistance
8. sensitivity
9. motor
10. commutator
11. armature

Review Questions

30-1. The equal and opposite forces acting on a current loop in a magnetic field form what is called a *couple*. Show that the resultant torque on such a couple is the product of one of the forces and the perpendicular distance between their lines of action.

30-2. Why is it necessary to provide a *radial* magnetic field for the coil of a galvanometer?

30-3. A coil of wire is suspended by a thread with the plane of the loop coinciding with the plane of the paper. If the coil is placed in a magnetic field directed from left to right, and if a clockwise current is sent through the coil, describe its motion.

30-4. How are the actions of a galvanometer and a motor similar? How are they different?

30-5. How does the core of a galvanometer coil affect the sensitivity of the instrument?

30-6. Explain the torque exerted on a bar magnet suspended in a magnetic field without

Summary and Review

referring to magnetic poles. Discuss, from an atomic standpoint, how the observed torque may arise from the same cause as the torque on a current loop.

30-7. Suppose the range of an ammeter is to be increased N-fold. Show that the shunt resistance that must be placed across the terminals of the ammeter is given by

$$R_s = \frac{R_a}{N-1}$$

where R_a is the ammeter resistance.

30-8. Suppose the range of a given voltmeter is to be increased N-fold. Show that the multiplier resistance that must be placed in series with the voltmeter is given by

$$R_m = (N-1)R_v$$

where R_v is the voltmeter resistance.

30-9. Show by diagrams how an ammeter and a voltmeter should be connected in a circuit. Compare the resistances of the two devices.

30-10. Discuss the error caused by the insertion of an ammeter into an electric circuit. How is this error minimized?

30-11. Discuss the error caused by the insertion of a voltmeter into a circuit. How is the error minimized?

30-12. A voltmeter is connected to a battery, and the reading is taken. An accurate resistance box is then placed in the circuit and adjusted until the voltmeter reading is one-half of its previous value. Show that the voltmeter resistance must be just equal to the added resistance. (This is called the *half-deflection method* for determining voltmeter resistance.)

30-13. Explain what happens when a voltmeter is erroneously placed in series in a circuit. What happens if an ammeter is mistakenly placed in parallel?

30-14. Prepare a short paper on the following topics:
  a. ballistic galvanometer
  b. ohmmeter
  c. dynamometer
  d. voltmeter

30-15. Plot a graph of the torque as a function of time for a single-loop dc motor.

Problems

Section 30-1—Force and Torque on a Loop

30-1. A rectangular loop of wire has an area of 30 cm^2 and is placed with its plane parallel to a 0.56-T magnetic field. What is the magnitude of the resultant torque if the loop carries a current of 15 A?
Ans. 0.0252 N · m

30-2. A coil of wire has 100 turns, each of area 20 cm^2. The coil is free to turn in a 4.0-T field. What current is required to produce a maximum torque of 2.30 N · m?

30-3. A rectangular loop of wire 6 cm wide and 10 cm long is placed with its plane parallel to a magnetic field of 0.08 T. What is the magnitude of the resultant torque on the loop if it carries a current of 14.0 A?
Ans. 6.72 × 10^{-3} N · m

30-4. A rectangular loop of wire has an area of 0.30 m^2. The plane of the loop makes an angle of 30° with a 0.75-T magnetic field. What is the torque on the loop if the current is 7.0 A?

30-5. Calculate the magnetic flux density required to give a coil of 100 turns a torque of 0.5 N · m when its plane is parallel to the field. The dimension of each turn is 84 cm^2, and the current is 9.0 A. **Ans. 66.1 mT**

30-6. What current is required to produce a maximum torque of 0.8 N · m on a solenoid having 800 turns of area 0.4 m^2? The flux density is 3.0 mT. What is the position of the solenoid in the field?

30-7. The axis of a solenoid that has 750 turns of wire makes an angle of 34° with a 5-mT field. What is the current if the torque is 4.0 N · m at that angle? The area of each turn of wire is 0.25 m². **Ans. 7.63 A**

Sections 30-3 to 30-5—Galvanometers, Voltmeters, and Ammeters

30-8. A galvanometer coil 50 mm × 120 mm is mounted in a constant radial 0.2-T magnetic field. If the coil has 600 turns, what current is required to develop a torque of 3.6×10^{-5} N m?

30-9. A galvanometer has a sensitivity of 20 μA per scale division. What current is required to give full-scale deflection with 25 divisions on either side of the central equilibrium position? **Ans. 500 μA**

30-10. A galvanometer has a sensitivity of 15 μA per scale division. How many scale divisions will be covered by the deflecting needle when the current is 60 μA?

30-11. A certain voltmeter draws 0.02 mA for full-scale deflection at 50 V. (a) What is the resistance of the voltmeter? (b) What is the resistance per volt? **Ans. 2.50 MΩ, 50 kΩ/V**

*30-12. For the voltmeter in Prob. 30-11, what multiplier resistance must be used to convert this voltmeter to an instrument that reads 100 mV full scale?

*30-13. The coil of a galvanometer will burn out if a current of more than 40 mA is sent through it. If the coil resistance is 0.5 Ω, what shunt resistance should be added to permit the measurement of 4.00 A? **Ans. 5.05 mΩ**

*30-14. A current of only 90 μA will produce full-scale deflection of a voltmeter that is designed to read 50 mV full scale. (a) What is the resistance of the meter? (b) What multiplier resistance is required to permit the measurement of 100 mV full scale?

*30-15. An ammeter that has a resistance of 0.10 Ω is connected in a circuit and indicates a current of 10 A at full scale. A shunt having a resistance of 0.01 Ω is then connected across the terminals of the water. What new circuit current is needed to produce full-scale deflection of the ammeter? **Ans. 110 A**

Challenge Problems

30-16. A wire of length 12 cm is made into a loop and placed into a 3.0-T magnetic field. What is the largest torque the wire can experience for a current of 6 A?

30-17. A solenoid has 600 turns of area 20 cm² and carries a current of 3.8 A. The axis of the solenoid makes an angle of 30° with an unknown magnetic field. If the resultant torque on the solenoid is 1.25 N m, what is the magnetic flux density? **Ans. 548 mT**

30-18. A flat coil of wire has 150 turns of radius 4.0 cm, and the current is 5.0 A. What angle does the plane of the coil make with a 1.2-T magnetic field if the resultant torque is 0.60 N · m?

30-19. A circular loop consisting of 500 turns carries a current of 10 A in a 0.25-T magnetic field. The area of each turn is 0.2 m². Calculate the torque when the loop makes the following angles with the field: 0°, 30°, 45°, 60°, and 90°. **Ans. 250, 217, 177, 125, and 0 N m.**

30-20. A solenoid of 100 turns has a cross-sectional area of 0.25 m² and carries a current of 10 A. What torque is required to hold the solenoid at an angle of 30° with a 40-mT magnetic field?

30-21. A solenoid consists of 400 turns of wire, each of radius 60 mm. What angle does the axis of the solenoid make with the magnetic flux if the current through the wire is 6 A, the flux density is 46 mT, and the resulting torque is 0.80 N m? **Ans. 39.8°**

*30-22. A galvanometer having an internal resistance of 35 Ω requires 1.0 mA for full-scale deflection for a current of 10 mA. What multiplier resistance is needed to convert this device to a voltmeter that reads a maximum of 30 V?

Summary and Review

*30-23. The internal resistance of a galvanometer is 20 Ω, and it reads full scale for a current of 10 mA. Calculate the multiplier resistance required to convert this galvanometer into a voltmeter with a range of 50 V. What is the total resistance of the resulting meter?   **Ans. 4980 Ω, 5000 Ω**

*30-24. What shunt resistance is needed to convert the galvanometer of Prob. 30-22 to an ammeter reading 10 mA full scale?

*30-25. A certain voltmeter reads 150 V full scale. The galvanometer coil has a resistance of 50 Ω and produces a full-scale deflection on 20 mV. Find the multiplier resistance of the voltmeter.   **Ans. 374,950 Ω**

*30-26. A galvanometer has a coil resistance of 50 Ω and a current sensitivity of 1 mA (full scale). What shunt resistance is needed to convert this galvanometer to an ammeter reading 2.0 A full scale?

*30-27. A laboratory ammeter has a resistance of 0.01 Ω and reads 5.0 A full scale. What shunt resistance is needed to increase the range of the ammeter tenfold?
                                Ans. 0.00111 Ω

*30-28. A commercial 3-V voltmeter requires a current of 0.02 mA to produce full-scale deflection. How can it be converted to an instrument with a range of 150 V?

Critical Thinking Questions

30-29. Consider the rectangular, 4- by 6-cm loop of 60 turns of wire, in Fig. 30-11. The plane of the loop makes an angle of 40° with a 1.6-T magnetic field **B** directed along the x-axis. The loop is free to rotate about the y-axis, and the clockwise current in the coil is 6.0 A. What is the torque, and will it turn toward the x-axis or toward the z-axis?
                                Ans. 1.06 N m, toward z-axis

*30-30. A voltmeter of range 150 V and total resistance 15,000 Ω is connected in series with another voltmeter of range 100 V and total resistance 20,000 Ω. What will each meter read when they are connected across a 120-V battery of negligible internal resistance?

*30-31. The magnetic moment μ is a vector quantity whose magnitude for a coil of N turns of area A is given by NIA, where I is the current in the coil. The direction of the magnetic moment is perpendicular to the plane of the coil in the direction given by the right-hand thumb rule (see Fig. 29-19). If such a coil is placed into a uniform **B** field, show that the torque has a magnitude given by:

$$\tau = \mu B \sin \theta$$

This is sometimes written as a vector (cross) product $\tau = \mu \times B.$

*30-32. Verify the answer obtained for Prob. 30-19 by applying the formula derived in the previous question. Do not confuse the angle

Figure 30-11

θ with the angle α that the plane of the loop makes with the **B** field.

*30-33. The internal wiring of a three-scale voltmeter is shown in Fig. 30-12. The galvanometer has an internal resistance of 40 Ω, and a current of 1.00 mA will produce full-scale deflection. Find the resistances R_1, R_2, and R_3 for the use of the voltmeter for 10, 50, and 100 V.

Ans. 9960 Ω, 49,960 Ω, 99,960 Ω

Figure 30-12

Electromagnetic Induction

We have seen that an electric field can produce a magnetic field. In this chapter, you will learn that the reverse is also true: a magnetic field can give rise to an electric field. An electric current is *generated* by a conductor that moves relative to a magnetic field. A rotating coil in a magnetic field *induces* an alternating emf, which produces an *alternating current* (ac). This process is called **electromagnetic induction** and it is the operating principle behind many electrical devices. For example, electric ac generators and transformers use electromagnetic induction to produce and distribute electric power economically.

31-1 FARADAY'S LAW

Faraday discovered that when magnetic flux lines are cut by a conductor, an emf is produced between the end points of the conductor. For example, an electric current is induced in the conductor of Fig. 31-1a as it moved downward across the flux lines. (The lowercase symbol i will be used for induced currents and for varying currents.) The faster the movement, the more pronounced the galvanometer deflection. When the conductor is moved upward across the flux lines, a similar observation is made, except that the current is reversed (see Fig. 31-1b). If no flux lines are crossed, for example, the conductor is moved parallel to the field, no current is induced.

Electromagnetic Induction

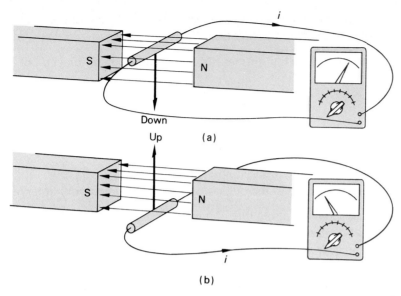

Figure 31-1 When magnetic flux lines are cut by a conductor, an electric current is induced.

Suppose that a number of conductors are moved through a magnetic field, as illustrated by dropping a coil of *N* turns across the flux lines in Fig. 31-2. The magnitude of the induced current is directly proportional to the number of coils and to the rate of motion. Evidently, *an emf is induced by the relative motion between the conductor and the magnetic field.* The same effect is observed when the coil is held stationary and the magnet is moved upward.

Summarizing what we have learned from these experiments, we can state that:

1. Relative motion between a conductor and a magnetic field induces an emf in the conductor.
2. The direction of the induced emf depends on the direction of motion of the conductor with respect to the field.
3. The magnitude of the emf is directly proportional to the rate at which magnetic flux lines are cut by the conductor.

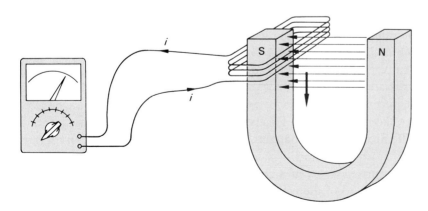

Figure 31-2 Induced emf in a coil is proportional to the number of turns of wire passing through the field.

4. The magnitude of the emf is directly proportional to the number of turns of the conductor crossing the flux lines.

A quantitative relationship for computing the induced emf in a coil of N turns is given by

$$\mathscr{E} = -N\frac{\Delta\phi}{\Delta t}$$ 31-1

where $\mathscr{E}$ = average induced emf
$\Delta\phi$ = change in magnetic flux occurring during time interval Δt

A magnetic flux changing at the rate of one weber per second will induce an emf of one volt for each turn of the conductor. The negative sign in Eq. (31-1) means that the induced emf is in such a direction as to oppose the change that produced it, as will be explained in Sec. 31-3.

Now let us discuss how magnetic flux ϕ linking a conductor may change. In the simple case of a straight wire moving through lines of flux, $\Delta\phi/\Delta t$ represents the rate at which the flux linked by the conductor changes. A continuous circuit is necessary for an induced current to exist, however, and more often we are interested in the emf induced in a loop or coil of wire.

Recall that the magnetic flux ϕ passing through a loop of effective area A is given by

$$\phi = BA$$ 31-2

where B is the magnetic flux density. When B is in *teslas (webers per square meter)* and A is in *square meters*, ϕ is expressed in webers.

A *change in flux ϕ* can occur in two principle ways:

1. By changing the flux density B going through a constant loop area A:

$$\Delta\phi = (\Delta B)A$$ 31-3

2. By changing the effective area A in a magnetic field of constant flux density B:

$$\Delta\phi = B(\Delta A)$$ 31-4

Two examples of changing flux density through a constant, stationary coil area are given in Fig. 31-3. In Fig. 31-3a, the north pole of a magnet is moved through a circular coil. The changing flux density induces a current in the coil, as indicated by the galvanometer. In Fig. 31-3b no current is induced in coil B as long as the current in coil A is constant. By quickly varying the resistance in the left circuit, however, the magnetic flux density reaching coil B can be increased or decreased. While the flux density is changing, a current is induced in the coil on the right.

Note that when the north (N) pole of the magnet is moved into the coil in Fig. 31-3a, the current flows in a clockwise direction as viewed toward the magnet. Therefore, the end of the *coil* near the N pole of the magnet becomes an N pole also (from the right-hand-thumb rule of Chap. 30). The magnet and the coil will experience a force of repulsion, making it necessary to exert a force to bring them together. If the magnet is removed from the coil, a force of attraction will exist

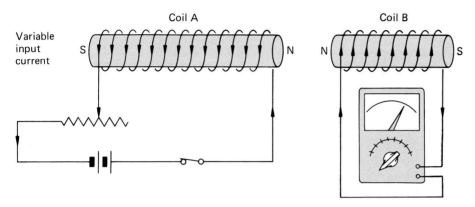

Figure 31-3 (a) Inducing a current by moving a magnet into a coil. (b) A changing current in coil A induces a current in coil B.

that makes it necessary to exert a force to separate them. We will see in Sec. 31-3 that such forces are a natural consequence of the conservation of energy.

EXAMPLE 31-1

A coil of wire having an area of 10^{-3} m² is placed in a region of constant flux density equal to 1.5 T. In a time interval of 0.001 s, the flux density is reduced to 1.0 T. If the coil consists of 50 turns of wire, what is the induced emf?

Solution

The change in flux density is

$$\Delta B = 1.5 \text{ T} - 1.0 \text{ T} = 0.5 \text{ T}$$

From Eq. (31-3), the change in flux is

$$\Delta \phi = (0.5 \text{ T})(10^{-3} \, m^2) = 5 \times 10^{-4} \text{ Wb}$$

Substituting into Eq. (31-1) yields

$$\mathscr{E} = -N\frac{\Delta\phi}{\Delta t}$$

$$= -50 \text{ turns} \frac{5 \times 10^{-4} \text{ Wb}}{0.001 \text{ s}} = -25 \text{ V}$$

The second general way in which the flux linking a conductor may change is by varying the effective area penetrated by the flux. The following example illustrates this point.

EXAMPLE 31-2

A square coil consisting of 80 turns of wire and having an area of 0.05 m² is placed perpendicular to a field of flux density 0.8 T. The coil is flipped until its plane is parallel to the field in a time of 0.2 s. What is the average induced emf?

Solution

The area penetrated by the flux varies from 0.05 m² to zero. Thus, the change in flux is

$$\Delta\phi = B(\Delta A)$$
$$= (0.8 \text{ T})(0.05 \text{ m}^2) = 4 \times 10^{-2} \text{ Wb}$$

The induced emf is

$$\mathcal{E} = -N\frac{\Delta\phi}{\Delta t} = (-80 \text{ turns})\frac{4 \times 10^{-2} \text{ Wb}}{0.2 \text{ s}}$$
$$= -16 \text{ V}$$

31-2 EMF INDUCED BY A MOVING WIRE

Another example of a changing area in a constant **B** field is illustrated in Fig. 31-4. Imagine that a moving conductor of length *l* slides along a stationary U-shaped conductor with a velocity *v*. The magnetic flux penetrating the loop

Figure 31-4   The emf induced by a wire moving perpendicular to a magnetic field.

increases as the area of the loop increases. Consequently, an emf is induced in the moving wire, and a current passes around the loop.

The origin of the emf can be understood by recalling that a moving charge in a magnetic field experiences a force given by

$$F = qvB$$

For example, in Fig. 31-4, free charges on the conductor are moved to the right through a magnetic field directed into the paper. The magnetic force **F** acting on the charges moves them through the length l of wire in a direction given by the right-hand-screw rule (away from the reader for conventional current). The work per unit of charge represents the induced emf, which is given by

$$\mathcal{E} = \frac{\text{work}}{q} = \frac{Fl}{q} = \frac{qvBl}{q}$$
$$= Blv \qquad\qquad\qquad \textbf{31-5}$$

If the velocity v of the moving wire is directed at an angle θ with the **B** field, a more general form is needed for Eq. (31-5):

$$\mathcal{E} = Blv \sin \theta \qquad\qquad\qquad \textbf{31-6}$$

EXAMPLE 31-3

A 0.2-m length of wire moves at a constant velocity of 4 m/s in a direction that is 40° with respect to a magnetic flux density of 0.5 T. Calculate the induced emf.

Solution

Direct substitution into Eq. (31-6) yields

$$\mathcal{E} = (0.5 \text{ T})(0.2 \text{ m})(4 \text{ m/s})(\sin 40°)$$
$$= 0.257 \text{ V}$$

The minus sign does not appear in Eq. (31-6) because the direction of the induced emf is the same as the direction of the magnetic force performing work on the moving charge.

31-3 LENZ'S LAW

Throughout the discussions of all physical phenomena, one guiding principle stands out above all the rest: the *principle of conservation of energy*. An emf cannot exist without a cause. Whenever an induced current produces heat or performs mechanical work, the necessary energy must come from the work done in inducing the current.

Recall the example discussed in Fig. 31-3a. The north pole of a magnet pushed into a coil induces a current that itself gives rise to another magnetic field. The second field produces a force that opposes the original force. Withdrawing the magnet creates a force that opposes the removal of the magnet. This is an illustration of **Lenz's law:**

> **Lenz's law** **An induced current will flow in such a direction that it will oppose by its magnetic field the motion of the magnetic field that is producing it.**

The more work that is done in moving the magnet into the coil, the greater will be the induced current and, hence, the greater the resisting force. We might have expected this result from the law of conservation of energy. To produce a larger current, we must perform a greater amount of work.

The direction of the current induced in a straight conductor moving through a magnetic field can be determined from Lenz's law. There is an easier method, however, as illustrated in Fig. 31-5. It is known as *Fleming's rule*, or the *right-hand rule:*

> **Fleming's rule** **If the thumb, forefinger, and middle finger of the right hand are held at right angles to each other, with the thumb pointing in the direction in which the wire is moving and the forefinger pointing in the field direction (N to S), the middle finger will point in the direction of induced conventional current.**

Fleming's rule is easy to apply and useful for studying the currents induced by a simple generator. Students sometimes remember the rule by memorizing *motion–flux–current.* These are the directions given by the thumb, forefinger, and middle finger, respectively.

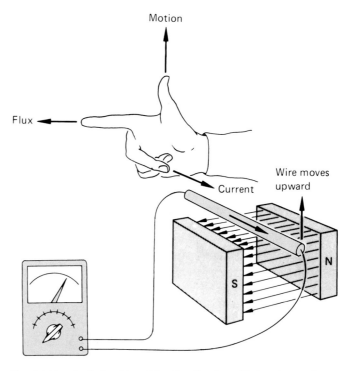

Figure 31-5 The right-hand rule for determining the direction of induced current.

31-4 THE AC GENERATOR

An electric generator converts mechanical energy into electric energy. We have seen that an emf is induced in a conductor when it experiences a change in flux linkage. When the conductor forms a complete circuit, an induced current can be detected. In a generator, a coil of wire is rotated in a magnetic field, and the induced current is transmitted by wires for long distances from its origin.

The construction of a simple generator is shown in Fig. 31-6. Essentially, there are three components: a **field magnet,** an **armature,** and **slip rings** with *brushes.* The field magnet may be a permanent magnet or an electromagnet. The armature for the generator in Fig. 31-6 consists of a single loop of wire suspended between the poles of the field magnet. A pair of slip rings is fused to each end of the loop; they rotate with the loop as it is turned in the magnetic field. Induced current is led away from the system by graphite brushes that ride on each slip ring. Mechanical energy is supplied to the generator by turning the armature in the magnetic field. Electric energy is generated in the form of an induced current.

The direction of the induced current must obey Fleming's rule of *motion–flux–current.* In Fig. 31-6, the downward motion of the left wire segment crosses a magnetic flux directed left to right. The induced current is, therefore, toward the slip rings. Similar reasoning shows that the current in the right segment of the loop, which is moving upward, will be away from the slip rings.

To understand the operation of an *ac generator,* let us follow the loop through a complete rotation, observing the current generated throughout the rotation. Figure 31-7 shows four positions of the rotating coil and the direction of the current delivered to the brushes in each case. Suppose that the loop is turned mechanically in a counterclockwise direction. In Fig. 31-7a the loop is horizontal, with side *M* facing the south (S) pole of the magnet. At this point, a maximum current is

Figure 31-6 The ac generator.

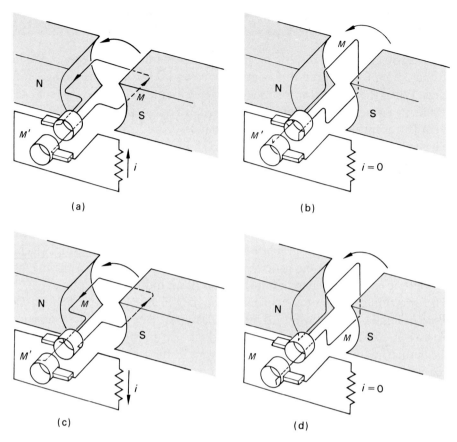

Figure 31-7 Production of an alternating current.

delivered in the direction shown. In Fig. 31-7b the loop is vertical, with side M facing upward. At this point, no flux lines are being cut, and the induced current drops to zero. When the loop becomes horizontal again, as in Fig. 31-7c, side M is now facing the north (N) pole of the magnet. Therefore, the current delivered to the slip ring M' has changed direction. An induced current flows through the external resistor in a direction opposite to that experienced earlier. In Fig. 31-7d the loop is vertical again, but now side M faces downward. No flux lines are cut, and the induced current again drops to zero. The loop next returns to horizontal as in Fig. 31-7a, and the cycle repeats itself. Thus, the current delivered by such a generator alternates periodically, the direction changing twice each rotation.

The emf generated in each segment of a rotating loop must obey the relation as given in Eq. 31-6:

$$\mathcal{E} = Blv \sin \theta$$

where v is the velocity of a moving wire segment of length l in a magnetic field of flux density **B.** The direction of the velocity v with respect to the **B** field at any instant is denoted by the angle θ. Let us consider the segment M of our rotating current loop when it reaches the position shown in Fig. 31-8. The *instantaneous* emf at that position is given by Eq. (31-6). If the loop rotates in a circle of radius r, the instantaneous velocity v can be found from

$$v = \omega r$$

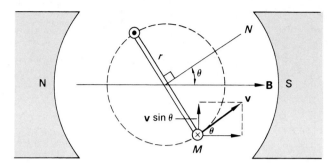

Figure 31-8 Calculating the induced emf.

where ω is the angular velocity in radians per second. Substituting into Eq. (31-6) gives the instantaneous emf

$$\mathcal{E} = Bl\omega r \sin \theta \qquad \text{31-7}$$

An identical emf is induced in the segment of wire opposite M, and no *net* emf is generated in the other segments. Hence, the total instantaneous emf is twice the value given by Eq. (31-7), or

$$\mathcal{E}_{\text{inst}} = 2Bl\omega r \sin \theta \qquad \text{31-8}$$

But the area A of the loop is

$$A = l \times 2r$$

and Eq. (31-8) can be further simplified to

$$\mathcal{E}_{\text{inst}} = NBA\omega \sin \theta \qquad \text{31-9}$$

where N is the number of turns of wire.

Equation (31-9) expresses an important principle relating to the study of alternating currents:

> **If the armature is rotating with a constant angular velocity in a constant magnetic field, the magnitude of the induced emf varies sinusoidally with respect to time.**

This fact is illustrated by Fig. 31-9. The emf varies from a maximum value when $\theta = 90°$ to a zero value when $\theta = 0°$. The maximum instantaneous emf is therefore

$$\mathcal{E}_{\text{max}} = NBA\omega \qquad \text{31-10}$$

since $\sin 90° = 1$. Stating Eq. (31-9) in terms of the maximum emf, we write

$$\mathcal{E}_{\text{inst}} = \mathcal{E}_{\text{max}} \sin \theta \qquad \text{31-11}$$

To see the explicit variation of generated emf with time, we should recall that

$$\theta = \omega t = 2\pi ft$$

where f is the number of rotations per second made by the loop. Thus, we can express Eq. (31-11) in the following form:

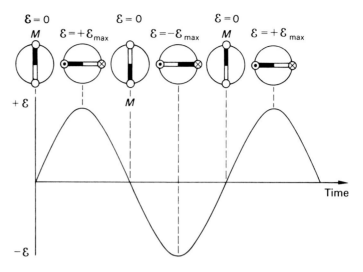

Figure 31-9 Sinusoidal variation of induced emf with time.

$$\mathscr{E}_{\text{inst}} = \mathscr{E}_{\text{max}} \sin 2\pi ft \qquad \text{31-12}$$

EXAMPLE 31-4

The armature of a simple ac generator consists of 100 turns of wire, each having an area of 0.2 m². The armature is turned with a frequency of 60 rev/s in a constant magnetic field of flux density 10^{-3} T. What is the maximum emf generated?

Solution
We must first calculate the angular velocity of the coil.

$$\omega = (2\pi \text{ rad/rev})(60 \text{ rev/s}) = 377 \text{ rad/s}$$

Substituting this value and other known parameters into Eq. (31-10), we obtain

$$\begin{aligned}
\mathscr{E}_{\text{max}} &= NBA\omega \\
&= (100 \text{ turns})(10^{-3} \text{ T})(0.2 \text{ m}^2)(377 \text{ rad/s}) \\
&= 7.54 \text{ V}
\end{aligned}$$

Since the induced current is proportional to the induced emf, from Ohm's law, the induced current will also vary sinusoidally according to

$$i_{\text{inst}} = i_{\text{max}} \sin 2\pi ft \qquad \text{31-13}$$

The maximum current occurs when the induced emf is a maximum. The sinusoidal variation is similar to that plotted in Fig. 31-9.

The SI unit for frequency is the **hertz** (Hz), which is defined as a cycle per second.

$$1 \text{ Hz} = 1 \text{ cycle/s}$$

Thus, a 60-cycle-per-second alternating current has a frequency of 60 Hz.

31-5 THE DC GENERATOR

A simple ac generator can be converted easily to a dc generator by substituting a split-ring commutator for the slip rings, as illustrated in Fig. 31-10. The operation is just the reverse of that discussed earlier for a dc motor (Chap. 30). In the motor, an electric current gives rise to an external torque. In the **dc generator,** an external torque generates an electric current. The **commutator** reverses the connections to the brushes twice per revolution. As a result, the current pulsates but never reverses direction. The emf of such a generator varies with time, as shown in Fig. 31-11. Note that the emf is always in the positive direction, but it rises to a maximum and falls to zero twice per complete rotation. Practical dc generators are designed with many coils in several planes so that the generated emf is large and nearly constant.

31-6 BACK EMF IN A MOTOR

In an electric motor, a magnetic torque turns a current-carrying loop in a constant magnetic field. We have just seen that a coil rotating in a magnetic field will induce an emf that opposes the cause that gave rise to it. This is true even if a current already exists in the loop. Thus, *every motor is also a generator.* According to Lenz's law, such an induced emf must oppose the current delivered to the motor. For this reason, the emf induced in a motor is called **back emf** or *counter emf.*

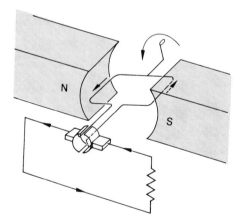

Figure 31-10   The dc generator.

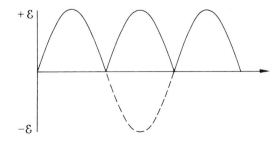

Figure 31-11   Pulsating emf produced by a dc generator.

The effect of a back emf is to reduce the net voltage delivered to the armature coils of the motor. Consider the circuit illustrated in Fig. 31-12. The net voltage delivered to the armature coils is equal to the applied voltage V less the induced voltage $\mathcal{E}_b$.

Applied voltage − induced voltage = net voltage

According to Ohm's law, the net voltage across the armature coils is equal to the product of the coil resistance R and the current I. Symbolically, we write

$$V - \mathcal{E}_b = IR \qquad\qquad \textbf{31-14}$$

Equation (31-14) tells us that the current through a circuit containing a motor is determined by the magnitude of the back emf. The magnitude of this induced emf, of course, depends on the speed of rotation of the armature. We can show this experimentally by connecting a motor, an ammeter, and a battery in series, as shown in Fig. 31-13. When the armature is rotating, a low current is indicated. The back emf reduces the effective voltage. If the motor is stalled by holding the armature stationary, the back emf will drop to zero. The increased net voltage results in a larger circuit and can cause the motor to overheat and even burn out.

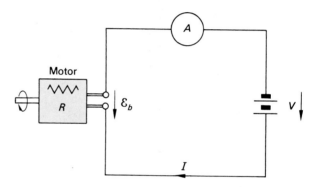

Figure 31-12   Back emf in a dc motor.

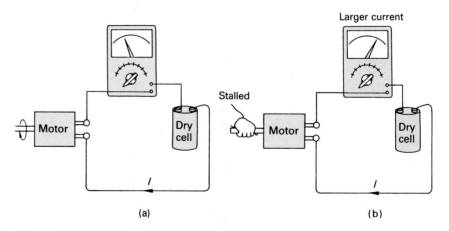

(a)   (b)

Figure 31-13   Demonstrating the existence of back emf in a dc motor.

31-7 TYPES OF MOTORS

DC motors are classified according to how the field coils and the armature are connected. When the armature coils and the field coils are connected in series, as shown in Fig. 31-14, the motor is said to be *series-wound*. In this type of motor, the current energizes both the field windings and the armature windings. When the armature turns slowly, the back emf is small and the current is large. Consequently, a large torque is developed at low speeds.

In a *shunt-wound* motor, the field windings and the armature windings are connected in parallel, as illustrated by Fig. 31-15. The entire voltage is applied across both windings. The primary advantage of a shunt-wound motor is that it produces more constant torque over a range of speeds. The starting torque is usually lower, however, than a similar series-wound motor.

In some applications, the field windings are in two parts, one connected in series with the armature and the other in parallel with it. Such a motor is called a *compound motor.* The torque produced by a compound motor varies between that of the series and shunt motors.

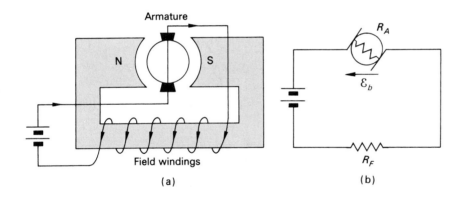

Figure 31-14   (a) The series-wound dc motor. (b) Schematic diagram.

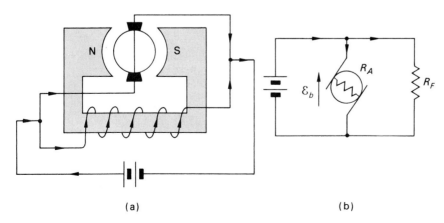

Figure 31-15   (a) The shunt-wound dc motor. (b) Schematic diagram.

In *permanent-magnet motors* no field current is necessary. These motors have torque characteristics similar to those of shunt-wound motors.

CONCEPTUAL EXAMPLE 31-5

A 120-V dc shunt motor has an armature resistance of 3 Ω and a field resistance of 260 Ω. When the motor is operating at full speed, the total current is 3 A. (a) What is the back emf of the motor at full speed? (b) Find the current in the motor at the moment its switch is turned on.

Solution (a)

To determine the back emf, we must see how the current is divided between the armature and field windings. The current I_F in the field windings can be found by writing the voltage equation for the outside loop in Fig. 31-15b:

$$V = I_F R_F$$

from which

$$I_F = \frac{V}{R_F} = \frac{120 \text{ V}}{260 \text{ } \Omega} = 0.46 \text{ A}$$

The current in the armature windings is therefore

$$I_A = 3 \text{ A} - 0.46 \text{ A} = 2.54 \text{ A}$$

Now the voltage equation for the loop containing the battery and the armature is

$$V - \mathscr{E}_b = I_A R_A$$

from which

$$
\begin{aligned}
\mathscr{E}_b &= V - I_A R_A \\
&= 120 \text{ V} - (2.54 \text{ A})(3 \text{ } \Omega) \\
&= 120 \text{ V} - 7.62 \text{ V} = 112.4 \text{ V}
\end{aligned}
$$

Solution (b)

At the instant the switch is turned on, the armature is not yet turning, and consequently $\mathscr{E}_b = 0$. In this case, the armature current is

$$I_A = \frac{V}{R_A} = \frac{120 \text{ V}}{3 \text{ } \Omega} = 40 \text{ A}$$

The field current is still 0.46 A, and the total starting current is

$$
\begin{aligned}
I &= I_A + I_F \\
&= 40 \text{ A} + 0.46 \text{ A} = 40.5 \text{ A}
\end{aligned}
$$

31-8 THE TRANSFORMER

It was noted earlier that a changing current in one wire loop will induce a current in a nearby loop. The induced current arises from the changing magnetic field associated with a changing current. Alternating current has a distinct advantage over direct current because of the inductive effect of a current that varies constantly in magnitude and direction. The most common application of this principle is offered by the *transformer*, a device that increases or decreases the voltage in an ac circuit.

A simple transformer is illustrated in Fig. 31-16. There are three essential parts: (1) a primary coil connected to an ac source, (2) a secondary coil, and (3) a soft iron core. As an alternating current is sent through the primary coil, magnetic flux lines move back and forth through the iron core, inducing an alternating current in the secondary coil.

The constantly changing magnetic flux is established throughout the core of the transformer and passes through both primary and secondary coils. The emf $\mathcal{E}_p$ induced in the primary coil is given by

$$\mathcal{E}_p = -N_p \frac{\Delta\phi}{\Delta t} \qquad\qquad \text{31-15}$$

where N_p = number of primary turns
$\Delta\phi/\Delta t$ = rate at which flux changes

Similarly, the emf $\mathcal{E}_s$ induced in the secondary coil is

$$\mathcal{E}_s = -N_s \frac{\Delta\phi}{\Delta t} \qquad\qquad \text{31-16}$$

where N_s is the number of secondary turns. Since the same flux changes at the same rate through each coil, we can divide Eq. (31-15) by Eq. (31-16) to obtain

$$\frac{\mathcal{E}_p}{\mathcal{E}_s} = \frac{N_p}{N_s} \qquad\qquad \text{31-17}$$

$$\frac{\text{Primary voltage}}{\text{Secondary voltage}} = \frac{\text{primary turns}}{\text{secondary turns}}$$

The induced voltage is in direct proportion to the number of turns. If the ratio of secondary turns N_s to primary turns N_p is varied, an input (primary) voltage can provide any desired output (secondary) voltage. For example, if there are 40

Figure 31-16   Transformer.

times as many turns in the secondary coil, an input voltage of 120 V will be increased to $40 \times 120 = 4800$ V in the secondary coil. A transformer that produces a larger output voltage is called a **step-up transformer.**

A **step-down transformer** can be constructed by making the number of primary turns greater than the number of secondary turns. Using a step-down transformer gives a lower output voltage.

The **efficiency of a transformer** is defined as the ratio of the power output to the power input. Recalling that electric power is equal to the product of voltage and current, we can write the efficiency E of a transformer as

$$E = \frac{\text{power output}}{\text{power input}} = \frac{\mathscr{E}_s i_s}{\mathscr{E}_p i_p} \qquad \text{31-18}$$

where i_p and i_s are the currents in the primary coil and the secondary coil, respectively. Most electric transformers are carefully designed for extremely high efficiencies, normally above 90 percent.

It is important to recognize that there is no power gain as a result of transformer action. When the voltage is stepped up, the current must be stepped down, so that the product $\mathscr{E}i$ does not increase. To see this more clearly, let us assume that a given transformer is 100 percent efficient. For this perfect transformer, Eq. (31-18) becomes

$$\mathscr{E}_s i_s = \mathscr{E}_p i_p$$

or

$$\frac{i_p}{i_s} = \frac{\mathscr{E}_s}{\mathscr{E}_p} \qquad \text{31-19}$$

This equation shows clearly the inverse relationship between current and induced voltage.

EXAMPLE 31-6

An ac generator that delivers 20 A at 6000 V is connected to a step-up transformer. What is the output current at 120,000 V if the transformer efficiency is 100 percent?

Solution
From Eq. (31-19),

$$i_s = \frac{\mathscr{E}_p i_p}{\mathscr{E}_s}$$

$$= \frac{(6000 \text{ V})(20 \text{ A})}{120,000 \text{ V}}$$

$$= 1 \text{ A}$$

Note in the preceding example that the current was reduced to 1 A, whereas the voltage was increased twentyfold. Since heat losses in transmission lines vary directly with the square of the current ($i^2 R$), this means that electric power can be transmitted for large distances without significant loss. When the electric power reaches its destination, step-down transformers are used to provide the desired current at lower voltages.

Summary and Review

Summary

Electromagnetic induction allows for the production of an electric current in a conducting wire. This is the basic operating principle behind many electrical devices. An understanding of the concepts summarized below is necessary for most applications involving the use of alternating current.

- A magnetic flux changing at the rate of 1 Wb/s will induce an emf of 1 V for each turn of a conductor. Symbolically,

$$\mathscr{E} = -N\frac{\Delta\phi}{\Delta t} \qquad \textit{Induced emf}$$

- Two principal ways in which the flux changes are

$$\Delta\phi = \Delta BA \qquad \Delta\phi = B\,\Delta A$$

- The induced emf due to a wire of length l moving with a velocity v at an angle θ with a field **B** is given by

$$\mathscr{E} = Blv \sin\theta \qquad \textit{Emf Due to Moving Wire}$$

- According to *Lenz's law*, the induced current must be in such a direction that it produces a magnetic force that opposes the force causing the motion.
- *Fleming's rule*: If the thumb, forefinger, and middle finger of the right hand are held at right angles to each other, with the thumb pointing in the direction in which the wire is moving and the forefinger pointing in the field direction (N to S), the middle finger will point in the direction of induced conventional current (*motion–flux–current*).

- The instantaneous emf generated by a coil of N turns moving with an angular velocity ω or frequency f is

$$\mathscr{E}_{inst} = NBA\omega \sin\omega t \qquad \mathscr{E}_{inst} = 2\pi f NBA \sin 2\pi ft$$

- The maximum emf occurs when the sine is zero. Thus,

$$\mathscr{E}_{max} = NBA\omega \qquad \mathscr{E}_{inst} = \mathscr{E}_{max} \sin 2\pi ft$$

- Since the induced current is proportional to $\mathscr{E}$, we also have

$$i_{inst} = i_{max} \sin 2\pi ft \qquad \textit{Instantaneous Current}$$

- The back emf in a motor is the induced voltage that causes a reduction in the net voltage delivered to a circuit.

Applied voltage − induced back emf = net voltage

$$V - \mathscr{E}_b = IR \qquad \mathscr{E}_b = V - IR$$

- For a transformer having N_p primary and N_s secondary turns,

$$\frac{\textit{Primary voltage}}{\textit{Secondary voltage}} = \frac{\textit{primary turns}}{\textit{secondary turns}} \qquad \frac{\mathscr{E}_p}{\mathscr{E}_s} = \frac{N_p}{N_s}$$

- The efficiency of a transformer is

$$E = \frac{\text{power output}}{\text{power input}} = \frac{\mathscr{E}_p i_p}{\mathscr{E}_s i_s} \qquad \textit{Transformer Efficiency}$$

Key Terms

1. induced emf
2. Lenz's law
3. ac generator
4. dc generator
5. field magnet
6. armature
7. slip rings
8. commutator
9. electromagnetic induction

(Continued)

10. back emf
11. shunt motor
12. series motor

13. compound motor
14. step-up transformer
15. step-down transformer

16. transformer efficiency

Review Questions

31-1. Discuss the various factors that influence the magnitude of an induced emf in a length of wire moving in a magnetic field.

31-2. A bar magnet is held in a vertical position with the north pole facing upward. If a closed-loop coil is dropped over the north end of the magnet, what is the direction of the induced current viewed from the top of the magnet?

31-3. A circular loop is suspended with its plane perpendicular to a magnetic field directed from left to right. The loop is removed from the field by moving it upward quickly. What is the direction of the induced current viewed along the field direction? Is a force required to remove the loop from the field?

31-4. An induction coil is essentially a transformer that operates on direct current. As shown in Fig. 31-17, the induction coil consists of a few primary turns wound around an iron core with a large number of secondary turns surrounding the primary. A battery current magnetizes the core so that it attracts the armature of the vibrator and opens the circuit periodically.

When the circuit is opened, the field collapses, and a large emf is induced in the secondary coil, producing a spark at the output terminals. What is the function of the capacitor C connected in parallel with the vibrator? Explain how an induction coil is used in the ignition system of an automobile.

31-5. Explain clearly how an ac generator can be converted to a dc generator. How would you proceed to convert an ac generator into an ac motor?

31-6. When the electric motor in a plant is starting, a worker notices that the lights are momentarily dimmed. Explain.

31-7. What type of dc motor should be purchased to operate a winch used to lift heavy objects? Why?

31-8. What type of motor should be used to operate an electric fan where uniform torque is desired at high speeds?

31-9. Explain how the existence of back emf in a motor helps keep its speed constant. *Hint:* What happens to $\mathscr{E}_b$ and I when the armature speed increases or decreases?

Figure 31-17 Induction coil.

31-10. There are three primary ways in which power is lost through the operation of a transformer: (a) wire resistance losses, (b) hysteresis losses, and (c) eddy-current losses. (*Eddy currents* are induced current loops that occur in the mass of a magnetic material resulting from a changing flux.) Explain how energy is wasted by these three processes.

31-11. Explain with the use of diagrams how transformers make it possible to transmit current economically from power installations to homes many miles away.

31-12. An ac generator produces 60 Hz alternating voltage. How many degrees of armature rotation will correspond to one-fourth of a cycle?

31-13. Why is it more economical for power companies to provide alternating current than direct current?

31-14. Prepare a brief report on the following topics and explain the part played by electromagnetic induction.
   a. the betatron
   b. induction coil
   c. the telephone
   d. eddy currents
   e. magnetohydrodynamic generator
   f. electric power transmission
   g. the universal motor
   h. induction motor
   i. synchronous motor

Problems

Section 31-2—Induced EMF

31-1. A coil of wire 8 cm in diameter has 50 turns and is placed in a B field of 1.8 T. If the B field is reduced to 0.6 T in 0.002 s, what is the induced emf? **Ans. −151 V**

31-2. A square coil of wire having 100 turns of area 0.044 m^2 is placed with its plane perpendicular to a constant B field of 4 mT. The coil is flipped to a position parallel with the field in a time of 0.3 s. What is the induced emf?

31-3. A coil of 300 turns moving perpendicular to the flux in a uniform magnetic field experiences a flux linkage of 0.23 mWb in 0.002 s. What is the induced emf?
Ans. −34.5 V

31-4. The magnetic flux linking a loop of wire changes from 5 to 2 mWb in 0.1 s. What is the average induced emf?

31-5. A coil of 120 turns is 90 mm in diameter and has its plane perpendicular to a 60-mT magnetic field produced by a nearby electromagnet. The current in the electromagnet is cut off, and as the field collapses, an emf of 6 V is induced in the coil. How long does it take for the field to disappear?
Ans. 7.63 ms

31-6. A coil of 56 turns has an area of 0.3 m^2. Its plane is perpendicular to a 7-mT magnetic field. If this field collapses to zero in 6 ms, what is the induced emf?

31-7. A wire 0.15 m long moves at a constant velocity of 4 m/s in a direction that is 36° with respect to a 0.4-T magnetic field. The axis of the wire is perpendicular to the magnetic flux lines. What is the induced emf? **Ans. 0.141 V**

31-8. A 0.2-m wire moves at an angle of 28° with an 8-mT magnetic field. The wire length is perpendicular to the flux. What velocity v is required to induce an emf of 60 mV?

Sections 31-4 and 31-5—Generators

31-9. The magnetic field in the air gap between the magnetic poles and the armature of an electric generator has a flux density of 0.7 T. The length of the wires on the armature is 0.5 m. How fast must these wires move to generate a maximum emf of 1.00 V in each armature wire? **Ans. 2.86 m/s**

31-10. A single loop of wire has a diameter of 60 mm and makes 200 rpm in a constant 4-mT magnetic field. What is the maximum emf generated?

31-11. The armature of a simple generator has 300 loops of diameter 20 cm in a constant 6-mT magnetic field. What must be the frequency of rotation in revolutions per second in order to induce a maximum emf of 7.00 V? **Ans. 19.7 rev/s**

31-12. An armature in an ac generator consists of 500 turns, each of area 60 cm². The armature is rotated at a frequency of 3600 rpm in a uniform 2-mT magnetic field. What is the frequency of the alternating emf? What is the maximum emf generated?

31-13. In Prob. 31-12, what is the instantaneous emf at the time when the plane of the coil makes an angle of 60° with the magnetic flux? **Ans. 1.13 V**

31-14. The armature of a simple ac generator has 100 turns of wire, each having a radius of 5.00 cm. The armature turns in a constant 0.06-T magnetic field. What must be the rotational frequency in rpm to generate a maximum voltage of 2.00 V?

31-15. A circular coil has 70 turns, each 50 mm in diameter. Assume that the coil rotates about an axis that is perpendicular to a magnetic field of 0.8 T. How many revolutions per second must the coil make to generate a maximum emf of 110 V? **Ans. 159 rev/s**

*31-16. The armature of an ac generator has 800 turns, each of area 0.25 m². The coil rotates at a constant 600 rpm in a 3-mT field. What is the maximum induced emf? What is the instantaneous emf 0.43 s after the coil passes a position of zero emf?

*31-17. A 300-Ω resistor is connected in series with an ac generator of negligible internal resistance. The armature of the generator has 200 turns of wire 30 cm in diameter, and it turns at 300 rpm in a constant 5-mT field. What is the instantaneous current through the resistor 0.377 s after the coil passes a position of zero emf? **Ans. 4.90 mA**

Section 31-6—Back EMF in a Motor

31-18. A 120-V dc motor draws a current of 3.00 A in operation and has a resistance of 8.00 Ω. What is the back emf when the motor is operating, and what is the starting current?

31-19. The armature coil of the starting motor in an automobile has a resistance of 0.05 Ω. The motor is driven by a 12-V battery, and the back emf at operating speed is 6.00 V. What is the starting current? What is the current at full speed? **Ans. 240 A, 120 A**

31-20. A 220-V dc motor draws a current of 10 A in operation and has an armature resistance of 0.4 Ω. What is the back emf when the motor is operating, and what is the starting current?

*31-21. A 120-V series-wound dc motor has a field resistance of 90 Ω and an armature resistance of 10 Ω. When operating at full speed, a back emf of 80 V is generated. What is the total resistance of the motor? What is the starting current? What is the operating current? **Ans. 100 Ω, 1.20 A, 0.400 A**

*31-22. The efficiency of the motor in Prob. 31-21 is the ratio of the power output to the power input. Determine the efficiency based on the known data.

Section 31-8—The Transformer

31-23. A step-up transformer has 400 secondary turns and only 100 primary turns. A 120-V alternating voltage is connected to the primary coil. What is the output voltage? **Ans. 480 V**

31-24. A step-down transformer is used to drop an alternating voltage from 10,000 to 500 V. What must be the ratio of secondary turns to primary turns? If the input current is 1.00 A and the transformer is 100 percent efficient, what is the output current?

31-25. A step-up transformer is 95 percent efficient and has 80 primary turns and 720 secondary turns. If the primary draws a current of 20 A at 120 V, what are the current and voltage for the secondary?

Ans. 2.11 A, 1080 V

31-26. A 25-W light bulb has a resistance of 8.0 Ω while burning. The light is powered from the secondary of a small transformer connected to a 120-V circuit. What must be the ratio of secondary turns to primary turns in this application? Assume 100 percent efficiency.

Challenge Problems

31-27. A 70-turn coil of wire has an area of 0.06 m^2 and is placed perpendicular to a constant 8-mT magnetic field. Calculate the induced emf if the coil flips 90° in 0.02 s.

Ans. 1.68 mV

31-28. A coil of area 0.2 m^2 has 80 turns of wire and is suspended with its plane perpendicular to a uniform magnetic field. What must be the flux density to produce an average emf of 2 V as the coil is flipped parallel to the field in 0.5 s?

31-29. The flux through a 200-turn coil changes from 0.06 to 0.025 Wb in 0.5 s. The coil is connected to an electric light, and the combined resistance is 2 Ω. What is the average induced emf, and what average current is delivered to the light filament?

Ans. 14.0 V, 7.00 A

31-30. A 90-mm length of wire moves with an upward velocity of 35 m/s between the poles of a magnet. The magnetic field is 80 mT directed to the right. If the resistance in the wire is 5.00 mΩ, what are the magnitude and direction of the induced current?

31-31. A generator develops an emf of 120 V and has a terminal potential difference of 115 V when the armature current is 25.0 A. What is the resistance of the armature?

Ans. 0.200 Ω

31-32. The coil of an ac generator rotates at a frequency of 60 Hz and develops a maximum emf of 170 V. The coil has 500 turns, each of area 4×10^{-3} m^2. What is the magnitude of the magnetic field in which the coil rotates?

31-33. A generator produces a maximum emf of 24 V when the armature rotates at 600 rpm. Assuming nothing else changes, what is the maximum emf when the armature rotates at 1800 rpm?

Ans. 72.0 V

*31-34. A shunt-wound motor connected across a 117-V line generates a back emf of 112 V when the armature current is 10 A. What is the armature resistance?

*31-35. A 110-V shunt-wound motor has a field resistance of 200 Ω connected in parallel with an armature resistance of 10 Ω. When the motor is operating at full speed, the back emf is 90 V. What is the starting current, and what is the operating current?

Ans. 11.6 A, 2.55 A

*31-36. A 120-V shunt motor has a field resistance of 160 Ω and an armature resistance of 1.00 Ω. When the motor is operating at full speed, it draws a current of 8.00 A. Find the starting current. What series resistance must be added to reduce the starting current to 30 A?

*31-37. A shunt generator has a field resistance of 400 Ω and an armature resistance of 2.00 Ω. The generator delivers a power of 4000 W to an external line at 120 V. What is the emf of the generator?

Ans. 187 V

Summary and Review

Critical Thinking Questions

31-38. A coil of wire has 10 loops, each of diameter D, placed inside a B field that varies at the rate of 2.5 mWb/s. If the induced emf is 4 mV, what is the diameter of the coil? What will the induced emf be if the diameter is doubled? What is the induced emf if the rate of change in the B field is doubled?

Ans. 45.2 cm, 16 mV, 8 mV

31-39. In Fig. 31-18a, the single loop of area 0.024 m^2 is connected to a 4-mΩ resistor. The magnet moves to the left through the center of the loop, causing the magnetic flux to increase at the rate of 2 mWb/s. What are the magnitude and direction of the current through the resistor? What if the magnet is then pulled back out of the loop with the same speed?

31-40. In Fig. 31-18b a changing B field is produced first by an increasing current through the loops and then by a decreasing current through the loops. In each case will the induced current be downward or upward on the near side of the single loop?

Ans. up, down

*31-41. A coil of 50 turns is rotated clockwise with a frequency of 60 Hz in a constant 3.0-mT magnetic field. The direction of the B field is along the positive x-axis, and the coil of area 0.070 m^2 rotates clockwise in the x-y plane. When the plane of the loop is parallel to the field, what is the direction of the current as viewed from the top (clockwise or counterclockwise)? What is the maximum induced emf? Starting time $t = 0$ when the emf is zero. At what later time will its emf first rise to 2.00 V?

*31-42. When the motor on a heat pump is first turned on, it momentarily draws 40.0 A. The current then immediately drops to a steady value of 12.0 A. If the motor operates on a 120-V power source, what is the back emf generated while the motor is running?

Ans. 84 V

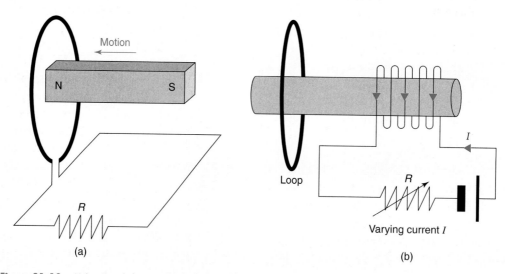

Figure 31-18 Using Lenz's law to determine the direction of the induced currents.

Alternating-Current Circuits

OBJECTIVES

After completing this chapter, you should be able to:

1. Determine the instantaneous current for changing and discharging a *capacitor* and for the growth and decay of current in an *inductor*.
2. Write and apply equations for calculating the *inductance* and *capacitance* for inductors and capacitors in an ac circuit.
3. Explain with diagrams the phase relationships for a circuit with (a) pure resistance, (b) pure capacitance, and (c) pure inductance.
4. Write and apply equations for calculating the *impedance*, the *phase angle*, and the *effective current* for a series ac circuit containing resistance, capacitance, and inductance.
5. Write and apply an equation for calculating the resonant frequency for an ac circuit.
6. Define and be able to determine the *power factor* for a series ac circuit.

About 99 percent of the energy generated in the United States is in ac form. There are reasons for the predominant use of ac circuits. A rotating coil in a magnetic field induces an alternating emf in an extremely efficient manner. Besides, the transformer provides a convenient method of transmitting ac currents over long distances with a minimal power loss.

The only element of importance in the dc circuit (besides a source of emf) was the resistor. Since alternating currents behave differently from direct currents, additional circuit elements become important. In addition to the normal resistance, electromagnetic induction and capacitance play important roles. In this chapter we present a few elementary aspects of alternating current in electric circuits.

32-1 THE CAPACITOR

In Chap. 26 we discussed the capacitor as an electrostatic device able to store change. Charging and discharging capacitors in an ac circuit provide an effective means of regulating and controlling the flow of charge. Before discussing the effects of **capacitance** in an ac circuit, however, it will be useful to describe the growth and decay of charge on a capacitor.

Consider the circuit illustrated in Fig. 32-1, containing only a capacitor and a resistor. When the switch is moved to S_1, the capacitor begins to be charged rapidly by the current i. As the potential difference Q/C between the capacitor plates rises, however, the rate of flow of charge to the capacitor decreases. At any instant, the

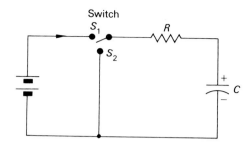

Figure 32-1 Circuit diagram illustrating a method for charging and discharging a capacitor.

iR drop through the resistor must equal the difference between the terminal voltage V_B of the battery and the back emf of the capacitor. Symbolically,

$$V_B - \frac{Q}{C} = iR \qquad \text{32-1}$$

where i = instantaneous current
Q = instantaneous charge on capacitor

Initially, the charge Q is zero, and the current i is a maximum. Thus, at time $t = 0$.

$$Q = 0 \qquad \text{and} \qquad i = \frac{V_B}{R} \qquad \text{32-2}$$

As the charge on the capacitor builds up, it produces a back emf Q/C opposing the flow of additional charge; the current i decreases. Both the increase in charge and the decrease in current are exponential functions, as shown by the curves in Fig. 32-2. If it were possible to continue charging indefinitely, the limits at $t = \infty$ would be

$$Q = CV_B \qquad \text{and} \qquad i = 0 \qquad \text{32-3}$$

The methods of calculus applied to Eq. (32-1) show that the instantaneous charge is given by

$$Q = CV_B(1 - e^{-t/RC}) \qquad \text{32-4}$$

and that the instantaneous current is given by

$$i = \frac{V_B}{R}e^{-t/RC} \qquad \text{32-5}$$

where t is the time. The logarithmic constant $e = 2.71828$ to six significant figures. Substitution of $t = 0$ and $t = \infty$ into the above relations will yield Eqs. (32-2) and (32-3), respectively.

The equations for computing instantaneous charge and current are simplified at the particular instant when $t = RC$. This time, usually denoted by τ, is called the *time constant* of the circuit.

$$\tau = RC \qquad \qquad \textit{Time Constant} \quad \text{32-6}$$

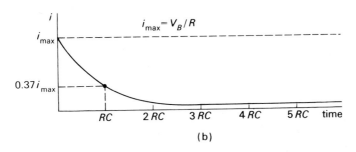

Figure 32-2 (a) The charge on a capacitor rises, approaching but never reaching its maximum value. (b) The current decreases, approaching zero as the charge builds to its maximum value.

We see from Eq. (32-4) that the charge Q rises to $1 - 1/e$ times its final value in one time constant:

$$Q = CV_B\left(1 - \frac{1}{e}\right) = 0.63CV_B \qquad \textbf{32-7}$$

In a capacitance circuit, the charge on a capacitor will rise to 63 percent of its maximum value after charging for a period of one time constant.

Substituting $\tau = RC$ into Eq. (32-5) shows that the current delivered to the capacitor decreases to $1/e$ times its initial value in one time constant:

$$i = \frac{V_B}{R}\frac{1}{e} = 0.37\frac{V_B}{R} \qquad \textbf{32-8}$$

In a capacitive circuit, the current delivered to a capacitor will decrease to 37 percent of its initial value after being charged for a period of one time constant.

Now let us consider the problem of a discharging capacitor. For practical reasons, *a capacitor is considered to be fully charged after a period of time equal to five time constants (5RC)*. If the switch in Fig. 32-1 has been in position S_1 for at least this long, it can be assumed that the maximum charge CV_B is on the capacitor. By moving the switch to position S_2, the voltage source is removed from the circuit, and a path is provided for discharge. In this case, the voltage Eq. (32-1) reduces to

$$-\frac{Q}{C} = iR \qquad \textbf{32-9}$$

Both the charge and the current decay along curves similar to that shown for the charging current in Fig. 32-2b. The instantaneous charge is found from

$$Q = CV_B e^{-t/RC} \qquad\qquad \textbf{32-10}$$

and the instantaneous current is given by

$$i = \frac{-V_B}{R} e^{-t/RC} \qquad\qquad \textbf{32-11}$$

The negative sign in the current equation indicates that the direction of i in the circuit has been reversed.

After discharging for one time constant, the charge and the current will have decayed to $1/e$ times their initial values. This can be shown by substituting τ into Eqs. (32-10) and (32-11).

In a capacitance circuit, the charge and the current will decay to 37 percent of their initial values after the capacitor has discharged for a length of time equal to one time constant.

The capacitor is considered to be fully discharged after five time constants ($5RC$).

CONCEPTUAL EXAMPLE 32-1

A 12-V battery having an internal resistance of 1.5 Ω is connected to a 4-μF capacitor through leads having a resistance of 0.5 Ω. (a) What is the initial current delivered to the capacitor? (b) How long will it take to charge the capacitor fully? (c) What is the value of the current after one time constant?

Solution (a)

Initially, there is no back emf from the capacitor. Hence, the current delivered to the circuit is the emf of the battery divided by the total resistance in the circuit:

$$i = \frac{\mathscr{E}_B}{R + r} = \frac{12 \text{ V}}{1.5 \ \Omega + 0.5 \ \Omega} = 6 \text{ A}$$

Solution (b)

The capacitor can be considered fully charged after a time.

$$t = 5RC = (5)(2 \ \Omega)(4 \ \mu\text{F}) = 40 \ \mu\text{s}$$

Solution (c)

After one time constant RC, the current will have decayed to 37 percent of its initial value. Thus,

$$i_\tau = (0.37)(6 \text{ A}) = 2.22 \text{ A}$$

In the preceding discussions, we simplified the approach by using direct currents. When an alternating voltage is impressed upon a capacitor, there are surges of charge into and out of the capacitor plates. Therefore, an alternating current is maintained in a circuit, even though there is no path between the capacitor plates. The effect of capacitance in an ac circuit will be discussed later.

32-2 THE INDUCTOR

Another important element in an ac circuit is the **inductor,** which consists of a continuous loop or coil of wire. (See Fig. 32-3). In Chap. 31 we showed that a change in magnetic flux in the region enclosed by such a coil will induce an emf in the coil. Until now, the flux changes originated from sources outside the coil itself. We now consider the emf induced in a coil as a result of changes in its *own* current. Regardless of how the flux change occurs, the induced emf must be given by

$$\mathscr{E} = -N\frac{\Delta\phi}{\Delta t} \qquad\qquad \textbf{32-12}$$

where N = number of turns
$\Delta\phi/\Delta t$ = rate at which flux changes

When the current through an inductor increases or decreases, a *self-induced* emf arises in the circuit that *opposes* the change. Consider the circuit illustrated in Fig. 32-3. When the switch is closed, the current rises from zero to its maximum value $i = V_B/R$. The inductor responds to this increasing current by setting up an induced back emf. Since the geometry of the inductor is fixed, the rate of change in flux, $\Delta\phi/\Delta t$, or the induced emf $\mathscr{E}$, is proportional to the rate of change in current, $\Delta i/\Delta t$. This proportionality is expressed in the equation

$$\mathscr{E} = -L\frac{\Delta i}{\Delta t} \qquad\qquad \textbf{32-13}$$

The proportionality constant L is called the **inductance** of the circuit. Solving explicitly for the inductance in Eq. (32-13), we write

$$L = -\frac{\mathscr{E}}{\Delta i/\Delta t} \qquad\qquad \textbf{32-14}$$

The unit of inductance is the **henry** (H).

USING A CALCULATOR

Some calculators have a graphing function. This allows you to see the ac output, for example, if you can enter your circuit information. Be sure to read your calculator's specific instruction manual to understand its features. Some calculators show both graphical and tabular information about your final plot. If yours has this feature, be sure to use the tabular information to check your answer.

Figure 32-3   The inductor.

A given inductance has an inductance of one henry (H) if an emf of one volt is induced by a current changing at the rate of one ampere per second.

$$1 \text{ H} = 1 \text{ V} \cdot \text{s/A}$$

The inductance of a coil depends on its geometry, the number of turns, the spacing of the turns, and the permeability of its core but not on voltage and current values. In this respect, the inductor is similar to capacitors and resistors.

We shall now consider the growth and decay of current in an inductive circuit. The circuit illustrated in Fig. 32-4 contains an inductor L, a resistor R, and a battery V_B. The switch is positioned so that the battery can be alternately connected and disconnected from the circuit. When the switch is moved to position S_1, a current begins to grow in the circuit. As the current rises, the induced emf $-L(\Delta i/\Delta t)$ is established in opposition to the battery voltage V_B. The net emf must equal the iR drop through the resistor. Thus,

$$V_B - L\frac{\Delta i}{\Delta t} = iR \qquad \text{32-15}$$

A mathematical analysis of Eq. (32-15) will show that the rise in current as a function of time is given by

$$i = \frac{V_B}{R}(1 - e^{-(R/L)t}) \qquad \text{32-16}$$

This equation shows that the current i is zero when $t = 0$ and has a maximum of V_B/R when $t = \infty$. The effect of inductance in a circuit is to delay the establishment of this maximum current. The rise and decay of current in an inductive current are shown in Fig. 32-5.

The time constant for an inductive circuit is

$$\tau = \frac{L}{R} \qquad \text{32-17}$$

τ is in *seconds* when L is in *henrys* and R is in *ohms*. Insertion of this value into Eq. (32-16) shows that:

In an inductive circuit, the current will rise to 63 percent of its final value in one time constant (L/R).

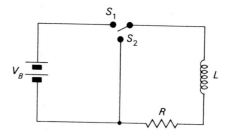

Figure 32-4 Circuit for studying inductance.

Figure 32-5 Rise and decay of current in an inductor.

After the current in Fig. 32-4 has attained a steady value, if the switch is moved to position S_2, the current will decay exponentially, as shown in Fig. 32-5. The equation for the decay is

$$i = \frac{V_B}{R} e^{-(R/L)t} \qquad \textbf{32-18}$$

Substitution of L/R into Eq. (32-18) shows that:

In an inductive circuit, the current decays to 37 percent of its initial value in one time constant (L/R).

Once again, for practical reasons the rise or decay time for an inductor is considered to be five time constants ($5L/R$).

32-3 ALTERNATING CURRENTS

Now that we are familiar with the basic elements in an ac circuit, it is necessary to understand more about alternating currents. The quantitative description of an alternating current is much more complicated than that for direct currents, whose magnitude and direction are constant. An alternating current flows back and forth in a circuit and has no "direction" in the sense direct current has. In addition, the magnitude varies sinusoidally with time, as we learned in our discussions of the ac generator.

The variation in emf or current for an ac circuit can be represented by a rotating vector or by a sine wave. These representations are compared in Fig. 32-6 on page 712. The vertical component of the rotating vector at any instant is the instantaneous magnitude of the voltage or the current. One complete revolution of the rotating vector or one complete sine wave on the curve represents one *cycle*. The number of complete cycles per second experienced by an alternating current is called its ***frequency*** and provides an important description of the current. The relationship between the instantaneous emf $\mathcal{E}$ or the instantaneous current i and the frequency was established in Chap. 31:

$$\mathcal{E} = \mathcal{E}_{\text{max}} \sin 2\pi ft \qquad \textbf{32-19}$$

$$i = i_{\text{max}} \sin 2\pi ft \qquad \textbf{32-20}$$

Note that the average value for the current in an ac circuit is zero since the magnitude alternates between i_{max} and $-i_{\text{max}}$. Even though there is no *net*

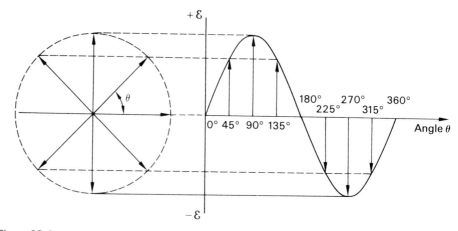

Figure 32-6 Rotating vector and its corresponding sine wave can be used to represent ac current or voltage.

current, however, charge is in motion, and electric energy can be released in the form of heat or useful work. The best method of measuring the effective strength of alternating currents is to find the dc value that will produce the same *heating* effect or develop the same *power* as the alternating current. This current value, called the ***effective current*** i_{eff}, is found to be 0.707 times the maximum current. A similar relation holds for the effective emf or voltage in an ac circuit. Thus,

$$i_{eff} = 0.707 i_{max} \qquad \text{32-21}$$

$$\mathscr{E}_{eff} = 0.707 \mathscr{E}_{max} \qquad \text{32-22}$$

One effective ampere is that alternating current that will develop the same power as one ampere of direct current.

One effective volt is that alternating voltage that will produce an effective current of one ampere through a resistance of one ohm.

AC meters are calibrated to show effective values. For example, if an ac meter measures household voltage to be 120 V at 10 A, Eqs. (32-12) and (32-22) will show that the maximum values of current and voltage are

$$i_{max} = \frac{10 \text{ A}}{0.707} = 14.14 \text{ A}$$

$$V_{max} = \frac{120 \text{ V}}{0.707} = 170 \text{ V}$$

Therefore, the house voltage actually varies between $+170$ and -170 V and the current ranges from $+14.14$ to -14.14 A. The usual frequency of voltage variation is 60 Hz.

32-4 PHASE RELATION IN AC CIRCUITS

In all dc circuits, the voltage and the current reach maximum and zero values at the same time and are said to be *in phase*. The effects of inductance and capacitance in ac circuits prevent the voltage and current from reaching maxima and minima at the same time. In other words, the current and voltage in most ac circuits are *out of phase*.

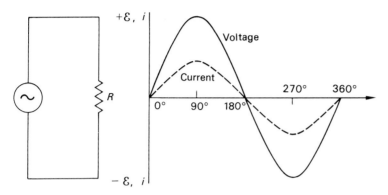

Figure 32-7   In a circuit containing pure resistance, the voltage and current are in phase.

To understand phase relations in an ac circuit, suppose we first consider a circuit containing a *pure* resistor in series with an ac generator, as in Fig. 32-7. This is an idealized circuit in which the inductive and capacitive effects are negligible. Many household devices, such as lights, heaters, and toasters, approximate a condition of pure resistance. In such devices, the instantaneous voltage V and current i are in phase. That is, variations in voltage will result in simultaneous variations of current. When the voltage is a maximum, the current is also a maximum. When the voltage is zero, the current is zero.

Next, we consider the phase relation between current and voltage across a *pure inductor*. The circuit illustrated in Fig. 32-8 on page 714 contains only an inductor in series with the ac generator. We have seen that the presence of inductance in a circuit whose current is changing at the rate $\Delta i / \Delta t$ results in a back emf

$$\mathcal{E} = -L \frac{\Delta i}{\Delta t}$$

which delays the current in reaching its maximum. The voltage reaches a maximum while the current is still at zero. When the voltage reaches a minimum, the current is at a maximum. In a circuit containing only inductance, the voltage is said to lead (occur before) the current by one-fourth of a cycle (or 90°). See the curve in Fig. 32-8.

In a circuit containing pure inductance, the voltage leads the current by 90°.

The effect of capacitance in an ac circuit is opposite to that of inductance. For the circuit shown in Fig. 32-9, the voltage must *lag behind* the current since the flow of charge to the capacitor is necessary to build up an opposing emf. When

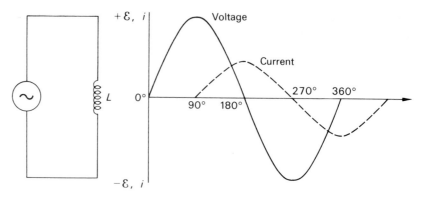

Figure 32-8 In a pure inductive circuit, the voltage leads the current by 90°.

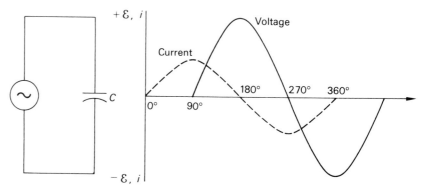

Figure 32-9 In a circuit containing only capacitance, the voltage lags the current by 90°.

the applied voltage is decreasing, charge flows from the capacitor. The rate of flow of this charge reaches a maximum when the applied voltage is zero.

In a circuit containing pure capacitance, the voltage lags the current by 90°.

This means that the variations in voltage occur one-fourth of a cycle later than the corresponding variations in current.

32-5 REACTANCE

In a dc circuit, the only opposition to current results from the material through which it passes. Resistive heat losses, which obey Ohm's law, also occur in ac circuits. In addition to normal resistance, however, one must contend with inductance and *capacitance*. Both inductors and capacitors impede the flow of an alternating current, and their effects must be considered along with the opposition of resistance.

The reactance of an ac circuit may be defined as its nonresistive opposition to the flow of alternating current.

We first consider the opposition to the flow of an alternating current through an inductor. Such opposition, called *inductive reactance,* arises from the self-induced back emf produced by a changing current. The magnitude of the inductive reactance X_L is determined by the inductance L of the inductor and by the frequency f of the alternating current and can be found from the formula

$$X_L = 2\pi f L \qquad \text{32-23}$$

The inductive reactance is measured in *ohms* when the inductance is in *henrys* and the frequency is in *hertz.*

The effective current i in an inductor is determined from its inductive reactance X_L and the effective voltage V by an equation analogous to Ohm's law:

$$V = iX_L \qquad \text{32-24}$$

EXAMPLE 32-2

A coil having an inductance of 0.5 H is connected to a 120-V, 60-Hz power source. If the resistance of the coil is neglected, what is the effective current through the coil?

Solution

The inductive reactance is

$$X_L = 2\pi f L = (2\pi)(60 \text{ Hz})(0.5 \text{ H}) = 188.4 \ \Omega$$

The current is found from Eq. (32-24):

$$i = \frac{V}{X_L} = \frac{120 \ \Omega}{188.4 \ \Omega} = 0.637 \text{ A}$$

Opposition to alternating current is also experienced because of the capacitance in a circuit. The *capacitive reactance* X_C is found from

$$X_C = \frac{1}{2\pi f C} \qquad \text{32-25}$$

where C = capacitance
$\quad f$ = frequency of alternating current

Capacitive reactance is expressed in *ohms* when C is in *farads* and f is in *hertz.* Once the capacitive reactance X_C of a capacitor is known, the effective current i can be found from

$$V = iX_C \qquad \text{32-26}$$

where V is the applied voltage.

32-6 THE SERIES AC CIRCUIT

In general, an ac circuit contains resistance, capacitance, and inductance in varying amounts. A series combination of these parameters is illustrated in Fig. 32-10. The total voltage drop in a dc circuit is the simple sum of the drop across each element in the circuit. In the ac circuit, however, the voltage and current are not in phase with each other. Recall that V_R is always in phase with the current but V_L leads the current by 90° and V_C lags the current by 90°. Clearly, if we are to determine the effective voltage V of the entire circuit, we must develop a means of treating phase differences.

This can best be accomplished by using a vector diagram, called the **phase diagram.** (See Fig. 32-11.) In this method, the effective values of V_R, V_L, and V_C are plotted as rotating vectors. The phase relationship is expressed in terms of the **phase angle** ϕ, which is a measure of how much the voltage leads the current in a particular circuit element. For example, in a pure resistor, the voltage and the current are in phase and $\phi = 0$. For an inductor, $\phi = +90°$, and in a capacitor $\phi = -90°$. The negative phase angle occurs when the voltage lags behind the current. Following this scheme, V_R appears as a vector along the x axis, V_L is represented by a vector pointing vertically upward, and V_C is directed downward.

The effective voltage V in an ac circuit can be defined as the vector sum of V_R, V_L, and V_C as they exist on the phase diagram. It can be seen from Fig. 32-11 that

Figure 32-10   Series ac circuit containing resistance, inductance, and capacitance.

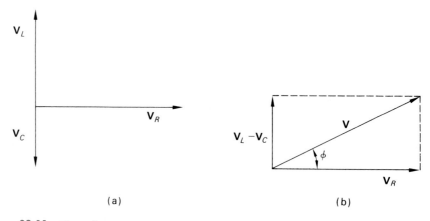

(a)   (b)

Figure 32-11   Phase diagram.

the magnitude of **V** is

$$V = \sqrt{V_R^2 + (V_L - V_C)^2} \qquad \text{32-27}$$

You should verify this equation by applying Pythagoras's theorem to the vector diagram.

Note from the phase diagram that a value of V_L that is greater than V_C results in a positive phase angle. In other words, if the circuit is inductive, the voltage leads the current. In a capacitive circuit, $X_C > X_L$, and a negative phase angle will result, indicating that the voltage lags the current. In any case, the magnitude of the phase angle can be found from

$$\tan \phi = \frac{V_L - V_C}{V_R} \qquad \text{32-28}$$

A more useful form of Eq. (32-27) can be found by recalling that

$$V_R = iR \qquad V_L = iX_L \qquad V_C = iX_C$$

Upon substitution, we find that

$$V = i\sqrt{R^2 + (X_L - X_C)^2} \qquad \text{32-29}$$

The quantity multiplied by the current in Eq. (32-29) is a measure of the combined opposition that the circuit offers to alternating current. It is called the **impedance** and is denoted by the symbol Z.

$$Z = \sqrt{R^2 + (X_L - X_C)^2} \qquad \text{32-30}$$

The higher the impedance in a circuit, the lower the circuit for a given voltage. Since R, X_L, and X_C are measured in *ohms*, the impedance is also expressed in ohms.

Therefore, the effective current i in an ac circuit is given by

$$i = \frac{V}{Z} \qquad \text{32-31}$$

where V = applied voltage
Z = impedance in circuit

It must be remembered that Z depends on the frequency of the alternating current as well as on the resistance, inductance, and capacitance.

Since the voltage across each element depends directly on resistance or reactance, an alternative phase diagram can be constructed by treating R, X_L, and X_C as vector quantities. Such a diagram can be used to compute the impedance, as indicated in Fig. 32-12. The phase angle ϕ in this representation can be found from

$$\tan \phi = \frac{X_L - X_C}{R} \qquad \text{32-32}$$

Of course, this angle is the same as that given by Eq. (32-28).

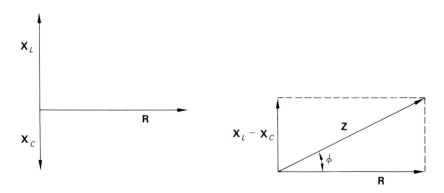

Figure 32-12 Impedance diagram.

CONCEPTUAL EXAMPLE 32-3

A 40-Ω resistor, a 0.4-H inductor, and a 10-μF capacitor are connected in series with an ac source that generates 120-V, 60-Hz alternating current. (a) Find the impedance of the circuit. (b) What is the phase angle? (c) Determine the effective current in the circuit.

Solution (a)

We must first compute X_L and X_C as follows:

$$X_L = 2\pi fL = (2\pi)(60 \text{ Hz})(0.4 \text{ H})$$
$$= 151 \ \Omega$$

$$X_C = \frac{1}{2\pi fC} = \frac{1}{(2\pi)(60 \text{ Hz})(10 \times 10^{-6} \text{ F})}$$
$$= 265 \ \Omega$$

The impedance of the circuit is

$$Z = \sqrt{R^2 + (X_L - X_C)^2}$$
$$= \sqrt{(40 \ \Omega)^2 + (151 \ \Omega - 265 \ \Omega)^2} = 121 \ \Omega$$

Solution (b)

From Eq. (32-32), the phase angle is

$$\tan \phi = \frac{151 \ \Omega - 265 \ \Omega}{40 \ \Omega} = -2.85$$
$$\phi = -71°$$

The negative sign is used to indicate that the phase angle is in the fourth quadrant.

Solution (c)

Finally, we can determine the effective current from the known impedance.

$$i = \frac{V}{Z} = \frac{120 \text{ V}}{121 \ \Omega} = 0.992 \text{ A}$$

The negative phase angle indicates that the voltage will lag behind this effective current. The circuit is more capacitive than it is inductive.

32-7 RESONANCE

Since inductance causes the current to lag behind the voltage and capacitance causes the current to lead the voltage, their combined effect is to cancel each other. The total reactance is given by $X_L - X_C$, and the impedance in a circuit is a minimum when $X_L = X_C$. When this is true, only the resistance R remains and the current will be a maximum. Setting $X_L = X_C$, we can write

$$2\pi f_r L = \frac{1}{2\pi f_r C}$$

and

$$f_r = \frac{1}{2\pi\sqrt{LC}} \qquad \text{32-33}$$

www.phy.ntnu.edu.tw
For a Java applet demonstrating the properties of an *RLC* circuit driven by a harmonic voltage source, see this Web site.

When the applied voltage has this frequency, called the **resonant frequency,** the current in the circuit will be a maximum. Additionally, it should be pointed out that since the current is limited only by resistance, it will be in phase with the voltage.

The antenna circuit in a radio receiver contains a variable capacitor that acts as a tuner. The capacitance is varied until the resonant frequency is equal to a particular signal frequency. The current peaks when this happens, and the receiver responds to the incoming signal.

32-8 THE POWER FACTOR

In ac circuits, no power is consumed because of capacitance or inductance. Energy is merely stored at one instant and released at another, causing the current and voltage to be out of phase. Whenever the current and voltage are in phase, the power P delivered is a maximum given by

$$P = iV$$

where i = effective current
V = effective voltage

This condition is satisfied when the ac circuit contains only resistance R or when the circuit is in resonance ($X_L = X_C$).

Normally, however, an ac circuit contains sufficient reactance to limit the effective power. In any case, the power delivered to the circuit is a function only of the component of the voltage V that is in phase with the current. From Fig. 32-11, this component is V_R, and we can write

$$V_R = V \cos \phi$$

where ϕ is the phase angle. Thus, the effective power consumed in an ac circuit is

$$P = iV \cos \phi \qquad \text{32-34}$$

The quantity $\cos \phi$ is called the **power factor** of the circuit. Note that $\cos \phi$ can vary from zero in a circuit containing pure reactance ($\phi = 90°$) to unity in a circuit containing only resistance ($\phi = 0°$).

Equation (32-30) and Fig. 32-12 show that the power factor can also be found from

$$\cos \phi = \frac{R}{Z} = \frac{R}{\sqrt{R^2 + (X_L - X_C)^2}} \qquad \textbf{32-35}$$

EXAMPLE 32-4

(a) What is the power factor for the circuit described in Example 32-3? (b) What power is absorbed in the circuit?

Solution (a)

The power factor is found in Eq. (32-35):

$$\cos \phi = \frac{R}{Z} = \frac{40 \ \Omega}{121 \ \Omega} = 0.33$$

Solution (b)

The power absorbed in the circuit is

$$P = iV \cos \phi = (0.992 \ \text{A})(120 \ \text{V})(0.33)$$
$$= 39.3 \ \text{W}$$

The power factor is sometimes expressed as a percentage instead of as a decimal. For example, the power factor of 0.33 in the preceding example could be expressed as 33 percent. Most commercial ac circuits have power factors from 80 to 90 percent because they usually contain more inductance than capacitance. Since this requires the electric power companies to furnish more current for a given power, the power companies extend a lower rate to users with power factors above 90 percent. Commercial users can improve their inductive power factors by adding capacitors, for instance.

Summary and Review

Summary

There are three principal elements in ac circuits: the *resistor*, the *capacitor*, and the *inductor*. A resistor is affected by ac current in the same manner as for dc circuits, and the current is determined by Ohm's law. The capacitor regulates and controls the flow of charge in an ac circuit; its opposition to the flow of electrons is called **capacitive reactance.** The inductor experiences a self-induced emf that adds **inductive reactance** to the circuit. The combined effect of all three elements in opposing electric current is called **impedance.** The major points to remember are summarized below.

- When a capacitor is being charged, the instantaneous values of the charge Q and the current i are found from

$$Q = CV_B(1 - e^{-t/RC}) \qquad i = \frac{V_B}{R}e^{-t/RC}$$

- The charge on the capacitor will rise to 63 percent of its maximum value as the current delivered to the capacitor decreases to 37 percent of its initial value during a period of one time constant τ.

$$\tau = RC \qquad \textit{Time Constant}$$

- When a capacitor is discharging, the instantaneous values of the charge and current are given by

$$Q = CV_Be^{-t/RC} \qquad i = \frac{-V_B}{R}e^{-t/RC}$$

Both the charge and the current decay to 37 percent of their initial values after discharging for one time constant.

- When alternating current passes through a coil of wire, an inductor, a self-induced emf arises to oppose the change. This emf is given by

$$\mathscr{E} = -L\frac{\Delta i}{\Delta t} \qquad L = -\frac{\mathscr{E}}{\Delta i/\Delta t}$$

This constant L is called the *inductance.* An inductance of one henry (H) exists if an emf of 1 V is induced by a current changing at the rate of 1 A/s.

- The rise and decay of current in an inductor are found from

$$i = \frac{V_B}{R}(1 - e^{-(R/L)t}) \qquad \textit{Current Rise}$$

$$i = \frac{V_B}{R}e^{-(R/L)t} \qquad \textit{Current Decay}$$

- In an inductive circuit, the current will rise to 63 percent of its maximum value or decay to 37 percent of its maximum in a period of one time constant. For an inductor, the time constant is

$$\tau = \frac{L}{R} \qquad \textit{Time Constant}$$

- Since alternating currents and voltages vary continuously, we speak of an *effective ampere* and an *effective volt* that are defined in terms of their maximum values as follows:

$$i_{\text{eff}} = 0.707i_{\text{max}} \qquad \mathscr{E}_{\text{eff}} = 0.707\mathscr{E}_{\text{max}}$$

- Both capacitors and inductors offer resistance to the flow of alternating current (called *reactance*), calculated from

$$X_L = 2\pi fL \qquad \textit{Inductive Reactance } X_L$$

$$X_C = \frac{1}{2\pi fC} \qquad \textit{Capacitive Reactance } X_C$$

The symbol f refers to the frequency of the alternating current in hertz. One hertz is one cycle per second.

Summary and Review

- The voltage, current, and resistance in a series ac circuit are studied with the use of phasor diagrams. Figure 32-12 illustrates such a diagram for X_C, X_L, and R. The resultant of these vectors is the effective resistance of the entire circuit called the *impedance Z*.

$$Z = \sqrt{R^2 + (X_L - X_C)^2} \quad \textit{Impedance}$$

- If we apply Ohm's law to each part of the circuit and then to the entire circuit, we obtain the following useful equations. First, the total voltage is given by

$$V = \sqrt{V_R^2 + (V_L - V_C)^2} \quad \textit{Voltage}$$

$$V_R = iR \qquad V_L = iX_L$$

$$V_C = iX_C \qquad V = iZ$$

$$V = i\sqrt{R^2 + (X_L - X_C)^2} \quad \textit{Ohm's Law}$$

- Because the voltage leads the current in an inductive circuit and lags the current in a capacitive circuit, the voltage and current maxima and minima usually do not coincide. The phase angle ϕ is given by

$$\tan \phi = \frac{V_L - V_C}{V_B}$$

or

$$\tan \phi = \frac{X_L - X_C}{R}$$

- The resonant frequency occurs when the net reactance is zero ($X_L = X_C$):

$$f_r = \frac{1}{2\pi\sqrt{LC}} \quad \textit{Resonant Frequency}$$

- No power is consumed because of capacitance or inductance. Since power is a function of the component of the impedance along the resistance axis, we can write

$$P = iV \cos \phi \quad \textit{Power Factor} = \cos \phi$$

$$\cos \phi = \frac{R}{Z}$$

$$\cos \phi = \frac{R}{\sqrt{R^2 + (X_L - X_C)^2}}$$

Key Terms

1. capacitance
2. inductor
3. inductance
4. henry
5. frequency
6. impedance
7. resonance
8. phase angle
9. effective current
10. effective voltage
11. phase diagram
12. capacitive reactance
13. inductive reactance
14. power factor
15. resonant frequency

Review Questions

32-1. Inductance in a circuit depends on its geometry and on the proximity of magnetic materials. Which of the following coils have higher inductances and why?
  a. closely or widely spaced turns
  b. long coil or short coil
  c. large cross section or small cross section
  d. iron core or air core

32-2. Show that the unit for the time constant L/R and RC for inductive and capacitive circuits is the second.

32-3. Sketch a curve of current versus time for (a) a circuit of high inductance and (b) a circuit of lower inductance.

32-4. Plot a curve of voltage versus time for (a) a charging capacitor and (b) a discharging capacitor. Compare these curves with those in the text for the current.

32-5. By plotting a curve of voltage as a function of time, show how the voltage changes in an inductive circuit. Compare these curves with those in the text for the current.

32-6. Should someone interested in establishing the breakdown voltage of a capacitor in an ac circuit be concerned with the average voltage, maximum voltage, or effective voltage?

32-7. A coil of wire is connected across the terminals of a 110-V dc battery. An ammeter connected in series with the coil indicates a current of 5 A. What happens to the current if an iron core is inserted into the coil? Now, disconnect the dc source, remove the iron core, and reconnect the system to a 110-V ac generator. The ammeter is adjusted to read ac effective amperes. Has the current decreased, increased, or stayed the same?

What happens to the current if the iron core is inserted? How do you account for your observations?

32-8. An incandescent lamp is connected in series with a 110-V ac generator and a variable capacitor. If the capacitance is increased, will the lamp glow more brightly or will it be dimmed? Explain. What would happen if the generator is replaced by a battery?

32-9. As the capacitance of a circuit increases, what happens to the resonant frequency of the circuit?

32-10. As the frequency is increased in an inductive circuit, what happens to the current in the circuit?

32-11. Inductive reactance depends *directly* on the frequency of the alternating current, whereas capacitive reactance varies *inversely* with the frequency. Both oppose the flow of charge in an ac circuit. Explain why their relationship to the frequency differs.

32-12. When a circuit is tuned to its resonant frequency, what is the power factor?

32-13. What is the power factor of a circuit containing (a) pure resistance, (b) pure inductance, (c) pure capacitance?

Problems

Section 32-1—The Capacitor

32-1. A series dc circuit consists of a 4-μF capacitor, a 5000-Ω resistor, and a 12-V battery. What is the time constant for this circuit?
Ans. 20 ms

32-2. What is the time constant for a series dc circuit containing a 6-μF capacitor and a 400-Ω resistor connected to a 20-V battery?

32-3. For the circuit described in Prob. 32-1, what are the initial current and the final current? How much time is needed to be assured that the capacitor is fully charged?
Ans. 2.40 mA, 0, 100 ms

32-4. For the circuit in Prob. 32-2, what is the maximum charge for the capacitor and how much time is required to fully charge the capacitor?

*32-5. An 8-μF capacitor is connected in series with a 600-Ω resistor and a 24-V battery.

After a time equal to one time constant, what are the charge on the capacitor and the current in the circuit?
Ans. 121 μC, 14.7 mA

*32-6. Assume that the fully charged capacitor of Prob. 32-5 is allowed to discharge. After one time constant, what are the current in the circuit and the charge on the capacitor?

*32-7. Assume that the 8-μF capacitor in Prob. 32-5 is fully charged and is then allowed to decay. Find the time when the current in the circuit will have decayed to 20 percent of its initial value.
Ans. 7.73 ms

*32-8. A 5-μF capacitor is connected in series with a 12-V source of emf and a 4000-Ω resistor. How much time is required to place 40 μC of charge on the capacitor?

Summary and Review

Section 32-2—The Inductor

32-9. A series dc circuit contains a 4-mH inductor and an 80-Ω resistor in series with a 12-V battery. What is the time constant for the circuit? What are the initial and final currents? **Ans. 50 μs, zero, 150 mA**

32-10. A 5-mH inductor, a 160-Ω resistor, and a 50-V battery are connected in series. What time is required for the current in the inductor to reach 63 percent of its steady-state value? What is the current at that instant?

32-11. What is the current in Prob. 32-9 after a time of one time constant? **Ans. 94.8 mA**

32-12. Assume that the inductor in Prob. 32-10 has reached its steady-state value. If the circuit is broken, how much time must pass before we can be sure that the current in the inductor is zero?

32-13. The current in a 25-mH solenoid increases from 0 to 2 A in a time of 0.1 s. What is the magnitude of the self-induced emf? **Ans. −500 mV**

32-14. A series dc circuit contains a 0.05-H inductor and a 40-Ω resistor in series with a 90-V source of emf. What are the maximum current in the circuit and the time constant?

*32-15. What is the instantaneous current in the inductor of Prob. 32-14 after a decay time of 1.0 ms? **Ans. 1.24 A**

*32-16. A 6-mH inductor, a 50-Ω resistor, and a 38-V battery are connected in series. At what time after the connection will the current reach an instantaneous value of 600 mA?

Section 32-3—Alternating Currents

32-17. An ac voltmeter, when placed across a 12-Ω resistor, reads 117 V. What are the maximum values for the voltage and current? **Ans. 165 V, 13.8 A**

32-18. In an ac circuit, the voltage and current reach maximum values of 120 V and 6.00 A. What are the effective ac values?

32-19. A capacitor has a maximum voltage rating of 500 V. What is the highest effective ac voltage that can be supplied to it without breakdown? **Ans. 354 V**

32-20. A certain appliance is supplied with an effective voltage of 220 V under an effective current of 20 A. What are the maximum and minimum values?

Section 32-5—Reactance

32-21. A 6-μF capacitor is connected to a 40-V 60-Hz ac line. What is the reactance? What is the effective ac current in the circuit containing pure capacitance? **Ans. 442 Ω, 90.5 mA**

32-22. A 2-H inductor of negligible resistance is connected to a 50-V 50-Hz ac line. What is the reactance? What is the effective ac current in the coil?

32-23. A 50-mH inductor of negligible resistance is connected to a 120-V 60-Hz ac line. What is the inductive reactance? What is the effective ac current in the circuit? **Ans. 18.8 Ω, 6.37 A**

32-24. A 6-μF capacitor is connected to a 24-V 50-Hz ac source. What is the current in the circuit?

32-25. A 3-μF capacitor connected to a 120-V ac line draws an effective current of 0.5 A. What is the frequency of the source? **Ans. 221 Hz**

32-26. Find the reactance of a 60-μF capacitor in a 600-Hz ac circuit. What is the reactance if the frequency is reduced to 200 Hz?

32-27. The frequency of an alternating current is 200 Hz, and the inductive reactance for a single inductor is 100 Ω. What is the inductance? **Ans. 79.6 mH**

*32-28. A 20-Ω resistor, a 2-μF capacitor, and a 0.70-H inductor are available. Each of these, in turn, is connected to a 120-V 60-Hz ac source as the only circuit element. What is the effective ac current in each case?

Section 32-6—The Series AC Circuit

*32-29. A 300-Ω resistor, a 3-μF capacitor, and a 4-H inductor are connected in series with a 90-V 50-Hz ac source. What is the net reactance of the circuit? What is the impedance? Ans. 196 Ω, 358 Ω

*32-30. What is the effective ac current delivered to the ac series circuit described in Prob. 32-29? What is the peak value for this current?

*32-31. A series ac circuit consists of a 100-Ω resistor, a 0.2-H inductor, and a 3-μF capacitor connected to a 110-V 60-Hz ac source. What are the inductive reactance, the capacitive reactance, and the impedance for the circuit? Ans. 75.4 Ω, 884 Ω, 815 Ω

*32-32. What are the phase angle and the power factor for the circuit described in Prob. 32-31?

*32-33. Assume that all the elements of Prob. 32-28 are connected in series with the given source. What is the effective current delivered to the circuit? Ans. 113 mA

*32-34. A series ac circuit contains a 12-mH inductor, an 8-μF capacitor, and a 40-Ω resistor connected to a 110-V 200-Hz ac line. What is the effective ac current in the circuit?

*32-35. When a 6-Ω resistor and a pure inductor are connected to a 100-V 60-Hz ac line, the effective current is 10 A. What is the inductance? What is the power loss through the resistor, and what is the power loss through the inductor? Ans. 24.5 mH, 600 W, 0

*32-36. A capacitor is in series with a resistance of 35 Ω and connected to a 220-V ac line. The reactance of the capacitor is 45 Ω. What is the effective ac current? What is the phase angle? What is the power factor?

*32-37. A coil having an inductance of 0.15 H and a resistance of 12 Ω is connected to a 110-V 25-Hz line. What is the effective current in the circuit? What is the power factor? What power is lost in the circuit? Ans. 4.16 A, 45.4 percent, 208 W

*32-38. What is the resonant frequency for the circuit described in Prob. 32-29?

*32-39. What is the resonant frequency for the circuit described in Prob. 32-34? Ans. 514 Hz

Challenge Problems

32-40. A 100-V battery is connected to a dc series circuit with a 60-mH inductor and a 50-Ω resistor. Calculate the maximum current in the circuit and the time constant.

32-41. If the circuit in Prob. 32-40 reaches a steady-state condition and is then broken, what will be the instantaneous current after one time constant? Ans. 736 mA

32-42. A resonant circuit has an inductance of 400 μH and a capacitance of 100 pF. What is the resonant frequency?

32-43. An LR dc circuit has a time constant of 2 ms. What is the inductance if the resistance is 2 kΩ? What is the instantaneous current 2 ms after the circuit is connected to a 12-V battery? Ans. 4.00 H, 3.79 mA

32-44. A series dc circuit consists of a 12-V battery, a 20-Ω resistor, and an unknown capacitor. The time constant is 40 ms. What is the capacitance? What is the maximum charge on the capacitor?

*32-45. A 2-H inductor having a resistance of 120 Ω is connected to a 30-V battery. How much

time is required for the current to reach 63 percent of its maximum value? What is the initial rate of current increase in amperes per second? What is the final current? Ans. 16.7 ms, 15 A/s, 250 mA

*32-46. Consider the circuit shown in Fig. 32-13. What is the impedance? What is the effective current? What is the power loss in the circuit?

Figure 32-13

Summary and Review

*32-47. A tuner circuit contains a 4-mH inductor and a variable capacitor. What must be the capacitance if the circuit is to resonate at a frequency of 800 Hz?   **Ans. 9.89 μF**

*32-48. An 8-μF capacitor is in series with a 40-Ω resistor and connected to a 117-V 60-Hz ac source. What is the impedance? What is the power factor? How much power is lost in the circuit?

*32-49. Someone wants to construct a circuit that has a resonant frequency of 950 kHz. If a coil in the circuit has an inductance of 3 mH, what capacitance should be added to the circuit?   **Ans. 9.36 pF**

*32-50. A 50-μF capacitor and a 70-Ω resistor are connected in series across a 120-V 60-Hz ac line. Determine the current in the circuit, the phase angle, and the total power loss.

*32-51. Refer to Prob. 32-50. What is the voltage across the resistor? What is the voltage across the capacitor? What inductance should be added to the circuit to reach resonance?   **Ans. 95.6 V, 72.5 V, 141 mH**

Critical Thinking Questions

*32-52. The resistance of an 8-H inductor is 200 Ω. If this inductor is suddenly connected across a potential difference of 50 V, what is the initial rate of increase of the current in A/s? [See Eq. (32-15).] What is the final steady current? At what rate is the current increasing after one time constant. At what time after connecting the source of emf, does the current equal one-half of its final value?

Ans. 250 mA, 6.25 A/s, 2.31 A/s, 27.7 ms

*32-53. In Chap. 29, we found that the magnetic flux density within a solenoid was given by:

$$B = \frac{\phi}{A} = \frac{\mu_0 NI}{l}$$

Here we have used l for the length of the solenoid in order to avoid confusion with the symbol L used for inductance. We also know that induced emf is found in two ways:

$$\mathcal{E} = -N\frac{\Delta\phi}{\Delta t} \quad \text{and} \quad \mathcal{E} = -L\frac{\Delta i}{\Delta t}$$

Prove the following relationships for a solenoid of length l having N turns of area A:

$$L = \frac{N\phi}{I} \quad \text{and} \quad L = \frac{\mu N^2 A}{l}$$

*32-54. An inductor consists of a coil 30 cm long with 300 turns of area 0.004 m^2. Find the inductance (see previous Prob. 32-53). A battery is connected, and the current builds from zero to 2.00 A in 0.001 s. What is the average induced emf in the coil?

Ans. 1.51 mH, −3.02 V

*32-55. An inductor, a resistor, and a capacitor are connected in series with a 60-Hz ac line. A voltmeter connected to each element in the circuit gives the following readings: $V_R = 60$ V, $V_L = 100$ V, and $V_C = 160$ V. What is the total voltage drop in the circuit? What is the phase angle?

*32-56. The antenna circuit in a radio receiver consists of a variable capacitor and a 9-mH coil. The resistance of the circuit is 40 Ω. A 980-kHz radio wave produces a potential difference of 0.2 mV across the circuit. Determine the capacitance required for resonance. What is the current at resonance?   **Ans. 2.93 pF, 5mA**

*32-57. A series *RLC* circuit has elements as follows: $L = 0.6$ H, $C = 3.5$ μF, and $R = 250$ Ω. The circuit is driven by a generator that produces a *maximum* emf of 150 V at 60 Hz. What is the effective ac current? What is the phase angle? What is the average power loss in the circuit?

Light and Illumination

OBJECTIVES

After completing this chapter, you should be able to:

1. Discuss the historical investigation into the nature of light and explain how light sometimes behaves as a wave and sometimes as particles.
2. Describe the broad classifications in the electromagnetic spectrum on the basis of frequency, wavelength, or energy.
3. Write and apply formulas for the relationship between velocity, wavelength, and frequency, and between energy and frequency for electromagnetic radiation.
4. Describe experiments that will result in a reasonable estimation of the speed of light.
5. Illustrate with drawings your understanding of the formation of shadows, labeling the *umbra* and *penumbra*.
6. Demonstrate your understanding of the concepts of *luminous flux, luminous intensity,* and *illumination,* and solve problems similar to those in the text.

An iron bar resting on a table is in thermal equilibrium with its surroundings. From its outward appearance one would never guess that it is active internally. All objects are continuously emitting radiant heat energy that is related to their temperature. The bar is in thermal equilibrium only because it is radiating and absorbing energy at the same rates. If the balance is upset by placing one end of the bar in a hot flame, the bar becomes more active internally and emits heat energy at a greater rate. As the heating continues to around 600°C, some of the radiation emitted from the bar becomes *visible;* that is, it affects our sense of sight. The color of the bar becomes a dull red, which turns brighter as more heat is supplied.

The radiant energy emitted by the object before this effect is visible consists of **electromagnetic waves** of longer wavelengths than red light. Such waves are referred to as **infrared rays,** meaning "beyond the red." If the temperature of the bar is increased to around 3000°C, it becomes white hot, indicating a further extension of the radiant energy into the **visible region.**

This example sets the stage for our discussion of light. The nature of light is no different fundamentally from the nature of other electromagnetic radiation, for example, heat, radio waves, or ultraviolet radiation. The characteristic that distinguishes light from the other radiation is its energy.

> Light **is electromagnetic radiation that is capable of affecting the sense of sight.**

The energy content of visible light varies from about 2.8×10^{-19} J to around 5.0×10^{-19} J.

33-1 WHAT IS LIGHT?

The answer to that question has been extremely elusive throughout the history of science. The long search for an answer provides an inspiring example of the scientific approach to the solution of a problem. Every hypothesis put forth to explain the nature of light was tested both by logic and by experimentation. The ancient philosophers' contention that visual rays are emitted from the eye to the seen object failed the test of both logic and experience.

By the latter part of the seventeenth century, two theories were advanced to explain the nature of light, the particle (corpuscular) theory and the wave theory. The principal advocate of the corpuscular theory was Sir Isaac Newton. The wave theory was upheld by Christian Huygens (1629–1695), a Dutch mathematician and scientist who was 13 years older than Newton. Each theory set out to explain the characteristics of light observed at that time. Three important characteristics can be summarized as follows:

1. *Rectilinear propagation:* Light travels in straight lines.
2. *Reflection:* When light is incident on a smooth surface, it turns back into the original medium.
3. *Refraction:* The path of light changes when it enters a transparent medium.

According to the corpuscular theory, tiny particles of unsubstantial mass were emitted by light sources such as the sun or a flame. These particles traveled outward from the source in straight lines at enormous speeds. When the particles entered the eye, the sense of sight was stimulated. Rectilinear propagation was easily explained in terms of particles. In fact, one of the strongest arguments for the corpuscular theory was based on this property. It was reasoned that particles cast sharp shadows, as illustrated in Fig. 33-1a, whereas waves can bend around edges. Such bending of waves, as shown in Fig. 33-1b, is called **diffraction.**

The sharp shadows formed by light beams indicated to Newton that light must consist of particles. Huygens, on the other hand, explained that the bending of

Figure 33-1 A strong argument for the particle theory of matter is the formation of sharp shadows. Waves were known to bend around obstacles in their path.

water waves and sound waves around obstacles is easily noticed because of the long wavelengths. He reasoned that if light were a wave with a short wavelength, it would appear to cast a sharp shadow because the amount of bending would be small.

It was also difficult to explain why particles traveling in straight lines from many directions could cross without impeding one another. In a paper published in 1690, Huygens wrote:

> If, furthermore, we pay attention to, and weigh up, the extraordinary speed with which light spreads in all directions, and also the fact that coming, as it does, from quite different, indeed from opposite, directions, the rays interpenetrate without impeding one another, then we may well understand that whenever we see a luminous object, this cannot be due to the transmission of matter which reaches us from the object, as for instance a projectile or an arrow flies through the air.

Huygens explained the propagation of light in terms of the motion of a disturbance through the distance between a source and the eye. He based his argument on a simple principle that is still useful today in describing the propagation of light. Suppose we drop a stone into a quiet pool of water. A disturbance is created that moves outward from the point of impact in a series of concentric waves. The disturbance continues even after the stone has struck the bottom of the pool. Such an example prompted Huygens to postulate that disturbances existing at all points along a moving wavefront at one instant can be considered sources for the wavefront at the next instant. Huygens's principle states:

Every point on an advancing wavefront can be considered a source of secondary waves called *wavelets*. The new position of the wavefront is the envelope of the wavelets emitted from all points of the wavefront in its previous position.

The application of this principle is illustrated in Fig. 33-2 for the common cases of a plane wave and a circular wave.

Huygens's principle was particularly successful in explaining reflection and refraction. Figure 33-3 shows how the principle can be used to explain the bending of light as it passes from air to water. When the plane waves strike the water surface at an angle, points *A, C,* and *E* become the sources of new wavelets. The

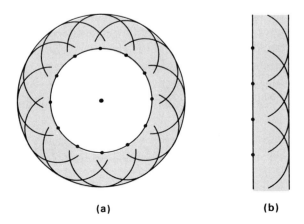

(a) **(b)**

Figure 33-2 Illustration of Huygens's principle (a) for a spherical wave and (b) for a plane wave.

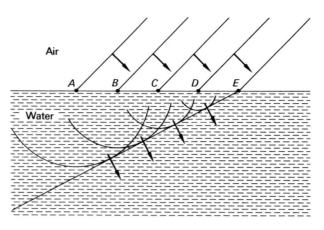

Figure 33-3 Huygens's explanation of refraction in terms of the wave theory.

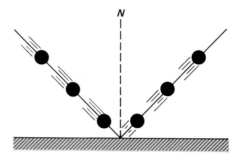

Figure 33-4 Explanation of reflection in terms of the particle theory of light.

envelope of these secondary wavelets indicates a change in direction. A similar construction can be made to explain reflection.

Reflection and refraction were also easily explained in terms of the particle theory. Figures 33-4 and 33-5 illustrate models that can be used to explain reflection and refraction on the basis of tiny corpuscles. Perfectly elastic particles of unsubstantial mass rebounding from an elastic surface could explain the regular reflection of light from smooth surfaces. Refraction could be analogous to the change in direction of a rolling ball as it encounters an incline. This explanation required that the particles of light travel faster in the refracted medium, whereas the wave theory required light to travel more slowly in the refracted medium. Newton recognized that if it could ever be shown that light travels more slowly in a material medium than it does in air, he would have to abandon the particle theory. It was not until the middle of the nineteenth century that Jean Foucault successfully demonstrated that light travels more slowly in water than it does in air.

33-2 THE PROPAGATION OF LIGHT

The discovery of interference and diffraction in 1801 and 1816 swung the debate solidly toward Huygens's wave theory. Clearly, interference and diffraction could be explained only in terms of a wave theory. However, there still remained one problem. All wave phenomena were thought to require the existence of a medium.

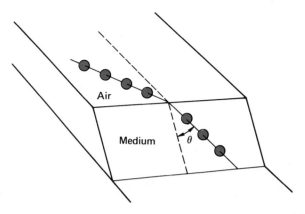

Figure 33-5 Refraction of light as it passes from air to another medium was explained by this mechanical example.

How, for example, could light waves travel through a vacuum if there were nothing to "vibrate"? Indeed, how could light reach the earth from the sun or other stars through millions of miles of empty space? To avoid this contradiction, physicists postulated the existence of a "light-carrying ether." This all-penetrating universal medium was thought to fill all the space between and within all material bodies. But what was the nature of this ether? Certainly it could not be a gas, solid, or liquid that obeyed the physical laws known at that time. And yet the wave theory could not be denied in light of the evidence of interference and diffraction. No other choice seemed possible except for the definition of ether as "that which carries light."

In 1865 a Scottish physicist, James Clerk Maxwell, set out to determine the properties of a medium that would carry light and also account for the transmission of heat and electric energy. His work demonstrated that *an accelerated charge can radiate electromagnetic waves into space.* Maxwell explained that the energy in an electromagnetic wave is equally divided between electric and magnetic fields that are mutually perpendicular. Both fields oscillate perpendicular to the direction of wave propagation, as shown in Fig. 33-6. Thus, a light wave would not have to depend on the vibration of matter. It could be propagated by oscillating transverse fields. Such a wave could "break off" from the region around an accelerating charge and fly off into space with the velocity of light. Maxwell's equations predicted that heat and electric action, as well as light, were propagated at the speed of light as electromagnetic disturbances.

Experimental confirmation of Maxwell's theory was achieved in 1885 by H. R. Hertz, who proved that radiation of electromagnetic energy can occur *at any frequency.* In other words, light, heat radiation, and radio waves are of the same nature, and they all travel at the speed of light (3×10^8 m/s). All types of radiation can be reflected, focused by lenses, polarized, and so forth. It seemed that the wave nature of light could no longer be doubted.

Confirmation of the electromagnetic theory paved the way for the eventual downfall of the "light-carrying ether" postulate. In 1887, A. A. Michelson, an American physicist, showed conclusively that the velocity of light is a constant, independent of the motion of the source. He could not establish any difference between the speed of light traveling in the direction of the earth's motion and traveling

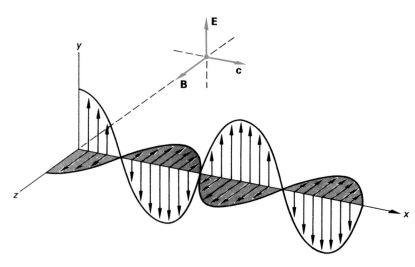

Figure 33-6   Electromagnetic theory holds that light is propagated as oscillating transverse fields. The energy is divided equally between electric **E** and magnetic **B** fields, which are mutually perpendicular.

opposite to the earth's motion. Those who are interested in a thorough discussion of the subject should look up the *Michelson–Morley experiment*. Einstein later interpreted Michelson's results to mean that the notion of ether must be abandoned in favor of a completely empty space.

33-3 THE ELECTROMAGNETIC SPECTRUM

Today, the electromagnetic spectrum is known to spread over a tremendous range of frequencies. A chart of the electromagnetic spectrum is presented in Fig. 33-7. The wavelength λ of electromagnetic radiation is related to its frequency f by the general equation

$$c = f\lambda \qquad \text{33-1}$$

where c is the velocity of light (3×10^8 m/s). In terms of wavelengths, the tiny segment of the electromagnetic spectrum referred to as the **visible region** lies between 0.00004 and 0.00007 cm.

Because of the small wavelengths of light radiation, it is more convenient to define smaller units of measure. The SI unit is the **nanometer** (nm).

One nanometer (1 nm) is defined as one-billionth of a meter.

$$1 \text{ nm} = 10^{-9} \text{ m} = 10^{-7} \text{ cm}$$

The visible region of the electromagnetic spectrum extends from 400 nm for violet light to approximately 700 nm for red light. Other older units are the *millimicron* (mμ), which is the same as a nanometer, and the *angstrom* (Å), which is 0.1 nm.

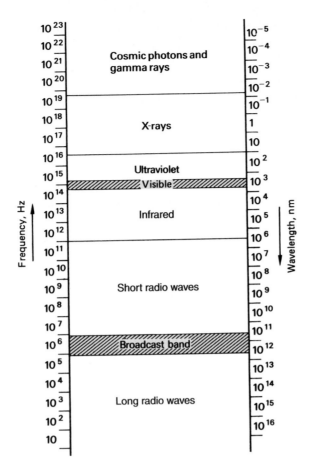

Figure 33-7   Electromagnetic spectrum.

EXAMPLE 33-1

The wavelength of yellow light from a sodium flame is 589 nm. Compute its frequency.

Solution

The frequency is found from Eq. (33-1).

$$f = \frac{c}{\lambda} = \frac{3 \times 10^8 \text{ m/s}}{589 \times 10^{-9} \text{ m}}$$

$$= 5.09 \times 10^{14} \text{ Hz}$$

Newton was the first to make detailed studies of the visible region by dispersing "white light" through a prism. In order of increasing wavelength, the spectral colors are violet (450 nm), blue (480 nm), green (520 nm), yellow (580 nm), orange (600 nm), and red (640 nm). Anyone who has seen a rainbow has seen the effects that different wavelengths of light have on the human eye.

The electromagnetic spectrum is continuous; there are no gaps between one form of radiation and another. The boundaries we set are purely arbitrary, depending upon our ability to sense one small portion directly and to discover and measure those portions outside the visible region.

The first discovery of radiation of wavelengths longer than those of red light was accomplished in 1800 by William Herschel. These waves are now known as thermal radiation and are referred to as *infrared waves.*

Shortly after the discovery of infrared waves, radiation of wavelengths shorter than visible light were noticed. These waves, now known as *ultraviolet waves,* were discovered in connection with their effect on certain chemical reactions.

How far the infrared region might extend in the direction of longer wavelengths was not known throughout most of the nineteenth century. Fortunately, Maxwell's electromagnetic theory opened the door for the discovery of many other classifications of radiation. The spectrum of electromagnetic waves is now conveniently divided into the eight major regions shown in Fig. 33-7: (1) long radio waves, (2) short radio waves, (3) the infrared region, (4) the visible region, (5) the ultraviolet region, (6) X rays, (7) gamma rays, and (8) cosmic photons.

33-4 THE QUANTUM THEORY

The work of Maxwell and of Hertz in establishing the electromagnetic nature of light waves was truly one of the most important events in the history of science. Not only was the wave nature of light explained, but the door was opened to an enormous range of electromagnetic waves. Remarkably, only two years after Hertz's verficiation of Maxwell's wave equations, the wave theory of light was again challenged. In 1887, Hertz noticed that an electric spark would jump more readily between two charged spheres when their surfaces were illuminated by the light from another spark. This phenomenon, known as the *photoelectric effect,* is demonstrated by the apparatus in Figure. 33-8. A beam of light strikes a metal surface *A* in an evacuated tube. Electrons ejected by the light are drawn to the collector *B* by external batteries. The flow of electrons is indicated by a device called an *ammeter.* The photoelectric effect defied explanation in terms of the wave theory. In fact, the ejection of electrons could be accounted for more easily in terms of the old particle theory. Still, there could be no doubt of the wave properties either. Science faced a remarkable paradox.

The photoelectric effect, along with several other experiments involving the emission and absorption of radiant energy, could not be accounted for purely in terms of Maxwell's electromagnetic wave theory. In an attempt to bring experimental observation into agreement with theory. Max Planck, a German physicist, published his *quantum hypothesis* in 1901. He found that the problems with the radiation theory lay with the assumption that energy is radiated continuously. It was postulated that electromagnetic energy is absorbed or emitted in discrete packets, or *quanta.* The energy content of these quanta, or *photons* as they were called, is proportional to the frequency of the radiation. Planck's equation can be written

$$E = hf \qquad \text{33-2}$$

Incident light

A

B

Batteries

Ammeter

Figure 33-8 The photoelectric effect.

where E = energy of photon

f = frequency of photon

h = proportionality factor called *Planck's constant*

$(6.626 \times 10^{-34} \text{ J} \cdot \text{s})$

In 1905, Einstein extended the idea proposed by Planck and postulated that the energy in a light beam does not spread continuously through space. By assuming that light energy is concentrated in small packets (photons) whose energy content is given by Planck's equation, Einstein was able to predict the photoelectric effect mathematically. At last, theory was reconciled with experimental observation.

Thus, it appears that light is *dualistic*. The wave theory is retained by considering the photon to have a frequency and an energy proportional to the frequency. The modern practice is to use the wave theory when studying the propagation of light. The corpuscular theory, on the other hand, is necessary to describe the interaction of light with matter. We may think of light as radiant energy transported in photons carried along by a wave field.

The origin of light photons was not understood until Niels Bohr in 1913 devised a model of the atom based on quantum ideas. Bohr postulated that electrons can move about the nucleus of an atom only in certain orbits or *discrete energy levels*, as shown in Fig. 33-9. Atoms were said to be *quantized*. If the atoms are somehow energized, as by heat, the orbital electrons may jump into a higher orbit. At some later time, these excited electrons will fall back to their original level, releasing as photons the energy that was originally absorbed. Although Bohr's model was not strictly correct, it provided a basis for understanding the emission and absorption of electromagnetic radiation in quantum units.

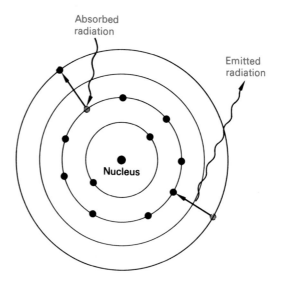

Figure 33-9 Bohr's theory of the atom.

33-5 THE SPEED OF LIGHT

The speed of light is probably the most important constant used in physics, and its accurate determination represents one of the most precise measurements accomplished by scientists. Its magnitude is so great (about 186,000 mi/s) that experimental measurements were entirely unsuccessful until the latter part of the seventeenth century. It was generally believed that the transmission of light must be instantaneous.

The first attempts to measure the speed of light experimentally were made in 1667 by Galileo. Briefly, his method consisted of stationing two observers in towers separated by a known distance. Signals were to be made at night with lanterns. The second experimenter was to uncover his lantern as soon as he received a light signal from the first experimenter. The speed of light could then be determined from the time required for the light to travel between the two towers. The experiment was inconclusive, and Galileo merely added to the prevailing opinion that light transmission was instantaneous. Such an experiment seems humorous to us only because of our present understanding of the true magnitude of the velocity of light.

Eight years later, a Danish astronomer, Olaus Roemer, made the first measurement of the speed of light. He based his calculations on irregularities in the predicted eclipses of one of the moons of the planet Jupiter. Roemer had been measuring the time interval between successive eclipses for several years. He noticed a consistent variation in his computations when measurements were made from different portions of the earth's orbit about the sun. Roemer correctly concluded that the irregularities in his measurements were due to the distance that light traveled to reach the earth. On the basis of his rather inaccurate measurements, Roemer calculated that the speed of light was 140,000 mi/s.

The first successful terrestrial measurement of the speed of light waves made by A. H. L. Fizeau, a French scientist. In 1849, his experiment involved the simple

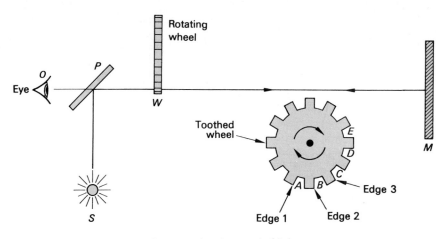

Figure 33-10 Fizeau's apparatus for measuring the speed of light.

computation of the time required for light to traverse a known distance. A schematic diagram of the apparatus is shown in Fig. 33-10. A rotating toothed wheel W allows light coming from the source to be interrupted, producing a series of short flashes. Let us assume first that the wheel is stationary and that light passes through one of the openings between the teeth. Light from the source S is reflected by the half-silvered plate glass P through the opening in the wheel and onto the plane mirror M located a few miles away. The reflected light returns through the opening and passes through the glass P and to the observer O.

Now suppose the toothed wheel is rotated at a constant frequency. Light from the source will pass through the opening A to the distant mirror and back to the wheel. The rotational frequency of the wheel is adjusted so that the returning light will just pass through the opening B. Under these conditions, light from the source passes by edge 1 onto the mirror and returns past edge 2. Thus, the time required for the light to travel the known distance is the same as the time for the wheel to rotate through the combined width of one opening and one tooth. The speed of light is therefore a function of the known frequency of rotation of the wheel and can be calculated. If the frequency is gradually increased, the light will eventually be blocked again until it reappears past edge 3. The speed of light calculated by Fizeau was 3.13×10^8 m/s. The error is attributed to inaccurate measurements.

In 1850, Foucault refined the apparatus developed by Fizeau by replacing the toothed wheel with a rotating mirror. He is remembered primarily for his measurements of the velocity of light in water. It was the first conclusive evidence that light travels more slowly in water than in air.

Probably no other scientist is more remembered for his work on measuring the speed of light than Albert A. Michelson (1852–1931). Using the Foucault method, he was able to obtain measurements that were extremely accurate. A schematic diagram of his apparatus is shown in Fig. 33-11. The light source, the eight-sided rotating mirror, and the telescope were located on Mt. Wilson, California. The reflecting mirrors were located approximately 22 miles away on the top of Mt. San Antonio. From the known distances and the time required for light to complete the trip, Michelson determined that the speed of light in air is 2.997×10^8 m/s. This value is accepted as one of the most precise measurements made using the Foucault principle.

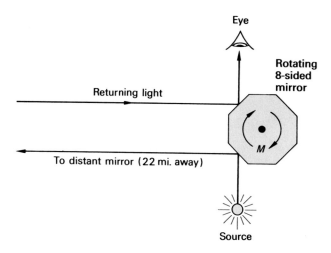

Figure 33-11 Michelson's method for measuring the speed of light in air.

Research concerning the velocity of light has continued until recent times when laser techniques have produced the value.

$$c = 2.99792457 \times 10^8 \text{ m/s}$$

which is accurate to about 12 parts in 100 million. This figure is based, however, on the older definition of the meter in terms of the krypton wavelength. The second can now be measured with an uncertainty of only 1 part in 10 trillion. As discussed in Chap. 2 on measurements, the meter was redefined in 1983 in terms of the unit of time. More precisely, the meter is established as the distance traveled by light in a time of $1/299,792,458$ s as measured by a cesium clock. This decision established the velocity of light, by definition, to be

$$c = 2.99792458 \times 10^8 \text{ m/s} \qquad \text{(exactly)} \qquad \text{33-3}$$

Useful approximations are 3×20^8 m/s and 186,000 mi/s. These values may be used in most physical calculations without fear of significant error.

33-6 LIGHT RAYS AND SHADOWS

One of the first properties of light to be studied was rectilinear propagation and the formation of shadows. Instinctively, we rely quite heavily on this property for estimating distances, directions, and shapes. The formation of sharp shadows on a sundial is used to estimate time. In this section we discuss how we can predict the formation of shadows.

According to Huygens's principle, every point on a moving wavefront can be considered a source for secondary wavelets. The wavefront at any time is the envelope of these wavelets. Thus, the light emitted in all directions by the point source of light in Fig. 33-12 can be represented by a series of spherical wavefronts moving away from the source at the speed of light. For our purposes, a *point source* of light is one whose dimensions are small in comparison with the distances studied. Note that the spherical wavefronts become essentially plane wavefronts in any

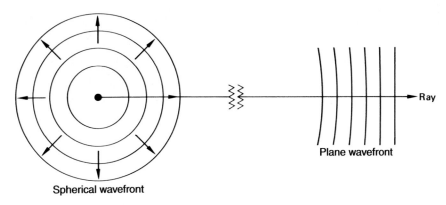

Figure 33-12 A ray is an imaginary line, drawn perpendicular to advancing wavefronts, that indicates the direction of light propagation.

specific direction at a long distance away from the source. An imaginary straight line drawn perpendicular to the wavefronts in the direction of the moving wavefronts is called a *ray*. There are, of course, an infinite number of rays starting from the point source.

Any dark-colored object absorbs light, but a black one absorbs nearly all the light it receives. Light that is not absorbed upon striking an object is either reflected or transmitted. If all the light incident upon an object is reflected or absorbed, the object is said to be *opaque*. Since light cannot pass through an opaque body, a shadow will be produced in the space behind the object. The shadow formed by a point source of light is illustrated in Fig. 33-13. Since light is propagated in straight lines, rays drawn from the source past the edges of the opaque object form a sharp shadow proportional to the shape of the object. Such a region in which no light has entered is called an *umbra.*

If the source of light is an extended one rather than a point, the shadow will consist of two portions, as shown in Fig. 33-14. The inner portion receives no light from the source and is therefore the *umbra*. The outer portion is called the *penumbra.* An observer within the penumbra would see a portion of the source but not all the source. An observer located outside both regions would see all the source. Solar and lunar eclipses can be studied by similar construction of shadows.

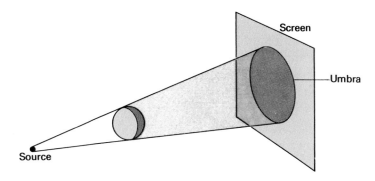

Figure 33-13 Shadow formed by a point source of light.

Figure 33-14   Shadows formed by an extended source of light.

33-7 LUMINOUS FLUX

Most sources of light emit electromagnetic energy distributed over many wavelengths. Electric power is supplied to a lamp, and radiation is emitted. The radiant energy emitted per unit of time by the lamp is called the radiant power or the **radiant flux.** Only a small portion of this radiant power is in the visible region: in the region between 400 and 700 nm. The sensation of sight depends only on the visible, or *luminous,* energy radiated per unit of time.

> **The luminous flux is that part of the total radiant power emitted from a light source that is capable of affecting the sense of sight.**

In a common incandescent light bulb, only about 10 percent of the energy radiated is luminous flux. The bulk of the radiant power is nonluminous.

The human eye is not equally sensitive to all colors. In other words, equal radiant power of different wavelengths does not produce equal brightness. A 40-W green light bulb appears brighter than a 40-W blue light bulb. A graph portraying the response of the eye to various wavelengths is shown in Fig. 33-15. Note that the sensitivity curve is bell-shaped around the center of the visible spectrum. Under normal conditions, the eye is most sensitive to yellow-green light of wavelength 555 nm. The sensitivity falls off rapidly for longer and shorter wavelengths.

If the unit chosen for luminous flux is to correspond to the sensual response of the human eye, a new unit must be defined. The watt (W) is not sufficient because the visual sensations are not the same for different colors. What is needed is a unit that measures *brightness.* Such a unit is the *lumen* (lm), which is determined by comparison with a standard source.

To understand the definition of a lumen in terms of the standard source, we must first develop the concept of a solid angle. A solid angle in **steradians** (sr) is defined the same way a plane angle is defined in radians. In Fig. 33-16 the angle θ in radians is

$$\theta = \frac{S}{R} \quad \text{rad} \tag{33-4}$$

Figure 33-15 Sensitivity curve.

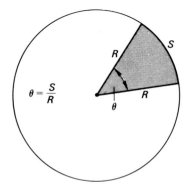

Figure 33-16 Definition of a plane angle θ expressed in radians.

where S is the arc length and R is the radius. The solid angle Ω is similarly defined by Fig. 33-17. It may be thought of as the opening from the tip of a cone that is subtended by a segment of area on the spherical surface.

> **One steradian (sr) is the solid angle subtended at the center of a sphere by an area A on its surface that is equal to the square of its radius R.**

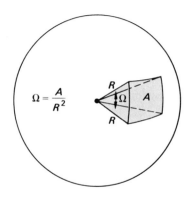

Figure 33-17 Definition of a solid angle Ω in steradians.

In general, the solid angle in steradians is given by

$$\Omega = \frac{A}{R^2} \quad \text{sr} \qquad \qquad 33\text{-}5$$

The steradian, like the radian, is a unitless quantity.

Just as there are 2π rad in a complete circle, it can be shown by Eq. (33-5) that there are 4π sr in a complete sphere.

$$\Omega = \frac{A}{R^2} = \frac{4\pi R^2}{R^2} = 4\pi \text{ sr}$$

Notice that the solid angle is independent of the distance from the source. There are 4π sr in a sphere, regardless of the length of its radius.

EXAMPLE 33-2

What solid angle is subtended at the center of an 8-m-diameter sphere by a 1.5-m^2 area on its surface?

Solution

Equation (33-5) yields

$$\Omega = \frac{1.5 \text{ m}^2}{(4 \text{ m})^2} = 0.0938 \text{ sr}$$

We are now in a position to clarify the definition of a unit that measures luminous flux. The **lumen** is defined by comparison with an internationally recognized standard source.

> **One lumen (lm) is the luminous flux (or visible radiant power) emitted from a $\frac{1}{60}$-cm^2 opening in a standard source and included within a solid angle of 1 sr.**

The standard source consists of a hollow enclosure maintained at the temperature of solidification of platinum, about 1773°C. In practice, it is more convenient to use standard incandescent lamps that have been rated by comparison with the standard.

Another convenient definition of the lumen utilizes the sensitivity curve (Fig. 33-15) as a basis for establishing luminous flux. By referring to the standard source, 1 lm is defined in terms of the radiant power of yellow-green light.

> **One lumen is equivalent to $\frac{1}{680}$ W of yellow-green light of wavelength 555 nm.**

To determine the luminous flux emitted by light of a different wavelength, the luminosity curve must be used to compensate for visual sensitivity.

EXAMPLE 33-3

A source of monochromatic red light (600 nm) produces a visible radiant power of 4 W. What is the luminous flux in lumens?

Solution

If the light were yellow-green (555 nm) instead of red, it would have a luminous flux F given by

$$F = (680 \text{ lm/W})(4 \text{ W}) = 2720 \text{ lm}$$

From the sensitivity curve, red light of wavelength 600 nm evokes about 59 percent of the response obtained with yellow-green light. Hence, the luminous flux issuing from the red light source is

$$F = (0.59)(2720 \text{ lm}) = 1600 \text{ lm}$$

The luminous flux is often calculated in the laboratory by determining the illumination it produces on a known surface area.

33-8 LUMINOUS INTENSITY

Light travels radially outward in straight lines from a source that is small in comparison with its surroundings. For such a source of light, the luminous flux included in a solid angle Ω remains the same at all distances from the source. Therefore, it is frequently more useful to speak of the *flux per unit solid angle* than simply to express the total flux. The physical quantity that expressed this relationship is called the **luminous intensity.**

The luminous intensity *I* of a source of light is the luminous flux *F* emitted per unit solid angle Ω.

$$I = \frac{F}{\Omega} \qquad\qquad 33\text{-}6$$

The unit for intensity is the *lumen per steradian* (lm/sr), called a *candela* (cd). The candela, or **candle** as it was sometimes called, originated when the international standard was defined in terms of the quantity of light emitted by the flame of a certain make of candle. This standard was found unsatisfactory, and eventually it was replaced by the platinum standard.

optics.org/home.ssi
This Web site is the home page of the Photonics Resources of Scientist and Engineers, where you can access the technical library and find out about research, industry news, events, and new products.

EXAMPLE 33-4

Most light sources have different luminous intensities in different directions. An **isotropic source** is one that emits light uniformly in all directions. What is the total luminous flux emitted by an isotropic source of intensity *I*?

Solution

From Eq. 33-6, the flux is given by

$$F = \Omega I$$

The total solid angle Ω for an isotropic source is 4π sr. Thus,

$$F = 4\pi I \qquad\qquad \textit{Isotropic Source} \qquad \textbf{33-7}$$

CONCEPTUAL EXAMPLE 33-5

A spotlight is equipped with a 40-cd bulb that concentrates a beam on a vertical wall. The beam covers an area of 9 m² on the wall, and the spotlight is located 20 m from the wall. Calculate the luminous intensity of the spotlight.

Solution

The total flux emitted by the 40-cd bulb is

$$F = 4\pi I = (4\pi)(40 \text{ cd}) = 160\pi \text{ lm}$$

This total flux is concentrated by reflectors and lenses into a solid angle given by

$$\Omega = \frac{A}{R^2} = \frac{9 \text{ m}^2}{(20 \text{ m})^2} = 0.0225 \text{ sr}$$

The intensity of the beam is now found from Eq. (33-6).

$$I = \frac{F}{\Omega} = \frac{160\pi \text{ lm}}{0.0225 \text{ sr}} = 2.23 \times 10^4 \text{ cd}$$

Note that the units of intensity (cd) and the units of flux (lm) are the same dimensionally. This is true because the solid angle in steradians is dimensionless.

33-9 ILLUMINATION

If the intensity of a source is increased, the luminous flux transmitted to each unit of surface area in the vicinity of the source is also increased. The surface appears brighter. In the measurement of light efficiency, the engineer is concerned with the density of luminous flux falling on a surface. We are therefore led to a discussion of the *illumination* of a surface.

> **The illumination *E* of a surface *A* is defined as the luminous flux *F* per unit area.**

$$E = \frac{F}{A} \qquad\qquad \textbf{33-8}$$

When the flux F is measured in lumens and the area A in square meters, the illumination E has the units of *lumens per square meter* or *lux* (lx). When A is expressed in square feet, E is expressed in *lumens per square foot*. The lumen per square foot is sometimes loosely referred to as the *footcandle*.

Direct application of Eq. 33-8 requires a knowledge of the luminous flux falling on a given surface. Unfortunately, the flux of common light sources is difficult to determine. For this reason, Eq. 33-8 is most often used to calculate the flux when A is known and E is computed from the measured intensity.

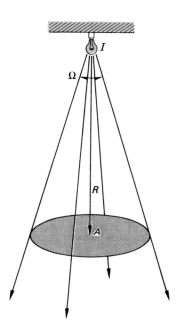

Figure 33-18   Computing the illumination of a surface perpendicular to the incident flux.

To see the relationship between intensity and illumination, let us consider a surface A at a distance R from a point source of intensity I, as shown in Fig. 33-18. The solid angle Ω subtended by the surface at the source is

$$\Omega = \frac{A}{R^2}$$

where the area A is perpendicular to the emitted light. If the luminous flux makes an angle θ with the normal to the surface, as shown in Fig. 33-19, we must consider the projected area $A \cos \theta$. This represents the effective area the flux "sees." Therefore, the solid angle, in general, can be found from

$$\Omega = \frac{A \cos \theta}{R^2}$$

Solving for the luminous flux F in Eq. (33-6), we obtain

$$F = I\Omega = \frac{IA \cos \theta}{R^2} \qquad\qquad \textbf{33-9}$$

We are now ready to express the illumination as a function of intensity. Substituting Eq. (33-9) into the defining equation for illumination gives the following:

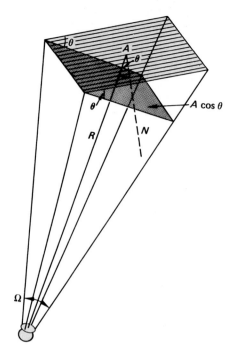

Figure 33-19 When a surface makes an angle θ with the incident flux, the illumination E is proportional to the component $A \cos \theta$ of the surface perpendicular to the flux.

$$E = \frac{F}{A} = \frac{IA \cos \theta}{AR^2}$$

or

$$E = \frac{I \cos \theta}{R^2} \qquad \text{33-10}$$

For the special case in which the surface is normal to the flux, $\theta = 0°$, and Eq. (33-10) is simplified to

$$E = \frac{I}{R^2} \qquad \text{\textit{Normal Surface}} \qquad \text{33-11}$$

You should verify that the units of *candela per square meter* are equivalent dimensionally to the units of *lumens per square meter* or *lux.*

EXAMPLE 33-6

A 100-W incandescent lamp has a luminous intensity of 125 cd. What is the illumination of a surface located 3 ft below the lamp?

Solution

By direct substitution into Eq. (33-11), we obtain

$$E = \frac{I}{R^2} = \frac{125 \text{ cd}}{(3 \text{ ft})^2} = 13.9 \text{ lm/ft}^2$$

EXAMPLE 33-7

A tungsten-filament lamp of intensity 300 cd is located 2.0 m away from a surface whose area is 0.25 m². The luminous flux makes an angle of 30° with the normal to the surface. (a) What is the illumination? (b) What is the luminous flux striking the surface? (Refer to Fig. 33-19.)

Solution (a)

The illumination is found directly from Eq. (33-10).

$$E = \frac{I \cos \theta}{R^2} = \frac{(300 \text{ cd})(\cos 30°)}{(2 \text{ m})^2} = 65 \text{ lx}$$

Solution (b)

The flux falling on the surface is found by solving for *F* in Eq. (33-8). Thus,

$$F = EA = (65 \text{ lx})(0.25 \text{ m}^2)$$
$$= 16.2 \text{ lm}$$

The equations above involving illumination and luminous intensity are mathematical formulations of the *inverse-square law*, which can be stated as follows:

The illumination of a surface is proportional to the luminous intensity of a point light source and is inversely proportional to the square of the distance.

If a light that is illuminating a surface is raised to twice the original height, the illumination will be only one-fourth as great. If the lamp distance is tripled, the illumination is reduced by one-ninth. This inverse-square relationship is illustrated in Fig. 33-20.

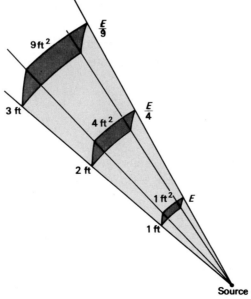

Figure 33-20 The luminous intensity *I* is constant for a given solid angle Ω. However, the illumination *E* (flux per unit area) decreases with the square of the distance from a point source.

Summary and Review

Summary

The investigation of the nature of light is still continuing, but experiments show that it sometimes behaves as particles and sometimes as a wave. Modern theory holds that light is electromagnetic radiation, and that its radiant energy is transported in photons carried along by a wave field. The main ideas and formulas presented in this chapter are summarized below.

- The wavelength λ of electromagnetic radiation is related to its frequency f by the general equation

$$c = f\lambda \qquad c = 3 \times 10^8 \text{ m/s}$$

- The range of wavelengths for visible light goes from 400 nm for violet to 700 nm for red.

$$1 \text{ nm} = 10^{-9} \text{ m} \qquad \begin{array}{l} \textit{The nanometer is} \\ \textit{used for wavelengths.} \end{array}$$

- The energy of light photons is proportional to the frequency.

$$E = hf \qquad E = \frac{hc}{\lambda} \qquad h = 6.626 \times 10^{-34} \text{ J} \cdot \text{s}$$

The constant h is *Planck's constant*.

- **Luminous flux** has been defined as that portion of the total radiant power emitted from a light source that is capable of affecting the sense of sight. Since visual brightness and sensitivity varies for different individuals, it was necessary to define luminous intensity in terms of a standard source and a well-defined solid angle (the steradian). By comparison to such standards, we are able to treat the illumination of surfaces, which is so important for the design of industrial workplaces.

- The **luminous intensity** of a light source is the luminous flux F per unit solid angle Ω. *Luminous flux* is the radiant power in the visible region. It is measured in *lumens*.

$$1 \text{ lm} = \frac{1}{680} \text{ W} \quad \text{for 555-nm light} \quad \textit{The Lumen}$$

$$\Omega = \frac{A}{R^2} \qquad \begin{array}{l} \textit{Solid Angle} \\ \textit{in Steradians} \end{array}$$

$$I = \frac{F}{\Omega} \qquad \begin{array}{l} \textit{Luminous Intensity} \\ \text{(1 cd = 1 lm/sr)} \end{array}$$

- For an isotropic source, one emitting light in all directions, the luminous flux is

$$F = 4\pi I \qquad \textit{Isotropic Source}$$

- The illumination E of a surface A is defined as the luminous flux per unit area.

$$E = \frac{F}{A} \qquad E = \frac{I \cos \theta}{R^2} \qquad \begin{array}{l} \textit{Illumination} \\ \text{(lm/m}^2\text{), lx} \end{array}$$

Key Terms

1. light
2. electromagnetic waves
3. visible region
4. nanometer
5. quantum theory
6. photons
7. light ray
8. umbra
9. penumbra
10. luminous flux
11. luminous intensity
12. steradian
13. illumination
14. isotropic source

Review Questions

33-1. What is meant by the dualistic nature of light? In what ways does light behave as particles? In what ways does light behave as a wave?

33-2. Explain how the energy of an electromagnetic wave depends on its frequency and how it depends on the wavelength.

33-3. When light enters glass from air, its energy in glass is the same as its energy in air. Is its frequency the same? What about its wavelength? Explain.

33-4. Microwave ovens, television, and radar utilize electromagnetic waves between infrared and radio waves. Compare the energy, frequency, and wavelengths of these waves with the energy, frequency, and wavelengths of visible radiation.

33-5. Review the definition of a radian, and discuss how the steradian for solid angles is similar to the radian for plane angles. How many radians are in a complete circle? How many steradians are in a complete sphere?

33-6. Draw a diagram illustrating a solar eclipse, labeling the umbra and penumbra regions. If you view a partial eclipse of the sun, are you standing in the umbra or the penumbra region?

33-7. Can you justify the following definition for a lumen? *A lumen is equal to the luminous flux falling on a surface on one square meter, all points of which are one meter from a uniform point source of one candela.*

33-8. An older unit for illumination was the *footcandle,* that illumination E received by a surface 1 ft^2 located at a distance of 1 ft from a 1-cd source of light. Explain how this definition is the same as that given in this text.

33-9. Describe the distribution of luminous flux from an incandescent lamp. Why is the lamp not an isotropic source?

33-10. Discuss the factors that will affect the illumination of a table in a machine shop.

33-11. Illumination is sometimes referred to as *flux density.* Explain why such a term might be appropriate.

33-12. *Photometry* is the science of measuring light. The intensity of a light source can be determined with the photometer, shown in Fig. 33-21. The luminous intensity I_x of an unknown source is found by visually comparing the unknown with a standard source of known intensity I_s. If the distances from each source are adjusted so

Figure 33-21 The grease-spot photometer is used to measure the intensity of an unknown source of light by comparing it to a standard source.

Summary and Review

that the grease spot is illuminated equally by each source, the unknown intensity I_x can be calculated from the inverse-square law. Derive the photometry equation

$$\frac{I_x}{r_x^2} = \frac{I_s}{r_s^2} \qquad \textit{Photometry Equation} \qquad 33\text{-}12$$

where r_s is the distance to the standard and r_x is the distance to the unknown source.

33-13. If two 40-W light bulbs are compared by using the photometer, will they necessarily be located at equal distances from the grease spot?

Problems

Section 33-3—The Electromagnetic Spectrum

33-1. An infrared spectrophotometer scans the wavelengths from 1 to 16 μm. Express this range in terms of the frequencies of the infrared rays.
Ans. 30.0×10^{13} to 1.88×10^{13}

33-2. What is the frequency of violet light of wavelength 410 nm?

33-3. A microwave radiator used in measuring automobile speeds emits a radiation of frequency 1.2×10^9 Hz. What is the wavelength? Ans. 250 mm

33-4. What is the range of frequencies for visible light?

33-5. If Plank's constant h is equal to 6.626×10^{-34} J s, what is the energy of light of wavelength 600 nm? Ans. 3.31×10^{-19}J

33-6. What is the frequency of light that has an energy of 5×10^{-19} J?

33-7. The frequency of yellow-green light is 5.41×10^{14} Hz. Express the wavelength of this light in nanometers and in angstroms.
Ans. 555 nm, 5550 A

33-8. What is the wavelength of light that has an energy of 7×10^{-19} J?

Section 33-5—The Speed of Light

33-9. The sun is approximately 93 million miles from Earth. How much time is required for the light emitted by the Sun to reach us on Earth? Ans. 8.33 min

33-10. If two experimenters in Galileo's experiment were separated by a distance of 5 km, how much time would have passed from the instant the lantern was opened until the light was observed?

33-11. The light reaching us from the nearest star, Alpha Centauri, requires 4.3 years to reach us. How far is this in miles? In kilometers?
Ans. 2.53×10^{13} mi, 4.07×10^{13} km

33-12. A spacecraft circling the moon at a distance of 384,000 km from the earth communicates by radio with a base on earth. How much time elapses between the sending and receiving of a signal?

33-13. An spacecraft sends a signal that requires 20 min to reach the earth. How far away is the spacecraft from earth?
Ans. 3.60×10^{11} m

Section 33-6—Light Rays and Shadows

33-14. The shadow formed on a screen 4 m away from a point source of light is 60 cm tall. What is the height of the object located 1 m from the source and 3 m from the shadow?

33-15. A point source of light is placed 15 cm from an upright 6-cm ruler. Calculate the length of the shadow formed by the ruler on a wall 40 cm from the ruler. Ans. 22.0 cm

33-16. How far must an 80-mm-diameter plate be placed in front of a point source of light if it is to form a shadow 400 mm in diameter at a distance of 2 m from the light source?

33-17. A source of light 40 mm in diameter shines through a pinhole in the tip of a cardboard box 2 m from the source. What is the diameter of the image formed on the bottom of the box if the height of the box is 60 mm?
Ans. 1.20 mm

*33-18. A lamp is covered with a box, and a 20-mm-long narrow slit is cut in the box so that light shines through. An object 30 mm tall blocks the light from the slit at a distance of 500 mm. Calculate the length of the umbra and penumbra formed on a screen located 1.50 m from the slit.

Section 33-9—Illumination

33-19. What is the solid angle subtended at the center of a 3.20-m-diameter sphere by a 0.5-m^2 area on its surface?   Ans. 0.195 sr

33-20. A solid angle of 0.080 sr is subtended at the center of a 9.00-cm-diameter sphere by surface area A on the sphere. What is this area?

33-21. An $8\frac{1}{2} \times 11$-cm sheet of metal is illuminated by a source of light located 1.3 m directly above it. What is the luminous flux falling on the metal if the source has an intensity of 200 cd. What is the total luminous flux emitted by the light source?
Ans. 1.11 lm, 2510 lm

33-22. A 40-W monochromatic source of yellow-green light (555 nm) illuminates a 0.5-m^2 surface at a distance of 1.0 m. What is the luminous intensity of the source, and how many lumens fall on the surface?

33-23. What is the illumination produced by a 200-cd source on a small surface 4.0 m away?   Ans. 12.5 lx

33-24. A lamp 2 m from a small surface produces an illumination of 100 lx on the surface. What is the intensity of the source?

33-25. A tabletop 1 m wide and 2 m long is located 4.0 m from a lamp. If 40 lm of flux fall on this surface, what is the illumination E of the surface?   Ans. 20.0 lx

33-26. What should be the location of the lamp in Prob. 33-25 in order to produce twice the illumination?

*33-27. A point source of light is placed at the center of a sphere 70 mm in diameter. A hole is cut in the surface of the sphere, allowing the flux to pass through a solid angle of 0.12 sr. What is the diameter of the opening?   Ans. 13.7 mm

Challenge Problems

33-28. When light of wavelength 550 nm passes from air into a thin glass plate and out again into the air, the frequency remains constant, but the speed through the glass is reduced to 2×10^8 m/s. What is the wavelength inside the glass?

33-29. A 30-cd standard light source is compared with a lamp of unknown intensity using a grease-spot photometer (refer to Fig. 33-21). The two light sources are placed 1 m apart, and the grease spot is moved toward the standard light. When the grease spot is 25 cm from the standard light source, the illumination is equal on both sides. Compute the unknown intensity.
Ans. 270 cd

33-30. Where should the grease spot in Prob. 33-29 be placed for the illumination by the unknown light source to be exactly twice the illumination of the standard source?

33-31. The illumination of a given surface is 80 lx when it is 3 m away from the light source. At what distance will the illumination be 20 lx?   Ans. 600 m

33-32. A light is suspended 9 m above a street and provides an illumination of 35 lx at a point directly below it. Determine the luminous intensity of the light.

*33-33. A 60-W monochromatic source of yellow-green light (555 nm) illuminates a 0.6 m^2 surface at a distance of 1.0 m. What is the solid angle subtended at the source? What is the luminous intensity of the source?
Ans. 0.60 sr, 68,000 cd

*33-34. At what distance from a wall will a 35-cd lamp provide the same illumination as an 80-cd lamp located 4.0 m from the wall?

*33-35. How much must a small lamp be lowered to double the illumination on an object that is 80 cm directly under it?
Ans. 23.4 cm

Summary and Review

*33-36. Compute the illumination of a given surface 140 cm from a 74-cd light source if the normal to the surface makes an angle of 38° with the flux.

*33-37. A circular tabletop is located 4 m below and 3 m to the left of a lamp that emits 1800 lm. What illumination is provided on the surface of the table? What is the area of the tabletop if 3 lm of flux falls on its surface? **Ans. 4.58 lx, 0.655 m²**

*33-38. What angle θ between the flux and a line drawn normal to a surface will cause the illumination of that surface to be reduced by one-half when the distance to the surface has not changed?

*33-39. In Michelson's measurements of the speed of light, as shown in Fig. 33-11, he obtained a value of 2.997×10^8 m/s. If the total light path was 35 km, what was the rotational frequency of the eight-sided mirror? **Ans. 1071 Hz**

*33-40. All the light from a spotlight is collected and focused on a screen of area 0.30 m². What must be the luminous intensity of the light for an illumination of 500 lx to be achieved?

*33-41. A 300-cd light is suspended 5 m above the left edge of a table. Find the illumination of a small piece of paper located a horizontal distance of 2.5 m from the edge of the table. **Ans. 8.59 lx**

Critical Thinking Questions

33-42. A certain radio station broadcasts at a frequency of 1150 kHz; a red beam of light has a frequency of 4.70×10^{14} Hz; and an ultraviolet ray has a frequency of 2.4×10^{16} Hz. Which has the highest wavelength? Which has the greatest energy? What are the wavelengths of each electromagnetic wave? **Ans. radio, ultraviolet, 261 m, 639 nm, 12.5 nm**

*33-43. An unknown light source A located 80 cm from a screen produces the same illumination as a standard 30-cd light source at point B located 30 cm from the screen. What is the luminous intensity of the unknown light source?

*33-44. The illumination of a surface 3.40 m directly below a light source is 20 lx. Find the intensity of the light source. At what distance below the light source will the illumination be doubled? Is the luminous flux also doubled at this location? **Ans. 231 cd, 2.40 m, no**

*33-45. The illumination of an isotropic source is E_A at a point A on a table 30 cm directly below the source. At what horizontal distance from A on the tabletop will the illumination be reduced by one-half?

*33-46. In Fizeau's experiment to calculate the speed of light, the plane mirror was located at a distance of 8630 m. He used a wheel containing 720 teeth (and voids). Every time the rotational speed of the wheel was increased by 24.2 rev/s, the light came through to his eye. What value did he obtain for the speed of light? **Ans. 3.01×10^8 m/s**

Reflection and Mirrors

OBJECTIVES

After completing this chapter, you should be able to:

1. Define and illustrate with drawings your understanding of the following terms: *virtual images, real images, converging mirror, diverging mirror, magnification, focal length,* and *spherical aberration.*
2. Use ray-tracing techniques to construct images formed by spherical mirrors.
3. Predict mathematically the nature, size, and location of images formed by spherical mirrors.
4. Determine the magnification and/or the *focal length* of spherical mirrors by mathematical and experimental methods.

The eye responds to light. Every object viewed is seen with light—either the light emitted by the object or light that is reflected from it. We now have a general understanding of the nature of light, and we have studied luminous objects and methods of measuring the light emitted from them.

Although all light can be traced to sources of energy, for example, the sun, an electric light bulb, or a burning candle, most of what we see in the physical world is a result of reflected light. In this chapter we treat the laws describing how light is turned back into its original medium as a result of striking a surface. Although this phenomenon, called *reflection,* can be interpreted in terms of Maxwell's electromagnetic wave theory, it is much simpler to describe it by tracing *rays.*

The ray treatment, generally referred to as **geometrical optics,** is based on the application of Huygens's principle. Remember that light rays are imaginary lines drawn perpendicular to advancing wavefronts in the direction of light propagation.

34-1 THE LAWS OF REFLECTION

When light strikes the boundary between two media, such as air and glass, one or more of three things can happen. As illustrated in Fig. 34-1, some of the light incident on a glass surface is reflected, and some passes into the glass. The light that enters the glass is partially absorbed and partially transmitted. The transmitted light usually undergoes a change in direction, called **refraction.** In this chapter we shall concern ourselves only with the phenomenon of reflection.

The reflection of light obeys the same general law of mechanics that governs other bouncing phenomena; that is, the angle of incidence equals the angle of reflection. For example, let us consider the pool table in Fig. 34-2a. To hit the black

Figure 34-1 When light strikes the boundary between two media, it may be reflected, refracted, or absorbed.

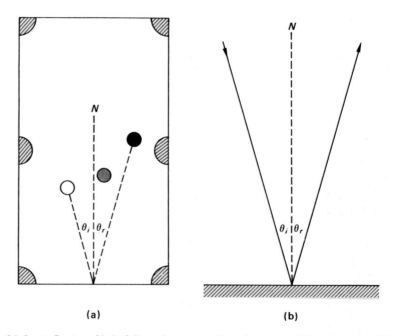

Figure 34-2 Reflection of light follows the same path as that expected for a bouncing billiard ball. The angle of incidence is equal to the angle of reflection.

ball on the right, we must aim at a point on the rail such that the incident angle θ_i is equal to the reflected angle θ_r. Similarly, light reflected from a smooth surface, as in Fig. 34-2b, has equal angles of incidence and reflection. The angles θ_i and θ_r are measured with respect to the normal to the surface. Two basic laws of reflection can be stated:

The angle of incidence is equal to the angle of reflection.

The incident ray, the reflected ray, and the normal to the surface all lie in the same plane.

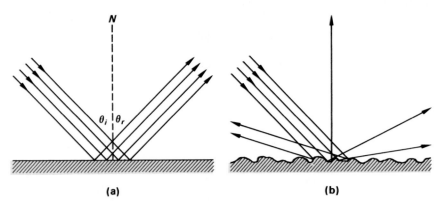

Figure 34-3   (a) Specular reflection and (b) diffuse reflection.

Light reflection from a smooth surface, in Fig. 34-3a, is called **regular** or **specular reflection.** Light striking the surface of a mirror or glass is specularly reflected. If all the incident light that strikes a surface were reflected in this manner, we could not see the surface. We would see only images of other objects. It is **diffuse reflection** (Fig. 34-3b) that enables us to see a surface. A rough or irregular surface will spread out and scatter the incident light, resulting in illumination of the surface. Reflection of light from brick, concrete, or newsprint provides an example of diffuse reflection.

34-2 PLANE MIRRORS

A highly polished surface that forms images by specular reflection of light is called a **mirror.** The mirrors hanging on the walls of our homes are usually flat, or **plane,** and we all have a certain familiarity with the images framed by plane mirrors. In every case, the image appears to be as far behind the mirror as the actual object is in front of it. As shown in Fig. 34-4, the images also appear right–left reversed. Anyone who has learned to tie a necktie or apply makeup by looking in a mirror is well aware of these effects.

To understand the formation of images by a plane mirror, let us first consider image I formed by rays emitted from a point O in Fig. 34-5. Four light rays are traced from the point source of light. The light ray OV is reflected back on itself at the mirror. Since the reflected light appears to have traveled the same distance

Figure 34-4   Images formed by mirrors are right-left reversed.

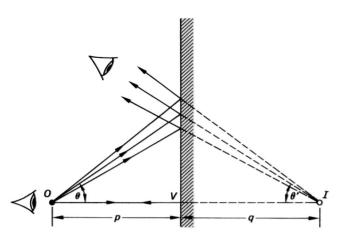

Figure 34-5 Constructing the image of a point object formed by a plane mirror.

as the incident light, the image will appear to be an equal distance behind the mirror when viewed along the normal to the reflecting surface. When the reflected light is viewed at an angle to the mirror, the same conclusion results: the image distance q is equal to the object distance p. This is true because the angle θ is equal to the angle θ' in the figure. Thus, we can say that:

The object distance is equal in magnitude to the image distance for a plane mirror.

$$p = q \qquad\qquad \textit{Plane Mirror} \quad \text{34-1}$$

We now consider the image formed by an extended object, as shown in Fig. 34-6. An extended object may be thought of as consisting of many point objects arranged according to the shape and size of the object. Each point on the object will have an image point located an equal distance behind the mirror. It follows that the image will have the same size and shape as the object. Right and left will be reversed, however, as discussed earlier.

Notice that the images formed by a plane mirror are, in truth, reflections of real objects. The images themselves are not real because no light passes through them.

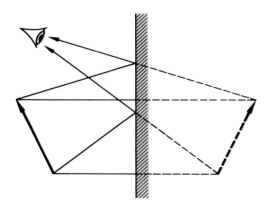

Figure 34-6 Image of an extended object.

These images that *appear* to the eye to be formed by rays of light but that in truth do not exist are called **virtual images**. A **real image** is an image formed by actual light rays.

> A virtual image **is one that seems to be formed by light coming from the image, but no light rays actually pass through it.**

> A real image **is formed by actual light rays that pass through it. Real images can be projected on a screen.**

Since virtual images are not formed by real light rays, they cannot be projected on a screen.

Real images cannot be formed by a plane mirror because the light reflected at a plane surface diverges. But if a plane mirror forms virtual images that do not physically exist, how can we see them? The full answer to this question must wait for a discussion of refraction and lenses. A preliminary answer is illustrated by Fig. 34-7, which also serves to demonstrate the two types of images. The eye makes use of the principle of refraction to reconverge the reflected light that *seems* to come from the virtual image. A *real* image is therefore projected on the retina of the eye. This image, which is formed by real, reflected light rays, is interpreted by the brain to have originated from a point behind the mirror. The brain is conditioned to the rectilinear propagation of light. It is fooled when light is somehow caused to change directions. Men who do not believe the brain can be conditioned to interpret images should try to tie someone's necktie without looking in a mirror. In this case, the real object seems less natural than its virtual image.

34-3 SPHERICAL MIRRORS

The same geometrical methods applied for the reflection of light from a plane mirror can be used for a curved mirror. The angle of incidence is still equal to the angle of reflection, but the normal to the surface changes at every point along the surface. A complicated relation between the object and its image results.

Most curved mirrors used in practical application are spherical. A **spherical mirror** is a mirror that may be thought of as a portion of a reflecting sphere. The

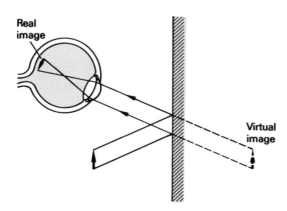

Figure 34-7 Image formed by a plane mirror is virtual. Such images appear to the eye to be located behind the mirror.

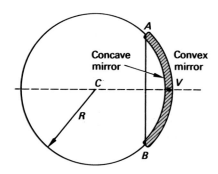

Figure 34-8 Definition of terms for spherical mirrors.

two kinds of spherical mirrors are illustrated in Fig. 34-8. If the inside of the spherical surface is the reflecting surface, the mirror is said to be **concave.** If the outside portion is the reflecting surface, the mirror is **convex.** In either case, R is the **radius of curvature,** and C is the *center of curvature* for the mirrors. The segment AB, often useful in optical problems, is called the **linear aperture** of the mirror. The dashed line CV that passes through the center of curvature and the topographical center, or *vertex,* of the mirror is known as the *axis* of the mirror.

Let us now examine the reflection of light from a spherical surface. As a simple case, suppose a beam of parallel light rays to be incident on a concave surface, as shown in Fig. 34-9. Since the mirror is perpendicular to the axis at its vertex V,

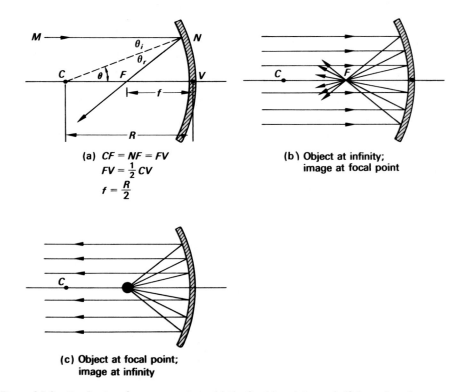

Figure 34-9 Focal point of a concave mirror: (a) The focal length is one-half the radius of curvature, (b) the object at infinity and the image at the focal point, and (c) the object at the focal point and the image at infinity.

a light ray *CV* is reflected back on itself. In fact, any light ray that proceeds along a radius of the mirror will be reflected back along itself. The parallel light ray *MN* is reflected so that the angle of incidence θ_i is equal to the angle of reflection θ_r. Both angles are measured with respect to the radius *CN*. The geometry of the reflection is such that the reflected ray passes through a point *F* on the axis halfway between the center of curvature *C* and the vertex *V*. The point, *F*, to which parallel light rays converge, is called the *focal point* of the mirror. The distance from *F* to *V* is called the **focal length** *f*. It is left as an exercise to show from Fig. 34-9a that

$$f = \frac{R}{2}$$

34-2

The focal length *f* of a concave mirror is one-half of its radius of curvature *R*.

All light rays from a distant object, such as the sun, will converge at the focal point *F*, as shown in Fig. 34-9b. For this reason, concave mirrors are frequently called **converging mirrors.** The focal point can be found experimentally by converging sunlight to a point on a piece of paper. The point along the axis of the mirror where the image formed on the paper is brightest will correspond to the focal point of the mirror.

Since light rays are reversible, a source of light placed at the focal point of a converging mirror will form its image at infinity. In other words, the emerging light beam will be parallel to the axis of the mirror, as shown in Fig. 34-9c.

A similar principle holds for a convex mirror, as illustrated in Fig. 34-10. Note that a parallel light beam incident on a convex surface diverges. The reflected light rays *appear* to come from a point *F* located behind the mirror, but no light rays actually pass through it. Even though the focal point is virtual, the distance *VF* is still called the **focal length** of the convex mirror. Since the actual light rays diverge when striking such a surface, a convex mirror is called a **diverging mirror.** Equation (34-2) also applies for a convex mirror. To be consistent with theory (to be developed later), however, the focal length *f* and the radius *R* must be reckoned as negative for diverging mirrors.

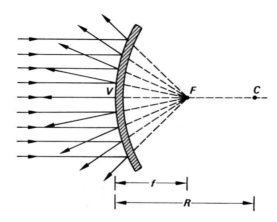

Figure 34-10 Focal point of a convex mirror.

34-4 IMAGES FORMED BY SPHERICAL MIRRORS

The best method for understanding the formation of images by mirrors is through geometrical optics, or *ray tracing*. This method consists of considering the reflection of a few rays diverging from some point of an object O that is *not* on the mirror axis. The point at which all these reflected rays will intersect determines the image location. We shall discuss three rays whose path can be traced easily. Each ray is illustrated for a converging (concave) mirror in Fig. 34-11 and for a diverging (convex) mirror in Fig. 34-12.

Ray 1 **A ray parallel to the mirror axis passes through the focal point of a concave mirror or seems to come from the focal point of a convex mirror.**

Ray 2 **A ray that passes through the focal point of a concave mirror or proceeds toward the focal point of a convex mirror is reflected parallel to the mirror axis.**

Ray 3 **A ray that proceeds along a radius of the mirror is reflected back along its original path.**

In any given situation, only two of these rays are necessary to locate the image of a point. By choosing rays from an extreme point of the object, the remainder of the image can usually be filled in by symmetry. In the figures, dashed lines are used to identify virtual rays and virtual images.

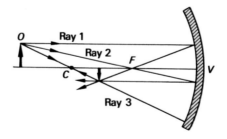

Figure 34-11 Principal rays for constructing the images formed by concave mirrors.

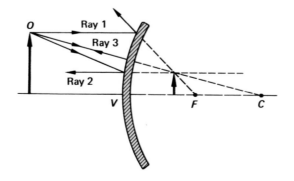

Figure 34-12 Principal rays for constructing the images formed by convex mirrors.

To illustrate the graphical method and at the same time visualize some of the possible images, let us consider several images formed by a concave mirror. Figure 34-13a illustrates the image formed by an object O that is located outside the center of curvature of the mirror. Note that the image is between the focal point F and the center of curvature C. The image is **real**, *inverted*, and *smaller* than the object.

In Fig. 34-13b the object O is located at the center of curvature C. The concave mirror forms an image at the center of curvature that is *real*, *inverted*, and the *same size* as the object.

In Fig. 34-13c the object O is located between C and F. Ray tracing shows that the image is located beyond the center of curvature. It is *real*, *inverted*, and *larger* than the object.

When the object is at the focal point F, all reflected rays are parallel (see Fig. 34-13d). Since the reflected rays will never intersect when extended in either direction, no image will be formed. (Some people prefer to say that the image distance is infinite.)

When the object is located inside the focal point F, as shown in Fig. 34-13e, the image *appears* to be behind the mirror. This can be seen by extending the reflected

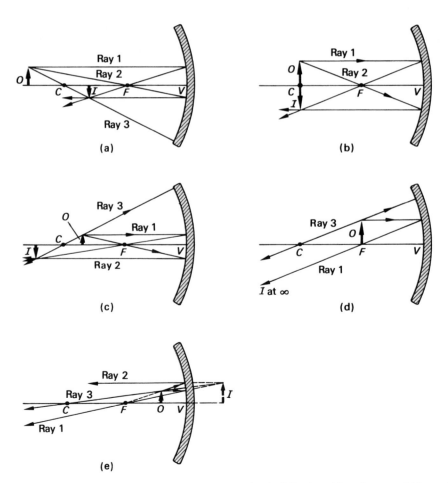

Figure 34-13 Images formed by a converging mirror for the following object distances: (a) beyond the center of curvature C, (b) at C, (c) between C and the focal length F, (d) at F, and (e) inside F.

rays to a point behind the mirror. Thus, the image is *virtual*. Notice also that the image is *enlarged* and erect (right side up). The **magnification** in this instance is the principle behind shaving mirrors and other mirrors that form enlarged virtual images.

On the other hand, all images formed by *convex* mirrors have the same characteristics. As was illustrated in Fig. 34-12, such images are *virtual, erect,* and *reduced in size.* This results in a wider field of view and accounts for many of the uses of convex mirrors. Automobile rear-view mirrors are usually convex to give maximum viewing capability. Some stores have large convex mirrors conveniently located to give them a panoramic view to help detect shoplifters.

34-5 THE MIRROR EQUATION

Now that we have a feel for the characteristics and formation of images, it will be useful to develop an analytical approach to image formation. Consider the reflection of light from a point object O, as illustrated in Fig. 34-14 for a concave mirror. The ray OV is incident along the axis of the mirror and is reflected back on itself. Ray OM is selected arbitrarily and proceeds toward the mirror at an angle α with the axis of the mirror. This ray is incident at an angle θ_i and reflected at an equal angle θ_r. The light rays reflected at M and V cross at the point I, forming an image of the object. The object distance p and image distance q are measured from the vertex of the mirror and indicated in the figure. The image at I is a *real* image since it is formed by actual light rays that pass through it.

Now, let us consider the image formed by an extended object OA, as shown in Fig. 34-15. The image of the point O is found to be I, as before. By tracing rays

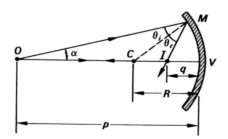

Figure 34-14   Converging mirror forms a point image of a point object.

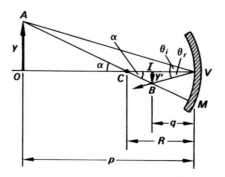

Figure 34-15   Deriving the mirror equation.

from the top of the arrow, we are able to draw the image of A at B. The ray AM passes through the center of curvature and is reflected back on itself. A ray AV that strikes the vertex of the mirror forms equal angles θ_i and θ_r. Rays VB and AM cross at B, forming an image of the top of the arrow at that point. The rest of the image IB can be constructed by tracing similar rays for corresponding points on the object OA. Notice that the image is *real* and *inverted*.

The following quantities are identified in Fig. 34-15.

$$
\begin{aligned}
\text{Object distance} &= OV = p \\
\text{Image distance} &= IV = q \\
\text{Radius of curvature} &= CV = R \\
\text{Object size} &= OA = y \\
\text{Image size} &= IB = y'
\end{aligned}
$$

We now attempt to relate these quantities. From the figure, it is noted that angles OCA and VCM are equal. Labeling this angle by α, we can write

$$
\tan \alpha = \frac{y}{p - R} = \frac{-y'}{R - q}
$$

from which

$$
\frac{-y'}{y} = \frac{R - q}{p - R} \qquad \text{34-3}
$$

The image size y' is negative because it is inverted in the figure. Similarly, angles θ_i and θ_r in the figure are equal, so that

$$
\tan \theta_i = \tan \theta_r \qquad \frac{y}{p} = \frac{-y'}{q} \qquad \text{34-4}
$$

Combining Eqs. (34-3) and (34-4), we have

$$
\frac{-y'}{y} = \frac{q}{p} = \frac{R - q}{p - R} \qquad \text{34-5}
$$

Rearranging terms, we obtain the important relation

$$
\frac{1}{p} + \frac{1}{q} = \frac{2}{R} \qquad \text{34-6}
$$

This relationship is known as the *mirror equation*. It is more often written in terms of the focal length f of the mirror, instead of the radius of curvature. Recalling that $f = R/2$, we can rewrite Eq. (34-6) as

$$
\frac{1}{p} + \frac{1}{q} = \frac{1}{f} \qquad \text{34-7}
$$

A similar derivation can be made for a convex mirror, and the same equations apply if the proper sign convention is adopted. Object and image distances, p and q, must be considered *positive* for real objects and images and *negative* for virtual objects and images. The radius of curvature R and focal length f must be considered *positive* for converging (concave) mirrors and *negative* for diverging (convex) mirrors.

EXAMPLE 34-1

What is the focal length of a converging mirror whose radius of curvature is 20 cm? What are the nature and location of an image formed by the mirror if an object is placed 15 cm from the vertex of the mirror?

Solution

The focal length is one-half of the radius of curvature, and the radius is positive for a converging mirror.

$$f = \frac{R}{2} = \frac{+20 \text{ cm}}{2} = +10 \text{ cm}$$

The location of the image is found from the mirror equation.

$$\frac{1}{p} + \frac{1}{q} = \frac{1}{f}$$

Solving explicitly for q yields

$$q = \frac{pf}{p - f}$$

from which

$$q = \frac{(15 \text{ cm})(10 \text{ cm})}{15 \text{ cm} - 10 \text{ cm}} = +30 \text{ cm}$$

Therefore, the image is real and located 30 cm from the mirror. Ray tracing similar to that of Fig. 34-13c shows that the image will also be inverted.

It is usually easier to solve the mirror equation explicitly for the unknown quantity than to substitute directly. You will find the following forms useful in most mirror problems:

$$p = \frac{qf}{q - f} \qquad q = \frac{pf}{p - f} \qquad f = \frac{pq}{p + q} \qquad \text{34-8}$$

The sign convention is summarized below:

1. The object distance p is positive for real objects and negative for virtual objects.
2. The image distance q is positive for real images and negative for virtual images.
3. The radius of curvature R and focal length f are positive for converging mirrors and negative for diverging mirrors.

This convention applies only to the *numerical* values substituted into Eq. (34-8). The quantities q, p, and f should maintain their signs unchanged until the substitution is made.

EXAMPLE 34-2

Find the position of the image if an object is located 4 cm from a convex mirror whose focal length is 6 cm.

Solution

In this case, $p = 4$ cm and $f = -6$ cm. The minus sign is necessary because a convex mirror is a diverging mirror.

$$q = \frac{pf}{p-f} = \frac{(4 \text{ cm})(-6 \text{ cm})}{4 \text{ cm} - (-6 \text{ cm})}$$

$$= \frac{-24 \text{ cm}^2}{4 \text{ cm} + 6 \text{ cm}} = -2.4 \text{ cm}$$

The image distance is negative, indicating that the image is virtual.

34-6 MAGNIFICATION

The images formed by spherical mirrors may be larger, smaller, or equal in size to the objects. The ratio of the image size to the object size is the *magnification M* of the mirror.

$$\text{Magnification} = \frac{\text{image size}}{\text{object size}} = \frac{y'}{y} \qquad \text{34-9}$$

The *size* refers to any linear dimension, height or width. Referring to Eq. (34-5) and to Fig. 34-15, we obtain the useful relation

$$M = \frac{y'}{y} = \frac{-q}{p} \qquad \text{34-10}$$

where q is the image distance and p is the object distance. A convenient feature of Eq. (34-10) is that *an inverted image will always have a negative magnification, and an erect image will have a positive magnification.*

EXAMPLE 34-3

A source of light 6 cm high is located 60 cm from a concave mirror whose focal length is 20 cm. Find the position, nature, and size of the image.

Solution

We first find the image distance q, as follows:

$$q = \frac{pf}{p-f} = \frac{(60 \text{ cm})(20 \text{ cm})}{60 \text{ cm} - 20 \text{ cm}}$$

$$= \frac{1200 \text{ cm}^2}{40 \text{ cm}} = 30 \text{ cm}$$

Since q is positive, the image is real. The image size is found from Eq. (34-10).

$$M = \frac{y'}{y} = -\frac{q}{p}$$

$$y' = -\frac{qy}{p} = -\frac{(30 \text{ cm})(6 \text{ cm})}{60 \text{ cm}}$$

$$= -3 \text{ cm}$$

The negative sign indicates that the image is inverted. Note that the magnification is $-\frac{1}{2}$.

Problem-Solving Strategy

Reflection and Mirrors

1. Read the problem carefully and draw a horizontal line representing the mirror axis. Indicate, by points on the mirror axis, the location of the radius R and focal length f of the mirror. (Remember that $f = R/2$.) Draw the concave or convex mirror as a curved line, and position the object as an erect arrow at its approximate location in front of the mirror.

2. Construct a geometric ray-tracing diagram to get a visual representation of the problem. A rough sketch should suffice, unless the problem requires a graphical solution.

3. Make a list of the given quantities, taking care to give the appropriate sign to each value. The radius and focal length are *positive* for *converging mirrors* and *negative* for *diverging mirrors*. Image distances q are *positive* when measured to *real images* and *negative* when measured to *virtual images*. The image size y' is *positive* for *erect images* and *negative* for *inverted images*.

4. Use the following equations to substitute and solve for the unknown quantities. Do not confuse the signs of operation (addition or subtraction) with the signs of substitution.

$$\frac{1}{p} + \frac{1}{q} = \frac{1}{f} \qquad M = \frac{y'}{y} = \frac{-q}{p}$$

$$R = \frac{f}{2}$$

5. It may be necessary to eliminate an unknown by solving both the mirror equation and the magnification equation simultaneously.

CONCEPTUAL EXAMPLE 34-4

How far should a pencil be held from a convex mirror to form an image one-half the size of the pencil? The radius of the mirror is 40 cm.

Solution

The focal length of the mirror is

$$f = \frac{R}{2} = \frac{-40 \text{ cm}}{2} = -20 \text{ cm}$$

The minus sign occurs because of the diverging mirror. Such a mirror always forms an erect image, reduced in size. (See Fig. 34-12.) The magnification in this case is $+\frac{1}{2}$. Thus,

$$M = -\frac{q}{p} = +\frac{1}{2}$$

$$q = -\frac{p}{2}$$

From the mirror equation, q is also

$$q = \frac{pf}{p - f}$$

Combining the two equations for q, we have

$$\frac{pf}{p-f} = -\frac{p}{2}$$

dividing by p gives

$$\frac{f}{p-f} = -\frac{1}{2}$$
$$2f = -p + f$$
$$p = -f$$
$$= -(-20 \text{ cm})$$
$$= 20 \text{ cm}$$

Thus, when an object is held at a distance equal to the focal length from a convex mirror, the image size is one-half of the object size.

34-7 SPHERICAL ABERRATION

In practice, spherical mirrors form reasonably sharp images as long as their apertures are small compared with their focal lengths. When large mirrors are used, however, some of the rays from objects strike near the outer edges and are focused to different points on the axis. This focusing defect, illustrated in Fig. 34-16, is known as *spherical aberration.*

A *parabolic mirror* does not exhibit this defect. Theoretically, parallel light rays incident on a parabolic reflector will focus at a single point on the mirror axis. (See Fig. 34-17.) A small source of light located at the focal point of a parabolic reflector is the principle used in many spotlights and search lights. The beam emitted from such a device is parallel to the axis of the reflector.

Figure 34-16   Spherical aberration.

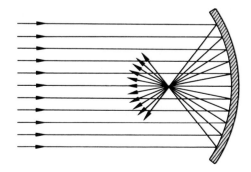

Figure 34-17   Parabolic reflector focuses all incident parallel light to the same point.

Summary and Review

Summary

In this chapter we have studied the reflective properties of converging and diverging spherical mirrors. The focal length and radius of curvature of such mirrors determine the nature and size of the images they form. Application of the formulas and ideas given in this chapter are necessary for understanding the operation and use of many technical instruments. The main concepts are summarized below.

- The formation of images by spherical mirrors can be visualized more easily with ray-tracing techniques. The three principal rays are listed below. You should refer to Fig. 34-18a for converging mirrors and to Fig. 34-18b for diverging mirrors.

  Ray 1   A ray parallel to the mirror axis passes through the focal point of a concave mirror or seems to come from the focal point of a convex mirror.

  Ray 2   A ray that passes through the focal point of a concave mirror or proceeds toward the focal point of a convex mirror is reflected parallel to the mirror axis.

  Ray 3   A ray that proceeds along a radius of the mirror is reflected back along its original path.

- Before listing the **mirror equations,** you should review what the symbols mean and the sign conventions.

R = radius of curvature,
    + for converging, − for diverging
f = focal length,
    + for converging, − for diverging
p = object distance,
    + for real object, − for virtual
q = image distance,
    + for real images, − for virtual
y = object size, + if erect, − if inverted
y' = image size, + if erect, − if inverted
M = magnification,
    + if both erect or both inverted

- The mirror equations can be applied to either converging (concave) or diverging (convex) spherical mirrors:

$$f = \frac{R}{2} \qquad M = \frac{y'}{y} = \frac{-q}{p}$$

$$\frac{1}{p} + \frac{1}{q} = \frac{1}{f} \qquad \textit{Mirror Equations}$$

- Alternative forms for the last equation are:

$$p = \frac{qf}{q - f} \qquad q = \frac{pf}{p - f} \qquad f = \frac{pq}{p + q}$$

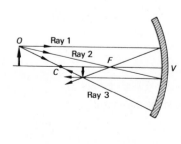

(a)
(b)

Figure 34-18   Ray tracing for (a) converging mirror and (b) a diverging mirror.

Key Terms

1. plane mirror
2. spherical mirror
3. linear aperture
4. focal length
5. real image
6. virtual image
7. diverging mirror
8. magnification
9. specular reflection
10. diffuse reflection
11. geometrical optics
12. radius of curvature
13. converging mirror
14. mirror equation
15. spherical aberration
16. parabolic mirror

Review Questions

34-1. Discuss the statement: One cannot "see" the surface of a perfect mirror.

34-2. Prove by a diagram that rays diverging from a point source of light appear to diverge from a virtual point after reflection from a plane surface.

34-3. Can an image of a real object be projected on a screen by a plane mirror? By a convex mirror? By a concave mirror?

34-4. State the laws of reflection and show how they may be demonstrated in a laboratory.

34-5. Use the mirror equation to show that the image of an infinitely distant object is formed at the focal point of a spherical mirror.

34-6. Use the mirror equation to show that the image of an object placed at the focal point of a concave mirror is located at infinity.

34-7. Use the mirror equation to show that, for a plane mirror, the image distance is equal in magnitude to the object distance. What is the magnification of a plane mirror?

34-8. In a concave shaving mirror, will greater magnification be achieved when the object is closer to the focal point or when it is closer to the vertex? Use diagrams to verify your conclusion.

34-9. Do objects moving closer to the vertex of a convex mirror form larger or smaller virtual images? Explain with diagrams.

34-10. Without looking at Fig. 34-13 in the text, construct the images formed by a concave mirror when the object is (a) beyond C, (b) at C, (c) between C and F, (d) at F, and (e) between F and V. Discuss the nature and relative size of each image.

34-11. Several small spherical mirrors are lying on a laboratory table. Describe how you would distinguish the diverging mirrors from the converging mirrors without touching them.

34-12. For real objects, is it possible to construct an inverted image by using a diverging mirror? What can you say about the magnification of diverging mirrors?

34-13. You wish to choose a shaving mirror that will give maximum magnification with an erect image. Does the focal length of the mirror play a part in determining its magnification? Explain.

34-14. Two concave spherical mirrors have the same focal length, but one has a larger linear aperture. Which forms the sharper image? Why?

34-15. Show how it is possible for a plane mirror to form a real image if light from the object is first converged by a concave mirror.

Problems

Section 34-1—Plane Mirrors

34-1. A man 1.80 m tall stands 1.2 m from a large plane mirror. How tall is his image? How far is he from his image? **Ans. 1.80 m, 1.20 m**

34-2. What is the shortest mirror length required to enable a 1.68-m-tall woman to see her entire image?

Summary and Review

*34-3. A plane mirror moves at a speed of 30 km/h away from a stationary person. How fast does this person's image appear to be moving in the opposite direction?

Ans. 60 km/h

*34-4. The optical lever is a sensitive measuring device that utilizes minute rotations of a plane mirror to measure small deflections. The device is illustrated in Fig. 34-19. When the mirror is in position 1, the light ray follows the path IVR_1. If the mirror is rotated through an angle θ to position 2, the ray will follow the path IVR_2. Show that the reflected beam turns through an angle 2θ, which is twice the angle through which the mirror itself turns.

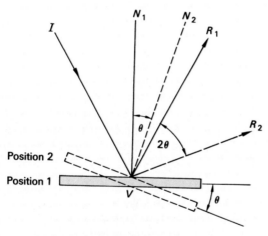

Figure 34-19 The optical lever.

Section 34-3—Spherical Mirrors

34-5. A light bulb 3 cm high is placed 20 cm in front of a concave mirror with a radius of curvature of 15 cm. Determine the nature, size, and location of the image formed. Sketch the ray-tracking diagram.

Ans. Real, $y' = -1.8$ cm, $q = +12$ cm

34-6. A spherical concave mirror has a focal length of 20 cm. What are the nature, size, and location of the image formed when a 6-cm-tall object is located 15 cm from this mirror?

34-7. A 8-cm-long pencil is placed 10 cm from a diverging mirror of radius 30 cm. Determine the nature, size, and location of the image formed. Sketch the ray-tracing diagram.

Ans. Virtual, $y' = +4.80$ cm, $q' = -6.00$ cm

34-8. A spherical convex mirror has a focal length 25 cm. What are the nature, size, and location of the image formed of a 5-cm-tall object located 30 cm from the mirror?

34-9. An object 5 cm tall is placed halfway between the focal point and the center of curvature of a concave spherical mirror of radius 30 cm. Determine the location and magnification of the image.

Ans. $q = +45$ cm, $M = -2.00$

34-10. A 4-cm-high source of light is placed in front of a spherical concave mirror that has a radius of 40 cm. Determine the nature, size, and location of the images formed for the following object distances: (a) 60 cm, (b) 40 cm, (c) 30 cm, (d) 20 cm, and (e) 10 cm. Draw the appropriate ray-tracing diagrams.

*34-11. At what distance from a concave spherical mirror of radius 30 cm must an object be placed to form an enlarged, inverted image located 60 cm from the mirror?

Ans. $p = 20$ cm

Section 34-6—Magnification

34-12. What is the magnification of an object if it is located 10 cm from a mirror and its image is erect and seems to be located 40 cm behind the mirror? Is this mirror diverging or converging?

34-13. A Christmas tree ornament has a silvered surface and a diameter of 3 in. What is the magnification of an object placed 6 in. from the surface of this ornament?

Ans. +0.111

34-14. What type of mirror is required to form an image on a screen 2 m away from the mirror when an object is placed 12 cm in front of the mirror? What is the magnification?

*34-15. A concave shaving mirror has a focal length of 520 mm. How far away from it should an object be placed for the image to be erect and twice its actual size?   Ans. 260 mm

*34-16. If a magnification of $+3$ is desired, how far should the mirror of Prob. 34-15 be placed from the face?

*34-17. An object is placed 12 cm from the surface of a spherical mirror. If an erect image is formed that is one-third the size of the object, what is the radius of the mirror? Is it converging or diverging?

   Ans. -6.00 cm, diverging

*34-18. A concave spherical mirror has a radius of 30 cm and forms an inverted image on a wall 90 cm away. What is the magnification?

Challenge Problems

34-19. What are the nature, size, and location of the image formed when a 6-cm-tall object is located 15 cm from a spherical concave mirror of focal length 20 cm?

   Ans. Virtual, erect, $q = -60$ cm, $y' = +24$ cm

34-20. An erect image has a magnification of $+0.6$. Is the mirror diverging or converging? What is the object distance if the image distance is -12 cm?

34-21. An object is located 50 cm from a converging mirror that has a radius of 40 cm. What are the image distance and the magnification?

   Ans. $q = +33.3$ cm, $M = -0.667$

34-22. What is the focal length of a diverging mirror if the image of an object located 200 mm from the mirror appears to be a distance of 120 mm behind the mirror?

34-23. A silver ball is 4.0 cm in diameter. Locate the image of a 6-cm-long object located 9 cm from the surface of the ball. What is the magnification?   Ans. -9.00 mm, $+0.100$

34-24. An object 80 mm tall is placed 400 mm in front of a diverging mirror of radius -600 mm. Determine the nature, size, and location of the image.

*34-25. An object 10 cm tall is located 20 cm from a spherical mirror. If an erect image 5 cm tall is formed, what is the focal length of the mirror?   Ans. $f = -20$ cm

*34-26. What is the magnification if the image of an object is located 15 cm from a diverging mirror of focal length -20 cm?

*34-27. An object is placed 200 mm from the vertex of a convex spherical mirror that has a radius of 400 mm. What is the magnification of the mirror?   Ans. $M = +\frac{1}{2}$

*34-28. A convex spherical mirror has a radius of -60 cm. How far away should an object be held if the image is to be one-third the size of the object?

*34-29. What should be the radius of curvature of a convex spherical mirror if it is to produce an image one-fourth as large as an object that is located 40 in. from the mirror?

   Ans. $q = -26.7$ in.

*34-30. A convex mirror has a focal length of -500 mm. If an object is placed 400 mm from the vertex, what is the magnification?

*34-31. A spherical mirror forms a real image 18 cm from the surface. The image is twice as large as the object. Find the location of the object and the focal length of the mirror.

   Ans. $p = 9.00$ cm, $f = 6.00$ cm

*34-32. A certain mirror placed 2 m from an object produces an erect image enlarged three times. Is the mirror diverging or converging? What is the radius of the mirror?

*34-33. The magnification of a mirror is -0.333. Where is the object located if its image is formed on a card 540 mm from the mirror? What is the focal length?

   Ans. $p = 1.62$ m, $f = +405$ mm

*34-34. What should be the radius of curvature of a concave mirror if it is to produce an image one-fourth as large as an object 50 cm away from the mirror?

*34-35. A spherical shaving mirror has a magnification of $+2.5$ when an object is located 15 cm from the surface. What is the focal length of the mirror?   Ans. $f = +25$ cm

Critical Thinking Questions

34-36. A baseball player is 6 ft tall and stands 30 ft in front of a plane mirror. The distance from the top of his cap to his eyes is 8 in. Draw a diagram showing location of the images formed of his feet and of the top of his cap. What is the minimum length of mirror required for him to see his entire image? If he walks 10 m closer to the mirror, what is the new separation of object and image? **Ans. 36 in., 40 ft**

*34-37. The diameter of the Moon is 3480 km, and it is 3.84×10^8 m away from Earth. A telescope on Earth utilizes a spherical mirror that has a radius of 8.00 m to form an image of the Moon. What is the diameter of the image formed? What is the magnification of the mirror?

*34-38. An image 60 mm long is formed on a wall located 2.3 m away from a source of light 20 mm high. What is the focal length of this mirror? Is it diverging or converging? What is the magnification?
Ans. 862 mm, converging, −3.00

*34-39. Derive an expression for calculating the focal length of a mirror in terms of the object distance p and the magnification M. Apply it to Prob. 34-25. Derive a similar relation for calculating the image distance q in terms of M and p. Apply it to Prob. 34-33.

*34-40. A concave mirror of radius 800 mm is placed 600 mm from a plane mirror that faces it. A source of light placed midway between the mirrors is shielded so that the light is first reflected from the concave surface. What are the position and magnification of the image formed after reflection from the plane mirror? (*Hint:* Treat the image formed by the first mirror as the object for the second mirror.)
Ans. 1.8 m behind plane mirror, +4.00

Refraction

Light travels in straight lines at a constant speed in a uniform medium. If the medium changes, the speed will also change and the light will travel in a straight line along a new path. The bending of a light ray as it passes obliquely from one medium to another is known as **refraction.** The principle of refraction is illustrated in Fig. 35-1 for a light wave entering water from the air. The angle θ_i that the incident beam makes with the normal to the surface is referred to as the *angle of incidence.* The angle θ_r between the refracted beam and the normal is called the *angle of refraction.*

Refraction explains familiar phenomena such as the apparent distortion of objects partially submerged in water. The stick appears to be bent at the surface

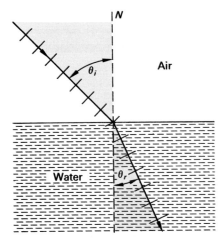

Figure 35-1   Refraction of a wavefront at the boundary between two media.

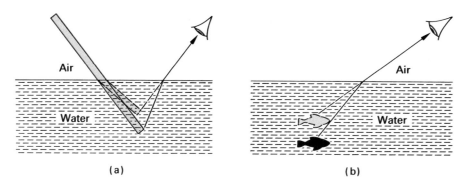

Figure 35-2 Refraction is responsible for the distortion of images. (a) The stick appears to be bent. (b) The fish seems to be closer to the surface than it actually is.

of the water in Fig. 35-2a, and the fish in Fig. 35-2b appears to be closer to the surface than it really is. In this chapter we study the properties of refractive media and develop equations to predict their effect on incident light rays.

35-1 INDEX OF REFRACTION

*inter*NET
CONNECTION

www.allexperts.com
If there are questions you can't answer, you might turn to the experts in a variety of fields at this Web site.

The velocity of light in a material substance is generally less than the free-space velocity of 3×10^8 m/s. In water the speed of light is almost 2.25×10^8 m/s, which is just about three-fourths of its velocity in air. Light travels about two-thirds as fast in glass, or around 2×10^8 m/s. The ratio of the velocity c of light in a vacuum to the velocity v of light in a particular medium is called the **index of refraction** n for that material.

The index of refraction n of a particular material is the ratio of the free-space velocity of light to the velocity of light through the material.

$$n = \frac{c}{v}$$

Index of Refraction 35-1

TABLE 35-1 Index of Refraction for Yellow Light of Wavelength 589 nm

Substance	n	Substance	n
Benzene	1.50	Glycerin	1.47
Carbon disulfide	1.63	Ice	1.31
Diamond	2.42	Quartz	1.54
Ethyl alcohol	1.36	Rock salt	1.54
Fluorite	1.43	Water	1.33
Glass		Zircon	1.92
Crown	1.52		
Flint	1.63		

The index of refraction is a unitless quantity, which is generally greater than unity. For water, $n = 1.33$, and for glass, $n = 1.5$. Table 35-1 lists the index of refraction for several common substances. Note that the values given apply for yellow light of wavelength 589 nm. The velocity of light in material substances is different for different wavelengths. This effect, known as **dispersion,** will be discussed in Sec. 35-4. When the wavelength of light is not specified, the index is usually assumed to correspond to that for yellow light.

EXAMPLE 35-1

Compute the velocity of yellow light in a diamond that has a refractive index of 2.42.

Solution
Solving for v in Eq. (35-1) gives

$$v = \frac{c}{n} = \frac{3 \times 10^8 \text{ m/s}}{2.42} = 1.24 \times 10^8 \text{ m/s}$$

It is the exceptionally large refractive index that provides one of the most positive tests for the identification of diamonds.

35-2 THE LAWS OF REFRACTION

Two basic laws of refraction have been known and observed since ancient times. These laws are stated as follows (refer to Fig. 35-3):

The incident ray, the refraction ray, and the normal to the surface all lie in the same plane.

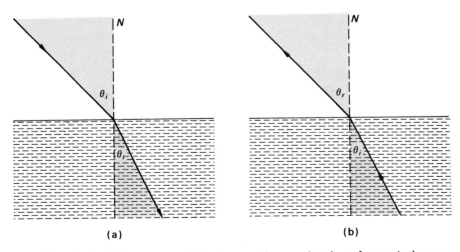

(a) (b)

Figure 35-3 (a) The incident ray, the refracted ray, and the normal to the surface are in the same plane. (b) Refracted rays are reversible.

The path of a ray refracted at the interface between two media is exactly reversible.

These two laws are demonstrated easily by observation and experiment. It is of much more importance, in a practical sense, however, to understand and predict the *degree* of bending that occurs.

To understand how a change in the velocity of light can alter its path through a medium, let us consider the mechanical analogy shown in Fig. 35-4. In Fig. 35-4a light incident on the glass plate is first bent toward the normal as it passes through the denser medium, and then it is bent away from the normal as it returns to the air. In Fig. 35-4b the action of wheels encountering a patch of sand resembles the behavior of light. As they approach the sand, one wheel strikes first and slows down. The other wheel continues at the same speed, causing the axle to assume a new angle. When both wheels are in the sand, the wheels again move in a straight line with uniform speed. The first wheel to enter the sand is also the first to leave, and it speeds up as it leaves the patch of sand. Thus, the axle swings around to its original direction. The path of the axle is analogous to the path of a wavefront.

The change in direction of light as it enters another medium can be analyzed with the help of the wavefront diagram in Fig. 35-5. A plane wave in a medium of refractive index n_1 meets at the plane surface of a medium whose index is n_2. The angle of incidence is labeled θ_2. In the figure, it is assumed that the second medium has a greater optical density than the first ($n_2 > n_1$). An example is light passing from air ($n_1 = 1$) to water ($n_2 = 1.33$). The line AB represents the wavefront at time $t = 0$ when it just comes into contact with medium 2. The line CD represents the same wavefront after the time t required to enter the second medium completely. Light travels from B to D in medium 1 in the same time t required for light to travel from A to C in medium 2. Assuming that the velocity v_2 in the second medium is smaller than the velocity v_1 in the first medium, the distance AC will be shorter than the distance BD. These lengths are given by

$$AC = v_2 t \qquad BD = v_1 t$$

(a) (b)

Figure 35-4 (a) The lateral displacement of light passing through glass. (b) A mechanical analogy.

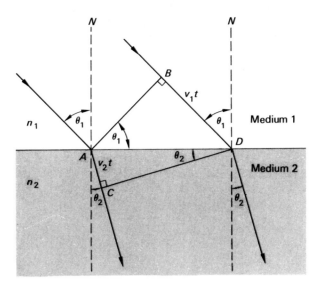

Figure 35-5 Deriving Snell's law.

It can be shown from geometry that angle BAD is equal to θ_1 and that angle ADC is equal to θ_2, as indicated in Fig. 35-5. The line AD forms a hypotenuse that is common to the two triangles ADB and ADC. From the figure,

$$\sin \theta_1 = \frac{v_1 t}{AD} \qquad \sin \theta_2 = \frac{v_2 t}{AD}$$

Dividing the first equation by the second, we obtain

$$\frac{\sin \theta_1}{\sin \theta_2} = \frac{v_1}{v_2} \qquad\qquad \textbf{35-2}$$

> **The ratio of the sine of the angle of incidence to the sine of the angle of refraction is equal to the ratio of the velocity of light in the incident medium to the velocity of light in the refracted medium.**

This rule was first discovered by the seventeenth-century Dutch astronomer Willebrord Snell and is called *Snell's law* in his honor. An alternative form for the law can be obtained by expressing the velocities v_1 and v_2 in terms of the indexes of refraction for the two media. Recall that

$$v_1 = \frac{c}{n_1} \qquad \text{and} \qquad v_2 = \frac{c}{n_2}$$

Utilizing these relations in Eq. (35-2), we write:

$$n_1 \sin \theta_1 = n_2 \sin \theta_2 \qquad\qquad \textbf{35-3}$$

Since the sine of an angle increases as the angle increases, we see that an increase in the index of refraction results in a decrease in the angle and vice versa.

EXAMPLE 35-2

Light passes at an angle of incidence of 35° from water into the air. What is the angle of refraction if the index of refraction for water is 1.33?

Solution

The angle θ_a can be found from Snell's law.

$$n_w \sin \theta_w = n_a \sin \theta_a$$
$$1.33 \sin 35° = 1.0 \sin \theta_a$$
$$\sin \theta_a = 1.33 \sin 35° = 0.763$$
$$\theta_a = 49.7°$$

The index of refraction *decreased* from 1.33 to 1.0, and so the angle *increased*.

EXAMPLE 35-3

A ray of light in water ($n_w = 1.33$) is incident upon a plate of glass ($n_g = 1.5$) at an angle of 40°. What is the angle of refraction into the glass? Refer to Fig. 35-6.

Figure 35-6

Solution

Applying Snell's law at the interface, we obtain

$$n_w \sin \theta_w = n_g \sin \theta_g$$
$$1.33 \sin 40° = 1.5 \sin \theta_g$$
$$\sin \theta_g = \frac{1.33}{1.5} \sin 40° = 0.57$$
$$\theta_g = 34.7°$$

This time *n* increased, and so θ decreased.

As an additional exercise, you should show that the ray is reversible.

35-3 WAVELENGTH AND REFRACTION

We have seen that light slows down when passing into a medium of greater *optical density*. What happens to the wavelength of light entering a new medium? In Fig. 35-7, light traveling in air at a velocity c encounters a medium through which it travels at the reduced speed v_m. Upon returning to the air, it again travels at the speed c of light in air. This does not violate the conservation of energy because the energy of a light wave is proportional to its frequency (see Sec. 33-4). The frequency f is the same inside the medium as it is outside the medium. That this is true will be realized if you consider that the frequency is the number of waves passing any point per unit of time. The same number of waves leaves the medium in a second as it enters the medium in a second. Thus, the frequency inside the medium cannot change. The velocity is related to the frequency and wavelength by

$$c = f\lambda_a \qquad \text{and} \qquad v_m = f\lambda_m \qquad\qquad \textbf{35-4}$$

where c and v_m are the speeds in air and inside the medium and λ_a and λ_m are the respective wavelengths. Since the velocity decreases inside the medium, the wavelength inside the medium must decrease proportionately for the frequency to remain constant. Dividing the first equation by the second in Eq. (35-4) yields

$$\frac{c}{v_m} = \frac{f\lambda_a}{f\lambda_m} = \frac{\lambda_a}{\lambda_m}$$

If we substitute $v_m = c/n_m$, we obtain

$$n_m = \frac{\lambda_a}{\lambda_m}$$

Therefore, the wavelength λ_m inside the medium is reduced by

$$\lambda_m = \frac{\lambda_a}{n_m} \qquad\qquad \textbf{35-5}$$

where n_m is the index of refraction of the medium and λ_a is the wavelength of the light in air.

***inter*NET**
CONNECTION

wigner.byu.edu
For a simulation of waves going from one material into another, see the applet at this Web site, at which you can change the angle of incidence, the index of refraction, and the color of the incoming light (i.e., its wavelength).

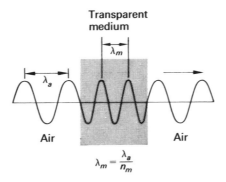

Figure 35-7 The wavelength of light is reduced when it enters a medium of greater optical density.

EXAMPLE 35-4

Monochromatic red light of wavelength 640 nm passes from air into a glass plate of refractive index 1.5. What is the wavelength of the light inside the medium?

Solution

Direct substitution into Eq. (35-5) yields

$$\lambda_m = \frac{\lambda_a}{n_m} = \frac{640 \text{ nm}}{1.5} = 427 \text{ nm}$$

As a summary of the relations discussed so far, we can write

$$\frac{\sin \theta_1}{\sin \theta_2} = \frac{v_1}{v_2} = \frac{n_2}{n_1} = \frac{\lambda_1}{\lambda_2} \qquad 35\text{-}6$$

where the subscripts 1 and 2 refer to different media. Here we see the relationship among all the important quantities affected by refraction.

35-4 DISPERSION

We have already mentioned that the velocity of light in different substances varies with different wavelengths. We defined the index of refraction as the ratio of the free-space velocity c to the velocity inside a medium.

$$n = \frac{c}{v_m}$$

The values given in Table 35-1 are valid strictly for monochromatic yellow light (589 nm). A different wavelength of light, such as blue light or red light, results in a slightly different index of refraction. Red light travels faster through a particular medium than blue light. This can be shown by passing white light through a glass prism, as in Fig. 35-8. Because of the different speeds inside the medium, the beam is *dispersed* into its component colors.

Dispersion **is the separation of light into its component wavelengths.**

From such an experiment we conclude that white light is actually a mixture of light, consisting of several colors. The projection of a dispersed beam is called a *spectrum.*

35-5 TOTAL INTERNAL REFLECTION

A fascinating phenomenon, known as ***total internal reflection,*** can occur when light passes obliquely from one medium to a medium with a lower optical density. To understand this phenomenon, consider a source of light submerged in

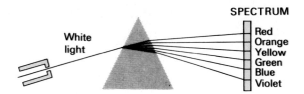

SPECTRUM
Red
Orange
Yellow
Green
Blue
Violet

White
light

Figure 35-8 Dispersion of light by a prism.

medium 1, as illustrated in Fig. 35-9. Consider the four rays A, B, C, and D, which diverge from the submerged source. Ray A passes into medium 2 normal to the interface. The angle of incidence and the angle of refraction are both zero for this special case. Ray B is incident at an angle θ_1 and refracted away from the normal at an angle θ_2. The angle θ_2 is greater than θ_1 because the index of refraction for medium 1 is greater than that for medium 2 ($n_1 > n_2$). As the angle of incidence θ_1 increases, the angle of refraction θ_2 also increases until the refracted ray C emerges tangent to the surface. The angle of incidence θ_c for which this occurs is known as the **critical angle.**

> The critical angle θ_c is the limiting angle of incidence in a denser medium that results in an angle of refraction of 90°.

A ray approaching the surface at an angle greater than the critical angle is reflected back inside medium 1. The ray D in Fig. 35-9 does not pass into the upper medium at all but is *totally internally reflected* at the interface. This type of reflection follows the same laws as any other reflection; in other words, the angle of incidence is equal to the angle of reflection. Total internal reflection can occur only when light is incident from a denser medium ($n_1 > n_2$).

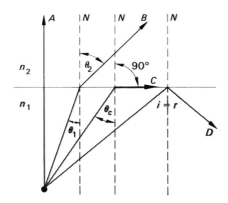

Figure 35-9 Critical angle of incidence.

The critical angle for two given media can be calculated from Snell's law.

$$n_1 \sin \theta_c = n_2 \sin \theta_2$$

where θ_c is the critical angle and $\theta_2 = 90°$. Simplifying, we write

$$n_1 \sin \theta_c = n_2(1)$$

or

$$\sin \theta_c = \frac{n_2}{n_1} \qquad \qquad \textit{Critical Angle} \qquad 35\text{-}7$$

Since $\sin \theta_c$ can never be greater than 1, n_1 must be greater than n_2.

EXAMPLE 35-5

What is the critical angle for a glass-to-air surface if the refractive index of the glass is 1.5?

Solution

Direct substitution yields

$$\sin \theta_c = \frac{n_a}{n_g} = \frac{1.0}{1.5} = 0.667$$
$$\theta_c = 42°$$

The fact that the critical angle for glass is 42° makes it possible to use 45° prisms in many optical instruments. Two such uses are illustrated in Fig. 35-10. In Fig. 35-10a a 90° reflection can be obtained with little loss of intensity. In Fig. 35-10b a 180° deflection is obtained. In each case, total internal reflection occurs because the angles of incidence are all 45° and therefore greater than the critical angle.

35-6 FIBER OPTICS AND APPLICATIONS

The application of fiber optics in communications has resulted in an information explosion. Optical fiber has a significantly larger bandwidth than copper wire, which means more information can be transmitted during a fixed time period. This increased information-carrying capability provides significant new possibilities, including interactive television and cable channel selections numbering in the thousands.

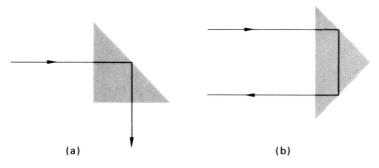

(a)   (b)

Figure 35-10 Right-angle prisms make use of the principle of total internal reflection to deviate the path of light.

Although it is easy to see how fiber optics is changing the world around us, it may come as a surprise that its ability to transmit information depends primarily on a single physical phenomenon, *total internal reflection*. As discussed in previous sections, total internal reflection results when light passing through one medium encounters a second medium of lower optical density. An optical fiber consists of two such mediums.

The structure of an optical fiber is illustrated in Fig. 35-11. Notice that the core of the fiber serves as the transmission medium, while the cladding acts to contain the transmitted signal. This means that the core must have a higher index of refraction than the cladding ($n_1 > n_2$).

Using the ray model of propagation, we can think of signals propagating through a fiber as distinct rays of light. As illustrated in Fig. 35-12, each ray has a fixed angle of incidence with the core/cladding boundary. As long as the beams maintain an angle of incidence equal to or greater than the fiber's critical angle, the signal will be confined to the fiber. (*Note:* Since an optical fiber is actually a cylindrical wave guide, light can propagate only at specific angles greater than the critical angle. These specific angles can be determined by solving the appropriate wave guide equations.)

As mentioned previously, fiber optics has played a significant role in the expansion of communication. At its basic level, optical fiber communications consist of sending information from a source to a destination by transmitting pulses of light. This is analogous to the historical practice of ship-to-ship communications using

Figure 35-11   The basic structure of an optical fiber.

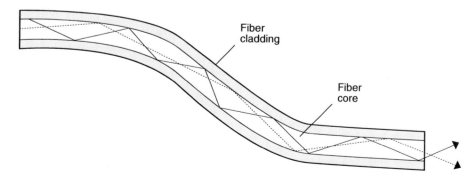

Figure 35-12   The ray model for propagation of light through an optical fiber. As long as curves are gentle enough to keep the angle of incidence greater than the critical angle, the transmitted light will experience total internal reflection.

Morse code. When one ship wished to communicate with another, a light source was flashed on and off in a sequence that both parties understood. This principle can be extended to fiber-optic communications, where numerous and highly efficient data coding schemes have been created. The encoding scheme chosen depends on the application and cost considerations.

Using optical fiber in communication systems has many advantages. Among the more important are immunity to electromagnetic interference, improved data security, higher transmission speed, and increased signal bandwidth. These advantages help make optical fiber the communications medium of choice for the foreseeable future.

While communications dominates the field of fiber optics, many other applications exist for optical fiber. Figure 35-13 illustrates a fluid level sensor. In this sensor, the behavior of light when traveling between different media is exploited. Notice that the fiber end has been specially shaped for this application so that when exposed to air (low fluid level) a majority of light transmitted down the fiber will be reflected back. When the fluid being sensed covers the fiber end, however, the index of the fluid more closely matches the index of the fiber core, thus reducing the amount of reflected light. This variation in reflected light is used to determine fluid level.

Another possible use for optical fiber is as a displacement sensor. This sensor exploits the spreading behavior of light as it leaves the end of a fiber. With this sensor, the distance between the fiber ends and the reflective plate in Fig. 35-14 can be measured. As the plate moves farther away from the fiber ends, less light is reflected from the transmitting fiber into the receiving fiber. Likewise, as the plate moves closer to the fiber ends, more light is reflected into the receiving fiber. This varying level of receiver intensity is used to determine the plate's position.

In medicine, fiber optics is making an impact in both diagnostics and treatment. For diagnostic purposes, a device known as a fiber-optic endoscope allows internal organs to be inspected visually. In this device, two optical fibers are used—one for illuminating the area of interest and the other for transmitting the image

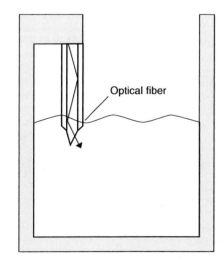

Figure 35-13 Fiber-optic fluid level sensor.

Figure 35-14 Fiber-optic displacement sensor.

being viewed. When connected to a video monitor, this device gives doctors the ability to see what previously required exploratory surgery.

In treatment, blocked arteries can be cleared using the LASTAC (laser enhanced transluminal angioplasty) system. In this system, laser light is transmitted through an optical fiber that has been inserted into a blocked artery. The laser vaporizes the impeding plaque, thus clearing the artery. With this treatment, more serious surgery can often be avoided.

35-7 IS SEEING THE SAME AS BELIEVING?

Because we are accustomed to light traveling in straight lines, refraction and total internal reflection often present us with pictures we do not believe. Atmospheric refraction accounts for many illusions that are referred to as *mirages*. Figure 35-15 provides two examples of such occurrences. In Fig. 35-15a a layer of hot air in contact with the heated ground is less dense than the cool layers of air above it. Consequently, light from distant objects is refracted upward, making them appear inverted.

At night, the situation is sometimes reversed: The cool layer of air is beneath warmer layers. The headlights of the car in Fig. 35-15b appear to be *looming* in the air. Many scientists believe that some of the unidentified flying objects (UFOs) reported for centuries can be explained in terms of atmospheric refraction.

Combinations of refraction and total internal reflection account for the bizarre photograph in Fig. 35-16. The picture was taken by an underwater camera looking upward at a girl sitting on the edge of a pool with her legs dangling in the

Figure 35-15 Atmospheric refraction accounts for the mirage in (a) and explains the phenomenon of *looming* in (b).

Figure 35-16 A photograph taken by an underwater camera presents a bizarre picture of a person sitting at the edge of a swimming pool. (From the film *Introduction to Optics*, Education Development Center.)

Figure 35-17 Combinations of refraction, reflection, and total internal reflection serve to deceive the eye.

water. An explanation is given in Fig. 35-17. The upper portion of the picture results from refraction at the surface. The inverted legs in the middle of the picture are due to total internal reflection at the surface of the water. The bottom of the picture represents the only undistorted image since the legs are viewed directly and in the same medium.

35-8 APPARENT DEPTH

Refraction causes an object submerged in a liquid of higher index of refraction to appear closer to the surface than it actually is. This shallowing effect is illustrated in Fig. 35-18. The object O appears to be at I because of the refraction of light from the object. The **apparent depth** is denoted by q, and the actual depth is denoted by p. Snell's law applied at the surface gives

$$\frac{\sin \theta_1}{\sin \theta_2} = \frac{n_2}{n_1} \qquad\qquad \textbf{35-8}$$

If we can relate the ratio of the indexes of refraction to actual and apparent depths, a useful relation can be obtained for predicting the apparent depths of submerged objects. From Fig. 35-18, it is noted that

$$\angle AOB = \theta_1 \qquad \text{and} \qquad \angle AIB = \theta_2$$

and that

$$\sin \theta_1 = \frac{d}{OA} \qquad \sin \theta_2 = \frac{d}{IA}$$

Using this information in Eq. (35-8), we obtain

$$\frac{\sin \theta_1}{\sin \theta_2} = \frac{d/OA}{d/IA} = \frac{IA}{OA} \qquad\qquad \textbf{35-9}$$

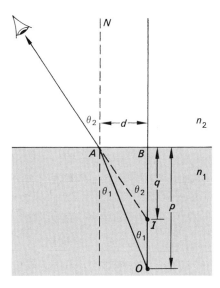

Figure 35-18 Relation between apparent depth and actual depth.

If we restrict ourselves to rays that are nearly vertical, the angles θ_1 and θ_2 will be small, so that the following approximations apply:

$$OA \approx p \qquad \text{and} \qquad IA \approx q$$

Applying these approximations to Eqs. (35-8) and (35-9), we can write

$$\frac{\sin \theta_1}{\sin \theta_2} = \frac{n_2}{n_1} = \frac{q}{p}$$

$$\frac{\text{Apparent depth } q}{\text{Actual depth } p} = \frac{n_2}{n_1}$$

35-10

EXAMPLE 35-6

A coin rests on the bottom of a container filled with water ($n_w = 1.33$). The apparent distance of the coin from the surface is 9 cm. How deep is the container?

Solution

In this problem, $n_1 = 1.33$ and $n_2 = 1.0$ are the indexes of refraction of water and air, respectively. The actual depth p is found by solving for p in Eq. (35-10).

$$p = \frac{qn_1}{n_2} = \frac{(9 \text{ cm})(1.33)}{1.0} = 12 \text{ cm}$$

The apparent depth is approximately three-fourths of the actual depth.

Summary and Review

Summary

Refraction has been defined as the bending of a light ray as it passes obliquely from one medium to another. We have seen that the degree of bending can be predicted based on either the change in velocity or the known index of refraction for each medium. The concepts of refraction, critical angle, dispersion, and internal reflection play important roles in the operation of many instruments. The major concepts covered in this chapter are summarized below.

- The index of refraction of a particular material is the ratio of the free-space velocity of light c to the velocity v of light through the medium.

$$n = \frac{c}{v} \qquad c = 3 \times 10^8 \text{ m/s} \qquad \begin{array}{c}\textit{Index of}\\ \textit{Refraction}\end{array}$$

- When light enters from medium 1 and is refracted into medium 2, Snell's law can be written in the following two forms (see Fig. 35-19):

$$n_1 \sin \theta_1 = n_2 \sin \theta_2 \qquad \frac{v_1}{v_2} = \frac{\sin \theta_1}{\sin \theta_2} \qquad \begin{array}{c}\textit{Snell's}\\ \textit{Law}\end{array}$$

- When light enters medium 2 from medium 1, its wavelength is changed by the fact that the index of refraction is different.

$$\frac{\lambda_1}{\lambda_2} = \frac{n_2}{n_1} \qquad \lambda_2 = \frac{n_1 \lambda_1}{n_2}$$

Figure 35-19   Snell's law.

- The critical angle θ_c is the maximum angle of incidence from one medium that will still produce refraction (at 90°) into a bordering medium. From the definition, we obtain

$$\sin \theta_c = \frac{n_2}{n_1} \qquad \textit{Critical Angle}$$

- Refraction causes an object in one medium to be observed at a different depth when viewed from above in another medium.

$$\frac{\text{Apparent depth } q}{\text{Actual depth } p} = \frac{n_2}{n_1}$$

Key Terms

1. refraction
2. optical density
3. Snell's law
4. dispersion
5. index of refraction
6. total internal reflection
7. critical angle
8. apparent depth

Summary and Review

Review Questions

35-1. State three laws of refraction and show how they can be demonstrated in the laboratory.

35-2. Is the index of refraction a constant for a particular medium? Explain.

35-3. Explain how the day is lengthened by atmospheric refraction.

35-4. A coin is placed on the bottom of a bucket so that it is just out of sight when viewed at an angle from the top. Show by the use of diagrams why the coin becomes visible if the bucket is filled with water.

35-5. Will objects of higher optical density have greater or smaller critical angles when surrounded by air?

35-6. On the basis of topics discussed in this chapter, explain why a diamond is much more brilliant than a glass replica.

35-7. Explain why right-angle prisms are more efficient reflectors than mirrored surfaces.

35-8. Why are colors observed in the light from a diamond?

35-9. A child stands waist deep in a swimming pool that has a uniform depth throughout. Why does it appear to him that he is standing in the deepest part of the pool?

35-10. The wavelength λ of a certain source of radiation is increased to 2λ. If the index of refraction was measured originally to be 1.5, what will it be when the wavelength is doubled?

35-11. What is the critical angle for an irregularly shaped piece of glass submerged in a liquid of the same index of refraction? Why would the glass be invisible in this case?

Problems

Section 35-1—Index of Refraction

35-1. The speed of light through a certain medium is 1.6×10^8 m/s in a transparent medium. What is the index of refraction in that medium. **Ans. 1.88**

35-2. If the speed of light is to be reduced by one-third, what must be the index of refraction for the medium through which the light travels?

35-3. Compute the speed of light in (a) crown glass, (b) diamond, (c) water, and (d) ethyl alcohol. **Ans. 1.97×10^8 m/s, 1.24×10^8 m/s, 2.26×10^8 m/s, 2.21×10^8 m/s**

35-4. If light travels at 2.1×10^8 m/s in a transparent medium, what is the index of refraction in that medium?

Section 35-2—The Laws of Refraction

35-5. Light is incident at an angle of 37° from air to flint glass ($n = 1.6$). What is the angle of refraction into the glass? **Ans. 22.1°**

35-6. A beam of light makes an angle of 60° with the surface of water. What is the angle of refraction into the water?

35-7. Light passes from water ($n = 1.33$) to air. The beam emerges into air at an angle of 32° with the horizontal water surface. What is the angle of incidence inside the water? **Ans. 39.6°**

35-8. Light in air is incident at 60° and is refracted into an unknown medium at an angle of 40°. What is the index of refraction for the unknown medium?

35-9. Light strikes from medium A into medium B at an angle of 35° with the horizontal boundary. If the angle of refraction is also 35°, what is the relative index of refraction between the two media? **Ans. 1.43**

35-10. Light incident from air at 45° is refracted into a transparent medium at an angle of 34°. What is the index of refraction for the material?

*35-11. A ray of light originating in air (Fig. 35-20) is incident on water ($n = 1.33$) at an angle

Figure 35-20

of 60°. It then passes through the water entering glass ($n = 1.50$) and finally emerges back into air again. Compute the angle of emergence. **Ans. 60°**

*35-12. Prove that, no matter how many parallel layers of different media are traversed by light, the entrance angle and the final emergent angle will be equal as long as the initial and final media are the same.

Section 35-3—Wavelength and Refraction

35-13. The wavelength of sodium light is 589 nm in air. Find its wavelength in glycerine.
Ans. 401 mm

35-14. The wavelength decreases by 25 percent as it goes from air to an unknown medium. What is the index of refraction for that medium?

35-15. A beam of light has a wavelength of 600 nm in air. What is the wavelength of this light as its passes into glass ($n = 1.50$)?
Ans. 400 mm

35-16. Red light (520 nm) changes to blue light (478 nm) when it passes into a liquid. What is the index of refraction for the liquid? What is the velocity of the light in the liquid?

*35-17. A ray of monochromatic light of wavelength 400 nm in medium A is incident at 30° at the boundary of another medium B. If the ray is refracted at an angle of 50°, what is its wavelength in medium B?
Ans. 613 nm

Section 35-5—Total Internal Reflection

35-18. What is the critical angle for light moving from quartz ($n = 1.54$) to water ($n = 1.33$)?

35-19. The critical angle for a given medium relative to air is 40°. What is the index of refraction for the medium? **Ans. 1.56**

35-20. If the critical angle of incidence for a liquid to air surface is 46°, what is the index of refraction for the liquid?

35-21. What is the critical angle relative to air for (a) diamond, (b) water, and (c) ethyl alcohol? **Ans. 24.4°, 48.8°, 47.3°**

35-22. What is the critical angle for flint glass immersed in ethyl alcohol?

*35-23. A right-angle prism like the one shown in Fig. 35-10a is submerged in water. What is the minimum index of refraction needed for the material to achieve total internal reflection? **Ans. 1.88**

Challenge Problems

35-24. The angle of incidence is 30°, and the angle of refraction is 26.3°. If the incident medium is water, what might the refractive medium be?

35-25. The speed of light in an unknown medium is 2.40×10^8 m/s. If the wavelength of light in this unknown medium is 400 nm, what is the wavelength in air? **Ans. 500 nm**

35-26. A ray of light strikes a pane of glass at an angle of 30° with the glass surface. If the angle of refraction is also 30°, what is the index of refraction for the glass?

35-27. A beam of light is incident on a plane surface separating two media of indexes 1.6 and 1.4. The angle of incidence is 30° in the medium of higher index. What is the angle of refraction? **Ans. 34.8°**

35-28. In going from glass ($n = 1.50$) to water ($n = 1.33$), what is the critical angle for total internal reflection?

Summary and Review

35-29. Light of wavelength 650 nm in a particular glass has a speed of 1.7×10^8 m/s. What is the index of refraction for this glass? What is the wavelength of this light in air?
Ans. 1.76, 1146 nm

35-30. The critical angle for a certain substance is $38°$ when it is surrounded by air. What is the index of refraction of the substance?

35-31. The water in a swimming pool is 2 m deep. How deep does it appear to a person looking vertically down?
Ans. 1.50 m

35-32. A plate of glass ($n = 1.50$) is placed over a coin on a table. The coin appears to be 3 cm below the top of the glass plate. What is the thickness of the glass plate?

Critical Thinking Questions

*35-33. Consider a horizontal ray of light striking one edge of an equilateral prism of glass ($n = 1.50$) as shown in Fig. 35-21. At what angle θ will the ray emerge from the other side?
Ans. 77.1°

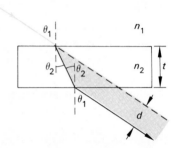

Figure 35-22

light passes through glass surrounded by air. The angle of incidence θ_1 is $40°$, and the glass ($n = 1.50$) is 2 cm thick.
Ans. 5.59 mm

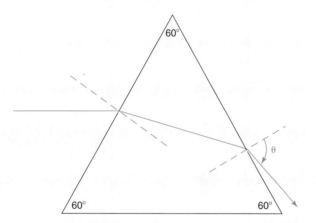

Figure 35-21

*35-34. What is the minimum angle of incidence at the first face of the prism in Fig. 35-21 such that the beam is refracted into air at the second face? (Larger angles do not produce total internal reflection at the second face.)

*35-35. Light passing through a plate of transparent material of thickness t suffers a lateral displacement d, as shown in Fig. 35-22. Compute the lateral displacement if the

*35-36. A rectangularly shaped block of glass ($n = 1.54$) is submerged completely in water ($n = 1.33$). A beam of light traveling in the water strikes a vertical side of the glass block at an angle of incidence θ_1 and is refracted into the glass, where it continues to the top surface of the block. What is the minimum angle θ_1 at the side such that the light does not go out of the glass at the top?

*35-37. Prove that the lateral displacement in Fig. 35-22 can be calculated from

$$d = t \sin \theta_1 \left(1 - \frac{n_1 \cos \theta_1}{n_2 \cos \theta_2}\right)$$

Use this relationship to verify the answer to Question 35-35.

Lenses and Optical Instruments

OBJECTIVES

After completing this chapter, you should be able to:

1. Determine mathematically or experimentally the focal length of a lens and state whether it is converging or diverging.
2. Apply the lensmaker's equation to solve for unknown parameters related to the construction of lenses.
3. Use ray-tracing techniques to construct images formed by diverging and converging lenses for various object locations.
4. Predict mathematically or determine experimentally the nature, size, and location of images formed by converging and diverging lenses.

A ***lens*** is a transparent object that alters the shape of a wavefront passing through it. Lenses are usually constructed of glass and shaped so that refracted light will form images similar to those discussed for mirrors. Anyone who has examined objects through a magnifying glass, observed distant objects through a telescope, or experimented in photography knows something of the effects lenses have on light. In this chapter we study the images formed by lenses and discuss their application.

36-1 SIMPLE LENSES

The simplest way of understanding how a lens works is to consider the refraction of light by prisms, as illustrated in Fig. 36-1 on page 794. When Snell's law is applied to each surface of a prism, light is bent toward the normal when entering a prism and away from the normal on leaving. The effect, in either case, is to cause the light beam to be deviated toward the base of the prism. The light rays remain parallel because both the entrance and emergent surfaces are planes forming equal angles with all rays passing the prism. Thus, a prism merely alters the direction of a wavefront.

Suppose we place two prisms base to base, as shown in Fig. 36-2a on page 794. Light incident from the left will converge, but it will not come to a focus. To focus the light rays to a point, the extreme rays must be deviated more than the central rays. This is accomplished by grinding the surfaces so that they have a uniformly curved cross section, as indicated in Fig. 36-2b. A lens that brings a parallel beam of light to a point focus in this fashion is called a ***converging lens.***

> A ***converging lens*** **is one that refracts and converges parallel light to a point focus beyond the lens.**

(a)   (b)

Figure 36-1   Parallel rays of light are bent toward the base of a prism and remain parallel.

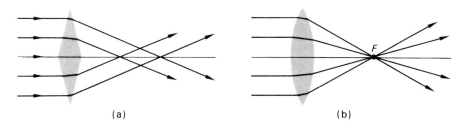

(a)   (b)

Figure 36-2   (a) Two prisms placed base to base will converge rays but will not bring all rays to a common focus. (b) A converging lens can be constructed by curving the surfaces uniformly.

The curved surfaces of lenses may be of any regular shape, such as spherical, cylindrical, or parabolic. Since spherical surfaces are easier to make, most lenses are constructed with two spherical surfaces. The line joining the centers of the two spheres is known as the *axis* of the lens. Three examples of converging lenses are shown in Fig. 36-3: *double convex, plano-convex,* and *converging meniscus.* Note that converging lenses are thicker in the middle than at the edge.

A second type of lens can be constructed by making the edges thicker than the middle, as shown in Fig. 36-4. Parallel light rays passing through such a lens bend toward the thicker part, causing the beam to diverge. Projection of the refracted light rays shows that the light appears to come from a virtual focal point in front of the lens.

A diverging lens is one that refracts and diverges parallel light from a point located in front of the lens.

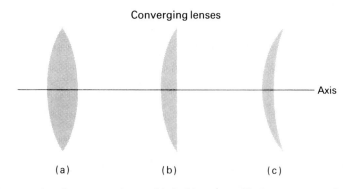

Converging lenses

(a)   (b)   (c)   Axis

Figure 36-3   Examples of converging lenses: (a) double convex, (b) plano-convex, and (c) converging meniscus.

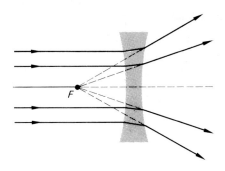

Figure 36-4 A diverging lens refracts light so that it appears to come from a point on the same side of the lens as the incident light.

Examples of **diverging lenses** are double concave, plano-concave, and diverging meniscus. See Fig. 36-5.

36-2 FOCAL LENGTH AND THE LENSMAKER'S EQUATION

A lens is regarded as "thin" if the thickness is small in comparison with the other dimensions involved. As with mirrors, image formation by thin lenses is a function of the focal length; however, there are important differences. One obvious difference is that light may pass *through* a lens in two directions. This results in two focal points for each lens, as shown in Fig. 36-6 for a converging lens and in Fig. 36-7 for a diverging lens. The former has a *real focus F,* and the latter has a **virtual focus F'.** The distance between the optical center of a lens and the focus on either side of the lens is the *focal length f.*

> **The focal length *f* of a lens is reckoned as the distance from the optical center of the lens to either focus.**

Since light rays are reversible, a source of light placed at either focus of a converging lens results in a parallel light beam. This can be seen by reversing the direction of the rays illustrated in Fig. 36-6.

Diverging lenses

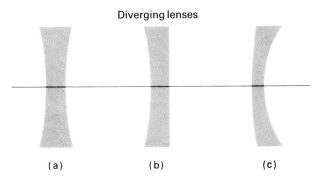

(a)   (b)   (c)

Figure 36-5 Examples of diverging lenses: (a) double concave, (b) plano-concave, and (c) diverging meniscus.

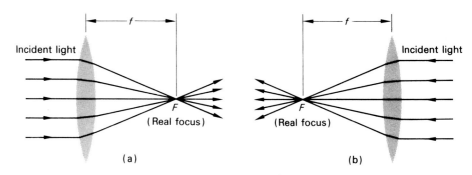

Figure 36-6 Demonstrating the focal length of a converging lens. The focal point is real because actual light rays pass through it.

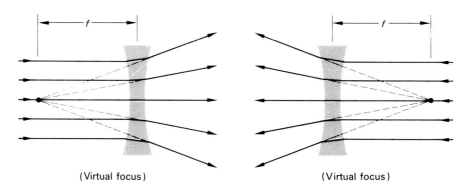

Figure 36-7 Demonstrating the virtual focal points of a diverging lens.

The focal length f of a lens is not equal to one-half the radius of curvature, as for spherical mirrors; it depends on the index of refraction n of the material from which it is made. It also is determined by the radii of curvature R_1 and R_2 of its surfaces, as defined in Fig. 36-8a. For thin lenses, these quantities are related by the equation

$$\frac{1}{f} = (n - 1)\left(\frac{1}{R_1} + \frac{1}{R_2}\right)$$ 36-1

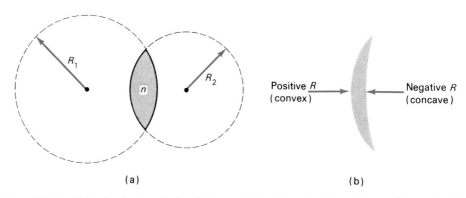

Figure 36-8 (a) The focal point of a lens is determined by the radii of its surfaces and the index of refraction. (b) The sign convention for the radius of a lens surface.

Because Eq. (36-1) involves the construction parameters for a lens, it is referred to as the **lensmaker's equation.** It applies equally to converging and diverging lenses if the following sign convention is followed:

1. The radius of curvature (either R_1 or R_2) is considered positive if the surface is curved outward (convex) and negative if the surface is curved inward (concave). (Refer to Fig. 36-8b).
2. The focal length f of a converging lens is considered positive, and the focal length of a diverging lens is considered negative.

EXAMPLE 36-1

A lensmaker plans to construct a plano-concave lens out of glass with an index of refraction of 1.5. What should the radius of its curved surface be if the desired focal length is -30 cm?

Solution

The radius of curvature R_1 for a plane surface is infinity. The radius R_2 of the concave surface is found from Eq. (36-1):

$$\frac{1}{f} = (n - 1)\left(\frac{1}{\infty} + \frac{1}{R_2}\right) = (n - 1)\frac{1}{R_2}$$

$$R_2 = (n - 1)\, f = (1.5 - 1.0)(-30 \text{ cm})$$

$$= -15 \text{ cm}$$

By convention, the minus sign shows that the curved surface is concave.

EXAMPLE 36-2

A **meniscus lens** has a convex surface with a radius of curvature of 10 cm and a concave surface with a radius of -15 cm. If the lens is constructed from glass with an index of refraction of 1.52, what will its focal length be?

Solution

Substitution into the lensmaker's equation yields

$$\frac{1}{f} = (n - 1)\left(\frac{1}{R_1} + \frac{1}{R_2}\right)$$

$$= (1.52 - 1)\left(\frac{1}{10 \text{ cm}} - \frac{1}{15 \text{ cm}}\right)$$

$$= 0.52\left(\frac{3 - 2}{30 \text{ cm}}\right) = \frac{0.52}{30 \text{ cm}}$$

$$f = \frac{30 \text{ cm}}{0.52} = 57.7 \text{ cm}$$

The fact that the focal length is positive indicates that it is a *converging* meniscus lens.

To understand how images are formed by lenses, we now introduce ray-tracing methods similar to those discussed for spherical mirrors. The method consists of tracing two or more rays from a chosen point on the object and using the point of intersection as the image of that point. The entire deviation of a ray passing through a thin lens can be considered to take place at a plane through the center of the lens. In Sec. 36-2 it was noted that a lens has two focal points. We define the *first focal point* F_1 as the one located on the same side of the lens as the incident light. The *second focal point* F_2 is located on the opposite, or far, side of the lens. With these definitions in mind, there are three principal rays that can be traced easily through a lens. These rays are illustrated in Fig. 36-9 for a converging lens and in Fig. 36-10 for a diverging lens:

Ray 1 A ray parallel to the axis passes through the second focal point F_2 of a converging lens or appears to come from the first focal point F_1 of a diverging lens.

Ray 2 A ray that passes through the first focal point F_1 of a converging lens or proceeds toward the second focal point F_2 of a diverging lens is refracted parallel to the lens axis.

Ray 3 A ray that passes through the geometrical center of a lens will not be deviated.

Figure 36-9   Principal rays for image construction using a converging lens.

Figure 36-10   Principal rays for constructing images formed by diverging lenses.

The intersection of any two of these rays (or their extensions) from an object point represents the image of that point. Since a real image produced by a lens is formed by rays of light that actually pass through the lens, *a real image is always formed on the side of the lens opposite the object. A virtual image will appear to be on the same side of the lens as the object.*

To illustrate the graphical method and, at the same time, to understand the various images formed by lenses, we shall consider several examples. Images formed by a converging lens are shown in the object locations described in Fig. 36-11a to e:

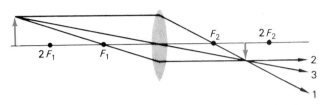

(a) Object located at a distance beyond twice the focal length. A real, inverted, and diminished image is formed between F_2 and $2F_2$ on the opposite side of the lens.

(b) Object at a distance equal to twice the focal length. A real, inverted image the same size as the object is located at $2F_2$ on the opposite side of the lens.

(c) Object located at a distance between one and two focal lengths from the lens. A real, inverted, and enlarged image is formed beyond $2F_2$ on the opposite side of the lens.

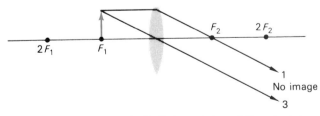

(d) Object at the first focal point F_1. No image is formed. The refracted rays are parallel.

Figure 36-11 Image construction is shown for the following object distances: (a) beyond $2F_1$, (b) at $2F_1$, (c) between $2F_1$, and F_1, and (d) at F_1.

(e) Object located inside the first focal point. A virtual, erect, and enlarged image is formed on the same side of the lens as the object.

Figure 36-11 (*cont.*)   (e) inside F_1.

Notice that the images formed by a *convex* lens are similar to those formed by *concave* mirrors. This is true because they both converge light. Since concave lenses diverge light, we would expect them to form images similar to those formed by a diverging mirror (convex mirror). Figure 36-12 demonstrates this similarity.

> **Images of real objects formed by diverging lenses are always virtual, erect, and diminished in size.**

To avoid confusion, one should identify both lenses and mirrors as either converging or diverging. Diverging lenses are often used to reduce or neutralize the effect of converging lenses.

36-4 THE LENS EQUATION AND MAGNIFICATION

The characteristics, size, and location of images can also be determined analytically from the *lens equation*. This important relation can be deduced by applying plane geometry to Fig. 36-13. The derivation is similar to the one used to derive the mirror equation, and the final form is exactly the same. The lens equation can be written

$$\frac{1}{p} + \frac{1}{q} = \frac{1}{f} \qquad \text{36-2}$$

Figure 36-12   Images formed by diverging lenses are always virtual, erect, and diminished in size.

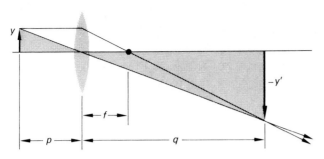

Figure 36-13   Deriving the lens equation and the magnification.

where p = object distance
q = image distance
f = focal length of lens

The same sign conventions established for mirrors can be used in the lens equation if converging and diverging lenses are compared with converging and diverging mirrors. This convention is summarized as follows:

1. The object distance p and the image distance q are considered positive for real objects and images and negative for virtual objects and images.
2. The focal length f is considered positive for converging lenses and negative for diverging lenses.

The following alternative forms of the lens equation are useful in solving optical problems:

$$p = \frac{fq}{q - f} \qquad q = \frac{fp}{p - f}$$

$$f = \frac{qp}{p + q} \qquad\qquad \text{36-3}$$

You should verify each of these forms by solving the lens equation explicitly for each parameter in the equation.

The **magnification** of a lens is also derived from Fig. 36-13 and has the same form as discussed for mirrors. Recalling that the magnification M is defined as the ratio of image size y' to the object size y, we can write

$$M = \frac{y'}{y} = -\frac{q}{p} \qquad\qquad \text{36-4}$$

where q is the image distance and p is the object distance. A *positive magnification indicates that the image is erect, whereas a negative magnification occurs only when the image is inverted.*

EXAMPLE 36-3

An object 4 cm high is located 10 cm from a thin converging lens having a focal length of 20 cm. What are the nature, size, and location of the image?

Solution

This situation corresponds to that illustrated in Fig. 36-11e. The image distance is found from

$$q = \frac{pf}{p - f} = \frac{(10 \text{ cm})(20 \text{ cm})}{10 \text{ cm} - 20 \text{ cm}}$$
$$= \frac{200 \text{ cm}^2}{-10 \text{ cm}} = -20 \text{ cm}$$

The minus sign indicates that the image is virtual. The magnification relation [Eq. (36−4)] now allows us to compute the image size.

$$M = \frac{y'}{y} = -\frac{q}{p}$$
$$y' = \frac{-qy}{p} = \frac{-(-20 \text{ cm})(4 \text{ cm})}{10 \text{ cm}}$$
$$= +8 \text{ cm}$$

The positive sign indicates that the image is erect. This example illustrates the principle of a magnifying glass. A converging lens held closer to an object than its focal point produces an erect and enlarged virtual image.

EXAMPLE 36-4

A diverging meniscus lens has a focal length of −16 cm. If the lens is held 10 cm from an object, where is the image located? What is the magnification of the lens?

Solution

Direct substitution yields

$$q = \frac{pf}{p - f} = \frac{(10 \text{ cm})(-16 \text{ cm})}{10 \text{ cm} - (-16 \text{ cm})}$$
$$= \frac{-160}{10 + 16} = -6.15 \text{ cm}$$

The minus sign again indicates that the image is virtual. The magnification is

$$M = -\frac{q}{p} = \frac{-(-6.15 \text{ cm})}{10 \text{ cm}} = +0.615$$

The positive magnification means that the image is erect.

Problem-Solving Strategy

Lenses and Optical Instruments

1. Read the problem carefully and draw a horizontal line representing the lens axis. Indicate the location of points equal to f and $2f$ on each side of a vertical line representing the concave or convex lens. Represent the object as an erect arrow at its approximate location to the left of the lens.

2. Construct a geometric ray-tracing diagram to get a visual representation of the problem. A rough sketch should suffice, unless the problem requires a graphical solution. Remember that converging and diverging lenses and mirrors form similar images, except that the virtual images are formed on the *opposite* side for mirrors and on the *same* side for lenses. Real images are formed by the actual light rays, and virtual images are formed where the light only appears to originate, for example, behind the mirror or on the same side of the lens as the incident light.

3. Make a list of the given quantities, taking care to place the appropriate sign to each value. The radius and focal length are *positive* for *converging* mirrors and *negative* for *diverging* mirrors. Image distances q are *positive* when measured to *real* images and *negative* when measured to *virtual* images. The image size y' is *positive* for erect images and *negative* for inverted images.

4. Use the following equations to substitute and solve for the unknown quantities. Do not confuse the signs of operation (addition or subtraction) with the signs of substitution.

$$\frac{1}{p} + \frac{1}{q} = \frac{1}{f} \qquad M = \frac{y'}{y} = \frac{-q}{p}$$

5. It may be necessary to eliminate an unknown by solving both the mirror equation and the magnification equation simultaneously.

36-5 COMBINATIONS OF LENSES

When light passes through two or more lenses, the combined action can be determined by considering the image that would be formed by the first lens as the object for the second lens, and so on. Consider, for example, the lens arrangement illustrated in Fig. 36-14. Lens 1 forms a real, inverted image I_1 of the object O. By considering this intermediate image as a real object for lens 2, the final image I_2 is seen to be real, erect, and enlarged. The lens equation can be applied successively to the two lenses to determine the location of the final image analytically.

The total magnification produced by a system of lenses is the product of the magnifications produced by each lens in the system. The truth of this can be seen from Fig. 36-14. The magnifications in this case are

$$M_1 = \frac{y'_1}{y_1} \qquad M_2 = \frac{y'_2}{y_2}$$

Since $y'_1 = y_2$, the product $M_1 M_2$ yields

$$\frac{y'_1}{y_1}\frac{y'_2}{y_2} = \frac{y'_2}{y_1}$$

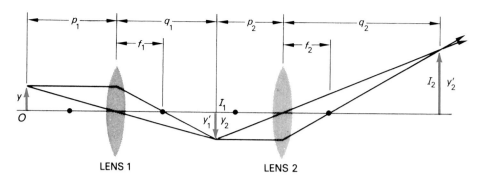

Figure 36-14   The microscope.

But y_2'/y_1 is the overall magnification M. In general, we can write

$$M = M_1 M_2 \qquad\qquad 36\text{-}5$$

Applications of the above principles are found for the microscope, the **telescope,** and other optical instruments.

36-6 THE COMPOUND MICROSCOPE

A compound **microscope** consists of two converging lenses, arranged as shown in Fig. 36-15. The left lens is of short focal length and is called the **objective lens.** This lens has a large magnification and forms a real, inverted image of the object being studied. The image is further enlarged by the eyepiece, which forms a virtual final image. The total magnification achieved is the product of the magnifications of the objective lens and the **eyepiece.**

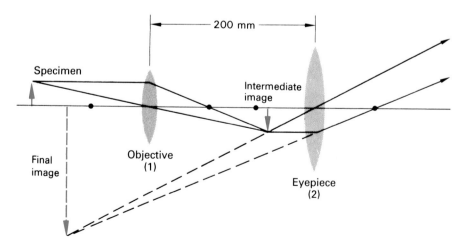

Figure 36-15   Combinations of lenses.

CONCEPTUAL EXAMPLE 36-5

In a compound microscope, the objective lens has a focal length of 8 mm, and the eyepiece has a focal length of 40 mm. The distance between the two lenses is 200 mm, and the final image appears to be at a distance of 250 mm from the eyepiece. (a) How far is the object from the objective lens? (b) What is the total magnification? (Refer to Fig. 36-15.)

Solution (a)

We begin by labeling the objective lens as 1 and the eyepiece as 2. Since more information is given about parameters that affect the second lens, we shall compute the position of the intermediate image first. From the lens equation, the distance p_2 of this image from the second lens is

$$p_2 = \frac{f_2 q_2}{q_2 - f_2} = \frac{(40 \text{ mm})(-250 \text{ mm})}{-250 \text{ mm} - 40 \text{ mm}}$$
$$= 34.5 \text{ mm}$$

The minus sign was used for the image distance q_2 because it was measured to a virtual image. Now, that p_2 is known, we can compute the image distance q_1 for the first image.

$$q_1 = 200 \text{ mm} - 34.5 \text{ mm} = 165.5 \text{ mm}$$

Thus, the object distance p_1 must be

$$p_1 = \frac{q_1 f_1}{q_1 - f_1} = \frac{(165.5 \text{ mm})(8 \text{ mm})}{165.5 \text{ mm} - 8 \text{ mm}}$$
$$= 8.41 \text{ mm}$$

Solution (b)

The total magnification is the product of the individual magnifications:

$$M_1 = \frac{-q_1}{p_1} \qquad M_2 = \frac{-q_2}{p_2}$$

Applying Eq. (36-5), we obtain

$$M = M_1 M_2 = \frac{+q_1 q_2}{p_1 p_2}$$
$$= \frac{(165.5 \text{ mm})(-250 \text{ mm})}{(8.41 \text{ mm})(34.5 \text{ mm})} = -143$$

The negative magnification indicates that the final image is inverted. This microscope would be rated 143×, and the object under study should be placed 8.41 mm from the objective lens.

Holographic Memory The CD-ROM and DVD medium is a thin disk of metallized polycarbonate that stores data as pits and ridges which can be read by reflecting a laser beam off this changing surface. However, these disks are essentially two-dimensional, in that data is stored only on the disk. A new technology could produce optical data in three dimensions—a hologram. In experiments done at CalTech, crystals of lithium niobate doped with iron and manganese atoms could be excited by different kinds of light and could store the energy until exposed to red or ultraviolet light. The stored pattern, created by a coordinated pair of red lasers in the presence of ultraviolet light, was held by the electrons of both iron and manganese atoms. IBM is very interested in this technology, because it could potentially lead to 3-D optical read-write memory devices.

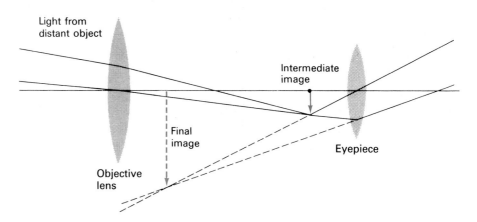

Figure 36-16 The refracting telescope.

36-7 TELESCOPE

The optical system of a refracting telescope is essentially the same as that of a microscope. Both instruments use an eyepiece, or *ocular,* to enlarge the image produced by an objective lens, but a telescope is used to examine large, distant objects, and a microscope is used for small, nearby objects.

The refracting telescope is illustrated in Fig. 36-16. The objective lens forms a real, inverted, and diminished image of the distant object. As with the microscope, the eyepiece forms an enlarged and virtual final image of the distant object.

A telescope image is usually smaller than the object being observed. Linear magnification is therefore not a meaningful way to describe the effectiveness of a given telescope. A better measure would be to compare the size of the final image with the size of the object observed without the telescope. If the image seen by the eye is larger than it would be without the telescope, the effect will be to make the object appear closer to the eye than it actually is.

36-8 LENS ABERRATIONS

Spherical lenses often fail to produce perfect images because of defects inherent in their construction. Two of the most common defects are known as **spherical aberration** and **chromatic aberration.** Spherical aberration, as discussed earlier for mirrors, is the inability of a lens to focus all parallel rays to the same point. (See Fig. 36-17.)

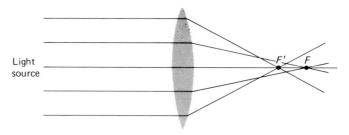

Figure 36-17 Spherical aberration.

Spherical aberration is a lens defect in which the extreme rays are brought to a focus nearer the lens than rays entering near the optical center of the lens.

This effect can be minimized by placing a *diaphragm* in front of the lens. The diaphragm blocks off the extreme rays, producing a sharper image along with a reduction in light intensity.

In Chap. 35 we discussed the fact that the index of refraction for a given transparent material varies with the wavelength of the light passing through it. Thus, if white light is incident upon a lens, the rays of the component colors are not focused at the same point. The defect, known as *chromatic aberration*, is illustrated in Fig. 36-18a, where blue light is shown to focus nearer the lens than red light.

Chromatic aberration is a lens defect that reflects its inability to focus light of different colors to the same point.

The remedy for this defect is the *achromatic lens,* illustrated in Fig. 36-18b. Such a lens can be constructed by combining a converging lens of crown glass ($n = 1.52$) with a diverging lens of flint glass ($n = 1.63$). These lenses are chosen and constructed so that the dispersion of one is equal and opposite to that of the other.

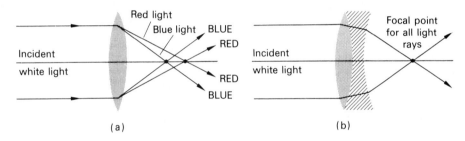

Figure 36-18 (a) Chromatic aberration and (b) an achromatic lens.

Summary and Review

Summary

A lens is a transparent device that converges or diverges light to or from a focal point. Lenses are used extensively in the design of many industrial instruments, and an understanding of how they form images is valuable. A summary of the major concepts discussed in this chapter is provided below.

- Image formation by thin lenses can be understood more easily through ray-tracing techniques as shown in Fig. 36-9 for converging lenses and in Fig. 36-10 for diverging lenses. Remember that the first focal point F_1 is the one on the same side of the lens as the incident light. The second focal point F_2 is on the far side.

  Ray 1 A ray parallel to the axis passes through the second focal point F_2 of a converging lens or appears to come from the first focal point F_1 of a diverging lens.

  Ray 2 A ray that passes through F_1 of a converging lens or proceeds toward F_2 of a diverging lens is refracted parallel to the lens axis.

  Ray 3 A ray that passes through the geometrical center of a lens will not be deviated.

- The *lensmaker's equation* is a relationship among the focal length, the radii of the two lens' surfaces, and the index of refraction of the lens material. The meaning of these parameters is seen from Fig. 36-8.

$$\frac{1}{f} = (n - 1)\left(\frac{1}{R_1} + \frac{1}{R_2}\right) \qquad \text{\textit{Lensmaker's Equation}}$$

R_1 or R_2 is positive if the outside surface is convex, negative if concave; f is considered positive for a converging lens and negative for a diverging lens.

- The equations for object and image locations and for the magnification are the same as for the mirror equations.

$$\frac{1}{p} + \frac{1}{q} = \frac{1}{f} \qquad\qquad p = \frac{qf}{q-f}$$

$$q = \frac{pf}{p-f} \qquad\qquad f = \frac{pq}{p+q}$$

$$\text{\textit{Magnification}} = \frac{\text{\textit{image size}}}{\text{\textit{object size}}} \qquad M = \frac{y'}{y} = \frac{-q}{p}$$

p or q is positive for real and negative for virtual; y or y' is positive if erect and negative if inverted.

Key Terms

1. lens
2. meniscus lens
3. lens equation
4. magnification
5. microscope
6. objective lens
7. eyepiece
8. telescope
9. converging lens
10. diverging lens
11. lensmaker's equation
12. spherical aberration
13. chromatic aberration
14. achromatic lens
15. diaphragm
16. virtual focus

Review Questions

36-1. Illustrate by diagrams the effect of a converging lens on a plane wavefront passing through it. What would the effect of a diverging lens be?

36-2. Explain why olives appear to be larger when viewed through their cylindrical glass container. Is the magnification due to the liquid, the glass, or both?

36-3. What happens to the focal length of a converging lens when it is immersed in water? What happens to the focal length of a diverging lens?

36-4. Distinguish between a *real* focus and a *virtual* focus. Which is on the same side of the lens as the incident light?

36-5. Distinguish between the first focal point and the second focal point as defined in the text.

36-6. The focal length of a converging lens is 20 cm. Choose a suitable scale and determine by graphical construction the nature, location, and magnification for the following object distances: (a) 15 cm, (b) 20 cm, (c) 30 cm, (d) 40 cm, (e) 60 cm, (f) infinity.

36-7. An object is moved from the surface of a lens to the focal point of the lens. Explain what happens to the image if the lens is (a) converging, (b) diverging.

36-8. Describe what happens to the magnification and location of an image as the object moves from infinity to the surface of (a) a converging lens, (b) a diverging lens.

36-9. Discuss the similarities and the differences between lenses and mirrors.

36-10. A camera uses a diaphragm to control the amount of light reaching the film. On a bright day, the diaphragm is almost closed, whereas on a dark day it must be opened wide to expose the film properly. Discuss the quality of the images produced in each case if the lens is not corrected for aberrations.

36-11. In a microscope the objective lens has a short focal length, whereas in a telescope the objective lens has a long focal length. Explain the reason for the difference in focal lengths.

36-12. Derive the lens equation with the help of Fig. 36-13.

36-13. Derive the magnification relation [Eq. (36-4)] with the help of Fig. 36-13.

36-14. Describe two methods you might use to determine the focal length of a double-concave lens.

36-15. Describe an experiment to determine the focal length of a double-concave lens.

36-16. Without referring to the text, write down the various sign conventions that must be applied when working with thin lenses.

36-17. According to convention, the object distance is considered negative when measured to a *virtual object*. Suggest some examples of virtual objects.

Problems

Section 36-2—Focal Length and the Lensmaker's Equation

(Assume $n=1.50$ unless told otherwise.)

36-1. A plano-convex lens is to be constructed out of glass so that it has a focal length of 40 cm. What is the radius of curvature of the curved surface?   **Ans. 20.0 cm**

36-2. If one uses a glass double-convex lens to obtain a focal length of 30 cm, what must be the curvature of each convex surface?

36-3. The curved surface of a plano-concave lens has a radius of -12 cm. What is the focal length if the lens is made from a material with a refractive index of 1.54?

           Ans. −22.2 cm

36-4. A converging meniscus lens has a concave surface with a radius of -20 cm and a convex surface with a radius of 12 cm. What is the focal length?

36-5. A converging lens such as the one shown in Fig. 36-8a is made of glass. The first surface has a radius of 15 cm, and the second surface has a radius of 10 cm. What is the focal length?   **Ans. 12.0 cm**

36-6. A meniscus lens has a convex surface with a radius of 20 cm and a concave surface with a radius of -30 cm. What is the focal length if the refractive index is 1.54?

36-7. A plano-convex lens is ground from crown glass ($n = 1.52$). What should be the radius of the curved surface if the desired focal length is to be 400 mm?   **Ans. 208 mm**

36-8. The magnitudes of the concave and convex surfaces of a glass lens are 200 and 600 mm, respectively. What is the focal length? Is it diverging or converging?

36-9. A plastic lens ($n = 1.54$) has a convex surface of radius 25 cm and a concave surface of -70 cm. What is the focal length? Is it diverging or converging?

Ans. 72.0 cm, converging

Section 36-3—Image Formation by Thin Lenses

36-10. A 7-cm-long pencil is placed 35 cm from a thin converging lens of focal length 25 cm. What are the nature, size, and location of the image formed?

36-11. An object 8 cm high is placed 30 cm from a thin converging lens of focal length 12 cm. What are the nature, size, and location of the image formed?

Ans. real, inverted, $y' = -5.33$ cm, $+20$ cm

36-12. A virtual, erect image appears to be located 40 cm in front of a lens of focal length 15 cm. What is the object distance?

36-13. A 50-mm-tall object is placed 12 cm from a converging lens of focal length 20 cm. What are the nature, size, and location of the image?

Ans. virtual, erect, $y' = 125$ mm, $q = -30$ cm

36-14. An object located 30 cm from a thin lens has a real, inverted image located a distance of 60 cm on the opposite side of the lens. What is the focal length of the lens?

36-15. A light source is 600 mm from a converging lens of focal length 180 mm. Construct the image using ray diagrams. What is the image distance? Is the image real or virtual?

Ans. 257 mm, real

36-16. A plano-convex lens is held 40 mm from a 6-cm object. What are the nature and location of the image formed if the focal length is 60 mm?

36-17. An object 6 cm high is held 4 cm from a diverging meniscus lens of focal length -24 cm. What are the nature, size, and location of the image?

Ans. virtual, $y' = 5.14$ cm, $q = -3.43$ cm

*36-18. The focal length of a converging lens is 200 mm. An object 60 mm high is mounted on a movable track so that the distance from the lens can be varied. Calculate the nature, size, and location of the image formed for the following object distances: (a) 150 mm, (b) 200 mm, (c) 300 mm, (d) 400 mm, (e) 600 mm.

36-19. An object 450 mm from a converging lens forms a real image 900 mm from the lens. What is the focal length of the lens?

Ans. $+30$ cm

Section 36-4—The Lens Equation and Magnification

36-20. An object is located 20 cm from a converging lens. If the magnification is -2, what is the image distance?

36-21. A pencil is held 20 cm from a diverging lens of focal length -10 cm. What is the magnification? **Ans. $+1/2$**

*36-22. A magnifying glass has a focal length of 27 cm. How close must this glass be held to an object to produce an erect image three times the size of the object?

*36-23. A magnifying glass held 40 mm from a specimen produces an erect image that is twice the object size. What is the focal length of the lens? **Ans. $+80$ mm**

*36-24. What is the magnification of a lens if the focal length is 40 cm and the object distance is 65 cm?

Challenge Problems

36-25. The radius of the curved surface in a plano-concave lens is 20 cm. What is the focal length if $n = 1.54$? **Ans. −37.4 cm**

36-26. A thin meniscus lens is formed with a concave surface of radius −40 cm and a convex surface of radius +30 cm. If the resulting focal length is 79.0 cm, what was the index of refraction of the transparent material?

36-27. A converging lens has a focal length of 20 cm. An object is placed 15.0 cm from the lens. Find the image distance and the nature of the image. **Ans. −60 cm, virtual**

36-28. How far from a source of light must a lens be placed if it is to form an image 800 mm from the lens? The focal length is 200 mm.

36-29. A source of light 36 cm from a lens projects an image on a screen 18.0 cm from the lens. What is the focal length of the lens? Is it converging or diverging? **Ans. 12.0 cm, converging**

36-30. What is the minimum film size needed to project the image of a student who is 2 m tall? Assume that the student is located 2.5 m from the camera lens and that the focal length is 55.0 mm?

*36-31. When parallel light strikes a lens, the light diverges, apparently coming from a point 80 mm behind the lens. How far from an object should this lens be held to form an image one-fourth the size of the object? **Ans. 240 mm**

*36-32. How far from a diverging lens should an object be placed so that its image will be one-fourth the size of the object? The focal length is −35 cm.

*36-33. The first surface of a thin lens has a convex radius of 20 cm. What should be the radius of the second surface to produce a converging lens of focal length 8.00 cm? **Ans. +5.00 cm**

*36-34. Two thin converging lenses are placed 60 cm apart and have the same axis. The first lens has a focal length 10 cm, and the second has a focal length of 15.0 cm. If an object 6.0 cm high is placed 20 cm in front of the first lens, what are the location and size of the final image? Is it real or virtual?

*36-35. A converging lens of focal length 25 cm is placed 50 cm in front of a diverging lens that has a focal length of −25 cm. If an object is placed 75 cm in front of the converging lens, what is the location of the final image? What is the total magnification? Is the image real or virtual? **Ans. $q_2 = -8.33$ cm in front of diverging lens, $M = -0.333$, virtual**

Critical Thinking Questions

*36-36. A camera consists of a converging lens of focal length 50 mm mounted in front of a light-sensitive film as shown in Fig. 36-19. When infinite objects are photographed, how far should the lens be from the film? What is the image distance when an object is photographed 500 mm from the lens? What is the magnification? **Ans. 50 mm, 55.5 mm, −0.111**

*36-37. An object is placed 30 cm from a screen. At what points between the object and the screen can a lens of focal length 5 cm be placed to obtain an image on the screen?

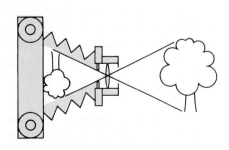

Figure 36-19 The camera.

Summary and Review

*36-38. A simple projector is illustrated in Fig. 36-20. The condenser provides even illumination of the film by the light source. The frame size of regular 8-mm film is 5 × 4 mm. An image is to be projected 600 × 480 mm on a screen located 6 m from the projection lens. What should be the focal length of the projection lens? How far should the film be from the lens?

Ans. $f = 49.6$ mm, $q = 50.0$ mm

*36-39. A telescope has an objective lens of focal length 900 mm and an eyepiece of focal length 50 mm. The telescope is used to examine a 30-cm-tall rabbit at a distance of 60 m. What is the distance between the lenses if the final image is 25 cm in front of the eyepiece? What is the apparent height of the rabbit as seen through the telescope?

*36-40. The Galilean telescope consists of a diverging lens as the eyepiece and a converging lens as the objective. The focal length of the objective is 30 cm, and the focal length of the eyepiece is −2.5 cm. An object 40 m away from the objective has a final image

Figure 36-20 The projector.

located 25 cm in front of the diverging lens. What is the separation of the lenses? What is the total magnification?

Ans. 27.5 cm, −0.068

*36-41. The focal length of the eyepiece of a particular microscope is 3.0 cm, and the focal length of the objective lens is 19 mm. The separation of the two lenses is 26.5 cm, and the final image formed by the eyepiece is at infinity. How far should the objective lens be placed from the specimen being studied?

Interference, Diffraction, and Polarization

OBJECTIVES

After completing this chapter, you should be able to:

1. Demonstrate by definition and drawings your understanding of the terms *constructive interference, destructive interference, diffraction, polarization,* and *resolving power.*

2. Describe Young's experiment and be able to use the results to predict the location of bright and dark fringes.

3. Discuss the use of a diffraction grating, derive the grating equation, and apply it to the solution of optical problems.

Light is dualistic in nature, sometimes exhibiting the properties of particles and sometimes those of waves. The demonstrative proof of the wave nature of light came with the discovery of interference and diffraction. Then polarization studies showed that, unlike sound waves, light waves are transverse instead of longitudinal.

In this chapter we study these phenomena and their significance in physical optics. We shall find that the neat geometrical-ray approach, which was so helpful in studying mirrors and lenses, must be discarded in favor of a more rigorous wave analysis.

37-1 DIFFRACTION

When light waves pass through an aperture or past the edge of an obstacle, they always bend to some degree into the region not directly exposed to the light source. This phenomenon is called *diffraction.*

Diffraction is the ability of waves to bend around obstacles placed in their path.

To understand this bending of waves, let us consider what happens when water waves strike a narrow opening. A plane-wave generator can be used in a ripple tank, as shown in Fig. 37-1 on page 814. The vibrating strip of metal serves as a source of waves at one end of a tray of water. The plane waves strike the barrier, spreading out into the region behind the gap. The diffracted waves appear to originate at the gap in accordance with Huygens's wave principle: *each point on a wavefront can be regarded as a new source of secondary waves.*

A similar experiment can be performed with light, but for the diffraction to be observable, the slit in the barrier must be narrow. In fact, diffraction is pronounced only when the dimensions of an opening or an obstacle are comparable to the

Mirrored Sunglasses
Mirrored sunglasses are made of several layers, each with specific light-absorbing or light-reflecting properties. The layer closest to the eye has a thin coating of silver or other metallic compound. This layer is thin enough to see through but thick enough to reflect light. The next layer is an ultraviolet-absorbing film. After that comes a polarizing filter, which eliminates all light that has bounced off the horizontal surface of roads or water. Even though the layers are embedded, they still perform their separate functions, just as if you were to add all three lenses together in a laboratory setting.

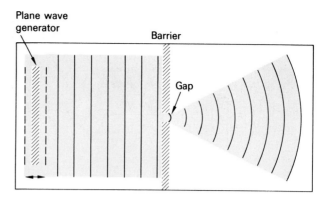

Figure 37-1 Schematic diagram illustrating the diffraction of plane water waves through a narrow gap.

wavelength of the waves striking it. This explains why diffraction of water waves and sound waves is often observed in nature and the diffraction of light is not.

37-2 YOUNG'S EXPERIMENT: INTERFERENCE

The first convincing evidence of diffraction was demonstrated by Thomas Young in 1801. A schematic diagram of Young's apparatus is shown in Fig. 37-2. Light from a monochromatic source falls on a slit A, which acts as a source of secondary waves. Two more slits S_1 and S_2 are parallel to A and equidistant from it. Light from A passes through both S_1 and S_2 and then on to a viewing screen. If light were not diffracted, the viewing screen would be completely dark. Instead, the screen is illuminated, as shown in Fig. 37-3. Even the point B located behind the barrier in direct line with slit A is illuminated. It is easy to see why this experiment caused early physicists to doubt that light consisted of particles traveling in straight lines. The results could be explained only in terms of the wave theory.

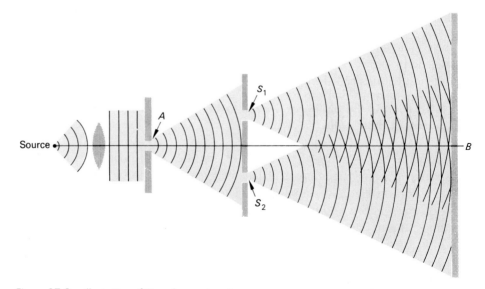

Figure 37-2 Illustration of Young's experiment.

Figure 37-3 Photograph of an interference pattern in Young's experiment (From F. A. Jenkins and H. E. White, *Fundamentals of Optics,* 4th ed., McGraw-Hill Company, New York, 1976. Reprinted by permission.)

The illumination of the screen in alternate bright and dark lines can also be explained in terms of the wave theory. To understand their origin, we recall the **superposition principle,** introduced in Chap. 21 for the study of interference in waves:

> **When two or more waves exist simultaneously in the same medium, the resultant amplitude at any point is the sum of the amplitudes of the composite waves at that point.**

Two waves are said to interfere *constructively* when the amplitude of the resultant wave is greater than the amplitudes of either component wave. *Destructive interference* occurs when the resultant amplitude is smaller.

In Young's experiment, light waves from slit A reach slits S_1 and S_2 at the same time and originate from a single source of one wavelength. Therefore, the secondary wavelets leaving slits S_1 and S_2 are *in phase*. The sources are said to be **coherent.** Figure 37-4 on page 816 shows how the light and dark bands are produced. The bright bands occur whenever the waves arriving from the two slits **interfere** constructively. Dark bands occur when destructive interference takes place. At point B in the center of the screen, light travels the same distances p_1 and p_2 from each slit. The difference in path length $\Delta p = 0$, and constructive interference results in a bright central band. At point C the difference in path lengths Δp causes the waves to interfere destructively. Another bright band occurs at point D when the waves from each slit again reinforce each other. The overall interference pattern is similar to that produced by the water waves in Fig. 37-5 on page 817.

Let us now consider the theoretical conditions necessary for the production of bright and dark bands. Consider light reaching the point D at a distance y from the central axis AB, as shown in Fig. 37-6 on page 817. The separation of the two slits is represented by d, and the screen is located at a distance x from the slits. The point D on the screen makes an angle θ with the axis of the system. The line S_2C is drawn so that the distances CD and p_2 are equal. As long as the distance x to the screen is much greater than the slit separation d, we may consider that p_1, p_2, and DA are all approximately perpendicular to the line S_2C. Therefore, the angle S_1S_2C is equal to the angle θ, and the difference Δp in path lengths of light coming from S_1 and S_2 is given by

$$\Delta p = p_1 - p_2 = d \sin \theta$$

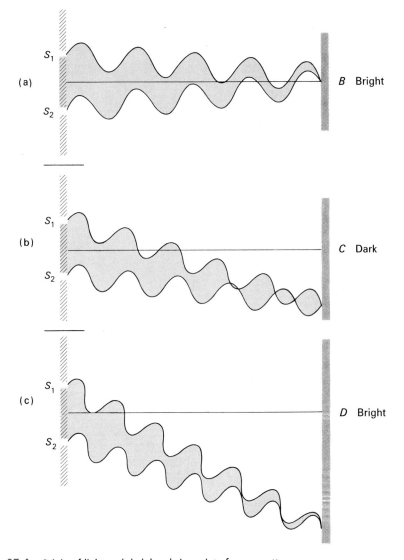

Figure 37-4 Origin of light and dark bands in an interference pattern.

Constructive interference will occur at D when this difference in path length is equal to

$$0, \lambda, 2\lambda, 3\lambda, \ldots, n\lambda$$

where λ is the wavelength of the light. Therefore, the conditions for bright fringes are given by

$$d \sin \theta = n\lambda \qquad n = 0, 1, 2, 3, \ldots \qquad \textbf{37-1}$$

where d is the slit separation and θ is the angle the fringe makes with the axis.

The conditions required for the formation of dark fringes at D will be satisfied when the path difference is

$$\frac{\lambda}{2}, \frac{3\lambda}{2}, \frac{5\lambda}{2}, \ldots$$

Figure 37-5 Interference pattern set up in water waves by two coherent sources. (From *PSCC Physics*, D. C. Heath and Company, Lexington, Mass., 1965.)

Under these conditions, destructive interference will cancel the waves. Thus, dark *fringes* will occur when

$$d \sin \theta = n\frac{\lambda}{2} \qquad n = 1, 3, 5, \ldots \qquad \textbf{37-2}$$

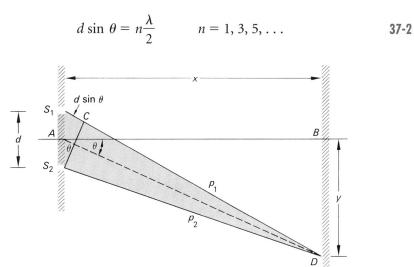

Figure 37-6 Theoretical interpretation of the double-slit experiment.

The preceding equations can be put into a more useful form by expressing them in terms of the measurable distances x and y. For small angles

$$\sin \theta \approx \tan \theta = \frac{y}{x}$$

Substitution of y/x for $\sin \theta$ in Eqs. (37-1) and (37-2) yields

Bright fringes: $$\frac{yd}{x} = n\lambda \qquad n = 0, 1, 2, \ldots \qquad \text{37-3}$$

Dark fringes: $$\frac{yd}{x} = n\frac{\lambda}{2} \qquad n = 1, 3, 5, \ldots \qquad \text{37-4}$$

In experiments designed to measure the wavelength of light, x and d are known initially. The distance y to any particular fringe can be measured and used to determine the wavelength.

EXAMPLE 37-1

In *Young's experiment,* the two slits are 0.04 mm apart, and the screen is located 2 m away from the slits. The third bright fringe from the center is displaced 8.3 cm from the central fringe. (a) Determine the wavelength of the incident light. (b) Where will the second dark fringe appear?

Solution (a)
For the third bright fringe, $n = 3$ in Eq. (37-3). Thus,

$$\frac{yd}{x} = 3\lambda$$

$$\lambda = \frac{yd}{3x} = \frac{(8.3 \times 10^{-2}\ \text{m})(4 \times 10^{-5}\ \text{m})}{3(2\ \text{m})}$$

$$= 5.53 \times 10^{-7}\ \text{m} = 553\ \text{nm}$$

Solution (b)
The displacement of the second dark fringe is found by setting $n = 3$ in Eq. (37-4).

$$\frac{yd}{x} = \frac{3\lambda}{2}$$

$$\lambda = \frac{3\lambda x}{2d} = \frac{(3)(5.53 \times 10^{-7}\ \text{m})(2\ \text{m})}{2(4 \times 10^{-5}\ \text{m})}$$

$$= 4.15 \times 10^{-2}\ \text{m}$$

www.laserprotector.com
www.diffraction.com
www.ion-optics.com
www.optometrics.com
For information on companies that manufacture diffraction gratings for a wide range of technical applications, see these sites.

37-3 THE DIFFRACTION GRATING

If many parallel slits similar to those in Young's experiment are spaced regularly and of the same width, a brighter and sharper diffraction pattern can be obtained. Such an arrangement is known as a *diffraction grating.* Gratings are made by ruling thousands of parallel grooves on a glass plate with a diamond point. The grooves act as opaque barriers to light, and the clear spaces form the slits. Most laboratory gratings are ruled from 10,000 to 30,000 lines per inch.

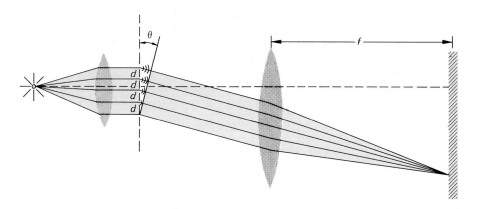

Figure 37-7 The diffraction grating.

A parallel beam of monochromatic light striking a diffraction grating, as shown in Fig. 37-7, is diffracted in a manner similar to that in Young's experiment. Only a few slits are shown in the figure, each separated by a distance *d*. Each slit acts as a source of secondary **Huygens's wavelets,** producing an interference pattern. A lens is used to focus the light from the slits on a screen.

The diffracted rays leave the slits in many directions. In the forward direction, the paths of all rays will be of the same length, setting up a condition for constructive interference. A central bright image of the source is formed on the screen. In certain other directions, the diffracted rays will also be in phase, giving rise to other bright images. One of these directions θ is illustrated in Fig. 37-7. The first bright line formed on either side of the central image is called the **first-order image.** The second bright line on either side of the central maximum is called the *second-order fringe,* and so forth. The condition for the formation of these bright fringes is the same as that derived for Young's experiment. (Refer to Fig. 37-8.) Therefore, we can write the grating equation as

$$d \sin \theta_n = n\lambda \qquad n = 1, 2, 3, \ldots \qquad \text{37-5}$$

Figure 37-8

where d = spacing of slits
 λ = wavelength of incident light
 θ_n = deviation angle for nth bright fringe

The first-order bright fringe occurs when $n = 1$. As illustrated in Fig. 37-9a, this image occurs when the paths of diffracted rays from each slit differ by an amount equal to one wavelength. The second-order image occurs when the paths differ by two wavelengths (Fig. 37-9b).

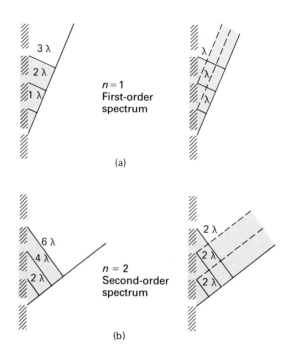

Figure 37-9

EXAMPLE 37-2

A diffraction grating having 20,000 lines per inch is illuminated by parallel light of wavelength 589 nm. What are the angles at which the first- and second-order bright fringes are formed?

Solution

The slit spacing d that corresponds to 20,000 lines per inch is

$$d = \frac{2.54 \text{ cm/in.}}{20,000 \text{ lines/in.}} = 1.27 \times 10^{-4} \text{ cm}$$

The first-order bright fringe occurs for $n = 1$, and the angle θ_1 is found from Eq. (37-5).

$$\sin \theta_1 = 1\frac{\lambda}{d} = \frac{5.89 \times 10^{-5} \text{ cm}}{1.27 \times 10^{-4} \text{ cm}} = 0.464$$

$$\theta_1 = 27.6°$$

The second-order fringe occurs when $n = 2$. Thus,

$$\sin \theta_2 = (2)(0.464) = 0.928$$
$$\theta_2 = 68.1°$$

The third-order fringe will not be formed because $\sin \theta_n$ cannot exceed 1.00. (The beam will not be deviated by an angle greater than 90°.)

37-4 RESOLVING POWER OF INSTRUMENTS

We have learned that light passing through a small opening or past an obstacle is diffracted so that the images formed are fuzzy. Interference fringes near the edges of images sometimes make it difficult to determine the exact shape of the source. Figure 37-10 shows the diffraction pattern formed by passing light through a small circular opening. Notice the large central maximum surrounded by dark and bright interference bands. This diffraction is of extreme importance in optical instruments because it sets the ultimate limit on the possible magnification.

To understand this limitation, consider light from the two sources in Fig. 37-11, passing through a small circular opening in an opaque barrier. In Fig.

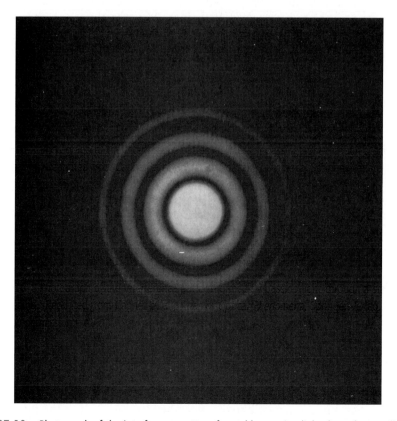

Figure 37-10   Photograph of the interference pattern formed by passing light through a small circular opening. (Reprinted from Cagnet et al., *Atlas of Optical Phenomena,* Springer-Verlag. Reprinted by permission.)

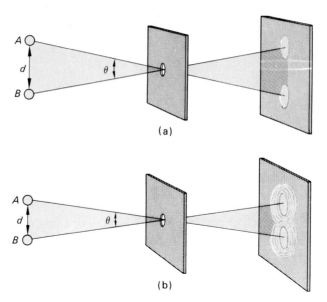

(a)

(b)

Figure 37-11 (a) The images of the sources A and B are easily distinguished. (b) As the sources are brought closer together, the images overlap, resulting in a confused image.

37-11a the images of the sources *A* and *B* are distinguished as separate images. The sources are said to be *resolved*. If they are brought closer together, however, as in Fig. 37-11b, their images overlap, resulting in a confused image. When the sources are so close together (or the opening is so small) that the separate images can no longer be distinguished, the sources are said to be *unresolved*.

The resolving power of an instrument is a measure of its ability to produce well-defined separate images.

A useful method of expressing the **resolving power** of an instrument is in terms of the angle θ subtended at the opening by the objects being resolved. (See Fig. 37-11.) The smallest angle θ_0 for which the images can be distinguished separately is a measure of the resolving power. No matter how perfectly a lens is constructed, the image of a point source of light will not be focused at a point. It will appear as only a tiny, bright dot with light and dark fringes around it. **Resolution** improves as the diameter of a lens is increased. It can be shown that for a given lens of diameter D (see Fig. 37-12) the optimum resolution occurs for the angular width θ_0, subtended by the radius of the central image in the diffraction pattern. In a telescope this limiting angle is

$$\theta_0 = \frac{1.22\lambda}{D} \qquad \text{37-6}$$

where λ is the wavelength of the light and D is the diameter of the objective lens.

Figure 37-13 illustrates the resolution of two point sources of light. In Fig. 37-13a the two images can still be distinguished; in Fig. 37-13b the images are at the limit of resolution. From the criterion used to obtain Eq. (37-6), two such images are just resolved when the central maximum of one pattern coincides with the first dark fringe of the other pattern.

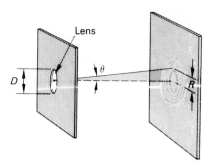

Figure 37-12 The limit of resolution.

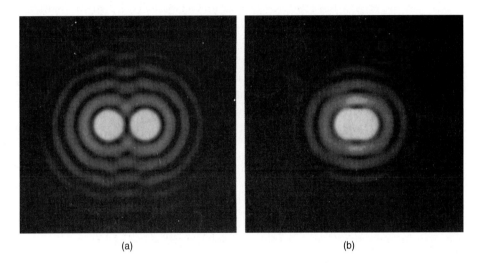

(a)   (b)

Figure 37-13 (a) Separate images formed from two point sources. (b) The images of two sources at the limit of resolution. (From Cagnet et al., *Atlas of Optical Phenomena*, Springer-Verlag. Reprinted by permission.)

To illustrate the topics we have been discussing in this section, consider a telescope. In Fig. 37-14 two objects are separated by a distance s_0 and are located at a distance p from the objective lens in the telescope. The angle between the objects, in radians, is approximately

$$\theta_0 = \frac{s_0}{p} \qquad\qquad \textbf{37-7}$$

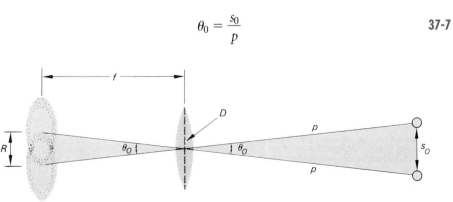

Figure 37-14 Resolution of two distant objects by a spherical lens.

The subscript 0 is used to show that θ_0 and s_0 represent the minimum conditions for resolution. Thus, we can rewrite Eq. (37-6) as

$$\theta_0 = 1.22\frac{\lambda}{D} = \frac{s_0}{p} \qquad \text{37-8}$$

Since s_0 represents the minimum separation of objects that can be resolved, this distance is also used to indicate the resolving power of an instrument.

CONCEPTUAL EXAMPLE 37-3

One of the largest refracting telescopes in the world is the 40-in.-diameter instrument at Yerkes Observatory in Wisconsin; its objective lens has a focal length of 19.8 m (65 ft). (a) What is the minimum separation of two features of the Moon's surface such that they are just resolved by this telescope? (b) What is the radius of the central maximum in the diffraction pattern set up by the objective lens? (For white light, the central wavelength of 500 nm can be used for computing the resolution. The Moon is 3.84×10^5 km from the Earth.)

Solution (a)

We must first convert all distances to meters, Thus,

$$p = 3.84 \times 10^5 \text{ km} = 3.84 \times 10^8 \text{ m}$$
$$\lambda = 500 \text{ nm} = 5 \times 10^{-7} \text{ m}$$
$$D = 40 \text{ in. } (2.54 \times 10^{-2} \text{ m/in.}) = 1.02 \text{ m}$$

The minimum separation s_0 is found by solving for s_0 in Eq. (37-8).

$$s_0 = 1.22\frac{\lambda p}{D}$$
$$= \frac{(1.22)(5 \times 10^{-7} \text{ m})(3.84 \times 10^8 \text{ m})}{1.02 \text{ m}} = 230 \text{ m}$$

Solution (b)

From Eq. (37-7) and Fig. 37-14, the radius R of the central maximum is

$$R = f\theta_0 = f\frac{s_0}{p}$$
$$= \frac{(19.8 \text{ m})(230 \text{ m})}{3.84 \times 10^8 \text{ m}} = 1.19 \times 10^{-5} \text{ m}$$

The 200-in. reflecting telescope at Mt. Palomar in California can distinguish features on the Moon 46.3 m (or 152 ft) apart.

37-5 POLARIZATION

All the phenomena discussed so far can be explained on the basis of either transverse or longitudinal waves. Interference and diffraction occur in sound waves, which are longitudinal, as well as in water waves, which are transverse. More exper-

interNET
CONNECTION

www.poc.com
www.modulationoptics.
com
www.mtinstruments.com
www.cvilaser.com
Manufacturers of polarizing lenses and other polarizing applications can be found at these sites.

imental evidence is needed to determine whether light waves are longitudinal or transverse. In this section we introduce a property of light waves that can be interpreted only in terms of transverse waves.

Let us first consider a mechanical example of transverse waves in a vibrating string. If the source of the wave causes each particle of the rope to vibrate up and down in a single plane, the waves are *plane-polarized.* If the rope is vibrated in such a manner that each particle moves in a random manner at all possible angles, the waves are *unpolarized.*

> **Polarization is the process by which the transverse oscillations of a wave motion are confined to a definite pattern.**

As an illustration of the polarization of a transverse wave, consider the rope passing through slotted frames, as shown in Fig. 37-15. The unpolarized vibrations pass through slot *A* and emerge polarized in the vertical plane. This frame is called the *polarizer.* Only waves with vertical vibrations can pass through the slot; all other vibrations will be blocked. The slotted frame *B* is called the *analyzer* because it can be used to test whether the incoming waves are plane-polarized. If the analyzer is rotated so that the slots in *B* are perpendicular to those in *A*, all the incoming waves are stopped. This can happen only if the waves reaching *B* are polarized in a plane perpendicular to the slots in *B*.

Polarization is a characteristic of *transverse* waves. Should the rope in our example be replaced by a spring, as in Fig. 37-16, the longitudinal waves would pass on through the slots, regardless of their orientation.

Now, let us consider light waves. In Chap. 33 we discussed the electromagnetic nature of light waves. Recall that such a wave consists of an oscillating electric field and an oscillating magnetic field perpendicular to each other and to the direction of propagation. Therefore, light waves consist of *oscillating fields* rather than of vibrating particles, as was the case for the waves in a rope. If it can be shown that these oscillations can be polarized, we can state conclusively that the oscillations are transverse.

A number of substances exhibit different indexes of refraction for light with different planes of polarization relative to their crystalline structure. Some exam-

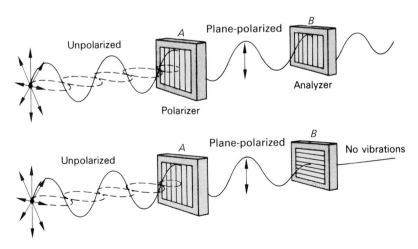

Figure 37-15   Mechanical analogy explaining the polarization of a transverse wave.

Figure 37-16   A longitudinal wave cannot be polarized.

ples are calcite, quartz, and tourmaline. Plates can be constructed from these materials that transmit light in only a single plane of oscillation. Thus, they can be used as polarizers for incident light whose oscillations are randomly oriented. Analogous to the vibrating rope passing through slots, two polarizing plates can be used to determine the transverse nature of light waves.

As shown in Fig. 37-17, light emitted by most sources is unpolarized. Upon passing through a tourmaline plate (the polarizer), the light beam emerges plane-polarized but with reduced intensity. Another plate serves as the analyzer. As this plate is rotated with respect to the polarizer, the intensity of the light passing through the system is gradually reduced until relatively no light is transmitted. It is therefore demonstrated that light waves are transverse rather than longitudinal.

Polarization has demonstrated that transverse nature of light in laboratories for many years, but the useful applications of the principle were not realized until the development of **Polaroid sheets.** These sheets are constructed by sandwiching a thin layer of iodosulfate crystals between two sheets of plastic. The crystals are aligned by a strong electric field. Two such sheets can be used to control the intensity of light. Photographers use Polaroid filters to vary the intensity and to reduce the glare from reflected light. Complex engineering studies can be made by examining stress patterns in certain plastic tool models. The light and dark fringes set up by polarized light give an indication of the areas of varying stress.

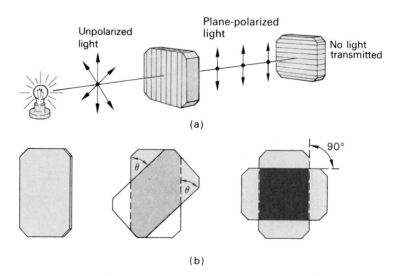

Figure 37-17   (a) Proof that light can be polarized. (b) The reduction of transmitted light intensity as the analyzer is rotated from 0° to 90°.

Summary and Review

Summary

In this chapter we have discussed several ways in which light behaves as a wave. The bending of light around obstacles placed in its path is called *diffraction*. A combination of diffraction and interference of light waves led to Young's experiment and to the diffraction grating. Modern industrial instruments use these concepts for a variety of applications. The major ideas presented in this chapter are summarized below.

- In Young's experiment, interference and diffraction account for the production of bright and dark fringes. The location of these fringes is given by the following equations (see Fig. 37-6):

Bright fringes: $\dfrac{yd}{x} = n\lambda$ $n = 0, 1, 2, 3, \ldots$

Dark fringes: $\dfrac{yd}{x} = n\dfrac{\lambda}{2}$ $n = 1, 3, 5, 7, \ldots$

- In a diffraction grating of slit separation d, the wavelengths of the nth order fringes are given by

$$d \sin \theta_n = n\lambda \qquad n = 1, 2, 3, \ldots$$

- The resolving power of an instrument is a measure of its ability to produce well-defined, separate images. The minimum conditions for resolution are illustrated in Fig. 37-14. For this situation the resolution equation is

$$\theta_0 = 1.22\dfrac{\lambda}{D} = \dfrac{s_0}{p} \qquad \textit{Resolving Power}$$

Key Terms

1. diffraction
2. interference
3. coherent
4. resolving power
5. resolution
6. polarization
7. polarizer
8. analyzer
9. Young's experiment
10. Huygens's principle
11. superposition principle
12. interference fringes
13. diffraction grating
14. order of an image
15. plane-polarized
16. Polaroid sheets

Review Questions

37-1. Radio waves and light waves are both electromagnetic radiations. Explain why radio waves can be received behind tall buildings whereas light cannot reach these areas.

37-2. Consider plane water waves striking a gap in a barrier, as in Fig. 37-1. Explain how the diffraction varies as (a) the slit width is decreased and (b) the wavelength of incoming waves is reduced.

37-3. Which effects in Young's experiment are due to diffraction and which are due to interference?

37-4. In a diffraction grating, how does the spacing of the lines affect the separation of the fringes in the interference pattern?

37-5. If white light were incident upon a diffraction grating, instead of monochromatic light, what would the resulting interference pattern look like?

37-6. In Young's experiment, what effect will reducing the wavelength of incident light have on the interference pattern?

37-7. What effect does increasing the aperture of a lens have on its resolving power? Will a larger wavelength of light result in increased resolution if other conditions are kept constant?

Summary and Review

37-8. The resolving power of some microscopes is increased by illuminating the object with ultraviolet light. Explain.

37-9. A polarizing plate absorbs about 50 percent of the intensity of an unpolarized beam. What accounts for this?

37-10. Assume that a beam of unpolarized light passes through a polarizer and an analyzer, as shown in Fig. 37-17. The analyzer has been rotated by an angle θ from the position for maximum transmission. Only the component of the amplitude A in the plane-polarized beam that lies along the axis is transmitted by the analyzer. This component is given by $A \cos \theta$. The intensity I of light is proportional to the square of the amplitude A. Show that the intensity I of the beam transmitted by the analyzer is given by

$$I = I_0 \cos^2 \theta \qquad \textbf{37-9}$$

where θ is the angle at which the analyzer is placed relative to the position for maximum transmitted intensity I_0.

37-11. Which of the following waves can be polarized: (a) X rays, (b) water waves, (c) sound waves, (d) radio waves?

37-12. When white light falls on a prism, it is dispersed, forming a spectrum of colors, with the red component receiving the least deviation. Compare this spectrum with that produced by a diffraction grating.

37-13. Give a strong argument for the wave theory of light. Counter this argument with reasons for considering that light consists of particles.

37-14. Suppose that you are using a diffraction grating with a spacing of 3000 lines per centimeter. Discuss the usefulness of this grating for examining (a) the infrared radiation of wavelength 3 nm and (b) the ultraviolet radiation of wavelength 100 nm.

Problems

Section 37-2—Young's Experiment; Interference

37-1. Light from a laser has a wavelength of 632 nm. Two rays from this source follow paths that differ in length. What is the minimum path difference required to cause (a) constructive interference and (b) destructive interference? **Ans. 632 nm, 316 nm**

37-2. Find the difference in path length required in Prob. 37-1 to provide the very next instances of constructive and destructive interference.

37-3. Monochromatic light illuminates two parallel slits 0.2 mm apart. On a screen 1.0 m from the slits, the first bright fringe is separated from the central fringe by 2.50 mm. What is the wavelength of the light? **Ans. 500 nm**

37-4. Monochromatic light from a sodium flame illuminates two slits separated by 1.0 mm. A viewing screen is 1.0 m from the slits, and the distance from the central bright fringe to the bright fringe nearest it is 0.589 mm. What is the frequency of the light?

37-5. Two slits 0.05 mm apart are illuminated by green light of wavelength 520 nm. A diffraction pattern is formed on a viewing screen 2.0 m away. What is the distance from the center of the screen to the first bright fringe? What is the distance to the third dark fringe? **Ans. 2.08 cm, 5.20 cm**

*37-6. For the situation described in Prob. 37-5, what is the separation of the two first-order bright fringes located on each side of the central band?

*37-7. Young's experiment is performed using monochromatic light of wavelength 500 nm. The slit separation is 1.20 mm, and the screen is 5.00 m away. How far apart are the bright fringes? **Ans. 2.08 mm**

*37-8. In Young's experiment it is noted that the second dark fringe appears at a distance of

2.5 cm from the central bright fringe. Assume that the slit separation is 60 μm and that the screen is 2.0 m away. What is the wavelength of the incident light?

Section 37-3—The Diffraction Grating

37-9. A diffraction grating having 300 lines per millimeter is illuminated by light of wavelength 589 nm. What are the angles at which the first and second-order bright fringes are formed? **Ans. 10.0°, 20.7°**

37-10. A diffraction grating has 250,000 lines per meter. What is the wavelength of incident light if the second-order bright fringe occurs at 12.6°?

37-11. A small sodium lamp emits light of wavelength 589 nm, which illuminates a grating marked with 6000 lines per centimeter. Calculate the angular deviation of the first- and second-order bright fringes? **Ans. 20.7°, 45.0°**

37-12. A parallel beam of light illuminates a diffraction grating with 6000 lines per centimeter. The second-order bright fringe is located 32.0 cm from the central image on a screen 50 cm from the grating. Calculate the wavelength of the light.

*37-13. The visible light spectrum ranges in wavelength from 400 to 700 nm. Find the angular width of the first-order spectrum produced by passing a white light through a grating marked with 20,000 lines per inch. **Ans. 15.0°**

*37-14. An infrared spectrophotometer uses gratings to disperse infrared light. One grating is ruled with 240 lines per millimeter. What is the maximum wavelength that can be studied with this grating?

Section 37-4—Resolving Power of Instruments

37-15. Light of wavelength 600 nm falls on a circular opening of diameter 0.32 mm. A diffraction pattern forms on a screen 80 cm away. What is the distance from the center of the pattern to the first dark fringe? **Ans. 1.83 mm**

37-16. The limiting angle of resolution for an objective lens in an optical instrument is 3×10^{-4} rad for a 650-nm light source. What is the diameter of the circular opening?

37-17. A certain radio telescope has a parabolic reflector that is 70 m in diameter. Radio waves from outer space have a wavelength of 21 cm. Calculate the theoretical limit of resolution for this telescope. **Ans. 3.66×10^{-4} rad**

37-18. Using a telescope with a 60-m-diameter lens, find how far apart two objects can be resolved if they are located a distance in space equal to that from the earth to the sun (93 million miles).

37-19. What is the angular limit of resolution of a person's eye when the diameter of the opening is 3 mm? Assume that the wavelength of the light is 500 nm. **Ans. 2.03×10^{-4} rad**

37-20. At what distance could the eyes described in Prob. 37-19 resolve wires in a door screen that are separated by 2.5 mm?

Challenge Problems

37-21. In Young's experiment a 600-nm light illuminates a slit located 2.0 m from a screen. The second bright fringe formed on the screen is 5 mm from the central maximum. What is the slit width? **Ans. 480 μm**

37-22. A transmission grating is ruled with 5000 lines per centimeter. For light of wavelength 550 nm, what is the angular deviation of the third-order bright fringe?

37-23. Monochromatic light passes through two slits separated by 0.24 mm. In the pattern formed on a screen 50 cm away, the distance between the first bright fringe of the left of the central maximum and the first fringe on the right is 2.04 mm. What is the wavelength of the light? **Ans. 490 nm**

37-24. A transmission grating ruled with 6000 lines per centimeter forms a second-order bright fringe at an angle of 53° from the central fringe. What is the wavelength of the incident light?

*37-25. If the separation of the two slits in Young's experiment is 0.10 mm and the distance to the screen is 50 cm, find the distance between the first dark fringe and the third bright fringe when the slits are illuminated with light of wavelength 600 nm.

Ans. 7.50 mm

*37-26. Light from a mercury-arc lamp is incident on a diffraction grating ruled with 7000 lines per inch. The spectrum consists of a yellow line (579 nm) and a blue line (436 nm). Compute the angular separation (in radians) of these lines in the third-order spectrum.

*37-27. A telescope will be used to resolve two points on a mountain 160 km away. If the separation of the points is 2.0 m, what is the minimum diameter for the objective lens? Assume that the light has an average wavelength of 500 nm.

Ans. 4.86 cm

Critical Thinking Questions

*37-28. A Michelson interferometer, as shown in Fig. 37-18, can be used to measure a small distance. The beam splitter partially reflects

Figure 37-18   The Michelson interferometer.

and partially transmits monochromatic light of wavelength λ from the source S. One mirror M_1 is fixed and another M_2 is movable. The light rays reaching the eye from each mirror differ, causing constructive and destructive interference patterns to move across the scope as the mirror M_2 is moved a distance x. Show that this distance is given by

$$x = m\frac{\lambda}{2}$$

where m is the number of dark fringes that cross an indicator line on the scope as the mirror moves a distance x.

37-29. A Michelson interferometer (see Prob. 37-28) is used to measure the advance of a small screw. How far has the screw advanced if krypton-86 light ($\lambda = 606$ nm) is used and 4000 fringes move across the field of view as the screw advances?

*37-30. A diffraction grating has 500 lines per millimeter ruled on its glass surface. White light passes through the grating and forms several spectra on a screen 1.0 m away. Is the deviation of colors with a grating different from those experienced for prisms? On the screen, what is the distance between the first-order blue line (400 nm) and the first-order red line (680 nm)? How many

complete spectra (400–700 nm) are possible for these conditions? **Ans. yes,**
$\Delta y = 15.9$ cm, two ($n = 2.85$)

*37-31. The taillights of an automobile are 1.25 m apart. Assume that the pupil of a person's eye has a diameter of 5 mm and that the light has an average wavelength of 604 nm. At night, on a long straight highway, how far away can the two taillights be resolved? Suppose you squint your eyes, forming a slit in which the limiting angle changes from $\theta_0 = 1.22\ \lambda/D$ to $\theta_0 = \lambda/d$. What is the new distance for resolution of the images?

*37-32. The intensity of unpolarized light is reduced by one-half when it passes through a polarizer. In the case of the plane-polarized light reaching the analyzer, the intensity I of the transmitted beam is given by

$$I = I_0 \cos^2 \theta$$

where I_0 is the maximum intensity transmitted and θ is the angle through which the analyzer has been rotated. Consider three Polaroid plates stacked so that the axis of each is turned 30° with respect to the spreading plate. By what percentage will the incident light be reduced in intensity when it passes through all three plates? **Ans. 28.1 percent**

Part **4**

Modern
Physics

38

Modern Physics and the Atom

OBJECTIVES

After completing this chapter, you should be able to:

1. Discuss Einstein's postulates and determine the relativistic changes in length, mass, and time.
2. Demonstrate your understanding of the photoelectric effect, the equivalence of mass and energy, the de Broglie wavelength, and the Bohr model of the atom.
3. Demonstrate your understanding of emission and absorption spectra and discuss with drawings the Balmer, Lyman, and Paschen spectral series.
4. Calculate the energy emitted or absorbed by the hydrogen atom when the electron moves to a higher or lower energy level.

By 1900, physical events had been observed that could not be explained satisfactorily by the laws of classical physics. Light, which was thought to be a wave phenomenon, was found to exhibit properties of particles as well. Newton's concepts of absolute mass, length, and time were found inadequate to describe certain physical events. The light produced by an electric spark in gases did not produce a continuous spectrum on passing through a prism or diffraction grating. These and other unexplained phenomena signaled the beginning of entirely new ways of viewing the world around us.

In 1905, Einstein's first paper on relativity was published; it was followed in 1916 by a second paper. They put classical physics into a new perspective. Einstein's relativity laid the basis for a universal physics that limited classical newtonian physics to situations involving speeds considerably less than the speed of light. The work of Einstein stimulated a rash of work by others, which has had immense ramifications; it established parameters for maximum energy use, space travel, modern electronics, chemical analysis, x rays, nuclear weaponry, and many other applications.

38-1 RELATIVITY

Einstein's two papers on relativity altered physics dramatically, but they are not generally understood by many people outside the scientific community. To understand relativity, you must put aside all your preconceived ideas and be willing to view physical events from a new perspective.

The special theory of relativity, published in 1905, is based on two postulates. The first states that every object is in motion compared with something else, that

Modern Physics and the Atom

there is no such thing as absolute rest. For example, picture a freight car moving along a railroad track at 40 mi/h. The cargo in relation to the freight car is stationary, but in relation to the earth it is moving at 40 mi/h. According to the first postulate, it is impossible to name anything that is at absolute rest; an object is at rest (or moving) only in relation to some specified reference point.

Einstein's first postulate also tells us that if we see something changing its position with respect to us, we have no way of knowing whether *it* is moving or *we* are moving. If you walk to a neighbor's house, it is just as correct according to this postulate to say that the house came to you. This sounds absurd to us because we are accustomed to using the earth as a frame of reference. Einstein's laws were designed to be completely independent of such a preferred reference frame. From the standpoint of physics, the first postulate is often restated as follows:

The laws of physics are the same for all frames of reference moving at a constant velocity with respect to one another.

Nineteenth-century physicists had suggested that there was a preferred frame of reference, the so-called luminiferous ether. This was the medium through which they thought electromagnetic waves propagated. However, experiments like the famous Michelson–Morley experiment of 1887 (discussed in Chap. 33) and others were unable to detect the existence of the ether. These experiments are the basis for Einstein's second revolutionary postulate:

The free-space velocity of light c is constant for all observers, independent of their state of motion.

To see why this second postulate was so revolutionary, let us consider a bus traveling at 50 km/h, as in Fig. 38-1. A person riding in the bus first throws a baseball

(a) Velocities relative to bus

(b) Velocities relative to ground

Figure 38-1 The velocity of the bus is 50 km/h to the right (north). Two balls thrown with equal speeds (20 km/h), one to the right and the other to the left, have different speeds relative to the ground. The speed of light is independent, however, of a frame of reference.

with a speed (relative to the person) of 20 km/h toward the front of the bus. Then a second ball is thrown with the same speed toward the rear of the bus. To the person who threw the balls, each ball traveled at 20 km/h, one toward the front of the bus and one toward the rear of the bus. But to an observer on the ground, the velocity of the bus adds to the velocity of the first ball, which appears to be traveling at 70 km/h in the same direction as the bus. And when the velocity of the bus is added to the second ball, the observer on the ground sees it traveling at 30 km/h *in the same direction as the bus.* But the speed of light does not change and will appear the same to the person on the bus and the observer on the ground. Light always travels at the same constant speed.

$$c = 3 \times 10^8 \text{ m/s}$$

whether it is traveling with the source or away from the source.

38-2 SIMULTANEOUS EVENTS: THE RELATIVITY OF TIME

A general understanding of *Einstein's postulates* does not require extensive mathematics. It requires only that we look at things in a fresh and different way. Our understanding of the world around us depends largely on our experience. The time of flight from Atlanta to New York may be listed as 2 h, and we would assume that an observer on the ground and another on the plane would each record the same time interval. Certainly, our experience seems to support such a conclusion. We shall see, however, that time intervals in one reference frame are in general not the same as those measured from a second frame that is in motion relative to the first. Thus, the observer aboard the plane will judge the trip to require less time than that reckoned by the person on the ground. Of course, the apparent discrepancy would not be noticeable at normal speeds.

Our measurements of time are usually based on the occurrence of simultaneous events. For example, the sweeping second hand on a stopwatch passes the point marked 0 just as a sprinter leaves the starting block. It later passes the point marked 10 just as the sprinter crosses the finish line. The judgment that the time interval was 10 s is based on the measurement of these simultaneous events. But Einstein's postulates force us to ask whether events judged to be simultaneous in one reference frame would also be judged as simultaneous in a moving frame of reference.

Einstein developed a thought experiment to illustrate the relativity of simultaneous measurements. Imagine that a boxcar is moving along a railroad track with uniform velocity as shown in Fig. 38-2. Two bolts of lightning strike the ends of the boxcar, leaving marks A_0 and B_0 on the boxcar and marks A and B on the ground. An observer P_0 is at the midpoint of the boxcar, and another observer P is on the ground halfway between points A and B. The events seen by each observer are light signals arriving from the lightning bolts. It must also be noted that the speed c of the light is constant for each observer and is not affected by the motion of the boxcar.

The light signals reaching the ground observer P are judged to arrive at the same time. The signals traveled the same distance, and the events are judged to occur simultaneously. Now consider the same events as observed by the person P_0 aboard

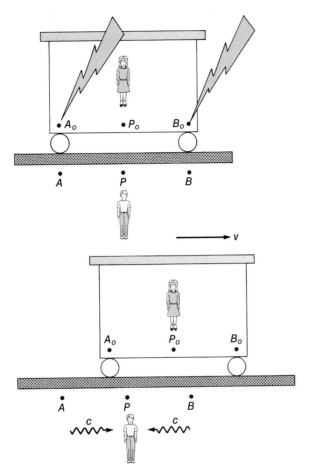

Figure 38-2 Two lightning bolts strike the ends of a boxcar, marking both car and track. The person at point P on the ground observes the events to be simultaneous. The person at point P_0 on the car judges that the bent B_0 occurred before the event at A_0.

the boxcar. By the time the light signal from B_0 has reached the point P on the ground, it has already passed point P_0 on the moving boxcar. The light from A_0 has not yet reached P_0. Thus, the judgment of the person on the boxcar is that events did not occur at the same time. Who is correct? Since according to the principle of relativity, there is no preferred frame of reference, we must conclude that each is correct from his or her own perspective.

38-3 RELATIVISTIC LENGTH, MASS, AND TIME

The relativity of simultaneous measurements has far-reaching consequences when material objects approach the speed of light. All physical measurements must account for relative motion. Time, length, and mass measurements will not be the same for all observers. A comparison of measurements made in different frames of reference is easier if we establish what is called a "proper" frame of reference. It might be noted that two events occurring in the *same* frame of reference are more fundamental than those same events as observed from another frame. We

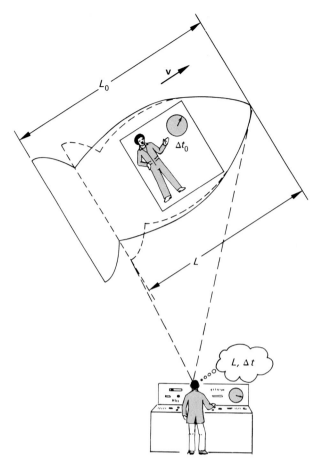

Figure 38-3 The length of objects and the duration of events are affected by relative motion. The person aboard the rocket measures the length L_0 and the time interval Δt; the person in the laboratory observes a shorter length L and records a longer time interval Δt.

will define a ***proper time interval*** Δt_0 as an interval between two events that occur at the same space point. For example, consider a rocket ship moving in space at a speed v relative to an observer in a laboratory on the earth. Refer to Fig. 38-3. A person aboard this ship measures *proper length* L_0, *proper mass* m_0, and *proper time intervals* Δt_0. The person on the ground who measures these same events that actually occur on the ship will obtain different values L, m, and Δt. Each observer is correct from his or her own viewpoint.

A series of relativistic equations can be developed to predict how measurements are affected by relative motion. In each case, the effect becomes more pronounced as the velocity v of objects approaches the limiting velocity c of light. If we find that the proper length of a rocket ship (see Fig. 38-3) is L_0, its length L when it is moving at a relative speed v will be given by

$$L = L_0 \sqrt{1 - \frac{v^2}{c^2}} \qquad \textit{Relativistic Contraction} \qquad \textbf{38-1}$$

Such foreshortening of length in the direction of motion is referred to as ***relativistic contraction***.

This means that the length L of a moving object is observed to be shorter by a factor of $\sqrt{1 - v^2/c^2}$ than its length at rest (its proper length). A study of the formula will reveal that the observed length L will be equal to the proper length L_0 when $v = 0$ (the object is at rest). It will begin to shorten as the velocity approaches c.

EXAMPLE 38-1

When a rocket ship is at rest with respect to us, its length is 100 m. How long do we measure its length to be when it moves by us at a speed of 2.4×10^8 m/s, or $0.8c$?

Solution

In this case the proper length L_0 is 100 m. Substitution into Eq. (38-1) gives

$$L = L_0 \sqrt{1 - \frac{v^2}{c^2}} = (100 \text{ m}) \sqrt{1 - \frac{(0.8c)^2}{c^2}}$$

$$= (100 \text{ m}) \sqrt{1 - \frac{0.64c^2}{c^2}} = (100 \text{ m}) \sqrt{1 - 0.64}$$

$$= (100 \text{ m}) \sqrt{0.36} = (100 \text{ m})(0.6) = 60 \text{ m}$$

As we have seen, time intervals are also affected by relative motion. A clock held by a person aboard the spaceship in Fig. 38-3 indicates a proper time interval Δt_0 that is shorter than that (Δt) observed from the laboratory on the earth. The time interval Δt is given by

$$\Delta t = \frac{\Delta t_0}{\sqrt{1 - v^2/c^2}} \qquad \text{Time Dilation} \qquad \textbf{38-2}$$

This slowing of time (longer time intervals) as a function of velocity is referred to as **time dilation.**

To be sure that you understand this equation, you should recognize that Δt and Δt_0 represent time *intervals,* or the time elapsed from the beginning until the end of an event. Consequently, a clock that is running more slowly will record longer time intervals. We can say that time has stopped when it becomes impossible to measure an event; in other words, the time interval is infinite. This is exactly what is predicted by the time-dilation equation in the limit where $v = c$.

$$\Delta t = \frac{\Delta t_0}{\sqrt{1 - c^2/c^2}} = \frac{\Delta t_0}{\sqrt{1 - 1}} = \frac{\Delta t_0}{0} = \infty$$

EXAMPLE 38-2

Suppose we observe a rocket ship moving past us at $0.8c$, as in the previous example. We measure the time between ticks on the rocket ship clock to be 1.67 s. What time between ticks is measured by the captain of the rocket ship?

Solution

In this case $\Delta t = 1.67$ s, and we are to find Δt_0 from Eq. (38-2). It will be helpful to recall from the previous example that $\sqrt{1 - v^2/c^2} = 0.6$ when $v = 0.8c$. Solving Eq. (38-2) for Δt_0, we have

$$\Delta t_0 = \Delta t \sqrt{1 - v^2/c^2} = (1.67 \text{ s})(0.6) = 1.00 \text{ s}$$

The interval measured by the captain is shorter than that measured by us.

Let us now consider yet another physical quantity that varies with relative velocity. For momentum to be conserved independent of a frame of reference, the mass of a body must vary in the same proportion as length and time. If the rest mass of an object is m_0, the mass m of a body moving with a speed v will be measured by

$$m = \frac{m_0}{\sqrt{1 - v^2/c^2}} \qquad \text{\textit{Relativistic Mass}} \qquad \textbf{38-3}$$

The mass m is referred to in this instance as the **relativistic mass.**

EXAMPLE 38-3

The rest mass of an electron is 9.1×10^{-31} kg. What is its relativistic mass if its velocity is $0.8c$?

Solution

In previous examples we found that $\sqrt{1 - v^2/c^2} = 0.6$ when $v = 0.8c$. Using this computation and substituting in Eq. (38-3), we obtain

$$m = \frac{m_0}{\sqrt{1 - v^2/c^2}} = \frac{9.1 \times 10^{-31} \text{ kg}}{0.6} = 1.52 \times 10^{-30} \text{ kg}$$

This represents a 67 percent increase in mass.

It should be clear from Eq. (38-3) that, if m_0 is not equal to zero, the value for the relativistic mass m approaches infinity as v approaches c. This would mean that an infinite force would be needed to accelerate a nonzero mass to the velocity of light. Apparently, the free-space velocity of light represents an upper limit for the speed of such masses. If the rest mass *is* zero, however, as for photons of light, the relativistic-mass equation *does* allow for $v = c$.

Startling as the predictions of Einstein's equations are, they are experimentally verified daily in the laboratory. Their conclusions are amazing to us only because we do not have direct experience with such fantastic speeds. The giant atom smashers, the betatrons, and many other devices for accelerating particles are in use all over the country to accelerate atomic and nuclear particles at speeds close to the speed of light. Protons in the big Brookhaven accelerator have been accelerated to within 99.948 percent of the speed of light. Their mass is shown to be increased precisely as relativity predicts. Electrons can now be accelerated to more than $0.9999999c$, + the speed of light, causing their mass to increase by more than 40,000 times the rest value.

38-4 MASS AND ENERGY

Before Einstein, physicists had always considered mass and energy as separate quantities that must be *conserved* separately. Now mass and energy must be considered as different ways of expressing the same quantity. If we say that mass can be converted to energy and energy to mass, we must recognize that mass and energy are the same thing expressed in different units. Einstein found the conversion factor to be equal to the square of the velocity of light.

$$E_0 = m_0 c^2 \qquad \text{38-4}$$

From the way the equation is written, it is easy to see that a tiny bit of mass corresponds to an enormous amount of energy. For example, an object whose rest mass m_0 is 1 kg has a rest energy E_0 of 9×10^{16} J.

A more general discussion of energy must take the effects of relativity into account. The expression for the total energy of a particle of rest mass m_0 and momentum $p = mv$ can be written

$$E = \sqrt{(m_0 c^2)^2 + p^2 c^2} \qquad \text{38-5}$$

Now, if we substitute for m_0 from the relativistic-mass relationship, Eq. (38-3), the total energy reduces to

$$E = mc^2 \qquad \text{38-6}$$

where m represents the relativistic mass. This is the most general form for the total energy of a particle.

Note that Eq. (38-5) reduces to $E_0 = m_0 c^2$ when the velocity is zero and hence $p = 0$. Furthermore, if we consider velocities considerably less than c, the equation simplifies to

$$E = \tfrac{1}{2} m_0 v^2 + m_0 c^2 \qquad \text{38-7}$$

Here we have the usual expression for kinetic energy with a new term added for the *rest energy.*

The more general expression for the kinetic energy of a particle must consider the effects of relativity. Recall that the kinetic energy E_k at speed v is defined as the work that must be done to accelerate a particle from rest to speed v. By the methods of calculus it can be shown that the relativistic kinetic energy of a particle is given by

$$E_K = (m - m_0)c^2 \qquad \text{38-8}$$

This represents the difference between the total energy of a particle and its rest-mass energy.

CONCEPTUAL EXAMPLE 38-4

An electron is accelerated to a speed of $0.8c$. Compare its relativistic kinetic energy with the value based on Newton's mechanics.

Solution

In the previous example we found the relativistic mass of an electron at this speed to be 1.52×10^{-30} kg. Since its rest mass is 9.1×10^{-31} kg, we can determine its kinetic energy by substitution into Eq. (38-8).

$$
\begin{aligned}
E_K &= (m - m_0)c^2 \\
&= (15.2 \times 10^{-31} \text{ kg} - 9.1 \times 10^{-31} \text{ kg})(3 \times 10^8)^2 \\
&= (6.1 \times 10^{-31} \text{ kg})(9 \times 10^{16}) \\
&= 5.49 \times 10^{-14} \text{ J}
\end{aligned}
$$

The newtonian value is based on $\frac{1}{2}m_0v^2$, where $v = 0.8c$.

$$
\begin{aligned}
\tfrac{1}{2}m_0v^2 &= \tfrac{1}{2}(9.1 \times 10^{-31} \text{ kg})(0.8c)^2 \\
&= (4.55 \times 10^{-31} \text{ kg})(0.64c^2) \\
&= (4.55 \times 10^{-31} \text{ kg})(0.64)(3 \times 10^8 \text{ m/s})^2 \\
&= 2.62 \times 10^{-14} \text{ J}
\end{aligned}
$$

The relativistic kinetic energy is more than twice its newtonian value.

38-5 QUANTUM THEORY AND THE PHOTOELECTRIC EFFECT

www.aip.org
This Web site, home of the American Institute of Physics, provides current information about employment, internship opportunities, and online physics journals.

Recall from Chap. 33 that the **photoelectric effect** led to the establishment of a dualistic theory of light. (See Fig. 38-4.) The ejection of electrons as a result of incident light could not be accounted for in terms of the existing electromagnetic theory.

In an attempt to bring experiment into agreement with theory, Maxwell Planck postulated that electromagnetic energy is absorbed or emitted in discrete packets, or *quanta*. The energy of such quanta, or *photons*, is proportional to the frequency of the radiation. **Planck's equation** can be written

$$ E = hf \qquad \text{38-9} $$

where h is the proportionality constant known as *Planck's constant*. Its value is

$$ h = 6.63 \times 10^{-34} \text{ J} \cdot \text{s} $$

Einstein used Planck's equation to explain the photoelectric effect. He reasoned that if light is emitted in photons of energy hf, it must also travel as photons. When a quantum of light strikes a metallic surface, it has an energy equal to hf. If all this energy is transferred to a single electron, the electron might be expected to leave the metal with energy hf. At least an amount of energy W is needed, however, to remove the electron from the metal. The term W is called the **work function** of the surface. Thus, the ejected electron leaves with a maximum kinetic energy given by

Incident light

A

e^- e^-

e^- e^- e^-

B $+$

Batteries

Ammeter

Figure 38-4   The photoelectric effect.

$$E_K = \tfrac{1}{2}mv^2_{\text{max}} = hf - W \qquad\qquad \textbf{38-10}$$

This is *Einstein's photoelectric equation.*

As the frequency of incident light is varied, the maximum energy of the ejected electron varies. The lowest frequency f_0 at which an electron is emitted occurs when $E_K = 0$. In this case,

$$f_0 = \frac{W}{h} \qquad\qquad \textbf{38-11}$$

The quantity f_0 is often called the **threshold frequency.**

EXAMPLE 38-5

Light of wavelength 650 nm is required to cause electrons to be ejected from the surface of a metal. What is the kinetic energy of the ejected electrons if the surface is bombarded with light of wavelength 450 nm?

Solution

The work function W of the surface is equal to the energy of the 650-nm light. Hence,

$$W = hf_0 = h\frac{c}{\lambda_0}$$

$$= 6.63 \times 10^{-34}\ \text{J} \cdot \text{s} \left(\frac{3 \times 10^8\ \text{m/s}}{650 \times 10^{-9}\ \text{m}} \right)$$

$$= 3.06 \times 10^{-19}\ \text{J}$$

The energy of the 450-nm light is

$$hf = \frac{hc}{\lambda} = \frac{(6.63 \times 10^{-34} \text{ J} \cdot \text{s})(3 \times 10^8 \text{ m/s})}{450 \times 10^{-9} \text{ m}}$$

$$= 4.42 \times 10^{-19} \text{ J}$$

From Einstein's photoelectric equation,

$$E_K = hf - W$$
$$= 4.42 \times 10^{-19} \text{ J} - 3.06 \times 10^{-19} \text{ J} = 1.36 \times 10^{-19} \text{ J}$$

38-6 WAVES AND PARTICLES

Electromagnetic radiation has a dual character in its interaction with matter. Sometimes it exhibits wave properties, as demonstrated by interference and diffraction. At other times, as in the photoelectric effect, it behaves like particles, which we have called *photons*. In 1924, Louis de Broglie was able to demonstrate this duality of matter by deriving a relationship for the wavelength of a particle.

This relationship can be seen by looking at two expressions for the energy of a photon. We have already seen from Planck's work that the energy of a photon can be expressed as a function of its wavelength λ.

$$E = hf = \frac{hc}{\lambda}$$

Another expression for the energy of a photon was given earlier in the section on relativity. Equation (38-5) stated that

$$E = \sqrt{(m_0 c^2)^2 + p^2 c^2}$$

This equation shows that photons have momentum $p = mv$ due to their relativistic mass. The rest mass m_0 of a photon is zero, however, so that Eq. (38-5) becomes

$$E = \sqrt{p^2 c^2} = pc \qquad \text{38-12}$$

Since we also know that $E = hc/\lambda$, we can write

$$\frac{hc}{\lambda} = pc$$

from which the wavelength of a photon is given by

$$\lambda = \frac{h}{p} \qquad \text{38-13}$$

de Broglie proposed that all objects have wavelengths related to their momentum—whether the objects are wavelike or particlelike. For example, the wavelength of an electron or any other particle is given by de Broglie's equation, which can be rewritten

$$\lambda = \frac{h}{mv} \qquad \text{\textit{de Broglie Wavelength}} \qquad \text{38-14}$$

EXAMPLE 38-6

What is the de Broglie wavelength of an electron that has a kinetic energy of 100 eV?

Solution

Recalling that 1 eV = 1.6×10^{-19} J, we see that 100 eV is equivalent to 1.6×10^{-17} J, which must be equal to $\frac{1}{2}mv^2$. Thus,

$$\tfrac{1}{2}mv^2 = 1.6 \times 10^{-17} \text{ J}$$

from which

$$v^2 = \frac{2(1.6 \times 10^{-17} \text{ J})}{m} = \frac{3.2 \times 10^{-17} \text{ J}}{9.1 \times 10^{-31} \text{ kg}} = 3.52 \times 10^{13} \text{ m}^2/\text{s}^2$$

The velocity of the electron is therefore

$$v = \sqrt{3.52 \times 10^{13} \text{ m}^2/\text{s}^2} = 5.93 \times 10^6 \text{ m/s}$$

Now the wavelength can be found from Eq. (38-14).

$$\lambda = \frac{h}{mv} = \frac{6.63 \times 10^{-34} \text{ J} \cdot \text{s}}{(9.1 \times 10^{-31} \text{ kg})(5.93 \times 10^6 \text{ m/s})}$$
$$= 1.23 \times 10^{-10} \text{ m} = 0.123 \text{ nm}$$

Note from de Broglie's equation that the higher the velocity of the particle, the shorter its wavelength. Remember that the relativistic mass must be used if the velocity of a given particle is large enough to warrant it.

38-7 THE RUTHERFORD ATOM

Early attempts to explain the structure of an atom were based on a model attributed to J. J. Thomson, the discoverer of the electron. In this model the electrons are implanted in a spherical space of positive charge, as illustrated in Fig. 38-5. Such a model explained the observed neutrality of atoms by postulating equal amounts of positive and negative charge.

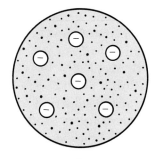

Figure 38-5 The Thomson model of the atom consisted of electrons implanted in a sphere of positive charge.

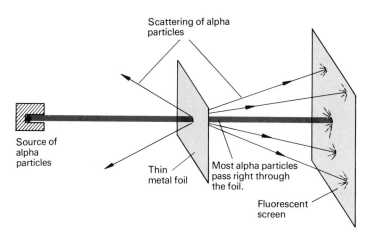

Figure 38-6 Rutherford scattering of alpha particles provided the first evidence for the atomic nucleus.

Thomson's model was short-lived, however, because of the work of Ernest Rutherford in 1911. He bombarded a thin metal foil with a stream of alpha particles, as illustrated in Fig. 38-6. An *alpha particle* is a tiny, positively charged particle emitted by a radioactive substance such as radium. Most of the positively charged particles penetrated the foil easily, as indicated by a flash of light when they struck the zinc sulfide screen, and a few were deflected slightly. Much to Rutherford's surprise, however, others were deflected at extreme angles. Some were even deflected backward. (Rutherford himself said that it was like firing a cannonball through tissue paper and having it come back and hit you in the face.) These extreme deflections could not be explained in terms of the Thomson model for the atom. If charge really was scattered through the atom, the electrical forces would be far too weak to repulse the alpha particles at the large angles actually observed.

Rutherford explained these results by assuming that all the positive charge of an atom is concentrated into a small region, called the *nucleus* of the atom. The electrons were assumed to be distributed in the space around this positive charge. With the atom consisting largely of empty space, the fact that most of the alpha particles passed right through the foil is explained easily. Furthermore, Rutherford found the large-angle scattering to be a consequence of electrostatic repulsion. Once the alpha particle has penetrated the electrons surrounding the positive charge, it is close to a large positive charge of great mass. Large electrostatic forces are therefore expected.

Rutherford has been credited with the discovery of the *nucleus* because he was able to develop formulas to predict the observed scattering of alpha particles. On the basis of his calculations, the diameter of the nucleus was estimated to be approximately one ten-thousandth of the diameter of the atom itself. The positive part of the atom was seen to be concentrated in the nucleus, approximately 10^{-5} nm in diameter. Rutherford believed that the electrons were grouped around the nucleus so that the diameter of the whole atom is about 0.1 nm.

38-8 ELECTRON ORBITS

An immediate difficulty accompanying the Rutherford atom was related to the stability of the atomic electrons. We know from Coulomb's law that the electrons should be attracted to the nucleus. One possible explanation is that the electrons are moving in circles around the nucleus much as planets revolve around the sun. The necessary centripetal force would be provided by coulomb attraction.

Consider, for example, the hydrogen atom, which consists of a single proton and a single electron. We might expect the electron to remain in a constant orbit about the nucleus, as illustrated in Fig. 38-7a. The charge of the electron is labeled $-e$, and the equal but opposite charge of the proton is labeled $+e$. At a distance r from the nucleus, the electrostatic force of attraction on the electron is given by applying Coulomb's law where $k = \frac{1}{4}\pi\epsilon_0$.

$$F_e = \frac{e^2}{4\pi\epsilon_0 r^2} \qquad\qquad \textbf{38-15}$$

where each charge has the magnitude e. For a stable orbit, this force must exactly equal the centripetal force, given by

$$F_c = \frac{mv^2}{r} \qquad\qquad \textbf{38-16}$$

where m is the mass of the electron traveling with a velocity v. Setting $F_e = F_c$, we have

$$\frac{e^2}{4\pi\epsilon_0 r^2} = \frac{mv^2}{r} \qquad\qquad \textbf{38-17}$$

Solving for the radius r, we obtain

$$r = \frac{e^2}{4\pi\epsilon_0 mv^2} \qquad\qquad \textbf{38-18}$$

According to classical theory, Eq. (38-18) should predict the orbital radius r of the electron as a function of its speed v.

The problem with this approach is that the electron must be continually accelerated under the influence of the electrostatic force. According to classical theory,

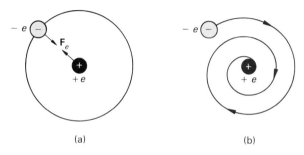

(a)   (b)

Figure 38-7   (a) A stable electron orbit in which the centripetal force is provided by the electrostatic force $\mathbf{F}_e$. (b) Instability due to electromagnetic radiation, which should cause the electron to lose energy and spiral into the nucleus.

an accelerated electron must radiate energy. The total energy of the electron would therefore decrease gradually, reducing the speed of the electron. As can be seen from Eq. (38-18), gradual reduction in electron speed v results in smaller and smaller orbits. Thus, the electron should spiral into the nucleus, as shown in Fig. 38-7b. The fact that it does not compels us to acknowledge a fundamental inconsistency in Rutherford's atom.

38-9 ATOMIC SPECTRA

All substances radiate electromagnetic waves when they are heated. Since each element is different, such emitted radiation can be expected to provide clues to atomic structure. These electromagnetic waves are analyzed by a **spectrometer**, which uses a prism or a diffraction grating to organize the radiation into a pattern called a *spectrum.* For an incandescent source of light, the spectrum is *continuous;* in other words, it contains all wavelengths and is similar to a rainbow. If, however, the source of light is a hot gas under low pressure, the spectrum of the emitted light consists of a series of bright lines separated by dark regions. Such spectra are called **line emission spectra.** The chemical composition of a vaporized material can be determined by comparing its spectrum with known spectra.

A line emission spectrum for hydrogen is shown in Fig. 38-8. The sequence of lines, called a **spectral series,** has a definite order, the lines becoming more and more crowded as the limit of the series is approached. Each line corresponds to a characteristic frequency or wavelength (color). The line of longest wavelength, 656.3 nm, is in the red and is labeled H_α. The others are labeled, in order, as H_β, H_γ, and so on.

It is also possible to obtain similar information from a gas or vapor in an unexcited state. When light is passed through a gas, certain discrete wavelengths are *absorbed.* These **absorption spectra** are like those produced by emission except that the characteristic wavelengths appear as *dark* lines on a light background. The

Figure 38-8 Line spectra for the Balmer series of the hydrogen atom.

absorption spectrum for hydrogen is compared with the emission spectrum in Fig. 38-8.

As early as 1884, Johann Jakob Balmer found a simple mathematical relationship for predicting the characteristic wavelengths of some of the lines in the hydrogen spectrum. His formula is

$$\frac{1}{\lambda} = R\left(\frac{1}{2^2} - \frac{1}{n^2}\right) \qquad \text{38-19}$$

where λ = wavelength
R = Rydberg constant
$n = 3, 4, 5, \ldots$

If λ is measured in meters, the value for R is

$$R = 1.097 \times 10^7 \text{ m}^{-1}$$

The series of wavelengths predicted by Eq. (38-19) is called the *Balmer series.*

EXAMPLE 38-7

Using Balmer's equation, determine the wavelength of the H_α line in the hydrogen spectrum. (The first line occurs when $n = 3$.)

Solution
Direct substitution yields

$$\frac{1}{\lambda} = 1.097 \times 10^7 \text{ m}^{-1}\left(\frac{1}{2^2} - \frac{1}{3^2}\right) = 1.524 \times 10^6 \text{ m}^{-1}$$

from which

$$\lambda = 656.3 \text{ nm}$$

Other characteristic wavelengths are found by setting $n = 4, 5, 6$, and so on. The limit of the series is found by setting $n = \infty$ in Eq. (38-19).

Since the discovery of Balmer's equation, several other series spectra have been discovered for hydrogen. In general, all these discoveries can be summarized by the single equation

$$\frac{1}{\lambda} = R\left(\frac{1}{l^2} - \frac{1}{n^2}\right) \qquad \text{38-20}$$

where l and n are integers with $n > l$. The series predicted by Balmer corresponds to $l = 2$ and $n = 3, 4, 5, \ldots$. The *Lyman series* is in the ultraviolet region and corresponds to $l = 1$, $n = 2, 3, 4, \ldots$; the *Paschen series* is in the infrared and corresponds to $l = 3$, $n = 4, 5, 6, \ldots$; and the *Brackett series,* also in the infrared, corresponds to $l = 4$, $n = 5, 6, 7, \ldots$.

38-10 THE BOHR ATOM

Observations of atomic spectra have indicated that atoms emit only a few rather definite frequencies. This fact does not agree with the Rutherford model, which predicts an unstable atom emitting radiant energy of all frequencies. Any theory of atomic structure must account for the regularities observed in atomic spectra.

The first theory to explain the line spectrum of the hydrogen atom satisfactorily was offered by Niels Bohr in 1913. He assumed, like Rutherford, that the electrons were in circular orbits about a dense, positively charged nucleus, but he decided that electromagnetic theory cannot be strictly applied on the atomic level. Thus, he avoided the problem of orbital instability due to emitted radiation. Bohr's first postulate is as follows:

> **An electron may exist only in those orbits where its angular momentum is an integral multiple of $h/2\pi$.**

Thus, contrary to classical prediction, electrons may be in certain, specified orbits without the emission of radiant energy.

The basis for Bohr's first postulate can be seen in terms of de Broglie wavelengths. The stable orbits are those in which an integral number of electron wavelengths can be fitted into the circumference of the Bohr orbit. Such orbits would allow for standing waves, as illustrated in Fig. 38-9 for four wavelengths. The conditions for such standing waves would be given by

$$n\lambda = 2\pi r \qquad n = 1, 2, 3, \ldots \qquad \text{38-21}$$

where r is the radius of an electron orbit that contains n wavelengths.

Since $\lambda = h/mv$, we can rewrite Eq. (38-21) as

$$n\frac{h}{mv} = 2\pi r$$

from which it can be shown that the angular momentum mvr is given by

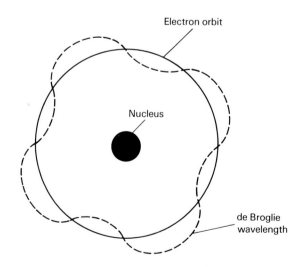

Figure 38-9 A stable electron orbit, showing a circumference equal to four de Broglie wavelengths.

$$mvr = \frac{nh}{2\pi} \qquad \textbf{38-22}$$

The number n, called the **principal quantum number,** may take on the values $n = 1, 2, 3, \ldots$.

A *second postulate* given by Bohr places even further restrictions on atomic theory by incorporating the quantum theory.

> **If an electron changes from one stable orbit to any other, it loses or gains energy in discrete quanta equal to the difference in energy between the initial and final states.**

In equation form, Bohr's second postulate can be written

$$hf = E_i - E_f$$

where hf = energy of emitted or absorbed photon
E_i = initial energy
E_f = final energy

Let us return to our discussion of the hydrogen atom to see whether Bohr's postulates will help to bring theory into harmony with observed spectra. Recall that application of Coulomb's law and Newton's law resulted in Eq. (38-18) for the radius r of the orbiting electron.

$$r = \frac{1}{4\pi\epsilon_0}\frac{e^2}{mv^2}$$

According to Bohr's theory,

$$mvr = \frac{nh}{2\pi}$$

Solving these two equations simultaneously for the radius r and for the velocity v, we obtain

$$r = n^2\frac{\epsilon_0 h^2}{\pi m e^2} \qquad \textbf{38-23}$$

$$v = \frac{e^2}{2\epsilon_0 nh} \qquad \textbf{38-24}$$

These equations predict the possible radii and velocities for the electron, where $n = 1, 2, 3, \ldots$.

We now derive an expression for the total energy of a hydrogen atom for any electron orbit.

$$E_T = E_K + E_P$$

The kinetic energy is found by substituting from Eq. (38-24),

$$E_K = \tfrac{1}{2}mv^2 = \frac{me^4}{8\epsilon_0^2 n^2 h^2} \qquad \textbf{38-25}$$

The potential energy of the atom for any orbit is

$$E_P = \frac{1}{4\pi\epsilon_0}\frac{e^2}{r} = -\frac{me^4}{4\epsilon_0^2 n^2 h^2} \qquad \textbf{38-26}$$

after substitution of r from Eq. (38-23). The potential energy is negative because outside work is necessary to remove the electron from the atom. Adding Eq. (38-25) to Eq. (38-26), we find the total energy to be

$$E_T = -\frac{me^4}{8\epsilon_0^2 n^2 h^2}$$

38-27

Returning to Bohr's second postulate, we are now in a position to predict the energy of an emitted or absorbed photon. Normally, the electron is in its ground state corresponding to $n = 1$. If the atom *absorbs* a photon, the electron may jump to one of the outer orbits. From an *excited* state, it will soon fall back to a lower orbit, *emitting* a photon in the process.

Suppose an electron is in an outer orbit of quantum number n_i and then returns to a lower orbit of quantum number n_f. The decrease in energy must be equal to the energy of the emitted photon.

$$E_i - E_f = hf$$

Substituting the total energy for each state from Eq. (38-27), we obtain

$$hf = E_i - E_f$$
$$= -\frac{me^4}{8\epsilon_0^2 n_i^2 h^2} + \frac{me^4}{8\epsilon_0^2 n_f^2 h^2}$$
$$= \frac{me^4}{8\epsilon_0^2 h^2}\left(\frac{1}{n_f^2} - \frac{1}{n_i^2}\right)$$

Dividing both sides of this relation by h and remembering that $f = c/\lambda$, we can write

$$\frac{1}{\lambda} = \frac{me^4}{8\epsilon_0^2 h^3 c}\left(\frac{1}{n_f^2} - \frac{1}{n_i^2}\right)$$

38-28

Substituting in the values, we obtain

$$\frac{me^4}{8\epsilon_0^2 h^3 c} = 1.097 \times 10^7 \ m^{-1}$$

which is equal to Rydberg's constant R. Therefore, Eq. (38-28) is simplified to

$$\frac{1}{\lambda} = R\left(\frac{1}{n_f^2} - \frac{1}{n_i^2}\right)$$

38-29

The above relationship is exactly the same in form as Eq. (38-20), which was established by Balmer and others from experimental data. Thus, the **Bohr atom** brings theory into harmony with observation.

EXAMPLE 38-8

Determine the wavelength of the photon emitted from a hydrogen atom when the electron jumps from the first excited state to the ground state.

Solution

In this case, $n_f = 1$ and $n_i = 2$. Thus,

$$\frac{1}{\lambda} = R\left(\frac{1}{n_f^2} - \frac{1}{n_i^2}\right)$$

$$= (1.097 \times 10^7 \text{ m}^{-1})\left(\frac{1}{1^2} - \frac{1}{2^2}\right) = 8.23 \times 10^6 \text{ m}^{-1}$$

from which

$$\lambda = 1.22 \times 10^{-7} \text{ m} = 122 \text{ nm}$$

38-11 ENERGY LEVELS

From Bohr's work we now have a picture of an atom in which orbital electrons can occupy a number of **energy levels.** The total energy at the nth level is given by Eq. (38-27).

$$E_n = -\frac{me^4}{8\epsilon_0^2 n^2 h^2} \qquad \textbf{38-30}$$

When the hydrogen atom is in its stable *ground* state, the quantum number n is equal to 1. The possible *excited* states are given for $n = 2, 3, 4, \ldots$.

EXAMPLE 38-9

Determine the energy of an electron in the ground state for a hydrogen atom.

Solution

The following constants are needed:

$$\epsilon_0 = 8.85 \times 10^{-12} \text{ C}^2/\text{N} \cdot \text{m}^2 \qquad h = 6.63 \times 10^{-34} \text{ J} \cdot \text{s}$$
$$m = 9.1 \times 10^{-31} \text{ kg} \qquad e = 1.6 \times 10^{-19} \text{ C}$$

Direct substitution into Eq. (38-30), with $n = 1$, yields

$$E_1 = -\frac{(9.1 \times 10^{-31} \text{ kg})(1.6 \times 10^{-19} \text{ C})^4}{8[8.85 \times 10^{-12}(\text{C}^2/\text{N} \cdot \text{m}^2)]^2(1)^2(6.63 \times 10^{-34} \text{ J} \cdot \text{s})^2}$$

$$= -2.17 \times 10^{-18} \text{ J}$$

A more convenient unit for measuring energy at the atomic level is the electronvolt (eV). Recall from Chap. 23 that an electronvolt is the energy acquired by an electron accelerated through a potential difference of 1 V. Consequently

$$1 \text{ eV} = 1.6 \times 10^{-19} \text{ J}$$

*inter*NET
CONNECTION

www2.slac.stanford.edu/vvc
A linear accelerator is a huge device used to find tiny subatomic particles. To find out more, visit this virtual site.

In applications in which a larger unit of energy is required, the megaelectronvolt (MeV) is more appropriate.

$$1 \text{ MeV} = 10^6 \text{ eV}$$

From the previous example, the energy of the ground-state electron can be expressed in electronvolts as follows:

$$E_1 = -2.17 \times 10^{-18} \text{ J} \frac{1 \text{ eV}}{1.6 \times 10^{-19} \text{ J}}$$
$$= -13.6 \text{ eV}$$

This fact can be used to write Eq. (38-30) in a simpler form:

$$E_n = \frac{-13.6 \text{ eV}}{n^2} \qquad\qquad \textbf{38-31}$$

Similar calculations will yield smaller negative values for the outer orbits. If the electron were to be entirely removed from the atom, a case where $n = \infty$, 13.6 eV of energy would be required. ($E_\infty = 0$.) Similarly, a photon of energy 13.6 eV could be emitted if an electron were to be captured by an ionized hydrogen atom and wind up in the ground state.

The atomic spectra observed for hydrogen is now understood in terms of energy levels. The Lyman series results from electrons returning from some excited state to the ground state, as seen in Fig. 38-10. The Balmer, Paschen, and Brackett series occur when the final states are orbits for which $n = 2$, $n = 3$, and $n = 4$, respectively.

An energy-level diagram for hydrogen is shown in Fig. 38-11. Such diagrams are often used to describe the various energy states of atoms.

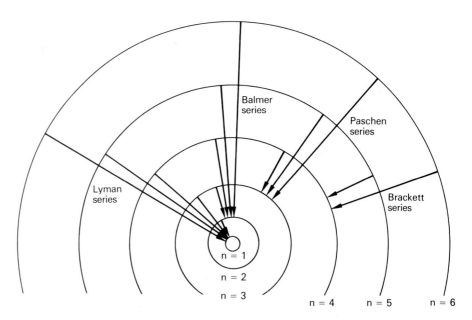

Figure 38-10   The Bohr atom and a description of the origin of the Lyman, Balmer, Paschen, and Brackett spectral series.

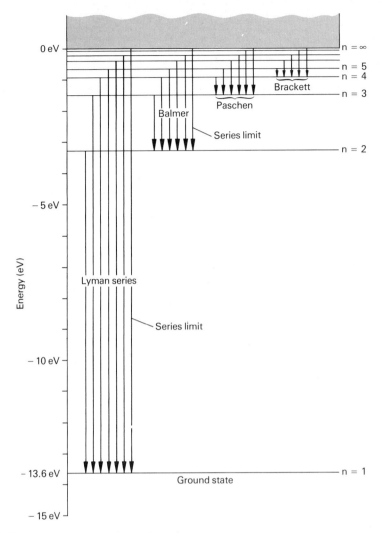

Figure 38-11   Energy-level diagram for the hydrogen atom.

38-12 LASERS AND LASER LIGHT

One of the most useful applications based on quantum physics and a study of the atom is the laser. The intense, highly focused, and coherent light emitted by these devices is responsible for many advances in science. In medicine, trained ophthalmologists can repair the retina of an eye with laser spot welds. By combining laser light with fiber optics (see Sec. 35-6), a revolution is already under way in the fields of electronics and communications. Powerful lasers have even been developed to drill small holes in diamonds.

The principle behind the operation of lasers is relatively easy to understand. It is merely an application of the quantum theory discussed earlier in this chapter for the energy levels of an atom. Basically there are three ways in which photons can interact with matter: (1) *absorption*, (2) *spontaneous emission*, and (3) *stimulated emission*. Each of these three events are described in Fig. 38-11.

(a) Absorption

(b) Spontaneous emission

(c) Stimulated emission

Figure 38-12   (a) An electron, originally in its ground state E_1, is boosted to the higher level E_2 because of the absorption of a photon. (b) Spontaneous emission occurs and a photo is emitted, $hf = E_2 - E_1$. (c) Stimulated emission occurs when an incident photon of energy hf causes the emission of a second photon of the same energy.

Absorption and emission were discussed in the previous section, where it was pointed out that the absorption of a photon can excite an atom by raising an electron to a higher energy level, as is illustrated by Fig. 38-12a. Such an electron is in an ***excited state,*** and eventually it will drop to its original level and undergo what is called *spontaneous emission* (see Fig. 38-12b). In each case the absorbed or the emitted photon has an energy given by

$$E_2 - E_1 = hf$$

Spontaneous emission accounts for the light we see from light bulbs and many other traditional sources of light. Although the light has a definite energy hf for a given photon, the total light emitted consists of many such photons of varying energies. Therefore, spontaneous emission produces light that is not directional and it cannot be focused sharply.

Stimulated emission provides the key to the operation and effectiveness of lasers. In fact, the word *laser* is an approximate acronym for "light amplification by stimulated emission of radiation." Suppose an atom is initially in an excited state E_2, as in Fig. 38-12c, and a second photon of energy $hf = E_2 - E_1$ is incident on the atom. Since the incident photon energy is the same as the excitation energy of the electron, there is an increased probability that the electron will drop to its lower energy level, thereby emitting a second photon of the same energy. Such stimulated emission, when coupled with the existence of the incident photon, has the effect of producing two photons with only one incident photon. Each photon has the same energy, direction, and polarization. These photons can, in turn, stimulate other atoms to emit similar photons. Thus, a chain reaction occurs in which many photons of light are emitted, forming the intense, coherent light characteristic of lasers.

Other factors are important for the effective operation of lasers. For example, for stimulated emission to occur, the atoms of the working substance must be in an excited state. At room temperature, most of the atoms have electrons in the

Figure 38-13 Population inversion occurs when applied external energy produces atoms with more electrons in the higher energy level than in the lower energy level.

Figure 38-14 A schematic diagram of the helium-neon laser.

ground state with only a few in the higher level. Such a normal population is shown in Fig. 38-13a. By introducing external energy in the form of heat, intense light, or electrical discharges, the population of electrons can be inverted. As in Fig. 38-13b, a population inversion places more electrons in the higher energy level than there are in the lower level. Conditions are therefore ripe for stimulating an avalanche of photons, all of the same frequency and highly directional.

The helium-neon laser illustrated in Fig. 38-14 is commonly found in physics laboratories. A high voltage is placed across a low-pressure mixture of helium and neon gas contained in a glass tube. Spontaneous emission occurs first, then a chain reaction of stimulated emissions. Both ends of the tube are silvered to form mirrors that reflect the photons back and forth, increasing the opportunities for additional stimulated emissions. One end is only partially silvered so that some of the photons can escape from the tube and form a laser beam.

38-13 MODERN ATOMIC THEORY

Although the Bohr atom remains a convenient way to describe the atom, a much more refined theory has been found necessary. The model of the electron as a point particle moving in a perfectly circular orbit does not explain many atomic phenomena. Additional quantum numbers have been established to describe the shape and orientation of the electron cloud about the nucleus, as well as the spin motion of the electrons. It has also been established that no two electrons of the same atom can exist in exactly the same state, even though they may exist at the same Bohr energy level. A more complete description of modern atomic theory can be found in textbooks in the field of atomic physics.

Summary and Review

Summary

The works of Einstein, Bohr, de Broglie, Balmer, and many others have led to a much clearer understanding of nature. We no longer view the world as though all phenomena can be seen, touched, and observed in traditional ways. Greater understanding of the atom has led to many industrial applications based on the principles discussed in this chapter. A summary of the major topics is given below.

- According to Einstein's equations of relativity, length, mass, and time are affected by relativistic speeds. The changes become more significant as the ratio of an object's velocity v to the free space velocity of light c becomes larger.

$$L = L_0 \sqrt{1 - \frac{v^2}{c^2}} \qquad \textit{Relativistic Contraction}$$

$$m = \frac{m_0}{\sqrt{1 - v^2/c^2}} \qquad \textit{Relativistic Mass}$$

$$\Delta t = \frac{\Delta t_0}{\sqrt{1 - v^2/c^2}} \qquad \textit{Time Dilation}$$

In the above equations, $c = 3 \times 10^8$ m/s.

- The total energy of a particle of rest mass m_0 and speed v can be written in either of the following forms:

$$E = mc^2 \qquad E = \sqrt{m_0^2 c^4 + p^2 c^2} \qquad \textit{Total Energy}$$

In these equations m is the relativistic mass as determined by the speed v, and p is the momentum mv.

- The *relativistic kinetic energy* is found from

$$E_K = (m - m_0)c^2 \qquad \textit{Relativistic Kinetic Energy}$$

- The quantum theory of electromagnetic radiation relates the energy of such radiation to its frequency f or wavelength λ.

$$E = hf \qquad E = \frac{hc}{\lambda} \qquad h = 6.63 \times 10^{-34} \text{ J} \cdot \text{s}$$

- In the photoelectric effect, the kinetic energy of the ejected electrons is the energy of the incident radiation hf less the work function of the surface W.

$$E_K = \tfrac{1}{2}mv^2 = hf - W \qquad \textit{Photoelectric Equation}$$

- The lowest frequency f_0 at which a photoelectron is ejected is the threshold frequency. It corresponds to the work-function energy W.

$$f_0 = \frac{W}{h} \qquad W = hf_0 \qquad \textit{Threshold Frequency}$$

- By combining wave theory with particle theory, de Broglie was able to give the following equation for the wavelength of any particle whose mass and velocity are known:

$$\lambda = \frac{h}{mv} \qquad h = 6.63 \times 10^{-34} \text{ J} \cdot \text{s} \qquad \begin{array}{l}\textit{de Broglie}\\\textit{Wavelength}\end{array}$$

- Bohr's first postulate states that the angular momentum of an electron in any orbit must be a multiple of $h/2\pi$. His second postulate states that the energy absorbed or emitted by an atom is in discrete amounts equal to the difference in energy levels of an electron. These concepts are given as equations below:

$$mvr = \frac{nh}{2\pi} \qquad hf = E_i - E_f \qquad \textit{Bohr's Postulates}$$

- Absorption and emission spectra for gases verify the discrete nature of radiation. The wavelength λ or frequency f that corresponds to a change in electron energy levels is given by

$$\frac{1}{\lambda} = R\left(\frac{1}{n_f^2} - \frac{1}{n_i^2}\right) \qquad f = Rc\left(\frac{1}{n_f^2} - \frac{1}{n_i^2}\right)$$

$$R = \frac{me^4}{8\epsilon_0^2 h^3 c} = 1.097 \times 10^7 \text{ m}^{-1}$$

Rydberg's Constant

- The total energy of a particular quantum state n for the hydrogen atom is given by:

$$E_n = -\frac{me^4}{8\epsilon_0^2 n^2 h^2} \qquad \text{or} \qquad E_n = -\frac{13.6 \text{ eV}}{n^2}$$

where $\epsilon_0 = 8.85 \times 10^{-12} \text{ C}^2/\text{N} \cdot \text{m}^2$
$e = 1.6 \times 10^{-19} \text{ C}$
$m = 9.1 \times 10^{-31} \text{ kg}$
$h = 6.63 \times 10^{-34} \text{ J} \cdot \text{s}$

Key Terms

1. relativistic mass
2. relativistic contraction
3. time dilation
4. Einstein's postulates
5. Planck's postulate
6. work function
7. photoelectric effect
8. spectrometer
9. line emission spectra
10. absorption spectra
11. spectral series
12. Bohr atom
13. principal quantum number
14. energy level
15. excited atom

Review Questions

38-1. An astronaut holds a clock of mass m and length L. Assume that the astronaut passes you at relativistic speed. Compare your measurements of m, L, and Δt with those made by the astronaut for the same clock.

38-2. Recalling Newton's second law of motion, what happens to the thrust requirements for propelling rockets to higher and higher relativistic speeds? Theoretically, what force would be required to achieve the velocity of light?

38-3. You are enclosed in a box with six opaque walls. Are there any experiments you can perform inside the box to prove that you are (a) moving with constant linear velocity, (b) accelerating, or (c) rotating with constant angular velocity?

38-4. Combine Eqs. (38-5) and (38-3) to obtain Einstein's equation for the total relativistic energy, $E = mc^2$.

38-5. Suppose you want photoelectrons to have a kinetic energy E_K and you know the work function W of the surface; how would you determine the required wavelength λ of the incident light?

38-6. Describe an experiment you might perform to determine the work function of a surface, assuming you have a light source of varying wavelength.

38-7. Sketch on a single diagram the Balmer series and the Lyman series for the hydrogen emission spectrum. What is meant by the *series limit?*

38-8. Explain clearly what is meant when we say that the energy of the ground state is -13.6 eV for the hydrogen atom. What is the significance of the minus sign?

38-9. Describe an experiment that will demonstrate (a) a line emission spectrum, (b) a line absorption spectrum, and (c) a continuous spectrum.

38-10. Hydrogen atoms in their ground state are bombarded by electrons that have been accelerated through a potential difference of 12.8 V. Which lines of the Lyman series will be emitted by the hydrogen atoms?

Summary and Review

Problems

Section 38-1—Relativity

38-1. A spaceship travels past an observer at a speed of 0.85c. A person aboard the spacecraft observes that it requires 6.0 s for him to walk the length of his cabin. What time would the observer record for the same event? **Ans. 11.4 s**

38-2. A rocket A moves past a lab B at a speed of 0.9c. A technician in the lab records 3.50 s for the time of an event that occurs on the rocket. What is the time as reckoned by a person aboard the rocket?

38-3. A blinking light on a spacecraft moves past an observer at 0.75c. The observer records that the light blinks at a frequency of 2.0 Hz. What is the actual frequency of the blinking light? **Ans. 3.02 Hz**

38-4. A particle on a table has a diameter of 2 mm when at rest. What must be the speed of an observer who measures the diameter as 1.69 mm?

38-5. A blue meterstick is aboard ship A, and a red meterstick is aboard ship B. If ship A moves past B at 0.85c, what will be the length of each meterstick as reckoned by a person aboard ship A?
Ans. L_B = 1.00 m, L_R = 52.7 cm

38-6. Three metersticks travel past an observer at speeds of 0.1c, 0.6c, and 0.9c. What lengths would be recorded by the observer?

38-7. What mass is required to run about 1 million 100-W light bulbs for 1 year? **Ans. 35.0 g**

38-8. Elementary particles called *mu-mesons* rain down through the atmosphere at 2.97×10^8 m/s. At rest the mu-meson would decay on average 2 μs after it came into existence. What is the lifetime of these particles from the viewpoint of an observer on earth?

Section 38-5—Quantum Theory and the Photoelectric Effect

38-9. The first photoelectrons are emitted from a copper surface when the wavelength of incident radiation is 282 nm. What is the threshold frequency for copper? What is the work function for a copper surface?
Ans. 1.06×10^{15} Hz, 4.40 eV

38-10. If the photoelectric work function of a material is 4.0 eV, what is the minimum frequency of light required to eject photoelectrons? What is the threshold frequency?

38-11. The energy E of a photon in joules is found from the product *hf*. Often we are given the wavelength of light and need to find its energy in electronvolts. Show that

$$E = \frac{1240}{\lambda}$$

such that if λ is in nanometers, E will be the energy in electronvolts.

38-12. Use the equation derived in Prob. 38-11 to verify that light of wavelength 490 nm has an energy of 2.53 eV. Also show that a photon with an energy of 2.10 eV has a wavelength of 590 nm.

*38-13. The threshold frequency for a certain metal is 2.5×10^{14} Hz. What is the work function? If light of wavelength 400 nm shines on this surface, what is the kinetic energy of ejected photoelectrons?
Ans. 1.04 eV, 2.07 eV

*38-14. When light of frequency 1.6×10^{15} Hz strikes a material surface, electrons just begin to leave the surface. What is the maximum kinetic energy of photoelectrons emitted from this surface when illuminated with light of frequency 2.0×10^{15} Hz?

*38-15. The work function of a nickel surface is 5.01 eV. If a nickel surface is illuminated by light of wavelength 200 nm, what is the kinetic energy of the ejected electrons?
Ans. 1.21 eV

*38-16. The stopping potential is a reverse voltage that just stops the electrons from being emitted in a photoelectric application. The stopping potential is therefore equal to the kinetic energy of ejected photoelectrons. Find the stopping potential for Prob. 38-13.

Section 38-6—Waves and Particles

38-17. What is the de Broglie wavelength of a proton ($m = 1.67 \times 10^{-27}$ kg) when it is moving with a speed of 2×10^7 m/s?

Ans. 1.99×10^{-14} m

38-18. The de Broglie wavelength of a particle is 3×10^{-14} m. What is the momentum of the particle?

38-19. Recalling formulas for kinetic energy and momentum, show that for nonrelativistic speeds, the momentum of a particle can be found from

$$p = \sqrt{2mE_k}$$

where E_k is the kinetic energy and m is the mass of the particle.

*38-20. Determine the kinetic energy of an electron if its de Broglie wavelength is 2×10^{-11} m.

*38-21. What is the de Broglie wavelength of the waves associated with an electron that has been accelerated through a potential difference of 160 V? **Ans. 9.71×10^{-11} m**

*38-22. The charge on a proton is $+1.6 \times 10^{-19}$ C, and its rest mass is 1.67×10^{27} kg. What is the de Broglie wavelength of a proton if it is accelerated from rest through a potential difference of 500 V?

Section 38-9—Atomic Spectra

38-23. Determine the wavelength of the first three spectral lines of atomic hydrogen in the Balmer series. **Ans. 656, 486, and 434 nm**

38-24. Find the wavelengths of the first three spectral lines of atomic hydrogen in the Paschen series.

38-25. Determine the radius of the $n = 4$ Bohr level of the classical Bohr hydrogen atom. **Ans. 847 nm**

38-26. What is the classical radius of the first Bohr orbit in the hydrogen atom?

38-27. Determine the wavelength of the photon emitted from a hydrogen atom when the electron jumps from the $n = 3$ Bohr level to ground level. **Ans. 103 nm**

*38-28. What is the maximum wavelength of an incident photon if it can ionize a hydrogen atom originally in its second excited state ($n = 3$)?

*38-29. What are the shortest and longest possible wavelengths in the Balmer series?

Ans. 364 and 656 nm

Challenge Problems

38-30. At a cost of 9 cents per kilowatt, what is the cost of the maximum energy to be released from a 1-kg mass?

38-31. An event that occurs on a spaceship traveling at $0.8c$ relative to the earth is observed by a person on the ship to last for 3 s. What time would be observed by a person on earth? How far will the person on earth judge that the spaceship has traveled during this event? **Ans. 5.00 s, 1.2×10^9 km**

38-32. When monochromatic light of wavelength 450 nm strikes a cathode, photoelectrons are emitted with a velocity of 4.8×10^5 m/s. What is the work function for the surface in electronvolts? What is the threshold frequency?

38-33. In the hydrogen atom, an electron falls from the $n = 5$ level to the $n = 2$ level and emits a photon in the Balmer series. What are the wavelength and energy of the emitted light? **Ans. 434 nm, 2.86 eV**

38-34. Calculate the frequency and the wavelength of the H_β line of the Balmer series. The transition is from the $n = 4$ level of the Bohr atom.

38-35. A spaceship A travels past another ship B with a relative velocity of 0.2c. Observer B determines that it takes a person on ship A exactly 3.96 s to perform a task. What time will be measured for the same event by observer A? **Ans. 3.88 s**

*38-36. The rest mass of an electron is 9.1×10^{-31} kg. What is the mass of an electron traveling at a speed of 2×10^8 m/s? What is the total energy of the electron? What is its relativistic kinetic energy?

Summary and Review

*38-37. What is the de Broglie wavelength of an electron with a kinetic energy of 50 MeV?
Ans. 0.174 pm

*38-38. The rest mass of a proton is 1.67×10^{-27} kg. What is the total energy of a proton that has been accelerated to a velocity of 2.5×10^8 m/s? What is its relativistic kinetic energy?

*38-39. Compute the mass and the speed of protons having a relativistic kinetic energy of 235 MeV. The rest mass of a proton is 1.67×10^{-27} kg.
Ans. 2.09×10^{-27} kg, 1.8×10^8 m/s

*38-40. How much work is required to accelerate a 1-kg mass from rest to a speed of $0.1c$? How much work is required to accelerate this mass from an initial speed of $0.3c$ to a final speed of $0.9c$? (Use the work-energy theorem.)

*38-41. A particle of mass m is traveling at $0.9c$. By what factor is its relativistic kinetic energy greater than its Newtonian kinetic energy?
Ans. 3.20

*38-42. What is the momentum of a 40-eV photon? What is the wavelength of an electron with the same momentum as this photon?

*38-43. When monochromatic light of wavelength 410 nm strikes a cathode, photoelectrons are emitted with a velocity of 4.0×10^5 m/s. What is the work function for the surface, and what is the threshold frequency?
Ans. 2.58 eV, 6.23×10^{14} Hz

*38-44. What is the velocity of a neutron ($m = 1.675 \times 10^{-27}$ kg) that has a de Broglie wavelength of 0.1 nm? What is its kinetic energy in electronvolts?

*38-45. What is the velocity of a particle that has a relativistic kinetic energy of twice its rest mass energy?
Ans. $0.943c$.

*38-46. Compute the relativistic mass and the speed of electrons that have a relativistic kinetic energy of 1.2 MeV.
Ans. 30.4×10^{-31} kg, 2.86×10^8 m/s

Critical Thinking Questions

38-47. A blue spacecraft is traveling at $0.8c$ relative to a red spacecraft. On the blue ship, a person moves a blue block a distance of 8 m in 3.0 s. On the red ship, a person moves a red block a distance of 4 m in 2.0 s. (a) What are the measurements of these four parameters from the viewpoint of the person on the blue ship? (b) What are the same measurements from the perspective of a person on the red ship?
Ans. Blue: $L_b = 8.00$ m, $t_b = 3.00$ s, $L_r = 2.40$ m, $t_r = 3.33$ s; Red: $L_b = 4.80$ m, $t_b = 5.00$ s, $L_r = 4.00$ m, $t_r = 2.00$ s

*38-48. Use the work-energy theorem to compare the work required to change relativistic speeds with values obtained from Newtonian physics. (a) The speed changes from $0.1c$ to $0.2c$. (b) The speed changes from $0.7c$ to $0.8c$.

*38-49. An electron in the hydrogen atom drops from the $n = 5$ level to the $n = 1$ level. What are the frequency, wavelength, and energy of the emitted photon? In which series does this photon occur? How much energy must be absorbed by the atom in order to kick the electron back up to the fifth level?
Ans. 95.0 nm, 3.16×10^{15} Hz, 2.09×10^{-18} J, Lyman, 2.09×10^{15} J

*38-50. In a photoelectric experiment shown in Fig. 38-15, a source of emf is connected in series with a galvanometer G. Light falling on the metal cathode produces photoelectrons. The source of emf is biased against the flow of electrons, retarding their motion. The potential difference V_0 just sufficient to stop the most energetic photoelectrons is called the *stopping potential*. Assume that a surface is illuminated with light of wavelength 450 nm, causing electrons to be ejected from the surface at a maximum speed of 6×10^5 m/s. What is the work function for the surface, and what is the stopping potential?

*38-51. In a photoelectric experiment, 400-nm light falls on a certain metal and photoelectrons are emitted. The potential required to stop the flow of electrons is 0.20 V. What is the energy of the incident photons? What is the work function? What is the threshold frequency?

Ans. 3.11 eV, 2.91 eV, 7.02 × 10¹⁴ Hz

Figure 38-15 The stopping potential V_0 is the potential difference that stops the transition of the most energetic photoelectrons ejected by the incident radiation.

Nuclear Physics and the Nucleus

The work of Rutherford and Bohr left us with a picture of the atom as a dense, positively charged nucleus surrounded by a cloud of electrons at distinct energy levels. From this point of view, the nucleus is the center of an atom, containing most of the atom's mass. The behavior of the atom is also affected by the nucleus because in the neutral atom the total number of positive charges in the nucleus must equal the number of electrons.

In this chapter, we look at the basic internal structure of the nucleus. We shall find that classical physics is not adequate to describe interactions at this level. Topics to be discussed include nuclear binding energy, radioactivity, and nuclear energy. The emphasis will be on providing a broad understanding of the atomic nucleus and its behavior.

Nuclear technology has grown enormously since its beginning in the early 1940s. The study of the atomic nucleus was once a subject reserved mainly for physicists, but today there are few people whose lives are not touched by some aspect of nuclear science. As patients we see the doctor use radioactive materials to diagnose a condition or treat it. As citizens, we are concerned with the promises and dangers of large-scale nuclear power production. More than ever before, technicians and engineers need a better understanding of the atomic nucleus and its potential.

39-1 THE ATOMIC NUCLEUS

All matter is composed of different combinations of *at least* three fundamental particles: protons, neutrons, and electrons. For example, a beryllium atom (Fig. 39-1) consists of a nucleus that contains four protons and five neutrons. The fact that beryllium is electrically neutral requires that four electrons surround the nucleus. The two inner electrons are at a different energy level ($n = 1$) than the outer two electrons ($n = 2$).

Rutherford's scattering experiments demonstrated that the nucleus contains most of the mass of an atom and that the nucleus is only about one ten-thousandth of the diameter of the atom. Thus, a typical atom with a diameter of 10^{-10} m (100 pm) would have a nucleus about 10^{-14} m (10 fm) in diameter. The prefixes *pico* (10^{-12}) and *femto* (10^{-15}) are useful for nuclear dimensions. Since the diameter of the atom is 10,000 times that of its nucleus, the atom, and therefore matter, consists largely of space that is almost empty.

Let us review what is known about the fundamental particles. The electron has a mass of 9.1×10^{-31} kg and a charge of $e = -1.6 \times 10^{-19}$ C. The proton is the nucleus of a hydrogen atom. It has a mass of 1.673×10^{-27} kg and a positive charge equal in magnitude to the charge of an electron ($+e$). Since the mass of an electron is extremely small, the mass of a photon is approximately the same as the mass of a hydrogen atom, which consists of one proton and one electron. The proton has a diameter of approximately 3 fm.

The other nuclear particle, the neutron, is present in the nuclei of all elements except hydrogen. It has a mass of 1.675×10^{-27} kg, which is slightly greater than that of the proton, but it has no charge. Thus, although neutrons contribute to the mass of a nucleus, they do not affect the net positive charge of the nucleus, which is due only to protons. The neutron also has a diameter of approximately 3 fm. Table 39-1 summarizes the data we have discussed for three fundamental particles.

From what we now know about the fundamental particles, it is clear that diagrams like Fig. 39-1 cannot be taken too seriously. Distances are not normally

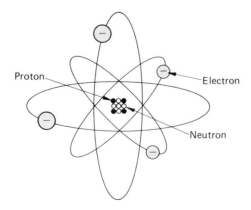

Figure 39-1 A model of a beryllium atom. The nucleus consists of four protons and five neutrons surrounded by four electrons. The positive charge of the protons is balanced exactly by the negatively charged electrons in the neutral electron.

TABLE 39-1 Fundamental Particles

Particle	Symbol	Mass, kg	Charge, C
Electron	e	9.1×10^{-31}	-1.6×10^{-19}
Proton	p	1.673×10^{-27}	$+1.6 \times 10^{-19}$
Neutron	n	1.675×10^{-27}	0

presented to scale in such schematic representations. Moreover, classical laws of physics often do not apply for the microworld of the nucleus.

A true understanding of atomic and nuclear events will require a new way of thinking. For example, one might ask what holds the nucleus together. Clearly, if Coulomb's electrostatic repulsion applies in the nucleus, it must be overcome by a much larger force. Both this much larger force and the electrostatic force are immense compared with the gravitational force. This third force is called the **nuclear force.**

The nuclear force is a strong, short-range force. If two nucleons (which are protons or neutrons) are separated by approximately 1 fm, a strong attractive force occurs that drops quickly to zero as their separation becomes larger. The force appears to be the same, or nearly the same, between two protons or two neutrons or between a neutron and a proton. If one nucleon is completely surrounded by other nucleons, its nuclear force field will be saturated, and it cannot exert any force on nucleons outside those surrounding it.

39-2 THE ELEMENTS

For many centuries, scientists have been studying the various elements found on the earth. A number of attempts have been made to organize the different elements according to their chemical and/or physical properties. The modern grouping of elements is the periodic table. One form of the periodic table is printed in Table 39-2.

Each element is assigned a number that distinguishes it from any other element. For example, the number for hydrogen is 1, the number for helium is 4, and the number for oxygen is 8. These numbers equal the number of protons in the nucleus of that element. The number is given the symbol Z and is called the **atomic number.**

The atomic number Z of an element is equal to the number of protons in the nucleus of an atom of that element.

The atomic number determines indirectly the chemical properties of an element because Z determines the number of electrons needed to balance the positive charge of the nucleus. The chemical nature of an atom depends on the number of electrons, in particular the outermost, or valence, electrons.

As the number of protons in a nucleus increases, so does the number of neutrons. In lighter elements, the increase is approximately one to one, but heavier elements may have more than $1\frac{1}{2}$ times more neutrons that protons. For example, oxygen has 8 protons and 8 neutrons, whereas uranium has 92 pro-

TABLE 39-2 The Periodic Table

Group

IA	IIA	IIIB	IVB	VB	VIB	VIIB	VIIIB			IB	IIB	IIIA	IVA	VA	VIA	VIIA	O
1 H 1.008																	**2 He** 4.0026
3 Li 6.941	**4 Be** 9.012											**5 B** 10.811	**6 C** 12.0111	**7 N** 14.0067	**8 O** 15.999	**9 F** 18.998	**10 Ne** 20.183
11 Na 22.990	**12 Mg** 24.305											**13 Al** 26.982	**14 Si** 28.086	**15 P** 30.9738	**16 S** 32.066	**17 Cl** 35.453	**18 Ar** 39.948
19 K 39.098	**20 Ca** 40.078	**21 Sc** 44.956	**22 Ti** 47.867	**23 V** 50.942	**24 Cr** 51.996	**25 Mn** 54.938	**26 Fe** 55.847	**27 Co** 58.933	**28 Ni** 58.71	**29 Cu** 63.54	**30 Zn** 65.39	**31 Ga** 69.723	**32 Ge** 72.61	**33 As** 74.9216	**34 Se** 78.96	**35 Br** 79.904	**36 Kr** 83.80
37 Rb 85.468	**38 Sr** 87.62	**39 Y** 88.906	**40 Zr** 91.224	**41 Nb** 92.906	**42 Mo** 95.94	**43 Tc** 97.907	**44 Ru** 101.07	**45 Rh** 102.906	**46 Pd** 106.4	**47 Ag** 196.967	**48 Cd** 112.411	**49 In** 114.82	**50 Sn** 118.71	**51 Sb** 121.75	**52 Te** 127.60	**53 I** 126.904	**54 Xe** 131.290
55 Cs 132.905	**56 Ba** 137.327	**57 La** 138.91	**72 Hf** 178.49	**73 Ta** 180.948	**74 W** 183.85	**75 Re** 186.2	**76 Os** 190.21	**77 Ir** 192.2	**78 Pt** 195.09	**79 Au** 196.967	**80 Hg** 200.59	**81 Tl** 204.383	**82 Pb** 207.19	**83 Bi** 208.980	**84 Po** 208.982	**85 At** 209.987	**86 Rn** 222.018
87 Fr 223.020	**88 Ra** 226.025	**89 Ac** 227.028	**104 Rf** 261.109	**105 Db** 262.114	**106 Sg** 263.119	**107 Bh** 264.121	**107 Hs** 265.131	**109 Mt** 266	**110 Uun*** (269)	**111 Uuu*** (272)	**112 Uub*** (277)		**114 Uuq*** (289)		**116 Uuh*** (289)		**118 Uuo*** (293)

Transition elements (groups IIIB–IIB)

Lanthanide series

58 Ce 140.12	**59 Pr** 140.908	**60 Nd** 144.24	**61 Pm** 144.913	**62 Sm** 150.36	**63 Eu** 151.96	**64 Gd** 157.25	**65 Tb** 158.925	**66 Dy** 162.50	**67 Ho** 164.930	**68 Er** 168.26	**69 Tm** 168.934	**70 Yb** 173.04	**71 Lu** 174.967

Actinide series

90 Th 238.03	**91 Pa** 231.036	**92 U** 238.03	**93 Np** 237.048	**94 Pu** 244.064	**95 Am** 243.061	**96 Cm** 247.070	**97 Bk** 247.070	**98 Cf** 251.080	**99 Es** 252.083	**100 Fm** 257.095	**101 Md** 258.099	**102 No** 259.101	**103 Lw** 260.105

Key
Atomic number → 6
Symbol → C
Atomic mass (weight) → 12.0111

Atomic weight values listed in parentheses are approximate.
*Indicates unnamed elements.

tons and 146 neutrons. The total number of nucleons in a nucleus is called the *mass number A.*

> **The mass number A of an element is equal to the total number of protons and neutrons in its nucleus.**

If we represent the number of neutrons by N, we can write the mass number A in terms of the atomic number Z and the number of neutrons.

$$A = Z + N \qquad \text{39-1}$$

Thus, the mass number of uranium is $92 + 146$, or 238.

A general way of describing the nucleus of a particular atom is to write the symbol for the element with its mass number and atomic number shown as follows:

$$\substack{\text{mass number} \\ \text{atomic number}} [\text{symbol}] = {}_{Z}^{A}X$$

For example, the uranium atom has the symbol ${}_{92}^{238}U$.

An alphabetical listing of all the elements is given in Table 39-3. The structures and symbols for the first four elements are shown in Fig. 39-2.

EXAMPLE 39-1

How many neutrons are in the nucleus of an atom of mercury ${}_{80}^{201}Hg$? The symbol shows that the atomic number is 80 and the mass number is 201. Thus, from Eq. (39-1), the number of neutrons is

$$N = A - Z = 201 - 80 = 121$$

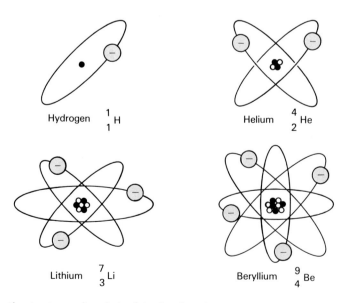

Figure 39-2 The structure and symbols of the first four elements.

TABLE 39-3 International Atomic Weights [Based on Carbon 12]*

Element	Symbol	Atomic Number	Atomic Weight
Actinium	Ac	89	(227)
Aluminum	Al	13	26.9815
Americium	Am	95	(243)
Antimony	Sb	51	121.75
Argon	Ar	18	39.948
Arsenic	As	33	74.9216
Astatine	At	85	(210)
Barium	Ba	56	137.34
Berkelium	Bk	97	(247)
Beryllium	Be	4	9.0122
Bismuth	Bi	83	208.980
Boron	B	5	10.811
Bromine	Br	35	79.909
Cadmium	Cd	48	112.40
Calcium	Ca	20	40.08
Californium	Cf	98	(251)
Carbon	C	6	12.01115
Cerium	Ce	58	140.12
Cesium	Cs	55	132.905
Chlorine	Cl	17	35.453
Chromium	Cr	24	51.996
Cobalt	Co	27	58.9332
Columbium	(see Niobium)		
Copper	Cu	29	63.546
Curium	Cm	96	(247)
Dysprosium	Dy	66	162.50
Einsteinium	Es	99	(254)
Erbium	Er	68	167.26
Europium	Eu	63	151.96
Fermium	Fm	100	(253)
Fluorine	F	9	18.9984
Francium	Fr	87	(223)
Gadolinium	Gd	64	157.25
Gallium	Ga	31	69.72
Germanium	Ge	32	72.59
Gold	Au	79	196.967
Hafnium	Hf	72	178.49
Helium	He	2	4.0026
Holmium	Ho	67	164.930
Hydrogen	H	1	1.00797
Indium	In	49	114.82
Iodine	I	53	126.9044
Iridium	Ir	77	192.2
Iron	Fe	26	55.847
Krypton	Kr	36	83.80
Lanthanum	La	57	138.91
Lawrencium	Lw	103	(257)
Lead	Pb	82	207.19
Lithium	Li	3	6.939
Lutetium	Lu	71	174.97
Magnesium	Mg	12	24.312
Manganese	Mn	25	54.9380
Mendelevium	Md	101	(256)
Mercury	Hg	80	200.59
Molybdenum	Mo	42	95.94
Neodymium	Nd	60	144.24
Neon	Ne	10	20.183
Neptunium	Np	93	(237)
Nickel	Ni	28	58.71
Niobium (Columbium)	Nb	41	92.906
Nitrogen	N	7	14.0067
Nobelium	No	102	(254)
Osmium	Os	76	190.2
Oxygen	O	8	15.9994
Palladium	Pd	46	106.4
Phosphorus	P	15	30.9738
Platinum	Pt	78	195.09
Plutonium	Pu	94	(244)
Polonium	Po	84	(210)
Potassium	K	19	39.102
Praseodymium	Pr	59	140.907
Promethium	Pm	61	(145)
Protactinium	Pa	91	(231)
Radium	Ra	88	(226)
Radon	Rn	86	(222)
Rhenium	Re	75	186.22
Rhodium	Rh	45	102.91
Rubidium	Rb	37	85.47
Ruthenium	Ru	44	101.07
Samarium	Sm	62	150.35
Scandium	Sc	21	44.956
Selenium	Se	34	78.96
Silicon	Si	14	28.086
Silver	Ag	47	107.870
Sodium	Na	11	22.9898
Strontium	Sr	38	87.62
Sulfer	S	16	32.064
Tantalum	Ta	73	180.948
Technetium	Tc	43	(99)
Tellurium	Te	52	127.60
Terbium	Tb	65	158.924
Thallium	Tl	81	204.37
Thorium	Th	90	232.038
Thulium	Tm	69	168.934
Tin	Sn	50	118.69
Titanium	Ti	22	47.90
Tungsten (Wolfram)	W	74	183.85
Unnilpentium	Unp	105	(260)
Unnilquadium	Unq	104	(261)
Uranium	U	92	238.03
Vanadium	V	23	50.942
Wolfram (Tungsten)	W	74	183.85
Xenon	Xe	54	131.30
Ytterbium	Yb	70	173.04
Yttrium	Y	39	88.905
Zinc	Zn	30	65.37
Zirconium	Zr	40	91.22

* Values in parentheses are mass numbers of longest-lived or best-known isotopes.

39-3 THE ATOMIC MASS UNIT

The small masses of nuclear particles call for an extremely small unit of mass. Scientists normally express atomic and nuclear masses in *atomic mass units* (u).

One atomic mass unit (1 u) is exactly equal to one-twelfth of the mass of the most abundant form of the carbon atom.

In terms of the kilogram, the atomic mass unit is

$$1 \text{ u} = 1.6606 \times 10^{-27} \text{ kg} \qquad \text{39-2}$$

The mass of a proton is 1.007276 u, and that of a neutron is 1.008665 u.

EXAMPLE 39-2

The periodic table shows the average atomic mass of barium to be 137.34 u. What is the average mass of the barium nucleus?

Solution

The mass of the nucleus is the atomic mass less the mass of the surrounding cloud of electrons. Thus, we must first determine the mass of an electron in atomic mass units.

$$m_e = 9.1 \times 10^{-31} \text{ kg} \frac{1 \text{ u}}{1.6606 \times 10^{-27} \text{ kg}}$$

$$= 5.5 \times 10^{-4} \text{ u}$$

Since the atomic number Z of barium is 56, there must be the same number of electrons. This total mass is

$$m_T = 56(5.5 \times 10^{-4} \text{ u}) = 3.08 \times 10^{-2} \text{ u}$$

The average atomic mass was given as 137.34 u. Thus, the nuclear mass is

$$137.34 \text{ u} - 0.0308 \text{ u}$$

or

$$137.31 \text{ u}$$

The masses given for atoms in the periodic table include the electron masses.

It should be remembered that mass can be equated to units of energy from Einstein's relation.

$$E = mc^2$$

Consequently, a mass of 1 u corresponds to an energy given by

$$(1 \text{ u})c^2 = (1.66 \times 10^{-27} \text{ kg})(3 \times 10^8 \text{ m/s})^2 = 1.49 \times 10^{-10} \text{ J}$$

In the more convenient units of electronvolts, we can write

$$(1 \text{ u})c^2 = 1.49 \times 10^{-10} \text{ J} \frac{1 \text{ eV}}{1.6 \times 10^{-19} \text{ J}}$$
$$= 9.31 \times 10^8 \text{ eV} = 931 \text{ MeV}$$

The conversion factor from mass units (u) to energy units (MeV) is therefore

$$c^2 = 931 \text{ MeV/u} \qquad \textbf{39-3}$$

As an exercise, you might verify that the electron and the proton have rest-mass energies of 0.511 and 938 MeV, respectively.

39-4 ISOTOPES

It is possible for two atoms of the same element to have nuclei containing different numbers of neutrons. Such atoms are called *isotopes.*

Isotopes are atoms that have the same atomic number Z but different mass numbers A.

For example, naturally occurring carbon is a mixture of two isotopes. The most abundant form, $^{12}_{6}\text{C}$, has six protons and six neutrons in its nucleus. Another form, $^{13}_{6}\text{C}$, has an extra neutron. Some elements have as many as 10 different isotopic forms.

Experimental verification of the existence of isotopes is accomplished with a *mass spectrometer.* This device, illustrated in Fig. 39-3, is used to separate the isotopes of an element. A source of singly ionized atoms of a particular element is positioned above the velocity selector. These ions are missing one electron and therefore have a charge of $+e$. They are propelled at varying speeds into the crossed **E** and **B** fields of the velocity selector. Ions with velocities sufficient to make the

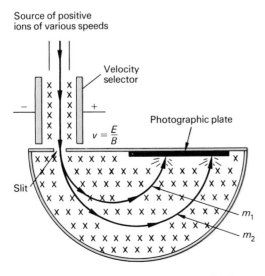

Figure 39-3 The mass spectrometer is used to separate isotopes of different mass; the crosses indicate that the direction of the magnetic field is into the paper.

magnetic force $\mathbf{F}_m$ equal and opposite to the electric force $\mathbf{F}_e$ will pass through the bottom slit undeflected. Recalling that $F_e = eE$ and $F_m = evB$, we write

$$evB = eE$$

from which

$$v = \frac{E}{B} \qquad\qquad \textbf{39-4}$$

Only ions with this velocity will pass through the slit at the bottom of the selector.

The fast-moving positive ions next pass into the lower region, where another $\mathbf{B}$ field causes them to experience a perpendicular magnetic force. The magnitude of this force will be constant and equal to evB, but its direction will always be at right angles to the velocity of the ion. The result is a circular path of radius R. The magnetic force provides the necessary centripetal force. In this case, we have

$$F_{\text{magnetic}} = F_{\text{centripetal}}$$

or

$$evB = \frac{mv^2}{R}$$

where m is the mass of the ion of charge e. Solving for R, we find that the radius of the semicircular path is given by

$$R = \frac{mv}{eB} \qquad\qquad \textbf{39-5}$$

Since v, e, and B are constant, Eq. (39-5) gives the radius as a function of the mass of given ions. Ions of different mass will strike the photographic plate at different positions because their semicircular paths are different. Wherever a beam of ions strikes the plate, a darkened line will be produced. The distance of a particular line from the slit is twice the radius in which that beam of ions moves. In this manner the mass can be determined from Eq. (39-5).

The mass spectrometer is used to separate and study isotopes. Most elements are mixtures of atoms with different mass numbers. For example, if an ion beam of pure lithium is injected into the mass spectrometer, two types of atoms are observed. The darker bank occurs because around 92 percent of the atoms have a mass of 7.016 u. The remaining 8 percent, producing a lighter band, are atoms with mass 6.015 u. These two isotopes of lithium are written $^{7}_{3}\text{Li}$ and $^{6}_{3}\text{Li}$.

Since some elements, for example, tin, have many different isotopic forms, it is not surprising that the average atomic mass for elements is often not very close to an integer. Average atomic mass is affected by the mass numbers and relative abundance of each isotopic form. For example, chlorine has an average atomic mass of 35.453 u, which results from a mixture of the two isotopes $^{35}_{17}\text{Cl}$ and $^{37}_{17}\text{Cl}$. The lighter chlorine isotope occurs about three times as often as the heavier one.

EXAMPLE 39-3

When chlorine is studied with the mass spectrometer, it is noted that an intense line occurs 24 cm from the entrance slit. Another lighter line appears at a distance of 25.37 cm. If the mass of the ions that form the first line is 34.980 u, what is the mass of the other isotope?

Solution

Since the given distances represent diameters, the radii of the two paths are

$$R_1 = 12 \text{ cm} \quad \text{and} \quad R_2 = 12.685 \text{ cm}$$

Now from Eq. (39-5),

$$R_1 = \frac{m_1 v}{eB} \quad \text{and} \quad R_2 = \frac{m_2 v}{eB}$$

Since e, v, and B are constant, we have

$$\frac{m_1}{m_2} = \frac{R_1}{R_2}$$

from which the mass m_2 is found to be

$$m_2 = \frac{m_1 R_2}{R_1} = \frac{(34.980 \text{ u})(12.685 \text{ cm})}{(12 \text{ cm})}$$
$$= 36.977 \text{ u}$$

39-5 THE MASS DEFECT AND BINDING ENERGY

One of the startling results that can be demonstrated with the mass spectrometer is that the mass of a nucleus is not exactly equal to the sum of the masses of its nucleons. Let us consider, for example, the helium atom, ^4_2He, which has two electrons about a nucleus containing two protons and two neutrons. The atomic mass is found from the periodic table to be 4.0026 u.

Now let us compare this value with the mass of all the individual particles that make up the atom:

$$2p = 2(1.007276 \text{ u}) = 2.014552 \text{ u}$$
$$2n = 2(1.008665 \text{ u}) = 2.017330 \text{ u}$$
$$2e = 2(0.00055 \text{ u}) = \underline{0.001100 \text{ u}}$$
$$\text{Total mass} = 4.032982 \text{ u}$$

The mass of the parts (4.0330 u) is apparently greater than the mass of the atom (4.0026 u).

$$m_{\text{parts}} - m_{\text{atom}} = 4.0330 \text{ u} - 4.0026 \text{ u}$$
$$= 0.0304 \text{ u}$$

pdg.lbl.gov
Have you ever wondered how physicists study subatomic particles? The Particle Adventure Web site is an introduction to particle physics (do not type www. in front of this site).

When protons and neutrons join to form a helium nucleus, the mass is decreased in the process. This difference is called the **mass defect.** A mass defect can be shown to exist for atoms of all elements.

> **The mass defect is defined as the difference between the rest mass of a nucleus and the sum of the rest masses of its constituent nucleons.**

We have seen from Einstein's work that mass and energy are equivalent. We might suppose, then, that the mass decrease in joining nucleons together will result in an energy decrease. Since energy is conserved, a decrease in the energy of the system means that energy must be released in joining the system together. In the case of helium, this energy would come from a mass of 0.0304 u and would be equal to

$$E = mc^2 = (0.0304 \text{ u})\frac{931 \text{ MeV}}{1 \text{ u}} = 28.3 \text{ MeV}$$

The total energy that would be released if we could build a nucleus from protons and neutrons is called the *binding energy* of the nucleus. As we have just seen, the binding energy of $^4_2$He is 28.3 MeV, as illustrated in Fig. 39-4a.

We can also reverse the above process and state that the **binding energy** is the energy required to break a nucleus apart into its constituent particles.

> **The binding energy of a nucleus is defined as the energy required to separate a nucleus into its constituent nucleons.**

In our example, an energy of 28.3 MeV must be supplied to $^4_2$He to separate the nucleus into two protons and two neutrons (Fig. 39-4b).

An isotope of atomic number Z and mass number A consists of Z protons, Z electrons, and $N = (A - Z)$ neutrons. If we neglect the binding energy of the elec-

(a) 28.3 MeV emitted   (b) 28.3 MeV absorbed

Figure 39-4   (a) When two protons and two neutrons are fused together to form a helium nucleus, energy is released. (b) The same amount of energy is required to break the nucleus apart into its constituent nucleons.

trons, a neutral isotope would have the same mass as Z neutral hydrogen atoms plus the mass of the neutrons. The masses of ^1_1H and m_n are

$$m_\text{H} = 1.007825 \text{ u} \qquad m_n = 1.008665 \text{ u} \qquad \textbf{39-6}$$

If we represent the atomic mass by M, the binding energy E_B can be approximated by

$$E_B = [(Zm_\text{H} + Nm_n) - M]c^2 \qquad \textit{Binding Energy} \qquad \textbf{39-7}$$

In applying this equation, we should remember that $N = A - Z$ and that $c^2 = 931$ MeV/u.

EXAMPLE 39-4

Determine the total binding energy and the binding energy per nucleon for the $^{14}_7\text{N}$ nucleus.

Solution
For nitrogen, $Z = 7$, $N = 7$, and $M = 14.003074$ u.

$$\begin{aligned}
E_B &= [(Zm_\text{H} + Nm_n) - M]c^2 \\
&= \{[7(1.007825 \text{ u}) + 7(1.008665 \text{ u})] - 14.003074 \text{ u}\}(931 \text{ MeV/u}) \\
&= (0.112356 \text{ u})(931 \text{ MeV/u}) = 104.6 \text{ MeV}
\end{aligned}$$

Since $^{14}_7\text{N}$ contains 14 nucleons, the binding energy per nucleon is

$$\frac{E_B}{A} = \frac{104.6 \text{ MeV}}{14 \text{ nucleons}} = 7.47 \text{ MeV/nucleon}$$

The atomic mass M used in Eq. (39-7) must be taken for the particular isotope of the element, not from the periodic table (Table 39-2) or from Table 39-3. These tables give the atomic masses of the naturally occurring mixture of isotopes for each element. The atomic mass of $^{12}_6\text{C}$, for example, is exactly 12.0000 u by definition. The periodic table gives a value of 12.01115 u because naturally occurring carbon contains small amounts of $^{13}_6\text{C}$ in addition to the more abundant $^{12}_6\text{C}$ atoms. The term **nuclide** is used to refer to a particular isotope that has a particular number of nuclear particles and, hence, a specific mass. The masses of several common nuclides are given in Table 39-4. These are the masses that should be used in determining mass defects and binding energies. Electron masses are included so that the given masses are basically the mass of the nucleus plus Z atomic electrons.

The binding energy per nucleon, as computed in Example 39-4, is an important way of comparing the nuclei of various elements. A plot of the binding energy per nucleon as a function of increasing mass number is shown in Fig. 39-5 for many stable nuclei. Note that the mass numbers toward the center (50 to 80) yield

TABLE 39-4 Atomic Masses for Several Nuclides			
Nuclide	Atomic Number	Mass Number	Atomic Mass
Hydrogen	1	1	1.007825 u
Deuterium	1	2	2.014102 u
Tritium	1	3	3.016049 u
Helium 3	2	3	3.016030 u
Helium 4	2	4	4.002603 u
Lithium 6	3	6	6.015126 u
Lithium 7	3	7	7.016003 u
Beryllium 9	4	9	9.012186 u
Boron 11	5	11	11.009305 u
Carbon 12	6	12	12.000000 u
Carbon 13	6	13	13.003354 u
Nitrogen 14	7	14	14.003074 u
Oxygen 16	8	16	15.994915 u
Oxygen 17	8	17	16.99913 u
Neon 20	10	20	19.992440 u
Copper 64	29	64	63.929759 u
Tin 120	50	120	119.902108 u
Gold 197	79	197	196.966541 u
Mercury 204	80	204	203.973865 u
Thallium 206	81	206	205.976104 u
Polonium 216	84	216	216.001922 u
Radon 222	86	222	222.017531 u
Radium 224	88	224	224.020218 u
Radium 226	88	226	226.025360 u
Thorium 233	90	233	233.041469 u
Thorium 234	90	234	234.043630 u
Protactinium 233	91	233	233.040130 u
Uranium 238	92	238	238.050786 u

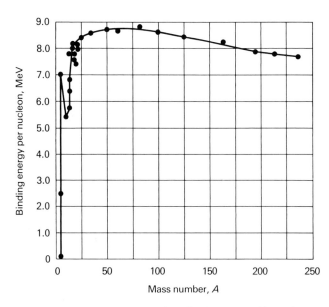

Figure 39-5 The average binding energy per nucleon for the most stable nucleus at each mass number.

the highest binding energy per nucleon. Elements ranging from $A = 50$ to $A = 80$ are the most stable.

39-6 RADIOACTIVITY

The strong nuclear force holds the nucleons tight in the nucleus, overcoming the coulomb repulsion of protons. The balance of forces is not always maintained, however, and sometimes particles or photons are emitted from the nuclei of atoms. Such unstable nuclei are said to be *radioactive* and have the property of **radioactivity.**

All naturally occurring elements with atomic numbers greater than 83 are radioactive. They are slowly decaying and disappearing from the earth. Uranium and radium are two of the better-known examples of naturally radioactive elements. A few other naturally occurring, lighter, and less active elements have also been discovered.

Unstable nuclei are also produced artificially as by-products of nuclear reactors, for study in laboratories, or for other purposes. In addition, some elements are made radioactive naturally by bombardment with high-energy photons.

There are three major forms of radioactive emission from atomic nuclei:

1. **Alpha particles** (α)   An alpha particle is the nucleus of a helium atom and consists of two protons and two neutrons. It has a charge of $+2e$ and a mass of 4.001506 u. Because of their positive charges and relatively low speeds ($\approx 0.1c$), alpha particles do not have great penetrating power.

2. **Beta particles** (β)   There are tow kinds of beta particles, a beta minus particle (β^-) and a beta plus particle (β^+). The beta minus particle is simply an electron of charge $-e$ and mass equal to 0.00055 u. A beta plus particle, also called a *positron,* has the same mass as an electron but the opposite charge ($+e$). These particles are generally emitted at speeds near the velocity of light. The beta minus particles are much more penetrating than alpha particles, but beta plus particles easily combine with electrons; then rapid annihilation of both the positrons and electrons occurs, with the emission of gamma rays.

3. **Gamma rays** (γ)   A gamma ray is a high-energy electromagnetic wave similar to heat and light but of much higher frequency. These rays have no charge or rest mass and are the most penetrating radiation emitted by radioactive elements.

To understand why these forms of radiation are emitted, it is helpful to look at nuclei that are relatively stable. Plotting the number of neutrons N versus the number of protons Z for these stable nuclei gives a rough graph like that shown in Fig. 39-6. Note that the light elements are stable when the ratio of Z to N is close to 1. More neutrons are required for stability in the heavier elements. The additional nuclear forces of the extra neutrons are needed to balance the higher electric forces that result as more protons are collected. Whenever a nucleus occurs that deviates very much from the line, it is unstable and will emit some form of radiation, thereby achieving stability.

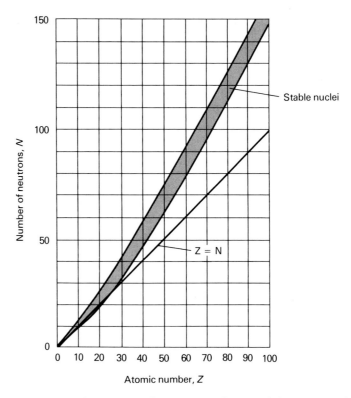

Figure 39-6 A comparison of the number of neutrons as a function of the atomic number. Notice that nuclei of higher Z have the greater proportion of neutrons.

39-7 RADIOACTIVE DECAY

Let us look at radioactive decay by alpha, beta, and gamma radiation and see what occurs during each process. The emission of an alpha particle $_2^4\alpha$ reduces the number of protons in the parent nucleus by 2 and the number of nucleons by 4. Symbolically, we write

$$_Z^A X \rightarrow {}_{Z-2}^{A-4} Y + {}_2^4\alpha + \text{energy} \tag{39-8}$$

The energy term results from the fact that the rest energy of the products is less than that of the parent atom. The difference in energy is carried away primarily by the kinetic energy imparted to the alpha particle. The recoil kinetic energy of the much more massive daughter atom is small by comparison.

EXAMPLE 39-5

Write the reaction that occurs when $_{88}^{226}$Ra decays by alpha emission.

Solution
Applying Eq. (39-8), we write

$$_{88}^{226}\text{Ra} \rightarrow {}_{86}^{222}\text{Rn} + {}_2^4\alpha + \text{energy}$$

Notice that the unstable element radium has been transformed into a new element, radon, which is closer to the stability line.

Next consider the emission of beta-minus particles from the nucleus. If beta-minus particles are electrons, how can an electron come from a nucleus containing only protons and neutrons? This can be answered, at least in part, by analogy to the Bohr atom. We have seen that photons, which do not exist in the atom, are emitted by atoms when they change from one state to another. Similarly, electrons, which do not exist in nuclei, can be emitted as a form of radiation when the nucleus changes from one state to another. When such a change does occur, the total charge must be conserved. This requires the conversion of a neutron into a proton and an electron.

$$_0^1n \rightarrow {}_1^1p + {}_{-1}^0e$$

Thus, in beta-minus emission, a neutron is replaced by a proton. The atomic number Z increases by 1, and the mass number is unchanged. Symbolically,

$$_Z^AX \rightarrow {}_{Z+1}^AY + {}_{-1}^0\beta + \text{energy} \qquad \text{39-9}$$

An example of beta emission is the decay of an isotope of neon into sodium:

$$_{10}^{23}Ne \rightarrow {}_{11}^{23}Na + {}_{-1}^0\beta + \text{energy}$$

The increase in Z is necessary to conserve charge.

Similarly, in positron (beta-plus) emission, a proton in the nucleus decays to a neutron and a positron.

$$_1^1p \rightarrow {}_0^1n + {}_{+1}^0e$$

The atomic number Z decreases by 1, and the mass number A is unchanged. Symbolically,

$$_Z^AX \rightarrow {}_{Z-1}^AY + {}_{+1}^0\beta + \text{energy} \qquad \text{39-10}$$

An example of positron emission is the decay of an isotope of nitrogen into an isotope of carbon:

$$_7^{13}N \rightarrow {}_6^{13}C + {}_{+1}^0\beta + \text{energy}$$

In both types of beta emission, the kinetic energy is shared mostly by the beta particle and another particle called a *neutrino*. The neutrino has no rest mass and no electric charge, but it can have both energy and momentum.

In gamma emission, the parent nucleus maintains the same atomic number Z and the same mass number A. The gamma photon simply carries away energy from an unstable nucleus. Frequently, a succession of alpha and beta decays is accompanied by gamma decays, which carry off excess energy.

The radioactive disintegration of $_{92}^{238}U$ is shown in Fig. 39-7 as a series of decays through a number of elements until it becomes a stable $_{82}^{206}Pb$ nucleus.

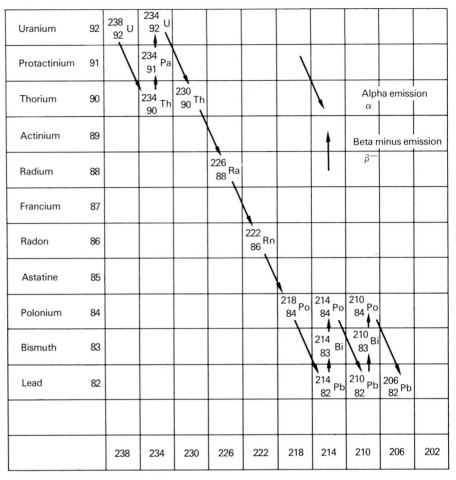

		238	234	230	226	222	218	214	210	206	202

Mass number, A

Figure 39-7 The uranium series of disintegration. Uranium decays, through a series of alpha- and beta-minus emissions, from ^{238}U to ^{206}Pb.

39-8 HALF-LIFE

A radioactive material continues to emit radiation until all the unstable atoms have decayed. The number of unstable nuclei decaying or disintegrating every second for a given isotope can be predicted on the basis of probability. This number is referred to as the **activity R,** given by

$$R = \frac{-\Delta N}{\Delta t}$$ 39-11

where N is the number of undecayed nuclei. The negative sign is included because N is decreasing with time. The units for R are inverse seconds (s^{-1}).

In practice, the activity in disintegrations per second is so large that a more convenient unit, the **curie** (Ci), is defined as follows.

> **One curie (Ci) is the activity of a radioactive material that decays at the rate of 3.7×10^{10} disintegrations per second.**

$$1 \text{ Ci} = 3.7 \times 10^{10} \text{ s}^{-1} \qquad \textbf{39-12}$$

The activity of 1 g of radium is slightly less than 1 Ci.

The random nature of nuclear decay means that the activity R at any time is directly proportional to the number of nuclei remaining; in other words, as the number of remaining nuclei decreases with time, the activity must also decrease with time. Therefore, if we plot the number of remaining nuclei as a function of time, as illustrated in Fig. 39-8, we see that radioactive decay is not linear. The time it takes for this curve to drop to one-half its original value is different for each radioactive isotope, and it is called the *half-life.*

> **The half-life $T_{1/2}$ of a radioactive isotope is the length of time in which one-half of its unstable nuclei will decay.**

For example, the half-life of radium 226 is 1620 yr; 1 g of this isotope will decay to 0.5 g in 1620 yr, to 0.25 g in 2(1620 yr), to 0.125 g in 3(1620 yr), and so on.

We can use this definition of the half-life to determine how many nuclei are present at a time t. If we start out at time $t = 0$ with a number N_0 of unstable nuclei, then after n half-lives have passed, there will be left a number of nuclei N given by

$$N = N_0 \left(\tfrac{1}{2}\right)^n \qquad \textbf{39-13}$$

The number n of half-lives in the period of time t is, of course, $t/T_{1/2}$. Thus, a more applicable form of the above relation is

$$N = N_0 \left(\tfrac{1}{2}\right)^{t/T_{1/2}}$$

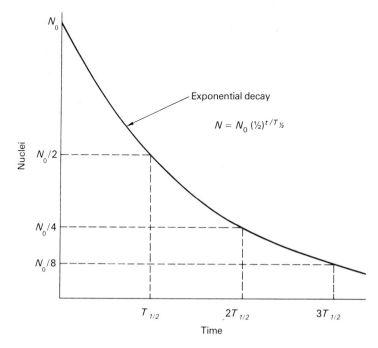

Figure 39-8   The radioactive decay curve, illustrating the half-life as the time $T_{1/2}$ required for one-half of the unstable nuclei, present at time $t = 0$, to decay.

Since the amount of radioactive material is determined by the number of nuclei present, an equation similar to Eq. (39-13) can be used to compute the mass of remaining radioactive material after a number of half-lives.

The same idea applied to the activity R of a radioactive sample yields the relation

$$R = R_0 \left(\tfrac{1}{2}\right)^{t/T_{1/2}} \qquad\qquad \text{39-14}$$

CONCEPTUAL EXAMPLE 39-6

The worst by-product of ordinary nuclear reactors is the radioactive isotope plutonium 239, which has a half-life of 24,400 yr. Suppose the initial activity of a sample containing 1.64×10^{20} $^{239}_{94}\text{Pu}$ nuclei is 4 mCi. (a) How many of these nuclei remain after 73,200 yr? (b) What will be the activity at that time?

Solution (a)

Substitution into Eq. (39-13) yields

$$N = N_0 \left(\tfrac{1}{2}\right)^{t/T_{1/2}} = 1.64 \times 10^{20} \left(\tfrac{1}{2}\right)^{73,200 \text{ yr}/24,400 \text{ yr}}$$
$$= 1.64 \times 10^{20} \left(\tfrac{1}{2}\right)^{3} = 1.64 \times 10^{20} \left(\tfrac{1}{8}\right)$$
$$= 2.05 \times 10^{19} \text{ nuclei}$$

Solution (b)

We obtain the remaining activity from Eq. (39-14).

$$R = R_0 \left(\tfrac{1}{2}\right)^{t/T_{1/2}} = 4 \text{ mCi} \left(\tfrac{1}{8}\right) = 0.5 \text{ mCi}$$

Both these calculations assume that no new $^{239}_{94}\text{Pu}$ nuclei are being created by other processes. It is easy to see from this example why disposal of some radioactive materials is such a difficult problem.

39-9 NUCLEAR REACTIONS

In a chemical reaction the atoms of two molecules react to form different molecules. In a *nuclear reaction,* nuclei, radiation, and/or nucleons collide to form different nuclei, radiation, and nucleons. If the colliding objects are charged, at least one of the colliding masses must be accelerated to a relatively high velocity. Normally the bombarding particle is light, for example, a proton $^{1}_{1}p$ or an alpha particle $^{4}_{2}\alpha$. These nuclear projectiles are accelerated with many different devices, for instance, Van de Graaff generators, cyclotrons, and linear accelerators.

In the nuclear reactions we shall study, several conservation laws must be observed, primarily *conservation of charge, conservation of nucleons,* and *conservation of mass-energy.*

Conservation of charge   **The total charge of a system can neither be increased nor decreased in a nuclear reaction.**

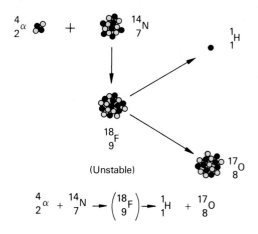

$$^4_2\alpha + ^{14}_7N \longrightarrow \left(^{18}_9F\right) \longrightarrow ^1_1H + ^{17}_8O$$

Figure 39-9   Striking a nitrogen-14 nucleus with an alpha particle.

Conservation of nucleons   **The total number of nucleons in the interaction must remain unchanged.**

Conservation of mass-energy   **The total mass-energy of a system must remain unchanged in a nuclear reaction.**

Now let us observe what happens when an alpha particle $^4_2\alpha$ strikes a nucleus in a sample of nitrogen gas $^{14}_7N$. (Refer to Fig. 39-9.) The first step is the entry of the alpha particle, which adds 2 protons and 2 neutrons to the nucleus. The atomic number Z is increased by 2, and the mass number A is increased by 4. The resulting nucleus is an *unstable* compound nucleus of fluorine $^{18}_9F$. This unstable nucleus quickly disintegrates into the final products, oxygen $^{17}_8O$ and hydrogen ^1_1H. The overall reaction can be written

$$^4_2\alpha + ^{14}_7N \rightarrow ^1_1H + ^{17}_8O \qquad \text{39-15}$$

Note how charge and nucleons are conserved in these reactions. There was a net charge of $+9e$ before the reaction and a net charge of $+9e$ after the reaction, and there are 18 nucleons before and after the reaction.

39-10 NUCLEAR FISSION

Before the discovery of the neutron in 1932, alpha particles and protons were the primary particles used to bombard atomic nuclei, but as charged particles they have the disadvantage of being repelled electrostatically by the nucleus. Consequently, large energies are required before nuclear reactions can occur.

Since neutrons have zero electric charge, they can easily penetrate the nucleus of an atom with no coulomb repulsion. Fast neutrons may pass completely through a nucleus or may cause it to disintegrate. Slow neutrons may be captured by a nucleus, creating an unstable isotope, which may disintegrate.

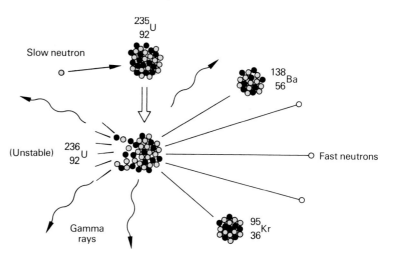

Figure 39-10 Nuclear fission of ^{235}U by capture of a slow neutron.

Whenever the absorption of an incoming neutron causes a nucleus to split into two smaller nuclei, the reaction is called *fission* and the product nuclei are called *fission fragments*.

Nuclear fission is the process by which heavy nuclei are split into two or more nuclei of intermediate mass numbers.

Whenever a slow neutron is captured by a uranium nucleus $^{235}_{92}$U, an unstable nucleus ($^{236}_{92}$U) is produced that may decay in several ways into smaller product nuclei (Fig. 39-10). Such fission reactions may produce fast neutrons, beta particles, and gamma rays in addition to the product nuclei. For this reason the products of a fission process, including fallout from a nuclear explosion, are highly radioactive.

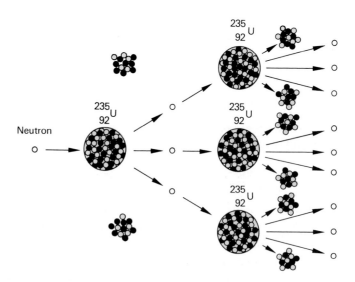

Figure 39-11 Nuclear chain reaction.

The fission fragments have a smaller mass number and therefore about 1 MeV more binding energy per nucleon (see Sec. 39-5). As a result, fission releases a large amount of energy. In the above example approximately 200 MeV per fission is produced.

Because each nuclear fission releases more neutrons, which may lead to additional fission, a ***chain reaction*** is possible. As seen in Fig. 39-11, the three neutrons released from the fission of $^{235}_{92}U$ produce three additional fissions. Thus, starting with one neutron, we have liberated nine after only two steps. If such a chain reaction is not controlled, it can lead to an explosion of enormous magnitude.

39-11 NUCLEAR REACTORS

A ***nuclear reactor*** is a device that controls the nuclear fission of radioactive material, producing new radioactive substances and large amounts of energy. These devices are used to furnish heat for electric power generation, propulsion, and industrial processes; to produce new elements or radioactive materials for a multitude of applications; and to supply neutrons for scientific experimentation.

A schematic diagram of a typical reactor is given in Fig. 39-12. The basic components are: (1) a *core* of nuclear fuel, (2) a *moderator* for slowing down fast neutrons, (3) *control rods* or other means for regulating the fission process, (4) a *heat exchanger* for removing heat generated in the core, and (5) *radiation shielding*. Steam produced by the reactor is used to drive a turbine, which generates electricity.

Figure 39-12   Schematic diagram of a nuclear reactor. Water heated under pressure in the reactor core is pumped into a heat exchanger, where it produces steam to operate a turbine.

The spent steam is changed to water in the condenser and pumped back to the heat exchanger for another cycle.

The essential ingredient in the reactor is the fissionable material, or nuclear fuel. About the only naturally occurring fissionable material is $^{235}_{92}U$, which constitutes about 0.7 percent of naturally available uranium. The remaining 99.3 percent is $^{238}_{92}U$. Fortunately, $^{238}_{92}U$ is a *fertile* material, meaning that it changes to a fissionable material when struck by neutrons. Plutonium $^{239}_{94}Pu$ produced in this manner can provide new fuel for the reactor.

The production of additional fuel as a part of the reactor's operation has led to the design of *breeder* reactors, in which there is a net increase in fissionable material. In other words, the reactor produces more fuel than it consumes. This does not violate the law of conservation of energy. It provides only for the production of fissionable material from fertile materials.

The fissionable fuel in most reactors depends on the availability of slow neutrons, which are more likely to produce fission. Fast neutrons liberated by the fission of nuclei must therefore be slowed down. For this reason reactor fuel is embedded in a suitable substance called a **moderator.** The function of this substance is to slow neutrons without capturing them.

Neutrons have a mass about the same as that of a hydrogen atom. It might be expected, then, that substances containing hydrogen atoms would be effective as moderators of neutrons. The neutron is analogous to a marble, which can be stopped by a collision with another marble but will merely bounce off a cannonball because of the great mass difference. Water (H_2O) and heavy water, containing ^2_1H instead of ^1_1H, are often used as moderators. Other suitable materials are graphite and beryllium.

To control the nuclear furnace, it is necessary to regulate the number of neutrons that initiate the fission process. Substances like boron and cadmium capture neutrons efficiently and are excellent control materials. A typical reactor has control rods that can be inserted into the reactor at variable distances. By adjusting the position of these rods, the activity of the nuclear furnace is controlled. A supplementary set of rods is available to allow the reactor to be shut down completely in an emergency.

39-12 NUCLEAR FUSION

In our earlier discussion on mass defect, we calculated that 28.3 MeV of energy is released in the formation of ^4_2He from its component nucleons. This joining together of light nuclei into a single heavier nucleus is called **nuclear fusion.** This process provides the fuel for stars like our own sun, and it is also the principle behind the hydrogen bomb. Many consider the fusion of hydrogen into helium as the ultimate fuel.

The use of nuclear fusion as a controlled source of energy is not without problems. It is still believed by most physicists that extremely high temperatures will be necessary to sustain nuclear fusion. The fusing nuclei would require millions of electronvolts of kinetic energy to overcome their coulombic repulsion. In the hydrogen bomb, this enormous energy is supplied by an atomic explosion, which then triggers the fusion process. The peaceful production of fusion by this method presents the problem of containment. The nuclear fuel would need to be so hot that it would instantly disintegrate any known substance. Present research methods involve containment by magnetic fields or rapid heating by powerful lasers. It is easy to see why the idea of "cold fusion" through an electrolytic process has generated so much excitement.

If the problems of fusion are ever solved, this energy source could provide a solution to our formidable problem of dwindling resources. The deuterium commonly found in seawater could provide us with an almost inexhaustible supply of fuel. It would represent more than a billion times the energy available in all our coal and oil reserves. In addition, it appears that fusion reactors would have much less of a problem with radioactive residue than that currently experienced with fission reactors.

Summary and Review

Summary

In this chapter we have studied the fundamental particles that make up the nuclei of atoms. Protons and neutrons are held together in the nucleus by strong nuclear forces that are active only within the nucleus. When such particles are joined together, the resulting mass is less that of the constituent parts. For massive nuclei it was also pointed out that energy results from tearing these nuclei apart. In either case, there is an enormous potential for useful energy. The major concepts that must be remembered in this chapter are as follows.

- The fundamental nuclear particles discussed in this chapter are summarized in the following table. The masses are given in atomic mass units (u) and the charge is in terms of the electronic charge $+e$ or $-e$, which is 1.6×10^{-19} C.

Fundamental Particles

Particle	Symbol	Mass, u	Charge
Electron	$_{-1}^{0}e$, $_{-1}^{0}\beta$	0.00055	$-e$
Proton	$_{1}^{1}p$, $_{1}^{1}$H	1.007276	$+e$
Neutron	$_{0}^{1}n$	1.008665	0
Positron	$_{+1}^{0}e$, $_{+1}^{0}\beta$	0.00055	$+e$
Alpha particle	$_{2}^{4}\alpha$, $_{2}^{4}$He	4.001506	$+2e$

The atomic masses of the various elements are given in the text.

- The atomic number Z of an element is the number of protons in its nucleus. The mass number A is the sum of the atomic number and the number of neutrons N. These numbers are used to write the nucleus symbol:

$$A = Z + N \qquad \text{Symbol: } _{Z}^{A}X$$

- One *atomic mass unit* (1 u) is equal to one-twelfth the mass of the most abundant carbon atom. Its value in kilograms is given below.

Also, since $E = mc^2$, we can write the conversion factor from mass to energy as c^2.

$$1 \text{ u} = 1.6606 \times 10^{-27} \text{ kg} \qquad c^2 = 931 \text{ MeV/u}$$

$$1 \text{ MeV} = 10^6 \text{ eV} = 1.6 \times 10^{-13} \text{ J}$$

In the mass spectrometer, the velocity v and the radius R of the singly ionized particles are

$$v = \frac{E}{B} \qquad R = \frac{mv}{eB} \qquad \textit{Mass Spectrometer}$$

- The *mass defect* is the difference between the rest mass of a nucleus and the sum of the rest masses of its nucleons. The *binding energy* is obtained by multiplying the mass defect by c^2.

$$E_B = [(Zm_H + Nm_n) - M]c^2 \qquad \textit{Binding Energy}$$

where $m_H = 1.007825$ u
$m_n = 1.008665$ u
$c^2 = 931$ MeV/u
$M = $ atomic mass
$N = A - Z$
$Z = $ atomic number

- Several general equations for radioactive decay are

$$_{Z}^{A}X \rightarrow {}_{Z-2}^{A-4}Y + {}_{2}^{4}\alpha + \text{energy} \qquad \textit{Alpha Decay}$$

$$_{Z}^{A}X \rightarrow {}_{Z+1}^{A}Y + {}_{-1}^{0}\beta + \text{energy} \qquad \textit{Beta-Minus Decay}$$

$$_{Z}^{A}X \rightarrow {}_{Z-1}^{A}Y + {}_{+1}^{0}\beta + \text{energy} \qquad \textit{Beta-Plus Decay}$$

- The *activity R* of a sample is the rate at which the radioactive nuclei decay. It is generally expressed in curies (Ci).

$$\text{One } \textit{curie} \text{ (1 Ci)} = 3.7 \times 10^{10}$$
disintegrations per second (s^{-1})

- The *half-life* of a sample is the time $T_{1/2}$ in which one-half the unstable nuclei will decay.
- The number of unstable nuclei remaining after a time t depends on the number n of half-lives that have passed. If N_0 nuclei exists at time $t = 0$, then a number N exists at time t. We have

$$N = N_0 \left(\tfrac{1}{2}\right)^n \qquad \text{where } n = \frac{t}{T_{1/2}}$$

- The activity R and mass m of the radioactive portion of a sample are found from similar relations:

$$R = R_0 \left(\tfrac{1}{2}\right)^n \qquad m = m_i \left(\tfrac{1}{2}\right)^n$$

- In any nuclear equation, the number of nucleons on the left side must equal the number of nucleons on the right side. Similarly, the net charge must be the same on each side.

Key Terms

1. nuclear force
2. nucleon
3. atomic number
4. mass number
5. atomic mass unit
6. isotopes
7. mass spectrometer
8. mass defect
9. binding energy
10. radioactivity
11. alpha particles
12. beta particles
13. gamma particles
14. half-life
15. activity
16. curie
17. nuclear fission
18. chain reaction
19. nuclear reactor
20. nuclear fusion
21. moderator

Review Questions

39-1. Write the symbol $_Z^A X$ for the most abundant isotopes of (a) cadmium, (b) silver, (c) gold, (d) polonium, (e) magnesium, and (f) radon.

39-2. From the curve describing the binding energy per nucleon (Fig. 39-5), would you expect the mass defect to be greater for chromium $_{24}^{52}Cr$ or uranium $_{92}^{238}U$? Why?

39-3. The binding energy is greater for the mass numbers in the central part of the periodic table. Discuss the significance of this in relation to nuclear fission and nuclear fusion. How do you explain the release of energy in both fusion and fission in view of the fact that one process brings nuclei together and the other tears them apart?

39-4. How is the stability of an isotope affected by the ratio of the mass number A to the atomic number Z? Does the element whose ratio is closest to 1 always appear to be the more stable?

39-5. Define and compare alpha particles, beta particles, and gamma rays. Which are likely to do the most damage to human tissue?

39-6. Given a source that emits alpha, beta, and gamma radiation, draw a diagram showing how you could demonstrate the charge and penetrating power of each type of radiation. Assume you have at your disposal a source of a magnetic field and several thin sheets of aluminum.

39-7. Describe and explain, step by step, the decay of $_{92}^{238}U$ to the stable isotope of lead, $_{82}^{206}Pb$. (Refer to Fig. 39-7.)

39-8. Write in the missing symbol, in the form $_Z^A X$, for the following nuclear disintegrations:

a. $_{90}^{234}Th \rightarrow _{91}^{234}Pa +$ _____
b. $_{15}^{32}P \rightarrow$ _____ $+ _{+0}^{1}e$
c. $_{94}^{239}Pu \rightarrow _{90}^{234}Th +$ _____
d. $_{92}^{238}U \rightarrow$ _____ $+ _2^4\alpha$

39-9. Write the missing symbol for the following nuclear reactions:

a. $_1^2H + _1^3H \rightarrow _2^4H +$ _____
b. $_{12}^{25}Mg +$ _____ $\rightarrow _{13}^{28}Al + _1^1H$
c. $_4^9Be + _2^4\alpha \rightarrow _6^{12}C +$ _____
d. $_1^1H +$ _____ $\rightarrow _6^{12}C + _2^4He$

Summary and Review

39-10. Explain the function of the following components of a nuclear reactor: (a) uranium, (b) radiation shielding, (c) moderator, (d) control rods, (e) heat exchanger, and (f) condenser.

39-11. Give examples to show how beta decay and alpha decay tend to bring unstable nuclei closer to the stability curve of Fig. 39-6.

39-12. Radon has a half-life of 3.8 days. Consider a sample of radon having a mass m and an activity R. What mass of radioactive radon remains after 3.8 days? Does this mean that the activity is reduced to one-half in the time of one half-life?

39-13. Radioactive carbon $^{14}_{7}C$ has a half-life of 5570 h. In a living organism, the relative concentration of this isotope is the same as it is in the atmosphere because of the interchange of materials between the organism and the air. When an organism dies, this interchange stops and radioactive decay begins without replacement from the living organism. Explain how this principle can be used to determine the age of fossil remains.

Problems

Refer to Table 39-4 for nuclidic masses.

Section 39-2—The Elements

39-1. How many neutrons are in the nucleus of $^{208}_{82}Pb$? How many protons? What is the ratio N/Z?   **Ans. 126, 82, 1.54**

39-2. The nucleus of a certain isotope contains 143 neutrons and 92 protons. Write the symbol for this nucleus.

39-3. From a stability curve, it is determined that the ratio of neutrons to protons for a cesium nucleus is 1.49. What is the mass number for this isotope of cesium?

Ans. 137

39-4. Most nuclei are nearly spherical in shape and have a radius that can be approximated by

$$r = r_0 A^{1/3} \qquad r_0 = 1.2 \times 10^{-15} \text{ m}$$

What is the approximate radius of the nucleus of a gold atom ($^{197}_{79}Au$)?

39-5. Study Table 39-4 for information on the several nuclides. Determine the ratio of N/Z for the following nuclides: beryllium-9, copper-64, and radium 224.

Ans. 1.25, 1.21, 1.55

Section 39-3—The Atomic Mass Unit

39-6. Find the mass in grams of a gold particle containing two million atomic mass units.

39-7. Consider a 2-kg cylinder of copper. What is the mass in atomic mass units? In megaelectronvolts? In joules? **Ans. 1.20×10^{27} u, 1.12×10^{30} MeV, 1.79×10^{17} J**

39-8. A certain nuclear reaction releases an energy of 5.5 MeV. How much mass (in atomic mass units) is required to produce this energy?

39-9. The periodic table gives the average mass of a silver atom as 107.842 u. What is the average mass of the silver nucleus?

Ans. 107.816 u

*39-10. Consider the mass spectrometer as illustrated by Fig. 39-3. A uniform magnetic field of 0.6 T is placed across both upper and lower sections of the spectrometer, and the electric field in the velocity selector is 120 V/m. A singly charged neon atom ($+1.6 \times 10^{-19}$ C) of mass 19.992 u passes through the velocity selector and into the spectrometer. What is the velocity of the neon atom as it emerges from the velocity selector?

*39-11. What is the radius of the circular path followed by the neon atom of Prob. 39-10?

Ans. 6.92 cm

Section 39-5—The Mass Defect and Binding Energy

*39-12. Calculate the mass defect and binding energy for the neon-20 atom ($^{20}_{10}$Ne).

*39-13. Calculate the binding energy and the binding energy per nucleon for tritium ($^{3}_{1}$H). How much energy in joules is required to tear the nucleus apart into its constituent nucleons? **Ans. 8.48 MeV, 2.83 MeV/nucleon, 1.36×10^{-12} J**

*39-14. Calculate the mass defect of $^{7}_{3}$Li. What is the binding energy per nucleon?

*39-15. Determine the binding energy per nucleon for carbon-12 ($^{12}_{6}$C). **Ans. 7.68 MeV/nucleon**

*39-16. What are the mass defect and the binding energy for a gold atom ($^{197}_{79}$Au)?

*39-17. Determine the binding energy per nucleon for tin-120 ($^{120}_{50}$Sn). **Ans. 8.50 MeV/nucleon**

Section 39-7—Radioactive Decay

39-18. The activity of a certain sample is rated as 2.8 Ci. How many nuclei will have disintegrated in a time of 1 min?

39-19. The cobalt nucleus ($^{60}_{27}$Co) emits gamma rays of approximately 1.2 MeV. How much mass is lost by the nucleus when it emits a gamma ray of this energy? **Ans. 0.00129 u**

39-20. The half-life of the radioactive isotope indium-109 is 4.30 h. If the activity of a sample is 1 mCi at the start, how much activity remains after 4.30, 8.60, and 12.9 h?

39-21. The initial activity of a sample containing 7.7×10^{11} bismuth-212 nuclei is 4.0 mCi. The half-life of this isotope is 60 min. How many bismuth-212 nuclei remain after

30 min? What is the activity at the end of that time? **Ans. 5.44×10^{11} nuclei, 2.83 mCi**

*39-22. Strontium-90 is produced in appreciable quantities in the atmosphere during a nuclear explosion. If this isotope has a half-life of 28 years, how long will it take for the initial activity to drop to one-fourth of its original activity?

*39-23. Consider a pure, 4.0-g sample of radioactive gallium-67. If the half-life is 78 h, how much time is required for 2.8 g of this sample to decay? **Ans. 135.5 h**

*39-24. If one-fifth of a pure radioactive sample remains after 10 h, what is the half-life?

Section 39-9—Nuclear Reactions

*39-25. Determine the minimum energy released in the nuclear reaction

$$^{19}_{9}F + ^{1}_{1}H \rightarrow ^{4}_{2}He + ^{16}_{8}O + \text{energy}$$

The atomic mass of $^{19}_{9}F$ is 18.998403 u.

Ans. 8.11 MeV

*39-26. Determine the approximate kinetic energy imparted to the alpha particle when radium-226 decays to form radon-222. Neglect the energy imparted to the radon nucleus.

*39-27. Find the energy involved in the production of two alpha particles in the reaction

$$^{7}_{3}Li - ^{1}_{1}H \rightarrow ^{4}_{2}He + ^{4}_{2}He + \text{energy}$$

Ans. 17.3 MeV

*39-28. Compute the kinetic energy released in the beta minus decay of thorium-233.

*39-29. What must be the energy of an alpha particle if it bombards a nitrogen-14 nucleus, producing $^{17}_{8}O$ and $^{1}_{1}H$? ($^{17}_{8}O$ = 16.999130 u)?

$$^{4}_{2}He + ^{14}_{7}N + \text{Energy} \rightarrow ^{17}_{8}O + ^{1}_{1}H$$

Ans. 1.19 MeV

Challenge Problems

*39-30. What is the average mass in kilograms of the nucleus of a boron-11 atom?

*39-31. What are the mass defect and the binding energy per nucleon for boron-11?
Ans. 0.0818 u, 6.92 megaelectronvolts per nucleon

*39-32. Find the binding energy per nucleon for thallium-206.

*39-33. Calculate the energy required to separate the nucleons in mercury-204.

Ans. 1.61 GeV

Summary and Review

*39-34. The half-life of a radioactive sample is 6.8 h. How much time passes before the activity drops to one-fifth of its initial value?

*39-35. How much energy is required to tear apart a deuterium atom? **Ans. 2.22 MeV**

*39-36. Plutonium-232 decays by alpha emission with a half-life of 30 min. How much of this substance remains after 4 h if the original sample had a mass of 4.0 g? Write the equation for the decay.

*39-37. If 32×10^9 atoms of a radioactive isotope are reduced to only 2×10^9 atoms in a time of 48 h, what is the half-life of this material? **Ans. 12.0 h**

*39-38. A certain radioactive isotope retains only 10 percent of its original activity after a time of 4 h. What is its half-life?

*39-39. When a ^6_3Li nucleus is struck by a proton, an alpha particle and a produce nucleus are released. Write the equation for this reaction. What is the net energy transfer in this case? **Ans. 4.02 MeV**

*39-40. Uranium-238 undergoes alpha decay. Write the equation for the reaction and calculate the disintegration energy.

*39-41. A 9-g sample of radioactive material has an initial activity of 5.0 Ci. Forty minutes later, the activity is only 3.0 Ci. What is the half-life? How much of the pure sample remains? **Ans. 54.3 min, 5.40 g**

Critical Thinking Questions

*39-42. Nuclear fusion is a process that can produce enormous energy without the harmful by-products of nuclear fission. Calculate the energy released in the following nuclear fusion reaction:

$$^3_2\text{H} + ^3_2\text{He} \rightarrow ^4_2\text{He} + ^1_1\text{H} + ^1_1\text{H}$$

Ans. 12.9 MeV

*39-43. Carbon-14 decays very slowly with a half-life of 5740 years. Carbon dating can be accomplished by seeing what fraction of carbon-14 remains, assuming that the decay process began with the death of a living organism. What would be the age of a chunk of charcoal if it was determined that the radioactive C-14 remaining was only 40 percent of what would be expected in a living organism?

*39-44. The velocity selector in a mass spectrometer has a magnetic field of 0.2 T perpendicular to an electric field of 50 kV/m. The same magnetic field is across the lower region. What is the velocity of singly charged lithium-7 atoms as they leave the selector? If the radius of the circular path

in the spectrometer is 9.10 cm, what is the atomic mass of the lithium atom? **Ans. 2.50×10^5 m/s, 7.014 u**

*39-45. A nuclear reactor operates at a power level of 2.0 MW. Assuming that approximately 200 MeV of energy is released for a single fission of U-235, how many fission processes are occurring each second in the reactor?

*39-46. Consider an experiment that bombards $^{14}_7\text{N}$ with an alpha particle. One of the two product nuclides is ^1_1H. The reaction is

$$^4_2\text{He} + ^{14}_7\text{N} \rightarrow ^A_Z\text{X} + ^1_1\text{H}$$

What is the product nuclide indicated by the symbol X? How much kinetic energy must the alpha particle have in order to produce the reaction? **Ans. 1.19 MeV**

*39-47. When passing a stream of ionized lithium atoms through a mass spectrometer, the radius of the path followed by ^7_3Li (7.0169 u) is 14.00 cm. A lighter line is formed by the ^6_3Li (6.0151 u). What is the radius of the path followed by the ^6_3Li isotopes?

Electronics

OBJECTIVES

After completing this chapter, you should be able to:

1. Explain with the use of diagrams the operation of a vacuum-tube diode, a cathode-ray tube, and an x-ray tube.
2. Distinguish, by diagrams and discussion, between N- and P-type semiconductors, describing how they can be produced commercially.
3. Explain with the use of diagrams how a transistor amplifier operates when it is connected with a common base, a common emitter, and a common collector.
4. Solve transistor application problems similar to those presented in this chapter.
5. Define and discuss applications of the *zener diode*, the *LED*, *photodiodes*, *silicon-controlled rectifiers*, and *integrated circuits*.

We are surrounded by electronics. The number of devices that are in some way electronic is constantly increasing. We now discuss some of the basic principles underlying this growing field of technology.

40-1 THERMIONIC EMISSION

The roots of the electronic revolution go back to the late nineteenth century, when, in 1883, Thomas Edison noticed that black deposits were forming inside the glass light bulbs he was experimenting with. Further experimentation showed that if a metal plate was also sealed inside the evacuated bulb and connected, as shown in Fig. 40-1, the galvanometer would indicate a current between the filament and the plate. This current flowed only when the positive terminal of the battery was connected to the plate. Several years later, this current was found to be the result of emission of electrons from the hot filament. The negatively charged electrons were attracted to the positive charge of the plate, generating an electric current between the plate and the filament.

The phenomenon that causes the emission of electrons from a heated filament is known as **thermionic emission.** It can be observed only in the absence of oxygen. A filament hot enough to cause thermionic emission will burn up in air.

Figure 40-1 The Edison effect.

40-2 VACUUM TUBES

The current flow observed by Edison occurred only when the plate (anode) was positive with respect to the filament (cathode). If the battery was reversed in the circuit, the current ceased. The negatively charged plate repelled the negatively charged electrons instead of attracting them, and the electrons that did escape the filament were attracted back to it by its positive charge.

Thermionic emission led to the development of the vacuum tube, which was the heart of all electronic devices until the 1950s. What made vacuum tubes so useful was the property that current could flow through them only in one direction—and that the current could be controlled.

A typical vacuum tube had a cathode heated by a separate heater filament and a plate anode. Tubes of this type, known as *diodes,* were the simplest form of the vacuum tube. A more sophisticated version, the *triode,* is shown in Fig. 40-2. The third element in the triode, the *grid,* was an electrode made up of a mesh of thin wires; electrons could pass easily through the many holes in it. When the tube was in operation, if a small negative voltage was applied to the grid, the electrons emit-

Figure 40-2

ted by the cathode were repelled by the grid and did not pass through it. The result was that no current flowed to the anode. If no voltage or a positive voltage was applied to the grid, however, the electrons from the cathode passed through the grid and current flowed between the cathode and the anode. Small voltages applied to the grid were able to control the flow of current in the tube. The small voltages were effectively amplified by this process.

These and similar tubes made possible the development of radio, television, computers, and numerous other modern devices. Now, the development and spread of *solid-state* technology utilizing *semiconductors* has made them virtually extinct. Two vacuum tubes, however, are still widely used, the **cathode-ray tube** (CRT) and the **x-ray tube.**

*inter*NET
CONNECTION

www.vacuum.org
This Web site, home of the American Vacuum Society, provides information on recommended practices, research, new materials, new devices, training, jobs, and topical conferences.

40-3 THE CATHODE-RAY TUBE

The cathode-ray tube is the key to electronically produced visual displays in oscilloscopes (Fig. 40-3), computers, and television sets. The tube consists of an *electron gun, deflection plates,* and the *screen.* These three parts are encased in a heavy-duty, evacuated glass case. The electron gun is in the long, thin part of the tube, as shown in Fig. 40-4. It is made up of a cathode and two grids and anodes. The anodes are shaped to allow a stream of electrons to pass through them into the larger section of the tube.

This beam of electrons passes a set of flat metal deflection plates, which can be charged like a capacitor, creating an electric field between them. Depending on the direction of the electric field, the electrons are deflected toward one plate or the other. In this way the beam can be deflected up or down or from side to side and aimed at any point on the screen at the front of the tube.

In a television set, the horizontal and vertical deflections are produced by magnetic fields rather than by electric fields. By sending alternating current through coils, it is possible to make the electron beam sweep back and forth across the tube. The screen itself is coated with **phosphor,** a substance that emits light when it is struck by high-energy electrons. Wherever the electron beam hits the screen, a bright spot appears. Wherever the picture should be light, the grid voltage is positive and

Figure 40-3

Figure 40-4   A cathode-ray tube using electrostatic deflection.

the beam is strong, resulting in a bright spot. Wherever the picture should be dark, the beam is retarded by making the grid negative and fewer electrons (or none) strike the screen, resulting in a dark spot. By covering the screen with closely spaced lines varying in brightness, the picture is produced on the screen. The lines are drawn at a rate of about 16,000 per second, and the entire screen is redrawn 60 times per second. Since both these speeds are much too great for the human eye to perceive, we see what appears to be a continuous moving image.

40-4 THE X-RAY TUBE

Almost all modern x-ray tubes (Fig. 40-5) are refinements of a tube designed in 1913 by William Coolidge, of the General Electric Company. As in other vacuum tubes, electrons are thermionically emitted by a hot cathode. In these tubes the cathode is cup-shaped to produce a narrow electron beam. The anode in the x-ray tube is held at a high voltage (20 to 100 kV), and the electrons reach it with a great deal of kinetic energy. The highly energetic electrons strike a metal target embedded in the anode. When the electrons collide with the metal atoms, many of the bombarding electrons knock inner electrons out of the metal atoms, leav-

Figure 40-5   Cathode x-ray tube.

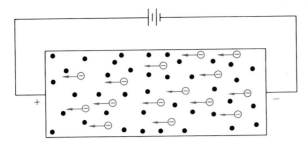

Figure 40-6 Flow of free electrons in a conductor.

ing holes in the inner shells. As outer electrons fall into these holes, they emit x-ray photons with relatively short wavelengths characteristic of the target metal.

40-5 SEMICONDUCTORS

Solid-state technology has completely replaced vacuum tubes in all but a few highly specialized applications. This has been possible because of the unique characteristics of **semiconductors.** These materials consists of a few elements and compounds that are not electrical insulators but do not exhibit the extremely high conductivity of true conductors either.

What makes something a conductor? The outer electrons of an atom, known as the **valence electrons,** are bound to the atom less tightly than the electrons that are closer to the nucleus. In a conductor they are really held loosely, and in the solid state many of them break away from the atoms and roam freely throughout the solid. Thus, they are called *free electrons.* When a battery or other means is used to place a potential difference across the solid, these free electrons are attracted to the positive potential and a current flows, as shown in Fig. 40-6.

Insulators, like rubber and glass, behave in just the opposite way. The outer valence electrons are held tightly to the atoms, and they do not move freely about the solid. Unless an extremely high voltage is applied to an insulator, no current flows through it.

Semiconductors fall midway between conductors and insulators. Their valence electrons are neither as free as those in a conductor nor as tightly bound as those in an insulator. The valence electrons are actually shared by the atoms in the semi-conductor. This process is known as *covalent bonding* and accounts for many of the properties of the semiconductors. Figure 40-7 shows how the electrons are

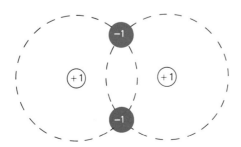

Figure 40-7 Two covalently bonded hydrogen atoms.

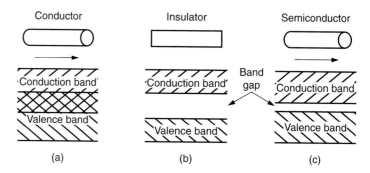

Figure 40-8 The band theory of conduction.

shared by the hydrogen atoms in the H$_2$ molecule. Covalent bonding permits electrons to move from atom to atom without being entirely free. The process is much like moving from person to person around your set in a square dance. You are always with a partner, yet you are traveling completely around the set.

Another way to look at the difference among conductors, insulators, and semiconductors is the *energy-band theory*. Since electrons can occupy a discrete number of energy levels, they can have only those energies that fall into *allowed bands*. The energy band in which the valence electrons normally move is called the *valence band*. Electrons that are free to move and conduct current are in the **conduction band.** As can be seen in Fig. 40-8, the conduction and valence bands overlap in conductors, are far apart in insulators, and are separated by a narrow energy gap in semiconductors. The overlap makes it easy for electrons to move into the conduction band in conductors. The large gap makes it almost impossible for them to do so in insulators, and the small gap makes it fairly easy in semiconductors.

Let us look at a real semiconductor. Because of its characteristics and availability, silicon is the most commonly used semiconductor. In its natural state silicon

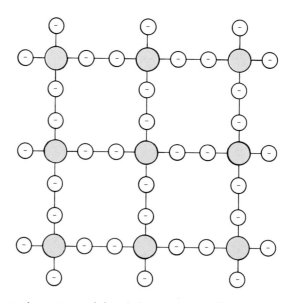

Figure 40-9 The atomic structure and shared electrons in pure silicon.

is a *crystal*. This means that, in a piece of silicon, the atoms are located at specific points in an ordered *lattice*. Figure 40-9 shows a diagram of the silicon lattice. There are four valence electrons surrounding each atom. In a perfect crystal each of these electrons would be shared with a neighboring atom. The crystal would be an insulator because no electrons would be available for current flow. However, any imperfection will cause the silicon to conduct some current.

The most common impurity occurs when an atom of another substance is placed at a point in the crystal lattice that would normally be occupied by a silicon atom. Two important types of impurities are *N-type donors* and *P-type acceptors.*

40-6 N-TYPE AND P-TYPE SEMICONDUCTORS

When an impurity atom has more than the four valence electrons needed to pair with the neighboring silicon atom, the extra unpaired electron is not covalently bonded and can move freely. Thus, the impurity donates extra negatively charged electrons to the crystal. This explains the names donor and N-type (for negative). A typical N-type impurity can be seen in Fig. 40-10.

Commercial **N-type semiconductor** material is made by adding small controlled quantities of a selected impurity to the silicon crystal. The intentional impurities are known as *dopants*. Phosphorous, arsenic, and antimony are common N-type dopants. Each has five valence electrons and therefore adds one free electron to the crystal. The P-type silicon is also produced commercially by doping. In this case the dopant has one less valence electron than silicon. Thus, P-type dopants have three valence electrons, and common P-type dopants are aluminum, boron, gallium, and indium. Any of these dopants yields a semiconductor like the one in Fig. 40-11.

Technology Today

Materials Used for Dopants, Semiconductors, and Microwave Devices

- GaAs (gallium arsenide) works better than silicon in microwave devices because it allows faster movement of electrons.

- Materials other than boron and arsenic are used as dopants.

- It is theoretically possible to make semiconductor devices from crystalline carbon.

- Crystal radio receivers were an early application of semiconductors.

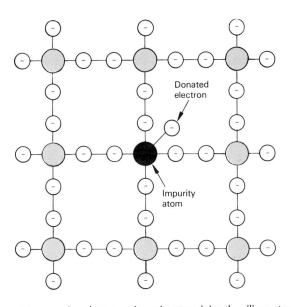

Figure 40-10   In an N-type semiconductor, an impurity atom joins the silicon structure, donating an extra electron.

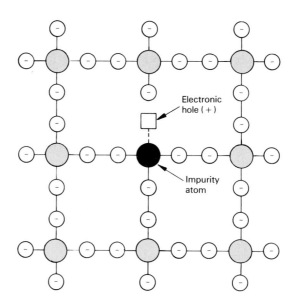

Figure 40-11 In a **P-type semiconductor,** the absence of a valence electron produces an electronic hole.

The *hole* marked in the figure is where the fourth valence electron would be if the spot were occupied by a silicon atom and not an impurity atom. When attracted by a positive potential, electrons near neighboring holes can jump over and fill the holes. Each time an electron moves in one direction, the hole moves in the opposite direction, to the atom from which the electron came. In this way a movement of electrons one way causes a movement of holes the other way. Since negative and positive charges always behave oppositely, it is just as if the holes were positive charges and the current can be treated as if it were a movement of positive charges (the holes).

howthingswork.virginia. edu

For more information on how audio amplifiers and computers work, see the How Things Work Web site, and go to earlier questions in Chap. 13.

40-7 PN JUNCTION

Some interesting effects result from the combination of the P-type and N-type materials. The most basic of these is formation of a **PN junction** by doping an N-type zone adjacent to a P-type zone. When the two zones are first formed, some of the free electrons from the N-type material jump past the junction and fill some of the holes in the P-type material. This leaves a thin layer of positively charged ions along the junction of the N side. These positive ions are the atoms that have lost an electron. Similarly, the atoms on the P side that have acquired electrons form a layer of negative ions. Once these layers are formed, the negative ions repel other electrons, preventing any more electrons from jumping over, and the positive ions prevent any holes from jumping over.

In this state the PN junction is rather stable. But things change when a battery is connected across the junction. We first look at what happens when the battery is connected with its negative terminal on the P side of the junction and its positive terminal on the N side, as in Fig. 40-12. The positive holes in the P-type material move toward the negative potential of the battery, and the negative electrons

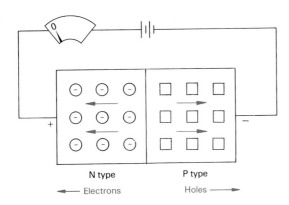

Figure 40-12   Reverse-biased PN junction.

move toward the positive potential of the battery. The result is that both holes and electrons move *away* from the junction and there is no current flow across the junction. When connected in this manner, the junction is said to be *reverse-biased*.

If the battery is connected across the junction in the opposite direction, positive to the P side and negative to the N side, the behavior is quite different. The electrons from the donor atoms in the N-type material are attracted to the positive potential and flow toward the junction. The holes flow toward the junction in the opposite direction toward the negative potential. In short, there is a current flow. This type of connection, shown in Fig. 40-13, is called *forward biasing* of the junction. The semiconductor device we have been discussing behaves just like the simplest vacuum tube discussed in Sec. 40-2; it allows current flow in only one direction. Because of this similarity, it is also known as a *diode*.

The significant advantages of the semiconductor diode over its vacuum-tube predecessor are shared by all other semiconductor devices. It is small and rugged. It does not require high voltage to operate, and it is more durable.

The semiconductor diode also exhibits another interesting characteristic. Its behavior is not predictable with Ohm's law. This is because the diode's resistance is not constant but varies with the applied bias. Figure 40-14 shows a graph of current versus voltage for an ordinary conductor and the same information for a

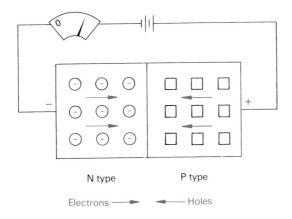

Figure 40-13   Forward-biased PN junction.

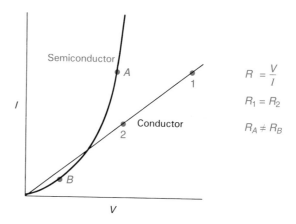

Figure 40-14 The relationship between voltage and current for a conductor and for a semiconductor.

forward-biased semiconductor diode. The conductor shows the straight line that results from the relation $V = IR$. The diode does not. Instead, the current rises slowly at the start and then rises rapidly. The slow rise at the beginning occurs because the forces resulting from the ion layers at the junction must be overcome. Once they are overcome, however, there is little resistance to current flow. Diodes do not show constant resistance. Each diode has its own characteristic curve, which depends on the material used to make it and the dopants used in it.

40-8 DIODE APPLICATIONS

We have already seen that a PN junction diode passes current only when forward-biased. This immediately makes it an excellent choice to convert alternating current to direct current (rectifying). Figure 40-15 shows a diode inserted in a circuit as a **half-wave rectifier.** (Notice the symbol representing the diode. The triangle can be seen as an arrow pointing in the direction of allowed current flow.) The output of this circuit is not a true, constant direct current but a current that flows in pulses only half of the time. The remainder of the current, which would normally flow the other way, is blocked.

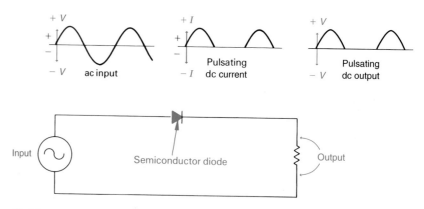

Figure 40-15 The semiconductor diode used as a half-wave rectifier.

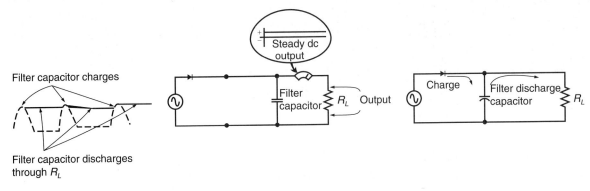

Filter capacitor charges

Filter capacitor discharges
through R_L

Figure 40-16 An electronic filtering circuit.

Often the pulsed output of the half-wave rectifier is not sufficient to meet the needs of a particular application. When this is the case, it is possible to *filter* the output of the rectifier to smooth out the pulses and make it look more like the continuous output of a battery. A simple but often used circuit for such filtering is shown on Fig. 40-16. A filter capacitor maintains the current through the load (R_L) when the pulsating voltage drops off. With a large capacitor, the discharge time constant is large and the load voltage is steady enough for many applications.

The half-wave rectifier is not the only means of accomplishing that purpose. More common is a *bridge rectifier,* which uses four diodes arranged as in Fig. 40-17. The ac input is connected to the bridge at points 1 and 3, and the output load is connected to points 2 and 4. When point 1 is positive and point 3 negative, the current flows as in Fig. 40-17a. Current entering the bridge at point 1 cannot pass through diode D_4, which is reverse-biased, but it can, and does, pass through diode D_1. It then leaves the bridge at point 2 and goes through the load R_L. It reenters the bridge at point 4, again facing two possible paths. D_4 remains reverse-biased and nonconducting, and so the current goes through D_3 and leaves at point 3. During the other half of the cycle, when the input polarity is reversed, the current flows in the path shown in Fig. 40-17b using the other pair of diodes, D_2 and D_4, in a similar manner. In this way the diode pairs that are conducting alternate according to the alternations in the input polarity. Such a rectifier is called a *full-wave rectifier* since all of the input signal is transmitted to the load.

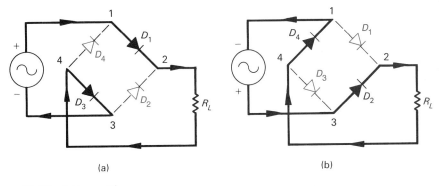

(a)   (b)

Figure 40-17 Bridge rectifier.

ac input   Half-wave rectifier   Full-wave rectifier

Figure 40-18   A comparison of waveforms.

A major disadvantage of the full-wave rectifier is its output form, shown in Fig. 40-18 in comparison with the ac input and the output of a half-wave rectifier. The figure shows that the full-wave rectifier leaves no gaps where the current output is zero. Instead, the current continuously varies between zero and its peak value in the forward direction. Thus, the output requires much less smoothing to make it look like the constant output of a battery, and the device does not in effect ignore half the source current.

40-9 THE ZENER DIODE

So far we have been discussing diodes as purely one-direction devices; when forward-biased, they conduct, and when reverse-biased, they do not. A look at a complete characteristic curve for a typical semiconductor diode (Fig. 40-19) shows that this is not quite the case. When the diode is reverse-biased, the current remains almost zero as the reverse voltage is increased until a critical voltage known as the *zener voltage* (or breakdown voltage) is reached. At that point the resistance of the diode abruptly drops, and the diode becomes an excellent conductor in the reverse direction.

This apparently strange effect is easily explained by examining the behavior of the electrons and holes at the junction in the diode. When the diode is reverse-biased, the electrons and holes all move away from the junction, leaving it nearly free of charges of either type. There are, however, always a few electrons and holes

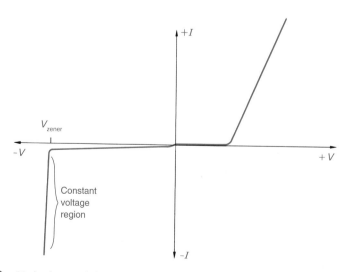

Figure 40-19   Diode characteristic curve.

created by random thermal motions in the crystal. These few created free holes and electrons flow away from the junction and constitute a nonzero *reverse current*. This reverse current, of a few nanoamperes, is always present when the diode is reverse-biased but can usually be ignored.

If the reverse-bias voltage is increased sufficiently, a point is reached where the electrons accelerated by this voltage have enough energy to knock more electrons loose from atoms with which they collide. These freed electrons knock more electrons loose, creating a large flow of electrons, known as the *avalanche effect*. Once the avalanche voltage is reached, the voltage across the diode does not increase to any extent, even though the current increases dramatically.

We might expect that a diode operated at such comparatively high current in its reverse direction would burn itself up. This is avoided by choosing rectifier diodes with an avalanche voltage rating greater than normal circuit potentials.

Zener diodes are manufactured using heavy doping, which causes the ions along the junction to be closer together. When reverse bias is applied, the electric field across the junction is more intense. Breakdown can occur at voltages of 4 V or less. Diodes that break down over a range of 4 to 6 V use both the avalanche effect and the Zener effect. The name **zener diode** is generally used for all breakdown diodes. Figure 40-20 shows the circuit symbol for a zener diode.

The tendency for a diode operating in the zener region of its characteristic curve to maintain a constant voltage regardless of current flow through it makes the zener diode especially useful in applications where a constant voltage at a particular point is critical, particularly in regulated power supplies. A regulated power supply is a voltage-current source with an output voltage that remains constant at a specified value, regardless of the output current.

Figure 40-21 shows a way of using two identical zener diodes connected back to back to limit the output of an ac source. When the two diodes are connected in this way, one of them is reverse-biased and voltage-limiting, while the other is forward-biased and conducting. When the ac source reverses its polarity, the diodes exchange roles and the conducting diode becomes the limiter and the limiter becomes the conductor. The peak output of this *voltage regulator* remains less than or equal to the zener voltage of the two diodes.

Figure 40-20 Zener diode symbol.

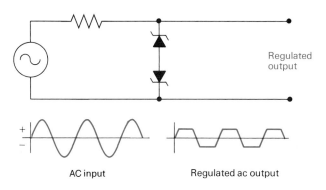

Figure 40-21 Zener diode regulator.

40-10 THE TRANSISTOR

A single PN junction device is a diode. The next step is the ***transistor,*** which consists of an N-type semiconductor section sandwiched between two sections of P-type semiconductor (called a *PNP transistor*) or a section of P-type semiconductor located between two sections of N-type semiconductor (called an *NPN transistor*). The transistor was invented in 1948 by Shockley, Bardeen, and Brattain at Bell Laboratories, and in the years since then it has completely revolutionized the field of electronics and many associated fields.

Let us consider a typical NPN junction transistor biased as shown in Fig. 40-22. The left section is made of heavily doped N-type semiconducting material and is called the *emitter*. The center section is thin and lightly doped with a P-type impurity. It is called the *base*. The right section is medium doped, N-type semiconducting material and is called the *collector*. Note that the emitter and base form an NP junction, and the base and collector form a PN junction.

The emitter-base NP junction in Fig. 40-22 is forward-biased, allowing electrons to flow easily from the N side to the P side. The base region is thin and has relatively few holes, allowing most of the electrons to diffuse right across it to the collector. But some of the electrons are trapped in the holes of the base region. If they are not removed from the base, the negative charge of the trapped electrons will repel the electrons from the emitter and rapidly shut off the electron current from the emitter to the collector. By removing or not removing this buildup of electrons from the base region, the transistor can be used as a switch.

When the electrons are removed from the base, there is an electron current from the base. The emitter current I_e is divided into base current I_b and a collector current I_c. That is,

$$I_e = I_b + I_c \qquad \textbf{40-1}$$

The ratio of the collector current to the emitter current is called α (alpha).

$$\alpha = \frac{I_c}{I_e} \qquad \textbf{40-2}$$

This number is a specification given by the manufacturer with each transistor. Since 95 to 99 percent of the electrons from the emitter travel through the base into the collector without being trapped, α typically ranges from 0.95 to 0.99.

Figure 40-22 An NPN transistor and its symbol. Arrows indicate the direction of electron currents. (Note that electron flow is opposite to conventional current.)

If we know α, we can calculate the relation of the small base current to the emitter current. Using Eq. (40-1), we can solve for I_b in terms of I_e and I_c.

$$I_b = I_e - I_c$$

But $I_c = \alpha I_e$, as computed from Eq. (40-2). Substitution gives

$$I_b = I_e - \alpha I_e$$

Finally, an expression for the base current of a transistor is

$$I_b = I_e(1 - \alpha) \qquad\qquad \textbf{40-3}$$

EXAMPLE 40-1

Given a transistor with $\alpha = 0.97$, calculate I_b and I_c when $I_e = 50$ mA and the transistor is connected as in Fig. 40-22.

Solution

Substitution into Eq. (40-3) gives

$$\begin{aligned}
I_b &= I_e(1 - \alpha) = (50 \text{ mA})(1 - 0.97) \\
&= (50 \text{ mA})(0.03) \\
&= 1.5 \text{ mA}
\end{aligned}$$

Now the collector current is found from Eq. (40-1).

$$\begin{aligned}
I_c &= I_e - I_b = 50 \text{ mA} - 1.5 \text{ mA} \\
&= 48.5 \text{ mA}
\end{aligned}$$

The PNP transistor is constructed just like the NPN transistor, except that the N-type material is sandwiched between two sections of P-type material. Because of this reversal of the positions of the two material types, the primary current carriers in the transistor are holes instead of electrons. Since holes are effectively positive charges, the biases are necessarily reversed. For the emitter-base junction to be forward-biased, the emitter voltage must be positive with respect to the base, and for the collector junction to be reverse-biased, the collector must be negative with respect to the base.

The biasing for a PNP transistor is shown in Fig. 40-23. The change in biasing does not affect the way the junctions operate. It merely accommodates the

Figure 40-23 A PNP transistor and its symbol. Arrows indicate electron flow.

rearrangement of materials and their properties. In operation, the holes flow from the emitter junction to the base and then most go into the collector.

40-11 TRANSISTOR AMPLIFICATION

In an amplification device we want to be able to input a varying signal with low-magnitude voltage or current characteristics and output a similarly varying signal with much higher voltage or current. A vacuum-tube triode is controlled by voltage. A small voltage on the grid of the triode controls a similar but much higher voltage in the plate circuit. The junction transistor, on the other hand, depends on the currents in it to control the amplification of the voltage or power in the output circuit.

Figure 40-24 shows a simple amplifier circuit utilizing the transistor. Because the transistor is a three-terminal device, one of its terminals must be common to both the input and output circuits. Any of the three terminals may be selected as the common terminal. In Fig. 40-24 the base is the common terminal.

As we discussed in Sec. 40-10, the emitter current and the collector current are almost equal. For this reason the ratio of output current to input current, called the *current gain*, is approximately 0.95 to 0.99 when the transistor is used in this configuration and is equal to α. In this case, collector current is considered as output and emitter current is considered as input.

Normally the voltage across the forward-biased region is small. This region represents a low resistance, and little voltage is needed to send a relatively high current through it. The second diode region, which is reverse-biased, represents a much higher resistance to the passage of current. Since voltage is the product of current and resistance, the resulting voltage must be high.

The ratio of output resistance to input resistance in a typical transistor is usually quite high. A typical input resistance might be 500 Ω, and a typical output resistance might be as much as 500 kΩ. This results in a ratio $R_{\text{out}}/R_{\text{in}}$ equal to 1000.

EXAMPLE 40-2

Assume that a transistor has an input resistance of 500 Ω and an output resistance of 400 kΩ. Determine the input and output voltages if the emitter current is 10 μA and $\alpha = 0.98$.

Figure 40-24 Common-base amplifier circuit.

Solution

The input voltage is found from Ohm's law.

$$V_{in} = I_e R_{in} = (10 \times 10^{-6} \text{ A})(500 \text{ }\Omega)$$
$$= 5 \times 10^{-3} \text{ V} = 5 \text{ mV}$$

To determine the output voltage, we must first determine the collector current. Recalling that $\alpha = I_c/I_e$, we can solve for the collector current as follows:

$$I_c = \alpha I_e = 0.98(10 \times 10^{-6} \text{ A}) = 9.8 \times 10^{-6} \text{ A}$$

Thus, the output voltage is

$$V_{out} = I_c R_{out} = (9.8 \times 10^{-6} \text{ A})(400 \times 10^{3} \text{ }\Omega)$$
$$= 3.92 \text{ V}$$

The **voltage amplification factor A_v** for a transistor is defined as the ratio of output voltage to input voltage.

$$A_v = \frac{V_{out}}{V_{in}} \qquad \text{\textit{Voltage Amplification Factor}} \qquad \textbf{40-4}$$

This ratio is also referred to as the *voltage gain*.

Along with an increase in voltage comes an increase in power. Recall that the power dissipated by a circuit component is equal to the voltage across it times the current through it. In any case, the power gain G is given by

$$G = \frac{\text{power out}}{\text{power in}} = \frac{V_{out} I_c}{V_{in} I_e} \qquad \textbf{40-5}$$

But since $\alpha = I_c/I_e$, we can write

$$G = \alpha \frac{V_{out}}{V_{in}} = \alpha A_v \qquad \text{\textit{Power Gain}} \qquad \textbf{40-6}$$

EXAMPLE 40-3

Determine the voltage amplification factor and the power gain for the transistor of Example 40-2.

Solution

In the previous example the input voltage was 5 mV, and the output voltage was 3.92 V. This gives an amplification factor of

$$A_v = \frac{V_{out}}{V_{in}} = \frac{3.92 \text{ V}}{5 \times 10^{-3} \text{ V}} = 784$$

Now, the power gain can be found from Eq. (40-6),

$$G = \alpha A_v = (0.98)(785) = 768$$

where we used $\alpha = 0.98$ from the previous example.

Figure 40-25 Common-emitter amplifier circuit.

Transistors can also be connected with the emitter as the common terminal. In other words, connected to the input and output circuits. A simple *common-emitter* amplifier circuit is shown in Fig. 40-25. As before, the input is across the emitter and base contacts, but the output in this case is from the emitter and collector terminals. The emitter junction is forward-biased, and the collector junction is reverse-biased. If the bias on the emitter junction is increased, the current flowing through the collector junction will increase as a result of the increase in current through the emitter junction. However, the base current will retain its usual low value. This low base current is also the current flowing in the input circuit. Thus, in this configuration we get a considerable current gain in the output circuit (collector circuit) over the current in the input circuit (base). We already know that the base current is given by

$$I_b = I_e(1 - \alpha)$$

and that the output, or collector, current is

$$I_c = \alpha I_e$$

These relations allow us to compute an expression for the current gain, or current amplification factor, A_i.

$$A_i = \frac{I_\text{out}}{I_\text{in}} = \frac{\alpha I_e}{(1 - \alpha)I_e} = \frac{\alpha}{1 - \alpha} \qquad \textbf{40-7}$$

A special symbol β is used to represent the current gain in the above form. Thus, for a transistor having a known α, the current gain can be found from

$$\beta = \frac{\alpha}{1 - \alpha} \qquad \textbf{40-8}$$

This is also known as the *base-collector current gain.*

EXAMPLE 40-4

In a common-emitter amplifier, the value of α is 0.96. What is the current gain?

Substitution into Eq. (40-8) yields

$$\beta = \frac{\alpha}{1 - \alpha} = \frac{0.96}{1 - 0.96} = 24$$

Typically, the values for β range from 20 to several hundred.

To determine the voltage gain we must again look at the effective input and output resistances. As in the common-base circuit, the input resistance is across the forward-biased emitter junction and is low. The output resistance is high but not quite as high as in the common-base circuit. The result is a voltage gain nearly as high as in the common-base circuit.

The power gain depends on the products of the current gain and voltage gain. While the voltage gain is slightly lower, the current gain β is much, much higher, and the power gain is also much higher.

The common-emitter circuit has one other interesting feature. When the incoming signal in Fig. 40-25 causes the base potential to become more positive, increasing the forward bias of the emitter junction, the collector potential moves in the opposite direction and becomes less positive. As a result, the common-emitter circuit has the property of changing the phase angle of the signal by 180° and acting as a *signal inverter.*

The final method of connecting a transistor is illustrated in Fig. 40-26. This is the common-collector arrangement. Looking at the same parameters as for the other two circuits, we get the following results. The input current is the base current, and the output current is the emitter current. The current gain is consequently

$$A_i = \frac{I_{\text{out}}}{I_{\text{in}}} = \frac{I_e}{I_e(1 - \alpha)}$$

from which

$$A_i = \frac{1}{1 - \alpha} \qquad \text{40-9}$$

Figure 40-26   Common-collector amplifier circuit.

TABLE 40-1 A Comparison of the Three Possible Transistor Amplifier Arrangements

	Common base	Common emitter	Common collector
Current gain	α (low)	β (high)	$\dfrac{1}{1-\alpha}$ (highest)
Voltage gain	Highest	High	Low
Power gain	Moderate	Highest	Lowest

In this case the base voltage is greater than the output voltage, and the net gain is less than 1, the lowest of the three arrangements. Related to this, the input resistance is high and the output resistance is low. Finally, because of the low voltage gain, the power gain is also the lowest of the three configurations. Table 40-1 compares the various gain characteristics of the three circuits.

There are two points to remember when using transistors in circuits: (1) Transistors require a dc voltage source for biasing before an ac signal can be applied to them for amplification or other modification. This dc voltage is often supplied by a regulated power supply using a diode bridge and zener diodes. (2) Since transistors are damaged easily by high temperature, the current must be rather low.

Transistor manufacturers publish data sheets and graphs of transistor characteristics. These documents are useful for circuit design and for verifying that none of the relevant parameters will be exceeded, destroying the transistor or other sensitive semiconductor circuit elements.

40-12 OTHER SEMICONDUCTOR DEVICES

Since the invention of the transistor, many other semiconductor devices have been developed. Some of them have interesting and unusual attributes that make possible many of the small electronic devices we enjoy today. One of the newest devices is the light-emitting diode (LED), used to provide visual displays in electronic devices. When arranged as shown in Fig. 40-27, a set of seven LEDs can be used to form numbers or letters. They are specially doped junction diodes that emit light due to the current flowing through them. The diodes are operated with a forward bias, allowing current to flow across the PN junction. At the junction, however, some of the electrons and holes combine to form electron-hole pairs. The electrons fill the holes, and the two are canceled out as current carriers. The energy of the recombined electrons and holes is released in the form of photons.

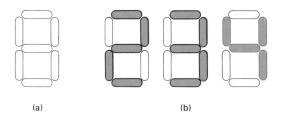

(a)   (b)

Figure 40-27 LED display. (a) Arrangement of LEDs. (b) Pattern of lighting to form the number 234.

Materials that behave this way, for example, gallium arsenide, gallium arsenide phosphide, and gallium phosphide, are used to make LEDs. The first two emit visible light, and the last emits infrared light.

Most visible-light LEDs emit red light, although some emit green and yellow light. Blue LEDs are difficult to make and are expensive. The advantages of LEDs are that they are small and rugged, they use little current and require low voltages, and their useful lifetimes are extremely long. Some LEDs may last as long as 100 years.

It is possible to combine an LED with an optical cavity resonator. The result is a laser diode that provides coherent light (all the waves are in phase). The cavity has two mirrored ends and a resonant frequency determined by its physical dimensions. The PN junction pumps light energy into the cavity at the resonant frequency, and the waves add in phase. The laser light escapes from the end, where the mirror is less than 100 percent reflective. Laser diodes are widely applied in medicine, measurements, and communications, and are used to read CDs and DVDs.

Similar to the LED but functioning in an inverse manner are photosensitive semiconductor devices. One of these is the photodiode. These diodes are run reverse-biased. When struck by light, the photons are absorbed by the atoms. This raises the energy level of the valence electrons sufficiently to cause them to jump into the conduction band. Thus, an electron-hole pair is created for each absorbed photon, and the electrons and holes are both swept away by the bias voltage, producing a current flow. Just as the light output of the LED is related to the current flow in the diode, the current flow in the photodiode is related to the intensity of the light striking the diode. The circuit symbols for the LED and the photodiode and a set of characteristic curves for a photodiode are shown in Fig. 40-28.

An interesting application of LEDs and light-sensitive diodes is in the construction of a type of semiconductor relay. A **relay** is a device for indirectly switching an electric current by using another current. Before the advent of semiconductor technology, relays were all electromechanical devices of various designs. One of the great advantages of a relay is that high, and consequently dangerous, currents can be switched without handling them directly. A relay for this purpose is found in

Figure 40-28 (a) Circuit symbols for the LED and photodiode. (b) Characteristic curves for a photodiode.

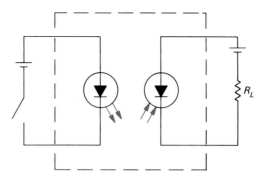

Figure 40-29 Simple LED-photodiode relay.

the starting circuit of a car. A small current in the starting switch activates the sole-noid switch; it actually turns on the high current that goes to the starting motor.

The optical semiconductor relay, which cannot be used for high-current appli-cations, is shown in Fig. 40-29. In operation, when a current is passed through the LED, it emits light. The light from the LED strikes the photodiode, which then conducts as long as the LED remains activated. In this way the photodiode acts as a switch turned on and off by the LED.

Another semiconductor device that may serve as a type of switch is the *silicon-controlled rectifier* (SCR) (Fig. 40-30). It is similar to a transistor with an extra layer of semiconductor material. Just as a transistor can be viewed as two diodes in series, the SCR can be viewed as three diodes in series. For the SCR in the figure they are, from left to right, an NP junction, a PN junction, and another NP junc-tion. In addition, the electrodes are renamed. The end electrodes are the *anode* and *cathode,* and the center electrode is called the *gate.*

When a low-bias voltage is applied between the anode and cathode, the center diode is reverse-biased and there is no current flow. In this state the device is said to be switched *off.* As the bias voltage is increased, a point is reached where the reverse-biased diode breaks down and the device begins to conduct. At this point the device has switched *on.* The point at which the SCR switches on, called the *breakover voltage,* is controlled by the positive bias applied to the gate. Once this point is reached, the gate loses control and the SCR must be disconnected in some way to restore control to the gate. The circuit symbol for the SCR and a set of characteristic curves for a typical SCR are shown in Fig. 40-31. Notice how the forward breakover voltage decreases with increasing gate current.

Two other simple semiconductor devices exhibit properties different from any others discussed up to this point, the *photoresistor* and the *thermistor.* The pho-

Figure 40-30 Silicon-controlled rectifier.

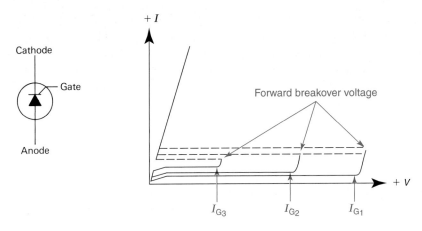

Figure 40-31 SCR symbol and the effect of gate current on breakover voltage.

toresistor is the key component of all modern light meters. In fact, almost all cameras that use batteries use a photoresistor to measure light intensity. Most photoresistors in use today are made of cadmium sulfide (CdS). Unlike the other semiconductor devices we have discussed, the photoresistor has no junction. It is a uniform crystal of CdS. When a voltage is applied across it, a current flows in proportion to the intensity of the light striking it. Conduction is not by free electrons but by electron-hole pairs, which are created when photons knock electrons loose from atoms in the crystal and into the conduction band. The more intense the light, the greater the number of photons hitting the crystal and the greater the conductance.

The last device to be discussed in this section is the *thermistor*. It takes advantage of the fact that the resistance of a semiconductor decreases as the temperature increases. The effect is opposite to, and much greater than, the change in the resistance of a conductor with temperature. This can be seen clearly in Fig. 40-32.

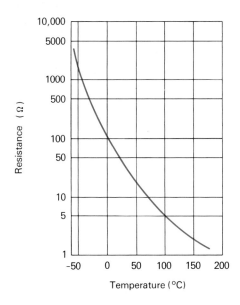

Figure 40-32 The characteristic curve of a typical thermistor.

The *y*-axis, showing the resistance, is logarithmic to accommodate the large changes in resistance. When a thermistor is maintained with a constant voltage across it, there will be large changes in the current that flows through the device as a result of changes in the temperature of the device. This means that a thermistor can be used to measure the temperature of its environment.

It also may be used as a safety device in complex solid-state components. The thermistor is placed in the parts of such components where high temperatures may be experienced. The circuit is then designed so that when the current through the thermistor rises to a critical value, the component shuts off, protecting the semiconductor devices within it from being destroyed by overheating. In this way the thermistor acts as a temperature-controlled fuse.

40-13 INTEGRATED CIRCUITS

As great as the electronic revolution the transistor caused in the 1950s and 1960s is an ongoing revolution due to the **integrated circuit** (IC). These devices combine numerous circuit elements (resistors, capacitors, transistors, diodes, and others) on tiny slices of ultrapure silicon. Each IC represents a complete circuit of a particular type, and ICs can be combined with each other and with other semiconductor devices to make large-scale components. They have also made possible certain devices like microprocessors that have created an electronic revolution of their own.

The most obvious attribute of the IC is its incredibly small size. A typical IC is 0.05-in. square and about 0.01 in. thick. This wafer may contain thousands of separate devices, enough to serve numerous complex purposes. It is protected by a container that often measures only a fraction of an inch in any direction.

Although the size of the IC is remarkable, size is not its only advantage over conventional devices. Because the entire circuit is fabricated at the same time, it is a highly reliable device, more so than circuits built up from individual parts. This characteristic is invaluable in devices such as guidance systems and pacemakers.

The individual devices that make up the IC function on the same principles as the semiconductor devices already discussed. They depend on the PN junction and the action of both electrons and holes for their various properties.

The silicon wafer that makes up the IC is known as the *substrate,* and the two processes of fabrication are *diffusion* and *etching.* **Diffusion,** as in the manufacture of junction-dependent devices, is the controlled injection of appropriate dopants into the silicon. **Etching** is the use of hydrofluoric acid to remove materials from the surface of the substrate to expose a region for further processing. These steps are combined with several steps that deposit materials on the substrate to make the necessary P-type and N-type regions forming diodes, transistors, and related devices.

Let us outline the steps needed to create an area of P-type silicon in the substrate. The first step is the oxidation of the top surface of the silicon to form a thin layer of silicon oxide. This oxide layer is impervious to most contaminants and serves to isolate and protect the substrate.

After the oxide layer comes the *photoresist* layer. The photoresist is a material that becomes acid-resistant when it is exposed to ultraviolet light. Exposure to ultraviolet light changes its chemical properties so that the areas that are exposed cannot be dissolved by acids.

The exposure to ultraviolet light is made after the coated substrate has been covered with a *mask*. This is a pattern that is opaque in the areas into which the dopants are eventually to be diffused. Masks are produced photographically by taking a picture of the appropriate pattern and then reducing it to the micro-miniature size needed for the IC.

After the placement of the mask and the exposure, the exposed photoresist is washed away with a solvent. This leaves a layer of exposed silicon oxide covering the area that will become the P-type region. Etching with hydrofluoric acid removes the oxide and leaves bare silicon substrate. The various steps in this process, in order, are shown in Fig. 40-33.

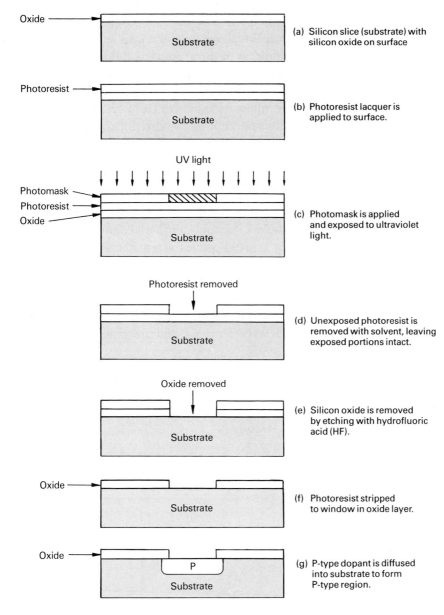

Figure 40-33 The steps in producing an integrated circuit.

Figure 40-34   The IC diode.

The bare silicon can now be doped with the desired dopant to produce a P-type region in the substrate. If the same process is repeated on an immediately adjacent area with an N-type dopant, a P-type region will border on an N-type region and create a PN junction. The addition of metal contacts on top of these regions gives a usable semiconductor diode like the one in Fig. 40-34.

Similar processes can be used to form all other necessary components in the IC. A P-type resistor on a P-type substrate is illustrated in Fig. 40-35. The upper N-type region is formed by diffusing enough N-type dopant into the already formed P-type region to convert it to an N-type region in that area. The plus sign indicates a high concentration of the N-type dopant. The purpose of the two N-type regions is to isolate the resistor from the substrate and the rest of the IC. The resistance is controlled by adjusting the length and cross-sectional area of the P-type conductor region and the amount of dopant diffused into the region.

A capacitor is formed similarly to a resistor. A diffused conducting layer acts as one of the plates of the capacitor, a metal contact acts as the other plate, and the oxide layer between them acts as the dielectric. A possible arrangement is shown in Fig. 40-36.

Of course ICs could not be the highly sophisticated circuits they are without the ability to contain transistors. Several types of transistors can be formed in ICs, and they can be formed in several ways. The basic and the preferred structures for IC transistors are shown in Fig. 40-37.

ICs are manufactured from circular silicon wafers about 6 in. in diameter. Each wafer is large enough to produce many ICs. After fabrication they are individually

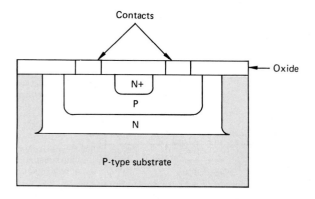

Figure 40-35   The IC resistor.

Figure 40-36 The IC capacitor.

tested, and defective circuits are marked. The wafer is then cut apart and the defective circuits are discarded. Since so many circuits are made at one time, the IC is one of the cheapest, smallest, and most reliable electronic devices produced. This fact has made pocket telephones, small computers, and similar high-quality electronic products as common and inexpensive as they are today, and it is hard even to imagine where the next step in the advance of electronics technology may lead.

(a) Basic structure

(b) Preferred structure

Figure 40-37 The IC transistor.

Summary and Review

Summary

Electronics is much too broad a subject to cover in one chapter. The intent of this unit of study was merely to describe many of the concepts important for today's industrial workers. The major ideas are summarized below:

- A *diode* is an electronic device consisting of two elements. In a vacuum tube these elements are the plate and the filament; in the semiconductor diode, the two elements are formed by a PN junction.
- A *semiconductor* is a solid or liquid material with a resistivity between that of a conductor and an insulator. An N-type semiconductor containing *donor* impurities and free electrons. A P-type semiconductor consists of *acceptor* atoms and electron holes. A *PN junction* is the union of a P-type crystal with an N-type crystal in such a way that current is conducted in only one direction.
- A *transistor* is a semiconductor device consisting of three or more electrodes, usually the emitter, the collector, and the base.
- For a common-base transistor amplifier

Emitter current = base current + collector current

$$I_e = I_b + I_c \qquad \alpha = \frac{I_c}{I_e} \qquad \text{\emph{Current Gain}}$$

- The *current gain* for other connections is

$$\beta = \frac{\alpha}{1 - \alpha} \qquad \text{\emph{Common-Emitter Amplifier}}$$

$$A_i = \frac{1}{1 - \alpha} \qquad \text{\emph{Common-Collector Amplifier}}$$

- The voltage gain A_v and the power gain G are given by

$$A_v = \frac{V_{\text{out}}}{V_{\text{in}}} \qquad \text{\emph{Voltage gain}}$$

$$G = \frac{\text{power out}}{\text{power in}} = \frac{V_{\text{out}} I_c}{V_{\text{in}} I_e}$$

$$G = \alpha \frac{V_{\text{out}}}{V_{\text{in}}} = \alpha A_v \qquad \text{\emph{Power Gain}}$$

Review Questions

40-1. In which case is there a larger current between anode and cathode in a vacuum-tube triode: when the grid is positive or negative?

40-2. Why is it necessary for vacuum tubes to be evacuated?

40-3. In terms of band theory, under what circumstances might an insulator be made to conduct? How can this be done?

40-4. In a semiconductor, why is the movement of a hole like that of a positive charge?

40-5. Why is doping necessary to make semiconductors conduct?

40-6. Explain the difference between P-type and N-type doping.

40-7. Discuss how a missing atom at a lattice site contributes to the conductivity of a semiconductor.

40-8. Which semiconductor device operates most like a check valve in a fluid system?

40-9. How could a change in the base bias be used to turn off a transistor so that no current would flow through it?

40-10. Transistors are sometimes mounted on finned metal heat sinks. Why is this necessary?

40-11. Which light source would be hotter, an LED or an incandescent light bulb? Why?

40-12. Which semiconductor device is best suited to ensuring that a specific point in a circuit is maintained at a specified voltage?

40-13. How does a photoresistor differ from a photodiode? Explain.

Problems

Section 40-10—The Transistor

40-1. Given an NPN transistor with a common-base connection and $\alpha = 0.98$, determine the base current and the collector current when the emitter current is 40 mA.

Ans. $I_b = 0.8$ mA, $I_c = 39.2$ mA

40-2. If the base current of a transistor with a common-base connection is 1.6 mA and the emitter current is 60 mA, what is α?

40-3. For a common-base amplifier, the input resistance is 800 Ω and an output resistance is 600 kΩ. (a) Determine the voltage gain if the emitter current is 12 mA and $\alpha = 0.97$. (b) What is the power gain?

Ans. (a) 727, (b) 706

40-4. The power gain is 800 for a common-base amplifier, and the voltage amplification factor is 840. Determine the collector current when the base current is 1.2 mA.

40-5. Determine the current gain in a common-emitter amplifier circuit when α is known to be 0.98.

Ans. 49

Challenge Problems

40-6. In the previous problem, what is the collector current if the emitter current is 20 μA? What is the base current?

40-7. A transistor with an effective input resistance of 400 Ω, an effective output resistance of 900 kΩ, and $\alpha = 0.96$ is connected in a common-base circuit. (a) What is the voltage gain when the input current I_e is 8 μA? (b) What is the power gain?

Ans. (a) 2160, (b) 2074

40-8. Calculate I_c and I_b for the conditions described in Prob. 40-7.

40-9. For a transistor with $I_e = 8$ μA and $\alpha = 0.97$ connected with a common emitter, calculate β, I_{in} and I_{out}.

Ans. 32.2, 0.24 μA, 7.76 μA

40-10. For a transistor with $\alpha = 0.99$ connected with a common collector, calculate the current gain A_i.

40-11. A transistor has $\beta = 99$; what is the value of α?

Ans. 0.99

40-12. For the transistor of Prob. 40-11, if $I_b = 0.1$ mA, what are the values of I_e and I_c?

Critical Thinking Questions

40-13. It is known that zener breakdown and avalanche breakdown are affected in opposite ways by temperature. From this fact, can you determine a good voltage rating for a zener diode when temperature stability is important?

40-14. A lab worker is directed to build the circuit shown in Fig. 40-2 but connects the diodes as shown in Fig. 40-38. Can you predict how the circuit will work?

Figure 40-38

40-15. Incandescent lamps have a large current flow when first turned on, and then the current drops as they heat up. Thus, they often burn out at the moment they are turned on. Is there a solution to this problem based on the information covered in this chapter?

*40-16. Incandescent lamps produce a lot of heat and are thus rather inefficient. Why can't they be replaced with LEDs that are more efficient?

*40-17. Write an equation that describes the current gain from base to emitter in terms of β and identify the amplifier configuration where it would be most useful.

*40-18. Suppose that there is an application in which the common-emitter amplifier would serve best but the phase reversal would not be acceptable. Can you think of a solution?

Appendix

SOLUTION TO QUADRATIC EQUATIONS

Often in solving physics problems, it is necessary to obtain the solution to a second-degree equation that has an unknown raised to the second power. For example, in kinematics, the position of a particle in a gravitational field varies with time according to the relation:

$$s = v_0 t + \tfrac{1}{2} a t^2$$

where s is the displacement, v_0 is the intitial velocity, a is the acceleration, and t is the time. Note that the appearance of t^2 means that there are two times at which the displacement may be the same. Such equations are called *quadratic equations*. Although there are several methods for solving these equations, perhaps the most useful technique for physics problems is to apply the quadratic formula.

> Given a quadratic equation of the form:
> $$ax^2 + bx + c = 0$$
> with a not equal to zero, the solutions are found from the quadratic formula:
> $$x = \frac{-b \pm \sqrt{b^2 - 4ac}}{2a}$$

EXAMPLE 1

Solve the following equation for x: $3x^2 = 12 + 5x$

Solution

First, we rewrite the equation in quadratic form ($ax^2 + bx + c = 0$):

$$3x^2 - 5x - 12 = 0$$

By inspection, we see that $a = 3$, $b = -5$, and $c = -12$. Now, we solve for x by substitution into the quadratic formula:

$$
\begin{aligned}
x &= \frac{b - \sqrt{b^2 - 4ac}}{2a} \\[2mm]
&= \frac{-(-5) \pm \sqrt{(-5)^2 - 4(3)(-12)}}{2(3)} \\[2mm]
&= \frac{+5 \pm \sqrt{(25) + (144)}}{2(3)} = \frac{+5 \pm \sqrt{169}}{6} = \frac{5 \pm 13}{6}
\end{aligned}
$$

To find the two solutions for x, use first the plus sign and then the minus:

First solution: $x = \dfrac{5 + 13}{6} = \dfrac{18}{6}$ or $x = +3$

Second solution: $x = \dfrac{5 - 13}{6} = \dfrac{-8}{6}$ or $x = -1.33$

The two answers are $x = +3$ and $x = -1.33$. Depending on the conditions of the problem, one of the solutions may be true mathematically but not possible physically. You should always try to interpret your results in light of the stated condtions.

EXAMPLE 2

A ball is thrown upward with an initial speed $v_0 = 20$ m/s. The acceleration due to gravity is $g = -9.8$ m/s^2. Given that the displacement $s = v_0 t + \frac{1}{2}at^2$, find the two times at which the displacement is 12 m above the release point.

Solution

First substitute the given values and write the equation with t as the unknown.

$$s = v_0 t + \tfrac{1}{2}at^2 \qquad \text{or} \qquad 12 = 20t + \tfrac{1}{2}(-9.8)t^2$$

Simplifying, and writing in quadratic form, we obtain:

$$4.9t^2 - 20t + 80 = 0$$

Now, we apply the quadratic formula to find the two solutions for a.

$$t = \frac{b - \sqrt{b^2 - 4ac}}{2a}$$

$$= \frac{-(-20) \pm \sqrt{(-20)^2 - 4(4.9)(12)}}{2(4.9)}$$

$$= \frac{+20 \pm \sqrt{400 - 235}}{9.8} = \frac{-20 \pm 12.8}{9.8}$$

Again, we find the two solutions by using first the plus and then the minus sign:

First solution: $t = \dfrac{20 + 12.8}{6} = \dfrac{32.8}{6}$ or $t = +3.35$ s

Second solution: $t = \dfrac{20 - 12.8}{9.8} = \dfrac{7.17}{9.8}$ or $t = +0.732$ s

The ball first reaches the height of 12 m at a time $t = 0.732$ s after release. It later returns to that same displacement at the time $t = 3.35$ s.

Index

A

Aberration
  chromatic, 806–807
  spherical, 758–759, 767, 806–807
Absolute humidity, 435
Absolute pressure, 333
Absolute temperature, 368–371
Absolute value, 7
Absolute zero, 369
Absorbers, ideal, 414
Absorption of photons, 855–857
Absorption spectra, 848–849
ac circuits
  impedance, 717–718
  phase diagrams, 716–717
  phase relations in, 713–714
  power factor, 719–720
  reactance, 714–715
  resonance, 719
  series, 716–718
ac generators, 689–692
Acceleration, 122–130
  angular, 244–246
  angular versus linear, 245
  average, 122–123
  centripetal, 214–216, 248
  constant, 123–125
  force and. *See* Newton's second law
  gravitational, 131–136
  linear versus angular, 245
  mass and. *See* Newton's second law
  restoring force and, 307–308
  sign convention for, 129–130
  in simple harmonic motion (SHM),
      312–314
  tangential, 248–249
  velocity and, 125–126
Acceptor atoms, 899–902
Achromatic lenses, 807
Action and reaction, law of, 74–75
Action-at-a-distance forces, 531
Action force, 74–75, 76–79
Activity (radioactive decay), 880–882
Actual mechanical advantage, 270
Addition of vectors
  component method, 63–65

Addition of vectors (*cont.*):
  parallelogram method, 55
  polygon method, 53–54
Addition rule, 7–8
Adiabatic processes, 448–450
Air
  dielectric constant, 580
  dielectric strength, 571–572, 580
  mass density, 327
  R-value, 408
  speed of sound in, 489–490
  thermal conductivity, 408
  vibrating air columns, 490–493
  weight density, 327
Algebra, 10–13
Allowed bands, 898
Alpha decay, 878–879
Alpha of transistors, 908–912
Alpha particles, 846, 877
Alternating currents, 711–714
Ammeters, dc, 672, 674–675
Ampere (unit), 593
Ampère, André Marie, 649
Ampere-hours, 633
Ampère's law, 655
Amplifiers
  common-base, 908–910
  common-emitter, 910–912
Amplitude, defined, 307
Analyzers (polarization), 825
Angle
  banking, 219–221
  critical, 781–782
  of incidence, 753–754, 773, 776–778
  phase, 716–718
  plane, 740–741
  radian (unit), 241, 742
  of reflection, 753–754, 773
  of refraction, 777–778
  shearing, 296–297
  solid, 740–742
  steradian (unit), 740–742
  of vectors, 53
Angle of repose, 91
Angles, 19
Angstrom (unit), 43, 732
Angular acceleration, 244–246

Angular deformations, 296–297
Angular displacement, 243–244
Angular momentum
  of electrons, 850–851
  mechanical, 256–259
Angular velocity, 243–244, 247–248, 310
Angular versus linear acceleration, 245
Angular versus linear velocity, 247–248
Anode of silicon-controlled rectifiers (SCRs), 914
Anode (plate) of tubes, 894–895, 896
Anodes, 628
  electrolysis and, 628–631
Antinodes in standing waves, 478, 490–493
Aphelion, 232
Apparent depth, 787–788
Archimedes, 337
Archimedes' principle, 337–341
Area expansion, 374–376
Armatures, 676, 689, 691, 692
Atmosphere (unit), 333
Atmospheric pressure, 333, 334–335
Atmospheric refraction, 785–786
Atomic mass, 427–429, 876
Atomic mass unit, 870–871
Atomic number, 866, 867, 876
Atomic spectra, 848–849, 854, 855
Atomic weights, 867, 869
Atoms
  acceptor, 899–902
  Bohr, 735, 850–853
  diameter, 865
  dipoles and, 578
  donor, 899–902
  electron energy levels in, 851–855
  electron orbits, 847–848
  electrons in, 515–516
  elements and, 866–868
  hydrogen, 848–849, 853–855
  ions of, 515, 628, 871–873
  isotopes, 871–873, 880–883
  magnetism and, 643–645
  neutrons in, 515–516
  nucleus, 515–516, 846, 865–866, 873–877

Atoms (cont.):
  protons in, 515–516, 865–866
  quantized, 735
  Rutherford, 846–848, 850
  Thomson, 845–846
Atwood machine, 160–161
Audible sound, 494–498
Avalanche (zener) effect, 905
Average acceleration, 122–123
Average speed, 121, 122
Average velocity, 122, 125
Avogadro's number, 428–429
Axis, 51
  of mirrors, 758–759
  of rotation, 101, 247
Axles, wheels and, 273–274

B

Back electromotive force (EMF)
  in motors, 693–694
Balmer, Johann Jakob, 849
Balmer series, 849, 854, 855
Band theory of conduction, 898
Banking angle, 219–221
Banking curves, 219–221
Bar magnets, 641–643
Bardeen, John, 906
Barometers, 334, 335
Base, 13
Base-collector current gain, 910–911
Base of transistors, 906, 907
Base units, 40–42
Batteries, 570
  capacity ratings, 633
  lead storage, 632–633
Beat frequencies, 499, 500
Bednorz, Georg, 606
Bel (unit), 496
Belt drives, 275–277
Bernoulli, Daniel, 345
Bernoulli's equation, 345–350
Beta-minus decay, 879
Beta-minus particles, 877
Beta of transistors, 910–911
Beta particles, 877
Beta-plus (positron) decay, 879
Beta-plus particles, 877
Bevel gears, 279
Bimetallic strips, 373–374
Binding energy, 873–877

Biological sciences, 2
Biot-Savart law, 655
Blackbody radiation, 414
Bohr, Niels, 735, 850
Bohr atom, 735, 850–853
Bohr's postulates, 850–853
Boiling, 434
Boiling point, 395, 434
Bonding, covalent, 897–898
Bore, 38
Boundary conditions, 478
Boyle, Robert, 422
Boyle's law, 421–424
Brackett series, 849, 854, 855
Brahe, Tycho, 232
Brattain, Walter Houser, 906
Breakover voltage, 914
Breeder reactors, 886
Bridge rectifiers, 903
British thermal unit (Btu), 385–387, 399
Brushes in generators, 689
Btu (British thermal unit), 385–387, 399
Bulk modulus, 298–299
Buoyant force, 337–341

C

Calipers, 45–47
Caloric theory of heat, 443–444
Caloric transfer, 384–385
Calorie (unit), 385–387, 399
Calorimeters, 392–393
Cams, 281
Candela (unit), 743
Cantoni, Marco, 606
Capacitance
  charge and, 570–573, 705–709
  defined, 571, 574
  farad (unit), 571
  potential difference and, 570–573
  voltage and, 570–573
Capacitive reactance, 715
Capacitors
  charge in, 570–573, 705–709
  current in, 705–708
  defined, 544, 570, 573
  dielectric constant and, 579–581
  dielectrics and, 577–581

Capacitors (*cont.*):
  electric fields in, 544–545
  IC, 916, 918
  in parallel, 581–585
  parallel-plate, 573–577
  phase relations in, 713–714
  potential energy in, 585
  in series, 581–585
  time constant, 706–708
Capacity rating of battery, 633
Carnot, Sadi, 454
Carnot cycle, 454
Cathode of silicon-controlled rectifiers (SCRs), 914
Cathode (filament) of tubes, 894–895, 896
Cathode-ray tubes (CRTs), 895–896
Cathodes, 628
  electrolysis and, 628–631
Celsius, Anders, 364
Celsius temperature scale, 364–367, 370–371
Center of curvature for mirrors, 758–759
Center of gravity, 111–112
Centi (c) prefix, 43
Centrifugal force, 218
Centrifugal governors, 222
Centripetal, defined, 215
Centripetal acceleration, 214–216, 248
Centripetal force, 216–219
Chain reactions, nuclear, 885
Change of phase, 394–399
Characteristic frequencies, 478–480
Charge. *See* Electric charge
Charles, Jacques, 424
Charles's law, 424, 425
Chromatic aberration, 806–807
Circuits. *See also* Capacitors; Inductors; Resistors
  ac. *See* ac circuits
  impedance, 717–718
  integrated circuits (ICs), 916–919
  Kirchhoff's laws for, 621
  parallel, 615–617
  phase relations in, 713–714
  power factor, 719–720
  reactance, 714–715

Circuits (*cont.*):
  resonance, 719
  series, 612–614, 716–718
Circular mil (unit), 603
Circular motion
  uniform, 213–234, 224
    centripetal acceleration and, 214–216
    entripetal force and, 216–219
    defined, 213–214
  in vertical circles, 223–225
Circular orbits, 228–232
Coefficient
  of area expansion, 375
  convection, 411–413
  of kinetic friction, 86
  of linear expansion, 372–374
  of performance, 460
  of static friction, 86, 91
  temperature coefficient of resistance, 604
  of volume expansion, 376
Coherent sources, 815
Cohesion of solids, 299
Coils. *See also* Inductors
  induced emf in, 709
  instantaneous current in, 692
  instantaneous emf in, 690–692
  magnetic fields of, 656
Cold fusion, 887
Collector of transistors, 906, 907
Combustion, heat of, 399
Common-base amplifiers, 908–910
Common-emitter amplifiers, 910–912
Communications system, fiber optics in, 782–784
Commutators, split-ring, 675
Completely elastic collisions, 202, 203
Completely inelastic collisions, 202, 203
Component method for adding vectors, 63–65
Components
  of force, 56–57
  of vectors, 60, 61
Compound microscopes, 804–805
Compound motors, 695

Compressibility, 298
Compression ratio, 458
Compression stroke, 456, 457
Compressions in sound waves, 487–488
Compressive stress, 291–292
Compressors, 461
Concave (converging) spherical mirrors, 758–759, 761, 762
Concurrent forces, 58, 63
Condensation, 397, 433–434, 469
Condensers, 461
Conduction
  band theory of, 898
  defined, 406
  thermal, 405–410
Conduction band, 898
Conductivity
  in liquids, 628
  of metals, 300
  thermal, 407–410
Conductors
  defined, 516, 897, 898
  examples of, 516
  magnetic force on current-carrying wires, 652–654
  superconductivity, 604–607
Conical pendulums, 221–223
Conservation
  of angular momentum, 258–259
  of charge, 882
  of electrical charge, 882–887
  of energy, 182–186, 445–446, 460
  of mass-energy, 883
  of mechanical energy, 182–186
  of momentum, 199–206, 258–259
  of nucleons, 883
  of thermal energy, 390–393
Constant acceleration, 123–125
Constant-pressure thermometers, 367, 368
Constant speed, 120–121
Constant-volume thermometers, 367–368
Constructive interference, 476, 815–818
Contact forces, 77, 530
Continuous spectra, 848–849

Control rods for reactors, 885
Convection, defined, 406
Convection, thermal, 405–406, 411–413
Convection coefficients, 411–413
Convection current, 411–412
Converging lenses, 793–794
Converging meniscus lenses, 794, 797
Converging (concave) spherical mirrors, 758–759, 761, 762
Conversion factors, 48
Converting units, 47–50
Convex (diverging) spherical mirrors, 758–759, 760, 767
Coolidge, William, 896
Copernicus, Nicolaus, 232
Core of nuclear fuel, 885
Corona discharge, 573
Corpuscular (particle) theory of light, 728–732
Cosine, 25, 26
Cosmic photons, 733, 734
Coulomb (unit), 521–522
Coulomb, Charles Augustine de, 521
Coulomb's law, 521–523, 530, 847
Covalent bonding, 897–898
Critical angle, 781–782
Critical (transition) temperature for superconductivity, 605
Critical temperature of gases, 431–432
CRTs (cathode-ray tubes), 895–896
Crystal lattice, 898–899
Curie (unit), 880–881
Current
  convection, 411–413
  electric. *See* Electric current
Current gain of transistors, 908–912
Curvature, radius of, for mirrors, 758, 763
Curves, banking, 219–221

D

dB (decibel), 496
dc ammeters, 672, 674–675
dc generators, 693
dc motors, 675–676
dc voltmeters, 672–674
de Broglie, Louis, 844
de Broglie wavelength, 844–845, 850–851
Deceleration, 124
Decibel (dB), 496
Decimal notation, 16
Deflection plates of cathode-ray tubes (CRTs), 895–896
Deformations
  angular, 296–297
  longitudinal, 293–296
Degrees, 19, 364
Density
  of fluids, 325–328
  linear, 471
  mass, 326–327
  relative, 328
  specific gravity and, 328
  weight, 326–327
Depth, apparent, 787–788
Depth gauges, 47
Derived units, 40, 41
Destructive interference, 476, 815–818
Dew point, 436
Diamagnetic materials, 648
Diamagnetism, 605
Diaphrams for lenses, 807
Dielectric constant, 579–581
Dielectric strength, 571–572, 580
Dielectrics, 577–581
Diesel engines, 458–459
Diffraction
  defined, 813–814
  introduced, 813–814
  of light, 728
  resolving power of optical instruments and, 821–824
  Young's (double-slit) experiment, 814–818
Diffraction gratings, 818–821
Diffuse reflection, 755
Diffusion, for ICs, 916
Dimensions, 49
Diodes
  applications, 902–904
  characteristics, 901–902
  defined, 901
  IC, 916, 918
Diodes (*cont.*):
  light-emitting diodes (LEDs), 912–914
  photodiodes, 913
  PN junctions as, 900–902
  vacuum tube, 894–895
  zener, 904–905
Dipoles, atomic, 578
Direction from zero, 7
Direction of vectors, 50–51
Dispersion of light, 780
Displacement, 39, 120–130
  angular, 243–244
  distance and, 121–122
  fiber optic sensors for, 784
  gravity and, 131–136
  sign convention for, 129–130
  in simple harmonic motion (SHM), 307–308
  vectors, 51–52
  work and, 172–174
Displacement vectors, 51–52
Distance, displacement and, 121–122
Diverging lenses, 794–795
Diverging meniscus lenses, 795
Diverging (convex) spherical mirrors, 758–759, 760, 767
Dividend, 9–10
Division rule, 9–10
Divisor, 9–10
Donor atoms, 899–902
Dopants in semiconductors, 899–902
Doppler effect, 500–504
Double concave lenses, 795
Double convex lenses, 794
Double-slit (Young's) experiment, 814–818
Ductility of metals, 300
Dynamic force, 56
Dynamics, defined, 3

E

Eccentricity of orbits, 232–233
Edison, Thomas, 893
Edison effect, 893
Effective current, 712
Effective electromotive force (EMF), 712
Effective voltage, 712

Efficiency
  defined, 268
  heat engines, 453–454, 455–456
  ideal engines, 455–456
  machines, 267–269, 270
  Otto cycle, 457–459
  simple machines, 267–269, 270
  transformers, 698
Einstein, Albert, 732, 735, 834–836, 840, 841, 842–844
Einstein's photoelectric equation, 842–844
Einstein's postulates of relativity, 834–836
Elastic and inelastic collisions, 202–206
Elastic bodies, 290
Elastic constants of metals, 295
Elastic limit, 292–293, 295–296
Elastic properties of matter, 290–293, 300
Elasticity
  modulus of, 293
  volume, 298–299
Electric charge
  alpha particles, 846, 877
  beta particles, 877
  capacitance and, 570–573, 705–709
  in capacitors, 705–709
  charge per unit area, 541–545
  conservation of, 882–887
  Coulomb (unit), 521–522
  Coulomb's law for, 521–523, 530, 847
  current and, 592–594
  electrification and, 512–515
  electromotive force (EMF) and, 595–597
  electrons, 562–564, 865, 866
  force per unit charge, 532–534
  Gauss's law for, 539–545
  ice-pail experiment, 543–544
  induced, 520–521
  kinetic energy and, 551, 564
  like, 514–516
  magnetic force on, 649–652
  Millikan's oil-drop experiment and, 562–564
  motion of, 592–594
  negative, 514–515, 557–559, 560

Electric charge (*cont.*):
  neutrons, 866
  positive, 514–515, 557–559
  positrons, 877
  potential and, 556–560, 564
  potential energy and, 551–560, 564
  protons, 866
  redistribution of, 519
  smallest unit of, 562–564
  transfer of, 515–516, 516
  unlike, 514–516
  work and, 560
  work per unit charge, 556–560, 564
Electric charge density, 541–545
Electric current
  alternating, 711–714
  ampere (unit), 593
  in capacitors, 705–708
  charge and, 592–594
  defined, 592–593
  direction of, 594–595
  effective, 712
  in inductors, 709–711
  instantaneous, in coils, 692
  Kirchhoff's first law for, 621–626
  magnetic fields and, 648–649
  magnetic force and, 652–654, 667–668
  magnetic torque and, 668–670
  Ohm's law and, 597–599
  power and, 601
  reverse, in zener diodes, 904–905
  reversing, through a source of emf, 620–621
  transient, 593, 595
Electric field intensity, 532–537
Electric field lines, 538–539, 557, 645–646
Electric fields
  in capacitors, 544–545
  defined, 531–532
  dielectric strength and, 571–572, 579–581
  equipotential lines for, 557
  Gauss's law for, 539–545
  potential gradient, 562
  resultant, 535–537
  vector sum of, 535–537

Electric fields (*cont.*):
  work done by, 560
Electric forces, Coulomb's law and, 521–523, 530, 847
Electric kinetic energy, 551, 564
Electric potential, 556–560, 564. *See also* Voltage
Electric potential difference, 560. *See also* Voltage
  capacitance and, 570–573
  terminal, 618–619
Electric potential energy, 551–560, 564
Electric power, 600–601
Electrification, 512–515
Electrolysis, 628–631
Electrolytes
  defined, 628
  electrolysis and, 628–631
Electromagnetic induction
  defined, 682
  Faraday's law and, 682–686
  Fleming's (right-hand) rule and, 688
  introduced, 682
  Lenz's law and, 687–688
Electromagnetic waves, 732–734
Electromotive force (EMF)
  back, in motors, 693–694
  effective, 712
  electric charge and, 595–597
  induced, 682–687, 709
  instantaneous, in loops, 690–692
  introduced, 595–597
  Kirchhoff's second law for, 621–626
  Ohm's law and, 597–599
  reversing current through a source of, 620–621
  terminal potential difference and, 618–619
  volt (unit), 556, 564
Electron gun of cathode-ray tubes (CRTs), 895–896
Electron orbits, 847–848, 851
Electrons
  angular momentum, 850–851
  in atoms, 515–516, 735
  charge, 562–564, 865, 866
  defined, 515
  discrete atomic energy levels of, 735

Electrons (*cont.*):
  energy levels in atoms, 851–855
  free, 516, 897
  kinetic energy, 843–844
  mass, 865, 866
  spin of, 643–645
  symbol, 866
  valence, 897
Electronvolt (unit), 564, 853
Electroplating, 631
Electroscopes
  defined, 512
  gold-leaf, 512, 517–519, 520–521
  pith-ball, 512–515
Electrostatics, defined, 512
Elements
  atomic mass, 427–429, 876
  atomic number, 866, 867, 876
  atomic weights, 867, 869
  atoms and, 866–868
  mass number, 866–868, 876
  periodic table of, 867
  symbols, 867
Elliptical orbits, 232–233
EMF. *See* Electromotive force (EMF)
Emissivity, 414–416
Emitter of transistors, 906, 907
Energy
  binding, 873–877
  conservation of, 182–186, 445–446, 460
  conservation of mass-energy, 883
  conservation of mechanical, 182–186
  defined, 172
  electromotive force (EMF) and, 595–597
  electrons in atoms, 515–516, 735
  electronvolt (unit), 564, 853
  friction forces and, 184–186
  internal energy function, 444–445
  introduced, 176–177
  kinetic. *See* Kinetic energy
  light photons, 734–735
  mass and, 841–842
  of periodic waves, 474–475

Energy (*cont.*):
  photons, 842–843, 844
  potential. *See* Potential energy
  quantum states of hydrogen atoms, 853–855
  thermal
    conservation of, 390–393
    emperature and, 360–367
  work-function, 842–843
Energy-bond theory, 898
Energy levels of electrons in atoms, 851–855
Engines
  Carnot, 454
  diesel, 458–459
  gasoline, 457–458
  heat, 452–459
  ideal, 455–456
  internal combustion, 456–459
  refrigeration, 459–461
Equilibrant force, 76
Equilibrium
  conditions for, 100–101
  defined, 73
  rotational, 107–110
  thermal, 360–362
  thermodynamic, 444–445
  translational, 75–76, 80–84, 107
Equipotential lines, 557
Etching for ICs, 916
Ether, light-carrying, 531, 730–732
Evaporation, 432, 434
Excited states of hydrogen atoms, 853–855
Exhaust stroke, 457
Expansion
  area, 374–376
  linear, 371–374
  volume, 376–378
  of water, 378–379
Expansion ratio, 459
Exponents, 9, 13–15
Eyepieces, 803–806

F

Factor, 9
Fahrenheit, Gabriel Daniel, 364
Fahrenheit temperature scale, 364–367, 370–371

Farad (unit), 571
Faraday, Michael, 538, 649
Faraday's ice-pail experiment, 543–544
Faraday's law, 682–686
Fast neutrons, 883–885, 886
Femto (f) prefix, 865
Ferromagnetic materials, 648
Fertile materials, 886
Fiber optics, 782–785
Field forces, 77
Field intensity
  electric, 532–537
  magnetic, 647
Field lines
  electric, 538–539, 557, 645–646
  magnetic, 643
Field magnets, 689
Fields
  concept of, 531–534
  electric. *See* Electric fields
  gravitational, 227–228, 530, 531–533
  magnetic. *See* Magnetic fields
Filament (cathode) of tubes, 894–895, 896
Filters for rectifiers, 903
Final velocity, 124–125, 126
First law of motion, 73–74
First law of thermodynamics, 445–446, 448, 460
First-order images, 819–821
Fission, nuclear, 883–886
Fizeau, Armand Hippolyte Louis, 736–737
Fizeau's apparatus, 736–737
Fleming's (right-hand) rule, 688
Flow rate, 341–343
Fluid flow, 341–343
Fluid pressure
  hydraulic presses and, 335–337
  introduced, 329–333
  measuring, 333–335
  Pascal's law and, 333–337
  velocity and, 343–345
Fluids
  Archimedes' principle for, 337–341
  barometers and, 334, 335
  Bernoulli's equation for, 345–350
  bulk moduli, 299

Fluids (*cont.*):
buoyant force, 337–341
density, 325–328
electric conduction in, 628
electrolysis and, 628–631
fiber optic sensors for, 784
gases turning to, 430–432
heat of fusion and heat of
vaporization, 395
index of refraction, 774
manometers and, 333–335
mass density, 327
normal boiling point, 434
pressure
hydraulic presses and, 335–
337
introduced, 329–333
measuring, 333–335
Pascal's law and, 333–337
and velocity, 343–345
R-values, 408
specific heat capacity, 390
speed of sound in, 489
thermal conductivities, 408
Torricelli's theorem for, 347–
348
venturi meters, 344, 348–349
volume expansion of, 376–377
weight density, 327
Flux
luminous, 740–743, 745
magnetic, induced emf and,
682–687
per unit solid angle, 743–744
radiant, 740
Flux density, magnetic, 645–648,
651–652, 658–659
Flux lines, magnetic, 643
Focal length
lenses, 795–797, 800–802
mirrors, 759
Focal point of mirrors, 759
Focal points of lenses, 795–797
Foot-pound (unit), 174
Foot-pound per second (unit),
187
Footcandle (unit), 745
Force
acceleration and. *See* Newton's
second law
action, 74–75, 76–79
action-at-a-distance, 531

Force (*cont.*):
buoyant, 337–341
centrifugal, 218
centripetal, 216–219
components of, 56–57
concurrent, 58, 63
contact, 77, 530
defined, 56
dynamic, 56
electric, Coulomb's law and,
521–523, 530, 847
electromotive (EMF). *See*
Electromotive force
(EMF)
equilibrant, 76
field, 77
force per unit charge, 532–534
of friction, energy and, 184–
186
of kinetic friction, 84–91
line of action of, 100–101
magnetic
on current-carrying wires,
652–654
law of, 642
on loops, 667–668
on moving charges, 649–652
mass and. *See* Newton's second
law
moment arm of, 101–102
moment of, 103–105
newton (unit), 56–57, 151, 152
normal, 84–91
nuclear, 866
per unit area. *See* Pressure
reaction, 74–75, 76–79
restoring, 306–308
resultant, 58, 62–63, 223
static, 56
of static friction, 84–86, 91
torque and, 103–107
vectors and, 56–57
work and, 172–174
Forced convection, 412–413
Forced vibrations, 493
Formulas, 10
Forward-biased PN junctions,
901
Foucault, Jean, 737
Franklin, Benjamin, 514–515
Free-body diagrams, 76–79
Free electrons, 516, 897

Free-falling bodies, 131–136
Free space
permeability of, 647–648
permittivity of, 540–541, 576–
580
speed of light in, 42, 732–734,
736–738, 834–836
Freezing (solidification), 397
Frequency
alternating currents, 711–714
beat, 499, 500
characteristic, 478–480
defined, 308
fundamental, 479, 491, 492
harmonic, 479, 490–493
overtone, 479, 491–493
of photons, 735, 852–853
resonant, 719
of rotation, 216
in simple harmonic motion
(SHM), 314–316
threshold, 843
wavelength and, 472–473, 732–
734
Friction
defined, 73
introduced, 84
kinetic
coefficient of, 86
force of, 84–91
static
coefficient of, 86, 91
force of, 84–86, 91
Friction force, energy and, 184–
186
Fulcrum, 271, 272
Full-scale deflection of meters,
672
Full-wave rectifiers, 903–904
Fundamental frequency, 479, 491,
492
Fundamental quantities, 40
Fusion, 394
latent heat of, 394–399
nuclear, 886–887
solid to liquid, 394
Fusion curve, 434–435

G

Galilei, Galileo, 131, 736
Gallium arsenide, 517

Galvanometers, 670–672, 674
Gamma decay, 879
Gamma rays, 733, 734, 877
Gas thermometers, 367–368
Gaseous phase, 394–399
Gases
    Boyle's law, 421–424
    Charles's law, 424, 425
    critical temperature, 431–432
    Gay-Lussac's law, 424–425
    general gas laws, 425–427
    heats of fusion and vaporiza-
        tion, 395
    ideal, 421, 429–430
    ideal gas law, 421, 429–430
    index of refraction, 774
    liquefaction of, 430–432
    mass density, 327
    R-values, 408
    speed of sound in, 489–490
    thermal conductivities, 408
    universal gas constant, 429–
        430
    vapor pressure and, 433–434,
        435–437
    vaporization and, 395–399,
        432, 434, 461
    weight density, 327
Gasoline engines, 457–458
Gate of silicon-controlled recti-
    fiers (SCRs), 914
Gauge pressure, 333
Gauss (unit), 646
Gaussian surfaces, 542–543
Gauss's law, 539–545
Gay-Lussac's law, 424–425
Gears, 277–278, 279
Generators
    ac, 689–692
    dc, 693
Geometrical optics. See Ray trac-
    ing
Geometry, 19–22
Geosynchronous orbits, 230–232
Germanium, 517
Giga (G) prefix, 43
Gold-leaf electroscopes, 512, 517–
    519, 520–521
Governors, centrifugal, 222
Graphs, 17–18
Gravitation, universal, 225–228,
    530

Gravitational acceleration, 131–136
Gravitational fields, 111, 227–
    228, 530, 531–533
Gravitational potential energy,
    176, 180–182
Gravity
    center of, 111–112
    specific, 328
    uniformly accelerated motion
        and, 131–136
    universal constant values, 225–
        228
Grid of tubes, 894–895, 895–896
Ground, 582
Ground state of hydrogen atoms,
    853–855
Guo, J. D., 606
Gyration, radius of, 251

H

Half-life, 880–882
Half-wave rectifiers, 902
Hardness of metals, 300
Harmonic frequencies, 479, 490–
    493
Hearing threshold, 495
Heat. See also Thermodynamics
    caloric theory of, 443–444
    change of phase and, 394–399
    of combustion, 399
    of condensation, 397
    defined, 362, 384
    latent heat of fusion, 394–399
    latent heat of vaporization,
        395–399, 461
    measurement of, 390–393
    methods of transfer
        introduced, 406
        thermal conduction, 405–410
        thermal convection, 405–
            406, 411–413
        thermal radiation, 405–406,
            413–416
    Prevost's law of heat exchange,
        415
    quantity of, 385–387, 399
    of solidification, 397
    specific heat capacity, 387–390
    of sublimation, 397
    thermal conductivity, 407–410
    work and, 443–444

Heat capacity, 388
Heat engines, 452–459
Heater filament of tubes, 894
Helical gears, 279
Helium-neon lasers, 857
Henry (unit), 709
Herschel, William, 734
Hertz (unit), 308, 472
Hertz, Heinrich Rudolph, 731, 734
High-temperature superconduc-
    tors, 607
Holes in semiconductors, 900
Hooke, Robert, 290
Hooke's law, 290–293, 296, 307
Horizontal projection, 137–143
Horsepower (unit), 187–188
Humidity
    absolute, 435
    relative, 435–436
Huygens, Christian, 728–732
Huygens's principle, 728–729,
    738–739
Hydraulic presses, 335–337
Hydrogen atoms, 848–849, 853–
    855
Hypotenuse, 21, 23, 26
Hypotheses, workable, 3–4
Hysteresis, 657–659
Hysteresis loops, 659

I

Ice
    index of refraction, 774
    specific heat capacity, 390
Ice-pail experiment, 543–544
Ice point (lower fixed point), 363
ICs (integrated circuits), 916–919
Ideal absorbers, 414
Ideal engines, 455–456
Ideal gas law, 421, 429–430
Ideal gases, 421, 429–430
Ideal mechanical advantage, 270
    belt drives, 275–277
    gears, 278
    hydraulic presses, 336–337
    inclined planes, 280–281
    levers, 271–272
    pulleys, 273–275
    screw jacks, 282
    wedges, 281
    wheels and axles, 273

Ideal radiators, 414
Illumination, 744–747
Image distance
    lenses, 800–802
    mirrors, 756, 762, 763
Image size
    lenses, 801
    mirrors, 763
Images
    formed by spherical mirrors,
        760–762
    formed by thin lenses, 798–800
    real, 756–757
    virtual, 757
Impedance of ac circuits, 717–718
Impulse
    defined, 197
    momentum and, 196–198
    newton-second (unit), 197
    pound-second (unit), 197
Impurities in semiconductors,
    899–900
In phase, defined, 472
In phase sources, 815
Incidence, angle of, 753–754, 773,
    776–778
Incident ray, 753–754, 775–778
Inclined planes, 278–282
Index of refraction, 774–775
Induced back electromotive force
    (EMF) in coils, 709
Induced charge, 520–521
Induced electromotive force
    (EMF), 682–687, 709
Induction
    electromagnetic. See
        Electromagnetic induction
    magnetic, 645, 647–648
Inductive reactance, 714–715
Inductors. See also Coils
    current in, 709–711
    induced emf in, 682–687, 709
    phase relations in, 713–714
    time constant, 710–711
Inelastic and elastic collisions,
    202–206
Inelastic bodies, 290
Inertia
    defined, 74, 131
    law of, 74
    moment of, 250–251
Infrared waves, 727, 733, 734

Infrasonic waves, 494
Instantaneous current in coils,
    692
Instantaneous electromotive force
    (EMF) in loops, 690–692
Instantaneous speed, 122
Instantaneous velocity, 122
Instruments, measuring, 45–47
Insulation R-values, 408, 410–411
Insulators
    defined, 516, 897, 898
    dielectric constant and dielec-
        tric strength, 580
    examples of, 516
Intake stroke, 456, 457
Integrated circuits (ICs), 916–919
Intensity
    electric field, 532–537
    luminous, 743–744
    magnetic field, 647
    sound, 494–498
Intensity level, sound, 495–497
Interference
    constructive, 476, 815–818
    destructive, 476, 815–818
    sound, 499
Internal combustion engines,
    456–459
Internal energy function, 444–
    445
Internal resistance, 618–619
International Bureau of Weights
    and Measures, 40–42
Inverse-square law for illumina-
    tion, 747
Inverters, signal, 911
Ions, 515, 628, 871–873
Isochoric processes, 450–451
Isothermal processes, 451–452
Isotopes, 871–873
    half-life, 880–882
Isotropic sources, 743–744

J

Joule (unit), 174, 385, 388, 389
Joule, Sir James Prescott, 385, 386
Junctions, defined, 622

K

Kamerlingh-Onnes, Heike, 604

Kelvin (unit), 389
Kelvin, Lord, 369
Kelvin temperature scale, 370–
    371
Kepler, Johannes, 232
Kepler's first law, 232
Kepler's second law, 232–233
Kepler's third law, 233
Kilo (k) prefix, 43
Kilocalorie (unit), 385–387
Kilogram (unit), 40–42, 379, 389
Kilogram-meter per second
    (unit), 197
Kilowatt (unit), 187–188
Kinetic energy
    electric, 551, 564
    electron orbits, 851
    electrons, 843–844
    of fluids, 345–346
    mechanical, 176, 177–179
    relativistic, 841–842
    rotational, 249–251
Kinetic friction
    coefficient of, 86
    force of, 84–91
Kirchhoff, Gustav, 621
Kirchhoff's laws, 621–626

L

Laser enhanced transluminal
    angioplasty (LASTAC),
    785
Lasers, 855–857
LASTAC (laser enhanced translu-
    minal angioplasty), 785
Latent heat of fusion, 394–399
Latent heat of vaporization, 395–
    399, 461
Lattice, crystal, 898–899
Law of action and reaction, 74–75
Law of areas, 232–233
Law of gravitation, universal,
    225–228
Law of inertia, 74
Law of magnetic force, 642
Law of orbits, 232–233
Law of periods, 233
Laws of motion
    Newton's first law, 73–74
    Newton's second law. See
        Newton's second law

Laws of motion (*cont.*):
Newton's third law, 74–75
Laws of reflection, 753–755
Laws of refraction, 775–778
Lead-acid storage batteries, 632–633
Leap seconds, 43
LEDs (light-emitting diodes), 912–914
Length
meter (unit), 40–42
mil (unit), 603
relativistic contraction, 838–839
Lens equation, 800–802
Lenses
achromatic, 807
chromatic aberration in, 806–807
combinations of, 803–806
converging, 793–794
converging meniscus, 794, 797
defined, 793
diaphragms for, 807
diverging, 794–795
diverging meniscus, 795
double concave, 795
double convex, 794
eyepieces, 803–806
focal length, 795–797, 800–802
focal points, 795–797
image distance, 800–802
image formation by thin lenses, 798–800
image size, 801
lens equation, 800–802
lensmaker's equation, 797
magnification, 800–802, 803–804
object distance, 800–802
object size, 801
objective, 804–805, 806
plano-concave, 795
plano-convex, 794, 795
resolving power of, 821–824
simple, 793–795
spherical aberration in, 806–807
thin, 795, 798–800
Lensmaker's equation, 797
Lenz's law, 687–688
Levers, 271–272

Light
defined, 728
diffraction, 728
dispersion of, 780
dualistic nature of, 735
electromagnetic spectrum and, 732–734
Huygens's principle, 728–729, 738–739
illumination, 744–747
isotropic sources, 743–744
lasers and, 855–857
lumen (unit), 740–741, 742–743
luminous flux, 740–743, 745
luminous intensity, 743–744
particle (corpuscular) theory, 728–732
photons, 734–735
point source of, 738–739
propagation of, 730–732
quantum theory, 734–735
rays, 738–739
rectilinear propagation, 728
reflection, 728–732
refraction, 728–732
shadows and, 738–739
spectrum, 780
speed of, 42, 732–734, 736–738, 834–836
visible, 732–733, 734, 780
wave theory, 729–732
Light-carrying ether, 531, 730–732
Light-emitting diodes (LEDs), 912–914
Like charges, 514–516
Limiting angle, 91
Line emission spectra, 848–849, 854, 855
Line of action of force, 100–101
Linear aperture of mirrors, 758
Linear density, 471
Linear expansion, 371–374
Linear media, 476
Linear speed, 216
Linear velocity, 247–248
Linear versus angular acceleration, 245
Linear versus angular velocity, 247–248
Linear versus rotational motion, 247–249

Liquefaction of gases, 430–432
Liquid phase, 394–399
Liquid storage tanks, 461
Liquids. *See* Fluids
Long radio waves, 733, 734
Longitudinal deformations, 293–296
Longitudinal strain, 293–296
Longitudinal stress, 293–296
Longitudinal waves, 469–470, 486–487, 488–489. *See also* Sound
Looming objects, 785
Loops
instantaneous emf in, 690–692
magnetic fields of, 655–656, 659
magnetic force on, 667–668
magnetic torque on, 668–670
Loudness, 498
Lower fixed point (ice point), 363
Lumen (unit), 740–741, 742–743
Luminous flux, 740–743, 745
Luminous intensity, 743–744
Lux (unit), 745–746
Lyman series, 849, 854, 855

M

Machines
efficiency of, 267–269, 270
mechanical advantage, 269–270
simple. *See* Simple machines
MAGLEV (magnetic levitation) vehicles, 607
Magnetic domains, 645
Magnetic field intensity, 647
Magnetic field lines, 643
Magnetic fields
of coils, 656
current and, 648–649
force, on moving charges, 649–652
introduced, 643
of loops, 655–656, 659
of solenoids, 656–657
of toroids, 657
of wires, 654–655
Magnetic flux, induced emf and, 682–687
Magnetic flux density, 645–648, 651–652, 658–659

Magnetic flux lines, 643
Magnetic force
    on current-carrying wires,
        652–654
    law of, 642
    on loops, 667–668
    on moving charges, 649–652
Magnetic induction, 645, 647–
    648
Magnetic levitation (MAGLEV)
    vehicles, 607
Magnetic poles, 641–643
Magnetic saturation, 645, 658–
    659
Magnetic torque
    on loops, 668–670
    on solenoids, 670
Magnetism
    diamagnetism, 605
    hysteresis and, 657–659
    introduced, 641–643
    modern theory of, 643–645
    residual, 658–659
    superconductivity and, 604–
        607
Magnets
    bar, 641–643
    defined, 641
    field, 689
    natural, 641
    superconducting, 606–607
Magnification
    of lenses, 800–802, 803–804
    of mirrors, 765
Magnitude, 7
    of physical quantities, 39
    of vectors, 51–53
Malleability of metals, 300
Manometers, 333–335
Masks for ICs, 917
Mass
    acceleration and. See Newton's
        second law
    alpha particles, 877
    atomic, 427–429, 876
    beta particles, 877
    electrons, 865, 866
    energy and, 841–842
    force and. See Newton's second
        law
    kilogram (unit), 40–42, 379,
        389

Mass (cont.):
    molecular, 427–429
    neutrons, 866
    newton (unit), 56–57, 151, 152
    positrons, 877
    protons, 865, 866
    relativistic, 837–838, 840–841
    rest, 841–842
    slug (unit), 153
    wavelength and, 844–845
    weight and, 153–155
Mass defect, 874
Mass density, 326–327
Mass-energy, conservation of, 883
Mass number, 866–868, 876, 877
Mass spectrometers, 848, 871–
    873
Mathematics, 6–29
    algebra, 10–13
    exponents, 9, 13–15
    geometry, 19–22
    graphs, 17–18
    physics and, 3–4
    radicals, 15
    roots, 15
    scientific notation, 16–17
    signed numbers, 7–10
    trigonometry, 23–26
Maxwell, James Clerk, 731, 734
Measuring instruments, 45–47
Mechanical advantage, 269–270
Mechanical energy, conservation
    of, 182–186
Mechanical waves, 468–469, 486–
    487. See also Sound
Medicine, fiber optics and, 784–
    785
Mega (M) prefix, 43
Megaelectronvolt (unit), 853–
    854, 871
Meissner-Ochsenfeld effect, 605
Melting point, 394
Mercury-in-glass thermometers,
    364–365, 367
Metals
    bimetallic strips, 373–374
    conductivity, 300
    as conductors, 516
    ductility, 300
    elastic constants, 295
    electroplating, 631
    hardness, 300

Metals (cont.):
    heat of fusion and heat of
        vaporization, 395
    malleability, 300
    mass density, 327
    R-values, 408
    resistivities, 602
    specific heat capacity, 390
    thermal conductivities, 408
    weight density, 327
Meter (unit), 40–42
Meters
    dc ammeters, 672, 674–675
    dc voltmeters, 672–674
    full-scale deflection, 672
    galvanometers, 670–672, 674
Metric system. See Système
    International d'Unités (SI)
Michelson, Albert Abraham, 731,
    737
Michelson-Morley experiment,
    731–732
Micro (μ) prefix, 43
Micrometer calipers, 45–47
Micrometer depth gauges, 47
Microscopes, compound, 804–805
Mil (unit), 603
Milli (m) prefix, 43
Millikan, Robert A., 562
Millikan's oil-drop experiment,
    562–564
Millimicron (unit), 732
Mirages, 785
Mirrors
    axis of, 758–759
    center of curvature, 758–759
    defined, 755
    focal length, 759
    focal point, 759
    image distance, 756, 762, 763
    image size, 763
    linear aperture of, 758
    magnification, 765
    object distance, 756, 762, 763
    object size, 763
    parabolic, 767
    plane, 755–757
    radius of curvature, 758, 763
    spherical. See Spherical mirrors
    vertex of, 758–759
Moderators for reactors, 885
Modes of oscillation, 478–480

Modulus
  bulk, 298–299
  of elasticity, 293
  of rigidity, 297
  shear, 296–297
Mole (quantity), 428–429
Molecular mass, 427–429
Moment arm of force, 101–102
Moment of force, 103–105
Moment of inertia, 250–251
Momentum
  angular, 256–259
    of electrons, 850–851
  conservation of, 199–206, 258–259
  defined, 197
  elastic and inelastic collisions, 202–206
  impulse and, 196–198
  kilogram-meter per second (unit), 197
  rest mass and, 841–842
  slug-foot per second (unit), 197
Motors
  back emf in, 693–694
  compound, 695
  dc, 675–676
  permanent-magnet, 696
  series-wound, 695
  shunt-wound, 695–696
  types of, 695–696
Müller, Karl Alex, 606
Multiplication rule, 9
Multiplier resistance, 672–674

N

N-type semiconductors, 899–900
Nano (n) prefix, 43
Nanometer (unit), 732
National Bureau of Standards, 43
Natural convection, 412–413
Natural magnets, 641
Negative charges, 514–515, 557–559, 560
Negative numbers, 7–10
Net (resultant) work, 174–176
Net work on fluids, 345–346
Neutrinos, 879
Neutrons
  in atoms, 515–516

Neutrons (cont.):
  charge, 866
  fast, 883–885, 886
  mass, 866
  slow, 883–885, 886
  symbol, 866
Newton (unit), 56–57, 151, 152
Newton, Sir Isaac, 728, 730, 733
Newton-meter (unit), 103
Newton-second (unit), 197
Newton's first law, 73–74
Newton's law of universal gravitation, 225–228, 530
Newton's second law
  applications, single-body problems, 155–164
  introduced, 150–153
  in rotation, 251–254
Newton's third law, 74–75
Newton's universal law of gravitation, 225–228, 530
Nodes in standing waves, 478, 490–493
Noncurrent forces, 100
Normal boiling point, 434
Normal force, 84–91
NPN transistors, 906–908
Nuclear chain reactions, 885
Nuclear fission, 883–886
Nuclear force, 866
Nuclear fusion, 886–887
Nuclear reactions, 882–887
Nuclear reactors, 885–886
Nucleons, conservation of, 883
Nucleus, 515–516, 846, 865–866, 873–877
Nuclides, 875

O

Object distance
  lenses, 800–802
  mirrors, 756, 762, 763
Object size
  lenses, 801
  mirrors, 763
Objective lenses, 804–805, 806
Oculars, 806
Oersted, Hans, 648, 649
Oersted's experiment, 648–649
Ohm (unit), 598
Ohm, Georg Simon, 598, 601

Ohm's law, 597–599
Opaque objects, 739
Optical fibers, 782–785
Optical instruments
  compound microscopes, 804–805
  resolving power of, 821–824
  telescopes, 806, 823–824
Orbits
  circular, 228–232
  eccentricity of, 232–233
  electron, 847–848
  elliptical, 232–233
  geosynchronous, 230–232
Oscillating fields, 825
Oscillation, modes of, 478–480
Ott, Hans, 606
Otto cycle, 457–459
Overtone frequencies, 479, 491–493
Oxidation, 629–630

P

P-type semiconductors, 899–900
P-V diagrams, 446–448
Pain threshold (sound), 495
Parabolic mirrors, 767
Parallel, 20
Parallel capacitors, 581–585
Parallel-plate capacitors, 573–577
Parallel resistors, 615–617
Parallelogram method for adding vectors, 55
Paramagnetic materials, 648
Particle (corpuscular) theory of light, 728–732
Particles and waves, 844–845
Pascal (unit), 294, 328
Pascal, Blaise, 333
Pascal's law, 333–337
Paschen series, 849, 854, 855
Pendulums
  conical, 221–223
  simple, 316–318
  torsion, 318–319
Penumbra, 739
Performance, coefficient of, 460
Perihelion, 232
Period
  defined, 308
  law of periods, 233

Period (*cont.*):
    of revolution, 216
    in simple harmonic motion
        (SHM), 314–316
    of simple pendulums, 317
    of torsion pendulums, 319
    wavelength and, 473
Periodic motion, 306–308. *See
    also* Simple harmonic
    motion (SHM)
Periodic table of elements, 867
Periodic waves
    energy, 474–475
    motion, 472–473
    power, 475
Permanent-magnet motors, 696
Permanently magnetized materi-
    als, 645
Permeability, 647–648
Permittivity, 540–541, 576–580
Perpendicular, 19–20
Phase angle, 716–718
Phase change, 394–399
Phase diagrams, 435, 716–717
Phase relations in ac circuits,
    713–714
Phosphors, 895
Photodiodes, 913
Photoelectric effect, 734, 842–844
Photons
    absorption of, 855–857
    cosmic, 733, 734
    energy of, 842–843, 844
    frequency of, 735, 852–853
    light, 734–735
    spontaneous emission of, 855–
        857
    stimulated emission of, 855–
        857
    wavelength of, 844–845, 852–
        853
Photoresist for ICs, 916
Photoresistors, 914–915
Physical quantities, 38–39
Physical sciences, 2–3
Physics
    defined, 2
    mathematics and, 3–4
    studying, 4–5
Pico (p) prefix, 43, 865
Pipes, sound in, 490–493
Piston displacement, 39

Pitch
    of screws, 282
    of sound, 498–499
Pith-ball electroscopes, 512–515
Planck, Maxwell, 734–735, 842
Planck's constant, 735, 842
Planck's equation, 735, 842–844
Plane angle, 740–741
Plane mirrors, 755–757
Plane-polarized waves, 825
Planetary gears, 279
Planetary motion, 232–233
Plano-concave lenses, 795
Plano-convex lenses, 794, 795
Plate (anode) of tubes, 894–895,
    896
PN junctions, 900–902
PNP transistors, 906, 907
Point source of light, 738–739
Polarization of waves, 824–826
Polarizers, 825
Poles, magnetic, 641–643
Polygon method for adding vec-
    tors, 53–54
Positive charges, 514–515, 557–
    559
Positron (beta-plus) decay, 879
Positrons, 877
Potential, electric, 556–560, 564.
    See also Voltage
Potential difference, electric, 560,
    570–573. *See also* Voltage
Potential energy
    in capacitors, 585
    electric, 551–560, 564
    electron orbits, 851
    of fluids, 345–346
    mechanical, 176, 180–182
Potential gradient, 562
Pound (unit), 56–57
Pound-second (unit), 197
Power
    defined, 172, 187
    electric, 600–601
    foot-pound per second (unit),
        187
    of optical instruments. *See*
        Magnification
    of periodic waves, 475
    radiant, 740
    resolving, 821–824
    rotational, and work, 254–256

Power (*cont.*):
    watt (unit), 187, 740
    work and, 187–188, 254–256
Power factor, 719–720
Power gain of transistors, 909
Power loss, 600
Power stroke, 456–457
Prefixes for units, 43–44, 865
Pressure
    absolute, 333
    atmospheric, 333, 334–335
    defined, 328
    fluid
        hydraulic presses and, 335–
            337
        introduced, 329–333
        measuring, 333–335
        Pascal's law and, 333–337
        and velocity, 343–345
    gauge, 333
    introduced, 328
    vapor, 433–434, 435–437
Prevost's law of heat exchange,
    415
Principal quantum number, 850–
    851
Prisms, 780, 781, 793–794
Problem-solving strategies
    acceleration, constant, 127
    Bernoulli's equation, 349
    calorimetry, 399
    circular motion, uniform, 224
    collisions, conservation of
        momentum, 204
    component method of vector
        addition, 64
    conservation
        of energy, 186
        of momentum, 204
    constant acceleration, 127
    converting units, 48
    Coulomb's law, 523–525
    electric fields, 538
    electric forces and Coulomb's
        law, 523–525
    electric potential and potential
        energy, 561
    energy
        conservation of, 186
        potential, and electric poten-
            tial, 561
    equilibrium

Problem-solving strategies (*cont.*):
  rotational, 107
  translational, 80
  expansion and temperature, 378
  fields, electric, 538
  fluid pressure, 333
  fluids
    Bernoulli's equation, 349
    rate of flow problems, 343
    at rest, 338–339
  forces, electric, and Coulomb's law, 523–525
  free-body diagrams, 77
  friction, 87
  heat, quantity of, 399
  laws of motion, Newton's second law, 158
  lenses and optical instruments, 803
  mirrors and reflection, 766
  momentum, conservation of, 204
  Newton's second law, 158
  optical instruments and lenses, 803
  polygon method of vector addition, 54
  potential, electric, and potential energy, 561
  potential energy and electric potential, 561
  projectile motion, 141
  quantity of heat, 399
  rate of flow problems, 343
  reflection and mirrors, 766
  resultant torque, 106
  rotation of rigid bodies, 257
  rotational equilibrium, 107
  second law of motion, 158
  temperature and expansion, 378
  torque, resultant, 106
  translational equilibrium, 80
  trigonometry, 27
  uniform circular motion, 224
  uniformly accelerated motion, 127
  units, converting, 48
  vector addition
    component method, 64
    polygon method, 54

Product, 9
Projectiles
  components of motion, 136–143
  defined, 137
  introduced, 136–137
Propagation
  of light, 730–732
  rectilinear, 728
Proper time interval, 837–838
Protons, 515–516, 865
  charge, 866
  mass, 866
  symbol, 866
Ptolemy, Claudius, 232
Pulleys, 273–275
Pythagorean theorem, 22, 23–24

Q

Quality of sound, 499
Quantity of heat, 385–387, 399
Quantum number, principal, 850–851
Quantum states of hydrogen atoms, 853–855
Quantum theory, 734–735, 842–844
Quotient, 9–10

R

R-values, 408, 410–411
Radial magnetic field, 671
Radian (unit), 241, 742
Radiant flux, 740
Radiant power, 740
Radiation
  blackbody, 414
  defined, 406
  electromagnetic, 730–732
  thermal, 405–406, 413–416
  visible. *See* Light
Radiators, ideal, 414
Radicals, 15
Radio waves, 733, 734
Radioactive decay, 878–882
Radioactivity, 877–882
Radius
  of curvature, for mirrors, 758, 763
  of electron orbits, 847–848

Radius (*cont.*):
  of gyration, 251
Rankine temperature scale, 370–371
Rarefactions in sound waves, 470, 487–488
Rate of flow, 341–343
Rate of radiation, 414
Ray tracing
  for optical fibers, 783
  for spherical mirrors, 760–762
  for thin lenses, 798–800
Rays. *See also* Waves
  gamma, 733, 734, 877
  incident, 754–755, 775–778
  light, 738–739
  reflected, 754–755
  refracted, 775–778
  X-rays, 733, 734
Reactance
  of ac circuits, 714–715
  capacitive, 715
  inductive, 714–715
Reaction force, 74–75, 76–79
Reactors, nuclear, 885–886
Real images, 756–757
Rectifiers
  bridge, 903
  filters, 903
  full-wave, 903–904
  half-wave, 902
  silicon-controlled rectifiers (SCRs), 914
Rectilinear propagation, 728
Reduction, 630
Reference circle, 308–310
Reference point, 308–309
Reflected ray, 754–755
Reflecting telescopes, 824
Reflection
  angle of, 753–754, 773
  defined, 753
  diffuse, 755
  laws of, 753–755
  of light, 728–732
  regular (specular), 755
  total internal, 780–782
Refracted ray, 775–778
Refracting telescopes, 806, 824
Refraction
  angle of, 777–778
  apparent depth and, 787–788

Refraction (*cont.*):
atmospheric, 785–786
defined, 753, 773
index of, 774–775
laws of, 775–778
of light, 728–732
principle of, 773–774
Snell's law, 777–778
velocity and, 776–778
wavelength and, 779–781
Refrigerants, 460–461
Refrigeration, 459–461
Regular (specular) reflection, 755
Relative density, 328
Relative humidity, 435–436
Relative permeability, 648
Relative permittivity, 580
Relativity
Einstein's postulates, 834–836
relativistic kinetic energy, 841–842
relativistic length, 838–839
relativistic mass, 837–838, 840–841
relativistic time, 837–838, 839
simultaneous events and, 836–837
special theory of, 834–836
Relays, semiconductor, 913–914
Residual magnetism, 658–659
Resistance
defined, 593, 597
internal, 618–619
multiplier, 672–674
Ohm (unit), 598
Ohm's law for, 597–599
power and, 601
shunt, 674–675
temperature coefficient of, 604
thermal, 408, 410–411
Wheatstone bridges and, 626–628
of wire, 601–603
Resistivity, 601–603
superconductivity and, 604–607
Resistors
IC, 916, 918
in parallel, 615–617
photoresistors, 914–915
in series, 612–614
thermistors, 915–916

Resolving power, 821–824
Resonance
of electric circuits, 719
sympathetic vibrations, 493
Resonant frequency, 719
Rest mass, 841–842
Restitution, coefficient of, 202–203
Restoring force, 306–308
Resultant electric field, 535–537
Resultant force, 58, 62–63, 223
Resultant of vectors, 62–63, 65
Resultant torque, 105–107
Resultant (net) work, 174–176
Retentivity (magnetism), 645
Reverse-biased PN junctions, 901
Reverse-biased zener diodes, 904–905
Reverse current in zener diodes, 904–905
Revolution, period of, 216
Revolutions per minute (unit), 216
Revolutions per second (unit), 216
Right angles, 19–20
Right-hand rules
Fleming's, 688
for magnetic fields, 654
for magnetic force, 649–650
Right triangles, 21–26
Rigidity, modulus of, 297
Roemer, Olaus, 736
Roots, 15
Rotation
axis of, 101, 247
frequency of, 216
Newton's second law in, 251–254
torque in, 253–254
Rotational equilibrium, 107–110
Rotational kinetic energy, 249–251
Rotational speed, 216
Rotational versus linear motion, 247–249
Rotational work and power, 254–256
Rowland, J. H., 658
Rowland's ring, 658
Rules (scales), 45–47
Rumford, Count, 384–385

Rutherford, Ernest, 846
Rutherford atom, 846–848, 850
Rutherford scattering, 846, 865
Rydberg's constant, 852

S

Satellites, 228–232
Saturated vapor pressure, 433–434, 435–437
Saturation, magnetic, 645, 658–659
Scalar quantities, 50
Scalene triangles, 21
Scales (rules), 45–47
Scattering, Rutherford, 846, 865
Schilling, Andreas, 606
Scientific notation, 16–17
Scientific theory, 3–4
Screen of cathode-ray tubes (CRTs), 895–896
Screw jacks, 282
SCRs (silicon-controlled rectifiers), 914
Second (unit), 43
Second law of motion. *See* Newton's second law
Second law of thermodynamics, 452–454
Second-order fringes, 819–821
Semiconductors, 895. *See also* Diodes; Transistors
conduction in, 898
crystal lattice in, 898–899
defined, 517, 897–898
dopants in, 899–900
examples of, 517
holes in, 900
impurities in, 899–900
N-type, 899–900
P-type, 899–900
photoresistors, 914–915
PN junctions, 900–902
relays, 913–914
silicon, 898–899
silicon-controlled rectifiers (SCRs), 914
thermistors, 915–916
Series ac circuits, 717–718
Series capacitors, 581–585
Series resistors, 612–614
Series-wound motors, 695

Shadows, 738–739
Shear modulus, 296–297
Shearing angle, 296–297
Shearing strain, 296–297
Shearing stress, 292, 296–297
SHM. *See* Simple harmonic motion (SHM)
Shockley, William, 906
Short radio waves, 733, 734
Shunt resistance, 674–675
Shunt-wound motors, 695–696
SI units. *See* Système International d'Unités (SI)
Signal inverters, 911
Signed numbers, 7–10
    for acceleration, 129–130
    for displacement, 129–130
    for velocity, 129–130
Significant figures, 44–45
Silicon, 517, 898–899
Silicon-controlled rectifiers (SCRs), 914
Simple harmonic motion (SHM)
    acceleration in, 312–314
    defined, 307
    displacement in, 307–308
    frequency in, 314–316
    period in, 314–316
    reference circle, 308–310
    simple pendulums and, 316–318
    torsion pendulums and, 318–319
    velocity in, 310–312
Simple lenses, 793–795
Simple machines
    belt drives, 275–277
    defined, 267
    efficiency and, 267–269, 270
    gears, 277–278, 279
    inclined planes, 278–282
    levers, 271–272
    mechanical advantage, 269–270
    pulleys, 273–275
    screw jacks, 282
    wedges, 281
    wheels and axles, 273–274
Simple pendulums, 316–318
Simultaneous events and relativity, 836–837
Sine, 25

Single-body problems, 155–164
Slip rings, 689, 690
Slow neutrons, 883–885, 886
Slug (unit), 153
Slug-foot per second (unit), 197
Snell, Willebrord, 777
Snell's law, 777–778
Solar system, 232–233
Solenoids, 656–657, 670
Solid angle, 740–742
Solid phase, 394–399
Solidification (freezing), 397
Solids
    area expansion of, 374–376
    cohesion of, 299
    dielectric constant and dielectric strength, 580
    heat of fusion and heat of vaporization, 395
    index of refraction, 774
    linear expansion of, 371–374
    mass density, 327
    R-values, 408
    resistivities, 602
    specific heat capacity, 390
    speed of sound in, 488–489
    thermal conductivities, 408
    volume expansion of, 376–378
    weight density, 327
Sound
    audible, 494–498
    beat frequencies, 499, 500
    defined, 487
    Doppler effect, 500–504
    forced vibrations, 493
    hearing threshold, 495
    infrasonic waves, 494
    intensity, 494–498
    intensity level, 495–497
    interference, 499
    loudness, 498
    pain threshold, 495
    in pipes, 490–493
    pitch, 498–499
    production of, 487–488
    quality, 499
    resonance (sympathetic vibration), 493
    speed of, 488–490
    standing waves, 476–478, 490–493
    ultrasonic waves, 494

Sound (*cont.*):
    vibrating air columns, 490–493
    vibrato effect, 499
Special theory of relativity, 834–836
Specific gravity, 328
Specific heat capacity, 387–390
Spectra
    absorption, 848–849
    atomic, 848–849, 854, 855
    continuous, 848–849
    defined, 848
    electromagnetic, 732–734
    hydrogen, 848–849, 854, 855
    line emission, 848–849, 854, 855
    visible, 780
Spectral series, 848–849, 854, 855
Spectrometers, mass, 848, 871–873
Specular (regular) reflection, 755
Speed. *See also* Velocity
    average, 121, 122
    banked curves and, 219–221
    constant, 120–121
    instantaneous, 122
    of light, 42, 732–734, 736–738, 834–836
    linear, 216
    rotational, 216
    of sound, 488–490
    velocity and, 120–122
    wave, calculating, 470–471
Speed ratio of belt drives, 276
Spherical aberration, 758–759, 767, 806–807
Spherical mirrors, 757–762
    convex (diverging), 758–759, 760, 767
    defined, 757–758
    images formed by, 760–762
    introduced, 757–759
    magnification, 765
    mirror equations, 762–765
    spherical aberration and, 767
Spin of electrons, 643–645
Split-ring commutators, 675
Spontaneous emission of photons, 855–857
Spring constant, 291–293

SQUIDs (superconducting quantum interference devices), 607

Standard, of physical quantities, 39

Standing waves, 476–478, 490–493

Static force, 56

Static friction, 84–86, 91

Statics, defined, 3

Steam, specific heat capacity of, 390

Steam point (upper fixed point), 363

Stefan-Boltzmann law, 414–416

Step-down transformers, 698

Step-up transformers, 698

Steradian (unit), 740–742

Stimulated emission of photons, 855–857

Strain, 291–293
longitudinal, 293–296
shearing, 296–297
volume, 298–299

Strategies for solving problems. *See* Problem-solving strategies

Streamline flow, 341–342

Stress, 291–293
compressive, 291–292
longitudinal, 293–296
shearing, 292, 296–297
tensile, 291
volume, 298–299

Sublimation, 397

Sublimation curve, 435

Substrates of ICs, 916

Subtraction of vectors, 66

Subtraction rule, 8–9

Superconducting quantum interference devices (SQUIDs), 607

Superconductivity, 604–607

Superposition principle for waves, 475–476, 815

Sympathetic vibrations (resonance), 493

Synchronous satellites, 230–232

Système International d'Unités (SI)
base units, 40–42
converting units, 47–50

Système International d'Unités (SI) (*cont.*):
derived units, 40, 41
introduced, 40–42
length (meter), 40–42
mass (kilogram), 40–42, 379, 389
prefixes, 43–44, 865
time (second), 42

T

Tangent, 25

Tangential acceleration, 248–249

Tangential velocity, 310

Telescopes, 806, 823–824

Temperature
absolute, 368–371
area expansion of, 374–376
Celsius scale, 364–367, 370–371
critical (transition), for superconductivity, 605
critical, of gases, 431–432
Fahrenheit scale, 364–367, 370–371
Kelvin scale, 370–371
linear expansion of, 371–374
lower fixed point (ice point), 363
measurement of, 362–367
quantity of heat versus, 385–387
Rankine scale, 370–371
thermal energy and, 360–367, 390–393
upper fixed point (steam point), 363
volume expansion of, 376–378

Temperature coefficient of resistance, 604

Temperature intervals, 364–367, 369–371

Tensile stress, 291

Tera (T) prefix, 43

Terminal potential difference, 618–619

Tesla (unit), 646

Thermal conduction, 405–410

Thermal conductivity, 407–410

Thermal convection, 405–406, 411–413

Thermal energy
conservation of, 390–393
temperature and, 360–367

Thermal equilibrium, 360–362

Thermal radiation, 405–406, 413–416

Thermal resistance, 408, 410–411

Thermionic emission, 893

Thermistors, 915–916

Thermodynamic equilibrium, 444–445

Thermodynamic state, 444–445

Thermodynamics
adiabatic processes, 448–450
Carnot cycle, 454
defined, 443
first law, 445–446, 448, 460
heat and work, 443–444
internal energy function, 444–445
isochoric processes, 450–451
isothermal processes, 451–452
Otto cycle, 457–459
P-V diagrams, 446–448
refrigeration, 459–461
second law, 452–454
throttling processes, 449–450

Thermometers
Celsius, 364–367, 370–371
constant-pressure, 367, 368
constant-volume, 367–368
defined, 362
Fahrenheit, 364–367, 370–371
gas, 367–368
introduced, 362–363
mercury-in-glass, 364–365, 367
requirements, 362–367
water equivalent of, 393

Thermostats, 373

Thin lenses, 795, 798–800

Third law of motion, 74–75

Thomson, Joseph John, 845

Thomson atom, 845–846

Threshold frequency, 843

Throttling processes, 449–450

Throttling valves, 449–450, 461

Time
leap seconds, 43
relativistic, 837–838, 839
second (unit), 43

Time constant
of capacitors, 706–708

Time constant (*cont.*):
of inductors, 710–711
Time dilation, relativistic, 839
Toroids, magnetic fields of, 657
Torque, 100, 103–107
force and, 103–107
magnetic
on loops, 668–670
on solenoids, 670
newton-meter (unit), 103
resultant, 105–107
in rotation, 253–254
transmission of, 275–278
Torricelli's theorem, 347–348
Torsion pendulums, 318–319
Total internal reflection, 780–782
Transfer of heat. *See* Heat, methods of transfer
Transformer oil, dielectric constant and dielectric strength, 580
Transformers, 697–698
Transient electric current, 593, 595
Transistors
alpha, 908–912
base, 906, 907
base-collector current gain, 910–911
beta, 910–911
collector, 906, 907
common-base amplifiers, 908–910
common-emitter amplifiers, 910–912
current gain, 908–912
defined, 906
emitter, 906, 907
IC, 916
NPN, 906–908
PNP, 906, 907
power gain, 909
voltage gain, 909
Transition (critical) temperature for superconductivity, 605
Translational equilibrium, 75–76, 80–84, 107
Translational motion, 241
Transmission of torque, 275–278
Transverse waves, 469, 471, 825–826
Triangles, 21–26

Trigonometry, 23–26, 59–63
Triodes, 894–895
Triple point of water, 370, 435
Tubes
cathode-ray tubes (CRTs), 895–896
vacuum, 894–895
X-ray, 896–897
Turbulent flow, 341
Turns ratio of transformers, 697

U

Ultimate strength, 293, 295–296
Ultrasonic waves, 494
Ultraviolet waves, 733, 734
Umbra, 739
Uniform circular motion, 213–234, 224
centripetal acceleration and, 214–216
centripetal force and, 216–219
defined, 213–214
Uniformly accelerated motion, 123–125
gravity and, 131–136
U.S. Customary System (USCS) of units, 40, 42
Units
converting, 47–50
metric (SI). *See* Système International d'Unités (SI)
prefixes for, 43–44, 865
U.S. Customary System (USCS) of units, 40, 42
Universal gas constant, 429–430
Universal gravitation, 225–228, 530
Unlike charges, 514–516
Unpolarized waves, 825
Upper fixed point (steam point), 363
USCS (U.S. Customary System) of units, 40, 42

V

Vacuum. *See* Free space
Vacuum tubes, 894–895
Valence band, 898
Valence electrons, 897
Valves, throttling, 449–450, 461
Vapor pressure, 433–434, 435–437

Vaporization, 395, 432, 434
latent heat of, 395–399, 461
Vaporization curve, 434–435
Vector quantities, 50
Vector sums of electric fields, 535–537
Vectors
addition of
component method, 63–65
parallelogram method, 55
polygon method, 53–54
angle of, 53
components of, 60, 61
direction of, 50–51
displacement, 51–52
force and, 56–57
introduced, 50
magnitude of, 51–53
resultant of, 62–63, 65
subtraction of, 66
trigonometry and, 59–63
Velocity. *See also* Speed
acceleration and, 125–126
angular, 243–244, 310
angular versus linear, 247–248
average, 122, 125
dimensions of, 49
elastic and inelastic collisions, 202–206
final, 124–125, 126
fluid pressure and, 343–345
gravity and, 131–136
instantaneous, 122
linear versus angular, 247–248
refraction and, 776–778
sign convention for, 129–130
in simple harmonic motion (SHM), 310–312
speed and, 120–122
tangential, 310
transverse waves, 471, 844–845
Venturi meters, 344, 348–349
Vernier calipers, 45–47
Vertex of mirrors, 758–759
Vertical projection, 138–143
Vibrations, forced and sympathetic (resonance), 493
Vibrato effect, 499
Virtual images, 757
Visible radiation. *See* Light
Visible spectrum, 780
Volt (unit), 556, 564

Voltage. *See also* Electric potential; Electric potential difference
  breakover, 914
  capacitance and, 570–573
  effective, 712
  Kirchhoff's second law for, 621–626
  Ohm's law and, 597–599
  power and, 600
  in series ac circuits, 716–718
  terminal, 618–619
  zener, 904–905
Voltage gain of transistors, 909
Voltage ratio of transformers, 697–698
Voltage regulators, 905
Voltmeters, 563
Voltmeters, dc, 672–674
Volume elasticity, 298–299
Volume expansion, 376–378
Volume strain, 298–299
Volume stress, 298–299

W

Water
  bulk modulus, 299
  expansion of, 378–379
  heat of fusion and heat of vaporization, 395
  ice
    index of refraction, 774
    specific heat capacity, 390
  index of refraction, 774
  mass density, 327
  R-value, 408
  steam, specific heat capacity, 390
  thermal conductivity, 408
  triple point, 370, 435
  vaporization curve, 434
  volume expansion coefficient, 376
  weight density, 327
Water equivalent of thermometers, 393
Watt (unit), 187, 740
Wave motion, 468–469
Wave theory of light, 729–732
Wavelength
  de Broglie, 844–845, 850–851
  defined, 472

Wavelength (*cont.*):
  frequency and, 472–473, 732–734
  mass and, 844–845
  period and, 473
  of photons, 844–845, 852–853
  refraction and, 779–781
  velocity and, 473, 844–845
  visible light, 732–733, 734, 780
Waves. *See also* Rays
  characteristic frequencies, 478–480
  constructive interference, 476, 815–818
  destructive interference, 476, 815–818
  electromagnetic, 732–734
  Huygens's principle, 728–729, 738–739
  infrared, 727, 733, 734
  infrasonic, 494
  isotropic sources, 743–744
  light. *See* Light
  longitudinal, 469–470, 486–487, 488–489. *See also* Sound
  mechanical, 468–469, 486–487. *See also* Sound
  particles and, 844–845
  periodic
    energy, 474–475
    motion, 472–473
    power, 475
  plane-polarized, 825
  polarization of, 824–826
  radio, 733, 734
  sound. *See* Sound
  standing, 476–478, 490–493
  superposition principle for, 475–476, 815
  transverse, 469, 471, 825–826
  ultrasonic, 494
  ultraviolet, 733, 734
  unpolarized, 825
  visible, 732–733, 734
Weber (unit), 646
Wedges, 281
Weight
  atomic, 867, 869
  defined, 56, 77
  gravitational fields and, 111, 227–228

Weight (*cont.*):
  mass and, 153–155
Weight density, 326–327
Wheatstone bridges, 626–628
Wheels and axles, 273–274
Wires
  induced emf in, 686–687
  magnetic fields of, 654–655
  magnetic force on, 652–654
  resistance of, 601–603
Work
  charge and, 560
  defined, 172
  by electric fields, 560
  electric power and, 600–601
  foot-pound (unit), 174
  force and, 172–174
  heat and, 443–444
  heat engines and, 452–459
  introduced, 172–174
  joule (unit), 174, 385, 388, 389
  kinetic energy and, 176, 177–179
  net, on fluids, 345–346
  per unit charge, 556–560, 564
  potential difference and, 560
  potential energy and, 176, 180–182
  power and, 187–188, 254–256
  resultant (net), 174–176
  rotational, and power, 254–256
Work-function energy, 842–843
Workable hypotheses, 3–4
Worm gears, 279

X

X-ray tubes, 896–897
X-rays, 733, 734

Y

Young, Thomas, 814
Young's (double-slit) experiment, 814–818
Young's modulus, 293–296, 488

Z

Zener diodes, 904–905
Zener (avalanche) effect, 905
Zener voltage, 904–905